Demystifying AI

As artificial intelligence advances at an exponential pace, understanding data science and machine learning has become increasingly essential. Yet, the wide range of available resources can be daunting, posing challenges for beginners. This second book builds on the foundation laid in the first, *Data Science and Machine Learning for Non-Programmers: Using SAS Enterprise Miner*, providing similar fundamental knowledge of data science and machine learning in an accessible way. It is specifically designed to cater to readers who prefer a hands-on guide using IBM SPSS Modeler, a widely popular software that does not require coding or programming skills. Like the first book, this volume helps learners from various non-technical fields gain practical insight into machine learning but shifts the focus to a different tool for those seeking alternatives to coding.

In this book, readers are guided through practical implementations using real datasets and IBM SPSS Modeler, a user-friendly data mining tool. The approach remains consistent with a focus on application, providing step-by-step instructions for all stages of the data mining process using two large datasets, ensuring continuity and reinforcing concepts in a cohesive project framework. This book also offers practical advice on presenting data mining results effectively, aiding readers in communicating insights clearly to stakeholders.

Together with the first book, this volume is a companion for beginners and experienced practitioners alike. It targets a broad audience, including students, lecturers, researchers, and industry professionals. It offers flexibility in learning pathways and deepens understanding of data science using easy-to-follow, software-based approaches.

Dothang Truong is the Associate Dean for the School of Graduate Studies and Professor of Aviation Data Science at Embry Riddle Aeronautical University, Daytona Beach, Florida. He has extensive teaching and research experience in machine learning, artificial intelligence, data analytics, aviation safety, air transportation management, and supply chain management. In 2022, Dr. Truong received the Frank Sorenson Award for the outstanding achievement of excellence in aviation research and scholarship. He is ranked #1 globally in the specialty of Traffic Management by ScholarGPS (2024) and recognized as a Highly Ranked Scholar—Prior Five Years, placing in the top 0.05% of scholars worldwide for exceptional scholarly productivity, impact, and quality.

Chapman & Hall/CRC
Data Mining and Knowledge Discovery Series
Series Editor: Vipin Kumar

Event Mining
Algorithms and Applications
Tao Li

Text Mining and Visualization
Case Studies Using Open-Source Tools
Markus Hofmann and Andrew Chisholm

Graph-Based Social Media Analysis
Ioannis Pitas

Data Mining
A Tutorial-Based Primer, Second Edition
Richard J. Roiger

Data Mining with R
Learning with Case Studies, Second Edition
Luís Torgo

Social Networks with Rich Edge Semantics
Quan Zheng and David Skillicorn

Large-Scale Machine Learning in the Earth Sciences
Ashok N. Srivastava, Ramakrishna Nemani, and Karsten Steinhaeuser

Data Science and Analytics with Python
Jesús Rogel-Salazar

Feature Engineering for Machine Learning and Data Analytics
Guozhu Dong and Huan Liu

Exploratory Data Analysis Using R
Ronald K. Pearson

Human Capital Systems, Analytics, and Data Mining
Robert C. Hughes

Industrial Applications of Machine Learning
Pedro Larrañaga et al

Automated Data Analysis Using Excel
Second Edition
Brian D. Bissett

Knowledge Guided Machine Learning: Accelerating Discovery using Scientific Knowledge and Data
Edited by Anuj Karpatne, Ramakrishnan Kannan, Vipin Kumar

Demystifying AI: Data Science and Machine Learning Using IBM SPSS Modeler
Dothang Truong

For more information about this series please visit: https://www.routledge.com/Chapman--HallCRC-Data-Mining-and-Knowledge-Discovery-Series/book-series/CHDAMINODIS

Demystifying AI
Data Science and Machine Learning Using IBM SPSS Modeler

Dothang Truong

CRC Press
Taylor & Francis Group
Boca Raton London New York

CRC Press is an imprint of the
Taylor & Francis Group, an **informa** business
A CHAPMAN & HALL BOOK

Designed cover image: Dothang Truong

First edition published 2026
by CRC Press
2385 NW Executive Center Drive, Suite 320, Boca Raton FL 33431

and by CRC Press
4 Park Square, Milton Park, Abingdon, Oxon, OX14 4RN

CRC Press is an imprint of Taylor & Francis Group, LLC

ISBN: 9781032740003 (hbk)
ISBN: 9781032742168 (pbk)
ISBN: 9781003468196 (ebk)

DOI: 10.1201/9781003468196

For Product Safety Concerns and Information please contact our EU representative: GPSR@taylorandfrancis.com
Taylor & Francis Verlag GmbH, Kaufingerstraße 24,
80331 München, Germany.

Support Material can be found at www.routledge.com/9781032740003

To my students,
May this book serve as a tool for your growth, and may your journey through data science and machine learning be filled with discovery and insight.

To the learners of today,
Who strive to unlock the power of data science and transform challenges into opportunities.

To the pioneers of data science and machine learning
For laying the foundation, and to all those who carry the torch, forward.

Contents

PART III Advanced Data Mining Methods

Preface

The fields of machine learning and artificial intelligence (AI) continue to transform industries and redefine our interactions with technology. As we move deeper into the AI era, leveraging machine learning algorithms to uncover meaningful patterns from rapidly growing datasets becomes increasingly essential. Reflecting on my journey, I vividly recall my first encounter with advanced AI, Tesla's full self-driving feature, and, more recently, the groundbreaking ChatGPT and other generative AI programs. These technologies reveal the immense potential of AI, demonstrating its ability to learn and adapt autonomously. It's an exhilarating time to be involved in this field, as AI tools continue to push the boundaries of what we once thought possible. At the core of these advancements lie data science and machine learning, fundamental components of AI programs.

WHY THIS BOOK?

Many people perceive AI applications as "black boxes"—unexplainable, overly complex, and difficult to understand. As a result, they may feel hesitant or even lack the confidence to begin learning about AI and machine learning and how to apply them in their fields. This is especially true for those without strong programming skills or a background in mathematics and computer science. It's crucial to "open the black box" and demystify AI by breaking down the foundational concepts of machine learning.

Learners from diverse fields, such as business, marketing, finance, aviation, social sciences, and the humanities, can often feel left behind amidst the emphasis on training data scientists who are skilled in programming and algorithm development. However, as software technologies advance, many applications now allow users to build machine learning models and perform data mining without writing a single line of code. These applications offer user-friendly interfaces that enable non-programmers to harness the power of machine learning.

The success of my first book, which introduced core concepts and practical skills in data science and machine learning using SAS Enterprise Miner, inspired me to continue this journey. That book focused on providing a strong foundation for learners without a technical background by using a powerful, easy-to-use tool that requires no coding. Building on that foundation, this second book is designed for learners from non-technical fields and experienced researchers alike, focusing on SPSS Modeler, a widely used data mining software that also requires no programming skills.

This book follows a similar approach by offering a structured and accessible way to understand machine learning fundamentals while providing hands-on experience with SPSS Modeler, which is popular among both academic and industry professionals. My goal is to offer learners a clear path to mastering AI concepts and practical skills, regardless of their technical expertise.

WHO SHOULD READ THIS BOOK?

This book is designed for students, lecturers, researchers, and industry professionals from a variety of backgrounds, including business, marketing, aviation, social sciences, and beyond. It is tailored for those seeking to understand the core concepts of data mining and machine learning, with practical examples that allow users to build models and analyze data without relying on programming. Whether you are just starting out or already have some experience in machine learning, this book aims to support your growth and help you unlock the potential of machine learning for your projects.

WHAT THIS BOOK COVERS

The content of this book closely follows the structure of the first book. Part I introduces foundational concepts of data mining and machine learning, presented without overwhelming mathematics, enabling learners to grasp the key principles and mechanisms behind various algorithms. Part

II covers widely used machine learning methods, while Part III explores more advanced techniques, such as random forest, gradient boosting, and Bayesian networks, to help readers build sophisticated models.

Throughout the chapters, readers will work with two large datasets, applying each step of the data mining process to a cohesive project. This book maintains a hands-on approach, guiding readers through practical implementations using SPSS Modeler. Step-by-step guidelines and screenshots are included to facilitate easy understanding, just as in the first book.

By using SPSS Modeler, readers will be able to perform data mining confidently, achieving accurate results without writing any code. This book also includes a chapter on presenting data mining results, providing examples to help users communicate their findings effectively to stakeholders.

HOW TO USE THIS BOOK

Like the first book, this one is designed to provide a flexible learning path. While each chapter builds upon the previous ones, readers can also navigate directly to topics or algorithms that interest them. The practical examples and exercises are based on two large datasets, allowing readers to apply learned concepts within the framework of a cohesive project.

DECLARATION OF LARGE LANGUAGE MODELS, GENERATIVE AI, AND AI-ASSISTED TECHNOLOGIES IN THE WRITING PROCESS

The content of this book was entirely developed and written by the author. During its preparation, the author used ChatGPT for editing purposes, including correcting grammar, typos, and sentence structure. Afterward, the author thoroughly reviewed, edited, and revised the content as necessary, verifying sources when needed. The author assumes full responsibility for the content of this book.

FINAL THOUGHTS

I am deeply grateful to everyone who supported my first book, and I hope this second book will continue to empower learners to explore machine learning in an accessible and practical way. The journey of discovery in data science and machine learning is endless, and I encourage you to dive into the examples, experiment with the tools, and enjoy the learning process.

Let's continue this journey together, unlocking the potential of machine learning, one model at a time.

Are you ready for the era of AI? Let's get started.

Dothang Truong, Ph.D.
Professor of Aviation Data Science

Part I

Introduction to Data Mining

1 Introduction to Data Mining and Data Science

LEARNING OUTCOMES

Upon completion of this chapter, you will be able to:

1. Define data mining.
2. Provide examples of data mining applications in real-life business practices.
3. Differentiate between reactive, proactive, and predictive strategies.
4. Explain the differences and overlaps between data science and data mining.
5. Describe the primary characteristics of data mining, how it differs from traditional statistical methods, and the potential advantages and disadvantages.
6. Discuss the evolution of artificial intelligence, data mining, machine learning, and deep learning.
7. Define big data and describe the primary characteristics of big data.
8. Explain the big data analytics process.
9. Identify ethical considerations in data mining.
10. Describe the Institutional Review Board requirements.
11. Implement the data de-identification process.

INTRODUCTION

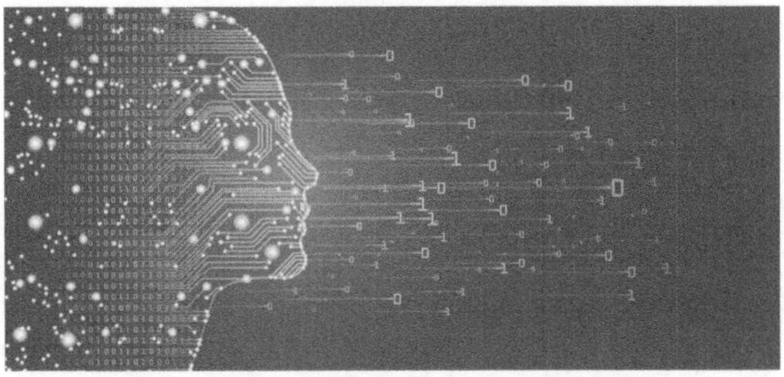

In recent years, artificial intelligence (AI) has experienced unprecedented growth, becoming a transformative force across various industries. This surge is largely attributed to advancements in data science and machine learning, which serve as the core components of AI systems. Data science provides the framework for collecting, processing, and analyzing vast amounts of data, while machine learning offers algorithms that enable systems to learn from the data and make informed decisions. Together, they empower AI to perform complex tasks, from natural language processing to predictive analytics.

The importance of data mining becomes evident when considering the role of data in AI development. Data mining involves extracting valuable insights from large datasets, uncovering patterns, and establishing relationships that might not be immediately apparent. These insights are crucial for training machine learning models. Without effective data mining techniques, the potential of AI systems would be significantly limited, as they would lack the rigorous, structured information needed for learning and adaptation.

In this chapter, we begin a journey into the world of AI, data mining, and machine learning, exploring fundamental concepts that serve as the foundation of these rapidly evolving fields. While the term **data mining** is not new, its significance continues to grow as it adapts to changing technologies and methodologies. Despite the tremendous amount of information and discussions on data

DOI: 10.1201/9781003468196-2

science, machine learning, and data mining we can find on the Internet, audiences often struggle to understand those concepts due to inconsistency and lack of clear explanations. In our modern world, we see many buzzwords bounding on the Internet and social media, often leading to misuse or misinterpretation. Thus, it becomes crucial to demystify these terms and understand their meanings accurately. Throughout this chapter, we will examine various concepts related to data mining, data science, machine learning, and big data. This comprehensive exploration lays the groundwork for deeper insights into data mining methods, which we will discuss in more detail in subsequent chapters.

Should you find yourself grappling with confusion after reading this chapter, don't worry. We will revisit these fundamental concepts with concrete examples of data mining and analysis results, providing clarity and context to reinforce your understanding.

Are you ready to embark on an exciting journey into the world of big data and AI? With curiosity and enthusiasm, let us start uncovering the mysteries of data mining and data science. Let's get started!

ARTIFICIAL INTELLIGENCE AND DATA MINING

Artificial Intelligence (AI) refers to the simulation of human intelligence in machines, allowing them to perform tasks that usually require human cognitive abilities. These tasks include learning, reasoning, problem-solving, perception, and language comprehension. AI systems use algorithms and data to recognize patterns, make decisions, and enhance their performance over time.

> **DEFINITION**
>
> **Artificial Intelligence**
> Artificial Intelligence (AI) is the simulation of human intelligence in machines, enabling them to perform tasks that typically require human cognition, such as learning, reasoning, problem-solving, perception, and language understanding.

AI systems require vast amounts of data to train effectively. A well-known example is Tesla's Full Self-Driving (FSD) system, which relies on extensive data collection to improve its autonomous capabilities. Similarly, smart assistants like Amazon Alexa, Apple Siri, and Google Assistant continuously process and analyze speech data to enhance their natural language understanding. Streaming services such as Netflix and Spotify rely on vast datasets of user preferences and viewing/listening habits to refine their recommendation algorithms. Facial recognition systems used in security and authentication, such as Apple's Face ID or airport biometric scanners, require massive image datasets to improve accuracy and reliability. Even healthcare AI systems, such as DeepMind Health, leverage vast amounts of medical

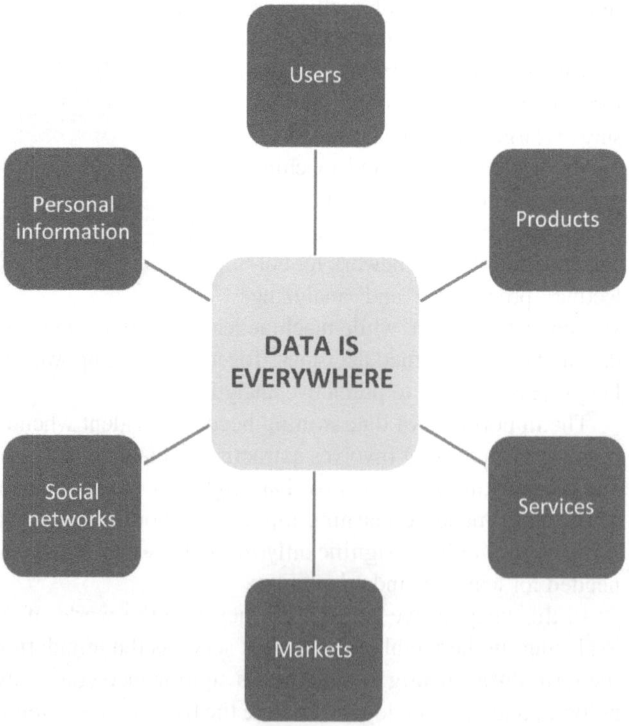

records, imaging scans, and diagnostic histories to assist in disease prediction and treatment planning. With AI becoming increasingly integrated into various industries, the question remains—where do all these data come from, and where do they exist?

We are living in a world of shared information, AI, social networks, wireless connectivity, and digital communications. In this world, data are streamed and collected continuously in real time. Data exist in many forms, including numbers, texts, images, voice, and videos. Big tech companies like Apple, Amazon, Meta (Facebook), and Tesla continue to collect and archive user data on their servers. Manufacturing companies collect data on their products, operations, supply chains, inventories, logistics, and retailers. Service providers, such as hotels, resorts, casinos, banks, and airlines, gather data on their markets and customers. Our normal lives are also captured constantly via security cameras installed almost everywhere. We also choose to share our own data by sharing information, thoughts, pictures, and videos on social networks. If we go to the Internet and google, we can find everything that we need. It is not an understatement that we are overwhelmed with the amount of data that we have access to, not to mention the proprietary data that are not accessible to the public. Remember, not seeing the data does not mean they don't exist. They are on the CLOUD, waiting to be used for AI-driven advancements.

Has Anyone Seen the TV Show *Person of Interest*?

With that vast amount of data stored in cloud-based servers, the question is how can we make use of the data to improve our lives, businesses, or customer experience? Let me start with one of my favorite TV shows, *Person of Interest*. It is a CBS sci-fi thriller show that aired from 2011 to 2016. The main character is Harold Finch, a math and computer genius. He received substantial government funding to develop an advanced machine learning algorithm and build an AI called *the Machine*. The main goal of this AI is to process and analyze all data being fed into it from various sources, including personal information, social networks, traffic data, business information, security information, government information, banking transactions, and others. Basically, the Machine receives all kinds of data that the government can provide, including numbers, texts, videos, images, and voices. The data are fed to the system in real time. So, you can imagine the volume of data the Machine receives every second. It certainly requires multiple supercomputers with very large bandwidth and very high-end servers to store that much data, not to mention the amount of energy needed to maintain the servers. Furthermore, computers with very high computing power are needed to process and analyze that huge amount of data promptly and effectively to produce practical outputs.

Finch developed a very robust and novel machine learning algorithm for the Machine, which took him several years to make it work properly. It was a tedious process with a lot of statistics, data analytics, computer programming, model training, and validation. But he finally succeeded. (Of course, we are never presented with the actual algorithm on the show.) Using the algorithm, the Machine is capable of discovering unknown patterns from the data to do one thing and one thing only, which is to predict a specific individual, a person of interest, that will be involved in crime or national security-related incidents. The prediction is not the name or address but the individual's social security number since it is a unique number assigned to each citizen. However, the Machine does not know whether this individual is a good guy or a bad guy; it only predicts that this person will be involved in an incident. It also does not know the reason or what caused the incident. In order to prevent the incident from happening, Finch hired an ex-Central Intelligence Agency (CIA) operative, John Reese, who has a set of special skills in hand-on-hand combat, weaponry, tactics, and covert operations. Reese's job is to find and follow that person of interest and intervene to stop the bad thing from happening. In each episode, the Machine makes a prediction, and the team rushes through time to follow the person of interest. Their goal is to determine whether the individual is a victim or perpetrator, enabling them to either save the person or prevent them from committing a harmful act.

While the synopsis may sound complicated, how the Machine functions is fairly straightforward. I presented that system in Figure 1.1 to demonstrate it. The input of the system is all data collected from various data sources in different forms, such as numbers, texts, voices, and images. The data are collected, stored, and fed to the cloud in real time. Imagine the amount of data the system has to process every second. The model uses machine learning algorithms to mine all of that data. It is a highly intelligent model that can learn, adapt, and improve itself, hence the name Artificial Intelligence. Of course, a lot of complicated algorithms are used in this model. The interesting thing is that the more data we collect and feed to the system, the more the model can learn and improve itself with new data via the model training process. In other words, the Machine learns from data (now you can see where the term *Machine Learning* comes from). Finally, the output of this system is the prediction of the target variable, more specifically, the person of interest represented by the social security number. It is a unique number assigned to one person, so there are no duplications.

The key takeaway from this series is that the Machine does not try to explain how the incident will happen because it has not yet happened, nor does it explain contributing factors based on any theory. It only analyzes the big data given to it and predicts what will happen in the near future. It does not know whether the person of interest is a victim or a perpetrator. Nonetheless, its prediction is never wrong; something terrible will happen involving that individual. Sounds fictional, right? Remember, it is a sci-fi show. Realistically, we should always account for the margin of error in prediction. In essence, it does not try to explain the incident but rather predicts it. Once Reese intervenes and stops the incident, then we will know what caused or may have caused the incident, but it is what we learn at the end of each episode. It is a great TV show, and if you are a fan of sci-fi thrillers, I strongly recommend watching it.

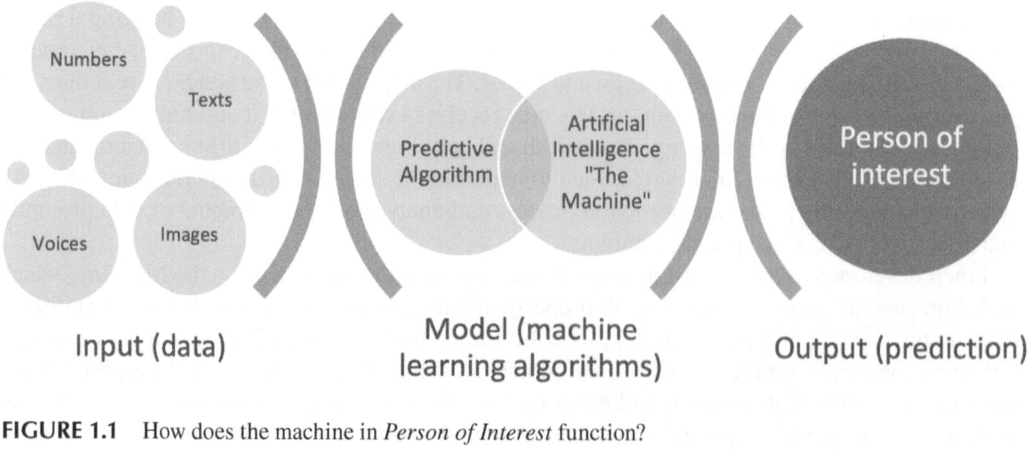

FIGURE 1.1 How does the machine in *Person of Interest* function?

Definitions of Data Mining

IBM defines data mining as "the process of finding anomalies, patterns, and correlations within large data sets to predict outcomes." Additionally, Nisbet et al. (2012) define data mining as the use of "machine learning algorithms to find faint patterns of relationship between data elements in large, noisy, and messy data sets, which can lead to actions to increase benefit in some form (diagnosis, profit, detection, etc.)."

In this book, I define **data mining** as a method of using machine learning algorithms to detect unknown *patterns* involving relationships among variables within large datasets to predict outcomes of interest, which can lead to informed business decisions. Figure 1.2 illustrates the concept of data mining. As you can see, what the Machine in the TV show *Person of Interest* does is nothing more than data mining. The futuristic part is that this AI can do it so well with the huge amount of data it has to process to make the prediction very effectively, so necessary mitigations can be implemented on time. In reality, that task requires significant resources to get it to work, and we have to account for the margin of error in prediction. There is no such thing as a perfect prediction in the real world.

> **DEFINITION**
>
> **Data mining**
> Data mining is a method of using machine learning algorithms to detect unknown patterns involving relationships among variables within large datasets to predict outcomes of interest, which can lead to informed business decisions.

In real life, small businesses, corporations, and governmental organizations can use data mining to analyze large datasets and detect unknown relationships between variables to make a prediction of outcomes, including performance, risk, revenue, customer satisfaction, and others. The predictive model and the relationships among variables can be used to provide the leaders with useful information about their operations and important business variables, based on which they can develop necessary business strategies and make informed decisions to enhance operational efficiency or mitigate risks. It may sound too scientific and may be hard to understand. Let me show you some examples of how data mining can be used in real-life scenarios.

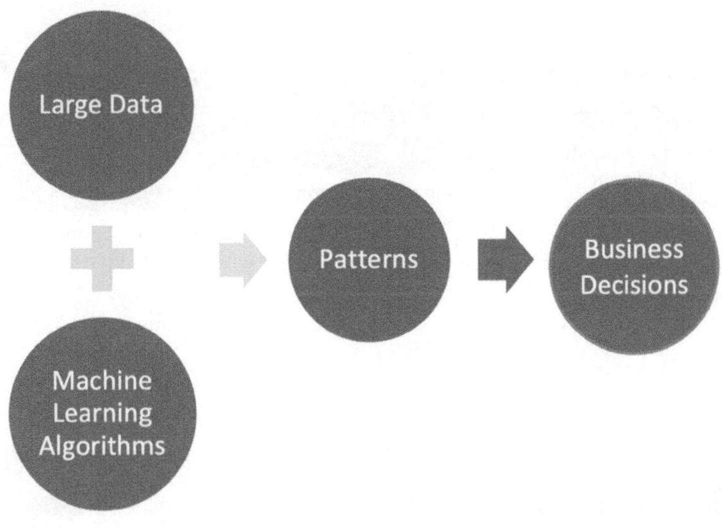

FIGURE 1.2 Data mining concept.

EVOLUTION OF ARTIFICIAL INTELLIGENCE, MACHINE LEARNING, AND DEEP LEARNING

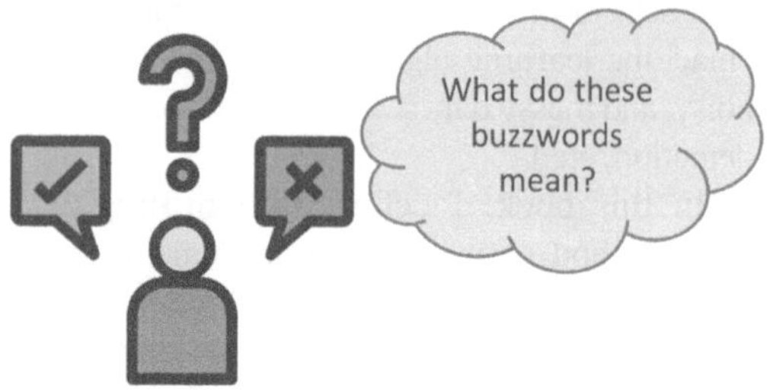

There are several buzzwords related to data mining and predictive modeling that we hear or read every day: **Artificial Intelligence, Machine Learning,** and **Deep Learning**. These terms are often overlapping and can be confusing, so I will briefly explain the differences among them. I will not go into great detail, as each is a field of study and a degree by itself. I will also explain the evolution of these fields of study over time (Figure 1.3).

First, **Artificial Intelligence (AI)**. While AI is a growing trend in research, it is not a new term. The term AI was created back in the 1950s to refer to software that uses complex logic to automate operations. AI has evolved significantly over time with advances in computer science and robotics. But AI is not just about building robots. A more current definition of AI is "a system that simulates human intelligence for solving problems or making decisions." AI can learn, predict, improve, and solve. Essentially, AI could be an application or solution that operates and acts with minimal human interference. Siri, Alexa, and Google Assistant are examples of AI virtual assistants that make our lives easier with smartphones, smart homes, and smart cars. Tesla's Full Self-Driving system is an AI that advances autonomous vehicles. ChatGPT is a generative AI that is capable of understanding natural language and performing various intelligent tasks based on users' demands. These applications are just the beginning of the AI era. We will soon see more amazing AI tools in all industries. Of course, as machines get smarter, concerns regarding ethics and social impacts will arise. However, that is a topic for a different book.

Machine learning is a core component of AI, and it started in the 1980s. Machine learning involves algorithms that learn patterns from data without explicit instructions. The data mining we have been discussing uses machine learning algorithms. These algorithms are developed based on the foundations of mathematics, statistics, and computer science. In the next chapters, I will describe specific machine learning algorithms that we can use in our research. In the past, a strong background in mathematics and coding skills were essential for using machine learning due to the extensive math involved, and only certain software, like MATLAB®, could handle the algorithms. Fortunately, in the last few years, more user-friendly applications have been created. As a result,

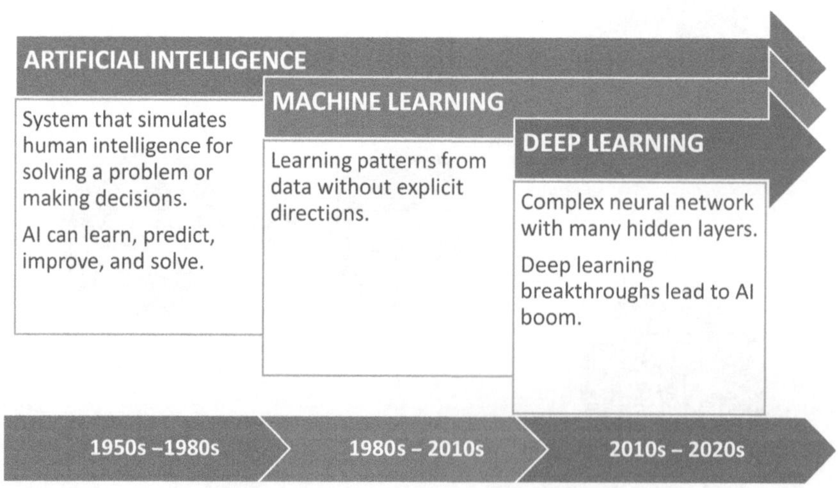

FIGURE 1.3 Evolution of artificial intelligence, machine learning, and deep learning.

machine learning is no longer just for mathematicians or computer scientists, it can also be used for applied research. While data scientists still use programming languages like Python or R for machine learning work, non-programmers now have more options to leverage this method for their research or businesses. Software applications such as SAS Enterprise Miner, IBM SPSS Modeler, Rapid Miner, KNIME, and Orange are easy-to-use data mining tools that allow us to analyze large datasets without writing codes.

Last but not least, **Deep Learning** is a buzzword we have started hearing more often in the past few years, even though it was introduced in the 2010s. Initially, it was not widely mentioned because only a few people understood the concept. Now, everyone talks about Deepfake, voice recognition, facial recognition on smartphones, and Tesla's Self-Driving Technology. In essence, deep learning is a complex neural network with many hidden layers. Some popular deep learning methods are convolutional neural networks and recurrent neural networks. I will not go into the technical details of this field, but it is a rapidly growing branch of machine learning. The technology that Apple uses for facial recognition and cameras is based on deep learning, as is Tesla's Self-Driving Technology. Deep learning is also used in natural language processing, art and music creation, and many other applications. You may have seen art and music created by AI that are indistinguishable from human creations. Deep learning remains an advanced subject, requiring a lot of data and complex algorithms to be performed correctly. Therefore, it is harder to find software that can handle deep learning without programming and coding.

MACHINE LEARNING: CAN MACHINES ACTUALLY LEARN?

The term "machine learning" is frequently used in books, papers, and online articles today. But one might wonder if machines can actually learn and what "machine learning" really means. It has become a buzzword, often used without a full understanding of its meaning. Instead of providing a technical definition, I'll explain it in simpler terms, as illustrated in Figure 1.4.

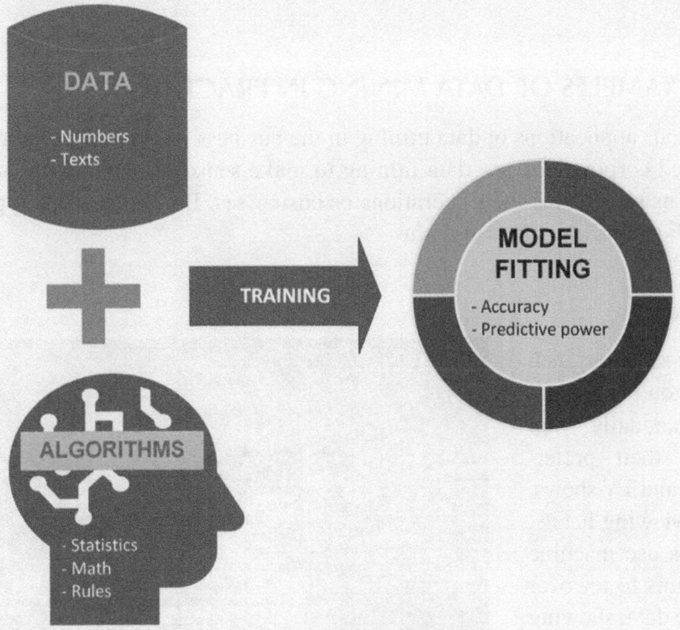

FIGURE 1.4 Machine learning: Can machines actually learn?

To develop a predictive model from data, we need a large dataset and a target variable. Next, we select one or more algorithms, which can be math-based, statistics-based, or rule-based. For each algorithm, a model is built by expressing relationships among variables using specific formulas and rules. The data scientist configures these settings based on data exploration and an understanding of the machine learning method. The next step is to fit that model to the actual dataset and evaluate its performance. If the model fits the data well, we have a well-fitting model. This process is known as training the model with data—in other words, the model is learning from the data, hence the term "machine learning."

If the model fit is poor and there is a significant difference between the model's predictions and the actual data (known as residuals), the data scientist needs to reconfigure the algorithm or try a different one, then retrain the model. This process is repeated until a good model fit is achieved. Essentially, the machine must continue to learn until it performs well, although we are the ones guiding its training. However, if we overtrain the model, it becomes overfit, meaning it will work well with the current dataset but may not generalize to other datasets. Therefore, it's important to avoid overtraining.

Over time, as we use the predictive model for actual predictions, adjustments are necessary to ensure it remains effective. By training the model with new data, we can improve its performance. Advanced programming can even automate this training process, allowing the model to continue learning from new data independently. When we reach this point, the machine can learn on its own without human intervention, resulting in true AI.

> We train the model with data, and the algorithm improves through this training. So, we can say that the machine does learn through the training we provide it.

I hope this explanation helps clarify the meaning of "machine learning."

PRACTICAL EXAMPLES OF DATA MINING IN PRACTICE

There are numerous applications of data mining in the business world that we experience every day. Organizations and corporations use data mining to make sense of the data they collect and make informed decisions to enhance their operations or businesses. Below are some typical examples of data mining applications in various domains.

Media Streaming

Many streaming services, such as Netflix, Amazon Prime, and Hulu, collect data daily from users regarding their preferences for movies and TV shows, as well as their viewing habits. These companies use machine learning algorithms to uncover patterns from the data, showing the relationship between user preferences and demographic information. The streaming application can then predict what types of TV shows or movies a user may want to watch and suggest similar content. For example, if I am a fan of action movies and have watched numerous Tom Cruise and Will Smith films, the next time I log in to Netflix, I will see suggestions for other action movies featuring these actors. The home page may also show several TV shows and movies in the same genre. Additionally, Netflix uses this information to decide which kinds of shows or movies to invest in to increase user subscriptions. Examples of popular Netflix TV shows and movies include *Stranger Things*, *House of Cards*, *The Crown*, *The Irishman*, and *Red Notice*.

Social Networks

Top social media platforms like Facebook and Twitter collect users' demographic information, product or service searches, and posts, photos, and communications shared on social media. Algorithms are developed to predict user behavior and their relationship with demographic information. The application suggests relevant sites, shows ads based on user preferences, and recommends potential friends or posts the user may want to see. For instance, I searched for new posts about Lexus vehicles on Facebook, and soon after, I began seeing more ads for Lexus vehicles for sale. I also received friend suggestions based on my friend database, allowing me to reconnect with long-lost friends. These companies use the data mining output to adjust or change their applications to better meet user demands and attract more users.

Online Shopping

Online shopping has gradually outperformed the traditional brick-and-mortar business model. Amazon is a dominant company in e-commerce and has achieved great success. Most traditional retailers have transitioned to e-commerce to improve profitability. Online retailers like Amazon, Best Buy, and Walmart use machine learning algorithms to predict customer shopping behavior. Their online platforms can suggest substitute products or new items based on customer demographics, past order history, or search history. For example, after searching for a smart outlet on Amazon, my Amazon app later suggested several options for smart outlets from different sellers. This convenience extends to predicting rising demand for specific products, allowing companies to prepare inventory to fulfill orders efficiently while reducing supply chain costs, inventory levels, and shipping times, thereby enhancing customer satisfaction.

Airline Industry

The airline industry is one of the largest and most competitive sectors in the United States. Airline executives use data mining to improve passenger satisfaction and operational efficiency. Airlines collect passenger demographic information and travel behavior data to uncover relationships, allowing them to predict future travel demand and offer passengers promotional deals based on travel history. For example, I have flown frequently with American Airlines. The airline app records my travel history and occasionally sends recommendations for discounted tickets to similar destinations. Additionally, I receive offers for frequent flyers, such as bonus reward points for specific trips or promotions for airline credit cards. Airlines also analyze factors like onboard services, Wi-Fi, cleanliness, legroom, food quality, flight delays, and cancellations to understand passenger loyalty. This enables them to improve operations and services to retain and attract loyal customers.

Banking and Finance

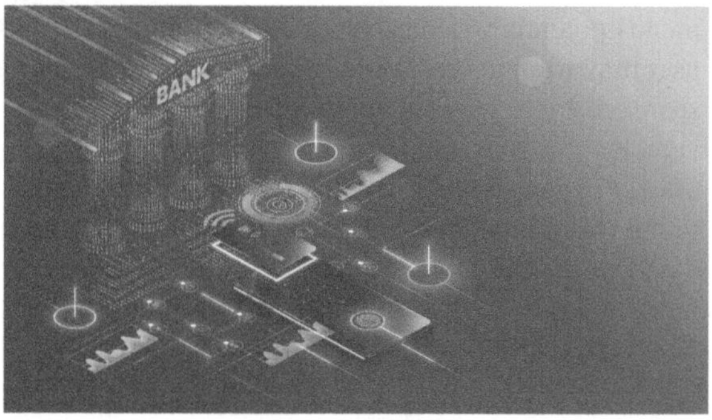

Banks and mortgage companies extensively use data mining to predict customer decisions and assess creditworthiness. Many of us have checking and savings accounts with banks, and we may decide to switch to another bank for various reasons. Banks use data mining to predict the likelihood of a customer closing an account, using demographics and banking history. Using these insights, bank managers can offer attractive deals to retain customers. Predictive modeling also helps banks offer appealing credit card deals or financing services. Additionally, mortgage companies use data mining to assess whether borrowers can repay their loans. They evaluate customer creditworthiness based on demographics and credit history to decide whether to lend, the loan amount, and the term. Data mining plays a key role in minimizing lenders' risks.

Healthcare

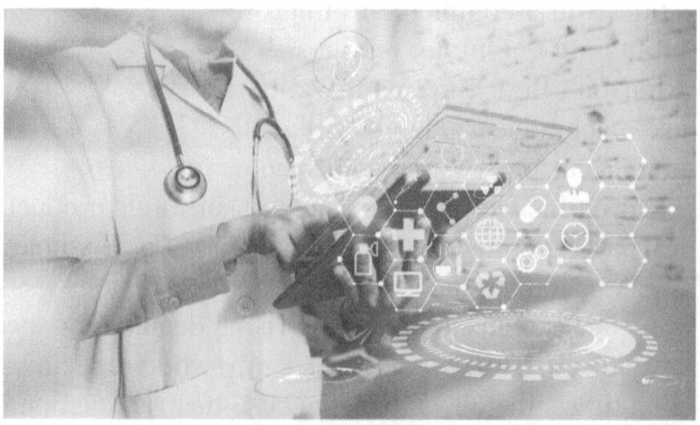

Data mining is vital in healthcare, helping providers improve service quality. Hospitals and doctors' offices collect patients' demographic information, health history, and treatment data. Predictive models can support patient treatment in several ways. For example, when a new patient arrives with certain symptoms, a doctor can enter the information into the system, which can suggest a preliminary diagnosis and treatment plan. The system can also predict the severity of the condition and suggest suitable treatments based on similar past cases. Furthermore, predictive modeling can determine if an issue is isolated or part of a larger pattern, enabling healthcare professionals to intervene and mitigate risks more effectively.

TRANSPORTATION

Traffic congestion is a significant issue in the United States and other developed countries, leading to delays, increased emissions, and wasted time. A large amount of traffic data is collected in real time, which can be used to predict the probability of traffic jams at specific locations and times. Authorities can use these predictions to reduce congestion and develop long-term transportation strategies, such as building new roads or promoting alternative transportation options. Accidents are also common and may lead to severe injuries or fatalities. Predictive models can be developed to anticipate accidents and their potential timing and location. Authorities can use these predictions to allocate resources effectively, ensuring that emergency medical technicians (EMTs) can reach the scene promptly. Remember, every second counts.

These are just a few examples of how data mining is being applied in practice to empower businesses and organizations to enhance operations, improve profitability, boost customer satisfaction, and ensure safety. There are numerous additional applications across domains such as education, training, manufacturing, engineering, aerospace, social science, and biology. Throughout this book, we will explore specific case studies across various sectors. Depending on your industry, you can determine how data mining can be applied. The reassuring aspect is that the same processes and algorithms can be used to analyze data; the difference lies in how you use the outcomes to make informed business decisions.

REACTIVE, PROACTIVE, AND PREDICTIVE STRATEGIES

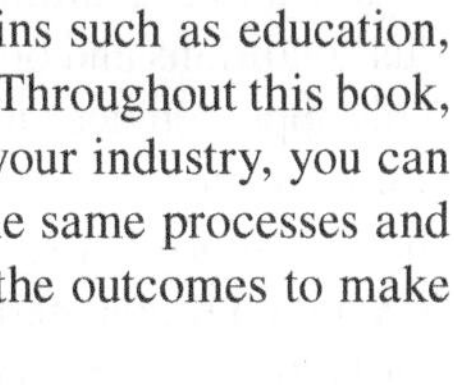

In order to use data mining effectively in businesses, it is essential to understand the primary purpose of data mining in predictive projects. I want to show you how the Federal Aviation Administration (FAA) differentiates between reactive, proactive, and predictive strategies. While it is meant for aviation safety strategies, I find this classification represents very well different types of research projects and where predictive research fits in. The FAA's primary responsibility is to ensure safety in air transportation. In aviation safety, the FAA categorizes three types of safety strategies, which are fundamentally different: reactive, proactive, and predictive (Figure 1.5). Aviation organizations can follow those strategies to enhance their operational safety.

The first type is the **Reactive strategy**, which focuses on examining events that have already happened and identifying the root causes of a specific aviation incident or accident. Based on the identified root causes, aviation organizations can form appropriate strategies to mitigate the risk. This strategy requires experts to investigate the incident very thoroughly, examine the black box,

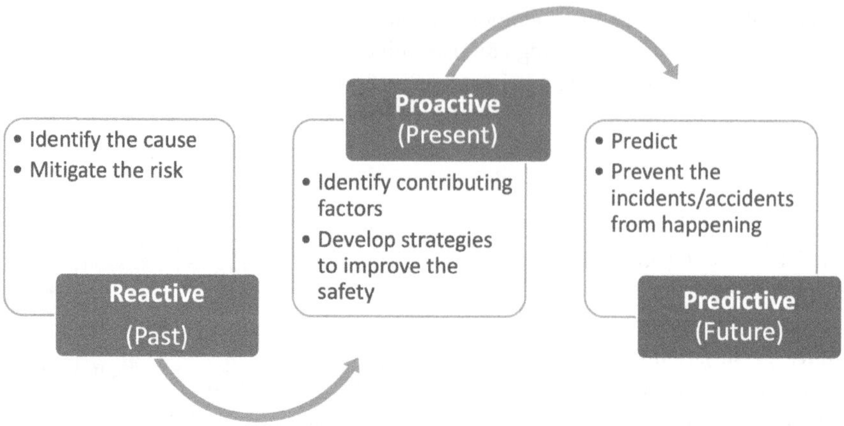

FIGURE 1.5 Reactive, proactive, and predictive safety strategies. (From Federal Aviation Administration.)

interview involved parties, and examine the scene of the incident. Depending on the scope and severity of the incident/accident, the investigation may take a long time to complete. For example, an airline accident may take months or even years to complete an investigation and confirm the causes. This prolonged period is necessary to ensure that every aspect of the incident is carefully examined, and the root causes are accurately identified. Despite the time-intensive nature of this strategy, the outcomes of such investigations are invaluable. The detailed findings provide critical insights that can significantly enhance safety protocols and training programs within the aviation industry. By understanding what went wrong and why, aviation organizations can make informed decisions to prevent similar incidents in the future, thereby improving overall safety standards.

The second type is the **Proactive strategy**, which looks into the present. In other words, aviation organizations examine contributing factors to an incident/accident from various aspects of hazardous conditions and organizational processes and see how they are related to the incident or accident. This approach does not attempt to find the causality of incidents but rather investigates the correlations between various contributing factors and the occurrence of aviation incidents. By understanding these correlations, aviation organizations can implement preventive measures to enhance safety. Several common methods are employed in the Proactive strategy, including statistical analysis, scientific theories, aeronautical scientific models, and engineering mechanisms. These methods help in identifying patterns and correlations between different variables that could contribute to potential hazards. For instance, statistical analysis can reveal trends in data that indicate an increased risk of certain types of incidents under specific conditions. Using these insights, aviation organizations can develop strategies and action plans that target the identified contributing factors. This proactive approach allows for the implementation of measures that can mitigate risks before they escalate into serious incidents. Moreover, organizations can design applications and systems that continuously monitor various parameters to detect potential incidents in real time. With the integration of sensors and data collection technologies, real-time monitoring becomes a vital component of the Proactive strategy. These technologies enable the detection of anomalies and potential incidents as they occur, providing authorities with the necessary information to make timely decisions. Although it may not always be possible to prevent incidents from happening entirely, immediate detection allows for swift action to minimize injuries or damages.

The third type is **Predictive strategy**, which looks into the future. In other words, aviation organizations analyze a large amount of data to predict new potential safety problems in the airspace system in the future. By mining big data, they can develop a predictive model that predicts an aviation incident or accident before it occurs. While the Reactive and Proactive strategies focus on understanding and addressing past and present incidents, respectively, the Predictive strategy aims to prevent incidents by forecasting them. This forward-looking approach does not attempt to explain how an incident will happen beforehand; rather, it concentrates on predicting the occurrence based on data patterns. The detailed explanation of how an incident might unfold can be analyzed later, but the primary goal is to predict what, where, and when an incident might occur, along with its probability. While the Reactive and Proactive strategies focus on understanding and addressing past and present incidents, respectively, the Predictive strategy aims to prevent incidents by forecasting them. This forward-looking approach does not attempt to explain how an incident will happen beforehand; rather, it concentrates on predicting the occurrence based on data patterns. The detailed explanation of how an incident might unfold can be analyzed later, but the primary goal is to predict what, where, and when an incident might occur, along with its probability.

To achieve this, aviation organizations employ data mining techniques to uncover previously unknown patterns and correlations within large datasets. These patterns enable the construction of predictive models that can provide early warnings of potential safety issues. For example, by analyzing trends in flight data, maintenance records, and environmental factors, a predictive model can identify conditions that are likely to lead to an incident. This strategy marks a significant advancement in aviation safety, as it allows for the prevention of incidents rather than merely mitigating their effects. By predicting incidents before they happen, aviation organizations can implement measures to avoid them entirely, thus ensuring a higher level of safety. However, the success of the Predictive strategy relies heavily on the availability of a large amount of high-quality data. Accurate predictions require comprehensive and detailed datasets, which can be obtained from various sources such as flight operations, maintenance logs, weather reports, and air traffic control data. The concept of predictive modeling in aviation can be likened to scenarios depicted in the TV show *Person of Interest*, where data mining is used to predict and prevent crimes before they occur. Similarly, in aviation, data mining and predictive modeling serve as crucial tools for identifying and preventing potential incidents, thereby enhancing the overall safety of the airspace system.

All three strategies are advantageous for businesses and organizations. However, in this book, our focus lies on the predictive strategy. Within business operations, predictive modeling plays a crucial role. It enables us to develop and validate predictive models that accurately predict a predetermined outcome (target variable). Additionally, we can identify significant predictors that contribute to these predictions. Subsequently, by extrapolating potential or forecasted future values of these predictors, we can anticipate outcomes in the future. Furthermore, we have the capability to conduct various what-if scenarios, modifying the values of specific predictors to observe how the outcomes change. Such scenarios empower decision-makers to determine the most effective strategies or decisions to tackle business problems with minimal uncertainty.

DATA MINING AND DATA SCIENCE

In recent years, analytics and broader fields have seen a surge in the popularity of the term **data science**. This surge has led to an increase in job openings for data scientists and the frequent appearance of titles like **data scientist** or **data science expert** on resumes. Despite its widespread use, **data science** often remains a buzzword—commonly spoken yet challenging to define clearly. This ambiguity largely stems from the lack of a universally accepted definition of data science. Before the rise of data science, the industry was buzzing with terms like **data mining, data analytics, business analytics,** and **big data**. It's important to note that in these terms, including **data science**, the emphasis is less on the **data** itself and more on what we do with it. We all know that we need data. The real value lies in the application of the data, not just its existence.

In this section, I aim to clarify the distinctions and connections between **data science** and **data mining**. While numerous comparisons are available online, they often lack consistency and, in some cases, contradict each other, leading to more confusion than clarification. It's worth mentioning that some of these comparisons inaccurately mix the concepts of data mining and data science, further muddying the waters. Therefore, it's crucial to approach these topics with caution and prioritize information from reliable sources.

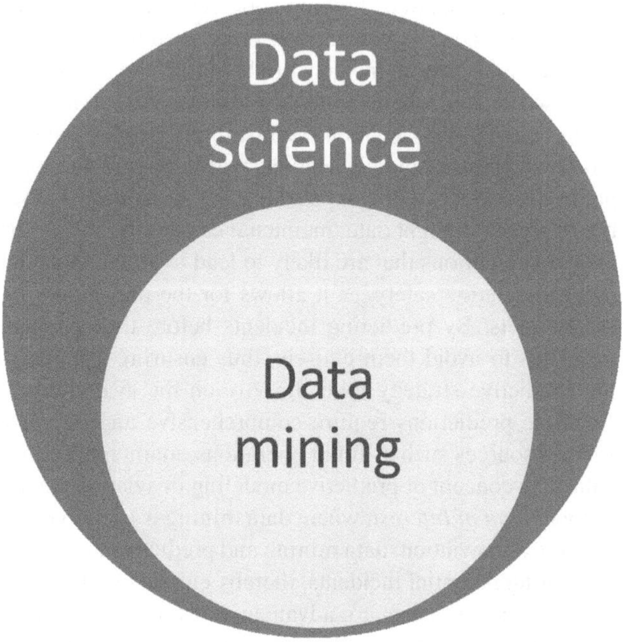

Data science is not a new term. Actually, an official definition of **data science** was introduced back in 1998 by Hayashi (1998). According to the author, data science is a "concept to unify statistics, data analysis and their related methods" in order to "understand and analyze actual phenomena" with data. In 2013, Dr. Vasant Dhar provided a definition of **data science** at a higher level in the Communications of ACM. He defined data science as "the study of the generalizable extraction of knowledge from data" (Dhar, 2013). It uses theories and techniques from various fields, including mathematics, statistics, computer science, and domain knowledge (Cao, 2017). **Data scientist** is a professional who "creates programming codes and combines them with statistical knowledge to create insights from data" (Davenport & Patil, 2012).

Thus, we can say that **data mining** and **data science** serve the same purposes of enabling us to explore and uncover patterns from large data to provide organizations and individuals with newly discovered information so they can make informed decisions. These terms overlap or can even be considered the same in some cases since both involve machine learning, big data, and analytics. A common understanding is that data mining is a method or interdisciplinary subfield, while data science is a discipline. From that perspective, we can consider data science broader than data mining, but data mining is an essential component of data science. Let's put it this way. Data mining covers methods of using machine learning to uncover patterns from large data to predict outcomes. Statistics, mathematics, and other rule-based methods are the core of data mining. On the contrary, data science is a larger and multidisciplinary field, focusing on capturing and extracting knowledge from data and communicating the outcomes. Data science consists of data mining as an essential part of analyzing data and other components such as data collection and management, data treatment, data visualization, computer programming, software engineering, and AI development.

> Data mining is an essential component of data science

Data scientists are often tasked with a variety of responsibilities that can include data collection and preparation, data mining, machine learning, computer programming, and software engineering. Consequently, their role may focus solely on data mining or require multidisciplinaries. If their primary responsibility is data mining, they concentrate on analyzing data and employing various machine learning methods to build predictive models. While computer programming skills can be advantageous, they are not strictly necessary because many data mining applications enable users to perform tasks without writing code (this book, for instance, focuses on machine learning for non-programmers). However, if the job includes additional responsibilities such as big data collection, data warehousing and management, developing an intelligence dashboard, creating complex deep learning algorithms, and deploying predictive models in practical applications, then programming skills become essential. This is why job qualifications and skill requirements for data scientist positions can vary significantly.

It's clear that as we move forward, the definitions of **data mining** and **data science** will continue to evolve and diversify. These fields are dynamic, with new dimensions and applications emerging regularly. To aid in understanding these distinctions, I've included a comparison in Figure 1.6. Since this book is tailored for individuals without a programming background, our focus will predominantly be on the data mining aspect of data science. This means I will cover the processes and methodologies used in machine learning to reveal hidden patterns within large datasets. However, this book will also touch upon essential concepts of data visualization and treatment, offering valuable knowledge for those keen to explore these areas.

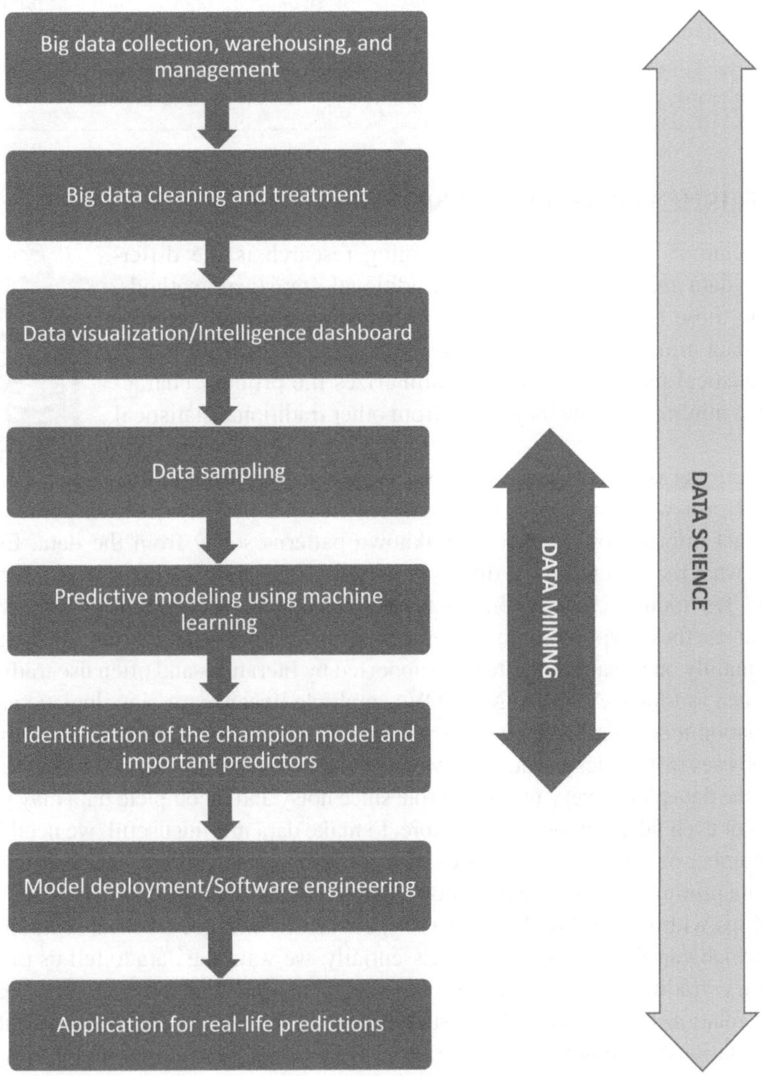

FIGURE 1.6 Data science and data mining.

Therefore, it is accurate to say that this book ventures into the domain of data science. In this context, the term **data scientists** will be used to describe specialists who employ machine learning techniques for data analysis. However, this book will not cover the roles of data scientists involved in data collection, warehousing, and management or those tasked with developing sophisticated visual intelligence dashboards, deploying predictive algorithms, and developing actual software applications. Our emphasis will be on unraveling the mysteries hidden within data through data mining techniques, providing a foundational understanding of how data science can be approached from a non-programmer's perspective.

In data mining, we still use data visualization, but primarily for data exploration purposes. This helps us select the appropriate machine learning algorithms and configurations.

Data visualization, as a component of Data Science, as shown in Figure 1.6, involves the use of advanced visualization tools, such as intelligence dashboards, which can capture and illustrate data patterns, trends, and relationships visually. These tools are equipped to handle real-time data feeds from various sources and can produce a wide array of visualizations, animations, and simulations. The purpose of such advanced visualization is not only exploratory but also to decisively support the decision-making process.

CHARACTERISTICS OF DATA MINING

One of the common confusions in data mining research is the difference between data mining methods and traditional statistical methods. Understanding those differences allows us to better understand the characteristics of data mining methods and know when to use them instead of traditional statistical methods. Table 1.1 summarizes the primary characteristics of data mining and how they differ from other traditional statistical methods.

First, data mining is a data-driven method. We use data mining to discover unknown or new patterns from large datasets. In other words, we use machine learning algorithms to extract information or uncover unknown patterns solely from the data. Essentially, we want to know what the data tell us. It differs from traditional statistical research, which tends to be theory-driven. Traditional research constructs relationships and hypotheses from existing theories, collects data to test these hypotheses, and confirms with evidence whether they are supported. These studies focus mainly on variables of interest, supported by literature, and often use traditional statistical methods such as t-tests, ANOVA, MANOVA, multiple linear regression, logistic regression, and structural equation modeling for testing correlations. In contrast, data mining does not rely on pre-existing hypotheses or theories because we want to find new patterns directly from the data. Hence, the quality of the data plays a very important role since noisy and incomplete data may provide meaningless results or even false outcomes. Therefore, to make data mining useful, we need large datasets with a large number of variables and records. We will discuss the characteristics of big data later.

Second, data mining research is exploratory; that is, we use data mining to explore new variables and relationships within data. We do not try to speculate in advance on what relationships are significant and which variables are important. Essentially, we want the data to tell us these outcomes. On the contrary, traditional statistical research is explanatory and tends to focus on confirming hypothesized relationships. Hence, those studies are limited to pre-identified variables of interest and their relationships based on the literature. A common question in data mining research is, "Do you discover any new variables, relationships, or patterns?" In other words, the findings of data mining research may surprise us, which makes them meaningful in predictive research.

TABLE 1.1
Differences between Data Mining and Traditional Statistical Methods

Items	Data Mining Methods	Traditional Statistical Methods
Objectives	*Data-driven*: The primary goal of data mining is to discover patterns, relationships, and insights in large datasets that may not be readily apparent. It focuses on extracting useful information from data, often with the aim of making predictions or finding hidden patterns.	*Theory-driven*: Traditional statistical methods aim to summarize and analyze data to make inferences about a population. They focus on understanding relationships between variables, testing hypotheses, and estimating parameters with a high degree of confidence. Theories are required to construct hypotheses in those models.
Exploratory vs. explanatory	*Exploratory:* Data mining techniques emphasize exploring large and complex datasets to identify new patterns and relationships. They can handle unstructured and semi-structured data and are capable of handling high-dimensional datasets.	*Explanatory*: Traditional statistical methods typically start with pre-defined hypotheses or research questions and use statistical tests to examine relationships between variables. As a result, they can provide an explanation for those relationships.
Data requirements	Data mining methods require large data for model training and validation. The datasets usually have many rows of data and many variables in different forms and structures.	Traditional statistical methods do not need many variables since they focus on proposed hypotheses. They also require a smaller sample size to work.
Assumptions	Data mining methods typically make fewer assumptions about the underlying data distribution and relationships between variables. They can handle noisy, incomplete, and inconsistent data.	Traditional statistical methods often assume specific data distributions (e.g., normality) and require adherence to certain statistical assumptions, such as linearity, homogeneity, and non-multicollinearity. Violation of these assumptions can affect the validity of results.
Model complexity	Data mining techniques can handle complex models, including decision trees, neural networks, and ensemble methods. They are capable of capturing complex, nonlinear relationships and interactions in the data.	Traditional statistical methods often use simpler models, such as correlations, t-test, ANOVA, linear regression, or logistic regression. These models are typically interpretable and easier to implement.
Interpretation	Data mining techniques prioritize discovering patterns and relationships, but the interpretability of the models can be a challenge. The focus is often on predictive accuracy rather than the explicit understanding of underlying mechanisms.	Traditional statistical methods emphasize interpretability, allowing researchers to understand the relationship between variables and make causal inferences. The emphasis is often on parameter estimation and hypothesis testing.

Of course, to discover new variables and relationships, we cannot limit ourselves to small datasets with limited variables. Therefore, the third characteristic of data mining is large data. Large data means datasets with a large number of variables and records in different forms and structures. Since we do not know what we may find, we need large data. As illustrated in the TV show *Person of Interest*, AI needs access to a huge amount of data, many of which are rather sensitive, to make accurate predictions. This is different from theory-driven, explanatory research, which usually uses smaller datasets with limited variables. When performing statistical analysis techniques such as ANOVA and multiple regression, or even structural equation modeling, we usually use a formula to find the minimum sample size needed to detect relationships at a particular significance level and expected test power. Typically, a sample size between 100 and 300 is sufficient, with models having 3–10 variables. On the contrary, in data mining research, we need to analyze data with hundreds of variables and millions of rows. Since data mining is data-driven, noise and missing values may negatively affect the quality of outcomes. Having larger data helps address these issues.

Due to the size of the data, data mining methods are not subject to strict statistical assumptions and can handle noisy and unstructured data. They usually do not require specific underlying data distributions. Many machine learning algorithms can perform well with missing values and outliers since they are rule-based rather than statistics-based. However, we still need to ensure data quality to obtain meaningful outcomes. The phrase "garbage in, garbage out" still applies here. On the contrary, theory-driven studies tend to use statistical techniques with a smaller sample size, which requires meeting strict statistical assumptions, such as normality, non-multicollinearity, linearity, and homogeneity. These requirements tend to create obstacles in the data analysis process and limit the inclusion of variables in the model.

Given the large data and variety of machine learning algorithms, data mining allows us to build complex predictive models with nonlinear relationships or interactions between variables. Since we do not know what the data may tell us, it is impossible to assume any type of relationship we are dealing with (linear or nonlinear) or how many variables will be in the final model. Advanced machine learning methods, such as decision trees, neural networks, or ensemble modeling, can model the complexity of the data. On the contrary, traditional statistical methods often use simpler models, such as t-tests, ANOVA, linear regression, or logistic regression. They rely on proposed hypotheses and may not be able to capture complex, nonlinear models.

Last but not least, interpretability is a key difference. Data mining can detect complex, nonlinear models with interactions of variables, often considered "black box" models because the relationships of variables are too complex to follow. Note that the model is not literally uninterpretable. Data mining outcomes allow us to examine those relationships using either trees or mathematical formulas. However, it requires a deeper understanding of the scientific background of the method. Note that the focus of data mining is not really to explain the underlying mechanism but rather to achieve accurate predictions. On the contrary, traditional statistical methods focus on explaining the relationship between variables and making causal inferences. The emphasis of these studies is often on parameter estimation and hypothesis testing, which makes their results easier to interpret.

The comparison highlights the strengths of data mining, but it is not without its weaknesses. The most obvious weakness is the tendency to overfit the model to the data, given the size of the data. Overfitting means training a predictive model with a lot of data to achieve very high accuracy. However, the model fits the training sample so well that it is not generalizable to other samples. When this model is run with another dataset, it achieves very low accuracy. Thus, the model is overfitted to the training sample and is not useful in real life since it lacks generalizability. Overfitting can usually be evaluated by partitioning the data randomly into training and validation samples. The training sample is used to train the model, and the validation sample is used to validate it. If the model fit metrics are similar between both samples, we have a reliable model without overfitting issues. We will delve deeper into overfitting in Chapter 6.

Additionally, some machine learning algorithms may require substantial computational power to complete the analysis. More specifically, neural network algorithms often require a very high number of iterations to search for the optimal solution. This search process demands a lot of computation, requiring high computing power and extended running time. An effective search algorithm and a high-power computer are needed to handle this. More importantly, in data mining, the most time-consuming part is data preparation and treatment. Typically, 80% of the time in data mining projects is dedicated to data cleaning and preparation.

These are general characteristics of data mining. When we cover each machine learning method, we will discuss its specific advantages and disadvantages. The rule of thumb is that *there is no perfect machine learning method, but there is the best method for a specific dataset.* While some methods are more complicated or advanced than others, all methods have pros and cons. The best method depends on the data. If there were a

There is no such thing as a perfect machine learning method. But there is the best method for a specific set of data.

perfect method that worked well for all types of data, we could always use that one for all projects. However, the reality is that we don't know which method will work best for our data until we train the model with the data. Hence, the common strategy is to use multiple methods and develop multiple predictive models with various configurations. By comparing these models, we can identify the best predictive model, also known as the champion model. We will discuss this step in detail in Chapter 6.

BIG DATA

I want to provide a quick introduction to big data. First, what is big data? A common-sense answer would be "a lot of data," which is not wrong. The AI in the TV show *Person of Interest* has to collect and process a huge amount of data in different types and formats coming from many sources in real time. Of course, it would require a supercomputer to store and process that amount of data promptly. Predictions would not be meaningful or useful if they took days to make.

> **DEFINITION**
>
> "Big data is a term that describes the large volume of data—both structured and unstructured—that inundates a business on a day-to-day basis. But it's not the amount of data that's important. It's what organizations do with the data that matters. Big data can be analyzed for insights that lead to better decisions and strategic business moves." (SAS)

SAS Institute defines big data as "large volumes of data—both structured and unstructured—that inundate a business on a day-to-day basis. But it's not the amount of data that's important. It's what organizations do with the data that matters. Big data can be analyzed for insights that lead to better decisions and strategic business moves."

Thus, big data is not only about the amount of data but also about what we do with it. Essentially, big data can be characterized by five dimensions: volume, velocity, variety, variability, and veracity (Figure 1.7).

- **VOLUME** pertains to the quantity of data, encompassing various sources such as business transactions, smart devices, industrial equipment, videos, social media, and more. Historically, managing such data would have posed challenges, but advancements in cloud computing and the availability of cost-effective storage solutions like data lakes and Hadoop have alleviated this burden.
- **VELOCITY** relates to the speed at which data are generated and transmitted. With the proliferation of the Internet of Things (IoTs), data stream into servers at an unprecedented rate, necessitating timely processing. RFID tags, sensors, and smart meters are driving the demand for handling these torrents of data in near real time.
- **VARIETY** refers to the diverse nature of data, which spans a wide array of formats. This encompasses structured, numeric data typically found in traditional databases, as well as unstructured data such as text documents, emails, videos, audio recordings, and financial transactions.

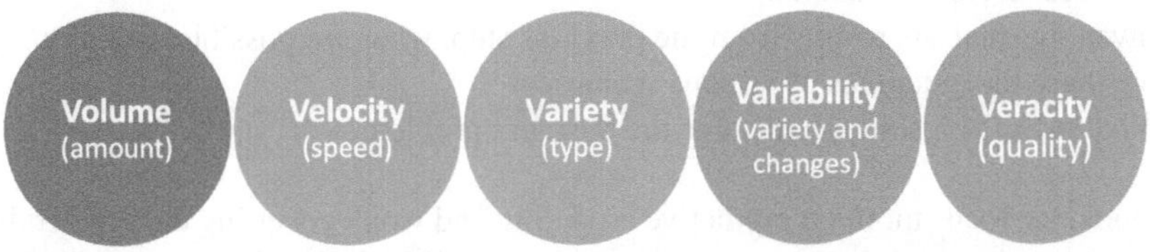

FIGURE 1.7 Big data dimensions.

FIGURE 1.8 Big data implementation process.

- **VARIABILITY** pertains to the diversity and fluctuations in data flows. In addition to the increasing speed and variety of data, data flows exhibit unpredictability due to frequent changes and significant variations. Businesses must stay attuned to trends in social media and effectively manage daily, seasonal, and event-triggered surges in data loads.
- **VERACITY** denotes data quality. Given the multitude of data sources, linking, matching, cleansing, and transforming data across systems can be challenging. Businesses must establish connections and discern relationships, hierarchies, and multiple data linkages; otherwise, their data may quickly become unmanageable.

Now we understand more about the complexity of big data. It is no longer just a very large dataset; it is a business framework for success. So how do we implement big data analytics? We need to follow the five steps presented in Figure 1.8. For each step, I list important questions we need to find answers to.

1. **Set a big data strategy.**
   - What is our goal?
   - What do we want to achieve with big data?
2. **Identify big data sources.**
   - What data are available?
   - Where do we get the data?
   - Do we have permission to access and use the data?
   - What data can we use for business purposes, and what restrictions do we have to follow?
3. **Access, manage, and store the data.**
   - How do we collect that huge amount of data from different sources?
   - Where do we store them?
   - Do we have enough infrastructure and computing power to categorize and store the data?
   - What equipment and protocols are needed to ensure the security of the data?
   - How do we access the data as needed without delay, lagging, or disruption?
4. **Analyze big data.**
   - What are our purposes?
   - What dataset do we analyze?
   - What variables do we use? How do we prepare and clean the data?
   - What analytic methods should we use?
   - How do we interpret the results?
5. **Make data-driven decisions.**
   - Given the analysis results from the previous step, what are possible solutions?
   - What are the outcomes of various scenarios?
   - What is the best solution to solve the business problem?

In this book, we focus more on predictive modeling and won't cover big data in detail. However, it is important to understand the concept and importance of big data and big data analytics. After all, we live in a world of data and digitization. Let me take you back to the TV show *Person of Interest* that I introduced at the beginning. The Machine possesses big data that streams to it in real time. It processes all this data to make predictions, which help save people's lives and, sometimes, save the world. This is certainly not an easy task.

In conclusion, we live in a world with an overwhelming amount of data. Some may say that too much data has created chaos in our society. But we cannot change this fact. We have to move forward and make the best of it. I believe that if we can find patterns and order in this chaos, we will succeed.

ETHICAL CONSIDERATION

When we work with data, especially a large amount of it, we naturally have to consider ethical questions. Data often includes personal, private, and sensitive information. While this information can help us develop good predictive algorithms, if it is leaked, it can harm those individuals. Accordingly, before we conduct data mining research, we must evaluate and address any ethical concerns to ensure no harm comes to human subjects.

ETHICAL PRINCIPLES AND DATA COLLECTION

Ethical consideration is an important topic in data mining because we must ensure compliance with requirements and restrictions regarding data access to protect human subjects and avoid violating privacy. Essentially, ethical considerations refer to the principles, guidelines, and moral values that govern the responsible and respectful use of data throughout the entire data mining process. According to the Belmont Report (U.S. Department of Health & Human Service, 1979), there are three basic ethical principles that all researchers or investigators must follow: (1) respect for persons, (2) beneficence, and (3) justice.

First, respect for persons means treating individuals as autonomous agents, and persons with diminished autonomy are entitled to protection. Second, beneficent actions ensure that the research does not harm persons and maximizes possible benefits while minimizing possible harms. Finally, justice ensures "fairness in distribution" to all participants. All researchers and investigators must comply with these ethical principles in their research projects.

Acquiring data is challenging, as few organizations are willing to share their data with the public. Some websites provide example datasets for practice purposes, but these could be fictional or have been de-identified, modified, or even reduced, making them unsuitable for research purposes. They may also lack validity or sufficient information on the variables, timeframe, or context needed to make a project meaningful.

There are limited organizations that provide de-identified data for public use, such as NASA, FAA, NTSB, BTS, the U.S. Census Bureau, and the Department of Education. These datasets have been carefully scrubbed to ensure no personal information or identification is revealed. Although these databases are valid and can be used for research, they are limited and may not provide the data needed for specific projects. An advantage of these publicly accessible data sources is that no permission or condition is required to use the data. Since the data have been cleaned and de-identified, researchers are allowed to publish their results on the Internet or in peer-reviewed journals without ethical concerns.

Private organizations usually have current and relevant data collected for business purposes, such as airlines, manufacturers, retailers, e-commerce companies, hotels, hospitals, social network providers, media providers, banks, and real estate companies. However, they consider the data proprietary and do not want to publish it for public use. They typically have in-house data analytics teams to analyze the data for business insights. Sometimes, these organizations may agree to share data for research purposes under restricted conditions, often through research grants or contracts.

In such cases, researchers need permission from the organization to access the data under the terms established by the organization. This permission is formalized through a non-disclosure agreement (NDA), which all parties should agree to and sign before sharing the data. Researchers must ensure they have permission before accessing and collecting the data and carefully review the terms and conditions before accepting them. Permission may include an expiration date for data access, the project's purpose, expectations for the final report submission, and whether the findings can be published in a peer-reviewed journal. Usually, permission to publish comes with restrictions, such as thorough de-identification of the data and non-disclosure of the data provider's identity in the report.

In some cases, government agencies or private parties agree to provide non-public data for research purposes but require the researcher to hold the publication of the results for a certain period. For example, a funding organization might provide the necessary data, which may contain time-sensitive information that should be kept from the public initially but can be publicly available after two years. Hence, the organization may require the researcher to hold the publication of the results for two years. After this expiration date, the researcher is free to publish the results, provided that the paper is approved by the data provider. This requirement is usually included in the contract or NDA signed by both the organization and the researcher.

> Ethics are very important. We cannot proceed with our research until we ensure that there is no harm to individuals.

When working with datasets provided by these organizations, we need to adhere to the three basic ethical principles mentioned earlier. We should begin by performing a systematic assessment of risks and benefits to the organization and individuals involved in the data. By understanding potential risks and harms to these parties, we can devise an appropriate action plan to avoid any harm. Revealing sensitive and private information can harm individuals and organizations alike. Additionally, misinformation or misinterpretation of the results poses a similar level of risk. Once we complete the risk and benefit analysis, we develop an action plan to ensure no harm is done to individuals or organizations.

Institutional Research Board (IRB) and the Review Process

An Institutional Research Board (IRB), alternatively known as an independent ethics committee (IEC), ethical review board (ERB), or research ethics board (REB), constitutes a committee tasked with the application of research ethics. Its role primarily involves scrutinizing proposed research methods to ascertain their ethical validity. These boards are officially authorized to either grant approval or withhold it, oversee, and evaluate research that involves human subjects. Frequently, they engage in a thorough risk-benefit analysis to determine the ethical suitability of conducting the research. The primary objective of the IRB is to ensure the implementation of necessary measures for safeguarding the rights and well-being of individuals participating as subjects in a research study.

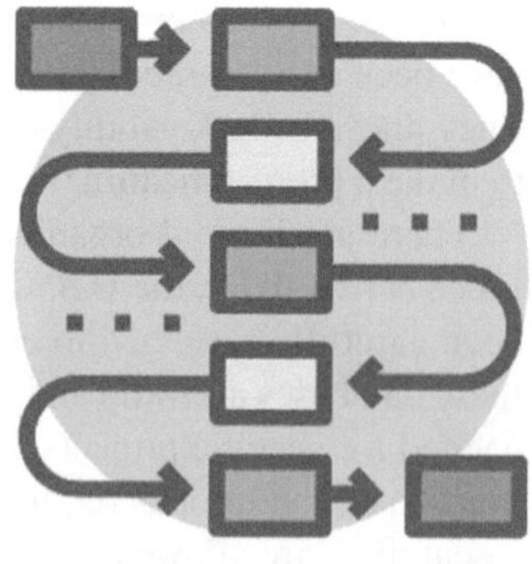

When a researcher conducts a study involving human subjects, they need to submit an application and get approved by two different IRBs: one at their own institution and another at the data provider's organization. For example, if I conduct a study involving individuals at an airline, I will need to submit applications to two different IRBs: one at the university where I work and another at the airline (which may use a different name than IRB). Both IRBs will review my application, and upon approval by both, I can proceed with my research.

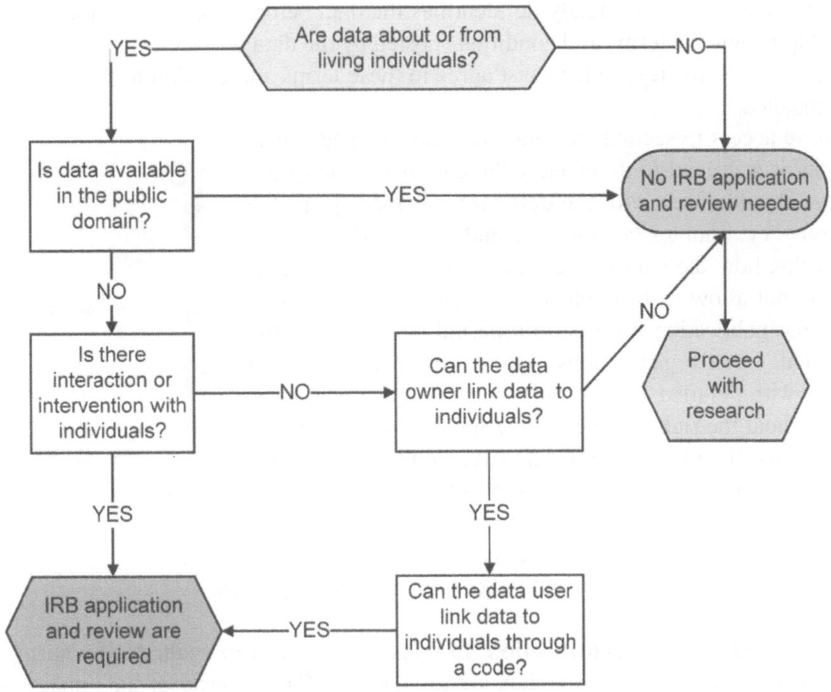

FIGURE 1.9 IRB decision process. (Adapted from the National Institute of Health, 2019.)

Each organization has different rules, processes, and requirements for IRB applications. The typical process includes the following steps:

1. **Application submission**: The researcher completes an application for IRB review, detailing the research purposes, methodology, potential use of human subjects, data collection process, experimental design, questionnaire, potential harms to human subjects, and efforts to avoid those harms.
2. **Initial review**: The IRB coordinator reviews the application and decides whether it should be reviewed by an IRB committee. If not, the researcher is informed of the exempt decision and can proceed with the research. If yes, an IRB committee, including experts in the field, will be formed.
3. **Committee review**: The IRB committee reviews, discusses, and votes on the application. They may request additional information or clarification from the researcher.
4. **Decision communication**: The IRB committee's decision is communicated to the researcher. The decision may include several options: exempt, approved, approved with conditions, or declined (resubmission required). Based on the decision, the researcher can either proceed with the research or must make changes and resubmit.

Figure 1.9 presents a diagram that explains the common IRB decision process, helping researchers determine whether IRB application and review are needed.

How to De-Identify Data?

Data de-identification is the process of removing or altering identifying information from a dataset to protect individuals' privacy and anonymity. It is typically unnecessary for publicly available data since the data provider already de-identifies the data before posting it online. Nonetheless, it is essential to review the terms and conditions posted by the data provider regarding how to use the data. In many cases, the researcher must agree to these terms and conditions before being granted access to the data.

If we have access to secondary data with sensitive and private information, it is crucial to de-identify the data before analyzing it. First, we should check with the data provider for any specific information they want to be removed and any conditions they have regarding how the data is used. For example, some organizations may not allow us to publish the findings or may require us to hold the publication for a certain period before submitting it to a journal. Other organizations may require us to submit the final report with detailed findings for their review, while others may want to hold the rights to the reports for various reasons. All these conditions should be discussed and agreed upon in advance. Additionally, any relevant privacy rules must be reviewed and discussed. Once we have a complete list of information that must be removed from the data, we can proceed with de-identification.

The de-identification process usually includes the following steps.

1. **Identify risks for re-identification**: In this step, we need to evaluate any harm to individuals or organizations if the data are re-identified. The severity of the impacts must be reviewed and recorded for risk assessment purposes. Additionally, we must assess the possibility that someone may be able to re-identify the data based on the information in the de-identified dataset. This means that sometimes the de-identified data may still contain some information that allows others to guess the hidden information, which is a serious issue.
2. **Develop and apply de-identification measures**: Once we know the risks of re-identification, we can develop and apply de-identification measures. The common approach is to remove personal identification and sensitive information or replace it with anonymous information or a temporary ID. While this sounds simple, we need to develop a clear and detailed protocol to avoid missing anything. The protocol should include the variables that should be removed or changed, the values used for replacement, and the location of those variables and data in the dataset. Since we are dealing with large datasets, some macros or programs can be written to automate the de-identification process to save time and avoid missing important data.
3. **Examine the data again**: Once we de-identify the data, it is vital to examine the new dataset to ensure the data are properly and completely de-identified. Since we are the ones who de-identified the data, there may be bias or oversights. Therefore, it is recommended to use external reviewers to examine the data. As they review the data for the first time, they are objective and can detect any issues we may have missed. This step is necessary to ensure the validity of the data de-identification.
4. **Document the method**: It is important to document the de-identification methods and processes carefully. We also need to document how we archive the original data and de-identified data and how we ensure the safety of the original data. This documentation can be used as evidence that we have followed a proper protocol to minimize any harm to human subjects.
5. **Keep the codes and guidelines for re-identification by authorized parties**: These codes and guidelines must be kept confidential and shared only with authorized parties, as they allow for the re-identification of the data.

These steps are essential for ensuring that the de-identification process effectively protects the privacy and anonymity of individuals involved in the data. It's always a good practice to stay updated with the latest guidelines and techniques in data privacy to enhance your de-identification protocols.

HOW IS THIS BOOK STRUCTURED?

In order to help you follow the chapters in this book easily, I present the general structure of each chapter in Table 1.2. Each chapter consists of multiple sections, details, and examples. Some information may be complicated or advanced, especially if you are new to data mining concepts. By understanding the general structure of these chapters, you should be able to navigate the introduction, details, guidelines, software references, and examples. I use this structure for most chapters, particularly those focusing on machine learning methods.

TABLE 1.2
General Structure of the Chapters

Sections	Description
Learning outcomes	A set of outcomes you are expected to learn and implement after completing the chapter.
Keywords	Primary terms used in the chapter.
Introduction	A brief introduction of the machine learning method and any important thing you need to pay attention to when reading the chapter.
Definitions of the method	Clear definitions of the method and related important terms.
Principle	The principle of the method is explained in detail to allow you to gain a fundamental understanding of the method and how it works.
Practical applications	Examples of how the method is applied in various domains.
Basic concepts	Basic concepts in the method and descriptions. These terms are used frequently in the chapter. You can refer to these concepts as you go through the chapter if you need a reminder of what they mean.
Process	The steps you need to follow to perform the method.
Advantages and disadvantages	Advantages and disadvantages of the method in comparison with other machine learning methods.
Detailed components	Detailed components of the method. I use figures, formulas, functions, and rules to explain them. These details help you understand how the method works and how to configure the model. The length of this section depends on the details provided in the chapter. Chapters covering important methods, including regression, decision trees, and neural networks, consist of more details. Chapters covering advanced methods, including random forest, gradient boosting, and Bayesian networks, are shorter because I want to avoid the complicated mathematical foundations of those methods. Instead, I focus on the analysis process, key takeaways, and common configurations.
Examples of predictive modeling	Specific examples of how to use the method with an actual dataset. The examples provide detailed steps with screenshots for property settings, how to use the software, explanations of decisions you can make, how to get the outcomes, which outcomes are important, and how to understand and interpret the results. This section is the most important section in each chapter, as it allows you to follow the steps and practice with the software and the data. In order to make it easy for you to follow through multiple steps of the data mining process, I use mainly two datasets throughout this book: the **Airline Recommendation** dataset and the **Boston Housing** dataset. Thus, you can see the whole picture and process of analyzing one big dataset using data mining.
Chapter summary	A summary of the method with key concepts and takeaways.
Chapter discussion questions	Important discussion questions covering all information and required content for the method. By answering those questions, you can test your understanding of the method in this chapter.
Exercises	Useful exercises that allow you to practice your understanding and skills in predictive modeling. Make sure to practice with the examples before doing these exercises. Following the steps in the examples should allow you to complete those exercises successfully.

Writing Tone and Style

In my research papers, I typically adopt a scholarly writing style, which is academic-focused, objective, and evidence-based. Each statement is meticulously supported by references through proper citations. Moreover, passive voice is frequently used to maintain a formal and authoritative tone.

However, for the purpose of this book, I have adopted a different writing strategy that primarily employs active voice to connect with readers and facilitate their understanding of the concepts and processes. In the initial seven chapters, I adopt a casual writing tone to infuse the topics with a sense of fun and engagement. By incorporating more real-life examples, I aim to explain the principles and concepts of machine learning in an accessible manner. Recognizing that machine learning can appear daunting, especially with its scientific terminologies and mathematical foundations, I assume that most readers do not possess a strong background in mathematics or computer science, and many may be non-programmers. Therefore, I strive to commence this complex subject with relatable statements, practical examples, and my personal experiences, all geared toward fostering interest and curiosity in the subject. Additionally, I have incorporated figures, tables, definition boxes, tips, and warnings to help readers follow the concepts and steps seamlessly.

As readers become more accustomed to terminologies, concepts, and processes, the writing tone and style will gradually shift toward a more scholarly approach starting from Chapter 8 onwards. The intention behind this transition is to foster the readers' comfort in reading and writing quality reports. By this book's conclusion, the readers should feel proficient in reading and writing various report types, including business reports, technical reports, and academic reports. It is essential to note that, as this book is not a research project or research reference book, I do not provide citations in the chapter content, as you regularly see in published articles, to avoid distraction while reading. However, a comprehensive list of references that served as the foundation for this book is available at the end, enabling readers to explore additional sources if desired.

Finally, in writing this book, I draw upon my extensive expertise and experience in teaching data mining in my classes and using machine learning in my research. As a result, some guidelines and rules of thumb are based on my perspectives and practical experience with these methods. It is important to recognize that the world of machine learning is dynamic, and information and guidelines may not always align across different books and sources. This variation arises from the fact that different authors possess their unique perspectives and experiences in using machine learning techniques. In several aspects of machine learning, particularly in model performance evaluation, there are no standardized cut-off values or thresholds. In this book, I provide various rules of thumb based on my experience with the methods, which have also been employed in my published articles, indicating acceptance within the academic community. Moreover, I endeavor to maintain consistency in terminologies, concepts, and steps throughout this book, facilitating a smoother understanding of the content for readers.

CHAPTER SUMMARY

Data science and data mining represent emerging fields of inquiry that empower us to analyze large datasets and use the findings to inform strategic business decisions. This chapter serves as a foundational exploration, introducing fundamental concepts within data science and data mining. We cover the evolutionary trajectory of AI and machine learning, examining their transformative impact. Furthermore, we explore how data mining applications are harnessed across various industries to strengthen safety measures and optimize operational efficiencies. We also look into the characteristics of big data and the ethical considerations related to data mining tasks. These fundamental concepts serve as essential prerequisites, laying the groundwork for a comprehensive understanding of the data mining process and machine learning methods.

Unlike traditional statistical analyses, which often rely on theoretical frameworks and limited sample sizes, data mining projects adopt an exploratory, data-driven approach, drawing insights from extensive datasets. Our objective is to discover unknown patterns concealed within the data. In other words, we have collected a large amount of data for our business, and we need to find out what useful insights the data can tell us about our business operations. The patterns we uncover from the data allow us to make informed decisions to improve our business operations.

Moreover, this chapter highlights the importance of the ethical considerations in data mining. Data mining involves the extraction of patterns, knowledge, and insights from large datasets, and while it provides valuable information, it also raises significant ethical concerns. Ethical considerations in data mining refer to the principles, guidelines, and moral values that govern the responsible and respectful use of data throughout the entire data mining process. Adhering to these ethical principles and guidelines ensures that data mining is conducted responsibly and with respect for individuals' rights and well-being. Upholding ethical standards not only fosters public trust in data science endeavors but also maximizes the beneficial impact of data-driven insights while mitigating potential harm. This chapter provides detailed steps for IRB application and de-identifying data, which ensures that the privacy and anonymity of individuals are protected throughout the research process.

CHAPTER DISCUSSION QUESTIONS

1. Define data mining and data science. What are the differences between them?
2. What are the primary differences between data mining methods and traditional statistic methods?
3. Describe the characteristics of data mining. What are the major challenges when using data mining?
4. Give a specific example of how data mining can be applied in various domains in real life.
5. Search on the Internet to find information on a business or organization using data mining to enhance their operations. Describe the purpose of their data mining projects, database, analytical methods, and findings.
6. Can machines learn? Why do we call it machine learning?
7. What is big data? What are the five dimensions of big data?
8. Describe the process of implementing big data analytics.
9. Describe the ethical principle in using secondary data. Why is ethical consideration important in data mining projects?
10. What are the main responsibilities of the Institutional Review Board (IRB)? Describe the IRB decision process.
11. Why do we need to de-identify data? Describe the data de-identification steps.

2 Data Mining Processes, Methods, and Software

LEARNING OUTCOMES

Upon completion of this chapter, you will be able to:

1. Define the data mining process.
2. Describe the CRISP-DM (Cross-Industry Standard Process for Data Mining) process, its steps, and when it should be used.
3. Describe the SEMMA (Sample-Explore-Modify-Model-Assess) process, its steps, and when it should be used.
4. Define the target variable and predictors.
5. Evaluate the scales of variables and their significance in data mining.
6. Compare the association model and classification model.
7. Differentiate between supervised learning and unsupervised learning.
8. Classify machine learning methods based on the target variable and study objectives.
9. List the primary characteristics of key machine learning methods.
10. Select the appropriate machine-learning method for a data mining project.
11. Compare different data mining software programs.

INTRODUCTION

In this chapter, we focus on the data mining process, exploring various machine learning methods, and explaining the art of selecting the most suitable method for our study purpose. Additionally, we explore various software options tailored for non-programmers, ensuring accessibility and ease of use in our data mining endeavors.

Throughout this journey, I introduce a number of data mining and machine learning terminologies. Do not worry if you find yourself confused or unable to grasp the concepts immediately. While these terminologies may initially appear unfamiliar, they will become increasingly familiar as you encounter them repeatedly throughout subsequent chapters. With time and exposure, you will gradually acquire mastery over these terms, gaining confidence and fluency in their usage. Trust in the learning process and rest assured that with persistence and engagement, clarity and comprehension will follow.

Recognizing the potential confusion caused by inconsistent terminology usage across different sources, I aim to provide clear and consistent definitions for the terms used throughout this book. This chapter serves as a reliable reference point for revisiting these definitions whenever needed, fostering a cohesive understanding of the concepts under discussion.

As the field of data mining evolves, it's common to encounter disparities in terminologies across various sources. However, I am confident that the terms I have chosen are used accurately and

DOI: 10.1201/9781003468196-3

consistently throughout this book. This consistency enhances your understanding of the methods, analysis processes, and results discussed in the forthcoming chapters, minimizing confusion and assuring clarity. Imagine the confusion that would arise if different terms were used interchangeably to denote the same concept—a scenario often encountered when we search online resources on data mining methods. To avoid such confusion, I adhere to a standardized terminology framework, ensuring cohesion and coherence of the content. Additionally, I provide references for equivalent terms, allowing you to seamlessly navigate other authors' studies. This approach not only enriches your comprehension but also empowers you to confidently navigate the complex landscape of data mining literature.

Furthermore, I compare various software options, focusing on those conducive to non-programmers. Contrary to popular belief, the landscape of data mining software is evolving, with an increasing range of programs designed to enable users to implement data mining and apply machine learning algorithms without requiring coding expertise. These user-friendly programs make data mining accessible to individuals from diverse backgrounds. Data mining and machine learning are no longer fields solely for computer engineering majors. While each software option has its own set of strengths and weaknesses, I provide the rationale for selecting SPSS Modeler as the primary tool for this book, while also offering insights into alternative options. By the conclusion of this book, I am confident that you will have refined your data mining skills using SPSS Modeler, equipped with the knowledge and proficiency to complete data mining projects effectively.

DATA MINING PROCESSES

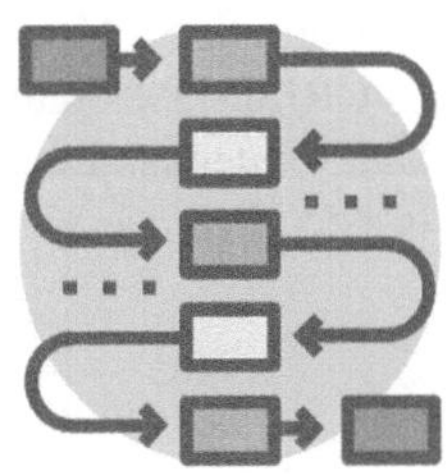

In Chapter 1, we defined data mining and discussed its primary characteristics. It is an exploratory, data-driven method to detect unknown patterns from large datasets and provide insights that enable informed business decisions. To properly perform data mining, we need to follow a process that guides us through specific steps to explore and analyze the data accurately. Since large datasets tend to be noisy, the analysis outcome may be affected by this noise, leading to less accurate results and potentially wrong decisions. In the business world, mispredictions can have substantial consequences. Imagine an airline sending a promotion package to the wrong passenger and losing another passenger, a bank executive opening a new branch in an incorrect location and losing money, or a false negative prediction in healthcare delaying a patient's treatment. These incorrect business decisions could be catastrophic for both the business and its customers.

We all know that any prediction has a margin of error. Our goal is to keep this margin within an acceptable range, which varies by organization and context. Depending on the organization's strategies, this could be 1%, 5%, or even 10%. For instance, in the service industry with high demand fluctuations, executives may accept a 10% error. On the other hand, in healthcare or aviation safety, a 1% error might be the maximum acceptable threshold due to the significant impact on outcomes.

So, how can we avoid or at least minimize such problems? The best practice is to follow a valid data mining process when analyzing the data. A good process guides us through essential steps to ensure the analysis is conducted correctly, avoiding common data mining mistakes. On the Internet or in other books, you may find various steps for the data mining process. Many of these are based on the authors' opinions and experiences, and some can be useful, especially for specific project types. However, in some cases, we might question whether such a process has been validated and can be generalized to other types of projects. Using an invalid process may result in questionable outcomes.

As stated in Chapter 1, data mining is an exploratory, data-driven method. This means the study's outcome is dependent on the quality of the data and the analysis process. It is imperative to conduct the analysis correctly to produce quality outcomes. Following incorrect steps could lead to invalid results, which, in turn, will lead to wrong business decisions. For example, if we train a model without first cleaning the data, the noise in the data might lead to messy outcomes that are neither interpretable nor meaningful. The prediction outcomes may be erratic or contradictory, making them useless for decision-making. In another scenario, we might overtrain the model with a lot of data, achieving high accuracy. However, without validating the model with a separate sample, it might not generalize to other datasets due to overfitting. When applied to new datasets, such a model may produce predictions with a high margin of error, rendering it invalid and unusable in the real world. Consequently, decision-makers might be misinformed and make wrong decisions based on these incorrect predictions. Therefore, it is crucial to follow a valid data mining process to ensure the quality of the prediction outcomes.

> A valid data mining process plays a vital role in producing an effective and usable predictive model.

In this book, I will introduce data mining processes that are validated, widely accepted, and used in both academia and industry. These include Cross-Industry Standard Process for Data Mining (CRISP-DM) and Sample-Explore-Modify-Model-Assess (SEMMA).

CRISP-DM PROCESS

CRISP-DM, which stands for Cross-Industry Standard Process for Data Mining, is based on the European Strategic Program on Research in Information Technology initiative (ESPRIT). It is more popular in the industry than in academia, and IBM has adopted and implemented this process in its software SPSS Modeler.

CRISP-DM breaks the process of data mining into six major steps: Business Understanding, Data Understanding, Data Preparation, Modeling, Evaluation, and Deployment (Figure 2.1). Below is a detailed overview of each step in the CRISP-DM process:

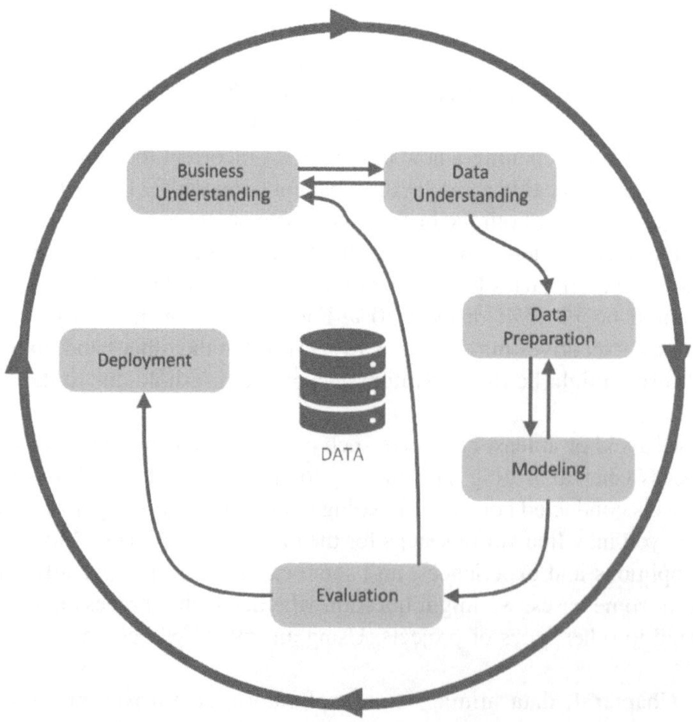

FIGURE 2.1 CRISP-DM process.

Business Understanding

- **Define the business problem or objective**: Clearly articulate the business goal you want to achieve through data mining.
- **Translate the business goal into a data mining problem**: Understand how the business goal can be addressed using data and analytics.
- **Identify success criteria**: Define measurable metrics that indicate the success of the project.

Data Understanding

- **Collect initial data**: Gather relevant data from various sources, including databases, files, and external systems.
- **Describe data**: Document the structure and content of the collected data, including variables, formats, and data quality issues.
- **Explore data**: Perform initial exploratory data analysis to identify patterns, trends, and potential issues.
- **Verify data quality**: Assess the quality, accuracy, and completeness of the data.

Data Preparation

- **Clean and preprocess data**: Address missing values, outliers, and inconsistencies in the data. Transform and format data for analysis.
- **Integrate data**: Combine data from different sources into a unified dataset.
- **Select relevant variables**: Choose the variables (features) that are most relevant to the analysis and discard unnecessary ones.
- **Create derived variables**: Generate new variables or features that might enhance the predictive power of the model.

Modeling

- **Select modeling techniques**: Choose appropriate algorithms or models based on the nature of the problem (classification, regression, clustering, etc.).
- **Build models**: Apply the selected algorithms to the prepared dataset to create predictive or descriptive models.
- **Assess model quality**: Evaluate the performance of the models using metrics such as accuracy, precision, recall, etc.
- **Fine-tune models**: Adjust model parameters to optimize performance or prevent overfitting.

Evaluation

- **Evaluate model performance**: Assess how well the models perform on unseen data using validation techniques (e.g., cross-validation).
- **Compare models**: Compare the performance of different models to select the best one for the business problem.
- **Review results with stakeholders**: Present the findings and insights to stakeholders and gather their feedback.

Deployment

- **Plan deployment**: Create a strategy for deploying the selected model in a real-world business environment.
- **Implement the model**: Integrate the model into the operational systems and processes.
- **Monitor model performance**: Continuously monitor the model's performance and update it as needed to maintain accuracy and relevance.
- **Document the process**: Document the entire data mining process, including decisions, methods, and outcomes.

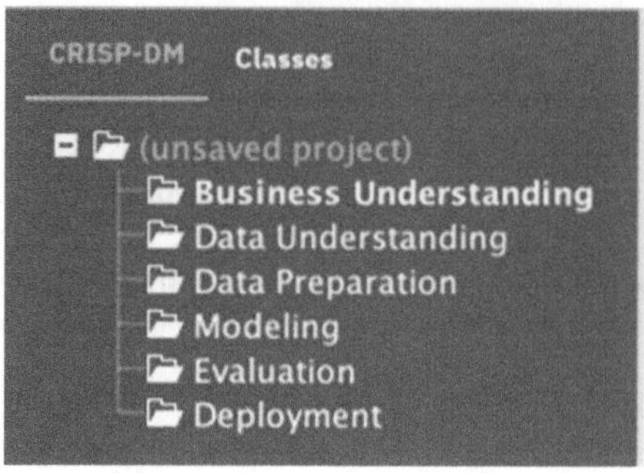

FIGURE 2.2 CRISP-DM view in SPSS Modeler.

FIGURE 2.3 Nodes pallet in SPSS Modeler.

Note that the sequence of the steps is not linear. Actually, the process requires moving back and forth between specific steps. The process diagram employs arrows to signify the primary and recurring interconnections between phases. The outer circle within the diagram serves as a representation of the iterative aspect inherent to data mining. The data mining process continues even after the deployment of a solution. Insights gained throughout this process have the potential to instigate fresh business inquiries, and future data mining endeavors stand to gain from the insights accumulated in prior iterations.

As we can see, CRISP-DM is very practical-oriented, which is why practitioners widely use it in the industry for real-life business problems. It should also be noted that CRISP-DM is built into the SPSS Modeler software. It is essentially an extension of the standard SPSS Modeler project tool (Figure 2.2). In fact, you can toggle between the CRISP-DM view and the standard Classes view to see your streams and output organized by type or by phases of CRISP-DM.

However, the nodes palette, which contains all data and modeling tools in the software, is not organized based on CRISP-DM. Specifically, the nodes palette organizes nodes into several groups, including Sources, Record Ops, Field Ops, Graphs, Modeling, Output, Export, etc. (Figure 2.3). These groups are not aligned with the CRISP-DM, meaning we cannot follow the CRISP-DM steps directly when looking for nodes in SPSS Modeler to build our models. This misalignment may cause further confusion for users.

SEMMA PROCESS

Another popular data mining process is SEMMA, which stands for Sample-Explore-Modify-Model-Assess. This process is proposed by SAS Institute and has been widely used in many industries, particularly by academic researchers. SEMMA is integrated into SAS Enterprise Miner software to guide users in conducting data mining. It provides a sequence of steps that should be followed to analyze large datasets and uncover new patterns (Figure 2.4). Some of these steps are very similar to those in CRISP-DM. Specifically, Data Understanding is similar to Explore, Data Preparation is similar to Modify, Modeling is similar to Model, and Evaluation is similar to Assess. However, two steps in CRISP-DM that differ from SEMMA are Business Understanding and Deployment.

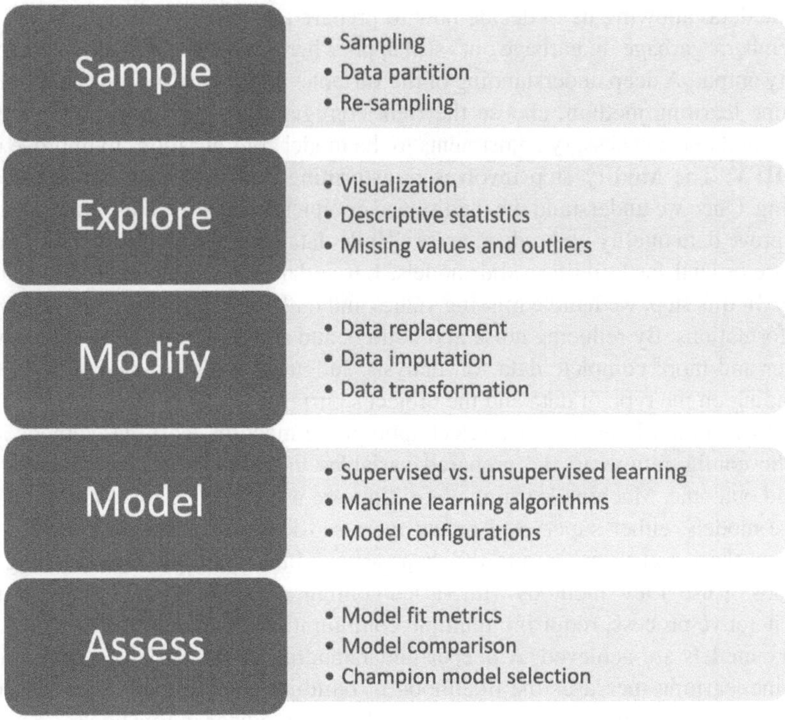

FIGURE 2.4 SEMMA process.

The following explains each step in the SEMMA process:

- **SAMPLE**: In the Sample step, we start with data sampling, selecting a portion of the whole dataset (sample) for analysis purposes. The sample should be large enough to contain sufficient information yet small enough to be used efficiently. If the original dataset is too large, we can decide to sample a part of it. For example, if we have a dataset with ten million data points, using the whole dataset may increase computing time. For data mining purposes, 10% of it, providing us with one million records, should be sufficient. However, if we do not sample correctly, our sample may be skewed from the original data, leading to sampling biases. In other words, our sample would not be representative of the full dataset, rendering the data mining results invalid. Therefore, we need to choose the right sampling method to minimize sampling biases and ensure a quality dataset for data mining.

  Additionally, this phase involves data partitioning, where the data are split into subsamples. The training sample is used to train the model, and the validation sample is used to validate the model. In some cases, a test sample may be used to test the final model. If we have an imbalanced dataset, our model may suffer from overfitting. In such cases, re-sampling may be needed to ensure the balance of the dataset. Details of different sampling methods will be covered later in Chapter 3.
- **EXPLORE**: In the Explore step, we explore and understand the data by discovering anticipated and unanticipated relationships between the variables and identifying abnormalities using data visualization. By using data visualization, we can produce multiple charts in different formats for various variables, gaining a deep understanding of the variables, scales, and values. Features such as descriptive statistics, missing value detection, and variable selection are used in this step. This step is crucial for understanding the variables

and the data, allowing us to decide how to prepare and modify the data before analysis. Remember, "garbage in, garbage out" still applies here. We aim for quality data to produce quality output. A deep understanding of the variables and data also helps us select the right machine learning method, choose the right configurations for the models, interpret the results, and make necessary adjustments to the model configurations to improve model fit.

- **MODIFY**: The Modify step involves transforming and preparing variables for model training. Once we understand the quality and completeness of the data, we can decide how to improve data quality and reduce noise. While data mining can handle noisy data, clean data are crucial for building valid models. Large datasets tend to be less structured and messy. In this step, we handle missing values and outliers and perform any necessary data transformations. By reducing noise and outliers and addressing missing values, we create cleaner and more complete data for analysis. Judgment calls are necessary in this step, depending on the type of data and the project's purpose.
- **MODEL**: In the Model step, we select appropriate machine learning methods and apply specific configurations to the prepared variables to train models that may provide the desired outcome. Machine learning algorithms are used in this step to build multiple predictive models, either supervised or unsupervised learning. Depending on the project's purpose, the target variable, and the data, we can build multiple models using different methods or use a few methods with varied configurations to produce more models. This is an iterative process, requiring multiple configurations and reconfigurations until satisfactory models are achieved. A deeper understanding of the variables, scales, data, and machine learning increases the likelihood of building good models. Detailed configurations for each machine learning method will be covered later in this book.
- **ASSESS**: The last step is Assess, where we evaluate model performance and compare models using appropriate metrics. Model fit metrics are chosen based on the study's purpose and the type of data mining model. Based on the model comparison, we select the champion model, which is the best model. We also examine the model's usefulness and usability and identify important predictors. The champion model, along with important predictors, is recommended to decision-makers to help them make informed decisions on using the tool to improve business operations.

> SEMMA is most appropriate for data mining projects. We use SEMMA in this book.

The SEMMA process provides a comprehensive framework for guiding the steps involved in importing, exploring, and analyzing data to yield high-quality outcomes. By constructing our project diagram in alignment with SEMMA's steps, we can systematically track our progress, document the results of each phase, and ultimately arrive at well-informed conclusions. This structured approach proves invaluable when troubleshooting errors or mispredictions during the modeling process. By retracing our steps and pinpointing where and why discrepancies occurred, we can make the necessary adjustments for improved accuracy and reliability.

While SEMMA is seamlessly integrated into the interface of SAS Enterprise Miner, making it an ideal fit for this software, its applicability extends beyond specific platforms. Widely accepted and utilized across both industry and academia, SEMMA streamlines the data mining process and facilitates logical and coherent reporting of findings. Therefore, we opt to adopt SEMMA as the guiding framework for this book, leveraging its established principles to navigate various data mining projects.

It's important to note that the SEMMA process concludes with the recommendation of the champion model and key predictors to decision-makers. Model deployment, however, falls outside

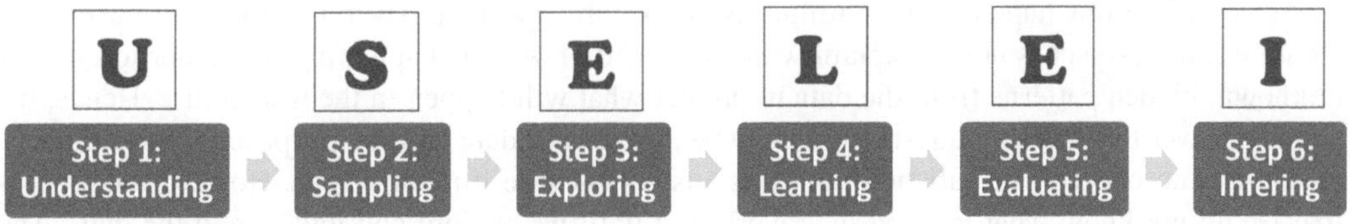

FIGURE 2.5 The USELEI process.

the scope of SEMMA and will be addressed in separate projects, focusing on the practical implementation of the tool and derived insights in real-world programs.

USELEI PROCESS

In some special cases, our business problem may require us to make causal inferences among variables from large datasets. This is common in risk analysis projects where we want to know if a change in one variable may lead to a change in another variable. In such cases, the expected outcome is the probability of the target variable based on the change in one predictor. To discover such a causal network, we should use a specific method, namely Bayesian Networks. At the end of this process, causal inference must be made, assuming we have prior domain knowledge and causal assumptions.

I developed a new process for this type of project since there isn't one in the literature. I named it the Understanding-Sampling-Exploring-Learning-Evaluating-Inferring (USELEI) process (Figure 2.5). This process is particularly useful for causal data mining. I will get into the details of this process and the Bayesian Network method in Chapter 16.

HOW TO CATEGORIZE DATA MINING METHODS?

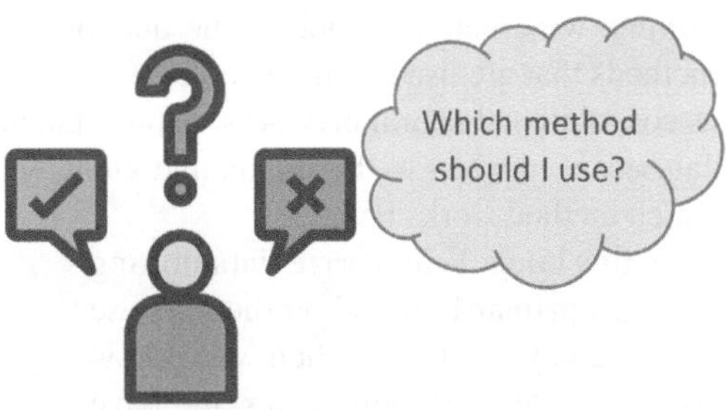

For someone new to data mining, the common question is, "Which method should I use to analyze my data?" This is the question that instructors tend to receive when teaching statistics courses. In this case, students usually expect a clear and concrete answer so they can use that exact method to complete their class assignments or projects. I face the same question in my data mining course, in which my students usually expect me to tell them exactly what method to use and which variables to include in the analysis. However, things are not that simple with data mining since data mining projects are exploratory rather than confirmatory. We aim to discover unknown patterns from data. Remember? We want to know what data will tell us about the relationships between variables.

The choice is not as straightforward and clear as black and white. We need to consider many factors, and even then, trials and errors are required. But no one says data mining is easy. That is why the data scientist job is very attractive because it requires a specific set of skills. I remembered that Netflix used to host a data mining competition in which all teams were given the same dataset and a target variable. The team that selected the right method and configured the right algorithm to produce the best prediction would win the competition. So, you see, the difficulties in data mining allow us to tell a great data scientist from a mediocre one.

As described in Chapter 1, data mining is an exploratory, data-driven method. The purpose of a data mining project is not to explain what happened or what is happening but rather to uncover unknown, hidden patterns from the data to predict what will happen in the near future. Hence, the simple answer to the above question is, "It depends on your data and the purpose of your project." Of course, no one will be satisfied with this answer because it is too vague. However, the fact is that we do not know what method may work best until we explore and understand the data. And even then, given the large volume and noise in the data, it is very hard to speculate which method would produce the best predictive model. But it is the reality of this type of project when we, as data scientists, are expected to explore and discover new patterns from the data, not to confirm what we already know or think we know.

Data mining is a very broad area that consists of numerous methods, which may cause confusion for data scientists. Hence, it is important that we determine how to select the appropriate method for the purpose of our project. In many cases, we select multiple methods for our analysis, train multiple models using those methods, and select the best one. As mentioned in the previous section, this best model is called *the champion model.* The reason we need to use multiple methods is due to the exploratory nature of data mining. Since we cannot speculate which method can produce the best model, developing models using only one method with one set of configurations will likely lead to an inferior model that fails to make a good prediction for our business operations. And we may not even know that it is inferior because we do not have anything to compare it with. As with other business activities, in order to find the best one, we have to have a baseline and other options for comparison.

There are several ways to categorize data mining methods: by data type or by the purpose of analysis. Data types typically include quantitative and qualitative. Quantitative data are numbers using interval, ratio, or ordinal scales. Qualitative data are text, voices, images, and videos. We also have mixed data, which consists of both quantitative and qualitative data. Naturally, the majority of data mining methods are quantitative since there are many methods of analyzing data quantitatively. On the other hand, qualitative methods are much scarcer due to the nature of qualitative data. Text mining is a typical method that we use to analyze textual data. Even when we use text mining, we eventually apply some quantitative methods to make sense of the results. Advanced methods that are used to analyze voices, images, and videos are generally quantitative-based, such as convolutional neural networks, natural language processing (NLP), or voice recognition. As you can see, even when we have data and know what type of data we have, it is still hard to determine which method works best.

> There is no such thing as a perfect data mining method, but there is the best method for a specific set of data.

In this book, I categorize data mining methods primarily based on the purpose of the analysis. Thus, when you know your variables and their scales and have a clear purpose for data analysis, you should be able to determine appropriate methods that you can use to develop your predictive models. However, note that each method has its own advantages and disadvantages, and since data mining is a data-driven approach, there is never a certainty in which method works best. In other words, we cannot speculate which algorithm will work best for the given data before running the data analysis. Basically, we explore multiple options and determine the best method based on the analysis results, which is the primary goal of data mining. My recommendation is to select several methods and configure various models to see which model produces the best prediction. Even if we choose to use only one method, we should develop several models with different configurations and compare them to choose the best one. Again, that model is called *the champion model.* Please keep in mind that the term *best model* here is relative; that is, it is not the true best one because we have to test all possible models to find the true best one. The best model refers to the one with the highest prediction accuracy and predictive power among the available models we construct. Hence, it is the relative best.

COMMON TERMS USED IN THIS BOOK

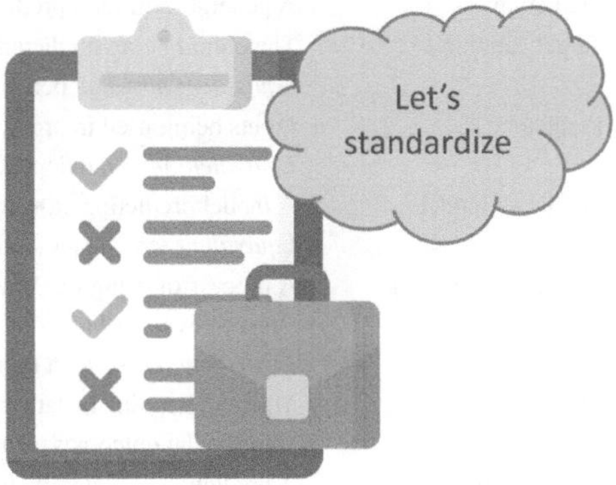

Before we get into the details of method categorization, it's important to understand the various terms used in data mining books and papers. Some of these terms are used interchangeably, while others might have different meanings in different contexts. For example, some authors use "regression" to refer to a model with a continuous target variable, while others use "prediction" for the same model. I think none of them is the best choice. First, "prediction" is a more general term referring to the act of predicting, so it's unclear why it should only apply to continuous target variables. Furthermore, "regression" seems specific to regression analysis methods, so it doesn't make much sense to use this term to refer to methods like decision trees or neural networks. Additionally, the logistic regression method handles binary or categorical target variables but is not considered a regression model, adding to the confusion. To complicate matters further, we have the term "predictive modeling," commonly referring to the data mining process in general. Therefore, we need to differentiate between predictive models with different types of target variables.

I have also noted that the terms used for variables vary. Some authors use traditional statistical analysis terms, that is, "dependent variable" for the output and "independent variable" for the input. Other authors use "predictor" for the input and "target variable" for the output. While the former is technically correct, it is often used in traditional statistical research. The latter seems to be more popular in data mining research as it captures the prediction intention. In data mining literature, authors often use "features" to refer to variables, leading to terms such as "feature selection" and "feature importance" instead of "variable selection" and "variable importance." Furthermore, SPSS Modeler uses the term "fields" in many places, including node settings, outputs, and guidelines. Note that "fields" are the same as variables.

> SPSS Modeler uses the term "fields" instead of "variables" in most contexts, including node settings, outputs, and guidelines, which could lead to confusion. It is important to note that "fields" and "variables" are synonymous in this context.
>
> To avoid misunderstanding, I will use the term "variables" throughout the chapter content, while "fields" will be explicitly used in the software instructions and demonstrations.

It's hard to say which terms work better than others since they are all technically correct. The challenge is that authors tend to use the terms they prefer, and sometimes they don't use them consistently, causing a great deal of confusion for readers. The key is to be consistent so that readers understand what is being discussed. For that purpose, in this book, I standardize the terms and use them consistently, as defined in Table 2.1. Thus, when I use terms such as "association model," "classification model," or "predictors," you will know exactly what I am referring to without confusion.

TABLE 2.1
Standardized Terms Used in this Book and Equivalent Terms

Terms	Description
Prediction	A general term for a predicting act
Target variable	The output being predicted *Equivalent terms*: dependent variable, output, target
Predictors	Inputs being used in predicting the output *Equivalent terms*: independent variables, input variables, features
Predictive model	A model predicting either a continuous or categorical target variable *Equivalent terms*: prediction model, data mining model, machine learning model
Predictive modeling	A process of using machine learning algorithms to develop a model that predicts future or unseen events or outcomes based on data *Equivalent terms:* data mining process, machine learning process
Machine learning	Machine learning is data analytic methods that detect unknown patterns from large data to predict the outcomes of interest
Association model	A predictive model with an interval target variable *Equivalent terms*: regression model
Classification model	A predictive model with a binary or nominal target variable
Supervised learning	A machine learning approach for structured data, where a model is trained on labeled data to make predictions of a pre-defined target variable
Class	A class refers to a distinct category or label that represents a specific group or type of data instances in a dataset, which is often the target. *Equivalent terms*: category, label, level
Unsupervised learning	A machine learning approach for unstructured data, where a model is trained on unlabeled data without a specific target variable, focusing on discovering hidden patterns or structures within the data
Training sample	A sample of data used to train the model
Validation sample	A sample of data used to validate the model
Test sample	A sample of data used to perform the final test of the model
Model fit	A measure of how well the model fist the data *Equivalent terms*: model performance, goodness of fit
Model fit metrics	Criteria used to evaluate model fit *Equivalent terms*: evaluation metrics, goodness-of-fit indices, evaluation criteria
Overfitting	A situation in which the model is overtrained to the training sample and not generalized to other datasets
Model reliability	The model's ability to produce consistent results with different runs using different datasets (internal consistency)
Model validity	The model's ability to consistently produces an accurate prediction of the target variable (accuracy of the model)
Predictive power	The model's ability to accurately capture and represent the underlying patterns, relationships, or trends present in the data and generalize the results to other datasets
Prediction accuracy	A measure of how well the model performs in terms of accurately predicting the target variable
Champion model	The model with the best model fit among compared models.
Variable importance	Relative importance of each predictor in determining the prediction of the target variable *Equivalent terms*: feature importance
Variable scale	A characteristic that describes the nature and properties of a variable; it indicates the type of values that a variable can take and the operations that can be performed on it *Equivalent terms*: level of measurement, scale of measurement, measurement scale, measurement of variable

DATASETS USED IN THIS BOOK

I will use two primary datasets in this book to demonstrate how to develop both the association model and classification model.

- **Association model**: Boston housing dataset; target variable: *medium housing price* (interval).
- **Classification model**: Airline recommendation dataset; target variable: *recommendation of the airline* (binary: Yes—recommend; No—not recommend).

To help you follow the steps in the data mining process, I use these two datasets throughout all chapters to demonstrate how to complete each step. This approach allows you to connect those steps together and see the whole picture of a data mining project. We will discuss the details of these datasets in the next few chapters. I am confident this approach works better than using a different dataset for each chapter.

VARIABLE MEASUREMENT SCALES

Another topic that requires consistency is the measurement scales for variables. The scale of variable measurement is a key factor in determining the methods used to analyze data. Essentially, it defines and categorizes variables or numbers within a dataset. This concept is also known as the "variable scale," "measurement scale," "scales of measurement," or "measurements of variables." Understanding the differences between these scales is crucial for selecting the appropriate analytical method. In this book, I refer to it as the variable scale.

In statistics, we typically use two types of scales: continuous (metric) and categorical (non-metric). Within these types, there are four specific scales: interval and ratio (continuous) and ordinal and nominal (categorical). Table 2.2 describes these four scales and provides some examples.

TABLE 2.2
Variable Measurement Scales

Types	Traditional Scales	Characteristics	Examples	SPSS Modeler Scales
Continuous (metric)	Interval	Numeric, rank order, no true zero values, difference between values is the same	F degree (0 degree is not true zero)	Continuous
	Ratio	Numeric, rank order, having true zero values, difference between values is the same	Age, distance, population, accident rate	Continuous
Categorical (non-metric)	Ordinal	Non-numeric classes, rank order, difference between responses is inconsistent	*Usage*: never; sometimes; often *Agreement*: disagree, somewhat disagree, somewhat agree, agree	Ordinal
	Nominal	Non-numeric classes, no rank order, difference between responses is meaningless	*Region*: north, south, east, west *Travel purpose*: business, pleasure	Nominal
		Binary: Nominal scale with only two levels	*Decision*: yes, no *Answer*: true, false	Flag

However, SPSS Modeler has a slightly different scale setup with several options. In this book, I focus on the three most relevant levels: continuous, ordinal, and nominal. As you may notice, SPSS Modeler uses different terminology from the traditional names. To avoid confusion, I include the SPSS scales with their corresponding names in the same table, so you know which ones to use when working with the software.

DEFINITION

Variable scale is a characteristic that describes the nature and properties of a variable; it indicates the type of values a variable can take and the operations that can be performed on those values.

Since SPSS Modeler uses some unique terminology, I list the interchangeable terms below for clarification

- Variables = Fields = Features
- Continuous = Interval
- Binary = Flag

Interval and Ratio

"Interval" and "ratio" scales are both continuous scales represented by numerical values, which can be ranked in order and have consistent gaps (intervals) between levels across the measurements. To avoid confusion with the interval scale, I use the term "gaps" instead. The key difference between these two scales is that the ratio scale has a true zero value, while the interval scale does not. For example, Fahrenheit degrees are considered "interval" because 0°F does not mean the absence of temperature; it represents a specific value indicating very cold conditions. In contrast, "ratio" variables, such as age, distance, or scores, have true zero values since they represent 0 years, 0 distance, and 0 points.

As you can see, ratio variables are more common than interval variables, as interval variables are often found in scientific fields rather than in everyday life or business activities. However, you may notice that the term "interval" is more frequently used than "ratio" on the Internet and in many materials. One possible explanation is that authors often use "interval" to refer to the continuous scale. Perhaps for this reason, SPSS Modeler does not distinguish between the two, offering only one scale called "continuous" for both cases. This should not affect the analysis results since these two scales are used in the same way during analysis. The distinction is only important when interpreting the results. Accordingly, in this book, I will use the term "interval" for both scales to simplify the content and maintain consistency. Just remember that the SPSS Modeler software uses "continuous" for this scale.

Ordinal

On the other hand, non-metric scales, such as "ordinal" and "nominal," are represented by non-numeric classes. The ordinal scale has a ranking order, but the gaps between levels can be inconsistent. Examples include levels of usage (never, sometimes, often) or levels of agreement (disagree, somewhat disagree, somewhat agree, agree). These categories reflect participants' responses to questions about their preferences or perceptions. Although they are non-numerical, they can still be ranked in order.

Nominal

Finally, the nominal scale is represented by non-metric classes that lack any rank order, making the differences between levels meaningless. For example, the "region" variable might have levels such as north, south, east, and west. These levels do not have any inherent relationship to one another and cannot be ranked. They are primarily used to categorize variables of interest in a qualitative manner.

SPSS Modeler includes a "nominal" scale but also has a special case known as the "flag" scale. The flag scale is essentially a binary scale, which is a subtype of the nominal scale with only two levels. For instance, a "decision" can be either Yes or No, or an "answer" can be True or False. This scale is frequently used to predict risks or decisions, which is why the SPSS Modeler has a dedicated scale for it.

Likert Scale

The Likert scale is a widely used measurement tool, typically consisting of either 5 or 7 points, designed to measure people's perceptions, behaviors, or attitudes. It always includes a middle point, as shown in Figure 2.6. Common examples include measuring "satisfaction" (very dissatisfied, dissatisfied, neutral, satisfied, very satisfied) or "quality" (very poor, poor, fair, good, excellent). The gaps between levels are generally consistent and considered equal. This scale is extensively used in business management and social science research.

A common question arises: should Likert scale variables be categorized as "interval" or "ordinal?" Based on the way responses are coded, Likert scales appear to be "ordinal." However, many researchers choose to treat Likert scale variables as "interval" variables. The primary reason is that this classification allows the use of parametric statistical techniques, which require metric variables. While there are non-parametric techniques available, they are often limited and less robust. For instance, when researchers are dealing with multiple hypotheses involving complex relationships among variables, they often rely on multiple regression analysis or structural equation modeling (SEM) methods to test those hypotheses. These are parametric methods and require metric variables. Since there are no equivalent non-parametric methods, researchers often treat Likert scale variables as interval scales, enabling the use of these parametric methods.

The decision to treat Likert-scale variables as interval data often receives criticism because the Likert scale is not a true interval scale. Nonetheless, researchers in fields such as marketing,

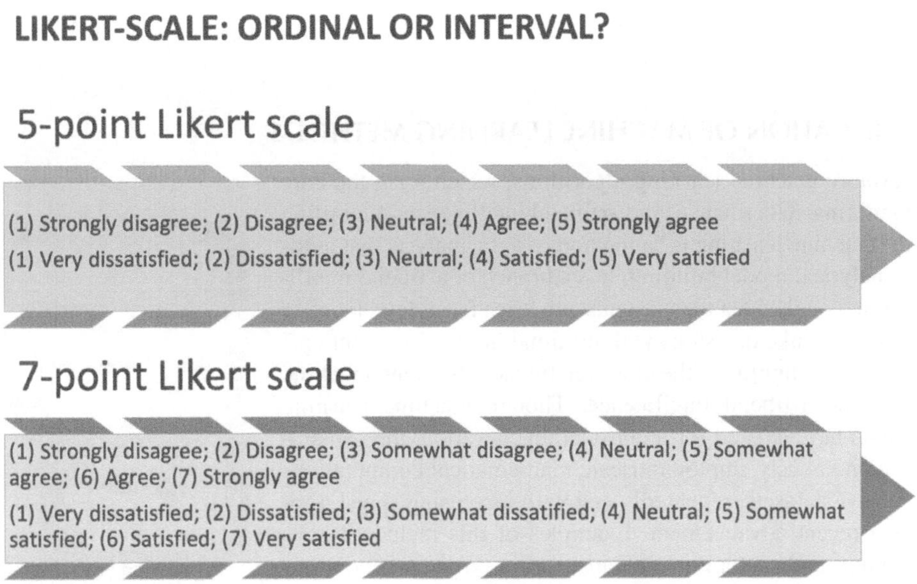

FIGURE 2.6 Likert scale.

management, and social behaviors frequently categorize Likert scale data as "interval" in their studies. Although this approach has limitations—such as the restricted number of levels (5 or 7) and challenges in meeting test assumptions—it allows researchers to obtain meaningful and practical analysis results. The crucial factor is ensuring that the gaps between levels are equal throughout the measurement. This consistency distinguishes continuous scales from categorical scales. For example, when designing a survey questionnaire, the gap between "Very dissatisfied" and "Somewhat dissatisfied" should be equal to the gap between "Somewhat satisfied," "Very satisfied," and so on.

Another reason researchers prefer treating Likert scale variables as "interval" is that it simplifies result interpretation. Since Likert scale data are coded numerically (e.g., 1–5 or 1–7), the statistical outputs—such as correlation coefficients, model fit, and explained variance—are easier to interpret as metric values. If we were to treat these variables as "ordinal," the interpretation process would resemble that used for nominal variables, which is more complex because the responses are non-metric. Chapter 7 provides specific examples of this when discussing logistic regression. In fact, ordinal regression functions similarly to logistic or multinomial regression.

> In data mining, it's advisable to treat Likert scale variables as "interval" data unless there is a specific reason to classify them as "ordinal" and a well-defined plan for analysis and result interpretation. To justify this approach, it's essential to provide evidence that the gaps between levels are equal.

Ultimately, how we treat the Likert scale depends on the study's purpose. My rule of thumb is to treat Likert scale variables as "interval" unless there is a compelling reason to do otherwise and a clear plan for data analysis and result interpretation. In such cases, it's essential to provide evidence that the gaps between levels are equal and test the assumptions required for the parametric methods used.

CLASSIFICATION OF MACHINE LEARNING METHODS

Let's explore machine learning algorithms, which form the core of data mining, data science, and artificial intelligence. According to SAS, machine learning is "a method of data analysis that automates analytical model building. It is a branch of artificial intelligence based on the idea that systems can learn from data, identify patterns, and make decisions with minimal human intervention." This definition highlights the close relationship between machine learning and artificial intelligence. Though machine learning algorithms have existed for a considerable duration, the capability to autonomously employ intricate mathematical computations on extensive datasets, repeatedly and with increasing speed, represents a recent advancement. Examples of this include Tesla's self-driving and autopilot systems; online recommendations from Amazon, Facebook, and Netflix; fraud detection by banks;

AI assistants like Siri, Alexa, and Google Assistant; and generative AI systems such as ChatGPT, Copilot, and Google Germini.

In this book, we define machine learning methods as "data analytic techniques that detect unknown patterns in large datasets to predict outcomes of interest." This definition aligns with the concept of data mining introduced in Chapter 1 and the methods discussed throughout this book. The methods we cover will enable us to train models and build predictive algorithms with the accuracy and predictive power required for making real-world predictions.

> **DEFINITION**
>
> Machine learning methods are data analytic techniques that detect unknown patterns in large data to predict the outcomes of interest.

Supervised vs. Unsupervised Learning

Machine learning algorithms can be broadly categorized into "supervised learning" and "unsupervised learning." "Supervised learning" is best suited for structured data and requires a target variable. Models are developed to predict this target variable based on numerous predictors. In contrast, "unsupervised learning" is more appropriate for less structured data, where there is no specific target variable; instead, all variables are treated as input variables. The primary goal of unsupervised learning is to detect the intrinsic structure of the data. The rule of thumb for deciding whether to use supervised or unsupervised learning is to check if your dataset has a target variable and whether your research aims to predict it.

> **DEFINITION**
>
> Supervised learning: A machine learning approach for structured data, where a model is trained on labeled data to make predictions of a pre-defined target variable.
>
> Unsupervised learning: A machine learning approach for unstructured data, where a model is trained on unlabeled data without a specific target variable, focusing on discovering hidden patterns or structures within the data.

Typically, there are more supervised learning methods available than unsupervised ones. Examples of supervised learning methods include regression and logistic regression, decision trees, random forests, gradient boosting, memory-based reasoning, support vector machines, neural networks, and Bayesian networks. The appropriate method for your research depends on the study's purpose and the scale of the target variable. Methods like regression and logistic regression share foundations with traditional statistical techniques, but in data mining, they are applied to much larger datasets with less stringent assumptions, such as the requirement for normal distributions. It's also worth noting that random forest and gradient boosting are ensemble methods based on decision trees, where multiple trees are constructed to improve predictive accuracy. We'll examine these machine learning methods and their characteristics in depth later. It's common practice to use multiple methods to build various predictive models and compare them to identify the best-performing model.

> Look for a target variable in the dataset. If your dataset contains a target variable and your goal is to predict it, use supervised learning. Use unsupervised learning if there is no target variable and you aim to explore patterns or relationships within the data.

Examples of unsupervised learning algorithms include cluster analysis, principal component analysis, and multidimensional scaling. These methods group cases based on similarity or detect underlying factors in the data. Like their supervised counterparts, the foundations of these methods are rooted in multivariate statistics, but they are applied to much larger datasets in data mining.

Figure 2.7 provides a diagram that categorizes these methods into supervised and unsupervised learning categories. While we won't cover every method in this book—some are very advanced—we will focus on the methods commonly used in practice. Don't worry, the methods covered range from standard to advanced, equipping you to build predictive models for any project.

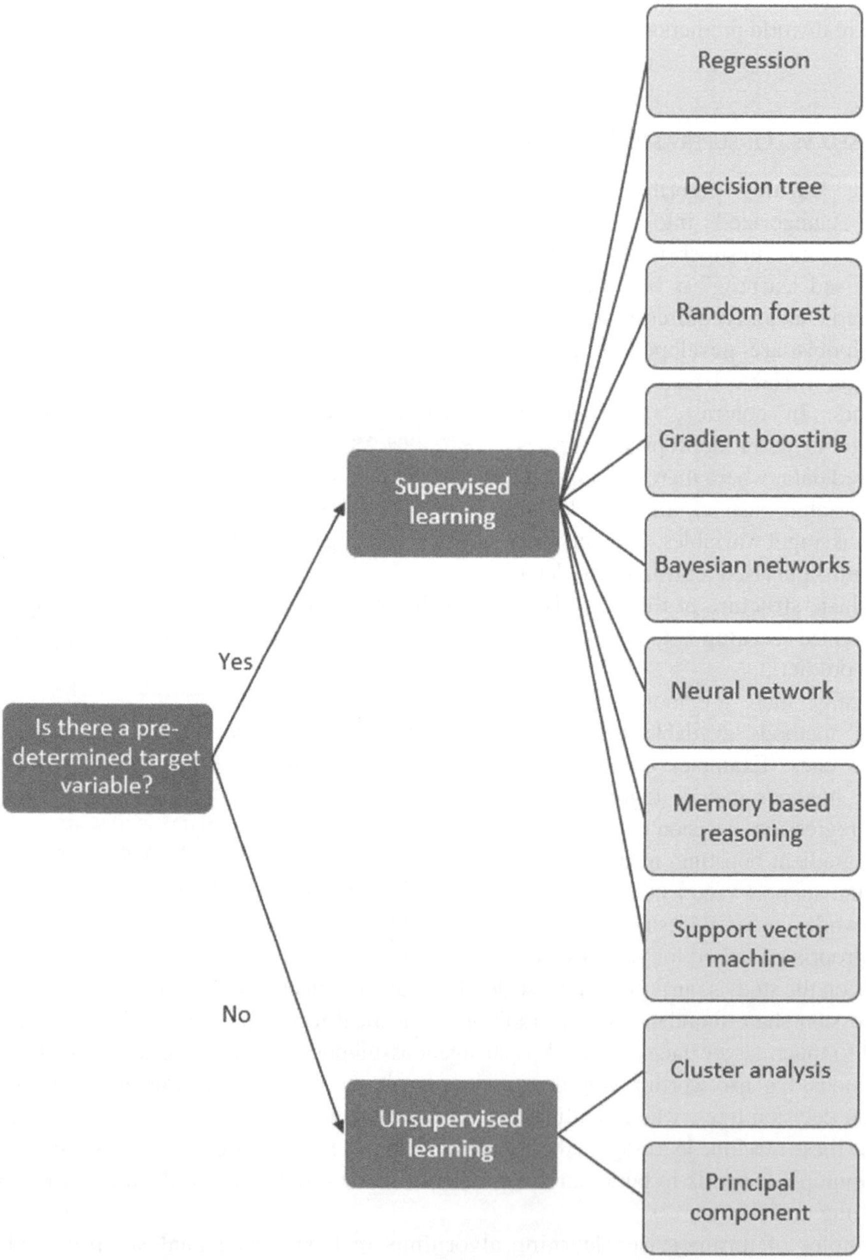

FIGURE 2.7 Classification of Machine Learning methods

Descriptions and Characteristics of Machine Learning Methods

Table 2.3 provides brief descriptions of primary supervised learning algorithms, and Table 2.4 describes some primary unsupervised learning algorithms. Note that those are the methods that are actually covered in this book. Don't worry if some of them do not make much sense to you. These tables just give us a quick introduction to those methods. The characteristics, advantages, disadvantages, and the modeling and analysis process of each algorithm will be described in the subsequent chapters. We will have one chapter for each method.

TABLE 2.3
Supervised Learning Methods

Machine Learning Algorithms	Description	Characteristics
Linear regression	Linear regression is a prediction method used when the target variable is continuous. The relationship between the target variable and predictors is linear.	• Statistic-based • Linear relationships • Multiple assumptions
Classification regression (logistic regression)	Classification regression predicts the probability of a binary or nominal target variable based on a function of multiple predictors. If the target variable is binary, it is called logistic regression; if the target variable is categorical, it is called multinomial logit regression. Note that SPSS Modeler has both models in the Logistic node: **Binomial option** for the binary target variable and **Multinomial option** for the nominal target variable. Hence, I use the term "logistic regression" for both cases.	• Statistic-based • Linear relationships between predictors and odds ratio of the target variable
Decision tree	Decision tree is a logical rule-based method that illustrates a hierarchical structure of variables, encompassing the root node (also known as the parent node) along with its subsequent child nodes. The root node is subdivided into two or more branches, employing the optimal condition for distinguishing individuals within each class. Decision tree can be used for both association and classification models.	• Rule-based • Can detect nonlinear relationships
Neural network	A neural network model is characterized by its composition of several layers, with each layer comprising individual units or neurons. Within a layer, each unit receives input from the preceding layer and computes its own output. The resulting outputs from neurons within one layer then serve as inputs for the subsequent layer in the sequence of layers.	• Math-based • Can detect nonlinear relationships • Can handle complex model structure
Gradient boosting	Gradient boosting is an ensemble learning method that employs a sequence of decision trees, collectively constructing a unified predictive model. The process of gradient boosting commences with an initial model and subsequently refines it by utilizing the residuals from the preceding model as the new target.	• Rule-based • Can detect nonlinear relationships • Can handle complex model structure
Random forest	Random forest method is an ensemble learning technique used in machine learning and statistics. It constructs a multitude of decision trees during the training process, where each tree is built from a different random subset of the training data and features. This diversity in trees helps reduce overfitting and increases the model's robustness. In the case of classification models, the random forest output corresponds to the mode of the classes predicted by the individual trees. When dealing with association models, the random forest output represents the average prediction derived from the individual trees.	• Rule-based • Can detect nonlinear relationships • Can handle complex model structure

(Continued)

TABLE 2.3 *(Continued)*
Supervised Learning Methods

Machine Learning Algorithms	Description	Characteristics
Bayesian networks	Bayesian network is a probabilistic method founded on Bayes' theorem, enabling the prediction of future events through the consideration of prior evidence. It employs a directed acyclic graph to visually illustrate causal connections between predictors and the target variable. Additionally, a Bayesian network is characterized by a collection of conditional probabilities, facilitating causal inference and probabilistic reasoning. It serves as a valuable tool for modeling and understanding complex systems by representing and analyzing relationships among variables in a probabilistic manner.	• Math-based • Can detect nonlinear relationships • Can handle complex model structure • Can infer causation

TABLE 2.4
Unsupervised Learning Methods

Machine Learning Algorithms	Description	Characteristics
Cluster analysis	Cluster analysis is an effective pattern discovery method. This method groups cases into clusters (or segments) based on their characteristics, aggregating them according to their similarities. Identified clusters can be applied to new data to classify new cases. Two common methods in cluster analysis include hierarchical and non-hierarchical clustering.	• Statistics based • Grouping cases into segments
Principal Components	Principal component analysis is a method widely used to reduce the dimensionality of data while retaining its essential information. It accomplishes this by transforming the original correlated variables into a new set of uncorrelated variables called principal components. These components are linear combinations of the original variables and are ordered by the amount of variance they capture. Principal component analysis helps uncover underlying patterns, simplifies data interpretation, and aids in visualizing high-dimensional data by focusing on the most significant sources of variation.	• Statistics based • Data reduction (underlying constructs)

HOW TO CHOOSE THE RIGHT MACHINE LEARNING METHODS? SELECTION GUIDELINES

Selecting the right machine learning methods is one of the most challenging decisions in data mining projects. Using the wrong methods can lead to unusable or, worse, incorrect outcomes, resulting in flawed business decisions. It's important to note that different authors and resources often offer inconsistent or even contradictory advice on this matter, making it difficult to determine how and why to choose the appropriate methods. To help you navigate this process, I will provide specific criteria and steps to guide you in selecting the right methods. Additionally, I will present scenarios and examples to make it easier to follow.

Guidelines To Choose The Right Machine Learning Methods

Decision Order →

Pre-determined target variable?								
YES	SUPERVISED LEARNING METHODS							
	Categorical target variable?	Interval target variable?	Linear relationships?	Nonlinear relationships?	Interrelationships among predictors?	Causal inference?	Model complexity?	Methods
	No	Yes	Yes	No	No	No	No	Linear regression
	Yes	No	Yes	No	No	No	No	Classification regression
	Yes	Yes	Yes	Yes	Yes	No	No	Decision tree
	Yes	Yes	Yes	Yes	Yes	No	Yes	Neural Network
	Yes	Yes	Yes	Yes	Yes	No	Yes	Random Forest
	Yes	Yes	Yes	Yes	Yes	No	Yes	Gradient Boosting
	Yes	Yes	Yes	Yes	Yes	Yes	Yes	Bayesian Networks

NO	UNSUPERVISED LEARNING METHODS				
	Data reduction?	Social or market segmentation?	Group variables into a small number of variables (underlying factors)?	Group observations to distinct clusters or segments by similarities?	Methods
	Yes	No	Yes	No	Principal Components
	No	Yes	No	Yes	Cluster Analysis

FIGURE 2.8 Guidelines on how to select the right machine learning methods.

The selection of methods should be driven by the purpose of the study and the variable scales involved. Specifically, you need to understand the target variable, the predictors, and the specific objectives of the analysis. Figure 2.8 outlines various decision scenarios that can help you choose the appropriate machine learning method for your project. To make an informed decision, you should consider the purpose of the analysis, the type and size of the data, and the scales of the variables. By moving from left to right in the decision table and answering the questions in sequence, you can systematically determine the most suitable methods.

I use two examples below as a demonstration of how to use the guidelines for method selection.

CASE 1

EXAMPLE

Suppose we want to predict passengers' recommendations of an airline based on passenger survey data. The "recommendation" is a non-metric target variable with two levels: Yes and No. Using this variable as the starting point, we begin with the first column on the left:

- **Is there a pre-determined target variable?** Yes, the target variable is the "recommendation." Therefore, we should focus on supervised learning methods only.
- **Is it categorical?** Yes, it is binary. Therefore, we should skip linear regression and consider logistic regression.
- **Are we expecting linear or nonlinear relationships?** If we are unsure, we could consider both possibilities. Logistic regression remains a viable option, but we must be mindful of its limitations (e.g., while logistic regression technically shows

nonlinear relationships in the form of a logarithm curve, the relationship between the odds ratio of the target variable and predictors is still linear).

- **Do we want to examine interrelationships among predictors?** If the answer is yes, we might need to consider other methods. However, if we are unsure, we can retain logistic regression. Let's assume, we do **not** want to examine interrelationships to keep the model simpler.
- **Do we want to make causal inferences?** Assuming the answer is No, Bayesian networks can still be an option even without intending to make causal inferences, so we keep this open.
- **Final question: Do we want model complexity?** Note that model complexity means the method can produce a complex structural model, but it also means the model is hard to understand and interpret. If the answer is No, because we prefer an easy-to-interpret model, we should skip neural networks, random forests, gradient boosting, and Bayesian networks. This leaves us with logistic regression and decision trees as the final options.

Conclusion: In this scenario, we should use supervised learning and focus on two methods: logistic regression and decision trees. To find the best predictive model, we need to build multiple models using these two methods with various configurations. By comparing the models, we can identify the champion model.

CASE 2

Suppose we want to identify different groups of citizens based on their shopping behaviors during the COVID-19 pandemic. The goal is to determine how many distinct groups exist in society based on common shopping behaviors. Let's use the guidelines to select the right machine learning method:

- **Is there a pre-determined target variable?** No. We do not aim to predict a specific decision or behavior, so we should focus on unsupervised learning methods in the second half of the table.
- **Is data reduction the purpose of this study?** No. Thus, we skip the Principal Component method.
- **Is social or market segmentation the purpose of this study?** Yes. This confirms that Cluster analysis is a possible choice.
- **Do we intend to group variables into a smaller set of variables?** No. Therefore, Cluster analysis remains an option.
- **Do we want to group observations (citizens) into distinct clusters or segments based on their similarities in certain perspectives?** Yes, we want to segment them based on common shopping behaviors during the pandemic. Thus, the final choice is Cluster analysis.

Conclusion: As demonstrated, the guidelines are straightforward and easy to follow without requiring complex diagrams with boxes and arrows. In some cases, you may end up with several possible choices, while in others, there may be only one. When multiple methods are available, you should develop multiple models with different configurations for each of them to find the best one. Even when only one choice is available, it's still advisable to build several models using different configurations to identify the champion model.

ENSEMBLE

In data mining, an important approach worth mentioning is "ensemble modeling." "Ensemble" models combine predictions from multiple individual models to create a single, consensus prediction. This practice is common in data mining, particularly when individual models yield differing results. An ensemble model typically offers greater accuracy than any of the individual models that compose it. While we cannot guarantee that the ensemble model will outperform all others, using an ensemble approach allows us to explore the potential of a combined model and compare its performance against individual models. This approach is highly valuable in data mining. We will get into the details of ensemble modeling in Chapter 10.

DATA MINING SOFTWARE COMPARISON

As mentioned earlier, machine learning algorithms are typically constructed and configured using specialized software. As the field of machine learning continues to expand, data mining software also evolves, offering more advanced features and capabilities. In this chapter, I will review and compare some of the top software programs used in data mining, focusing on those widely adopted in industry and academia. These programs have been validated and are known to provide robust machine learning algorithms and configurations for analysis.

There is a long list of software claiming to perform data mining or big data analytics. However, it's important to be cautious with programs designed primarily for learning purposes. While these tools may be easy to use, they often work best for small-scale projects and offer limited algorithms and features. When dealing with large datasets and requiring multiple configurations for models, the limitations of such programs become apparent. Moreover, new software programs on the market should be validated, as they may contain errors in their algorithms.

Furthermore, gaining skills and experience with well-recognized software adds value to your resume, as many companies and organizations already use these tools. These software vendors often provide certification programs, training courses, workshops, and extensive technical support, making them a reliable choice for professionals in the field.

What Are the Most Popular Data Mining Software Programs on the Market?

To answer this question, I decided to consult two generative AIs—ChatGPT and Bing AI. The reason for this approach is that my initial search on Google yielded inconsistent information, with different websites listing various programs, making it difficult to compile a comprehensive list. Generative AIs excel at synthesizing information from the Internet and can produce a more reliable list of programs for review. Furthermore, by comparing the results from two different sources, we can ensure the reliability of the outcomes.

Table 2.5 shows the lists generated by ChatGPT and Bing AI, sorted alphabetically. The order does not indicate the superiority of any program over others. I've shaded the programs that do not appear on both lists for easy comparison. Interestingly, the two lists share eight common programs: IBM SPSS Modeler, KNIME, Microsoft SQL Server Analysis Services (SSAS), Oracle Data Mining (ODM), Orange, RapidMiner, SAS Enterprise Miner, and Weka. It's worth noting that most of these programs do not require programming skills, making them suitable for non-programmers.

TABLE 2.5
Lists of Popular Data Mining Programs Provided by ChatGPT and Bing AI

ChatGPT	Bing AI
IBM SPSS Modeler	Alteryx Analytics
KNIME	H2O
Microsoft SQL Server Analysis Services (SSAS)	IBM SPSS Modeler
Oracle Data Mining (ODM)	KNIME
Orange	Microsoft SSAS
Python	ODM
Rapid Miner	Orange
SAS Enterprise Miner	Rapid Miner
Tenson Flow	SAS Enterprise Miner
Weka	Weka

This reflects an emerging trend in the market, with more companies adopting non-coding programs in the field of data mining.

Let's compare some common data mining programs on the market, including both coding and non-coding options. This book primarily targets audiences interested in non-coding programs, so the focus will be on those. However, I will also discuss common coding software, such as R and Python, for comparison. More specialized programming tools like MATLAB® are not included, as they are better suited for students with computer science backgrounds.

In this book, I will cover five major data mining software programs: SPSS Modeler, SAS Enterprise Miner, RapidMiner, R, and Python. These are among the top data mining tools available today. SAS Enterprise Miner and Python are highly regarded in both industry and academia, with a particular dominance in the industry. Many organizations rely on SAS and Python for their data analysis needs, and employees are often required to obtain certifications to prove their proficiency. Both SAS and Python offer extensive certification programs recognized by numerous companies across various industries.

On the other hand, SPSS Modeler, RapidMiner, and R are more popular within academic circles. Instructors and researchers frequently use these tools for teaching and research purposes, which is why we see a higher number of journal publications utilizing them.

I will outline the primary criteria and compare these software programs, highlighting their pros and cons so that you can see both sides of each option (Table 2.6).

TABLE 2.6
Data Mining Software Comparison

Criteria	SPSS Modeler	SAS Enterprise Miner	Rapid Miner	R	Python
Common Users	Academia & Industry	Industry & Academia	Academia	Academia	Industry & Academia
Certification programs	Moderate, offered by IBM	Extensive, offered by SAS	Moderate, offered by Rapid Miner Academy	Extensive, offered by third parties	Extensive, offered by third parties
Coding and user interface	User-friendly, no coding required	User-friendly, no coding required	User-friendly, no coding required	Programming language, coding required	Programming language, coding required

(Continued)

TABLE 2.6 *(Continued)*
Data Mining Software Comparison

Criteria	SPSS Modeler	SAS Enterprise Miner	Rapid Miner	R	Python
Workspace diagram	Yes, less intuitive, getting crowded with many models	Yes, intuitive, easy to view and follow	Yes, not intuitive, getting crowded with many models	No workspace diagram	No workspace diagram
Algorithms	Robust ML algorithms	Robust ML algorithms, robust text mining	Robust ML algorithms	Improved ML algorithms	Robust ML algorithms
Connections of nodes and steps	Easy, the sequence of nodes is not intuitive	Intuitive, connection is easy	Not intuitive, connections are confusing, easy to make mistakes	Coding required	Coding required
Configurations	Easy using the property window	Easy using the property panel, too many choices	Easy using the property window	Coding required	Coding required
Variable importance feature	Built-in, easy to use and interpret	Built-in, easy to use and interpret	Requiring extra steps	Requiring extra steps	Requiring extra steps
Outputs graphs	Requiring extra steps	Included in the output window	Requiring extra steps	Coding required	Coding required
Performance evaluation	Included in the output window	Included in the output window	Requiring extra steps	Coding required	Coding required
Pricing	Expensive professional version, free education version	Expensive professional version, free education version	Free education version	Open source, free	Open source, free
Challenges	Confusing diagram and model comparison, confusing result outputs	Cloud-based, slow, cumbersome file and folder management, not designed for data preparation	Confusing diagram and model comparison, confusing terms	Steep learning curve, cumbersome visualization	Steep learning curve, cumbersome visualization

User Interface

One important criterion for comparing data mining software is the user interface and the level of coding required. For students, instructors, and researchers in fields like math, science, and engineering—who often have a strong foundation in programming languages—coding is not an issue and may even be preferred. However, for users in business management or social sciences, coding can be a significant obstacle. Learning a new programming language is time-consuming and comes with a steep learning curve. Not everyone is inclined toward coding, but that shouldn't prevent them from engaging in data mining.

From this perspective, SPSS Modeler, SAS Enterprise Miner, and RapidMiner offer more user-friendly interfaces that require minimal coding. These programs allow users to perform data mining with simple point-and-click operations. However, they also provide flexibility, enabling users to integrate or write code if desired, including compatibility with Python through various libraries and tools. This flexibility is particularly valuable; non-programmers can build predictive models without coding yet still convert their algorithms into code for integration into other platforms.

On the other hand, R and Python are programming languages that require coding for every step of the data mining process, including visualization. For newcomers to programming, this presents a steep learning curve, requiring significant practice before any meaningful analysis can be performed. Additionally, in programming, even a small mistake like a missing comma can result in errors, so precision is crucial. However, for those proficient in programming, R and Python offer considerable flexibility and extensive resources for algorithm development. Python, in particular, has a rich library of codes that can be accessed for modeling purposes. These programs are widely accepted in the industry and are often required skills for data scientists.

Workspace

The next item is the workspace diagram. Essentially, the data mining process involves many steps and models, so tracking these steps and models in our analysis is not easy. A workspace diagram is a visual tool that allows us to incorporate all steps and models into diagrams. Typically, we create such a workspace diagram by dragging and dropping nodes into the workspace and connecting them as appropriate. Steps like importing data files, sampling, partitioning, exploring, transforming data, modeling, and assessing can be done easily with this diagram. I usually build many models using different algorithms or the same algorithms with different configurations, so I find this tool extremely useful. I can easily compare, track steps, make modifications, run a part of the diagram, and review the final results. Data mining is an exploratory and iterative process, so it is unlikely to get the desired result on the first try. The diagram makes the work easier to handle.

Among the software programs, SPSS Modeler, SAS Enterprise Miner, and RapidMiner all provide a workspace diagram. On the other hand, R and Python lack this feature due to the programming language nature of these programs. Among the three programs with this feature, I have tried all three and found that the diagram in SAS Enterprise Miner is the easiest to use. I have built many models using SAS Enterprise Miner diagrams, including some fairly complicated ones, and I have never encountered any confusion or difficulties. Additionally, this software integrates its nodes in the SEMMA process, making it easier to find the node you need. SPSS Modeler is not too bad, but the required sequence of nodes is not very logical, which makes it hard to find the right one. Furthermore, the way it handles multiple models and outputs could be more intuitive, and if the steps are not followed correctly, vague error messages may appear. Also, you need additional nodes to collect the results, unlike in SAS Enterprise Miner, which produces results within each node. While SPSS Modeler allows us to associate each step in the analysis with the CRISP-DM process, this process does not guide us through the analysis; it is used primarily for reference purposes. That means you still need to understand the overall data mining process and select the right nodes yourself. Finally, RapidMiner's diagram shares similar challenges, and figuring out how to put multiple models in the diagram is a challenging task. We also need to connect inputs separately from outputs when connecting two steps. Moreover, I found that RapidMiner uses very unusual names for operators in the diagram, which could be confusing to new users. For example, they use "label" instead of "target variable," "example" instead of "dataset," etc. In the diagram, they use acronyms for these names, which make it even harder to follow. As a matter of fact, there are so many unusual operator names that they had to create a glossary to provide definitions for them.

Machine Learning Algorithms

Regarding the machine learning algorithms provided by these software programs, I find that SAS Enterprise Miner, RapidMiner, and Python offer more robust algorithms that allow for complex modeling with various levels of configuration. Nonetheless, SPSS Modeler and R are continuously improving their algorithms, and we have begun to see more features in these programs. SPSS Modeler also offers certain features that provide more configuration options than SAS Enterprise Miner, particularly in advanced methods. While SAS Enterprise Miner tends to build in the configurations for these advanced methods to make them easier for users to implement, this approach can limit the options available for fine-tuning the models.

It's also important to note that all of these software programs provide text analytics tools, but SAS Text Miner is easier to use when dealing with textual data. The steps for text parsing, text filtering, text topic, and text clustering are well-organized and seamlessly connected. On the other hand, Python offers a versatile collection of NLP tools and libraries that enable developers to handle various NLP tasks, including sentiment analysis, part-of-speech (POS) tagging, document classification, topic modeling, word vectors, and more.

Pricing

Pricing is a sensitive topic. R and Python are fully open-source and free to use. RapidMiner is also open-source and free for academic use, though it requires a paid license for commercial purposes. SPSS Modeler is not free and is quite expensive for enterprises, though IBM offers a free version for educational use. SAS Enterprise Miner is also costly for enterprises, but universities with campus SAS licenses can access it for free, and SAS OnDemand for Academics offers a free cloud-based version for teachers, researchers, and students, though it currently only works with Windows OS.

Challenges

Each program has its challenges. As mentioned earlier, R and Python are programming languages, so there is a steep learning curve for new users. Both lack workspace diagrams and visualization can be cumbersome with these programs. SPSS Modeler and RapidMiner, while offering workspace diagrams, have a confusing design, making model comparison difficult, and they feature confusing node and operator names. In both SPSS Modeler and RapidMiner, viewing output graphs requires extra steps, especially with SPSS Modeler. SAS Enterprise Miner has an advantage here as all results, whether in table or graph formats, are presented within the same node.

However, SAS Enterprise Miner, particularly the free cloud-based version, also has its weaknesses. First, it requires a strong Internet connection to work, meaning the software's performance is subject to Internet speed, and it cannot be used offline. Second, the way SAS handles folders and files can be challenging for some users. Unlike other software such as SPSS Modeler and RapidMiner, SAS manages folders through libraries. This means that folders containing files must be connected to a library, and SAS Enterprise Miner must be connected to the same library to access the files in that folder. This process is cumbersome and requires several steps to open a file. However, this method is efficient for managing data in SAS projects, especially when dealing with multiple folders and numerous files. SAS Enterprise Miner is also not designed to upload files, change file names, or move files between folders; these actions must be done using SAS Studio, which is free. It will take some time to get used to, but the process becomes easier with understanding. From this perspective, SPSS Modeler and RapidMiner are better choices, as they allow users to import datasets from various sources, including their own computers, with ease. They do not require an Internet connection, allowing for offline use. Furthermore, if you are already familiar with SPSS Statistics, you'll find that SPSS Modeler integrates well with SPSS datasets, making it easier to navigate, even though there are some differences.

It's also important to note that data mining software is not designed for easy data preparation or modification. These programs are dedicated to data mining, and they intentionally limit features that could lead to accidental changes to the data. This limitation is critical because altering data or variable names in these programs could result in invalid outputs, with no easy way to trace the errors. Therefore, data preparation should be done before using the data mining software. Typically, each vendor offers other tools for data preparation, but if you want to avoid learning another software, Microsoft Excel is a good tool for coding, merging, and cleaning data. After downloading raw datasets, Excel can be used to recode data, merge data from various sources, and clean the data. Excel offers many useful features for these tasks, and most users are familiar with it. Once the dataset is cleaned, it can be converted into SAS or SPSS format and then used for modeling in the data mining software.

Conclusion

In conclusion, no data mining software is perfect; each has its pros and cons and serves different audiences. The choice of software should be based on your needs, background, and the purpose of the project. Programming software like Python and R are favorable for science and engineering students and researchers due to their versatility and adaptability, and they are also required skills for some data scientist roles. However, data mining can still be effectively conducted without programming skills using software like SPSS Modeler, SAS Enterprise Miner, and RapidMiner. As the data mining software market continues to advance, students in business management and social sciences should take advantage of these technologies. Skills acquired with these software programs enable effective data mining without a programming background. In this book, I chose SPSS Modeler for its robustness and flexibility with datasets. Additionally, the education version is free and does not require an Internet connection.

SPSS MODELER INTERFACE

In this section, I will describe the software interface and its primary components. You will also find detailed guidelines provided by IBM. I highly recommend installing the software and exploring it to familiarize yourself with how the interface works. Although SPSS Modeler is part of the SPSS family, its interface is quite different from that of the SPSS Statistics program, so there will be a learning curve at the beginning. You can always return to this chapter for additional information and assistance regarding the software interface, nodes, and streams.

Stream Building

Using SPSS Modeler for data mining entails processing data through a sequence of nodes, referred to as a "stream." Each node represents a specific operation applied to the data, while the connections between nodes define the flow of data. A typical data stream is used to import data into SPSS Modeler, guide it through various transformations, and finally direct the processed data to a destination, such as a table or viewer.

The main workspace in SPSS Modeler is called the "stream canvas," where users build and manage these data streams. Streams are created by designing diagrams of data operations directly on the canvas. Each operation is represented by an icon or node, and the nodes are linked in a sequence that visually represents the flow of data through each step (https://www.ibm.com/docs/en/spss-modeler).

For instance, imagine you need to open a data source, create a new field, select records based on the values in that new field, and display the results in a table. In this scenario, your data stream would consist of four nodes.

Variable File node: Opens and imports the data from the data source.

Derive node: Adds a new field based on a calculation or transformation.

Select node: Filters records based on criteria related to the new field.

Table node: Displays the filtered results in a table format.

This sequence of nodes, or stream, allows you to easily manage and manipulate data within SPSS Modeler, ensuring that each operation is clearly defined and linked within the workflow. SPSS Modeler also allows users to handle multiple streams concurrently, either within the same stream canvas or by opening a new one. All active streams are organized in the Streams Manager, located in the upper right corner of the SPSS Modeler window. Figure 2.9 illustrates examples of streams and their associated nodes.

Data mining projects often involve cleaning and transforming data, building multiple models, and comparing results, which can lead to confusion and make it difficult to keep track of your work. The software interface in SPSS Modeler allows you to work with diagrams of data streams, helping to organize and manage the steps in your analysis. The typical steps for building a data stream include:

- **Add nodes** to the stream canvas.
- **Connect the nodes** to form a stream.

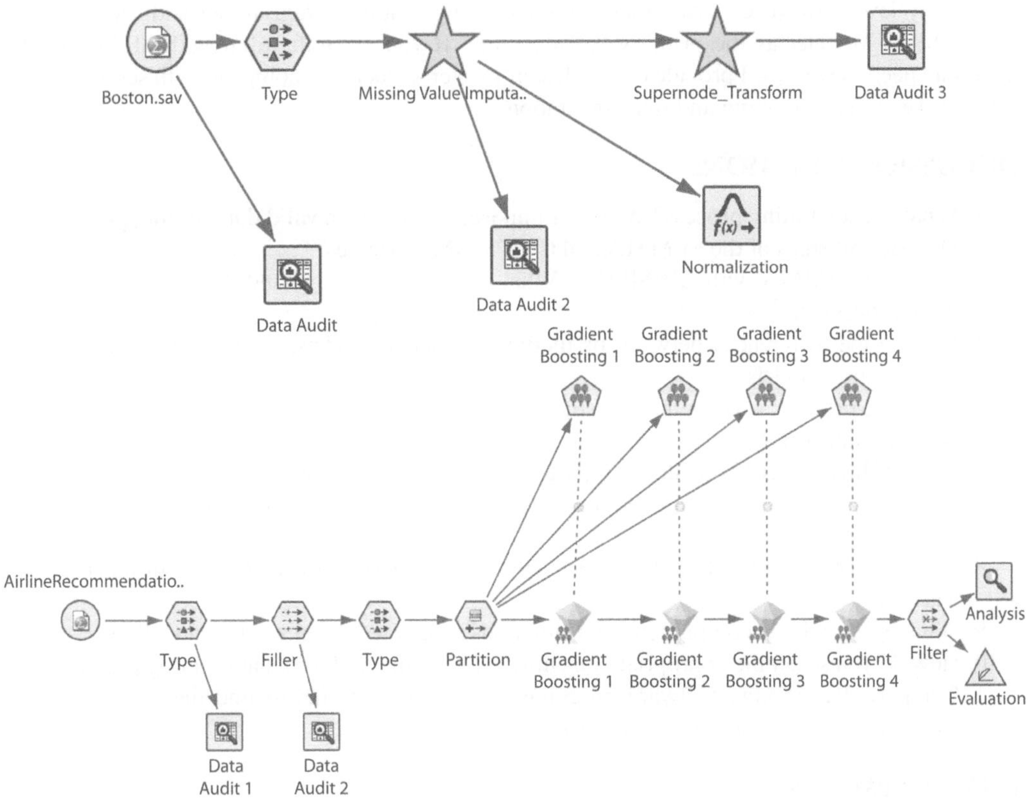

FIGURE 2.9 Examples of nodes and streams in SPSS Modeler.

- **Specify options** for any node or the entire stream.
- **Run the stream** to execute the analysis.

In the following pages, I provide detailed guidelines on how to work with nodes and streams in SPSS Modeler. As you progress through the subsequent chapters, you can always refer back to these guidelines if you need assistance with building data streams.

CHAPTER SUMMARY

In the field of advanced analytics, data mining demands a strong grasp of key terminologies, statistical principles, variable scaling, and the overall analytical process. This chapter explores two primary data mining processes widely adopted in both academic and industrial settings: SEMMA and CRISP-DM. While each methodology has its own strengths, our focus is primarily on the SEMMA process. Known for its robustness and ease of implementation, SEMMA aligns well with the objectives of data mining projects, providing a structured approach that yields clear and actionable insights. However, it is important to acknowledge that CRISP-DM remains a respected framework, valued for its practicality and effectiveness in industrial applications. Although we emphasize SEMMA, we recognize the continued relevance and applicability of CRISP-DM in certain contexts.

To maintain consistency throughout this book, we define key terminologies essential for understanding data mining methodologies and their associated programs. Additionally, we differentiate between supervised and unsupervised learning, providing a comprehensive list of key machine learning methods along with their descriptions and characteristics. This equips readers with the knowledge to select the appropriate method based on their study objectives, target variables, variable scales, and data size.

This chapter also includes a comparative analysis of five popular data mining software programs, highlighting their respective advantages and disadvantages. It is important to note that no software program is without flaws, and the choice of software ultimately depends on individual needs and proficiency levels. However, given that this book caters to a non-programmer audience, we have chosen SPSS Modeler as the primary software tool. If you are unfamiliar with SPSS Modeler, there's no need to worry—I provide detailed, step-by-step guidelines, complete with screenshots, to ensure ease of understanding and implementation.

DISCUSSION QUESTIONS

1. What is data mining process? Why is it important to follow a valid data mining process?
2. Describe all steps of the SEMMA and CRIPS-DM processes.
3. Compare SEMMA with CRISP-DM. What are the similarities and differences between these processes?
4. Explain the differences between prediction, predictive models, association models, and classification models.
5. What are primary variable measurement scales? When should we treat Likert scale variables as "interval"?
6. Can machines learn? Why do we refer to it as "machine learning?"
7. Differentiate between supervised learning and unsupervised learning? Provide examples of each type.
8. What are common machine learning methods for supervised learning and unsupervised learning?
9. What is ensemble modeling? In what situations should we use ensemble modeling?
10. How do you select the appropriate machine learning method for a data mining project?
11. Compare data mining software programs. Search the Internet to find the most current usage statistics for data mining software.

CHAPTER EXERCISES

Exercise 1: Add the Data File in SPSS Modeler and Identify the Target Variable

Follow the steps given below:

1. Open SPSS Modeler.
2. In the File menu, select New Stream.
3. Go to File/Save Stream As.
4. Select a location for your stream and name it Airline_Explore.
5. From the Sources group in the Node Pallet, double-click the Statistics File node to add it to the stream.
6. Double-click the Statistics File node to open the setting window for this node.
7. Click the ellipsis button (three-dot button) and select the AirlineRecommendation.sav file from your computer.

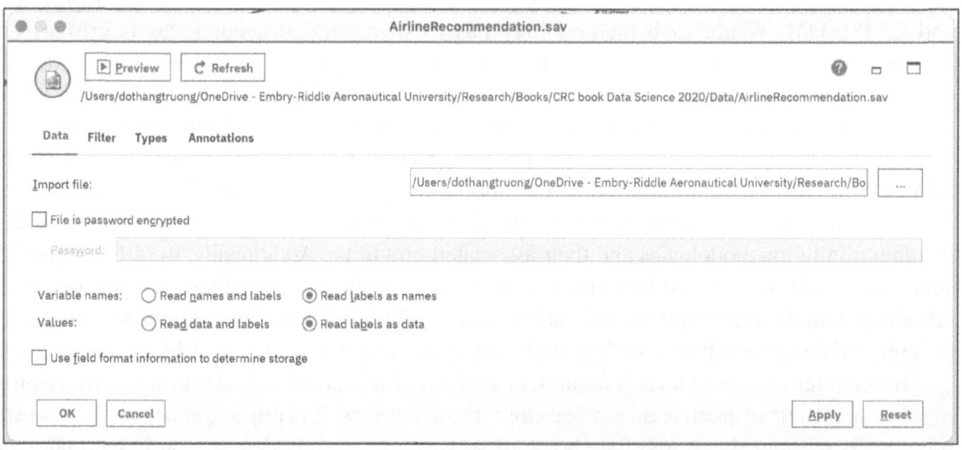

8. Click Preview to review the first ten rows of the data.

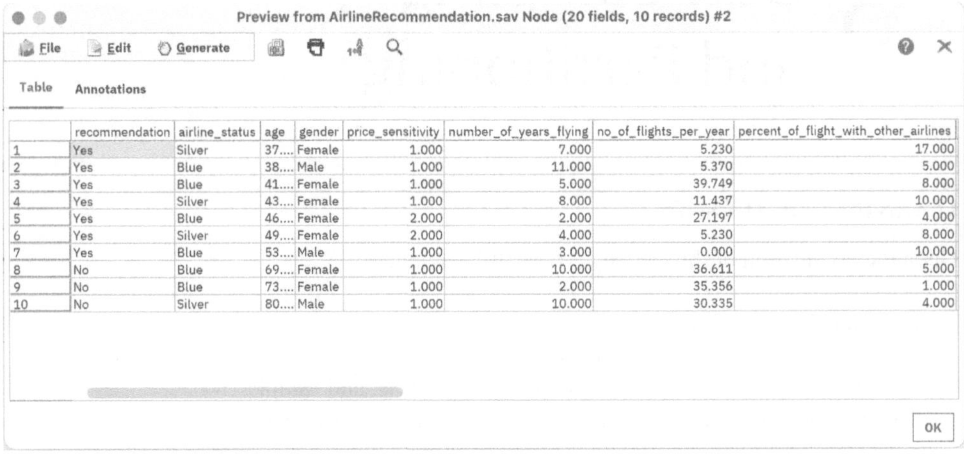

	recommendation	airline_status	age	gender	price_sensitivity	number_of_years_flying	no_of_flights_per_year	percent_of_flight_with_other_airlines
1	Yes	Silver	37....	Female	1.000	7.000	5.230	17.000
2	Yes	Blue	38....	Male	1.000	11.000	5.370	5.000
3	Yes	Blue	41....	Female	1.000	5.000	39.749	8.000
4	Yes	Silver	43....	Female	1.000	8.000	11.437	10.000
5	Yes	Blue	46....	Female	2.000	2.000	27.197	4.000
6	Yes	Silver	49....	Female	2.000	4.000	5.230	8.000
7	Yes	Blue	53....	Male	1.000	3.000	0.000	10.000
8	No	Blue	69....	Female	1.000	10.000	36.611	5.000
9	No	Blue	73....	Female	1.000	2.000	35.356	1.000
10	No	Silver	80....	Male	1.000	10.000	30.335	4.000

9. Click Apply and OK.
10. From the Field Ops group, add the Type node to the stream and connect it to the data node. We use this node to view and change the scales and roles of fields.
11. Double-click the Type node to open the setting window.
12. Check the measurement scales of all fields as below.
13. For the field *recommendation*, change the role to T*arget*.

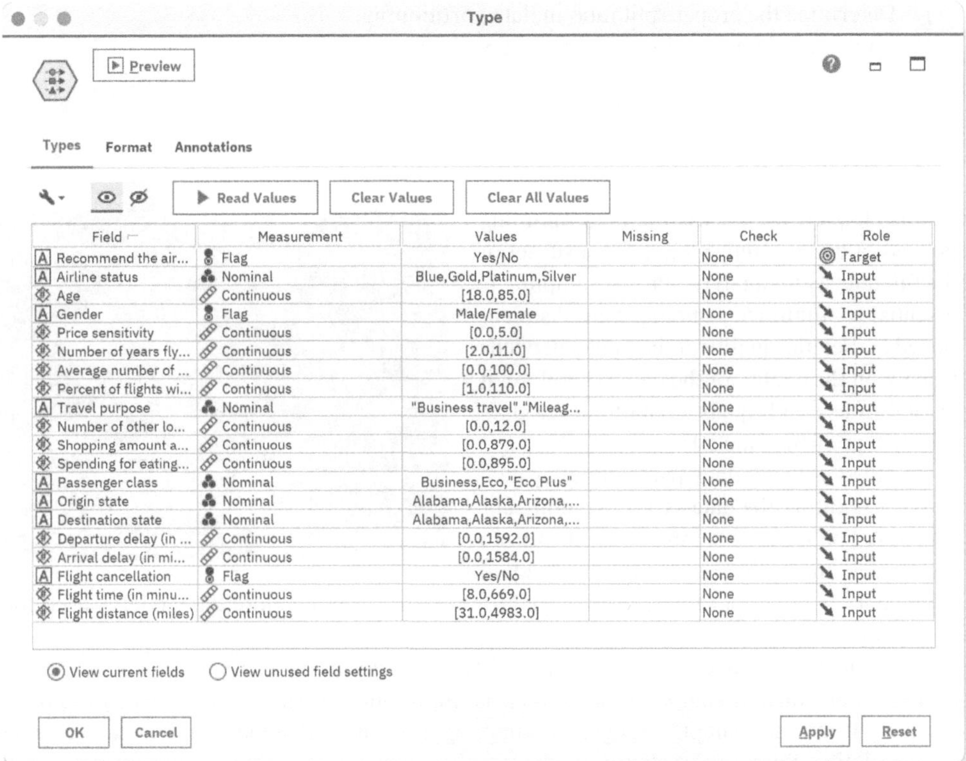

Field	Measurement	Values	Missing	Check	Role
Recommend the air...	Flag	Yes/No		None	Target
Airline status	Nominal	Blue,Gold,Platinum,Silver		None	Input
Age	Continuous	[18.0,85.0]		None	Input
Gender	Flag	Male/Female		None	Input
Price sensitivity	Continuous	[0.0,5.0]		None	Input
Number of years fly...	Continuous	[2.0,11.0]		None	Input
Average number of ...	Continuous	[0.0,100.0]		None	Input
Percent of flights wi...	Continuous	[1.0,110.0]		None	Input
Travel purpose	Nominal	"Business travel","Mileag...		None	Input
Number of other lo...	Continuous	[0.0,12.0]		None	Input
Shopping amount a...	Continuous	[0.0,879.0]		None	Input
Spending for eating...	Continuous	[0.0,895.0]		None	Input
Passenger class	Nominal	Business,Eco,"Eco Plus"		None	Input
Origin state	Nominal	Alabama,Alaska,Arizona,...		None	Input
Destination state	Nominal	Alabama,Alaska,Arizona,...		None	Input
Departure delay (in ...	Continuous	[0.0,1592.0]		None	Input
Arrival delay (in mi...	Continuous	[0.0,1584.0]		None	Input
Flight cancellation	Flag	Yes/No		None	Input
Flight time (in minu...	Continuous	[8.0,669.0]		None	Input
Flight distance (miles)	Continuous	[31.0,4983.0]		None	Input

14. Click Apply and OK.

3 Data Sampling and Partitioning

LEARNING OUTCOMES

Upon the completion of this chapter, you will be able to:

1. Define data sampling.
2. Explain the importance of data sampling in data mining.
3. Differentiate between population, sampling frame, and sample.
4. Compare probability sampling and non-probability sampling.
5. Identify sampling bias and understand its implications.
6. Evaluate and select the appropriate sampling method for various data mining projects.
7. Sample a subset of the full dataset using SPSS Modeler.
8. Determine the appropriate sample size for a data mining project.
9. Describe the challenges of an imbalanced dataset and overfitting.
10. Evaluate various strategies to address the imbalanced dataset.
11. Define oversampling and undersampling and understand their applications.
12. Describe SMOTE technique and its strengths and weaknesses.
13. Explain the importance of data partitioning in model validation.
14. Determine the proper split ratio in data partitioning.
15. Implement data partitioning using SPSS Modeler.

INTRODUCTION

In this chapter, we discuss the data sampling process, a crucial step in selecting a subset of the full dataset for analysis. Proper sampling is essential to minimize sampling bias. As mentioned earlier, the quality of the model depends heavily on the quality of the data. Therefore, it is vital to use high-quality data to develop an effective predictive model.

We also explore data partitioning, which involves splitting the dataset into several subsamples. The training sample is used to train the model, the validation sample is used to validate it, and the test sample is employed for the final testing before the model is used for actual predictions.

This chapter introduces key concepts and terminology in data sampling, which are necessary to understand various sampling methods. It is important to differentiate between the population, sampling frame, and sample. Our goal in sampling is to ensure that the sample accurately represents the population. Since our predictive model is trained using the training sample, we must ensure that the model is generalizable to other datasets. In other words, the model should be capable of making accurate predictions with any new dataset, thereby demonstrating its usability.

DOI: 10.1201/9781003468196-4

SAMPLING THE DATA

> **DEFINITION**
>
> Sampling is the process of selecting a portion of the entire dataset (sample) for analysis purposes.

Sampling is essentially the process of selecting a subset of the entire dataset (sample) for analysis purposes. The goal of sampling is to gather representative and reliable information from a population, especially when it is impractical or impossible to collect data from every individual or element within that population. Sampling is a crucial step in any research project to ensure that the data are sufficient and representative. Proper sampling allows us to create a quality dataset that is adequately sized and representative of the target population, enabling us to produce a valid and generalizable model.

To better understand the sampling process and how to sample data effectively, I will describe the concepts of population, sampling frame, sampling strategy, and sample size from the perspective of data mining. These concepts are deeply discussed in research methodology books that focus on survey, experiment, and interview designs, which are common methods for collecting primary data. However, in data mining studies, we apply these concepts slightly differently because we work with secondary data—data that has been collected prior to the project. Typically, such data are collected by marketing companies, industrial organizations, or governmental agencies. Given that these databases are often very large, it is not practical to analyze the full dataset. Therefore, it is essential to have a strategy to correctly sample a subset of the large dataset for analysis.

Figure 3.1 illustrates the differences between the population, sampling frame, and sample. We will discuss the details of these concepts, their interrelationships, and the various sampling methods available.

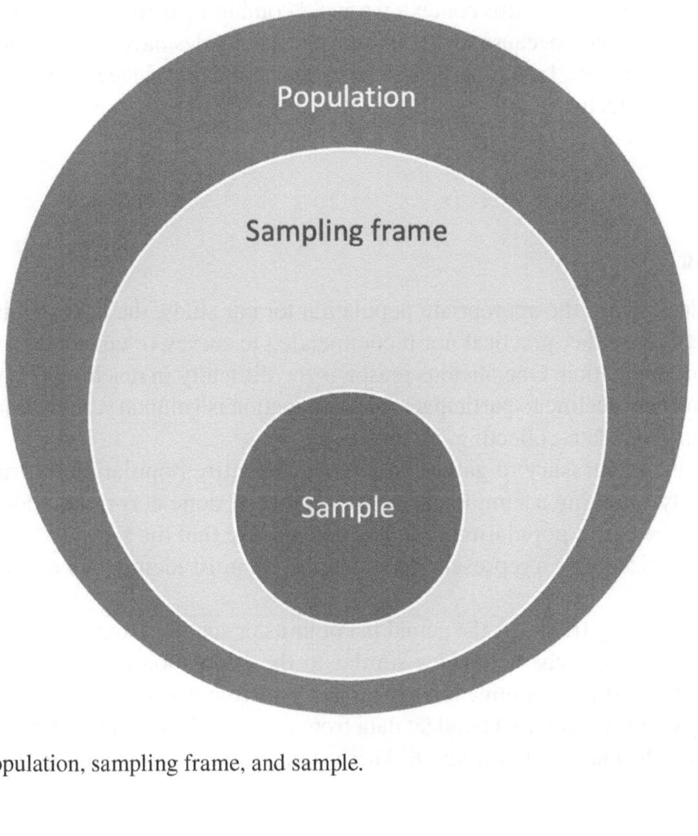

FIGURE 3.1 Population, sampling frame, and sample.

Population

"Population" refers to the primary target audience from which we want to examine data patterns and draw conclusions. The population in our study could consist of individuals, objects, cases, or specific industry sectors. The target population should be large enough to ensure the significance of our findings, yet focused enough to maintain the feasibility of the study.

> **DEFINITION**
>
> **Population** is the primary target audience from which we aim to examine the patterns of the data and draw conclusions.
>
> **Sampling frame** is the actual list of units or sources from which a sample is drawn.
>
> **Sample** is a subset of the population.
>
> **Sampling method** refers to the procedure that we use to select participants and collect the data.

For example, consider a project aimed at predicting passengers' airline recommendations. In this case, U.S. airline passengers represent an appropriate target population, covering U.S. residents who use air transportation for travel. Targeting worldwide residents would make data collection challenging due to the vastness of the population. Moreover, when focusing on residents of all countries, confounding variables—such as cultural, political, educational, or economic factors specific to each country—could affect the target variables. The absence of data on these variables could compromise the validity of our findings. By narrowing the focus to the U.S. market, we can more readily access data and draw valid conclusions about U.S. air travelers' decisions. We may choose to include residents of all ages or focus on those aged 18 and older, depending on the study's scope. In this book, I use a passenger survey dataset focused on predicting airline recommendations, specifically targeting adult passengers, as they are the ones who typically purchase tickets.

Another example of a population is all highway transportation incidents in Canada. By concentrating on this population, we can build predictive models to assess the risk of injuries in traffic incidents within that country. In this context, we avoid combining different modes of transportation, such as air, road, and water, because incidents across these modes have vastly different characteristics and contributing factors. Including such diverse data could introduce noise, leading to an invalid predictive model. By focusing solely on highway transportation, we can develop an effective and reliable model that accurately predicts highway incidents.

Sampling Frame

Once we have determined the appropriate population for our study, the next step is to establish the sampling frame. It is neither practical nor recommended to survey or collect data from every individual within our population. One obvious reason is the difficulty in reaching every individual, and the fact that some may decline to participate, as participation is voluntary. Individuals have the right to refuse requests from data-collecting organizations.

Moreover, it is unnecessary to gather data from the entire population to draw valid conclusions. Statistically, studying a sample of the population, if done correctly, allows us to generalize the results to the entire population. The key is to ensure that the sample is representative and of adequate size. To achieve a representative sample, we must identify an appropriate sampling frame.

Simply put, "sampling frame" is the actual list of units or sources from which a sample is drawn. This frame should possess characteristics similar to the population to ensure representativeness. For example, in the airline recommendation project, where our population consists of U.S. adult residents who travel by air, we might collect data from residents in six states: New York, Minnesota, Florida, Texas, California, and Washington. These states serve as our sampling frame, representing different regions—Northeast, Southeast, Midwest, Southwest, and Northwest—thereby ensuring the sample's representativeness.

However, if we were to collect data only from residents in California, a state in the West, our sample would likely be underrepresented. This is because residents of California may have unique characteristics and behaviors specific to the Western region, making them unrepresentative of the entire U.S. population. Such a discrepancy between the sampling frame and the population would introduce sampling bias, leading to analysis results that do not accurately reflect the population. This is why it is crucial to carefully identify the sampling frame before proceeding with data sampling.

Sample

"Sample" is a small subset of a population that we use for data analysis to make inferences about the entire population. For our conclusions about the population to be valid, the sample must be representative. To achieve this, we need a well-defined sampling frame and must draw the sample from that frame.

For example, in the airline recommendation project, we selected a sampling frame consisting of six states (New York, Minnesota, Florida, Texas, California, and Washington). We then collect data from qualified residents in those states. The qualifications require that residents must be 18 years or older and have used air transportation for travel in the past. This ensures that we focus on individuals who have used air travel and are capable of purchasing tickets. Their airline recommendations are valuable for predicting travel demand. A sample is considered representative if it shares the same attributes and characteristics as the population. Consequently, the analysis results based on this sample can be used to make inferences about the population.

Sampling Methods

Even with a well-constructed sampling frame, we may still end up with a biased sample if we do not use an appropriate sampling strategy. "Sampling method" refers to the procedure we use to select participants and collect data. For instance, when an organization conducts a survey, they send invitations to residents within the sampling frame (the six states mentioned above) and request their responses. However, how they select those participants is crucial to ensuring a representative sample.

In data mining, if the full dataset has been collected by another organization, such as an airline or a government agency, data scientists must decide how to select individuals from that dataset for the sample. This process is known as the sampling method. Choosing the right sampling method is essential to ensuring the quality of the data. In the next section, I will describe common sampling methods in detail.

BASIC CONCEPTS IN DATA SAMPLING

Data sampling is a fundamental technique in statistics and data mining used to make inferences about a larger population based on a smaller subset of data, known as a sample. The sample is carefully selected to ensure it accurately reflects the characteristics and diversity of the population. By analyzing the sample, data analysts can draw conclusions or make predictions about the entire population. Table 3.1 presents basic concepts in data sampling that are essential for accurately understanding and applying sampling methods.

TABLE 3.1
Basic Concepts in Data Sampling

Terms	Description
Population	Population is the primary target audience from which we want to examine the patterns of the data and draw conclusions.
Sampling frame	Sampling frame is the actual list of units or sources from which a sample is drawn.
Sample	Sample is a subset of the population. We use the sample for data analysis and make inferences about the population.
Sampling method	Sampling method is the procedure that we use to select participants and collect the data.
Sample size	Sample size refers to the number of individuals or subjects included in the sample. It should be determined based on statistical considerations.
Representative sample	Representative sample is a sample that accurately reflects the characteristics and diversity of the population.
Sampling bias	Sampling bias occurs when the selected sample does not accurately represent the population, leading to skewed or inaccurate results.
Probability sampling	Probability sampling is the method of selecting individuals randomly in such as way each individual in the sampling frame has an equal probability of being chosen.
Non-probability sampling	Non-probability sampling is the method of selecting individuals based on non-random criteria, and not every individual in the sampling frame has a chance of being included.
Margin of error	The margin of error is a measure of the uncertainty or variability in the estimates obtained from the sample.
Confidence level	The confidence level represents the degree of certainty or confidence we have in the estimates obtained from the sample.
Imbalanced dataset	An imbalanced dataset refers to a dataset where the number of samples in different classes significantly varies, resulting in unequal representation of the classes. In an imbalanced dataset, one class (the minority class) typically has significantly fewer instances than the other classes (majority classes).
Oversampling	Oversampling is a technique that balances the data by incrementing the size of the minority (rare events) to the same size as the majority.
Undersampling	Undersampling is a technique that balances the data by reducing the size of the majority class to the same size as the minority class.

SAMPLING METHODS

There are many different sampling methods, typically categorized into two primary types: probability sampling and non-probability sampling.

- **Probability sampling** involves the random selection of participants. This method allows for strong statistical inferences about the entire population.
- **Non-probability sampling** involves the non-random selection of participants based on convenience or other criteria. While this method is easier and more cost-effective, it raises concerns about sampling bias.

Before discussing these types of sampling methods, it's important to understand what sampling bias is and why minimizing it is crucial. Sampling bias occurs when a sample is collected in such a way that certain members of a population are systematically more likely to be selected than others. When sampling bias occurs, the findings from the data analysis of this sample cannot be generalized to the entire population. In other words, sampling bias poses a threat to external validity, limiting the generalizability of the findings. In such cases, conclusions can only be drawn about the specific sample rather than the intended population.

Sampling bias occurs when some members of a population are systematically more likely to be selected in a sample than others, which limits the representativeness of the sample.

Sampling bias is a serious issue since it poses a threat to external validity and limits the generalizability of the findings.

Although it can be challenging to completely eliminate sampling bias, efforts should be made to minimize it. Some sampling methods are more effective than others at reducing sampling bias.

Probability Sampling Methods

Probability sampling is recommended because it allows us to minimize sampling bias.

Probability sampling is the recommended method because it ensures data quality and minimizes sampling bias. In this approach, individuals in the sampling frame are selected randomly, giving each person an equal probability of being chosen. This method is effective in producing an unbiased representation of the total population. There are four common probability sampling methods used in research, which I will briefly describe with examples. Figure 3.2 provides a visual illustration of these methods.

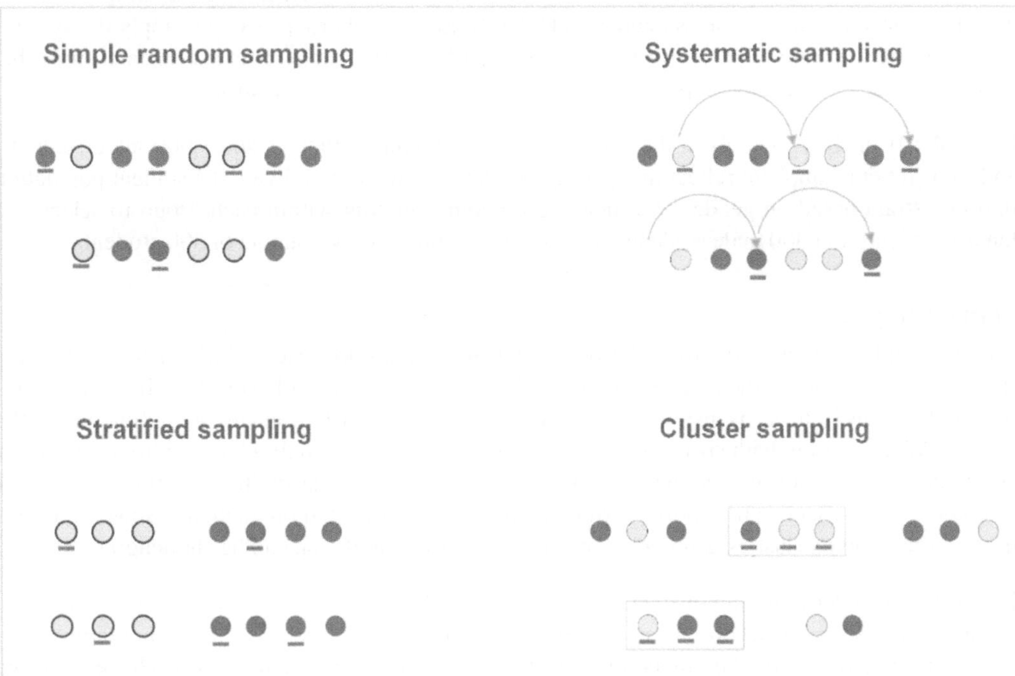

FIGURE 3.2 Probability sampling methods.

Simple Random Sampling

This method ensures that every member of the population has an equal chance of being selected. A common approach is to assign a number to each individual in the sampling frame and then use a random number generator (such as the "RAND()" function in Excel) to select the required number of participants. This way, the selection is entirely based on chance, eliminating sampling bias.

Example: Suppose we want to select a simple random sample of 5,000 high school students in a county. We would assign a number to each student in the database, from 1 to 5,000. Then, we use a random number generator to select 500 random numbers. The students corresponding to those 500 numbers will be chosen for the sample.

Systematic Sampling

Systematic sampling is a form of random sampling where a random generator is not employed. Instead, each individual is assigned a number, and we select these numbers at consistent intervals. It's crucial to emphasize that we must ensure the absence of any hidden patterns in the list that could potentially bias the sample. For instance, if the database arranges students by grade and lists them in ascending order of test scores, there exists a risk that our chosen intervals might inadvertently miss students with lower scores, leading to a sample that is skewed toward higher-scoring individuals.

Example: Continuing with our high school students, if we have a total of 5,000 students in the county, we first list them alphabetically. Then, we select every 10th student on the list, resulting in a sample of 500 individuals.

Stratified Sampling

Stratified sampling involves dividing the population into subgroups, or strata, based on relevant characteristics (e.g., gender, age range, grade, score range). After determining the sample size, we calculate how many individuals should be selected from each subgroup. We then apply the systematic sampling method within each subgroup to select the participants. This approach ensures that every subgroup is adequately represented, allowing for more precise conclusions.

Example: If the high schools in the county have 2,000 female students and 3,000 male students, and we want our sample to reflect this gender distribution, we would divide the student population into two strata based on gender. We then use random sampling within each group to select 200 female students and 300 male students, resulting in a representative sample of 500 students.

Cluster Sampling

Cluster sampling involves dividing the population into subgroups, each of which should possess similar characteristics to the entire population. These subgroups are called clusters. Instead of sampling individuals from each cluster, we randomly select several clusters and use all individuals in the clusters. When dealing with sizable clusters, it is possible to sample individuals within each cluster using one of the previously mentioned techniques, a method known as multistage sampling. Cluster sampling proves effective in handling extensive and widely dispersed populations. However, ensuring that the sampled clusters genuinely represent the entire population can be challenging.

Example: Consider high school students across a state with 20 high schools. If we don't have the resources to visit every school, we could assume that the schools have similar demographics and characteristics. We would then randomly select eight schools to form our clusters. However, if the schools differ significantly in demographics, some clusters might not be representative of the population, leading to a potentially biased sample. This is the primary disadvantage of cluster sampling.

Non-Probability Sampling Methods

In non-probability sampling, individuals are selected based on non-random criteria, meaning not every individual in the sampling frame has an equal chance of being included. Non-probability sampling is easier and more cost-effective to conduct, but it carries a higher risk of sampling bias, resulting in a less representative sample. Consequently, the inferences we can make about the population are weaker and more limited than those from probability samples. Therefore, if we opt to use non-probability sampling for any reason, we should still aim to make the sample as representative of the population as possible.

DEFINITION

Non-probability sampling is a method of selecting individuals based on non-random criteria, where not every individual in the sampling frame has an equal chance of being included.

Non-probability sampling methods are often used in exploratory and qualitative research with smaller sample sizes, such as interviews. Given the difficulty of randomly selecting a large sample for interview purposes, researchers often resort to non-probability sampling. In these types of studies, the goal is not to make broad generalizations about a large population but to develop an initial understanding of a smaller or under-researched population. Non-probability sampling is also common in experimental studies, where the focus is on examining the causal effect of a treatment on outcomes of interest, rather than studying a large population. Since experiments often require participants to spend extended periods in a controlled environment, such as a lab or classroom, it is challenging to recruit a large random sample. Researchers must rely on volunteers, leading to the use of non-probability sampling.

In data mining research, where we work with large datasets and aim to produce predictive models that are generalizable to other datasets, I do not recommend using non-probability sampling. With the amount of data typically available, we should be able to select a large sample and use probability sampling. However, I will briefly describe some common non-probability sampling methods so you can understand what they are and how they work.

Convenience Sampling

Convenience sampling involves selecting individuals who are readily accessible to the researcher. While it offers a straightforward and cost-effective means of collecting initial data, it falls short in producing results that can be generalized to the broader population. For instance, in a scenario where we're researching opinions on employee services within our organization, asking only colleagues working at the same level to complete a survey is a convenient approach. However, this limits the sample's representativeness as it doesn't encompass all employees at all levels within the organization.

Voluntary Response Sampling

Voluntary response sampling relies predominantly on accessibility. In this context, individuals volunteer to participate rather than being actively selected and contacted by the researcher. It's important to note that voluntary response samples tend to carry at least some degree of bias, as certain individuals are naturally more inclined to volunteer than others, potentially leading to non-representative results. For example, we send out a survey to all employees at the organization regarding health insurance benefits, and a lot of them decide to

> Non-probability sampling is easier and more cost-effective, but it carries a higher risk of sampling bias.
>
> In data mining research, probability sampling methods should be used to ensure that findings are generalizable.

complete it. While this approach can provide us with some insights into the subject, it's worth noting that the respondents are more likely to hold strong opinions about the services. Therefore, we cannot definitively assert that their viewpoints accurately reflect those of all employees.

Purposive Sampling

Purposive sampling, also known as judgment sampling, requires us to use our expertise to select a sample that is most useful for the purposes of the research. It is often employed in qualitative research, especially when the researcher seeks in-depth insights into a specific phenomenon rather than making statistical generalizations. It is particularly useful when dealing with a small and specific population. To ensure the effectiveness of purposive sampling, it's essential to establish clear criteria and a rational basis for inclusion. For example, if our goal is to gain a deeper understanding of the opinions and experiences of disabled employees within the organization, we deliberately select several employees with diverse support needs. This deliberate selection allows us to gather a wide range of data regarding their experiences with employee services.

Snowball Sampling

Snowball sampling involves recruiting participants through other participants. The researcher contacts a few people they know and asks them to refer others who might also participate. As more people are contacted, the sample size "snowballs," hence the term. For example, if we are researching the experiences of the LGBTQ+ community in our city, and there is no list of all LGBTQ+ individuals, probability sampling is not feasible. We might start by interviewing one person who then refers us to others in the community.

As you can see, non-probability sampling methods are easier and cheaper to implement, but they come with a higher risk of sampling bias and may result in a sample that is not representative of the population. Therefore, the findings cannot be generalized to a larger population. In data mining research, where we typically work with larger datasets, probability sampling should be used to ensure the data are representative. This approach allows us to develop predictive models that are more generalizable and applicable to other datasets within the same population.

SECONDARY DATA AND SAMPLE REPRESENTATIVENESS

When we collect data ourselves, we can select the sampling frame and apply one of the sampling methods described above to ensure the sample's representativeness. However, in many data mining projects, we rely on data collected by other organizations or parties—referred to as secondary data. In these cases, we treat the entire dataset as the sampling frame and use one of the sampling methods to draw a sample from it. But since we often don't know how the original data were sampled or collected, how can we be sure that our sample is representative?

To address this uncertainty, I recommend an effective approach: analyzing the demographic information of the sample and comparing it with that of the target population. By demonstrating that our sample shares similar demographic attributes and characteristics with the population, we can confidently conclude that our sample is representative, even if we are unaware of the original data collection methods. This approach is also useful when we collect data ourselves. Even if we choose the sampling frame and use probability sampling, there are still uncertainties in the data collection process that may introduce sampling bias—factors beyond our control that can affect representativeness. For example, the individuals from whom we collect data may not be as representative as we anticipated. Therefore, to ensure representativeness, it is prudent to compare the demographic information of the sample with that of the population.

Table 3.2 provides an example of comparing demographics between the sample and the target population. In most cases, demographic information about the population can be obtained from online

TABLE 3.2
Example of a Comparison between Sample and Target Population

Demographic Characteristics	Sample (%)	Population (%)
Gender		
• Male	45	48
• Female	43	42
• Other	12	10
Age ranges		
• 18–30	20	18
• 31–40	38	40
• 41–50	30	28
• >50	12	14
Income ranges		
• <$40,000	38	40
• 40,000–99,999	29	30
• 100,000–200,000	24	20
• >200,000	9	10
Education		
• High school	23	20
• Bachelor	48	50
• Graduate degree	29	30
Marital status		
• Married	43	45
• Single	57	55
Region		
• Northeast	15	17
• Southeast	23	21
• Southwest	17	18
• Midwest	19	20
• West	26	24

sources such as the U.S. Census Bureau or other organizations' publications. As shown in the table, while the characteristics of the sample and population may not be identical, the differences are minimal, and the data distributions are very similar. This comparison indicates the similarity between the sample and the population. In other words, our sample is representative of the population, and the findings from our sample can be confidently applied to the broader population.

> Comparing the demographics of the sample with those of the target population helps demonstrate the sample's representativeness.

SAMPLE SIZE

What Is Sample Size, and Why Is It Important?

Sample size refers to the number of individuals or subjects included in a sample. In data mining, sample size is a crucial consideration, as an adequate sample is vital to ensuring the quality and reliability of the findings.

Below are some key reasons highlighting the importance of sample size:

DEFINITION

Sample size refers to the number of individuals or subjects that are included in the sample.

1. **Representative patterns**: Data mining aims to discover meaningful patterns and relationships within data. A larger sample size provides a more representative subset of the population, increasing the likelihood of capturing accurate and reliable patterns. This leads to more precise and representative results of the population being studied.
2. **Increased generalizability**: A larger sample size enhances the generalizability of research findings. When the sample is more diverse and includes a greater number of individuals or cases, the results are more likely to apply to the broader population from which the sample is drawn, thereby improving the external validity of the study.
3. **Reliable model building**: Data mining often involves building predictive models or classifiers. A larger sample size allows for more robust and accurate model building, as the model can capture a wider range of variations and complexities in the data, leading to more reliable and generalizable predictions.
4. **Increased statistical power**: Statistical power refers to the ability to detect true effects or relationships within the data. A larger sample size increases the statistical power of a study, making it more likely to detect significant differences or associations if they exist. This is particularly important for drawing meaningful conclusions from the data, as data mining algorithms can detect patterns and relationships that might remain hidden in smaller samples, leading to more robust findings.
5. **Enhanced accuracy and confidence**: A larger sample size reduces variability and uncertainty in predictions. Confidence intervals become narrower, providing a more accurate range within which the true population parameter is likely to fall. This enhances the accuracy and reliability of the study's findings.
6. **Subgroup analysis**: With a larger sample size, researchers can conduct more robust subgroup analyses, investigating effects or relationships within specific subgroups of the population. Adequate sample sizes within each subgroup ensure more reliable conclusions about those particular groups.
7. **Increased stability**: A larger sample size provides more stability to the results, reducing the influence of random fluctuations or outliers. Outliers or extreme values have a lesser impact when the sample size is large, ensuring more robust and reliable findings.
8. **Accounting for variability**: Many real-world phenomena exhibit inherent variability. A larger sample size allows for capturing a broader range of variability, providing a more comprehensive understanding of the phenomenon under investigation.
9. **Handling complex and rare patterns**: In certain cases, data mining seeks to uncover complex or rare patterns that occur infrequently. A larger sample size increases the chances of encountering such patterns, making it easier to identify and understand these less common phenomena.
10. **Validating results**: Data mining often involves testing and validating discovered patterns or models on unseen data. A larger sample size increases confidence in the generalizability and reliability of the results during the validation process.

Sample Size for Traditional Statistical Research

Given the importance of sample size, the obvious question is: How large should the sample size be for a project?

Typically, in traditional research, there are two primary ways to determine the appropriate sample size: based on the population size and the statistical techniques used for analysis. First, the sample size must be sufficient to adequately represent the population. In this context, the sample size is determined by considering the population size, confidence level, and margin of error. Qualtrics provides a useful sample size calculator for this purpose (https://www.qualtrics.com/blog/calculating-sample-size/). The mathematical formula for calculating sample size is also included below as a reference.

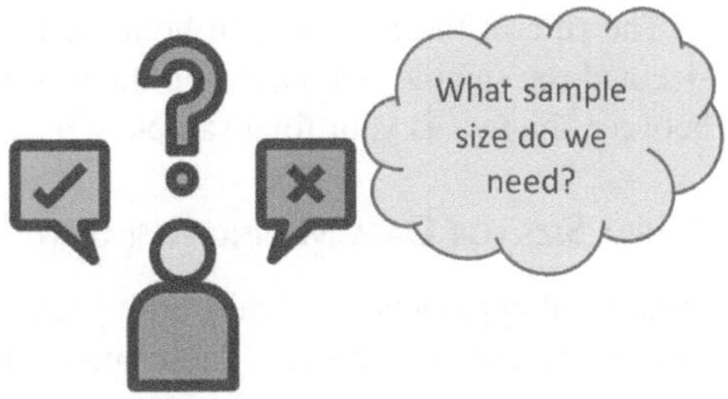

$$Sample\ size = \frac{(z\_score)^2 \times StdDev \times (1 - StdDev)}{(Margin\ of\ error)^2}$$

The confidence level indicates how confident you are in the results, expressed as the percentage of times that different samples (if repeated samples were drawn) would produce the same result. The common value for the confidence level is 95%. The margin of error is a statistical measurement that accounts for the difference between the actual and projected results. A typical value for the margin of error is 5%.

For example, using a 95% confidence level, a 5% margin of error, and a population size of 100,000, the calculated sample size is 383. If we increase the population size to 100 million, the required sample size is still 385. This is because the population size is used to calculate the z-score, which saturates once the population reaches a certain threshold. Therefore, even if our population size is 300 million, the needed sample size remains 385. This approach may raise further questions, such as whether the population size matters at all. This leads us to the second approach.

In the second approach, the sample size must be sufficient to ensure adequate statistical power and effect size for the analysis. The formula used to calculate this sample size varies depending on the statistical technique selected. I won't discuss the details of these methods, as there are numerous formulas for calculating sample size based on different statistical methods. However, I highlyrecommend the following two resources:

- **Sample size calculators by Daniel Soper**: This site offers a variety of statistical calculators, including those for sample size, effect size, and beta (Type II error rate). It also provides the formulas and academic references used to develop these calculators, ensuring their validity. The most useful calculators include those for multiple regression, hierarchical multiple regression, and structural equation modeling (see https://www.danielsoper.com/statcalc/).
- **G*Power software**: This open-source software application allows users to calculate sample size for various statistical methods, including t-tests, ANOVA, MANOVA, repeated measures, logistic regression, and multiple regression. It is widely used in academic research. However, some features require careful reading of the manuals to avoid miscalculations (see https://www.psychologie.hhu.de/arbeitsgruppen/allgemeine-psychologie-und-arbeitspsychologie/gpower).

The key to using these sample size calculators is to determine the primary parameters you expect, such as effect size, desired test power, significance level, number of independent variables (predictors), and number of dependent variables. Once you select the appropriate statistical method and expected values for these parameters, the tools will calculate the minimum sample size required to meet these statistical requirements. I won't go into detail about these parameters, as they are covered extensively in statistical literature.

Calculate two sample sizes: one based on the population size and the other based on the test power and effect size. Then, select the higher number between the two as your final sample size.

The rule of thumb is to use both methods: calculate the sample size based on the population size and the sample size based on the test power and effect size. Then, choose the higher number between the two as your final sample size.

Sample Size for Data Mining Research

Traditional approaches to determining sample size may not be as applicable in data mining studies. These approaches primarily focus on traditional statistical research, whereas many machine learning methods are not based on statistics. Additionally, these traditional calculations often result in sample sizes ranging from 50 to 500 responses, which, while sufficient for traditional statistical analysis, do not adequately capture the scope of data mining studies that focus on uncovering unknown patterns within large datasets.

As discussed in Chapter 1, when we refer to data size, we consider both the number of variables (variety) and the number of records (volume). To detect meaningful patterns in terms of relationships among variables without a predefined theoretical foundation, we need a large number of variables in the model. Limited variables restrict the number of patterns (relationships) that can be discovered. Moreover, excluding important input variables can result in a predictive model that fits the data well from a modeling perspective but lacks practical implications.

When more variables are included in the model, the explanatory power of the model often decreases. I will explore the concept of explanatory power in more detail in Chapter 6, but essentially, it refers to the model's ability to explain the variance in the target variable based on the variance in the predictors. As we add more variables, the model's explanatory power may diminish. To maintain explanatory power with a large number of variables, we require more data points for training the model, which means a larger sample size. Additionally, data partitioning for validation purposes further increases the need for a larger sample size.

So, how large a sample size is sufficient for data mining projects? This is a challenging question, as there is no definitive answer. Some authors suggest that "the more, the better," while others claim that 10,000 data points are sufficient. I found no consistent guidelines on this topic. The required sample size also depends on the data source, as we cannot use more data than what is available. For instance, some data providers may offer millions of records from hourly operational data, while others, such as those collecting accident data, may only provide thousands of records. Ultimately, we need enough data to train and validate models and ensure the generalizability of the results. However, we must also consider the computing power needed to process very large datasets, as training complex models, such as neural networks, with vast amounts of data can be time-consuming. Based on my experience with data mining projects, I will offer my recommendations.

Imbalance Ratio

Another important factor in sampling is the imbalance ratio (IR), which is the rate of the events we want to predict. For example, if our target variable is flight delays, the imbalance ratio is the proportion of delayed flights to the total number of flights. The events in this case are delayed flights, while non-events are on-time flights. A balanced dataset is crucial for developing a valid and generalized predictive model. An imbalanced dataset, where the imbalance ratio is too low or too high, can lead to inaccurate predictions. For instance, if the imbalance ratio is 0.1, meaning 10% of the data are events and 90% are non-events, the model may inaccurately predict that all cases are non-events, resulting in a high rate of false negatives. Although the overall prediction accuracy may appear high (90%), the model fails to predict the events (flight delays), rendering it invalid and unusable. Conversely, if the imbalance ratio is too high, with a large number of events compared to non-events, the model may have a high rate of false positives, leading to similar issues. One common solution for an imbalanced dataset is to resample the data to achieve a more manageable response ratio, a topic we will explore later in this chapter.

In conclusion, when selecting a sample size for data mining, it is essential to consider both the number of records and the imbalance ratio. For example, a sample size of 100,000 records with a imbalance ratio of 0.4–0.6 is generally sufficient, whereas a sample size of 1 million cases with a imbalance ratio of 0.03 may still be inadequate. Ideally, a imbalance ratio of 0.5 represents a perfect balance between events and non-events. However, achieving this perfect balance in practice is rare. My recommendation is to aim for a sample size of at least 100,000 cases with a imbalance ratio between 0.4 and 0.6.

> For data mining projects aim to obtain a sample size of at least 100,000 records with a imbalance ratio of 0.4–0.6.

EXAMPLE: SAMPLING

In this section, we will learn how to sample data using SPSS Modeler. Suppose you have a very large dataset for data mining purposes and do not intend to use all the data points. The main reason for this is that such a large dataset may be unnecessary for modeling and could slow down the computing process, affecting model training time. For example, if you have a dataset with 10 million data points but decide to use 100,000 data points for modeling, how do you sample these 100,000 data points from the total 10 million without sacrificing data quality and generalizability? I will demonstrate how to do this using SPSS Modeler.

The first step is to choose the appropriate sampling method. As previously mentioned, probability sampling methods are recommended for data mining projects. SPSS Modeler provides several sampling methods to choose from. In most cases, when dealing with a balanced dataset, random sampling is the preferred method unless there is a specific reason to select another method. Random sampling allows you to draw a sample randomly while maintaining the imbalance ratio. For example, if your full dataset has a imbalance ratio of 0.45 and you want to sample 10% of the data, random sampling will provide you with 10% of the data while preserving the 0.45 imbalance ratio.

Once you have selected the appropriate sampling method, drawing a sample from the full dataset is quite straightforward. Most data mining software applications have built-in sampling features that make this process easy. I will now demonstrate how to perform data sampling in SPSS Modeler. Suppose you decide to use the random sampling method. In SPSS Modeler, there are two ways to perform sampling: when creating a new data source or by using the Sample node. I will show you how to use the Sample node, as it offers more configuration options, allowing you to customize your sampling process and adjust the settings as needed.

> From this chapter onward, I will primarily use field names rather than field labels in the examples and software instructions to avoid potential confusion caused by lengthy labels. Although the outputs may still display labels, using field names provides a more concise way to explain the analysis steps and results. Please refer to Table 3.1 for the corresponding field names and labels.

We will use the example dataset, **AirlineRecommendation**, which we explored in detail in Chapter 3. For the sampling step, we will focus on the target variable, **recommendation**, which indicates whether passengers would recommend the airline for air travel. This is a binary variable with two levels: **Yes** (recommend the airline—events) and **No** (not recommend the airline—non-events). The total dataset contains 124,464 data points, with a imbalance ratio of 0.475 (or 47.5%), which is considered balanced. Our goal is to sample 50% of the data using random sampling.

Follow the steps below to set up the Sample node:

1. Open SPSS Modeler.
2. Create a new stream and name it "Airline_Sample."
3. From the **Sources** group in the Node Pallet, double-click the **Statistics File** node to add it to the stream.
4. Double-click the **Statistics File** node to open the settings window for this node.
5. Click the ellipsis button (three-dot button) and select the **AirlineRecommendation.sav** file from your computer.
6. From the **Field Ops** group, add a **Type** node to the stream and connect it to the data node. This node is used to view and modify the measurement scales and roles of fields.
7. Double-click the **Type** node to open its settings window.
8. Check the measurement scales of all fields as listed below. For the field **recommendation,** change the role to **Target**.

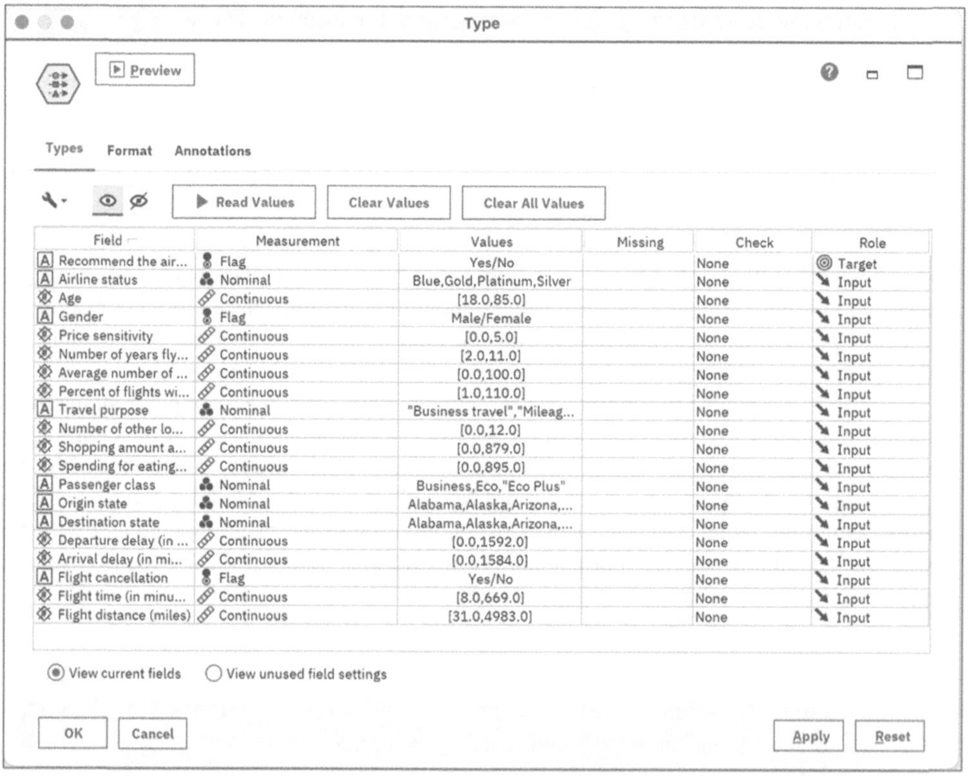

Field	Measurement	Values	Missing	Check	Role
Recommend the air...	Flag	Yes/No		None	Target
Airline status	Nominal	Blue,Gold,Platinum,Silver		None	Input
Age	Continuous	[18.0,85.0]		None	Input
Gender	Flag	Male/Female		None	Input
Price sensitivity	Continuous	[0.0,5.0]		None	Input
Number of years fly...	Continuous	[2.0,11.0]		None	Input
Average number of ...	Continuous	[0.0,100.0]		None	Input
Percent of flights wi...	Continuous	[1.0,110.0]		None	Input
Travel purpose	Nominal	"Business travel","Mileag...		None	Input
Number of other lo...	Continuous	[0.0,12.0]		None	Input
Shopping amount a...	Continuous	[0.0,879.0]		None	Input
Spending for eating...	Continuous	[0.0,895.0]		None	Input
Passenger class	Nominal	Business,Eco,"Eco Plus"		None	Input
Origin state	Nominal	Alabama,Alaska,Arizona,...		None	Input
Destination state	Nominal	Alabama,Alaska,Arizona,...		None	Input
Departure delay (in ...	Continuous	[0.0,1592.0]		None	Input
Arrival delay (in mi...	Continuous	[0.0,1584.0]		None	Input
Flight cancellation	Flag	Yes/No		None	Input
Flight time (in minu...	Continuous	[8.0,669.0]		None	Input
Flight distance (miles)	Continuous	[31.0,4983.0]		None	Input

9. From the **Records** group in the Node Pallet, add the **Sample** node to the stream and connect it to the **Type** node.

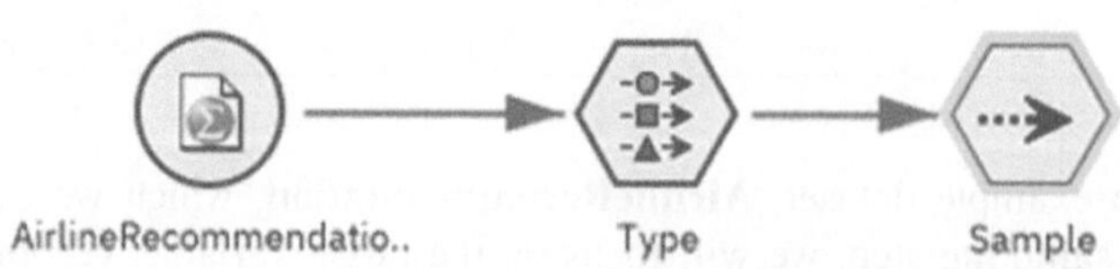

10. Double-click the **Sample** node to open its settings window.

11. Under the Settings tab, select **Complex** for the Sample method. Choose **Random** for the Sample type and **Proportions** for the Sample units. To sample 50% of the data, set the **Fixed** value to **0.5** for the Sample size.

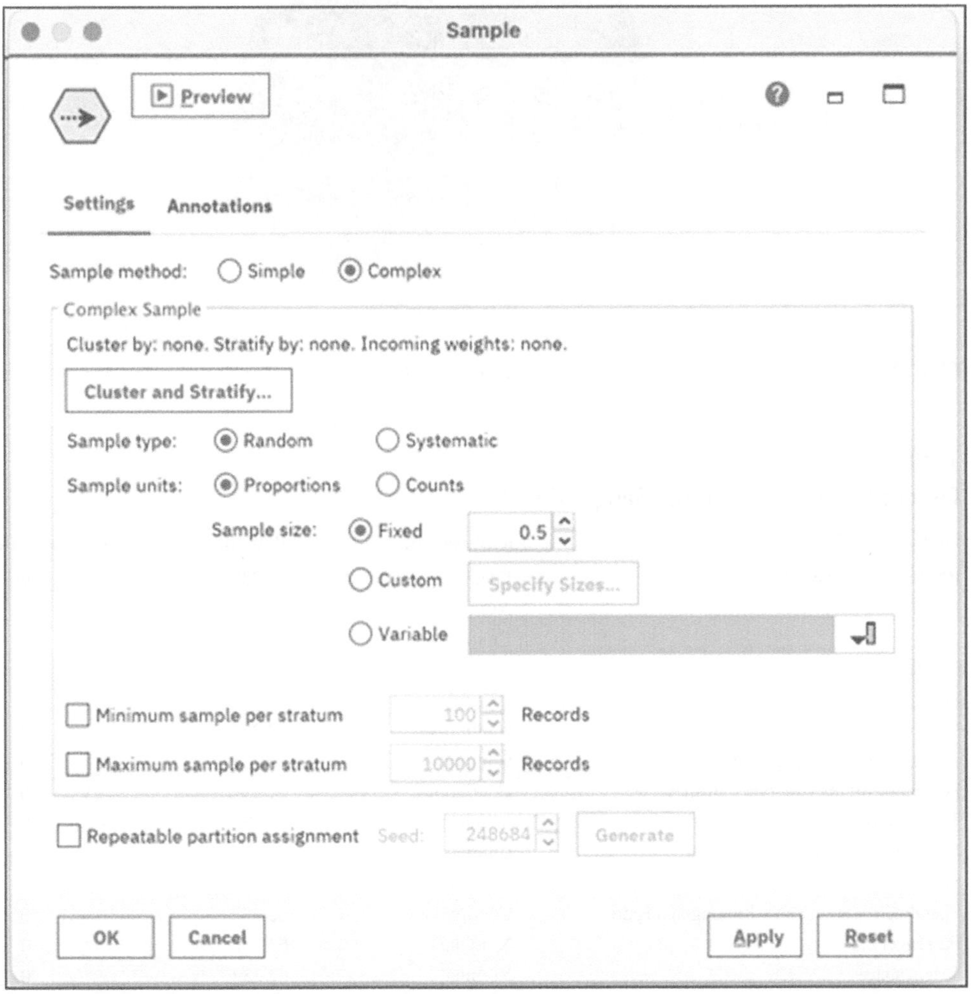

12. Click **Apply**, and then click **OK**.

This process will prompt SPSS Modeler to randomly sample 50% of the data, as intended. To view the sampled data, add a **Data Audit** node to the Sample node and run it. As mentioned earlier, the **Data Audit** node will generate graphical results along with descriptive statistics for our sampled data.

Figure 3.3 presents the sampling results for the target variable, recommendation. Table 3.3 provides the descriptive statistics of the sampled data. For this example, I have selected only the field name, measurement scale, valid records, and null value (missing data). Some key outputs to focus on are as follows:

> From this chapter onward, instead of displaying screenshots of numeric results, I will convert the Data Audit outputs into a table format for easier viewing.

- Imbalance ratio remains consistent after sampling: recommended: 52.2%; not recommended: 47.8%.
- Sample data size: 62,232 valid records.

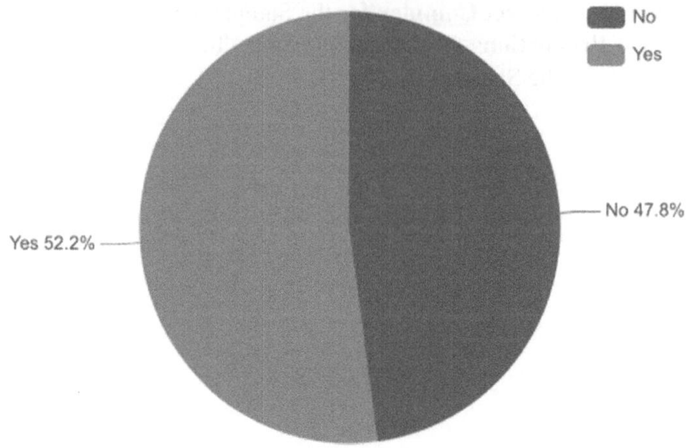

FIGURE 3.3 Results of the sampling for the **recommendation** field.

TABLE 3.3
Descriptive Statistics of the Sampled Data

Field	Measurement	Valid Records	Null Value (Missing)
Recommend the airline	Flag	62,232	0
Airline status	Nominal	62,232	0
Age	Continuous	62,232	0
Gender	Flag	62,232	0
Price sensitivity	Ordinal	62,232	0
Number of years flying	Continuous	62,232	0
Average number of flights per year	Continuous	62,232	0
Percent of flights with other airlines	Continuous	62,232	0
Travel purpose	Nominal	62,232	0
Number of other loyalty cards	Continuous	62,232	0
Shopping amount at airport	Continuous	62,232	0
Spending for eating and drinking at airport	Continuous	62,232	0
Passenger class	Nominal	62,232	0
Origin state	Nominal	62,232	0
Destination state	Nominal	62,232	0
Departure delay (in minutes)	Continuous	61,101	1,131
Arrival delay (in minutes)	Continuous	60,908	1,324
Flight cancellation	Flag	62,232	0
Flight time (in minutes)	Continuous	60,908	1,324
Flight distance (miles)	Continuous	62,232	0

As you can see, the software randomly selects 50% of the data for the sample, maintaining nearly the same imbalance ratio as the full dataset (52.46% recommended and 47.54% not recommended). This confirms that the sampling process has been completed correctly.

IMBALANCED DATASET

As previously discussed, a balanced dataset is crucial for producing a valid and useful predictive model. An imbalanced dataset poses a high risk of prediction error. A balanced dataset typically has a imbalance ratio close to 0.5, meaning there is a similar number of events (positives) and

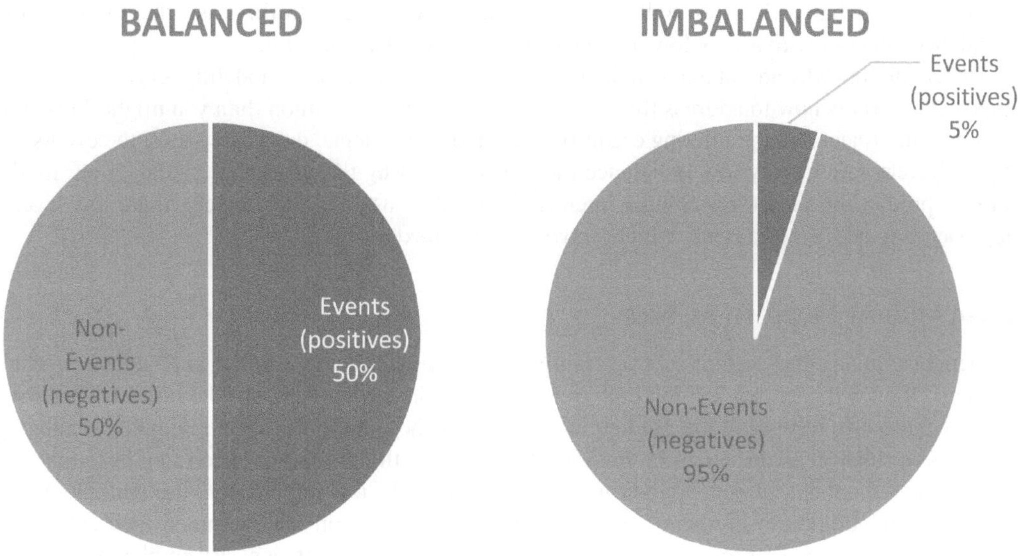

FIGURE 3.4 Balanced vs. imbalanced datasets.

non-events (negatives). Figure 3.4 illustrates the difference between a balanced dataset and an imbalanced dataset.

In the balanced dataset, 50% of the data represents events, and 50% represents non-events, resulting in a imbalance ratio of 50%. Conversely, the imbalanced dataset consists of 5% events and 95% non-events, leading to a imbalance ratio of 5%. In the context of the imbalanced dataset, the events are referred to as the **minority class** (comprising only 5% of the data), while the non-events are referred to as the **majority class** (comprising 95% of the data).

Be cautious when building predictive models for imbalanced data, as the overall prediction accuracy can be misleading.

Why does a highly imbalanced dataset become a problem in predictive modeling? Consider a dataset where 5% of the data represent events (positives) and 95% represent non-events (negatives). When we train our model with this dataset, the imbalance causes the algorithm to treat all cases as non-events because it aims to achieve the highest prediction accuracy. Remember, the algorithm cannot understand the meaning of the variables or data; it only processes the numbers and tries to fit them into a model. In this scenario, the model might achieve an overall prediction accuracy of 95%, which seems very high. However, upon closer inspection, we realize that the model fails to predict any events (positives), rendering it practically useless, especially if our goal is to predict those specific events.

Why does this happen? When we break down the predictions into positive and negative categories, we find that the true positive rate is 0%, while the false negative rate is 100%. This occurs because the algorithm classifies all cases as non-events (negatives) to maximize overall accuracy. As a result, the overall prediction accuracy can be misleading. To accurately evaluate model performance, we must also consider other metrics, such as specificity, sensitivity, F1-score, Gini coefficient, ROC curve, and Lift chart, which we will discuss in detail in Chapter 6. In this case, due to the highly imbalanced dataset with a very low imbalance ratio, the model is invalid.

Understanding why an imbalanced dataset poses a significant risk to modeling is crucial. It's also important to know how to address this issue because it is more common than you might think. For example, in projects like predicting credit fraud, aviation accidents, or cyberattacks, the events are typically very rare (a very low imbalance ratio), yet predicting these risks is critical. If our model fails to predict the target variable of interest, it becomes useless. Fortunately, there are several approaches to address this issue, which we will explore next.

Using Multiple Model Fit Metrics

As demonstrated in the example above, training a good model with an imbalanced dataset is challenging. Often, a model may exhibit high overall prediction accuracy but still be invalid. Therefore, it is crucial not to rely solely on prediction accuracy or the misclassification rate when evaluating model performance, as these can be misleading. Note that the misclassification rate is simply one minus the prediction accuracy. To thoroughly evaluate models, it is important to use multiple model fit metrics. In addition to prediction accuracy and misclassification rate, common metrics include sensitivity, specificity, F1-score, Lift chart, ROC curve, and Gini coefficient. By employing multiple metrics, we can assess the model's performance from various perspectives, providing a comprehensive view of its effectiveness. We will explore these metrics in detail in Chapter 6.

> Misclassification rate = 1– prediction accuracy

Data Partitioning and Model Validation

One primary concern in predictive modeling is overfitting, where a model is overly trained on the training data, leading to high prediction accuracy for that specific dataset but poor performance on new data. This lack of generalizability renders the model unusable. To mitigate this risk, in addition to using multiple model fit metrics, it is essential to partition the data into training and validation samples. The training sample is used to build the model, while the validation sample is used to test its performance. This approach helps detect any issues with the model by comparing results from the training and validation samples. Model validation is especially crucial when dealing with imbalanced datasets. We will discuss the details of data partitioning later in this chapter.

Cost-Sensitive Learning

Another approach to addressing imbalanced datasets is cost-sensitive learning (CSL). Rather than comparing models based solely on prediction accuracy or misclassification rate, CSL considers the costs associated with misclassification. The goal is to develop a model that minimizes the total cost of misclassification. To achieve this, a payoff table, or decision matrix, is created for the target profile, outlining the costs of misclassification for each class.

CSL is a widely used method in data mining to enhance the performance of predictive models with imbalanced datasets. This subset of machine learning takes into account the unequal costs of misclassification (false negatives or false positives). By defining these costs for all classes, we can train and select the model that minimizes the total cost.

In scenarios with an imbalanced dataset and a very low or high imbalance ratio, the cost of false negatives is typically much higher than the cost of false positives. For example, in predicting serious health issues, false negatives can lead to delayed treatment and severe consequences. When comparing models, the one with the lowest total cost is preferred over the one with the lowest misclassification rate. CSL thus optimizes the model's performance by considering the costs associated with different types of classification errors, effectively addressing class imbalance issues and improving the model's applicability in real-world situations where the consequences of errors vary significantly.

Alternatively, if the focus is on maximizing profits rather than minimizing costs, a profit matrix can be defined with detailed profits for accurate predictions. In this case, the model with the highest total profit is selected. Throughout this book, we will focus on examples that emphasize cost considerations. I will get into the details of CSL and how to apply it in Chapter 6.

Resampling—Oversampling and Undersampling

One common approach to addressing an imbalanced dataset is to resample the data to create a balanced dataset. This approach should be considered if there is a risk to model validity due to the imbalanced data. In other words, if the previous methods do not effectively resolve data imbalance issues and there is a concern that our models may be invalid, then resampling becomes a viable option.

> **DEFINITION**
>
> **Oversampling**: Balance the data by incrementing the size of the minority class (rare events) to match the size of the majority class.
>
> **Undersampling**: Balance the data by reducing the size of the majority class to match the size of the minority class.

Resampling involves modifying the imbalanced dataset to balance events and non-events. Two primary methods for accomplishing this are oversampling and undersampling. Figure 3.5 illustrates the differences between these two methods.

- **Oversampling (Upsampling)**: This technique balances the data by increasing the size of the minority class (rare events) to match the size of the majority class. In oversampling, we increase the number of records in the minority class within the training sample by duplicating cases. The primary advantage of this method is that it preserves all the information from the original dataset, allowing us to retain all records from both the minority and majority classes. However, the duplication of data makes models more prone to overfitting, as they may overly learn from the same data points. Additionally, oversampling can result in a very large dataset, which could increase computing time.

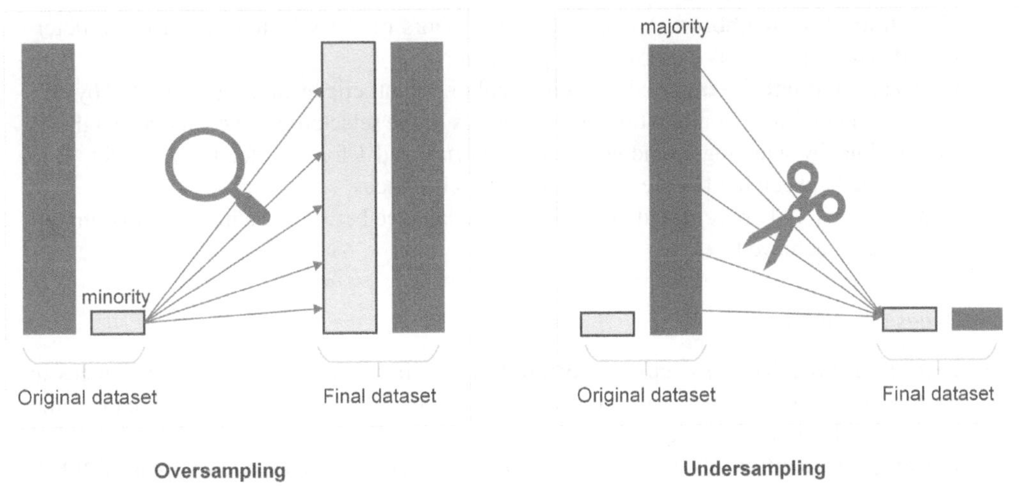

FIGURE 3.5 Oversampling and undersampling.

The most common technique for oversampling is **random oversampling**, which randomly replicates minority class records. While this technique is straightforward, it often leads to overfitting due to repeated exposure to the same data points. A more advanced technique is the **Synthetic Minority Oversampling Technique (SMOTE)**, which uses the k-nearest neighbor method to create synthetic cases that are close in the feature space, reducing the risk of overfitting.

- **Undersampling (Downsampling)**: This technique balances the data by reducing the size of the majority class to match the size of the minority class. A common method for undersampling is **stratified sampling**. The primary advantage of this approach is improved training time, as it results in a smaller dataset. However, by removing many records from the majority class, we risk losing critical information that could be important for the model.

It's important to note that resampling is not random sampling, which can raise concerns about sampling bias. As previously discussed, it is crucial to address potential sampling bias by comparing the demographics of the sample with the population to ensure the sample's representativeness.

Synthetic Minority Over-Sampling Technique (SMOTE)

SMOTE is a widely used method in machine learning to address the problem of data imbalance. The SMOTE algorithm, proposed by Chawla et al. (2002), effectively creates synthetic examples for the minority class, thereby balancing the class distribution. This approach helps in training machine learning models that can better recognize and generalize patterns related to the minority class. However, while SMOTE can be beneficial, it may not always lead to improved performance, and its effectiveness depends on the specific dataset and problem at hand.

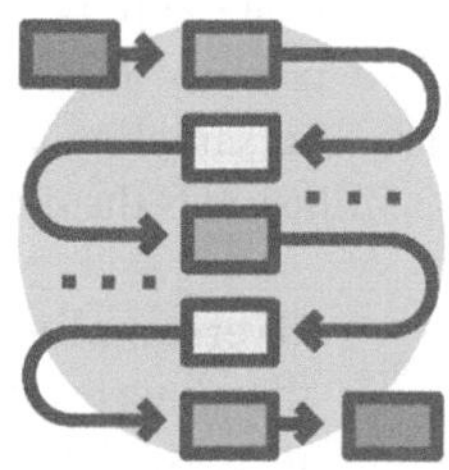

Overview of How SMOTE Works

- **Identify minority class instances**: The first step in SMOTE is to identify instances belonging to the minority class, as the algorithm is specifically designed to address this imbalance.
- **Select a minority instance**: A random instance from the minority class is chosen as the starting point for generating synthetic samples.
- **Find k-nearest neighbors**: The k-nearest neighbors of the selected instance are determined. The value of k is a user-defined parameter.
- **Generate synthetic instances**: For each neighbor, a synthetic instance is generated by creating a random linear combination of the features of the selected instance and its neighbor. This is done by selecting a random value between 0 and 1 for each feature and using it to interpolate between the feature values of the two instances.
- **Repeat**: Steps 2–4 are repeated until the desired balance between the minority and majority classes is achieved.

Advantages of SMOTE

- **Increased minority class representation**: SMOTE increases the number of instances in the minority class by generating synthetic samples, leading to a more balanced dataset and improving the model's ability to learn patterns in the minority class.
- **Better generalization**: By augmenting the minority class with synthetic samples, SMOTE can help the model generalize better and avoid overfitting on the majority class.

- **Improved model performance**: In scenarios where class imbalance negatively impacts model performance, applying SMOTE can enhance overall model performance, particularly in terms of recall and sensitivity for the minority class.
- **No information loss**: SMOTE does not result in the loss of information from the original dataset. It addresses the imbalance issue without discarding any existing data points.

Disadvantages of SMOTE

- **Risk of overfitting**: While SMOTE aims to address class imbalance, it may introduce noise and lead to overfitting, especially if the synthetic samples are not representative of the true underlying distribution of the minority class.
- **Sensitivity to noise**: SMOTE may be sensitive to noisy samples in the dataset. If the minority class contains outliers or noisy data points, the synthetic samples generated by SMOTE might not accurately represent the class.
- **Potential misclassification**: The introduction of synthetic samples can sometimes result in misclassification errors, as the model may struggle to differentiate between genuine and synthetic instances, particularly if the synthetic samples are not well-distributed.
- **Computational complexity**: Generating synthetic samples involves additional computational overhead, especially for large datasets, which can increase training times and resource requirements.
- **Dependency on k-nearest neighbors (k-NN)**: SMOTE relies on the k-NN algorithm to generate synthetic samples. The choice of the k parameter can impact the quality of synthetic samples, and an inappropriate choice may lead to suboptimal results.

In summary, while SMOTE is a valuable tool for addressing data imbalance, its application should be done carefully, considering the specific characteristics of the dataset and the potential impact on model performance. It is essential to evaluate the trade-offs and monitor the model's behavior when using SMOTE in a data mining project.

EXAMPLE: RESAMPLING DATA

I will demonstrate how to perform data resampling using SPSS Modeler. For this example, let's consider the **Credit Card Fraud** dataset. The dataset is highly imbalanced, with a imbalance ratio of 0.2%, meaning that only 0.2% of the records are fraud cases, while 99.8% are not (Table 3.4).

EXAMPLE

- **Dataset:** CreditFraud.sav
- **Target variable:** Class (Flag: True-1; False-0)

We apply three different resampling approaches: undersampling, oversampling, and SMOTE.

TABLE 3.4
Hypothetical Imbalanced Dataset

Variable	Formatted Value	Frequency Count	Percent
Class	True (1)	492	0.2
Class	False (0)	284,315	99.8

Undersampling Strategy

Before running the software, we need to calculate the factor that will be used to set the record balancing directives. In undersampling, the goal is to reduce the size of the majority class to match the size of the minority class. The factor is calculated as follows:

$$Factor = \frac{Minority\ Class\ Size}{Majority\ Class\ Size} = \frac{True}{False} = \frac{492}{284{,}315} = 0.00173$$

Thus, in this case, we use a factor of 0.0018 to balance the dataset.

Follow the steps below to perform the undersampling strategy.

1. Open SPSS Modeler.
2. Create a new stream and name it "Credit_Fraud."
3. From the **Sources** group in the Node Pallet, double-click the **Statistics File** node to add it to the stream.
4. Double-click the **Statistics File** node to open its settings window.
5. Click the ellipsis button (three-dot button) and select the **CreditFraud.sav** file from your computer.
6. From the **Field Ops** group, add a **Type** node to the stream and connect it to the data node. This node allows you to view and change the measurement scales and roles of fields.
7. Double-click the **Type** node to open its settings window.
8. Check the measurement scales of all fields. For the field **Class**, change the role to **Target**.
9. From the Records group in the Node Pallet, add a **Balancing** node to the stream and connect it to the **Type** node. Rename it to **Undersampling**.
10. Double-click the **Undersampling** node to open its settings window.
11. Under the Settings tab, set **Condition**: **Class=0; Factor=0.002** (this setting reduces the number of negative cases to match the size of the positive cases, balancing the data).

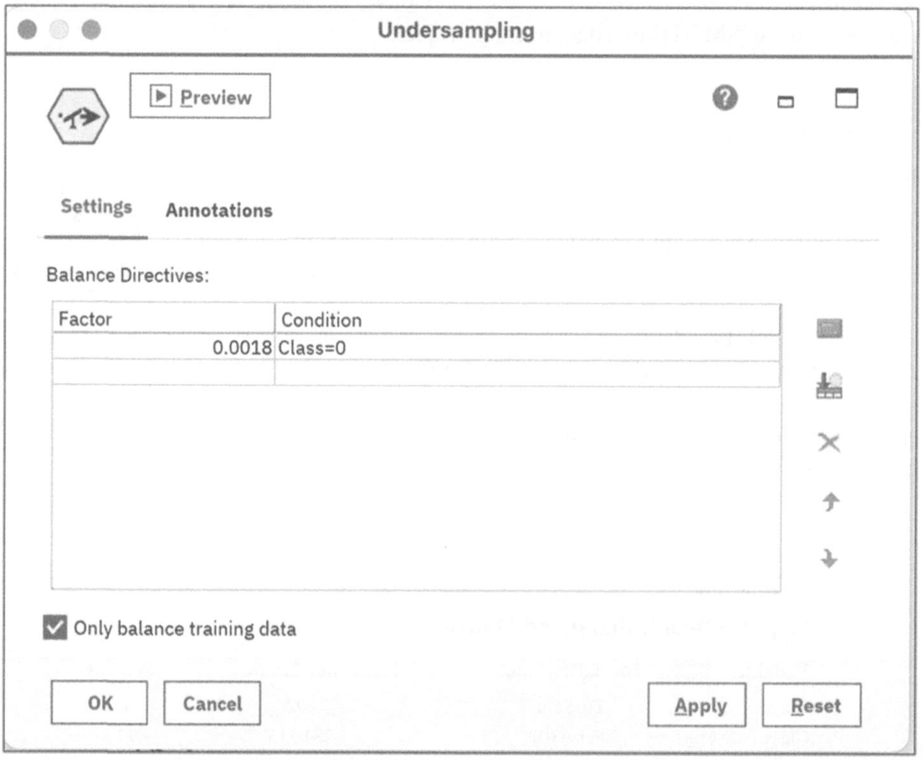

12. Click **Apply**, and then click **OK**.

TABLE 3.5
Results of the Undersampling

Data	Field	Formatted Value	Frequency Count	Percent
Sample	Class	True (1)	492	50.62
	Class	False (0)	480	49.38

You could use the **Data Audit** node to obtain the full statistical results. However, in this case, I used the **Distribution Chart** node to view the resampling results (Table 3.5). As you can see, the Balance node produces a balanced dataset with 50.62% fraud cases and 49.38% non-fraud cases, which is very close to the ideal 50% response, indicating a balanced dataset. The total number of records is now 972 cases. The output of this node is the newly balanced dataset, which can be used to train models. However, it's important to note that the dataset is now much smaller, which may result in information loss. This example demonstrates how to balance data using undersampling in SPSS Modeler.

Oversampling Strategy

In oversampling, the goal is to increase the size of the minority class to match the size of the majority class. The factor is calculated as follows:

$$Factor = \frac{Majority\ Class\ Size}{Minority\ Class\ Size} = \frac{False}{True} = \frac{284315}{492} = 577.88$$

Accordingly, we use a factor of 580 to balance the dataset.

Follow the same steps (1–8) as with the undersampling approach. Then, continue with the following steps:

9. From the **Records** group in the Node Pallet, add a second **Balancing** node to the stream and connect it to the Type node. Rename it to **Oversampling**.
10. Double-click the **Oversampling** node to open its settings window.
11. Under the Settings tab, set **Condition**: **Class=1; Factor=580** (this setting increases the number of positive cases to match the size of the negative cases, balancing the dada).

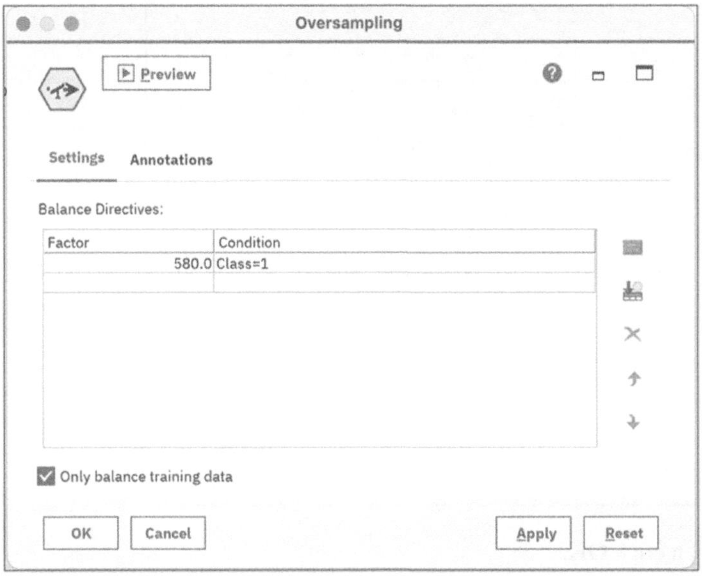

12. Click Apply, and then click OK.

TABLE 3.6
Results of the Oversampling

Data	Field	Formatted Value	Frequency Count	Percent
Sample	Class	True (1)	285,360	50.09
	Class	False (0)	284,315	49.91

Table 3.6 shows the resampling results. As you can see, we now have a balanced dataset with 50.09% fraud cases and 49.91% non-fraud cases. The total number of records is now 569,675, which is significantly larger than the original sample size. The output of this node is the new balanced dataset, which can be used to train the models. However, because many duplicated positive cases have been added, the data may be prone to overfitting. Additionally, the increased dataset size may require more time and computing power during the modeling step.

SMOTE

As mentioned earlier, SMOTE is a more advanced resampling technique that uses the k-nearest neighbor method to select cases that are close in the feature space. By augmenting the minority class with synthetic samples, SMOTE can enhance model generalizability and reduce the risk of overfitting on the majority class.

Follow the same steps (1–8) as with the undersampling approach. Then, continue with the following steps.

9. From the **Records** group in the Node Pallet, add a **SMOTE** node to the stream and connect it to the Type node. Rename it to **Oversampling**.
10. Open the settings window of this node.
11. Under the Settings tab, select **Class** for Target Field, set **Over Sample Ratio** to **Auto**, select **Regular** for **Algorithm Kind**, and set **K Neighbour** to **5** (for **Sample Rules**).

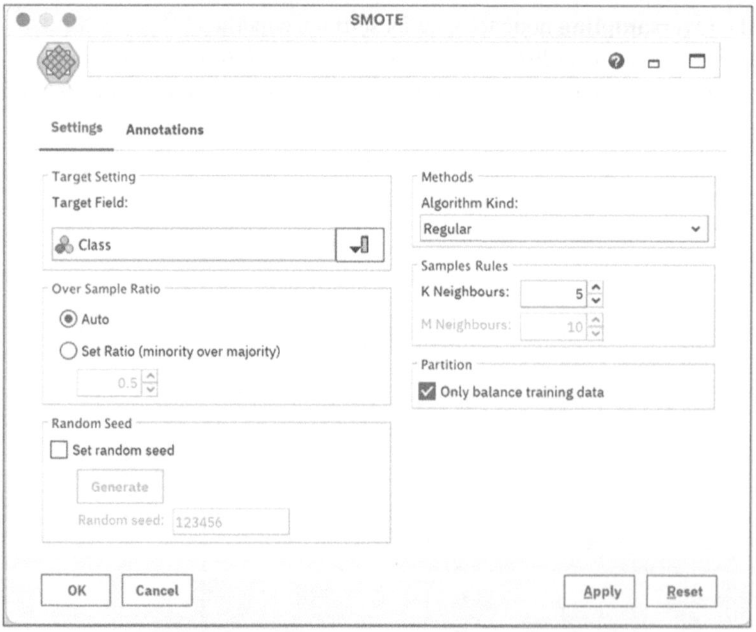

12. Click **Apply**, and then click **OK**.

TABLE 3.7
SMOTE Results

Data	Field	Formatted Value	Frequency Count	Percent
Sample	Class	True (1)	284,315	50
	Class	False (0)	284,315	50

Table 3.7 shows the SMOTE results, indicating a balanced dataset with 50% fraud cases and 50% non-fraud cases. The total number of records is now 568,630. Although we still end up with a much larger dataset, the data are more generalizable compared to the simple oversampling technique. Consequently, the models are less prone to overfitting.

DATA PARTITION

In data mining, a popular strategy for assessing the quality of model generalization is to partition the data source. As mentioned earlier, overfitting can occur in data mining when a model is overtrained to achieve high accuracy. Such a model may perform poorly with a new dataset, raising concerns about its generalizability. In other words, the model works well with the training sample but is likely to fail when applied to different datasets.

To avoid this issue, we partition the data into multiple datasets. The **training sample** is used for preliminary model fitting. The remaining data are reserved for empirical validation and are typically split into two parts: the **test sample** and the **validation sample**. The test sample is used to evaluate the model and prevent overfitting. By comparing the results between the training and test samples, we can determine whether the model is generalizable. Finally, the validation sample is used for a final assessment of the model.

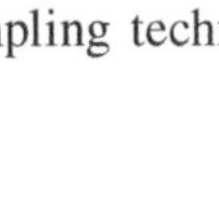

- While most software programs consistently name the first partition sample **Train**, the names for the second and third samples can vary, which may cause confusion.
- Keep in mind that the purposes of the second and third samples are the same; the differences lie only in their naming. Table 3.8 illustrates the differences between SPSS Modeler and SAS Enterprise Miner.
- Since we are using SPSS Modeler in this book, I will use the names provided by this software to ensure consistency and make it easier to follow.

TABLE 3.8
Sample Name Differences between SPSS Modeler and SAS Enterprise Miner

Software	First Partition Sample	Second Partition Sample	Third Partition Sample
SPSS Modeler	Train	Test	Validation
SAS Enterprise Miner	Train	Validation	Test

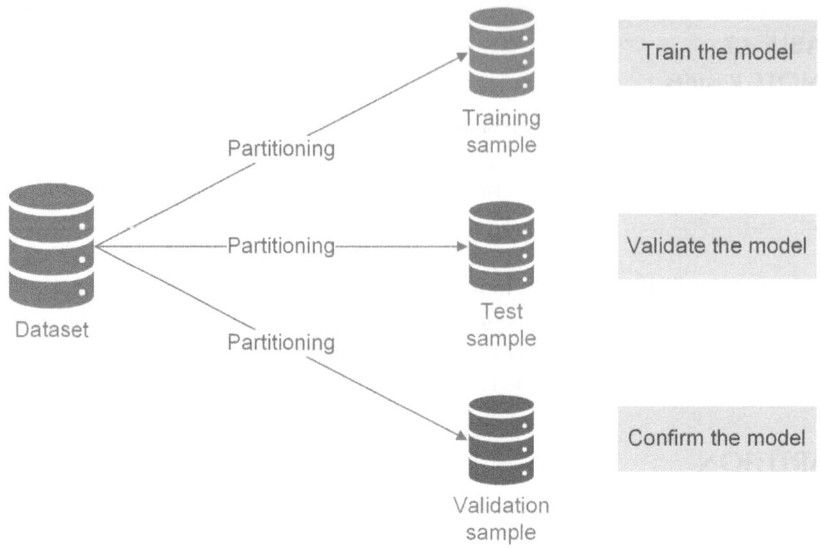

FIGURE 3.6 Data partitioning.

Figure 3.6 illustrates how data partitioning works. Why do we need three subsamples for data mining purposes?

First, the **training sample** is used to train the model and identify one with a good model fit. Next, we apply this model to the **test sample** to evaluate and validate its performance. If the model does not validate successfully, we reconfigure the model, retrain it, and test the new version. This process is iterative and may require multiple trials and errors until we find a well-validated model. Finally, we use the **validation sample** to confirm the generalizability of this final model.

Split Ratio

A common question in data partitioning is about the split ratio—how much of the data should be allocated to the training sample versus the validation and test samples? For example, a 60–40 split ratio typically means that 60% of the data go to the training sample, while the remaining 40% are divided between the validation and test samples. In this case, the 40% is usually split evenly, with 20% of the data used for testing and another 20% for final validation.

There is no standard rule for the split ratio. The literature suggests various ratios, ranging from 50–50, 60–40, 70–30, 80–20, to even 90–10. To determine the appropriate split ratio, it is essential to consider the size of the dataset. The rule of thumb is that the three samples must reflect data trends similar to those of the original dataset. Otherwise, sampling bias may occur, leading to a biased dataset and, ultimately, an invalid predictive model. It is also crucial to ensure that each sample has enough data points for modeling purposes. Typically, more data are allocated to the training sample to achieve a good model fit, but the validation and test samples should also have sufficient data points to evaluate and validate the model adequately. Therefore, data size is a critical factor.

For example, if the original dataset contains 100,000 data points, a 60–40 split ratio could be used, resulting in 60,000 records for training, 20,000 for testing, and 20,000 for final validation. This split could also be referred to as 60–20–20 for clarity. If the dataset has more than 1 million data points, a split ratio of 80–20 (or 80–10–10) may be appropriate, providing 800,000 records for training, 100,000 records for testing, and 100,000 records for final validation. On the contrary, if the sample size is small, with fewer than 1,000 data points, a 60–40 split ratio should be used without a validation sample. This ensures that the test sample contains enough data to evaluate the model effectively.

TABLE 3.9
Recommended Split Ratio Based on the Sample Size

Sample Size	Recommended Split Ratio
Fewer than 1,000 data points	60–40 with no test sample (train: 60%; validation: 40%)
1000–1 million data points	60–20–20 (train: 60%; test: 20%; validation: 20%)
1–10 million data points	80–10–10 (train: 80%; test: 10%; validation: 10%)
More than 10 million data points	90–5–5 (train: 90%; test: 5%; validation: 5%)

Table 3.9 provides my recommended split ratios for different data sizes. Please note that these are general recommendations, and depending on the specific machine learning method used, adjustments to the split ratio may be necessary.

EXAMPLE: DATA PARTITIONING

This section demonstrates how to partition data in SPSS Modeler using the Airline Recommendation dataset. The software sets default values of 50–50–0 for three subsamples. Since we have more than 100,000 data points, we will partition our dataset into three subsamples using a split ratio of 60–40. Specifically, 60% of the data will be used for training, 20% for testing, and 20% for final validation.

EXAMPLE

- **Dataset:** AirlineRecommendation.sav
- **Target variable:** Recommend (Flag: Yes/No)

Follow the steps below to partition the data.

1. Open the stream "Airline_Sample."
2. We should have the data node and Type node created before.
3. From the Field Ops group in the Node Pallet, add a **Partition** node to the stream and connect it to the **Type** node.

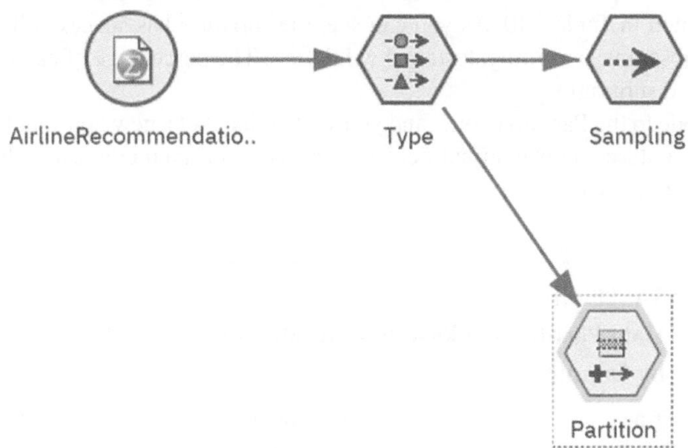

4. Open the settings window of this node.
5. Under the Settings tab, select **Train, test, and validation** for **Partition**.

6. Set the **Training partition size** to **60,** the **Testing partition size** to **20,** and the **Validation partition size** to **20**.

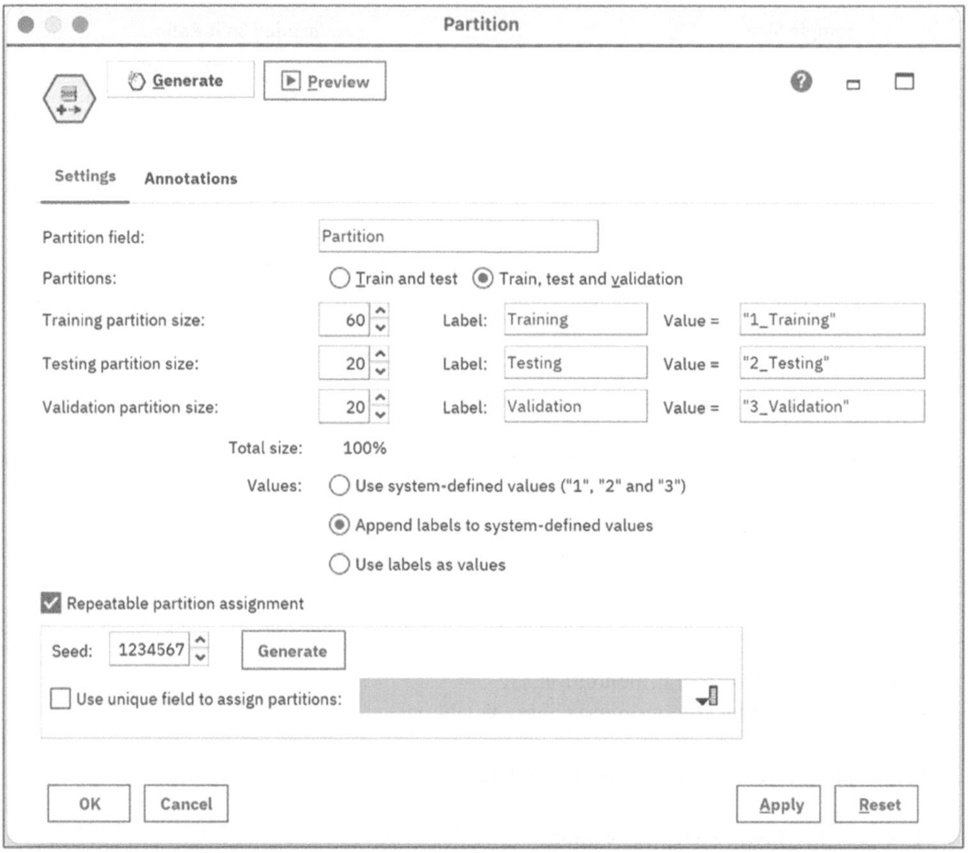

7. Click **Apply**, and then click **OK**.

SPSS Modeler will partition the data according to the size and split ratio we predefined. The results are presented in Table 3.10. As you can see, the software has successfully split the original data into three subsamples: training, test, and validation. The percentage of each subsample is very close to the desired split ratio.

Add a Type node to the Partition node, and you will notice that a new nominal field, Partition, has been added to the dataset. This nominal field enables us to train a model using the training sample and then test it with the test sample.

TABLE 3.10
Data Partitioning Results—Number of Observations by Sample

Type	Number of Observations	Percent
Training sample	74.747	60.6
Test sample	24,862	19.98
Validation sample	24,855	19.97

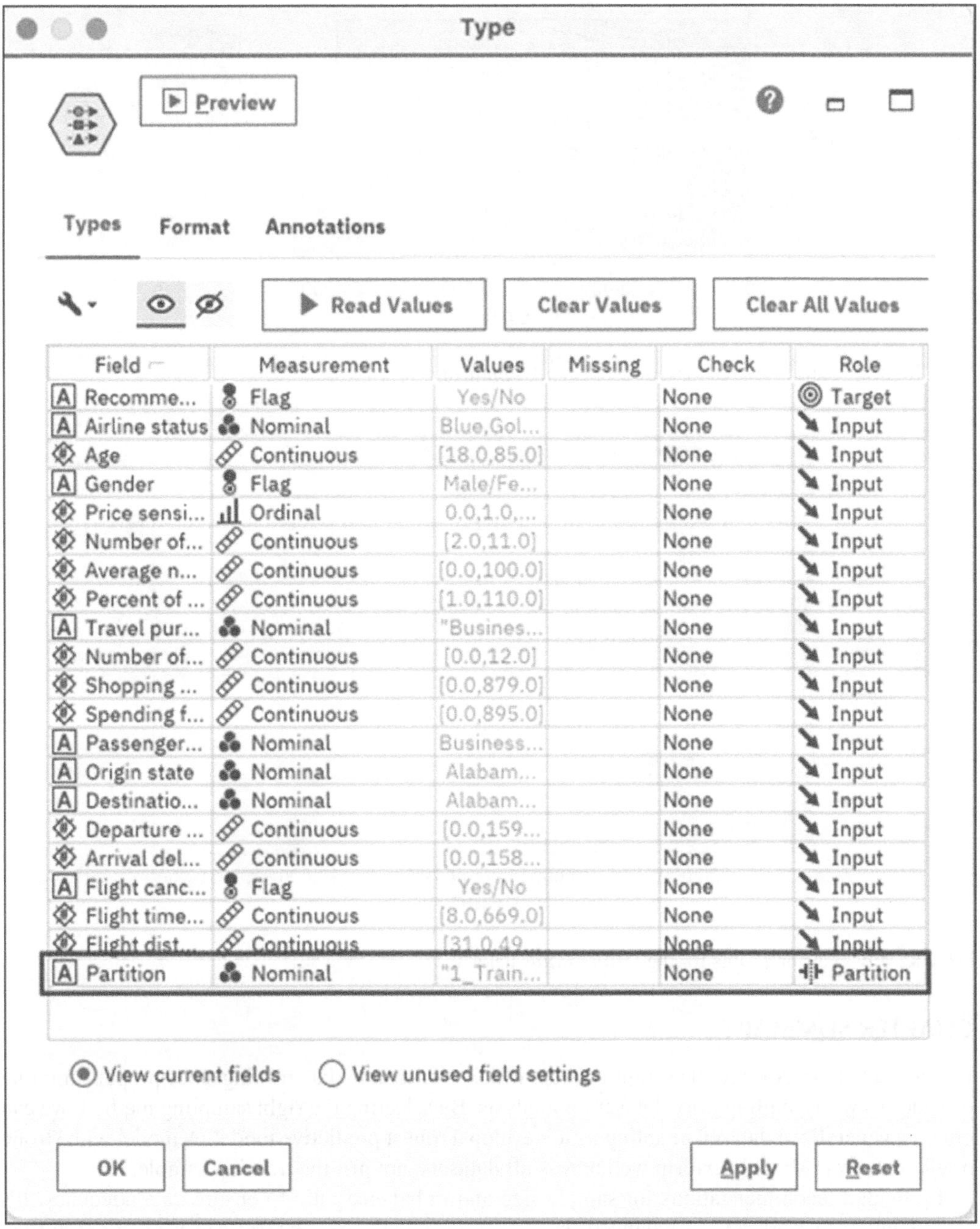

One important thing to keep in mind is that the partitioning process is done randomly while preserving all the characteristics of the original dataset. Figures 3.7 and 3.8 show the partition results for **recommendation** and **airline status**. As you can see, the imbalance ratio for **recommendation** was maintained at 52.5% across all three subsamples, just as in the original dataset. Similarly, the ratio across the four levels of **airline status** remains consistent across the subsamples, mirroring the original dataset. This indicates that our partition samples are representative of the full dataset. In other words, it is safe to train and test our model using these subsamples.

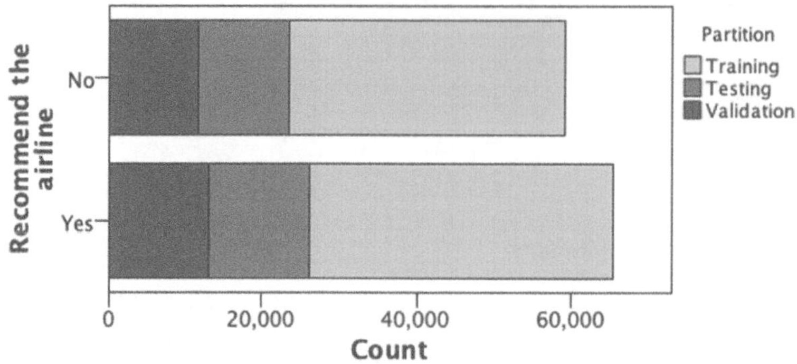

FIGURE 3.7 Patrion results for the *recommendation* field.

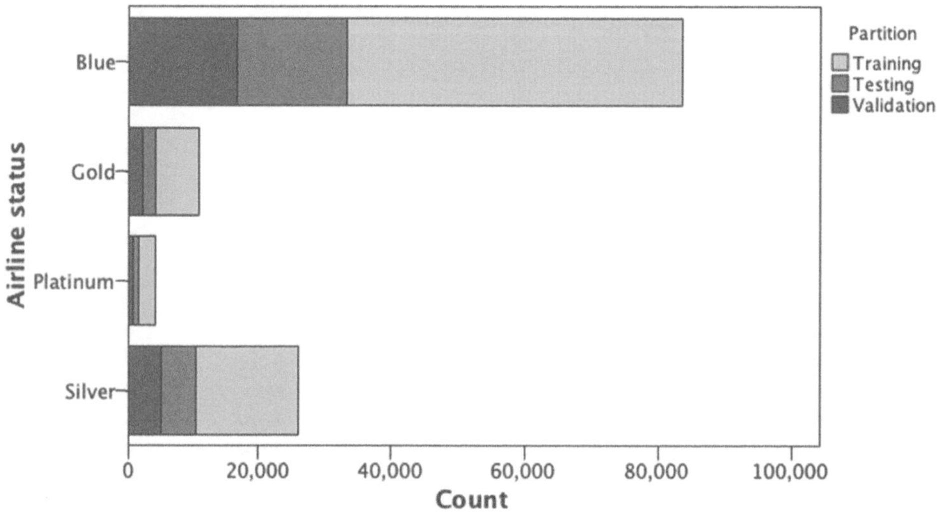

FIGURE 3.8 Patrion results for the *airline status* field.

CHAPTER SUMMARY

In this chapter, we covered the sampling step, which is crucial in data mining since proper sampling provides us with a high-quality dataset for analysis. By selecting the right sampling method, we can obtain a generalized dataset, enabling us to develop a robust predictive model. A model with strong predictive power should perform well across all datasets, not just the training sample.

I provided recommendations for sample size and imbalance ratio to ensure data adequacy. It's also essential to work with a balanced dataset, as an imbalanced dataset can lead to overfitting, resulting in an invalid and unusable predictive model. To address data imbalance, several strategies can be employed, including using multiple model fit criteria, partitioning data, resampling, and CSL.

Resampling is a common technique used to balance an imbalanced dataset. We explored three different resampling methods: undersampling, oversampling, and SMOTE. Additionally, data partitioning is an important step to divide the dataset into subsamples. Typically, the training sample is used to train the model, the test sample is used to evaluate the model, and the validation sample is used to confirm the final model's generalizability.

I also provided recommendations for the split ratio based on different sample sizes. This chapter included demonstrations on how to use SPSS Modeler for data sampling, resampling, and data partitioning.

CHAPTER DISCUSSION QUESTIONS

1. What is the purpose of sampling, and why is it important in data mining?
2. Describe the differences between population, sampling frame, and sample. How are they related? Provide real-life examples of a population, sampling frame, and sample.
3. What is sampling bias? Provide some examples of sampling bias.
4. Compare probability sampling and non-probability sampling. Which methods are more effective in minimizing sampling biases?
5. Provide an example of a data mining project in your business. Choose the appropriate sampling method for this project and justify your decision.
6. What are the methods for calculating the required sample size? Use Daniel Soper's sample size calculator and G*Power to calculate a sample size for your project.
7. What is overfitting, and how does it affect the validity of a predictive model?
8. Describe the main concerns with an imbalanced dataset. What are the primary strategies to address an imbalanced dataset?
9. What are the differences between oversampling and undersampling? Provide specific examples.
10. Describe the SMOTE technique. What are the advantages and disadvantages of this resampling technique?
11. What is the purpose of data partitioning? How do we determine the appropriate split ratio in data partitioning?

CHAPTER EXERCISES

Exercise 1: Data Partitioning for the Boston Housing Dataset

1. Upload the **Boston** dataset to your computer.
2. Open the dataset in SPSS Modeler.
3. Create a new stream **Partition** and add the data source to the workspace. Add a Type node and connect it to the data source.
4. Add a Partition node to the stream and connect it to the Type node.
5. Partition the data using the split ratio 60–40: training sample: 60%; test sample: 40%; validation sample: 0%.
6. Run the Partition node and report the results.

Note: We will continue using the **Boston** dataset in the next chapters. Assume that data partitioning has already been completed. Refer to this exercise if you need to redo the data partitioning.

Exercise 2: Data Sampling for the Try Buy Dataset

1. Upload the **try_buy** dataset to your computer.
2. Open the dataset in SPSS Modeler. Change the Role of BuyReturn to Target, and change the Level to Flag (Binary).
3. Create a new stream named **Sample** and add the data source to the workspace. Add a Type node and connect it to the data source.
4. Add a Sample node to the stream and connect it to the Type node.
5. Sample 80% of the full dataset using two different sampling methods:
   i. Systematic
   ii. Stratified
6. Run the Sample node and report the results.

Exercise 3: Data Resampling for the Try Buy Dataset

1. Upload the **try_buy** dataset to your computer
2. Open the dataset in SPSS Modeler.
3. Open the diagram named **Sampling** created in Exercise 2. Run the Sample node and report the imbalance ratio for the target variable, BuyReturn. Is the dataset imbalanced? Why?
4. Perform resampling to balance the data using the following three techniques.
   - Undersampling: Use a Balance node and calculate the required factor.
   - Oversampling: Use a Balance node and calculate the required factor.
   - SMOTE: Use a SMOTE node and set k-Neighbours to 5.
5. Report the results of each resampling technique.
6. Evaluate the new sample: Is the data balanced in the new sample? What is the new sample size? What are the potential concerns with the new data?

Exercise 4: Data Sampling and Partitioning for the Passenger Dataset

1. Upload the **Passengers** dataset to your computer.
2. Open the dataset in SPSS Modeler. Change the Role of **return** to Target, and change the Level to Flag (binary).
3. Create a new stream named **Sample**, and add the data source to the workspace. Add a Type node and connect it to the data source.
4. Add a Sample node and connect it to the Type node.
5. Sample 20% of the full dataset using the Random method.
6. Run the Sample node and report the results.
7. Add a Partition node and connect it to the Sample node.
8. Partition the data using the split ratio 60–40: train sample: 60%; test sample: 20%; validation sample: 20%.
9. Run the Data Partition node and report the results.

Note: We will continue to use the **Passenger** dataset in the next chapters. Assume that the sampling and data partitioning have already been completed. Refer to this exercise if you need to redo the sampling and data partitioning.

4 Data Visualization and Exploration

LEARNING OUTCOMES

Upon the completion of this chapter, you will be able to:

1. Explain the importance of data exploration in the data mining process.
2. Describe the primary components of a dataset.
3. Discuss the advantages of visualization in data exploration.
4. Select the appropriate chart types for categorical and interval variables.
5. Create various charts using graph tools in SPSS Modeler.
6. Compare bar charts and histograms.
7. Explain the applications of line charts, pie charts, box plots, and scatter plots.
8. Compare multi-plots with other graphical tools.
9. Interpret the chart results for data mining purposes.
10. Create descriptive statistics for the data.
11. Interpret descriptive statistics for data mining purposes.

INTRODUCTION

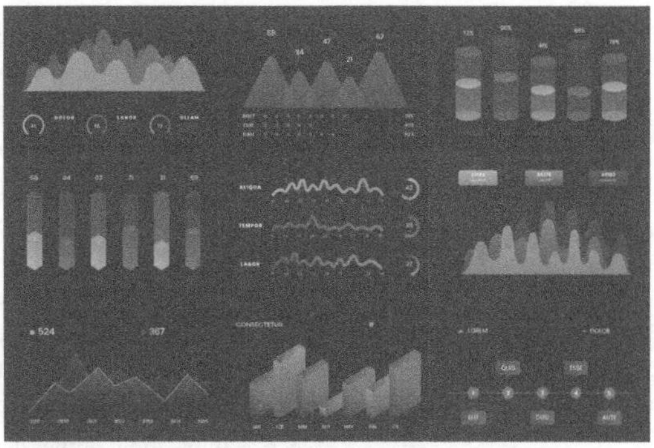

Data exploration is a crucial step in data mining because it provides a comprehensive understanding of the data, focusing on parameters, characteristics, trends, and potential concerns. This understanding enables us to take appropriate actions to clean, prepare, and transform the data as needed. It also guides us in selecting the right machine learning methods and analytical configurations to develop a robust predictive model. During this step, we use visualization and descriptive statistics to explore and understand the data, the variables, and any potential issues, such as missing values or outliers. Visualization is a powerful tool for data exploration, and there are many visualization techniques available. I will demonstrate these tools with examples.

DATA UNDERSTANDING

Before conducting any data analysis, it is essential to understand the data, its parameters, and its quality. The quality of the inputs directly influences the quality of the outputs—hence the saying, "Garbage in, garbage out." This step is crucial for any type of data analysis, but it is particularly important for data mining, where we often deal with large, noisy datasets. When faced with a dataset that contains many variables and records, gaining a comprehensive understanding can be challenging due to the data's sheer size. Without a thorough understanding, we risk using invalid or irrelevant variables as predictors during modeling, or we may overlook important variables.

DOI: 10.1201/9781003468196-5

Additionally, it can be difficult to select the appropriate machine learning method without understanding the data's characteristics and limitations. A poor model fit might also result from excessive missing values or outliers in the dataset.

To address these challenges, it is vital to thoroughly explore the data, focusing on the type of information available, the variables present, the measurement scales of those variables, and any emerging trends. As discussed in Chapter 2, the selection of a machine learning method depends on the data, the scale of the target variable, and the purpose of the project.

Data exploration is one of the most critical steps in data mining because it allows us to assess the data's structure and identify potential issues. Based on this understanding, we can determine the necessary strategies for data modification and transformation. A good predictive model requires quality inputs—the model is only as good as the data it's built upon. For example, imagine you are an airport executive responsible for predicting the likelihood of flight delays or cancellations at your airport on specific days or times. If your predictive model, based on poor inputs such as missing values or excessive noise, inaccurately predicts a low likelihood of delays (a false negative), you might decide not to take any preventive action. As a result, you could face significant chaos at the airport due to unexpected flight cancellations—an outcome that could have been mitigated with better data and a more accurate prediction. Although other factors can contribute to poor predictions, poor data quality is often a primary cause. Remember, "garbage in, garbage out." Table 4.1 highlights some key reasons why data exploration is vital in data mining.

TABLE 4.1
Importance of Data Exploration in Data Mining Projects

Items	Importance of Data Exploration
Understanding data characteristics	Data exploration is crucial for gaining insights into the characteristics of a dataset. It allows us to understand the data's distribution, range, variability, and potential quality issues. By exploring the data, we can identify outliers, missing values, data inconsistencies, or other anomalies that may impact the quality and validity of the analysis.
Variable selection	Understanding the data helps us identify relevant variables for the data mining task. By analyzing the relationships between variables, their distributions, and their potential impact on the target variable, we can make informed decisions about which variables to include or exclude in the modeling process.
Data preprocessing and cleaning	Data exploration is also essential for identifying and addressing data quality issues. It enables us to detect and handle missing values, outliers, duplicates, or inconsistent data. Based on insights gained during the exploration phase, we can perform data treatment tasks such as imputation, outlier detection, or data transformation. Effective data cleaning and preparation lead to improved model performance and more reliable results.
Identifying patterns and relationships	Exploring the data allows us to identify potential patterns, trends, or relationships between variables. By visualizing the data, applying statistical measures, or conducting exploratory data analysis, we can uncover meaningful insights and understand the underlying structure of the data. These insights guide the selection of appropriate machine learning methods and help improve the predictive power of the model.
Requirements and limitations	Data exploration helps us identify any violations of assumptions required by specific machine learning algorithms. This awareness enables us to make informed decisions about the suitability of certain techniques and avoid potential pitfalls or biases in the modeling process.
New discoveries	Data exploration can inspire the generation of new discoveries. By observing patterns or trends in the data, we may identify new relationships to investigate further. These initial insights can guide subsequent modeling and analysis, leading to new discoveries and research directions.

There are two primary ways to explore data: visualization and statistical exploration. Visualization allows us to observe data trends and patterns through various types of graphs, making it easy to extract useful information. This method is user-friendly and can be incorporated into business intelligence dashboards for managers or executives. Patterns and trends can be quickly identified through visualization. Many visualization tools are available on the market, such as Tableau, Power BI, and SAS Viya. These applications enable the addition and customization of charts and graphs to meet specific needs. For big data, these tools can process and display data in real time, allowing users to monitor patterns of variables of interest over time. These dynamic visualization tools provide up-to-date information.

In this book, I focus primarily on the charts used to explore data for analysis purposes. These charts are deterministic, meaning they do not allow for interaction or real-time adjustments. Therefore, separate charts must be created for different variables and analysis purposes. Nevertheless, they provide essential information that helps us understand, process, and develop good predictive models. In this context, data visualization is considered a step in the data mining process. More extensive data visualization tools can be discussed in another book.

> It is recommended to use both data visualization and statistical exploration.

For data mining purposes, visualization alone may not be sufficient. Statistical exploration offers valuable quantitative information, such as descriptive statistics of variables, which help us understand the characteristics of the data. Additionally, statistical exploration provides preliminary insights into correlations between the target variable and predictors. By combining visualization with statistical exploration, we can leverage the strengths of both methods. In this book, I will demonstrate how to use both approaches to explore and understand data effectively.

DATA DESCRIPTION

One important thing to remember in data exploration is the need to understand the variables and their attributes before creating charts and graphs. This understanding is crucial because it helps us determine the appropriate types of graphs to use and how to interpret the outputs. Therefore, before running any visualizations, it's essential to first gather basic information about the dataset, including the number of variables, their descriptions, and their measurement scales. This information will guide us in selecting the most suitable graphs to make sense of the data.

In this chapter, I will use the airline recommendation dataset as an example to demonstrate various types of visualizations using SPSS Modeler. First, let's examine the parameters and characteristics of this dataset.

Case Study—Airline Recommendation

In this case study, a large-scale survey was conducted with passengers from multiple airlines across the United States flying a specific airline from one origin to a destination. They were asked to complete a survey about their flight experience and whether they would recommend that airline to others. The survey collected demographic information and details about their flight experience, including the origin and destination states, number of previous flights, travel purposes, spending at the airport, and price sensitivity. The passenger survey data were then consolidated with airline status and flight variables, such as flight distance, flight time, departure and arrival delays, and flight cancellations.

The goal of this project is to predict the target variable: passengers' airline recommendations. We aim to identify which model best predicts this target variable and which variables are significant predictors. These predictions would help airline executives develop strategies to improve operations and services, thereby attracting more passengers.

Given the number of variables in the dataset, it's essential to first examine those variables, their roles, and their scales. Table 4.2 describes the variable names, descriptions, and scales in this dataset.

> Variable name requirements:
> - Clarity and Conciseness: Ensure the variable name is clear and concise.
> - Length: Keep it short.
> - Special Characters: Avoid using special characters.
> - Spaces: Do not use blank spaces.

- **Variable names**: Since variable names are used in actual analyses and results, they should be both clear and concise. A variable name should be descriptive enough to allow easy recognition of the variable, yet brief enough to prevent it from being cut off in the output. Many software applications impose length limits on variable names. I strongly recommend avoiding the use of blank spaces or special characters (e.g., @#$%^&*) in variable names. Some software programs, like SAS Enterprise Miner and RapidMiner, do not permit spaces or special characters and will generate error messages if these issues are detected in variable names. The same requirements apply to programming languages such as R or Python.

TABLE 4.2
Data Description

Variable Name	Label/Description	Scale	Role
recommend	Recommend the airline	Flag (Binary) (Yes/No)	Target
airline_st	Airline status	Nominal	Input
age	Age	Continuous	Input
gender	Gender	Flag (Binary) (Male/Female)	Input
price_sensitivity[a]	Price sensitivity	Continuous	Input
no_years_flying	Number of years flying	Continuous	Input
no_flt_pa	Average number of flights per year	Continuous	Input
pct_ flt_other_airline	Percent of flights with other airlines	Continuous	Input
type_of_travel	Travel purpose	Nominal	Input
no_other_lty_cards	Number of other loyalty cards	Continuous	Input
shopping_amt_airpt	Shopping amount at airport	Continuous	Input
eat_drink_airpt	Spending for eating and drinking at airport	Continuous	Input
Class	Passenger class	Nominal	Input
origin_state	Origin state	Nominal	Input
destin_state	Destination state	Nominal	Input
depart_delay_in_minutes	Departure delay (in minutes)	Continuous	Input
arrival_delay_in_minutes	Arrival delay (in minutes)	Continuous	Input
flt_cancel	Flight cancellation	Flag (Binary) (Yes/No)	Input
flt_time	Flight time (in minutes)	Continuous	Input
flt_distance	Flight distance (miles)	Continuous	Input

[a] Likert scale, treated as Continuous scale.

While some software, like SPSS Modeler, may be more flexible with variable names, integrating models with platforms like Python can still lead to errors if naming conventions are not followed. Therefore, it is best to adhere to correct naming practices. To separate words in a variable name, use an underscore (_) instead of a space.

- **Variable labels/descriptions**: The details of a variable can be provided in its description, often referred to as "labels" in some applications. The description or label offers users a more detailed explanation of the variables. As shown in Table 4.2, the labels clearly describe each variable and include measurement units, such as years, minutes, or miles. Labels are especially helpful when variable names are short or use acronyms, as they provide clarity and context. While there are no strict limits on variable descriptions, it is advisable to keep them clear and concise to avoid confusion or misinterpretation.
- In software applications with user interfaces (UI), such as SAS Enterprise Miner, SPSS Modeler, or RapidMiner, labels can often be displayed alongside variable names. This feature allows users to refer to the label for a full understanding of the variable when the name alone is not sufficiently descriptive. Additionally, labels can be used instead of variable names in outputs for reporting purposes. However, lengthy labels can pose challenges in reports by enlarging output tables, making them difficult to fit within the report. They can also create difficulties when presenting graphs. Therefore, it's important to choose labels carefully.
- **Measurement scales**: Measurement scales are crucial for analysis, particularly when selecting the appropriate machine learning algorithms during the modeling step. As discussed in Chapter 2, variables can have different scales, including nominal, flag (or binary), ordinal, or continuous. In Table 4.2, the dataset includes one flag (binary) target variable, two flag (binary) predictors, five nominal predictors, and 11 continuous predictors. Note that the **price_sensitivity** variable uses a Likert scale, and we assume the intervals between levels are equal. Therefore, it is treated as a continuous variable (see Chapter 2 for a more detailed discussion on the Likert scale).
- **Role**: The role of a variable is another critical piece of information. Variables generally have two primary roles: **Target** or **Input**. The target variable is the one we intend to predict, while input variables (also known as predictors) are used to make that prediction. In our example, **recommendation** is the target variable with a flag (binary) scale, while the other variables serve as predictors. Additionally, there are other roles, such as **Record ID** (a record identifier), **Partition** (used to partition the data), **Split** (a model is built for each possible value of the field), **Frequency** (a frequency weighting factor), and **None** (the variable is ignored). For now, we will focus primarily on the Target and Input roles. The Partition role will be used later in the modeling process.

Add Data to SPSS Modeler

In this book, we will use the **Statistics File** node to add data files (.sav) to the stream. I assume you are familiar with SPSS Statistics and have experience working with .sav files. The Statistics File node allows us to read .sav files from our local computer hard drive. All data files used in this book are provided in .sav format.

Follow the steps below to add the **AirlineRecommendation.sav** file to the stream:

1. Open SPSS Modeler.
2. Open the "Airline_Explore" stream created before.
3. From the **Sources** group in the Node Pallet, double-click the **Statistics File** node to add it to the stream.
4. Double-click the **Statistics File** node to open its settings window.

5. Click the ellipsis button (three-dot button) and select the **AirlineRecommendation.sav** file from your computer.

6. Click **Preview** to review the first ten rows of the data.

Reminder: To make it easier to follow the steps in using SPSS Modeler and avoid confusion, I refer to **"variables"** in the chapter content and **"fields"** in the software instructions and demonstrations. This distinction is important because SPSS Modeler primarily uses the term **"fields"** rather than **"variables"** in most instances.

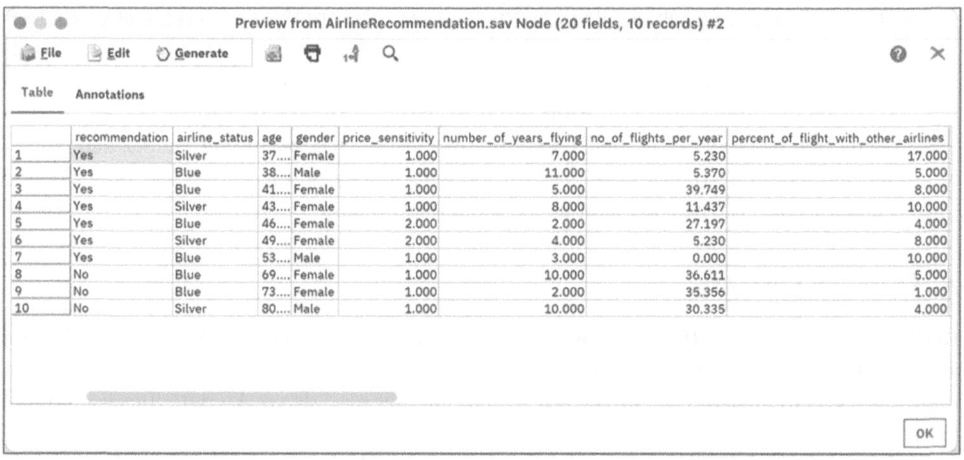

	recommendation	airline_status	age	gender	price_sensitivity	number_of_years_flying	no_of_flights_per_year	percent_of_flight_with_other_airlines
1	Yes	Silver	37....	Female	1.000	7.000	5.230	17.000
2	Yes	Blue	38....	Male	1.000	11.000	5.370	5.000
3	Yes	Blue	41....	Female	1.000	5.000	39.749	8.000
4	Yes	Silver	43....	Female	1.000	8.000	11.437	10.000
5	Yes	Blue	46....	Female	2.000	2.000	27.197	4.000
6	Yes	Silver	49....	Female	2.000	4.000	5.230	8.000
7	Yes	Blue	53....	Male	1.000	3.000	0.000	10.000
8	No	Blue	69....	Female	1.000	10.000	36.611	5.000
9	No	Blue	73....	Female	1.000	2.000	35.356	1.000
10	No	Silver	80....	Male	1.000	10.000	30.335	4.000

7. Click **Apply** and **OK**.
8. From the Field Ops group, add a **Type** node to the stream and connect it to the data node. This node allows us to view and modify the measurement scales and roles of fields.
9. Double-click the **Type** node to open its settings window.
10. Review the measurement scales of all fields as below. For the **recommendation**, change the role to **Target**.

Field	Measurement	Values	Missing	Check	Role
Recommend the air...	Flag	Yes/No		None	Target
Airline status	Nominal	Blue,Gold,Platinum,Silver		None	Input
Age	Continuous	[18.0,85.0]		None	Input
Gender	Flag	Male/Female		None	Input
Price sensitivity	Continuous	[0.0,5.0]		None	Input
Number of years fly...	Continuous	[2.0,11.0]		None	Input
Average number of ...	Continuous	[0.0,100.0]		None	Input
Percent of flights wi...	Continuous	[1.0,110.0]		None	Input
Travel purpose	Nominal	"Business travel","Mileag...		None	Input
Number of other lo...	Continuous	[0.0,12.0]		None	Input
Shopping amount a...	Continuous	[0.0,879.0]		None	Input
Spending for eating...	Continuous	[0.0,895.0]		None	Input
Passenger class	Nominal	Business,Eco,"Eco Plus"		None	Input
Origin state	Nominal	Alabama,Alaska,Arizona,...		None	Input
Destination state	Nominal	Alabama,Alaska,Arizona,...		None	Input
Departure delay (in ...	Continuous	[0.0,1592.0]		None	Input
Arrival delay (in mi...	Continuous	[0.0,1584.0]		None	Input
Flight cancellation	Flag	Yes/No		None	Input
Flight time (in minu...	Continuous	[8.0,669.0]		None	Input
Flight distance (miles)	Continuous	[31.0,4983.0]		None	Input

11. Click **Apply** and **OK**.

DATA VISUALIZATION

With the vast amount of data collected daily, visualization has become a powerful tool for reviewing data trends and patterns without the need to delve into raw data or rely solely on descriptive statistics. Data visualization simplifies the process, making it easy for users to understand the data, even if they lack a background in data analytics or statistics. In many industries, organizations develop business intelligence dashboards that allow users to quickly select key variables or key performance indicators (KPIs) and generate charts representing data trends or patterns. Operators use these tools to monitor daily operations, while decision-makers leverage visualization to form strategies and make business decisions. However, it's crucial to know what to look for (KPIs or metrics) and how to interpret the outcomes.

Visual analytics applications such as Tableau, Spotfire, SAS Visual Analytics, and Microsoft Power BI offer many features for users. Regardless of the tool, several common chart types are typically used for visualization. In this book, I will cover some of these common chart types, explain how to create them, and demonstrate how to interpret the results. I will show you how to create charts using SPSS Modeler, though you can use other tools you're comfortable with, such as SPSS Statistics or MS Excel. As previously mentioned, in this book, data visualization is just one step in the data mining process. Therefore, we will use these charts in a deterministic mode rather than the full dynamic mode often seen in intelligent dashboards.

I will use several primary visualization tools in SPSS Modeler: Graphboard, Distribution, Histogram, Plot, Multiplot, and Time Plot. These tools are all listed under the **Graphs** group in the Node Palette. My favorite node for visualization is the **Graphboard** node, as it is the most robust and versatile. It allows us to create different types of graphs based on the variables and their scales. The Graphboard node can automatically detect the appropriate chart types based on the number of variables and their scales. For example, if we select one nominal variable, the node will suggest using bar charts, plot charts, etc. Figure 4.1 shows an example where we select one nominal variable (recommendation) and one continuous variable (price sensitivity), displaying the available chart options for this selection. This feature helps us avoid using inappropriate chart types for our variables.

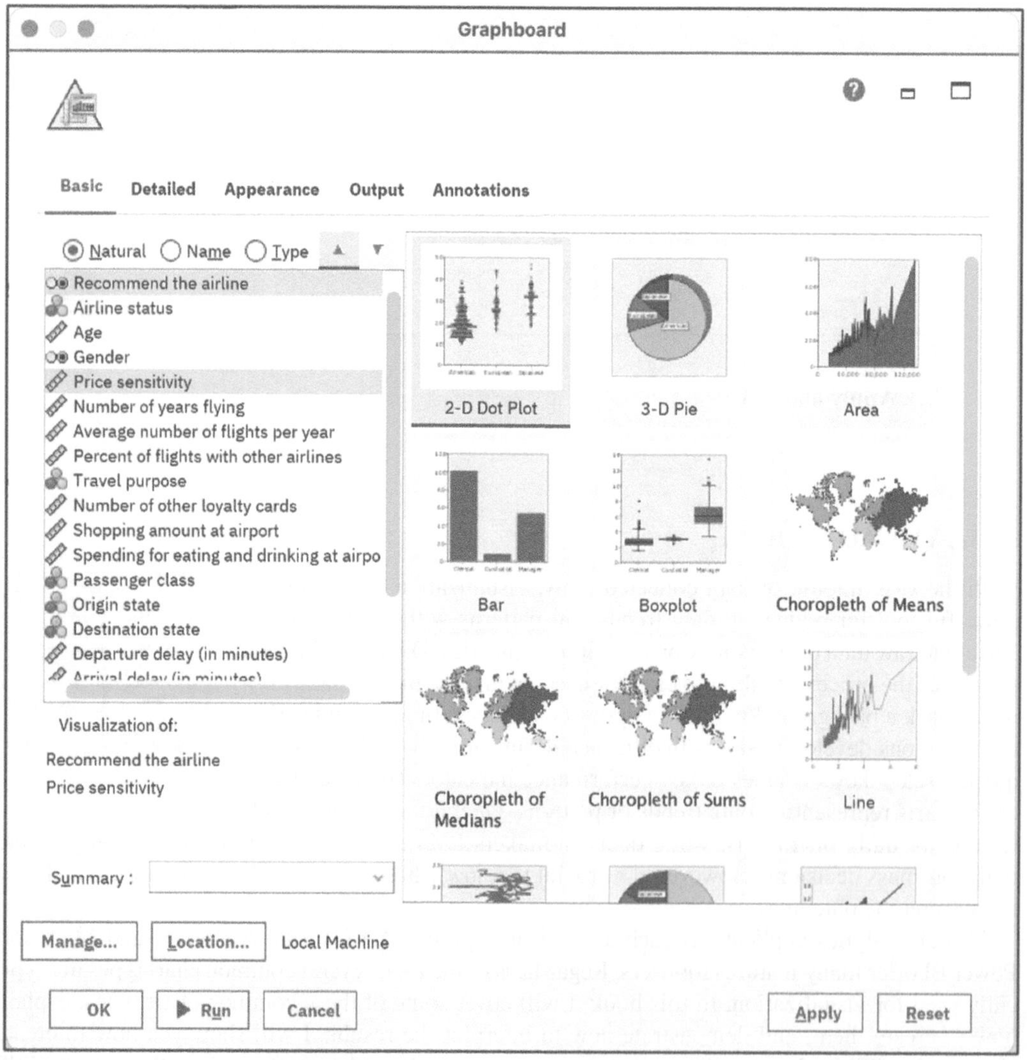

FIGURE 4.1 Example of the Grapboard setting.

TABLE 4.3
Differences between Bar Charts and Histograms

Items	Bar Charts	Histograms
Data type	Bar charts are used to display categorical data. The categories or groups are displayed on the x-axis, with the height or length of the bars representing the frequencies or percentages associated with each category.	Histograms are suitable for continuous data. The data are divided into intervals or bins, and the height of the bars represents the frequency of data points within each interval.
X-axis representation	In a bar chart, the x-axis represents discrete categories or groups. Each category is usually represented by a separate bar.	In a histogram, the x-axis represents the range of values or intervals for the continuous variable being measured. The intervals are continuous and adjacent, forming a continuous scale.
Data presentation	Bar charts are effective for comparing data across different categories, showing distinct categories and comparing their frequencies.	Histograms display the distribution of values within a continuous variable, providing insights into the shape, central tendency, and spread of the data.
Bar placement	In a bar chart, the bars are usually separated from each other with gaps to emphasize the distinction between categories.	In a histogram, the bars are typically placed adjacent to each other without gaps, as the intervals are continuous and form a continuous scale.
Axis labels	Bar charts use a label for each category on the x-axis to indicate a specific group or category.	Histograms use labels on both the x-axis and y-axis. The x-axis label represents the measured variable, while the y-axis label represents the frequency of data points within each interval.

Bar Charts and Histograms

Bar charts and histograms are among the most common graphical tools used in data visualization. Both use bars to present results, either vertically or horizontally. So, what are the differences between them? Table 4.3 summarizes some key distinctions between these two graphical tools. Bar charts are ideal for categorical variables, while histograms are used for continuous variables. We will explore the details of these chart types and discuss when to use each one.

Bar Charts

Bar charts are used for categorical variables to present the frequency or percentage for each category or level. They are straightforward to understand and interpret. Percentages are particularly effective for comparison purposes, while frequencies can show the number of cases in each category.

As mentioned earlier, the **Graphboard** node is the most robust and versatile option, allowing us to create various types of graphs with ease. However, SPSS Modeler also provides a specialized node for creating bar charts, the **Distribution** node. In this example, I will first use the Graphboard node to create bar charts for our AirlineRecommendation dataset. Then, I'll demonstrate how to use the Distribution node for the same purpose. Moving forward, I will primarily use the Graphboard node for creating charts, as it is a more efficient approach. This method allows us to choose and customize charts as needed, though it requires creating one chart at a time. There is, however, a quicker way to generate charts for all variables in the dataset, which I will demonstrate later.

Follow the steps below in SPSS Modeler to add and set up the Graphboard node.

> Bar charts are used for categorical variables.
>
> Histograms are used for continuous variables.

1. Open the "Airline_Explore" stream. We should have the data node and the Type node created before.
2. From the **Graphs** group in the Node Pallete, add a **Graphboard** node to the stream and connect it to the **Type** node.

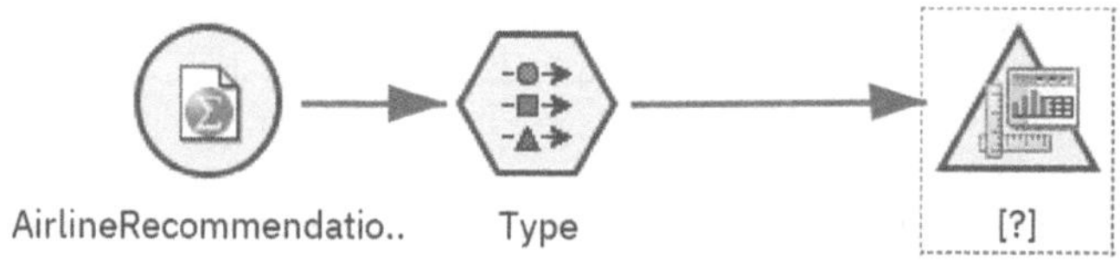

3. Open the settings window of the **Graphboard** node.
4. Select the **Recommendation the airline** field on the left, and then choose the Bar of Counts chat option on the right.

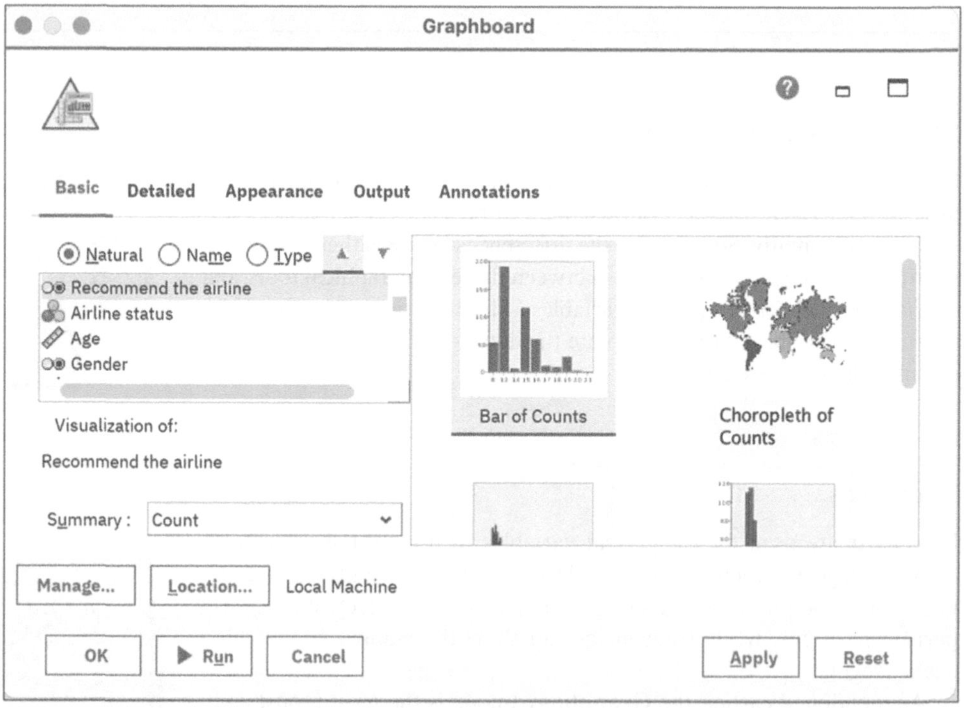

5. Under the **Appearance** tab, select **Use all data.**
6. Click **Apply** and **Run.**

The output will display the same bar chart as shown in Figure 4.2.

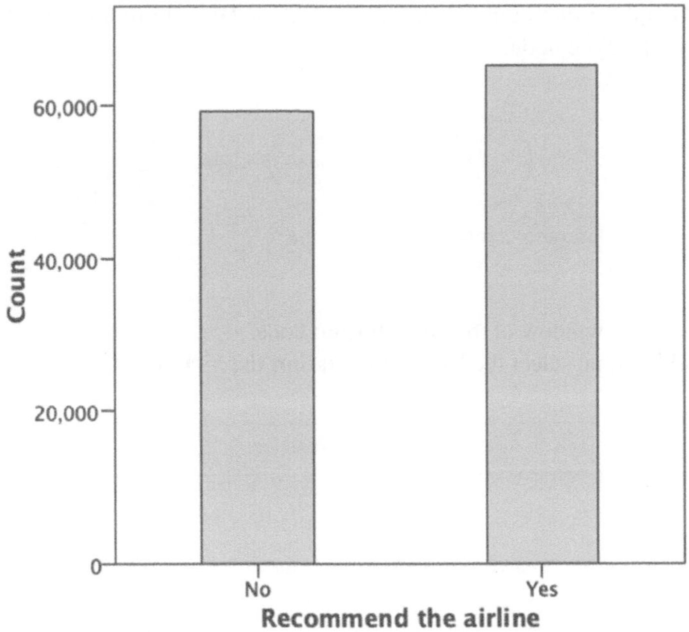

FIGURE 4.2 Bar chart for **Recommendation** (Graphboard node).

To view the full chart data, hover the pointer over the chart, and it will display the data for each category. Alternatively, you can go to Edit and select Copy Graph Data. This allows you to paste the raw data used to construct the chart into your document, as shown in Table 4.4. Since this chart does not have an option to display percentages, you can perform the calculations in Excel to obtain that result.

> Generally, graph tools in SPSS Modeler show only count numbers, not percentages. To get the percentage for each category, copy and paste the graph data into Excel and manually calculate the percentage values.

We can also create a bar chart using the Distribution node. This node is specifically designed to construct a bar chart and is therefore suitable only for nominal variables.

Follow the steps below:

1. Open the "Airline_Explore" stream. We should have the data node and the Type node created before.

TABLE 4.4
Chart Data for the Recommendation Target Variable

Recommend the Airline	Count	Percentage
Yes	65,297	52.46
No	59,167	47.54

2. From the **Graphs** group in the Node Pallete, add the **Distribution** node to the stream and connect it to the Type node.

3. Open the settings window of the **Distribution** node.
4. For the **Field** option, select the **Recommendation the airline** field.

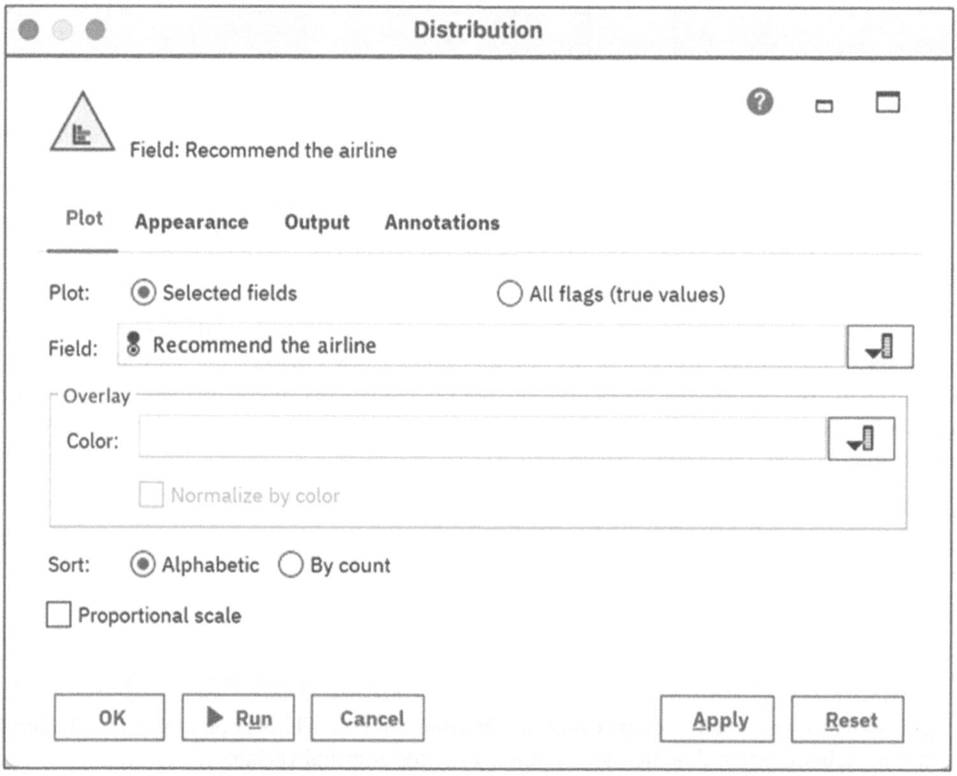

5. Click **Apply** and **Run**.

The output on the **Table** tab displays both the number of responses and the percentage for the two levels of the **recommendation** variable.

Table Graph Annotations

Value	Proportion	%	Count
No		47.54	59167
Yes		52.46	65297

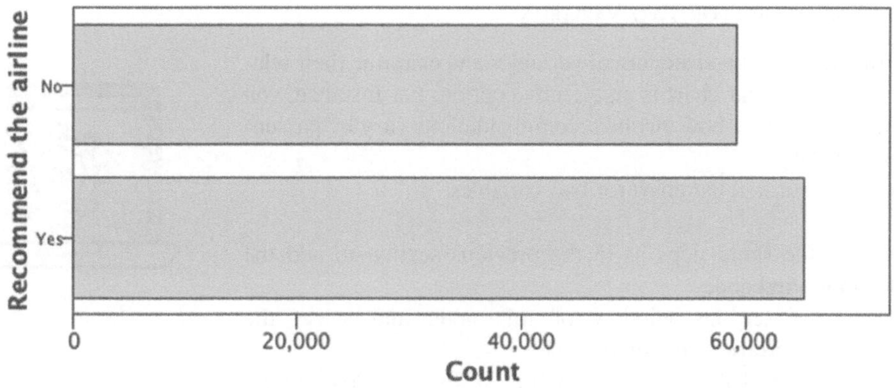

FIGURE 4.3 Bar chart for **Recommendation** (distribution node).

To display the bar chart, click on the **Graph** tab. Figure 4.3 shows the bar charts for the target variable, **recommendation**, presented in frequency. Note that in these charts, the label is used instead of the variable name, which is why you see **Recommend the airline** instead of **recommendation**. You can click the icon ( ) to toggle between the label and variable name. Since the target variable is binary (Yes/No), the charts show the frequencies and percentages for these two levels. As we can see, 52.46% of passengers recommend the airline, while 47.54% do not recommend it.

Graph Formatting

To customize graph properties such as colors, format, titles, axes, values, or legends, you can use the graph edit mode. This mode allows you to adjust the layout and visual aspects of your graph. To enter edit mode, go to *View > Edit Mode* from the menu or click the corresponding toolbar icon. Once in edit mode, you can make the following modifications:

- Edit and format text.
- Adjust the color, transparency, and patterns of frames and graphic elements.
- Modify the color of borders and lines.
- Rotate and alter the shape and aspect ratio of point elements.
- Resize graphic elements like bars and points.
- Adjust margins and padding to change the space around items.
- Set number formatting options.
- Change axis and scale settings.
- Sort, exclude, and collapse categories on a categorical axis.
- Set the orientation of panels.
- Apply transformations to the coordinate system.
- Modify statistics, graphic element types, and collision modifiers.
- Reposition the legend.
- Apply different visualization stylesheets.

The next section provides guidelines on setting up overlays and aesthetics for visualizations in SPSS Modeler. It's important to experiment with different features and settings to find the best options for your data visualization.

Stacked Bar Charts for Two Variables

If you want to graph two categorical variables and examine their relationship, a stacked bar chart is an effective option. For instance, you might want to analyze how airline recommendations vary by passengers' airline status.

To create a stacked bar chart for two variables:

1. Follow the same steps as in the previous section to add the **Graphboard** node.
2. Open the settings window of this node and select the **Recommendation the airline** field.
3. On the **Detailed** tab, select **Airline status** for Color under Optional Aesthetics.

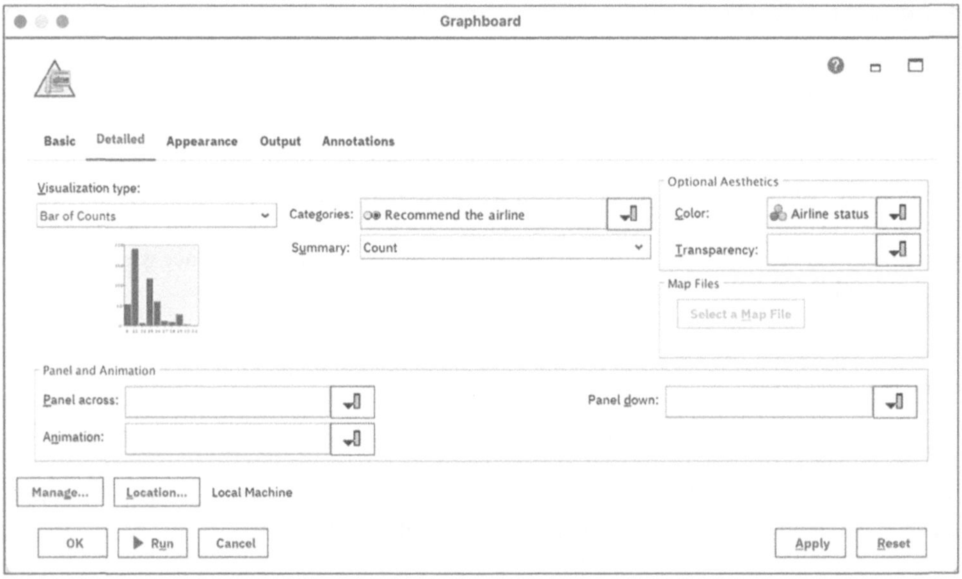

4. Under the **Appearance** tab, select **Use all data**.
5. Click **Apply** and **Run.**

Figure 4.4 shows the stacked bar charts for two variables: **Recommendation** and **Airline status**. These bar charts display the percentage breakdown of the target variable by the categories of airline status.

From the chart, we can interpret that among passengers who recommend the airline, the majority are in the Blue class (30%), followed by the Silver class (15%). However, a significantly larger percentage of Blue class passengers do not recommend the airline (40%), while only about 5% of Silver class passengers do not recommend it. In other words, Blue class passengers are more likely not to recommend the airline, whereas Silver class passengers are more likely to recommend it.

3D Bar Charts For Three Variables

When dealing with more than two variables, a 3D bar chart can be an effective option for visualizing the relationships among three variables. For instance, if you want to analyze the number of years flying by airline status and gender, a 3D bar chart can provide valuable insights.

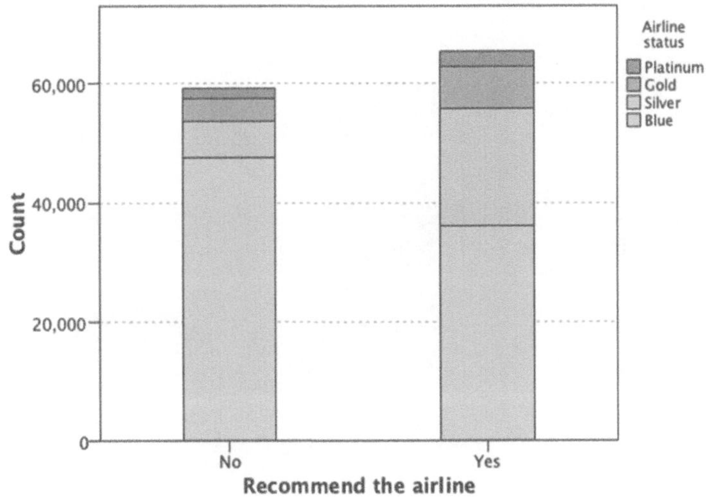

FIGURE 4.4 Stacked bar charts for two variables.

To create a 3D bar chart for three variables:

1. Follow the same steps as in the previous section to add the Graphboard node.
2. Open the settings window of this node.
3. On the **Basic** tab, select three fields: **Number of years flying, Airline status,** and **Gender** (hold the Control button to select multiple fields).
4. Select the 3D bar chart type.
5. On the **Detailed** tab, assign the fields as follows: **Airline status** for the **X** field, **Number of years flying** for the **Y** field, and **Gender** for the **Z** field.

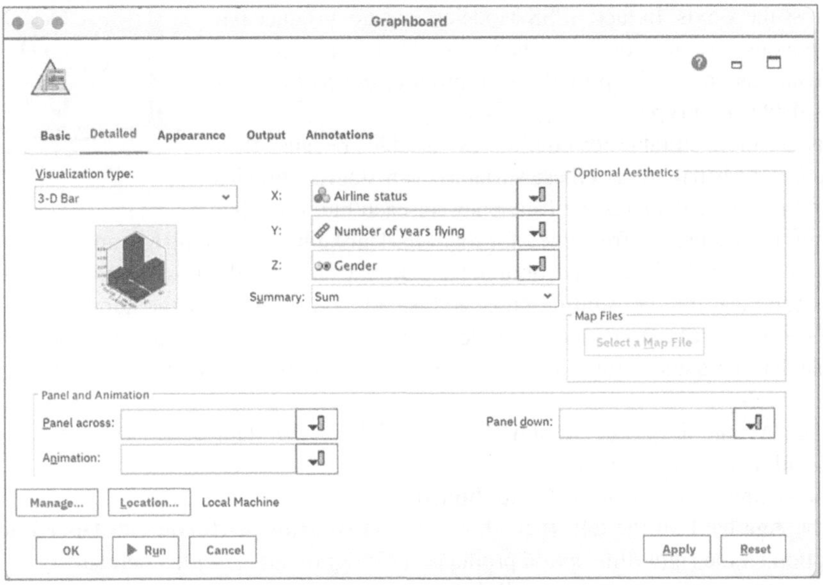

6. Click **Apply** and **Run.**

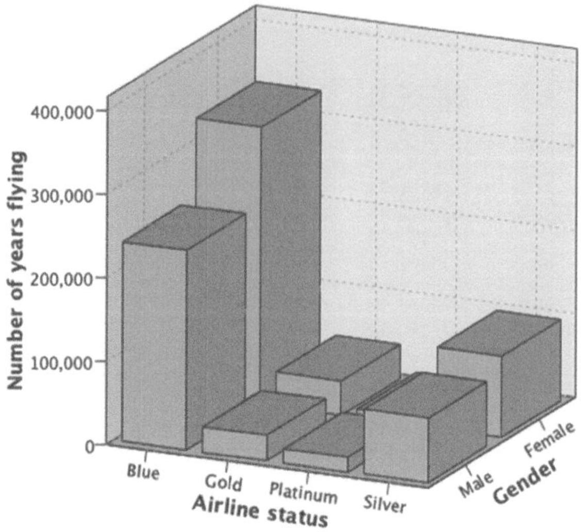

FIGURE 4.5 3D bar charts for three variables.

Figure 4.5 shows the 3D bar charts for these three variables. The chart displays the comparison between airline statuses and gender in terms of the number of years flying. For example, the majority of passengers are in the Blue class, but female passengers in this category tend to have more years of flying experience than male passengers. In other airline statuses, the difference between the two genders is minimal. In some cases, 3D bar charts allow for comparisons of a field across the categories of other fields.

Histograms

For continuous variables, bar charts may not be the best choice. Given the continuous scale, bar charts can become very large and noisy, with data spread out across the x-axis. In fact, SPSS Modeler does not produce bar charts for continuous variables. Instead, when a variable has a continuous scale, the software automatically provides histogram or dot plot options as the only available chart types.

Histograms are more suitable for continuous variables because they present the frequency distribution of these variables. A histogram divides the values into bins and calculates the frequency for each bin. Each bar in the histogram represents the frequency for that bin. Unlike bar charts, histograms have no gaps between bars, as the variable's scale is continuous. While it's technically possible to define the bins manually, this can be a time-consuming process. Instead, you can specify the number of bins you need, and the software will automatically divide the data into bins and generate the histogram.

For example, let's create a histogram for the **Age** variable. Follow these steps in SPSS Modeler:

1. Open the "Airline_Explore" stream. We should have the data node, Type node, and Graphboard node created before.
2. Open the settings window of the **Graphboard** node.
3. Select the **Age** field on the left, then choose the **Histogram with Normal Distribution** chart option on the right (this option produces a histogram along with a bell curve).
4. On the **Appearance** tab, select **Use all data**.
5. Click **Apply** and **Run**.

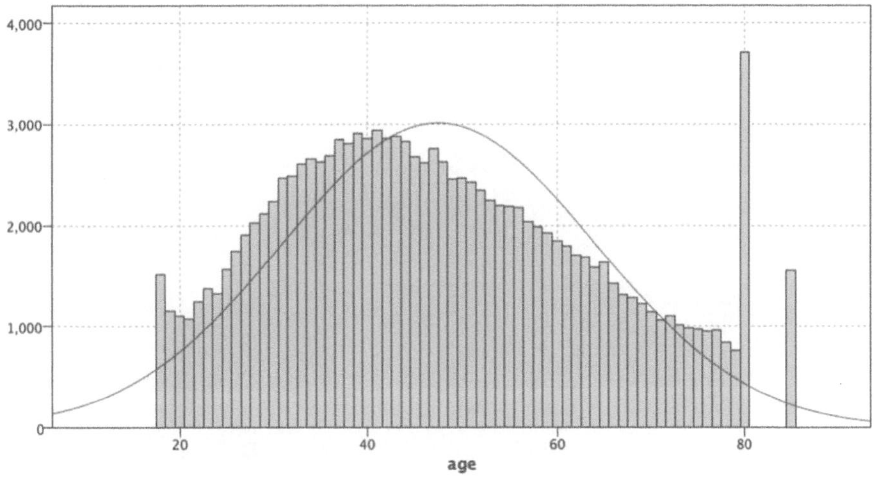

FIGURE 4.6 Histogram for **Age** (number of bins determined by the software).

The above steps will create a histogram with a large number of bins, automatically determined by the software (as shown in Figure 4.6). This chart displays the count for each age range, giving an idea of the age distribution, which is close to a normal distribution.

However, with many bins, the histogram may be a bit hard to interpret. Suppose we want a histogram for **Age** with five bins to make the chart easier to interpret. To do this, we need to use the **Histogram** node.

Follow these steps to add and set up the Histogram node:

1. Open the "Airline_Explore" stream. We should have the data node and Type node created before.
2. Add a **Histogram** node to the stream and connect it to the **Type** node.
3. Open the settings window of the **Histogram** node.
4. On the **Plot** tab, Select the **Age** field.
5. On the **Options** tab, set **No. of Bins** to 5.
6. Check the box **Show normal curve**.

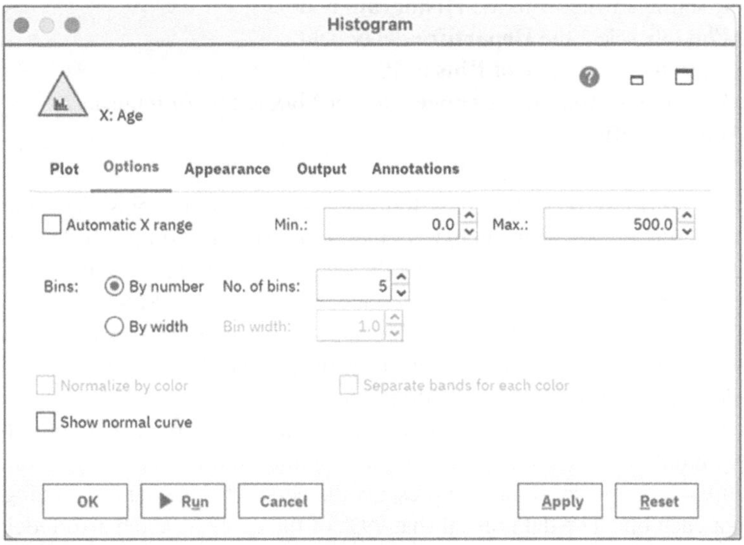

7. Click **Apply** and **Run**

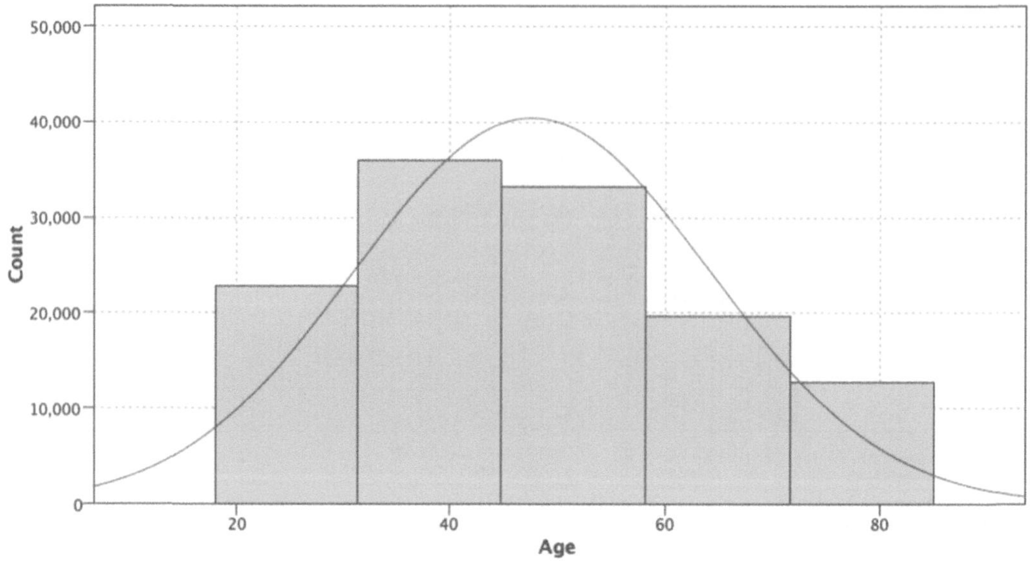

FIGURE 4.7 Histogram for **Age** with five bins.

Figure 4.7 shows the histogram for **Age** with five bins. The chart is now much easier to interpret. It indicates that most passengers are between the ages of 31 and 49, followed by those in the 49–58 range. In contrast, there are fewer older passengers between 72 and 85 years. Additionally, the normal curve suggests that the data points are fairly close to a normal distribution.

Let's explore another example using a histogram. This time, we'll create a diagram for **Departure delay** with ten bins.

1. Open the "Airline_Explore" stream. We should have the data node and Type node created before.
2. Add a **Histogram** node to the stream and connect it to the **Type** node.
3. Open the settings window of the **Histogram** node.
4. On the **Plot** tab, select the **Departure delay** field.
5. On the **Options** tab, set **No. of Bins** to 10.
6. Uncheck the box **Automatix X range**, and set **Max** to 500 (minutes).
7. Click **Apply** and **Run**

As shown in Figure 4.8, the software will divide the data into ten bins. To obtain more detailed information about the ranges, you can extract the graph data. To do this, go to the **Edit** menu and select **Copy Graph Data**. Table 4.5 presents the detailed graph data, including the average delay minutes for each bin and the corresponding number of cases. Since SPSS Modeler does not calculate percentages in charts, I calculated the percentages using Excel to provide an overview of the data. As shown, 97% of the cases have a delay time of less than 100 minutes, followed by 3% of the cases with a delay time between 101 and 200 minutes.

Since the data are extremely right-skewed, let's examine the trend more closely by displaying a histogram with only three bins. By adjusting the number of bins to 3, we generate the histogram shown in Figure 4.9. Similarly, Table 4.6 presents the detailed chart data, including the ranges and percentages for each bin. The data reveal that 99% of the cases have departure delay times of less than 116 minutes.

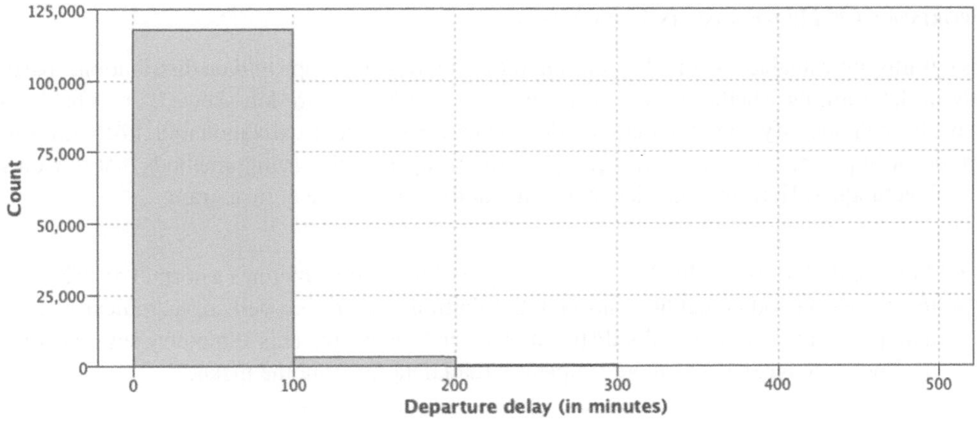

FIGURE 4.8 Histogram for **Departure delay** with five bins.

TABLE 4.5
Graph Data for Departure Delay with 5 Bins

Departure Delay (in Minutes)	Range	Count	Percent
50	0–100	118,083	97
150	101–200	3,247	3
250	201–300	635	1
350	301–400	197	0
450	401–500	59	0

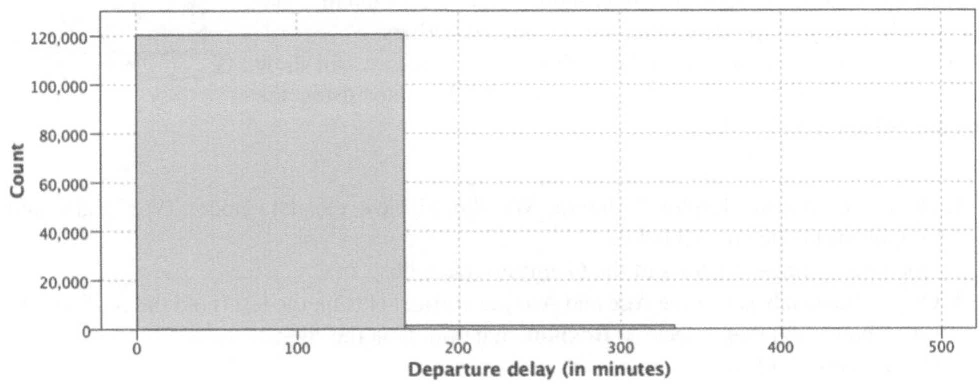

FIGURE 4.9 Histogram for **Departure delay** with three bins.

TABLE 4.6
Graph Data for Departure Delay with 3 Bins

Departure Delay (in Minutes)	Range	Count	Percent
83	0–166	120,751	99
250	167–333	1,312	1
417	334–500	158	0

Importance Of Histograms in Data Mining

Histograms are valuable tools in data mining for evaluating the shape of data distributions, particularly in determining whether they follow a normal distribution, are left-skewed, or right-skewed. They also help identify potential outliers. Beyond their role as a visualization tool, histograms have significant implications in data mining, especially in supervised learning methods that rely on statistical techniques. Here are some key points about the importance of histograms:

- **Normal distribution check**: The linear regression method assumes a normal distribution to produce a good model fit. A normal distribution appears as a bell curve, indicating that data points are symmetrically distributed around the mean. This symmetry suggests that data near the mean occur more frequently than data far from the mean.
- **Skewness detection**: If a histogram reveals skewed data—whether left or right—it may be necessary to consider data transformation efforts to normalize the distribution. Proper normalization can improve the performance of statistical models.
- **Outlier Identification**: Histograms can also help identify outliers. The presence of extensive outliers can indicate noise in the dataset, potentially affecting the predictive power of the model. By examining the extent of these outliers, we can make informed decisions about whether to retain or remove cases with extreme outliers.

Box Plots

In addition to histograms, another useful graphical tool for indicating data skewness and identifying potential outliers is the box plot, also known as the whisker plot. A box plot graphically represents the five-number summary of a dataset, which includes the minimum, lower quartile (Q1), median, upper quartile (Q3), and maximum values. Additionally, box plots highlight any observations that are considered outliers.

Let's create two box plots related to airline status. The first will show the age distribution by airline status. To create a box plot using the Graphboard node, follow these steps:

1. Open the "Airline_Explore" stream. We should have the data node, Type node, and Graphboard node created before.
2. Open the settings windows of the **Graphboard** node.
3. On the **Basic** tab, select the **Age** and **Airline status** fields on the left (hold the Shift key to select both), and then select the **Boxplot** chat option on the right.
4. Click Apply and Run.

As shown in Figure 4.10, the median and the interquartile range (the box) are centered, indicating the symmetry of the data. The interquartile range appears consistent across the four airline statuses. Notably, there is only one outlier for the Silver passenger category (the data point outside the whisker).

If we follow the same steps but select the **Number of years flying** and **Airline status** fields, we generate another box plot, as presented in Figure 4.11. This plot displays the distribution of the number of flights per year by airline status. You may observe that the median and the box are slightly off-center, suggesting asymmetry in the data. The box for Blue passengers is slightly larger than that for other statuses, indicating a difference in the variance of the data.

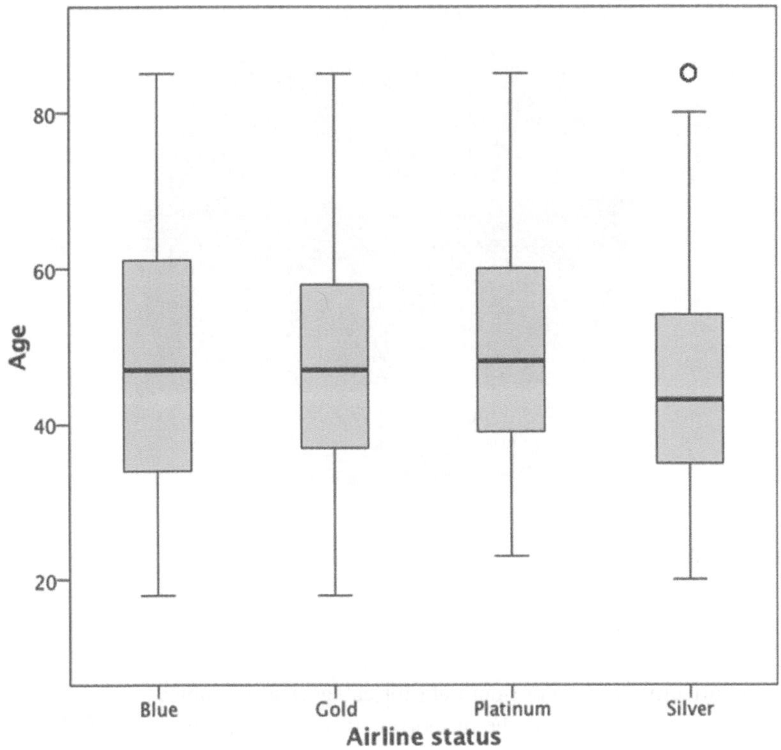

FIGURE 4.10 Box plots for **Age** by **Airline status**.

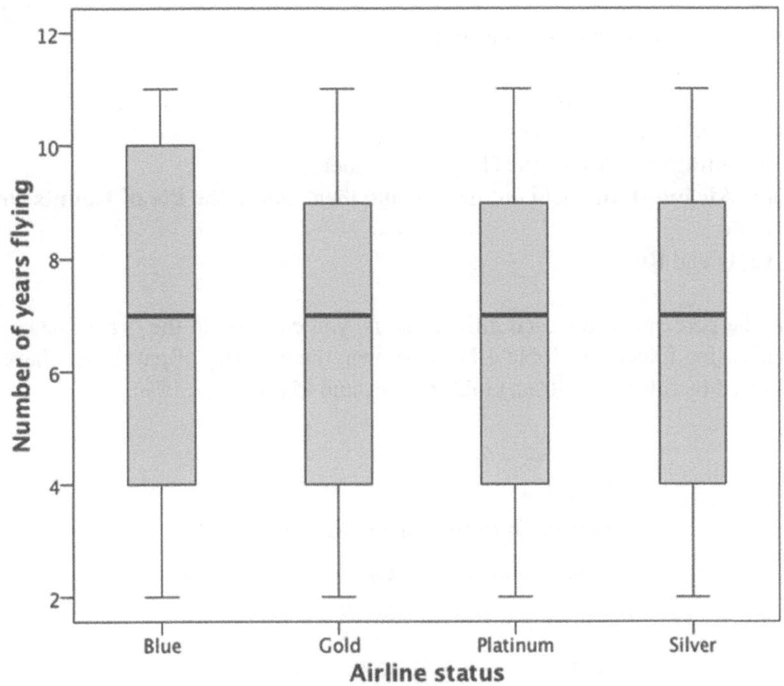

FIGURE 4.11 Box plots for **Age** and **Average number of years flying** by **Airline status**.

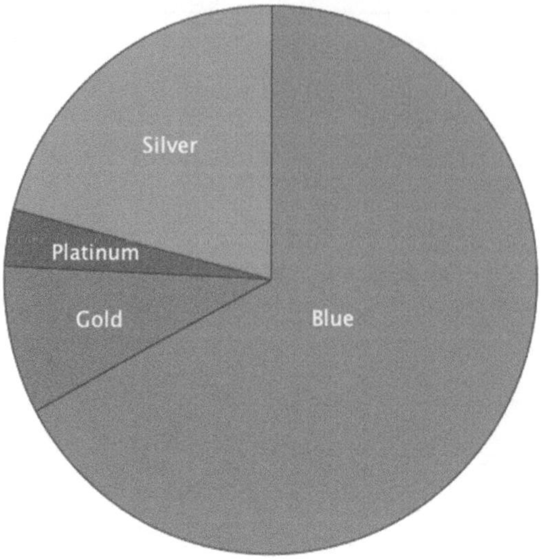

FIGURE 4.12 Pie chart for **Airline status**.

Pie Charts

Pie charts are an effective way to present data for categorical variables as percentages, visually representing the proportion of each category as a slice of a pie. This format is easy to understand and interpret. For example, Figure 4.12 shows a pie chart illustrating the percentage distribution of passenger airline status. The chart indicates that the majority of passengers have Blue status, followed by Silver, Gold, and Platinum statuses.

Follow the steps below to create a pie chart:

1. Open the "Airline_Explore" stream. We should have the data node, Type node, and Graphboard node created before.
2. Open the settings window of the Graphboard node.
3. Select the **Airline status** field on the left, and then choose the **Pie of Counts** chat option on the right.
4. Click **Apply** and **Run**.

To display the percentage for each airline status, you can obtain the graph data and calculate the percentages using Excel (see Table 4.7). As shown, the majority of passengers have Blue status (67.08%), followed by Silver (20.86%), Gold (8.71%), and Platinum (3.35%).

TABLE 4.7
Graph Data for Airline Status

Airline Status	Count	Percent
Blue	83,490	67.08
Silver	25,967	20.86
Gold	10,837	8.71
Platinum	4,170	3.35

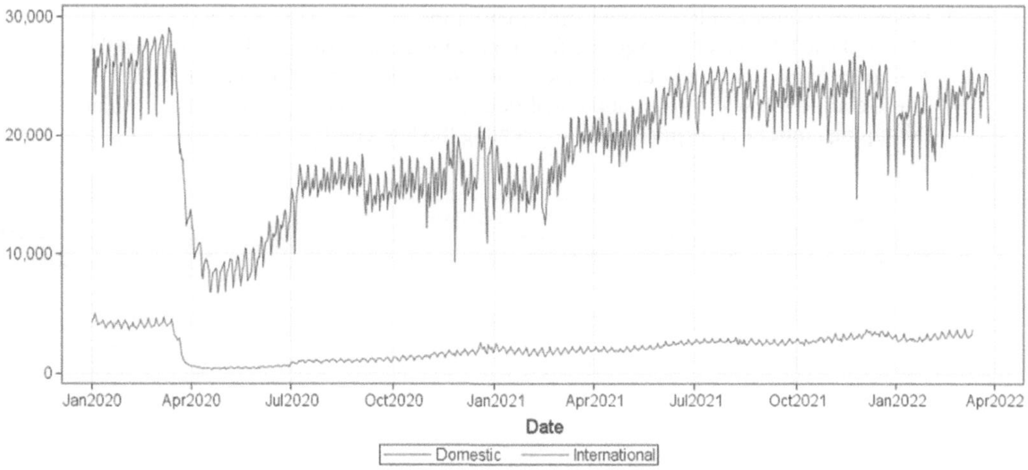

FIGURE 4.13 Line chart for domestic and international flights over time.

Line Charts—Trend Analysis

When analyzing data collected over time, line charts are an excellent choice for visualizing trends in events or incidents. A line chart connects a series of data points with a line, presenting sequential values that help identify trends. In this chart type, the x-axis represents a sequential progression of values.

For instance, Figure 4.13 illustrates the trend of domestic and international flights from January 1, 2020, to March 1, 2022. As observed, when the COVID-19 pandemic began in early 2020, the number of flights decreased significantly in both categories. Moving into late 2021, as vaccination efforts expanded nationwide, travel demand surged, leading to an increase in daily flights. This upward trend continued into 2022, with more flights recorded in the first half of the year. Line charts enable us to analyze data trends over time, monitor changes in events or incidents, and detect noise within the data. In this example, two variables are plotted on the same chart for comparison. It is evident that international flights are consistently lower than domestic flights, with less fluctuation over time.

It's important to note that line charts require two variables: one for the x-axis and one for the y-axis. In this example, time is represented on the x-axis, while the number of flights (domestic and international) is represented on the y-axis.

Since the **Airline recommendation** dataset is not organized by date or year, the application of line charts is somewhat limited based on the types of variables available. However, I will demonstrate two examples of line charts using this dataset.

Example 1 Line Chart for Shopping Amount by Age

Follow the steps below to create a line chart using the Graphboard node:

1. Open the "Airline_Explore" stream. We should have the data node, Type node, and Graphboard node created before.

2. Open the settings window of the Graphboard node.
3. On the **Basic** tab, select the **Age** and **Shopping amount at airport** fields on the left (hold the Control key to select both), and then select the Line chat option on the right.
4. The **Detailed** tab will display additional details, with the mean value of the Y variable (**Shopping amount at airport**) used for creating the line chart.

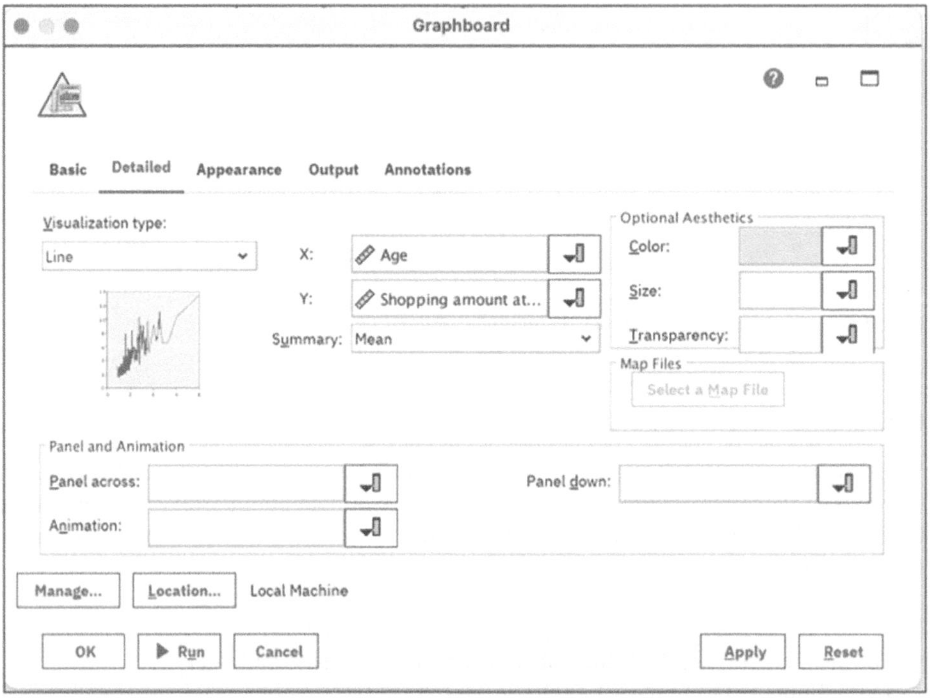

5. Click **Apply** and **Run.**

The chart is displayed in Figure 4.14. This line chart plots the values of the Y variable (shopping amount at the airport) against each value of the X variable (age) and connects them with a line. The figure indicates that passengers between the ages of 34 and 50 tend to spend the most money on shopping at airports. In contrast, younger passengers and those older than 70 years tend to spend less.

Example 2 Line Chart for Flight Time Against Flight Distance

In this second example, we use a line chart to illustrate the correlation between flight time (in minutes) and flight distance (in miles). To create this line chart, follow the steps below:

1. Open the "Airline_Explore" stream. We should have the data node, Type node, and Graphboard node created before.
2. Open the settings window of the Graphboard node.
3. On the **Basic** tab, select the **Flight time** and **Flight distance** fields on the left (hold the Control key to select both), and then select the Line chat option on the right.
4. On the **Detailed** tab, set X to **Flight distance** and Y to **Flight time**.
5. Click **Apply** and **Run.**

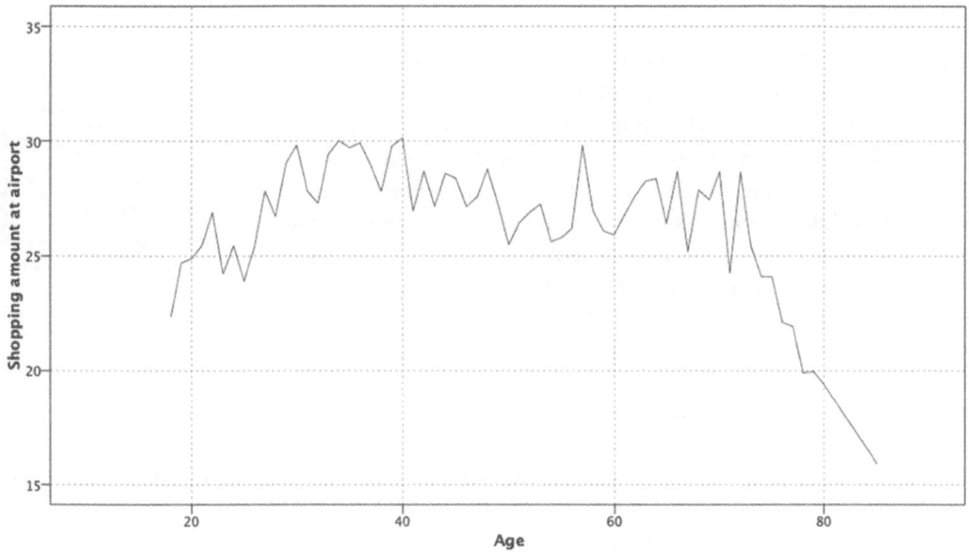

FIGURE 4.14 Line chart for the **Shopping amount at airport** by **Age**.

Figure 4.15 displays this line chart. The trend is linear and positive, indicating that longer distances correspond to longer flight times, which is expected. An interesting observation is the increased noise in the data for flights under 300 miles, compared to much less noise for distances above 3000 miles. This suggests that for flight distances of 3,000 miles or shorter, there is some fluctuation in flight time for reasons that warrant further investigation.

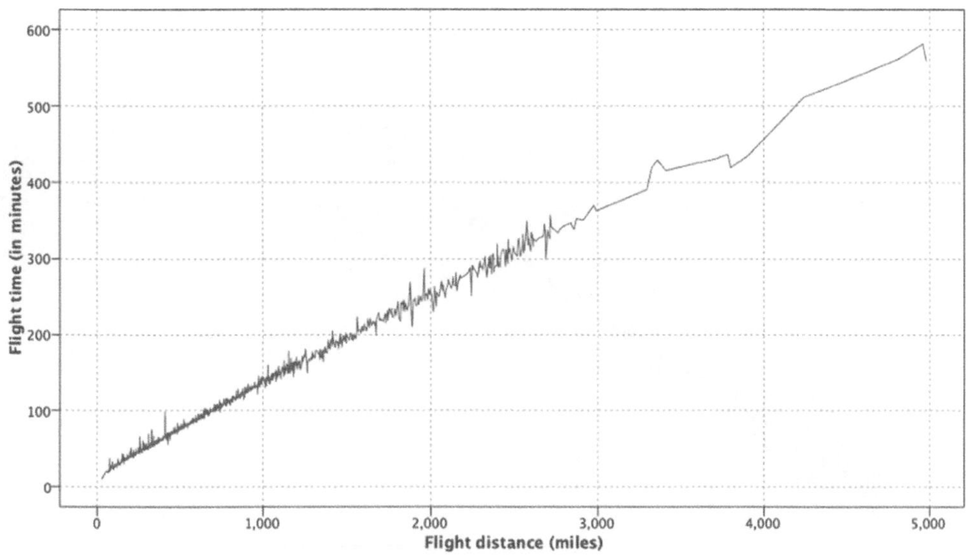

FIGURE 4.15 Line chart for **Flight time** by **Flight distance**.

Scatter Plots

When analyzing the relationship between two variables, a scatter plot is a highly effective graphical tool. Scatter plots display the two-dimensional distribution of data points, using markers to represent the data and showing how the points are dispersed across the chart.

For example, consider the relationship between departure delay time and arrival delay time. As shown in Figure 4.16, there is a positive correlation between these two variables: when a flight departs late, it tends to arrive late as well. The scatter plot reveals that this correlation is particularly strong for delay times of 500 minutes or less, as indicated by the higher density of data points within this range.

Follow the steps below to create that scatter plot:

1. Open the "Airline_Explore" stream. We should have the data node, Type node, and Graphboard node created before.
2. Open the settings window of the Graphboard node.
3. On the **Basic** tab, select the **Departure delay** and **Arrival delay** fields on the left (hold the Control key to select both), and then select the Line chart option on the right.
4. On the **Detailed** tab, set X to **Departure delay** and Y to **Arrival delay**.
5. Click **Apply** and **Run.**

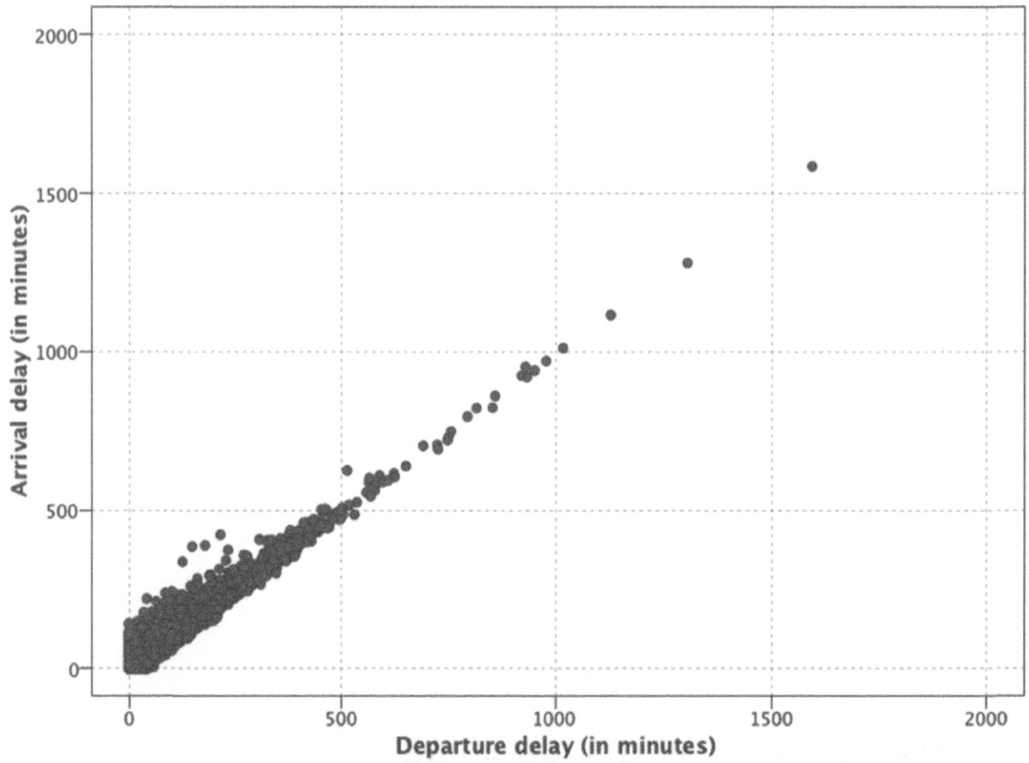

FIGURE 4.16 Scatter plot for **Arrival delay** by **Departure delay**.

Multiplot Node

A multiple Y fields against a single X field. This feature allows you to create multiple plots within a single graph, making it easier to compare different variables simultaneously. For instance, if you want to graph **Price sensitivity** and **Shopping amount at the airport** over **Age** in the same graph, you would typically need to create two separate graphs using the tools discussed earlier. However, by using the Multiplot node, you can combine these variables into a single graph.

Follow the steps below to create a Multiplot node:

1. Open the "Airline_Explore" stream. We should have the data node and Type node created before.
2. Add a Multiplot node to the stream and connect it to the Type node.
3. Open the settings window of the Multiplot node.
4. On the **Plot** tab, select the **Age** as the X field, and then select **Price sensitivity** and **Shopping amount at the airport** as the Y fields.

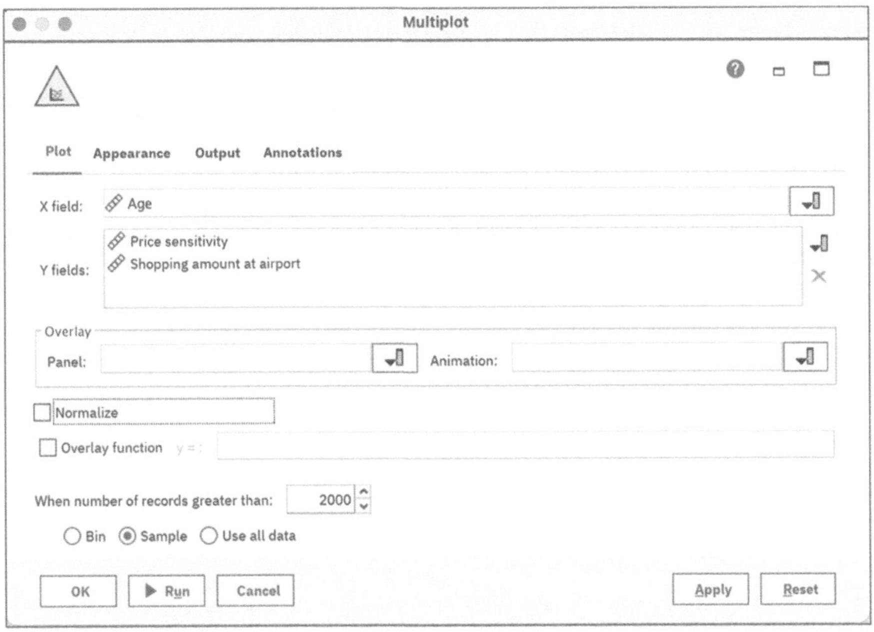

5. Select Sample (to reduce the running time, as this node will require more time to run with all data).
6. Click **Apply** and **Run**.

The multiplot chart is shown in Figure 4.17. In this chart, we can observe that the **Price sensitivity** information is not clearly presented, with the chart seeming to emphasize the **Shopping amount at the airport** instead. This issue arises because the two Y fields have different scales, making it difficult for the chart to display both fields in a meaningful way.

To resolve this issue, we need to normalize the data to ensure that the Y fields are on the same scale. In the settings window, check the box labeled Normalize and then run the Multiplot node again. The result is shown in Figure 4.18. As you can see, the data for the Y fields (**Shopping amount at airport** and **Price sensitivity**) have been normalized to the same scale, ranging from 0 to 1. This adjustment allows us to clearly see the patterns for both variables across different age ranges within the same chart.

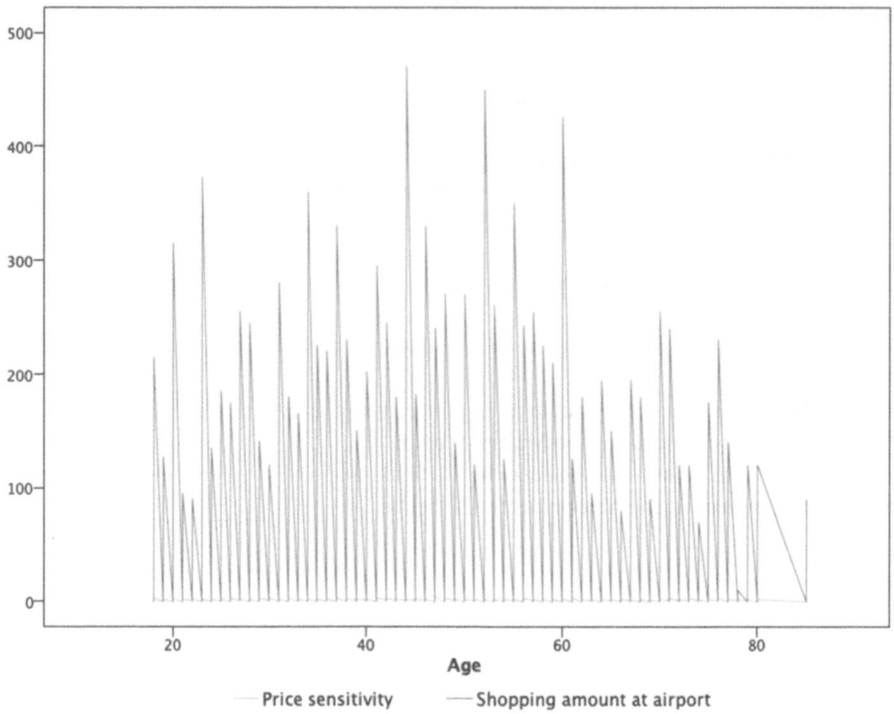

FIGURE 4.17 Multiplot node for **Price sensitivity** and **Shopping amount at the airport** over **Age** (non-normalized).

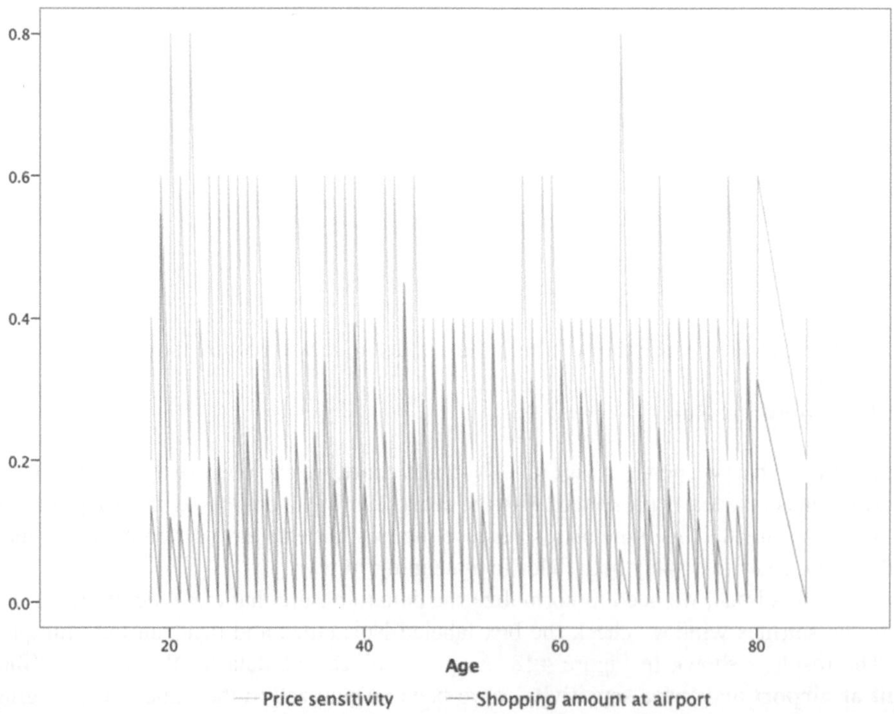

FIGURE 4.18 Multiplot node for **Price sensitivity** and **Shopping amount at the airport** over **Age** (normalized).

Quick Chart Explorer Using Data Audit Node

One question you may have is: What if we want to view the distributions of all variables in the dataset at once to examine their distributions and missing values? Instead of manually creating those charts one by one, as shown before, SPSS Modeler offers a way to generate all these charts simultaneously using the **Data Audit** node.

The **Data Audit** node provides a comprehensive review of the data. This information is presented in a matrix that is easy to understand, sortable, and conducive to generating detailed graphs along with a range of data preparation nodes.

- The **Audit** tab presents summary statistics, histograms, and distribution graphs, offering an initial insight into the data. Additionally, the report includes a storage icon positioned before the field name.
- The **Quality** tab details outliers, extremes, and missing values, and provides tools to address these issues.

This feature is particularly useful when you need to quickly visualize and assess the distribution of multiple variables without creating separate charts for each one. By running the Data Audit node, you can efficiently review the frequency distributions and identify any potential issues, such as skewness or outliers, across all variables in your dataset simultaneously.

> Data Audit node can create histograms and bar charts for all fields in the dataset within a single output window.

Follow the steps below to add and set up the Data Audit node.

1. Open the "Airline_Explore" stream. We should have the data node and Type node created before.
2. From the Output group in the Node Pallete, add the Data Audit node to the stream and connect it to the Type node.
3. Open the settings window of the Data Audit node.
4. On the **Settings** tab, keep the Default setting, which will include all fields in the analysis.
5. Check the box Advanced statistics.

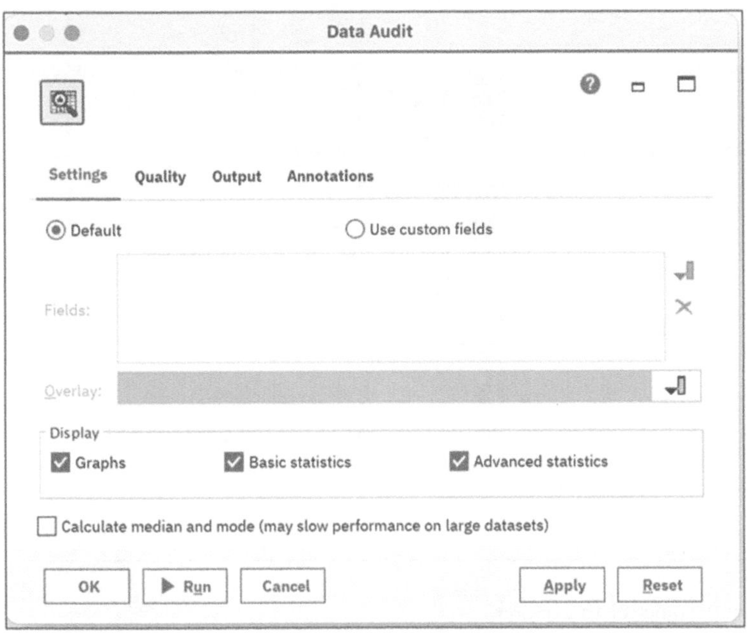

6. On the **Quality** tab, make sure to check two boxes for **Missing values: Count of records with valid values** and **Breakdown counts of records with invalid values**.
7. Click **Apply** and **Run.**

This approach is highly effective for viewing the distributions of all variables simultaneously. Figure 4.19 shows the results of these multi-charts for the **Airline recommendation** dataset. You can see all histograms and bar charts for the variables across multiple windows. Expand each window to view the detailed chart for that variable. Remember that bar charts are used for categorical variables, while histograms are used for continuous variables.

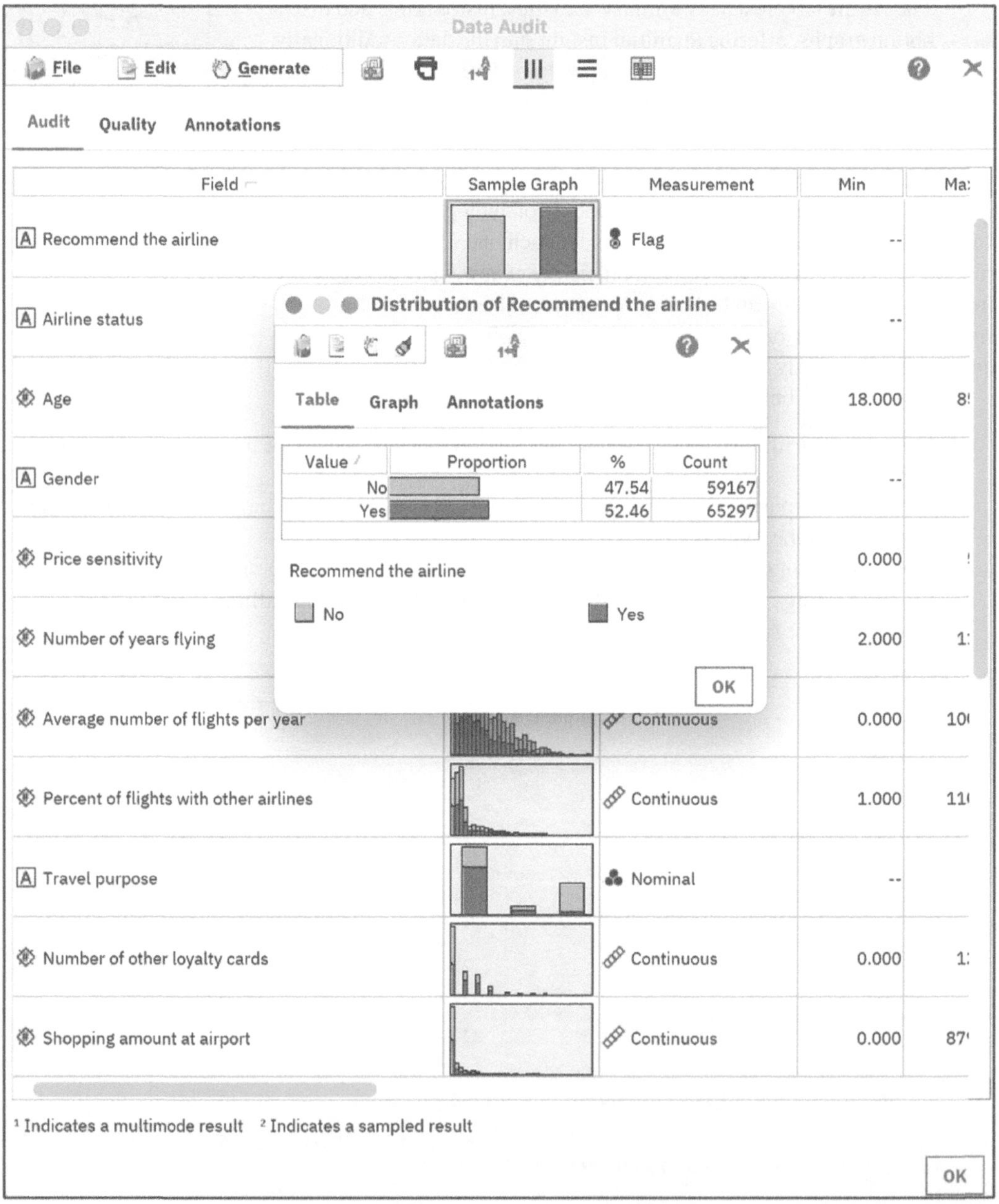

FIGURE 4.19 Results of the quick chart explorer.

STATISTICS EXPLORATION USING DATA AUDIT NODE

Data visualization offers a variety of graphical tools to help us visualize data trends and patterns. However, to select the appropriate machine learning method and determine the right configurations for predictive modeling, it's also essential to explore the statistical characteristics of our data. Specifically, we need to review the descriptive statistics of all variables to understand their attributes, identify any potential concerns, and decide if any data transformations are required in the Modify step. Key descriptive statistics to consider include:

Sample size

- **for categorical variables**: Frequency of each level, percentage, and missing values.
- **for interval variables**: Minimum, maximum, mean, standard deviation, skewness, kurtosis, and missing values.

These are the typical descriptive statistics covered in statistics courses. As a reminder, skewness and kurtosis provide insights into the distribution of data and whether they follow a normal distribution, which is a required assumption for the linear regression method.

For statistical exploration, we use the **Data Audit** node in SPSS Modeler. This multipurpose tool allows us to examine variable distributions and statistics within our datasets.

To illustrate, we'll run the same **Data Audit** node that was created earlier. In this example, to display the basic descriptive statistics, I selected only the "Basic statistics" option in the settings window. Figure 4.20 shows the output of the **Data Audit** node under the **Audit** tab. It includes measurements for all fields. For continuous variables, the output displays minimum, maximum, mean, standard deviation, skewness, and the number of valid cases. For categorical variables, it shows the number of unique levels and valid cases.

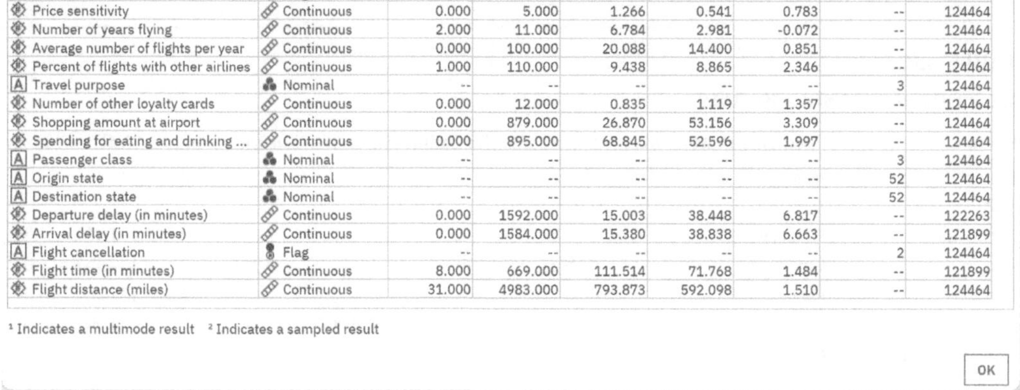

Field	Measurement	Min	Max	Mean	Std. Dev	Skewness	Unique	Valid
Recommend the airline	Flag	--	--	--	--	--	2	124464
Airline status	Nominal	--	--	--	--	--	4	124464
Age	Continuous	18.000	85.000	47.503	16.498	0.316	--	124464
Gender	Flag	--	--	--	--	--	2	124464
Price sensitivity	Continuous	0.000	5.000	1.266	0.541	0.783	--	124464
Number of years flying	Continuous	2.000	11.000	6.784	2.981	-0.072	--	124464
Average number of flights per year	Continuous	0.000	100.000	20.088	14.400	0.851	--	124464
Percent of flights with other airlines	Continuous	1.000	110.000	9.438	8.865	2.346	--	124464
Travel purpose	Nominal	--	--	--	--	--	3	124464
Number of other loyalty cards	Continuous	0.000	12.000	0.835	1.119	1.357	--	124464
Shopping amount at airport	Continuous	0.000	879.000	26.870	53.156	3.309	--	124464
Spending for eating and drinking ...	Continuous	0.000	895.000	68.845	52.596	1.997	--	124464
Passenger class	Nominal	--	--	--	--	--	3	124464
Origin state	Nominal	--	--	--	--	--	52	124464
Destination state	Nominal	--	--	--	--	--	52	124464
Departure delay (in minutes)	Continuous	0.000	1592.000	15.003	38.448	6.817	--	122263
Arrival delay (in minutes)	Continuous	0.000	1584.000	15.380	38.838	6.663	--	121899
Flight cancellation	Flag	--	--	--	--	--	2	124464
Flight time (in minutes)	Continuous	8.000	669.000	111.514	71.768	1.484	--	121899
Flight distance (miles)	Continuous	31.000	4983.000	793.873	592.098	1.510	--	124464

[1] Indicates a multimode result [2] Indicates a sampled result

FIGURE 4.20 Descriptive statistics for all fields from the Data Audit node.

Additional Information: Missing Values, Outliers, and Extreme Values

The **Audit** tab does not directly display invalid cases (missing values). While it's possible to calculate missing cases using Excel, the **Quality** tab conveniently presents missing values under the "Null values" column. This information can be copied and merged with the descriptive statistics. Additionally, the **Quality** tab highlights the number of outliers and extreme values for each field.

> Data Audit node can generate both basic and advanced descriptive statistics for all fields in the dataset, all within the same output window.

EXAMPLE: STATISTICS EXPLORATION

We use the Data Audit node to explore descriptive statistics for all fields in the **Airline recommendation** dataset. By combining the results from the **Audit** tab and **Quality** tab, we obtain full descriptive statistics for all fields, as presented in Table 4.7.

I summarize some key findings as follows:

- **Target variable distribution**: The frequency and percentage of the two levels of the target variable indicate a relatively balanced dataset, with 52.46% of the data being "Yes" (recommend) and 47.54% being "No" (not recommend).
- **Missing values**: There are no missing values in the categorical variables. However, among the continuous variables, there are missing values for three variables: **Arrival delay time**, **Departure delay time**, and **Flight time**. While other continuous variables do not seem to have missing values, a closer look at the **Price sensitivity** variable reveals something unusual. This variable has a minimum value of zero, which is unexpected because **Price sensitivity** is measured on a Likert scale (typically ranging from 1 to 5). Therefore, a value of zero indicates a missing value, and we will need to recode this value as missing. Handling these missing values will be discussed in Chapter 5.
- **Descriptive statistics for categorical variables**: The dataset includes variables with varying numbers of levels: three variables with two levels, two variables with three levels, one variable with four levels, and two variables with 52 levels (representing the 52 U.S. states).
- **Descriptive statistics for continuous variables**: Table 4.8 presents key statistics such as field name, minimum, maximum, mean, median, standard deviation, skewness, and kurtosis. These are standard descriptive statistics that most statistical software applications generate. It is noteworthy that only a few variables have distributions close to normal, with

TABLE 4.8
Descriptive Statistics Results

Field	Measurement	Min	Max	Mean	Std. Deviation	Skewness	Kurtosis	Unique	Valid	Missing
Recommend the airline	Flag	–	–	–	–	–	–	2	124,464	0
Airline status	Nominal	–	–	–	–	–	–	4	124,464	0
Age	Continuous	18	85	47.503	16.498	0.316	−0.711	–	124,464	0
Gender	Flag	–	–	–	–	–	–	2	124,464	0
Price sensitivity	Continuous	0	5	1.266	0.541	0.783	1.189	–	124,464	0
Number of years flying	Continuous	2	11	6.784	2.981	−0.072	−1.291	–	124,464	0
Average number of flights per year	Continuous	0	100	20.088	14.4	0.851	0.468	–	124,464	0

(Continued)

TABLE 4.8 *(Continued)*
Descriptive Statistics Results

Field	Measurement	Min	Max	Mean	Std. Deviation	Skewness	Kurtosis	Unique	Valid	Missing
Percent of flights with other airlines	Continuous	1	110	9.438	8.865	2.346	7.625	–	124,464	0
Travel purpose	Nominal	–	–	–	–	–	–	3	124,464	0
Number of other loyalty cards	Continuous	0	12	0.835	1.119	1.357	1.867	–	124,464	0
Shopping amount at the airport	Continuous	0	879	26.87	53.156	3.309	15.935	–	124,464	0
Spending for eating and drinking at the airport	Continuous	0	895	68.845	52.596	1.997	9.948	–	124,464	0
Passenger class	Nominal	–	–	–	–	–	–	3	124,464	0
Origin state	Nominal	–	–	–	–	–	–	52	124,464	0
Destination state	Nominal	–	–	–	–	–	–	52	124,464	0
Departure delay (in minutes)	Continuous	0	1,592	15.003	38.448	6.817	100.767	–	122,263	2201
Arrival delay (in minutes)	Continuous	0	1,584	15.38	38.838	6.663	95.221	–	121,899	2565
Flight cancellation	Flag	–	–	–	–	–	–	2	124,464	0
Flight time (in minutes)	Continuous	8	669	111.514	71.768	1.484	2.688	–	121,899	2565
Flight distance (miles)	Continuous	31	4,983	793.873	592.098	1.51	2.778	–	124,464	0

skewness and kurtosis values near zero. These include **Age**, **Number of flights per year**, **Number of loyalty cards**, **Price sensitivity**, and **Number of years flying**. Variables like **Flight distance** and **Flight time** are not perfectly normal but are close. Conversely, other variables exhibit high skewness and kurtosis values, indicating that their distributions are skewed and far from normal. Depending on the machine learning methods chosen for modeling, we may need to consider data normalization and transformation.

HIDDEN FEATURE IN SPSS MODELER— INTERACTIVE GRAPH TOOL

In the previous sections, we've explored various graph tools available in SPSS Modeler under the **Graphs** group in the **Node Pallet**. While I focused on the most commonly used tools, it's worth taking the time to practice with all of them. You may discover ways to effectively leverage these tools to create meaningful visualizations for your data. As previously mentioned, these tools are useful but not particularly interactive, requiring the addition of new nodes to create separate charts for different fields or chart types. Additionally, most graphs do not provide percentage values, so if you want to display percentages, you'll need to export the graph data to Excel and calculate those values manually. This process is not ideal, especially when other software, like SAS Enterprise Miner, allows you to switch between frequency and percentage within the same settings window easily.

But what if I told you that SPSS Modeler actually has a hidden feature that simplifies data visualization with a more interactive interface, and the graphs can indeed display percentage values? In this section, I'll reveal this hidden feature, which is not listed in the menu bar or **Node Pallet**. Even the SPSS Modeler guidelines do not provide instructions for this feature. It is integrated into a menu option called "**View Data**."

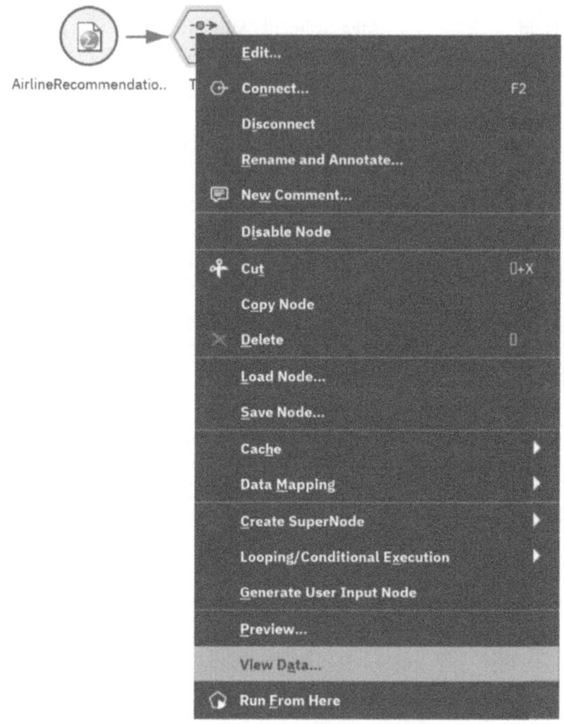

View Data is a menu option that allows you to view your data at any node, enabling you to observe changes in the dataset after sampling, partitioning, transformation, or modeling. You can also use it to view your original data. However, most users are unaware that this menu option includes a chart feature, offering an interactive way to create charts more efficiently.

Let's walk through an example. Suppose you have a stream with the **Airline Recommendation** data source connected to a **Type** node. If you right-click on the **Type** node (you can also right-click on the data source node), a drop-down menu will appear. Select **View Data** from this menu, which will open a **View Data** window (this feature runs on Java, so ensure your Java is up to date).

This window contains several tabs, including **Charts**, **Spreadsheet**, **Data Audit**, and **Dashboard**. I'll focus on the **Charts** tab because the other tabs are primarily for viewing data and statistics, which we've already covered. As shown in Figure 4.21, the **Charts** tab offers 28 different types of charts, including the ones we've discussed in this chapter, such as **Bar**, **Histogram**, **Pie**, **Scatter**, **Line**, **Box Plot**, and **Multi-Series** (similar to Multiplot). Additionally, it provides other useful tools like **Treemap**, **Q-Q Plot**, **Population**, and more.

I won't go over all of the chart types again since we've already covered several in detail. What's important to note is that SPSS Modeler provides this graph tool in a very interactive manner. Once you select the desired chart type, simply identify the fields on the left panel, as required for that chart

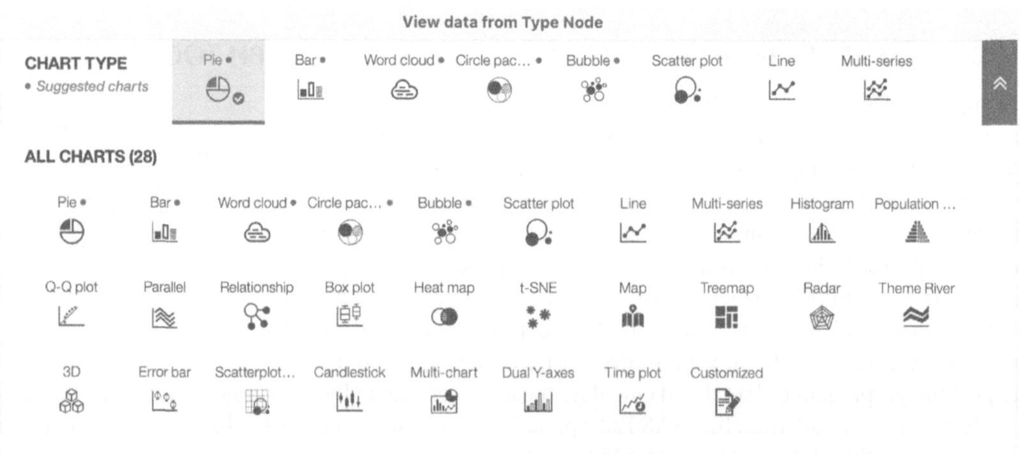

FIGURE 4.21 Chart options in the View Data menu.

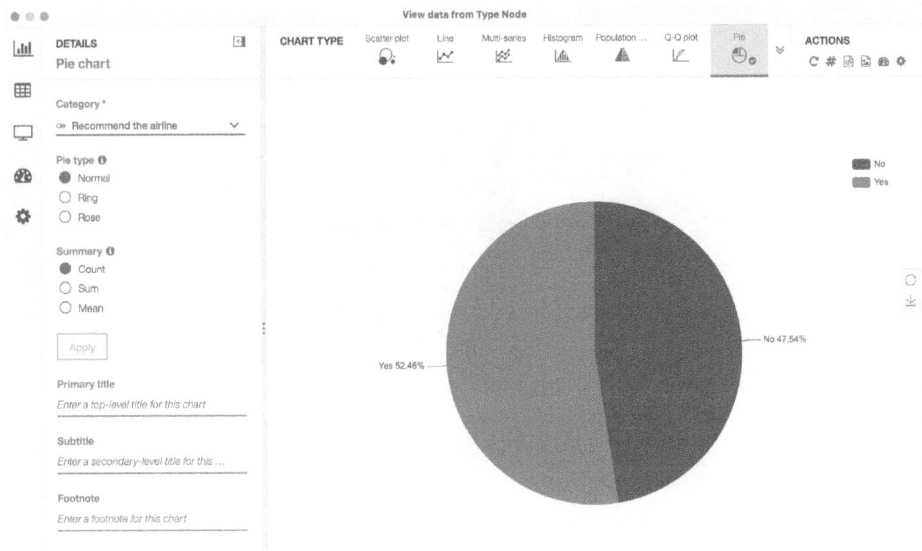

FIGURE 4.22 Interactive chart building interface.

type, and click **Apply**. The chart will then appear in the middle panel. For example, Figure 4.22 shows how easily we can build a pie chart for the **Recommend the airline** field. The pie chart displays percentage values for added convenience. You can change the pie chart type and see the changes reflected instantly. Additionally, you can click on a specific class of the variable (such as **Yes**) to focus on data for that level, and the chart will automatically adjust. This tool is highly interactive and dynamic, making it great for visualization demonstrations.

However, this tool is not without its shortcomings. First, it doesn't offer much customization for colors and fonts, so you'll need to accept the default settings. Additionally, only pie charts display percentages, while other charts do not. If you need percentage values for other chart types, you'll still have to calculate them in Excel. Moreover, you can't adjust the axis scales as you can with the other graph tools we've discussed. Finally, since this tool runs on Java, working with very large datasets may take some time, and in some cases, the interface might freeze.

Despite these shortcomings, I still believe this hidden feature is a valuable tool for data visualization. You should definitely give it a try and determine when it's appropriate to use this tool to create charts for your data mining projects. To demonstrate its capabilities, I've included several charts created with this tool, so you can compare them with the charts we made earlier. I've also added a treemap to this list, as it proves to be a better chart type for presenting data on **Original states** and **Destination states** compared to traditional bar charts, given the large number of levels in these variables. The Figures 4.23–4.31 are listed below for easy reference:

- Figure 4.23: Pie chart for **Airline status**.
- Figure 4.24: Stacked bar chart for **Passenger class** and **Gender**.
- Figure 4.25: Scatter plot for **Departure delay** and **Flight time**.
- Figure 4.26: Histogram for **Age**.
- Figure 4.27: Box plot for **Flight time** and **Airline status**.
- Figure 4.28: Multiplot for **Price sensitivity** and **Shopping amount at airport** over **Age**.
- Figure 4.29: 3D bar chart for **Shopping amount at airport** by **Gender** and **Travel purpose**.
- Figure 4.30: Line chart for **Arrival delay** by **Departure delay**.
- Figure 4.31: Treemap for **Original states**.

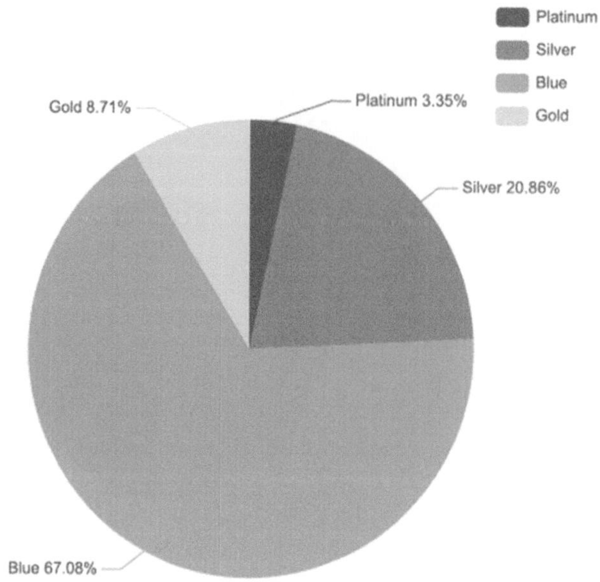

FIGURE 4.23 Pie chart for **Airline status**.

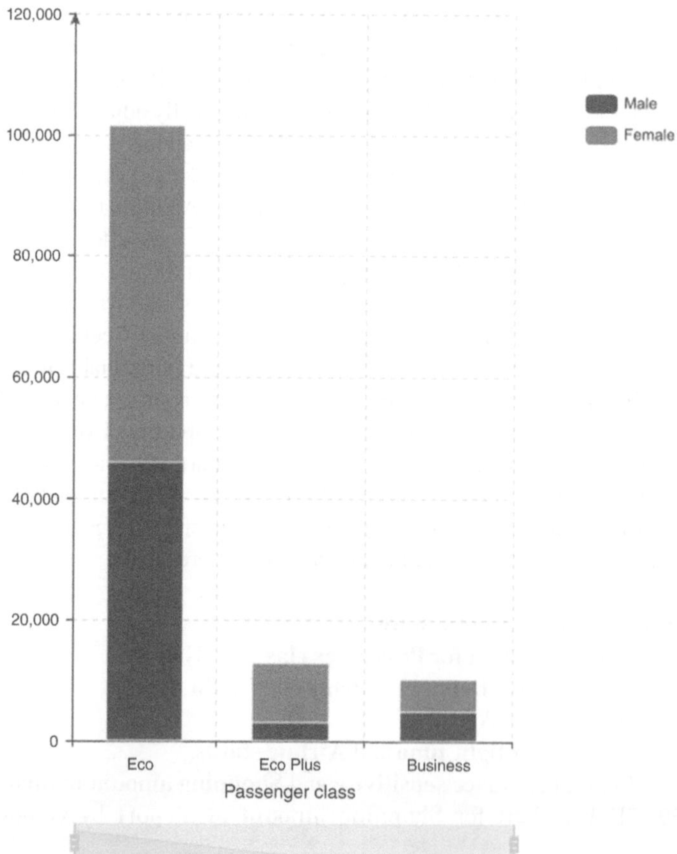

FIGURE 4.24 Stacked bar chart for **Passenger class** and **Gender**.

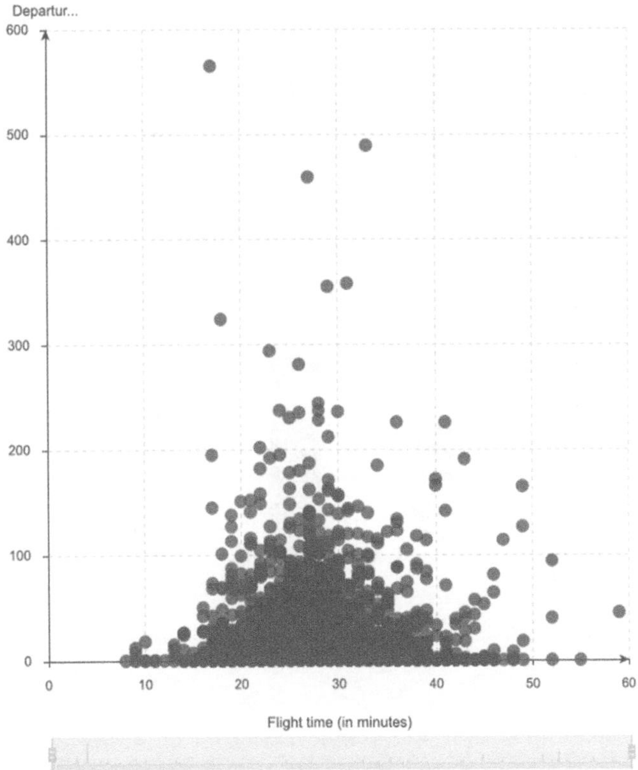

FIGURE 4.25 Scatter plot for **Departure delay** and **Flight time**.

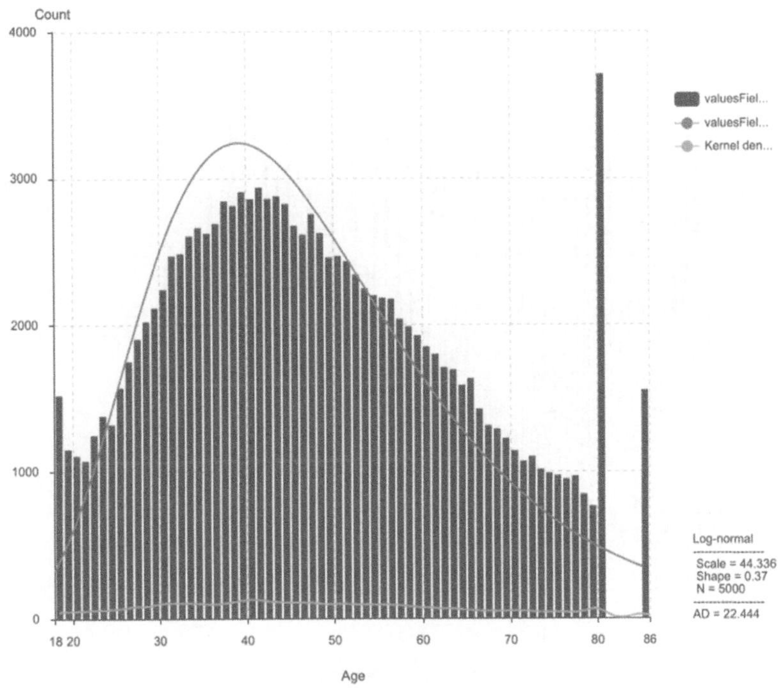

FIGURE 4.26 Histogram for **Age**.

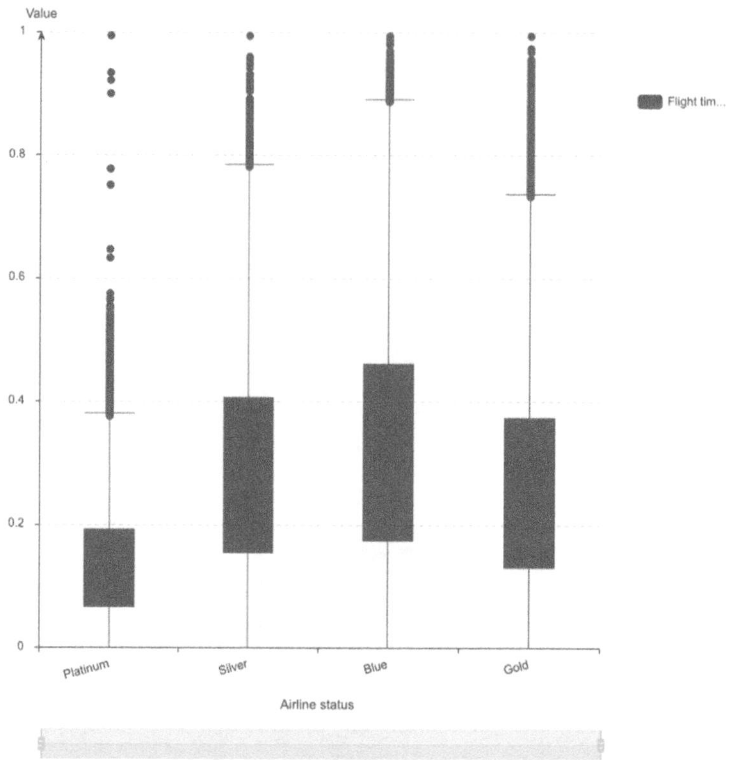

FIGURE 4.27 Box plot for **Flight time** and **Airline status**.

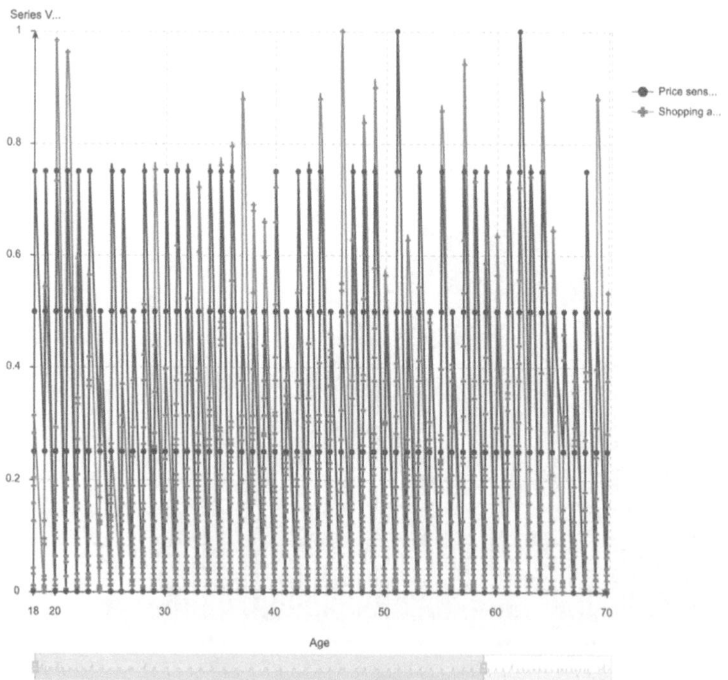

FIGURE 4.28 Multiplot for **Price sensitivity** and **Shopping amount at airport** over **Age**.

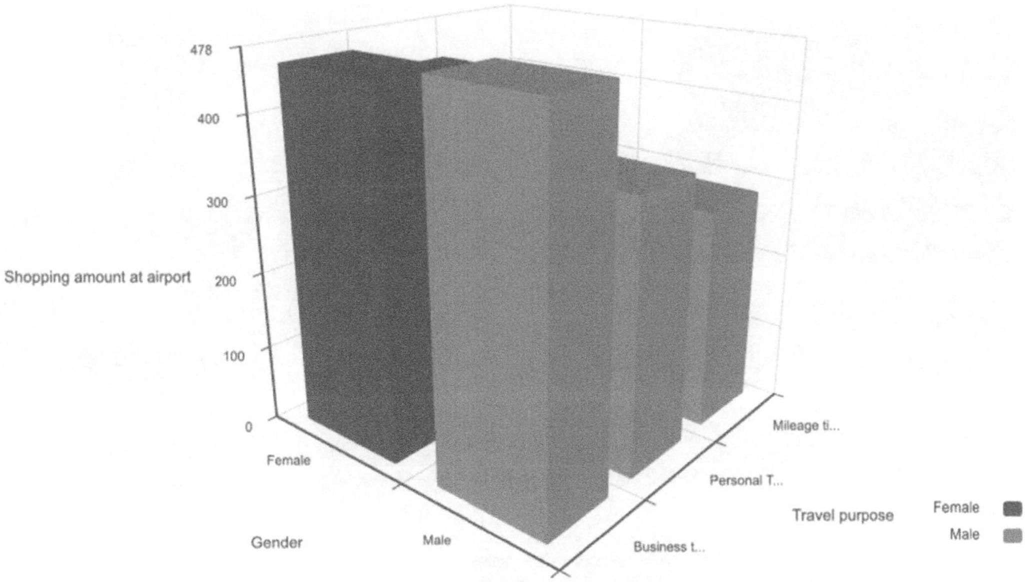

FIGURE 4.29 3D Bart chart for **Shopping amount at airport** by **Gender** and **Travel purpose**.

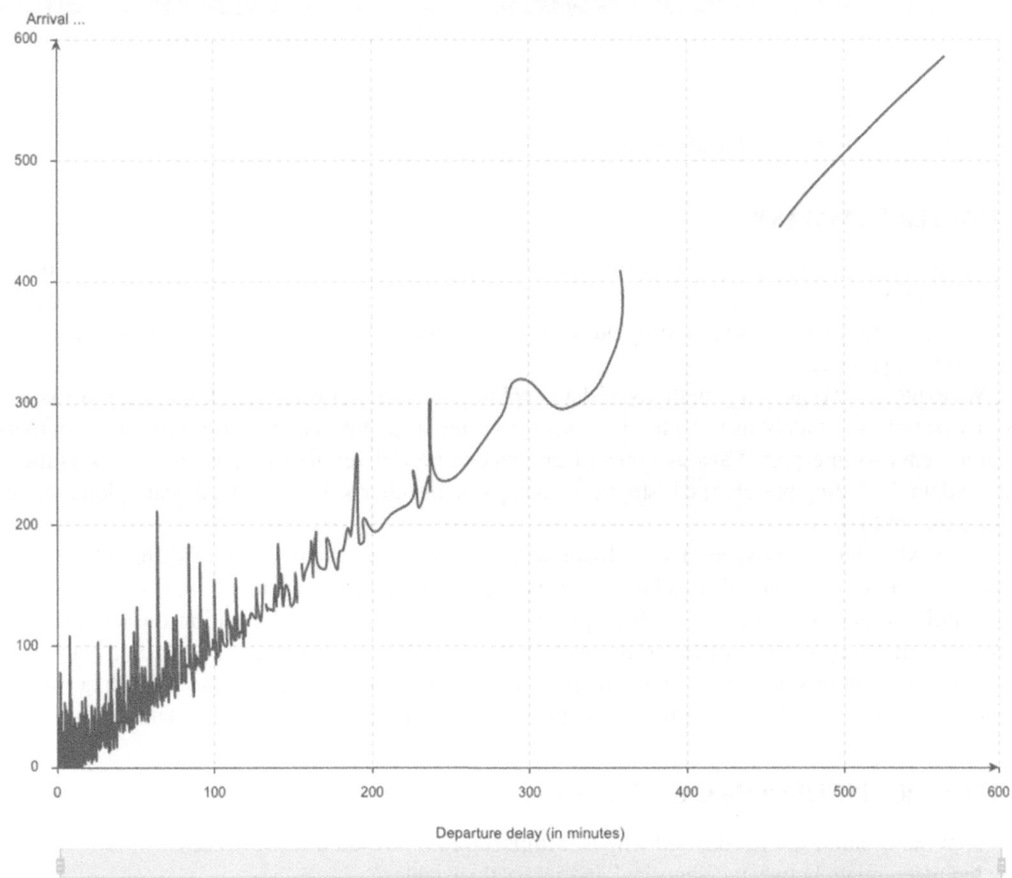

FIGURE 4.30 Line chart for **Arrival delay** by **Departure delay**.

FIGURE 4.31 Treemap for **Original states**.

CHAPTER SUMMARY

Data exploration is a crucial step in the data mining process because it enables us to thoroughly understand the data before analysis. This exploration provides insights into the variables, patterns, and trends within the dataset, guiding our decisions on which variables and methods to use during the modeling phase.

We explored two primary methods of data exploration: **visualization** and **descriptive statistics**. Visualization is a widely used technique for examining large datasets because it produces outputs that are easy to interpret. Various types of graphs can be utilized to create visual representations of the data, including bar charts, histograms, box plots, pie charts, line charts, density plots, scatter plots, and multiplots.

SPSS Modeler offers several visualization tools that allow us to generate different types of charts, tailored to the variables and the analysis objectives. Beyond visualization, it's also essential to examine descriptive statistics, which provide detailed statistical information on all variables, such as frequency, missing values, minimum and maximum values, standard deviation, skewness, and kurtosis. These statistical insights help us determine the subsequent steps in the data mining process, particularly whether we need to impute missing values or transform the data.

CHAPTER DISCUSSION QUESTIONS

1. What is data exploration, and why is it important in the data mining process?
2. Differentiate between categorical and interval variables.
3. Discuss the advantages of data visualization. Research and describe the primary software programs that provide visualization tools for big data.

4. What is the main difference between bar charts and histograms? When should we use histograms?
5. Describe the purpose of box plots. Provide some examples of their application.
6. Give examples of how pie charts and line charts can be used to explore data trends and patterns.
7. What is the purpose of scatter plots, and how can they be used in data visualization?
8. What is a multiplot, and in which scenarios should the multiplot node be used?
9. Describe the purpose of the Data Audit node. What are the primary advantages of this node in data exploration?
10. What are the applications of these visualization features in various industries?
11. Describe the primary descriptive statistics for continuous variables. How can descriptive statistics be used to improve data mining?

CHAPTER EXERCISES

Exercise 1: Graph Exploration for the Bank Dataset

1. Review the variable description of the **Bank** dataset (refer to Appendix 1).
2. Open the **Bank** dataset in SPSS Modeler.
3. Create a new stream named "Bank_Explore" and add the data source to the stream canvas. Add a Type node and connect it to the data source.
4. Plot the following pairs of fields against each other using the appropriate graph types. Create the graphs and explain the results:
   - **Income vs. Age**
   - **Saving balance vs. Credit card balance**
   - **IRA balance vs. Age**
   - **Line of credit balance vs. Income**
   - **Credit card purchase vs. Mortgage**
5. Use the Data Audit tool to create charts for all fields. Present the charts and explain the results.

Exercise 2: Graph Exploration for the Passengers Dataset

1. Review the variable description of the **Passengers** dataset (refer to Appendix 2).
2. Open the **Passengers** dataset in SPSS Modeler.
3. Create a new stream named "Passengers_Explore" and add the data source to the stream canvas. Add a Type node and connect it to the data source.
4. Create histograms for all continuous fields. Report and explain the results regarding the distributions of those fields.
5. Create bar charts for all flag (binary) and nominal fields. Report and explain the results.
6. Use the Data Audit tool to create charts for all fields. Present the charts and explain the results.
7. Identify fields with missing values.

Exercise 3: Statistics Exploration for the Passengers Dataset

1. Open the **Passengers** dataset in SPSS Modeler.
2. Open the stream "Passengers_Explore."
3. Add a Data Audit node to the stream and connect it to the Type node.
4. Set up the Data Audit node and run it to produce descriptive statistics for all fields. Identify fields with missing values and report the number of missing values.
5. Explain the descriptive statistics results. Identify any issues or concerns with the data. (Note: Fields using the Likert scale should not have zero values. Re-code zero values for these fields as missing values.)

Exercise 4: Graph Exploration for the Boston Dataset

1. Review the variable description of the **Boston** dataset (refer to Appendix 1).
2. Open the **Boston** dataset in SPSS Modeler.
3. Create a new stream named "Boston_Explore" and add the data source to the workspace. Add a Type node and connect it to the data source.
4. Create histograms for five fields: **Imv, AgePre40, CrimePC,** and **Tax**. Explain the results regarding the distributions of those fields.
5. Create a scatter plot for **LogDistance** by **Imv**. Explain the results. Then, create a density plot for the same fields and compare the results.
6. Use the Data Audit node to create histograms for all fields simultaneously. Explain the distributions.

Exercise 5: Graph Exploration for the Try Buy Dataset

1. Open the **try_buy** dataset in SPSS Modeler.
2. Create a new stream named "TuyBuy_Explore" and add the data source to the workspace. Add a Type node and connect it to the data source.
3. Create a stacked bar chart for **BuyReturn** by **Promotion Type**, using Promotion Type as the Optional Aesthetic. Explain the results.
4. Create pie charts for **Promotion Type** and **Market Segment**. Explain the results.

Exercise 6: MultiPlot for the Cardiac Dataset

1. Open the **cardiac** dataset in SPSS Modeler.
2. Create a new stream named "Cardiac_Explore" and add the data source to the workspace. Add a Type node and connect it to the data source.
3. Add a MultiPlot node to the workspace and connect it to the Type node.
4. Create a multiplot chart with **Albumin** and **Calcium** as the Y fields and **Age** as the X field. Present and explain the results.

Exercise 7: Statistics Exploration for the Boston Dataset

1. Open the **Boston** dataset in SPSS Modeler.
2. Open the stream "Boston_Explore."
3. Add a Data Audit node to the stream and connect it to the Type node.
4. Set up and run the Data Audit node to produce descriptive statistics for all fields. Identify any missing values and evaluate the normality of the data based on skewness and kurtosis values.
5. Create histograms for all fields using the Data Audit node. Evaluate the distributions and compare them with the descriptive statistics results from Step 4.

5 Data Modification

LEARNING OUTCOMES

Upon the completion of this chapter, you will be able to:

1. Explain the important role of data modification in the data mining process.
2. Provide examples of when data modification is needed.
3. Explain the advantages and disadvantages of various data modification techniques.
4. Compare replacement, imputation, and transformation techniques in data modification.
5. Select the appropriate data modification techniques for specific scenarios.
6. Evaluate various replacement methods and their applications.
7. Perform data replacement using SPSS Modeler.
8. Asses various imputation methods and their applications.
9. Execute data imputation using SPSS Modeler.
10. Evaluate different normalization methods and their applications.
11. Assess various discretization methods and their applications.
12. Perform data normalization and discretization using SPSS Modeler.

INTRODUCTION

In the previous chapter, we explored how to use data visualization to uncover patterns and trends in the data. We also examined descriptive statistics to gain detailed insights into the characteristics of each variable. This analysis allows us to assess the information within the data, identify missing or incorrect information, detect noise, and recognize any concerns that could impact the modeling process. But what should we do if we encounter issues with the data? The straightforward answer is that any concerns regarding data quality must be resolved before proceeding to the modeling step.

This chapter focuses on the next phase of the data mining process: data modification. Specifically, we will discuss the importance of data modification and explore various techniques for preparing and adjusting data. Key topics include replacement, imputation, and transformation. I will also demonstrate how to carry out data modification using SPSS Modeler. Keep in mind that our predictive models are only as good as the data we use to train them. By preparing and transforming data, we can address quality issues and develop an effective, accurate predictive model.

IMPORTANCE OF DATA MODIFICATION

DEFINITION

Data modification involves preparing and transforming the raw data to make it suitable for training our predictive models.

Data modification, also known as data preprocessing, is a critical step in machine learning that involves preparing and transforming the raw data to make it suitable for training our predictive models. The goal of data modification is to improve the quality and relevance of the

DOI: 10.1201/9781003468196-6

data, which enables us to perform more robust model training. As a result, we can achieve more accurate and useful predictive models.

There are many reasons why data modification is important in data mining. Following are some key points.

1. **Data quality improvement**: Techniques like data cleaning and replacement improve the quality of datasets by identifying and correcting errors, inconsistencies, and missing values. High-quality data are essential for valid and reliable predictive modeling, as it reduces the risk of misleading or invalid results. Remember, "garbage in, garbage out."
2. **Noise reduction**: Raw data often contain noise or irrelevant information that can negatively impact model accuracy. Techniques such as outlier detection and data replacement effectively reduce the impact of noisy data points, leading to more reliable and valid models.
3. **Handling missing data**: Missing data are a common issue in real-world datasets due to the data collection process. Imputation techniques address this problem by filling gaps with estimated values, ensuring that valuable information is not lost and minimizing bias in the results.
4. **Variable selection and dimensionality reduction**: These techniques identify relevant variables for analysis. Removing irrelevant or redundant variables reduces data complexity, enhances computational efficiency, and improves model performance by focusing on the most informative parameters.
5. **Normalization and standardization**: These techniques ensure that variables are on a comparable scale, which is particularly useful for algorithms sensitive to data scale. Normalizing and standardizing data enable fair comparisons and prevent certain features from dominating the analysis, improving model fit for specific algorithms.
6. **Handling heterogeneous data**: When data from different sources or formats need to be combined for analysis, data integration and aggregation harmonize the data. This integration provides a more comprehensive view of the problem, facilitates analysis, and reveals valuable patterns that may be hidden within individual datasets.
7. **Preparing data for specific algorithms**: Different machine learning algorithms have specific requirements and assumptions about input data. Data modification helps prepare the data to meet these requirements. For example, discretization may be necessary for algorithms requiring categorical inputs, while normalization is essential for multiple regression methods.

Overall, data modification is vital in data mining as it helps ensure the accuracy, reliability, and efficiency of predictive modeling. By enhancing data quality, reducing noise, handling missing data, selecting relevant features, and preparing the data for specific algorithms, data modification techniques allow us to produce an effective and accurate predictive model.

DATA MODIFICATION TECHNIQUES

There are various techniques available for data modification, which can be broadly categorized into three primary groups:

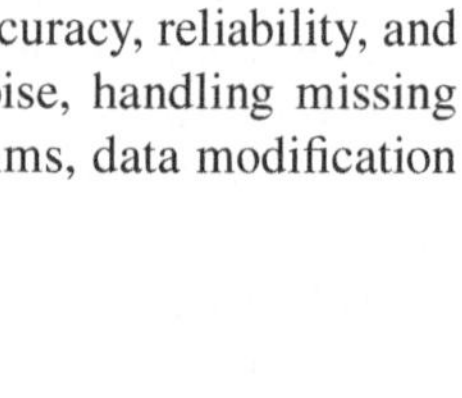

- **Data cleaning**: Data cleaning is the process of identifying and correcting errors in the data. It includes tasks such as removing duplicate entries, correcting typos, and addressing missing data.
- **Data transformation**: Data transformation is the process of changing the format or structure of the data. Several transformation practices we can use include converting categorical data to numerical data, normalizing the data, and aggregating the data.

- **Data reduction**: Data reduction involves reducing the size of the data without losing important information. Techniques such as dimensionality reduction, variable selection, and sampling are commonly used in data reduction.

Data modification is a crucial part of data mining, involving a complex process that requires careful planning and execution. However, the benefits of data modification are significant, including improved accuracy, efficiency, and interpretability of data mining results. When performing data modification, several key considerations should be kept in mind:

- **Use a variety of techniques**: No single technique is best for all datasets. It is important to use a variety of techniques to ensure that the data are properly modified.
- **Avoid overfitting**: Overfitting can occur if the data modification process is too aggressive, leading to data that no longer accurately represents the target population. This can result in inaccurate data mining outcomes.
- **Test the results**: It is essential to test the outcomes of the data modification process to ensure they are both accurate and useful. Cross-validation is a valuable technique for this purpose.

In this chapter, not all data modification techniques are covered. Sampling and resampling techniques were discussed in depth in Chapter 3. Additionally, dimensional reduction methods will be addressed in Chapters 12 and 13 when we explore unsupervised methods, including principal component analysis and cluster analysis. This chapter focuses on common techniques in data modification, particularly data cleaning and transformation. Specifically, we will cover three primary techniques: replacement, imputation, and transformation. I will explain the differences among these techniques and when to use each one. These techniques are straightforward and highly useful in most projects. We will also explore how to implement them using SPSS Modeler through specific examples.

REPLACEMENT, IMPUTATION, AND TRANSFORMATION

As mentioned earlier, this book covers three primary data modification techniques available in SPSS Modeler: replacement, imputation, and transformation. Table 5.1 provides a summary of the differences between these techniques and when they should be applied.

- **Replacement**: The replacement technique allows us to clean the data by re-coding or correcting data to ensure the accuracy of the analysis. In the field of data mining, replacement is often considered part of data transformation. However, in this book, we treat replacement as distinct from transformation to avoid confusion, as replacement is used to clean data or reassign values before handling missing values and transforming variables. The replacement task should be completed before imputation to ensure that the data are accurate and that the correct information and codes are in place for missing values.
- **Imputation**: Imputation is a common technique used to handle missing values. When faced with missing values, the instinct may be to remove them. However, removing missing values can lead to the loss of important information and potentially increase bias in the data. Therefore, researchers often choose to fill missing values with estimated ones. The challenge lies in selecting the appropriate imputation method to ensure data quality and avoid introducing bias. Imputation should be completed before transformation to ensure that all data points are available for the transformation process.

TABLE 5.1
Replacement, Imputation, and Transformation

	Replacement	Imputation	Transformation
What does it do?	Recode, correct, or reassign values for categorical or continuous variables in the data.	Replace missing values in datasets with estimated values.	Create new variables that are transformations of existing variables in datasets or create interaction variables. It can be used to standardize, normalize, convert, or aggregate data.
When to use this tool?	Replacement should precede the Imputation.	Imputation should precede the transformation.	Transformation should precede the machine learning algorithm to ensure data readiness for model training.

- **Transformation**: This technique enables the alteration of the data format or structure to meet the requirements of data analysis. It can be used to convert, normalize, or aggregate data. Certain machine learning algorithms, such as linear regression, perform better with standardized and normalized data. If the distribution of a variable is not normal, transformation can convert the data into a normal distribution. Additionally, transformation can be used to convert a continuous variable into a categorical one or to aggregate two variables into a single variable for analysis purposes. Transformation should be completed before building models with machine learning algorithms to ensure that the data are prepared and ready for model training.

In this chapter, we will focus on three primary data modification techniques: replacement, imputation, and transformation.

ADVANTAGES AND DISADVANTAGES OF DATA MODIFICATION

Data modification is an important step in the data mining process with many benefits. However, there are some challenges we may have to face when applying these techniques. In this section, we discuss the primary advantages and disadvantages of data modification.

Advantages of Data Modification

1. **Improved data quality**: Data cleaning enhances the quality and reliability of the dataset. They improve data quality by addressing errors, inconsistencies, and missing values, which leads to more accurate and reliable results.
2. **Enhanced model performance**: Dimensionality reduction and data transformation are useful techniques to identify the most relevant and informative variables for model training. These techniques allow us to reduce irrelevant or redundant features and create new meaningful variables, which improves the model performance, resulting in better predictive accuracy and generalization.
3. **Handling missing data**: Data imputation ensures that valuable information is not lost during analysis, reducing potential bias and improving the representativeness of the data.

4. **Efficient computation**: Techniques like dimensionality reduction and data sampling can significantly reduce the computational complexity of the analysis. They help reduce the number of variables or the size of the dataset, which significantly reduces the training time of machine learning models.
5. **Addressing data imbalance**: Data sampling techniques, including oversampling and undersampling, help address the data imbalance issue where one class is significantly underrepresented. By rebalancing the class distribution, model performance on minority classes can be improved, leading to more accurate predictions.

Disadvantages of Data Modification

1. **Information loss**: Changing data can result in information loss since removing certain variables or reducing the dimensionality of the data may discard valuable information that could have been useful for the analysis.
2. **Overfitting risks**: Oversampling can introduce synthetic or duplicated instances, which may lead to overfitting. Overfitting occurs when a model performs well on the training data but fails to generalize to new, unseen data.
3. **Subjectivity and bias**: Data modification often involves making subjective decisions about how to handle missing values, outliers, or feature selection. If not done correctly, these decisions could introduce bias into the analysis, potentially impacting the validity of the results.
4. **Increased complexity**: Data modification techniques add complexity to the data preprocessing stage, requiring additional time and effort. Selecting appropriate techniques and optimizing their parameters can be challenging, especially for large and complex datasets.
5. **Sensitivity to assumptions**: Some data modification techniques, such as imputation or transformation, rely on certain assumptions about the data distribution or relationships. If these assumptions are violated, the modified data may not accurately represent the underlying patterns, leading to biased results.

It is important to carefully consider the advantages and disadvantages of data modification techniques in the context of the specific project and dataset. The choice of techniques should be guided by the study purpose, requirements, and limitations of the analysis, as well as a thorough understanding of the data characteristics.

REPLACEMENT

Replacement is a vital data modification technique in data cleaning, especially when working with large datasets. It allows us to remove incorrect information or correct errors by reassigning values to specific variables, ensuring the accuracy and reliability of the data. In SPSS Modeler, the primary tool for this purpose is the Reclassify node. Below are detailed scenarios where the replacement technique is essential.

The text is generally well-written but contains a few grammatical errors and could benefit from slight refinements for clarity and readability. Below is a revised version:

DEFINITION

Replacement is a process of recoding, correcting, or reassigning values in the datasets to ensure all information is correct.

- **Remove incorrect information**: We tend to assume that collected data are always correct. However, you may be surprised that datasets often include errors due to flaws in data collection instruments,

data collection processes, or incorrectly entered information by participants. These errors can be detected during the data exploration step by reviewing data using visualization and descriptive statistics of variables, as described in Chapter 4. For example, when exploring age data using a histogram, we may notice some cases with an age of zero. Obviously, this is incorrect information since age cannot be zero. Similarly, we may see cases with ages under 18 years, even though the data description clearly indicated that only adults were surveyed. There are several reasons for these situations. First, the participants might have entered the age information incorrectly, making it unusable for analysis. Second, it could be a missing value (the participant decided to leave it blank) that was coded as zero by the software or survey instrument. In either case, these values should be treated as missing data rather than zero to avoid misanalyzing the data. Using the replacement technique, we can remove the incorrect information, and the software will treat those values as missing data.
- **Correct errors**: In some cases, when the incorrect information is obvious, we can make corrections in the dataset. For example, when exploring income information using a histogram, we might find several cases with negative salaries. After examining the survey forms, we may discover that the salary was entered incorrectly as a negative value when it should be positive. This incorrect information can lead to inaccurate predictions during the modeling step. In this case, we can use the replacement technique to correct the error by removing the negative sign. Another example is typos in categorical variables, which can create additional categories and cause misinformation. For instance, the job title "Manaer" should be treated the same as "Manager" due to the typo, but the software might consider them two different classes. Using the Replacement node, we can correct this data.
- **Reassign values**: One common application of the replacement technique is to reassign missing values. You may be surprised to find that data providers code missing values in various ways, depending on how they collect or archive data. SPSS Modeler recognizes several types of missing values: null (system missing values), empty strings, blanks, or user-defined missing values (e.g., 99 or −1). Therefore, if we notice missing values coded differently, we need to reassign those values with the proper format to ensure the software correctly detects missing values. In some cases, when extreme outliers are present, we may not want to use them in the analysis since they can bias the modeling. However, instead of removing the entire data row, we can replace those outliers with missing values.

These are some examples of how the replacement technique can be used in data mining. Depending on the issues you encounter in your project, you should be able to leverage this tool to clean your data. I will demonstrate some examples using SPSS Modeler.

SPSS Modeler provides several tools specifically designed for data replacement purposes:

- **Reclassify node**: This tool allows for the transformation of one set of categorical values into another, making it useful for collapsing categories or regrouping data for analysis.
- **Binning node**: The Binning node is used to transform numeric ranges into categorical sets using an automatic method, such as ranks or percentiles.
- **Filler node**: This node is employed to replace field values and adjust storage. You can choose to replace values based on specific conditions or replace all blanks or null values with a particular value. The Filler node is often used in conjunction with the Type node to handle missing values.
- **Derive node**: This tool can be used to manually classify numeric ranges into sets. For instance, if you need to collapse salary values into specific salary range categories, the Derive node allows you to define each category manually.
- **Derive node for conversion**: Additionally, the Derive node can convert a categorical field into numeric storage. For example, you could convert categorical values like "No" and "Yes" into numeric values, such as 0 and 1, respectively.

IMPUTATION FOR MISSING VALUES

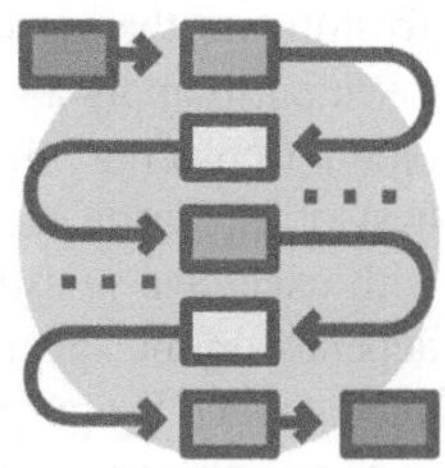

In data mining, missing values refer to the absence of data for certain variables or observations within a dataset. These gaps in data can arise due to various factors, such as data entry errors, incomplete data collection, or intentional omission by the data source. Missing values can be represented in several ways depending on the dataset format and context. SPSS Modeler can recognize different types of missing values, which include:

- **Blank or empty cells**: In most datasets, missing values are often represented by blank or empty cells. These cells contain no recorded value or information for the corresponding variable and observation.
- **Null or NA values**: Some datasets use specific codes or markers, such as "null" or "NA," to indicate missing values. These markers are typically assigned to denote the absence of data for a particular variable or observation. They are considered system-missing values because null values are intentionally used to signify missing data.
- **User-defined values**: In certain cases, missing values may be represented using specific placeholder values that are distinct from the valid values of the variable. For example, "–1" or "9999" might be used to denote missing values in a numeric variable.

DEFINITION

Missing values refer to the absence or lack of data for certain variables or observations in a dataset.

How Do Missing Values Impact Data Mining?

Handling missing values is an essential step in data mining, as they can create biases in the data and impact the quality and validity of the results. The following are some typical cases showing the impact of missing values in data mining.

1. **Bias in data analysis**: When missing values are present in a dataset, excluding or ignoring those observations can lead to biased results. If the missing values are not handled properly, the remaining data may not accurately represent the population, leading to inaccurate predictions and ungeneralized models.
2. **Reduced sample size**: Missing values can reduce the effective sample size needed for data mining. It results in decreased statistical power and may affect the performance of machine learning algorithms, especially those that rely on large amounts of data to learn meaningful patterns.
3. **Distorted relationships**: Missing values can introduce distortion or bias in the relationships between variables. Correlations and patterns within the data may be influenced by the absence of data points, leading to incorrect interpretations of the underlying relationships. This issue can impact the performance of predictive models, as they rely on accurate relationships between features.
4. **Algorithm compatibility**: Some machine learning algorithms, such as linear regression, cannot directly handle missing values. Other algorithms expect complete data as input and may fail to train the model or produce incorrect results if missing values are present.
5. **Loss of information**: Missing values represent lost information, such as relevant variables in the data. Depending on the extent and pattern of missing data, valuable information may be lost, potentially leading to decreased predictive accuracy.

How Do We Handle Missing Values?

To mitigate the issues caused by missing values, it is essential to apply appropriate techniques such as imputation (replacing missing values with estimated values) or exclusion (removing observations or variables with excessive missing data). The choice of technique depends on the specific dataset, the nature of the missing data, and the purpose of the project. While removing missing values can be done using any software program, the challenge lies in deciding whether such removal would lead to any severe issues in predictive modeling. I will focus on imputation, the most common strategy for handling missing values.

> **DEFINITION**
>
> Imputation is a process of estimating or filling in missing values in a dataset with estimated values.

Imputation is the process of estimating or filling in missing values in a dataset with estimated values. It is a technique used to ensure that the dataset is complete before conducting further analysis. There are many methods of imputation, and when choosing an imputation method, several important considerations should be taken into account:

- The choice of imputation method should be guided by the characteristics of the dataset, the nature of the missing data, and the purpose of the analysis. Different methods have different assumptions and may lead to varying results.
- Imputation introduces uncertainty into the dataset since the filled-in values are estimated based on available data. It is important to account for this uncertainty when performing subsequent modeling.
- Imputation should be performed cautiously, as it may introduce bias if the missing data are not completely random. There may be specific reasons behind the missing data, making them systematically missing rather than random. It is essential to understand the reasons for the missing data and consider potential biases in the imputed values.
- It is a good practice to document the imputation process, including the method used and any assumptions made, to ensure transparency and replicability of the analysis.

Overall, imputation is a widely used technique in data modification to handle missing values and ensure dataset completeness. The choice of imputation method should be carefully considered based on the characteristics of the missing data and the purpose of the analysis. It is important to understand the limitations and potential biases associated with imputation.

Table 5.2 presents some common imputation methods and their characteristics. For each method, I provide a brief description, the assumptions made, how the imputed value is calculated, and when

TABLE 5.2
Common Imputation Methods

Imputation Methods	Description	When to Use
Mean	It replaces missing values with the mean of the available values. It assumes that the missing values are similar to the average of the observed values. ***Imputed value = Mean (available data)***	***For continuous variables.*** It is a good choice for datasets that are normally distributed and that contain a small number of missing values.
Median	It replaces missing values with the median of the available values. It is less sensitive to outliers compared to mean imputation. ***Imputed value = Median (available data)***	***For continuous and ordinal variables.*** It is a good choice for datasets that are not normally distributed (skewed) and that contain a small number of missing values. It is also a good choice for ordinal data.

(Continued)

TABLE 5.2 (Continued)
Common Imputation Methods

Imputation Methods	Description	When to Use
Mid-range	It replaces the missing values with the midpoint between the minimum and maximum values of the available data for that variable. The midrange is a rough measure of central tendency. ***Imputed value = (max + min)/2***	***For continuous variables.*** It is a good choice for normally distributed datasets.
Distribution	It uses the distribution of the data to fill in missing values. This method is based on the assumption that the missing values are randomly distributed and that they can be estimated from the values of the observed data. The estimated values are determined based on the probability distribution of the non-missing observations.	***For continuous and categorical variables.*** It is a good choice if missing values are randomly distributed. This method typically does not change the distribution of the data very much.
Tree-based	Also known as predictive mean matching, it uses decision trees to impute missing values in a dataset. It is particularly useful for imputing missing values in variables with both continuous and categorical data types. Since the imputed value for each input variable relies on the values of other input variables, this method of imputation could potentially offer greater accuracy compared to a straightforward replacement of missing tree values with the variable's mean or median.	***For continuous and categorical variables.*** This method should be used for nonlinear relationships or multiple imputations (when multiple imputed datasets are created).
Tree imputation with surrogate split	It is an extension of tree-based imputation that incorporates surrogate variables when imputing missing values. Surrogate splits are additional split rules created in decision trees to handle missing values in predictor variables.	***For continuous and categorical variables.*** This method is a good choice if we have a high rate of missing values, complex dependencies, or interaction effects among predictors.
Mid-minimum spacing	It uses the median and the minimum value of the observed data to impute missing values. It specifies the proportion of the data to be contained in the spacing. It assumes that the missing values would fall within the range defined by the adjacent minimum values. There are two steps in calculating this statistic: (1) Trim the data using N% of the data (e.g., 10%) (2) The mid-minimum spacing value is calculated as the maximum plus the minimum (of the trimmed data) divided by two. ***Imputed value = [max(trimmed data) + min(trimmed data)]/2*** 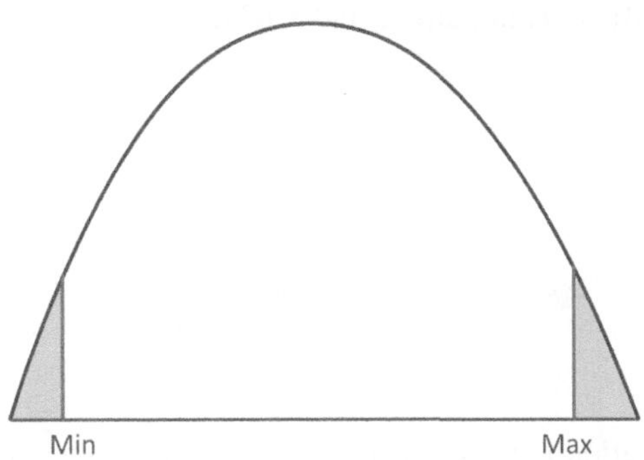	***For continuous variables.*** It is useful when the minimum values in a time series have meaningful significance or represent some kind of lower limit.

(Continued)

TABLE 5.2 *(Continued)*
Common Imputation Methods

Imputation Methods	Description	When to Use
Robust M-estimators of location	Robust M-estimators of location are statistical estimators used to estimate the central location or location parameters of a distribution. They are less sensitive to outliers. There are three primary M-estimators: (1) Tukey's Biweight; (2) Huber; (3) Andrew's Wave. 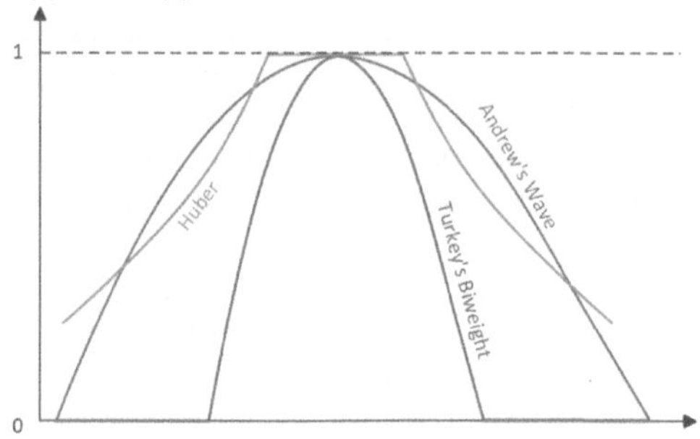 	***For continuous variables.*** They are particularly useful when dealing with datasets that contain outliers or data points that deviate significantly from the majority of the data.

it is appropriate to use that method. While you don't need to worry about calculating these values manually, as the software handles this task automatically, it is important to understand when to use each method. It is also worth noting that these methods are primarily used for continuous variables due to the need for quantitative calculations. Imputation methods for categorical variables are more limited due to their non-metric nature.

TRANSFORMATION

In data mining, **transformation** refers to the process of creating new variables derived from existing variables. This process can be used to standardize and normalize data, convert data into different formats, or aggregate data for analysis.

As previously mentioned, replacement and imputation are sometimes considered parts of the transformation process. However, since these methods have already been covered, this section will focus on two primary transformation techniques: **data normalization** and **variable discretization**.

NORMALIZATION

This book covers two primary data transformation tools: normalization and variable discretization.

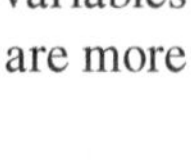

Data normalization, also known as variable scaling, is a preprocessing step in machine learning that adjusts the variables in a dataset to a similar range. It is one of the most commonly used transformation techniques. Before diving into why normalization is essential, it's important to understand the differences between un-normalized and normalized data.

TABLE 5.3
Normalized and Un-Normalized Data

Items	Un-Normalized Data	Normalized Data
Range	Un-normalized data can have different ranges across various variables. The values may span a wide range, making direct comparison or interpretation challenging.	Normalized data are transformed to have a consistent range. It typically falls within a predefined range, such as [0, 1] or [−1, 1], making it easier to compare values across features.
Bias	Un-normalized data may exhibit biases due to the inherent differences in ranges. Variables with larger values might dominate the analysis or disproportionately impact models.	Normalization helps to eliminate biases caused by different ranges. By scaling the data, all variables contribute more equally to the modeling process.
Outliers	Outliers in un-normalized data can significantly influence the overall modeling process.	Depending on the normalization method used, outliers may have a reduced impact on the overall data representation. Certain normalization techniques, like robust scaling, can handle outliers more effectively.
Interpretation	Interpreting the magnitude of un-normalized data may be challenging since the range of each variable varies.	Normalized data allows for a more intuitive interpretation and comparison. The values are transformed to a standardized range, enabling comparisons and assessments based on relative magnitudes.

Table 5.3 highlights these differences, while Figure 5.1 illustrates them through scatter plots and histograms.

Normalization provides great benefits in data mining but also poses some challenges. Following are some key advantages and disadvantages of data normalization.

Advantages of data normalization

- Improved performance of machine learning algorithms: By bringing data into a consistent range, normalization prevents certain variables from dominating others and ensures that all variables contribute equally to the learning process.
- Mitigation of the impact of outliers: Data normalization can reduce the effect of outliers by scaling the data, making it less sensitive to extreme values.
- Simplified interpretation and visualization: Normalizing data allows for easier interpretation and visualization of the data. When data are scaled to a common range, patterns and trends become more apparent, and it becomes simpler to compare and understand different variables.
- Accelerated convergence in optimization algorithms: Some machine learning methods require search optimization. Optimization algorithms can converge faster when working with normalized data.

Disadvantages of Data Normalization

- **Loss of original data distribution**: Normalization alters the original distribution of the data, compressing or stretching it to fit within a specific range. This can potentially lead to the loss of some information about the data distribution, which may be relevant in certain analysis scenarios.

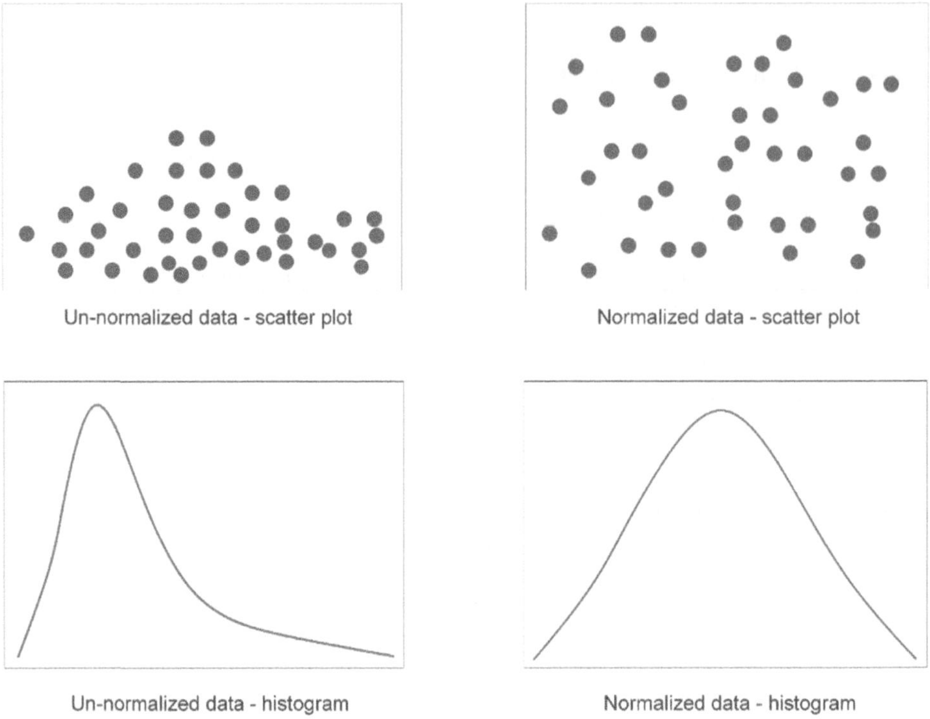

FIGURE 5.1 Un-normalized data and normalized data.

- **Increased complexity in interpretation**: While normalization can simplify interpretation in some cases, it can also introduce complexity. Transformed values need to be converted back to their original scale for meaningful interpretation, which can be challenging, especially when dealing with multiple normalization techniques or complex data transformations.
- **Sensitivity to outliers during normalization**: Although normalization can help mitigate the impact of outliers, certain normalization techniques, such as min-max normalization, are sensitive to extreme values. Outliers can disproportionately affect the normalized data, leading to potential distortions or loss of information.
- **Application-specific considerations**: The suitability of data normalization depends on the specific data and the machine learning methods. In some cases, maintaining the original scale of the data might be necessary for meaningful interpretation or compatibility with certain algorithms. Normalization should be carefully evaluated based on the requirements and characteristics of the data and the intended modeling techniques.

Table 5.4 presents common data normalization methods. When choosing a normalization method, several factors must be considered:

- **Purpose of normalization**: Determine whether the goal is to enhance the performance of your machine learning model or to make the data easier to visualize or interpret.
- **Type of data**: Consider whether you are working with numerical or categorical data. Note that normalization is applicable only to continuous variables.
- **Size of dataset**: For large datasets, it's essential to select a more efficient normalization method to handle the volume of data effectively.

TABLE 5.4
Common Data Normalization Methods and When to Use Them

Normalization Methods	Descriptions	When to Use
Min-max normalization (also known as rescaling)	Formula: $X' = \frac{X - X_{min}}{X_{max} - X_{min}}$ Range: [0, 1]	When you want to scale the data to a specific range, particularly when the distribution is known to be bounded.
Z-score normalization (also known as standardization)	Formula: $X' = \frac{X - X_{mean}}{SD}$	When you want to standardize the data to have a mean of 0 and a standard deviation (SD) of 1. This method is commonly used when the distribution of the data is unknown or not normally distributed.
Robust scaling (also known as median and MAD scaling)	Formula: $X' = \frac{X - X_{median}}{MAD}$	When you have data with outliers or a skewed distribution and you want to scale the data based on the median and median absolute deviation (MAD) instead of the mean and SD.
Log transformation	Formula: $X' = \log(X)$	When the data are highly skewed or contain exponential growth patterns, applying a logarithmic transformation can help normalize the distribution.
Power transformation	Formula: $X' = X^n$	When the data have a specific power-law relationship, a power transformation can help normalize the distribution. Common power values include $n = 0.5$ (square root), $n = 2$ (square), and $n = -1$ (inverse).

Note: X' = New value; X = Current value.

- **Algorithms used**: Some algorithms, such as k-nearest neighbors and artificial neural networks, are more sensitive to data scaling than others. Linear regression, for instance, performs best performed with normally distributed data. If you're using these algorithms, ensure that all variables are on the same scale by applying normalization.
- **Model performance**: Experiment with different normalization methods and evaluate which one yields the best performance for your model.

> Not all machine learning methods require normalization; in some cases, using the original scales may improve model performance. It's essential to assess whether normalization is necessary based on the specific method and data characteristics.
>
> Additionally, some advanced machine learning tools have built-in standardization and normalization features. In such cases, separate normalization may not be required, as the tool automatically handles the scaling of variables during the preprocessing stage.

Variable Discretization

Variable discretization is the process of converting a continuous variable into a categorical variable. This is a common preprocessing step in data mining when you need to transform a variable with a continuous scale into discrete categories. Discretization can be particularly useful when you want to capture nonlinear

> **DEFINITION**
>
> Variable discretization is a process of converting a variable with continuous scale (continuous variable) into discrete categories (a categorical variable).

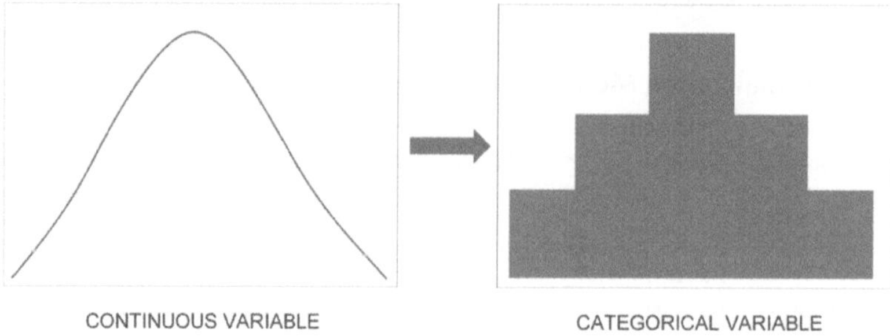

FIGURE 5.2 Data discretization.

relationships or when a specific data mining algorithm requires the variable to be treated as categorical. Figure 5.2 illustrates this concept, showing how the input (a continuous variable) is transformed into the output (a categorical variable with discrete levels).

A general approach to converting a continuous variable into categorical variables can be described as follows:

1. **Determine the number and range of categories**: Decide on the desired number of categories or groups you want to create for the new variables. This number represents the number of levels that you need for the categorical variable. This decision may depend on the specific dataset, the nature of the variable, and the purpose of the analysis. You can also consider using domain knowledge or statistical techniques to guide your decision.
2. **Partition the range of values**: Divide the range of values of the continuous variable into distinct bins. Each bin represents a category or level. The intervals should be defined to capture meaningful distinctions in the data while maintaining a balance between the number of categories and the information contained within each category.
3. **Assign categories to data points**: Map each data point from the original continuous variable to its corresponding category based on its value falling within the defined intervals. This mapping can be achieved using if-else statements, mathematical functions, or specialized functions available in data mining software.
4. **Encode categories into the new categorical variable**: Once the data points are assigned to categories, represent each category as a separate level of the new categorical variable. This step can be done differently depending on the software.

For example, when converting a continuous variable like age into a categorical variable, you might define categories such as **young, middle-aged,** and **elderly** based on specific age ranges. You would create three distinct bins based on the predefined ranges and assign continuous age values to these bins accordingly. Finally, the new categorical variable (e.g., **r_age**) would be created and encoded with three levels: **young**, **middle-aged**, and **elderly**.

It's important to note that converting continuous variables into categorical variables can result in some loss of information due to the discretization of the data. Additionally, the specific approach and the number of categories chosen can impact the results of your data mining efforts. It's advisable to experiment with different categorization schemes and evaluate their effects on model performance.

Table 5.5 describes common methods for variable discretization, with guidance on when to use each method based on the data's characteristics and the analysis purpose.

The selection of a discretization method depends on various factors, including the characteristics of the data, underlying data distributions, the presence of outliers, and the requirements of the machine learning method. Exploring the data using visualization, descriptive statistics, and domain expertise is crucial for understanding the data and choosing an appropriate method.

TABLE 5.5
Common Variable Discretization Methods and When to Use Them

Discretization Methods	Description	When to Use
Fixed-width/binning (equal-width binning)	This method divides the range of the continuous variable into a predetermined number of equal-width bins and assigns category labels based on which bin each value falls into.	When you want to perform a simple discretization without considering the distribution of the data. This method is less sensitive to outliers but may not capture the underlying patterns in the data.
Quantiles/binning (equal-frequency binning)	This method divides the data into bins based on equal-frequency or equal-sized subsets to ensure that each bin contains an equal number of data points. Then, it assigns category labels based on the bin each value belongs to.	When you want to ensure an equal representation of data points in each category and capture the distribution of the data. This method can be helpful for handling skewed data, but it may result in uneven bin sizes if the data contain outliers or extreme values.
Clustering[a]	This method applies a clustering algorithm like k-means to group similar values together and then assigns category labels based on the cluster assignments.	When you want to let the data determine the category boundaries based on the similarity between the records. This method is useful when the natural clusters in the data correspond to meaningful categories. However, you need to specify the number of clusters in advance.
Domain-specific thresholds	This method defines predefined thresholds based on domain knowledge or specific requirements and assigns category labels based on those thresholds.	When there are well-established thresholds or standards in the field that define categorical boundaries. This method is suitable when there are clear and meaningful cut-off points for categorization.

[a] We will get into the details of cluster analysis later in Chapter 14.

It's important to note that converting a continuous variable into a categorical one involves making decisions about how to group and label the data. These decisions can significantly impact subsequent modeling steps, so it's essential to carefully evaluate the implications and assess the trade-offs associated with each method. By doing so, you can ensure that the chosen method aligns with the goals of your analysis and enhances the overall model performance.

AUTOMATED DATA PREPARATION

The data modification approaches I described earlier require a deep understanding and careful execution to ensure accuracy. This process is often one of the most time-consuming steps in data mining.

You may wonder: Is there a quicker way to prepare and modify data without going through all those steps? The answer is yes. Many data mining software programs provide an automated tool that handles data preparation quickly and efficiently. For instance, SPSS Modeler offers a node called Automated Data Preparation (ADP). The ADP node manages the data preparation process on your behalf by examining your data, identifying issues, excluding problematic or less valuable variables, creating new variables when necessary, and enhancing performance using intelligent screening methods. This algorithm can operate fully automatically, making decisions and implementing fixes, or interactively, allowing you to preview modifications before they are applied and decide whether to accept or reject them according to your preferences.

Typically, the ADP node suggests data preparation measures that enhance the efficiency of subsequent model-building algorithms and improve their predictive capabilities. These measures may include transformations and variable selection. The target variable may also be subject to potential transformations. You can specify the priorities for the data preparation process as follows:

- **Balance speed and accuracy**: This option equally emphasizes both the processing speed of model-building algorithms and the accuracy of predictions.
- **Optimize for speed**: This option focuses on preparing the data to prioritize rapid processing by model-building algorithms. It is particularly useful for large datasets or when a quick response is desired.
- **Optimize for accuracy**: This option prioritizes the accuracy of predictions generated by model-building algorithms.
- **Custom analysis**: Choose this option if you wish to adjust the algorithm settings manually. Note that this setting is automatically activated if subsequent changes to the options on the Settings tab conflict with the objectives of other options.

The ADP node has both advantages and disadvantages. Automated tools can perform tasks quickly but offer limited control to researchers. This means you might not fully understand what modifications have been made to the data or how these changes could affect the quality of the data or the validity of the data mining results. Therefore, it's important to consider the nature of your data, the modeling methods, and potential trade-offs before deciding whether to use the ADP node. I will provide some examples of using the ADP node and compare it to other data modification methods.

Advantages of the ADP

- Automatic and efficient data preparation process.
- Optimization for different objectives: speed, accuracy, or a balance between the two.
- Fast and efficient data transformation and imputation.
- Automatic recoding of string values of categorical variables into numeric values, which is required by SPSS Modeler for certain analysis methods, such as logistic regression and linear regression. These methods require numeric values to run.

Disadvantages of the ADP

- Very limited control over the data modification process; only built-in features are available.
- Only simple imputation methods are offered (mean value or mode).
- Limited transformation methods are included: z-score normalization and min/max transformation.
- Advanced data modification methods cannot be applied.

> I find the ADP node particularly useful when running statistics-based models, such as linear regression, which requires converting categorical variables with string values into numeric format (a requirement specific to SPSS).
>
> I recommend using the Custom analysis option because it allows us to specify exactly which data modifications we want to implement.

EXAMPLES OF DATA MODIFICATION WITH AIRLINE RECOMMENDATION DATASET

The first example involves the **Airline Recommendation** dataset, where we apply two data modification methods, replacement and imputation, using SPSS Modeler.

- **Dataset**: AirlineRecommendation.sav
- **Target variable**: Recommend (Flag: Yes/No)

Before performing these modification tasks, it's important to explore the data to identify any issues. By using the Data Audit node, as described in Chapter 3, we can obtain the descriptive statistics for the fields in the dataset. Since this dataset contains two groups of fields—continuous and categorical—I present the descriptive statistics for each separately in Tables 5.6 and 5.7. Note that I do not provide the full descriptive statistics as shown in Chapter 3, as our focus here is on identifying missing values and other data issues.

TABLE 5.6
Brief Descriptive Statistics for Categorical Fields

Field Name	Role	Number of Levels	Missing
airline_status	INPUT	4	0
Class	INPUT	3	0
destination_state	INPUT	51	0
flight_cancelled	INPUT	2	0
Gender	INPUT	2	0
origin_state	INPUT	51	0
type_of_travel	INPUT	3	0
recommendation	TARGET	2	0

TABLE 5.7
Brief Descriptive Statistics for Continuous Fields

Field	Mean	Standard Deviation	Non-Missing	Missing	Min	Median	Max
Age	47.5	16.50	124,464	0	18	46	85
arrival_delay_in_minutes	15.38	38.84	121,899	2,565	0	0	1,584
departure_delay_in_minutes	15.00	38.45	122,263	2,201	0	0	1,592
eating_and_drinking_at_airport	68.84	52.60	124,464	0	0	60	895
flight_distance	793.87	592.10	124,464	0	31	630	4,983
flight_time_in_minutes	111.51	71.77	121,899	2,565	8	92	669
no_of_flights_per_year	20.09	14.40	124,464	0	0	17.36	100
no_of_other_loyalty_cards	0.84	1.12	124,464	0	0	0	12
number_of_years_flying	6.78	2.98	124,464	0	2	7	11
percent_of_flight_with_other_Air	9.44	8.87	124,464	0	1	7	110
price_sensitivity	1.27	0.54	124,464	0	0	1	5
shopping_amount_at_airport	26.87	53.16	124,464	0	0	0	879

The results show no missing values for the categorical fields (Table 5.6), indicating that these fields are in good shape. However, Table 5.7 reveals that three continuous fields—**arrival delay time, departure delay time,** and **flight time**—contain missing values. Additionally, there's an issue with the **price sensitivity** field. As mentioned in Chapter 3, this field uses a Likert scale ranging from 1 to 5. However, the descriptive statistics indicate a minimum value of 0, which is not valid. A likely explanation is that 0 was used to code missing values. Since SPSS Modeler represents missing values as blank spaces or null, it may have misinterpreted the zeros as actual data values. This could lead to inaccurate predictive modeling results.

To address this, we need to correct the coding of missing values using the replacement technique. Afterward, we will apply the imputation technique to handle the remaining missing values in the dataset.

Replacement Example for the Airline Recommendation

The purpose of this example is to use the replacement technique to re-code the zero values in the **price sensitivity** field as missing values, thereby correcting the information in the dataset. Since SPSS Modeler does not have a specific node designated for replacement, we will use the Filler node for this task. You might wonder: Why not use the Reclassify node, which can also re-code data? The main reason is that the Reclassify node is primarily intended for categorical fields. In this case, we are working with a continuous field, making the Filler node more appropriate. As mentioned earlier, the Filler node is particularly useful for replacing field values and is commonly used to handle missing values.

Follow the steps below to re-code the missing values using the Filler node:

1. Open SPSS Modeler.
2. Create a new stream and name it "**Airline_Modify**."
3. From the Sources group in the Node Pallet, double-click the Statistics File node to add it to the stream.
4. Double-click the Statistics File node to open its settings window.
5. Click the ellipsis button (three-dot button) and select the **AirlineRecommendation.sav** file from your computer.
6. From the **Field Ops** group, add a **Type** node to the stream and connect it to the data node. This node allows you to view and change the measurement scales and roles of fields.
7. Check the measurement scales of all fields. For the field **recommendation**, change the role to **Target**.
8. From the Field Ops group, add a **Filler** node to the stream and connect it to the Type node. Rename this node to **Missing value recoding**. Open the node settings.
9. Select **Price sensitivity** for **Fill in fields** and **Based on condition** under Replace.
10. For Condition, type "**Price sensitivity**"**=0** (this condition tells the software to select cases with zero values only).
11. For **Replace with**, type **undef** (*undef* returns the undefined value, which SPSS Modeler will interpret as blank or null).

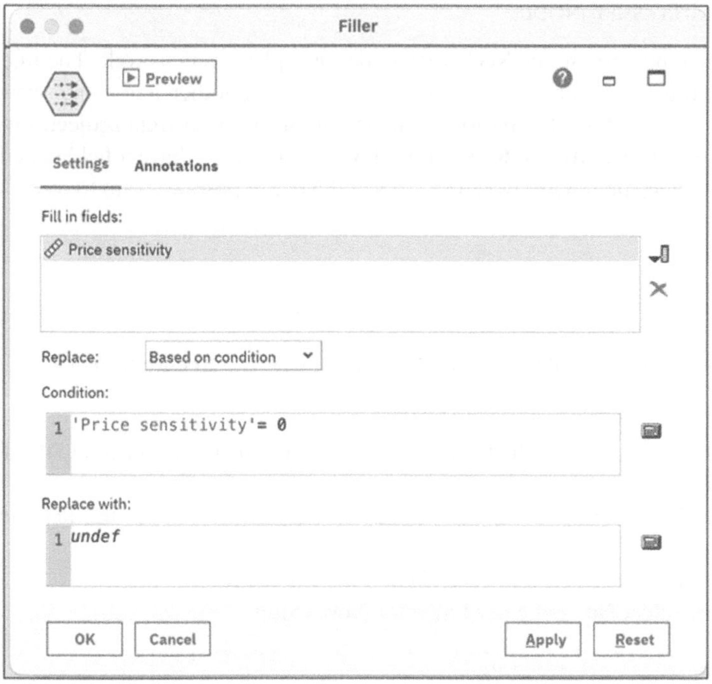

12. Click Apply, and then click OK.

The Filler node will replace all zero values in the **Price sensitivity** field with blanks, saving the modified information in the same field. As shown in Figure 5.3, the results indicate 3,976 missing values for **Price sensitivity**. These missing values represent the original zero values that have been successfully replaced. Thus, the replacement task has been completed successfully.

Audit Quality Annotations

Complete fields (%): 80% Complete records (%): 94.81%

Field	Measurement	Out...	Extre...	Act...	Impu...	Met...	% Complete	Valid Records	Null Value
Price sensitivity	Continuous	1571	181	None	Never	Fixed	96.806	120488	3976
Arrival delay (in minutes)	Continuous	1735	846	None	Never	Fixed	97.939	121899	2565
Flight time (in minutes)	Continuous	2017	102	None	Never	Fixed	97.939	121899	2565
Departure delay (in minutes)	Continuous	1730	819	None	Never	Fixed	98.232	122263	2201
Recommend the airline	Nominal	--	--	--	Never	Fixed	100	124464	0
Airline status	Nominal	--	--	--	Never	Fixed	100	124464	0
Age	Continuous	0	0	None	Never	Fixed	100	124464	0
Gender	Nominal	--	--	--	Never	Fixed	100	124464	0
Number of years flying	Continuous	0	0	None	Never	Fixed	100	124464	0
Average number of flights ...	Continuous	770	14	None	Never	Fixed	100	124464	0
Percent of flights with othe...	Continuous	2334	350	None	Never	Fixed	100	124464	0
Travel purpose	Nominal	--	--	--	Never	Fixed	100	124464	0
Number of other loyalty cards	Continuous	721	80	None	Never	Fixed	100	124464	0
Shopping amount at airport	Continuous	2074	700	None	Never	Fixed	100	124464	0
Spending for eating and dri...	Continuous	1355	307	None	Never	Fixed	100	124464	0
Passenger class	Nominal	--	--	--	Never	Fixed	100	124464	0
Origin state	Nominal	--	--	--	Never	Fixed	100	124464	0
Destination state	Nominal	--	--	--	Never	Fixed	100	124464	0
Flight cancellation	Nominal	--	--	--	Never	Fixed	100	124464	0
Flight distance (miles)	Continuous	1462	143	None	Never	Fixed	100	124464	0

FIGURE 5.3 Results of re-coding missing values for **Price sensitivity**.

Example of Using the Reclassify Node

I would like to demonstrate how to use the Reclassify node for replacement as well. The Reclassify node is primarily used to transform one set of categorical values into another, and it only works with categorical fields. Since we do not need to perform this operation for the current project, I will create a hypothetical scenario. In the **Airline Recommendation** dataset, we have a field called **flight cancellation**, which is a binary field with two levels: Yes and No. Suppose we want to re-code these values as follows:

- Yes -> True
- No -> False

Follow the steps below to implement the replacement task using the **Reclassify** node:

1. Open the "**Airline_Modify**" stream created before
2. From the **Field Ops** group, add the Reclassify node to the stream and connect it to the Type node. Open the node settings.
3. Select **Flight cancellation** for the **Reclassify field** and check the box **Existing field** under **Reclassify into.**
4. Under **Original value**, select **Yes**, and enter **True** for **New value.**
5. For **Original value**, select **No**, and enter **False** for **New value**.

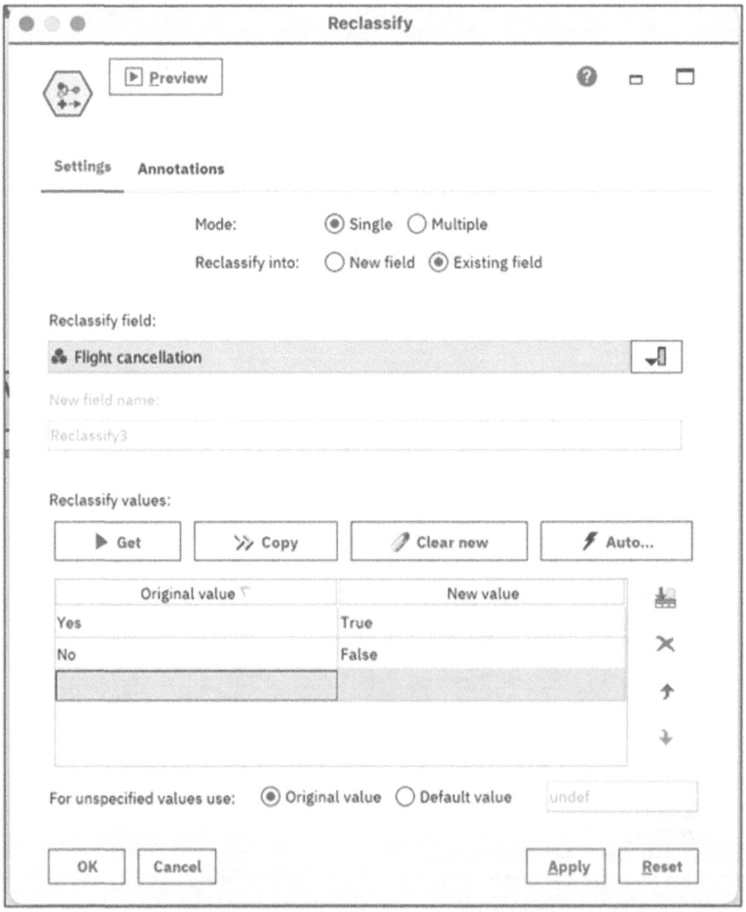

6. Click Apply, and then click OK.

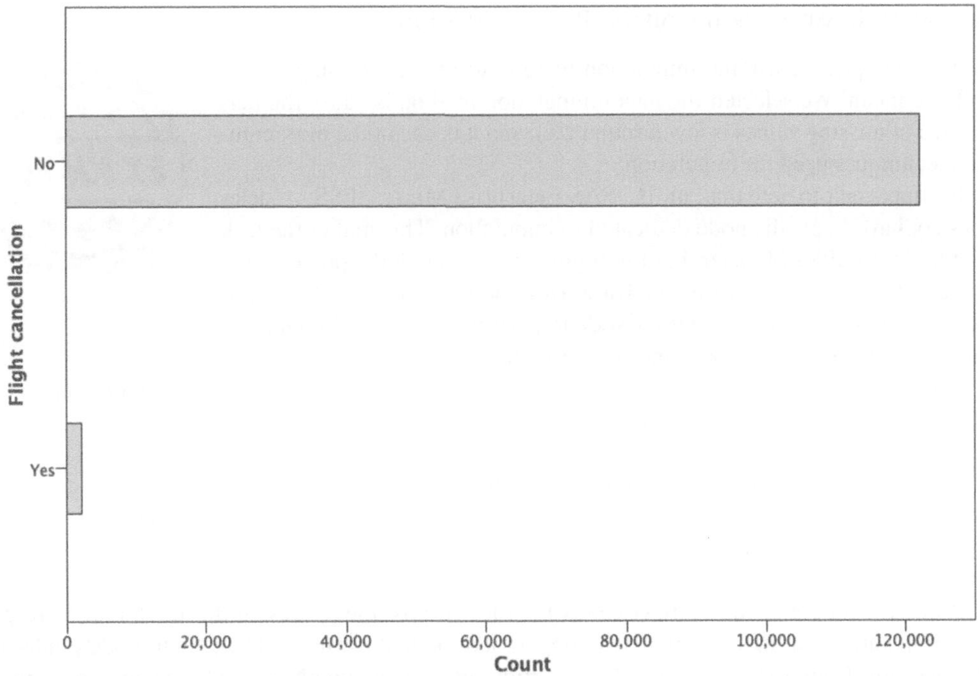

FIGURE 5.4 Flight cancellation values before replacement.

The Reclassify node will replace the values as intended. Figure 5.4 shows the bar chart for **Flight cancellation** before the replacement, and Figure 5.5 shows the chart after. As you can see, *Yes* has been replaced by *True* and *No* by *False*. The process is straightforward.

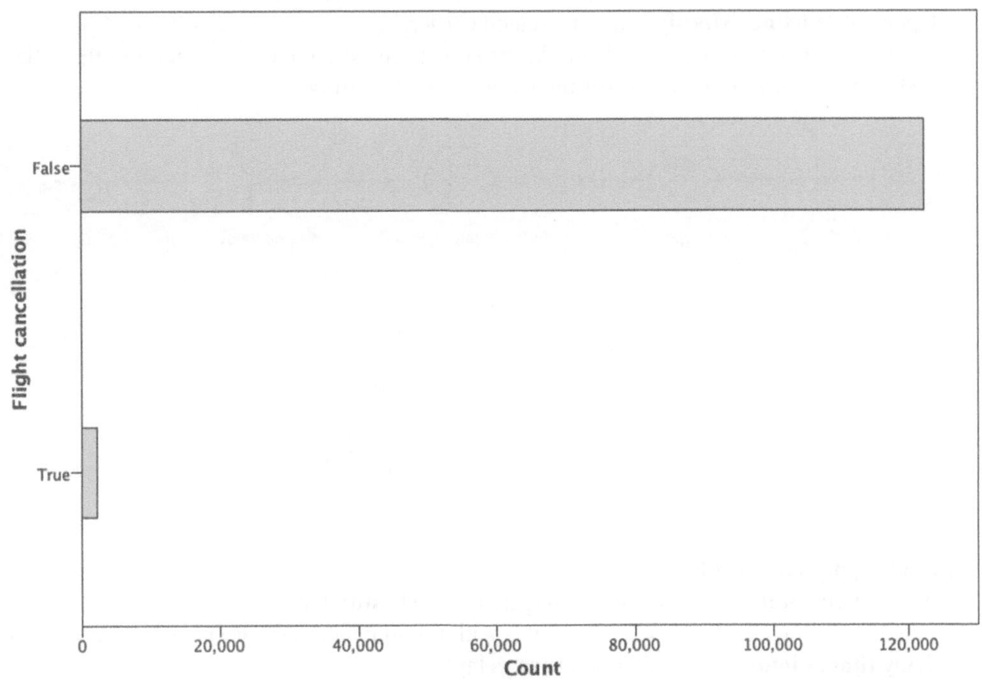

FIGURE 5.5 Flight cancellation values after replacement.

Imputation Example for the Airline Recommendation

In this example, we use the imputation method to replace missing values with the mean. We selected the mean imputation method because the percentage of missing values is low (around 2%), and it is one of the most common techniques used for imputation.

It's important to note that, unlike SAS Enterprise Miner, SPSS Modeler does not have a specific node dedicated to imputation. This makes the task a bit tricky in SPSS Modeler, but I will guide you through the process step by step. We will begin by using the **Data Audit** node to configure the imputation settings, then create a **SuperNode** to perform the actual imputation. But first, what is a SuperNode, and what does it do?

SuperNodes consolidate multiple nodes into a single node, effectively encapsulating parts of a data stream. This offers several advantages for data miners:

- It simplifies and organizes complex data streams.
- Nodes can be grouped into a SuperNode, customized for specific business needs.
- SuperNodes can be exported to libraries and reused across multiple data mining projects.

While this might sound more complex than it really is, the process is straightforward. To perform imputation, we first configure the imputation method using the Data Audit node (under the **Quality** tab). Then, we encapsulate these configurations in a SuperNode and execute it to complete the imputation.

Performing Imputation Using the SuperNode

Follow these steps to perform imputation using the SuperNode:

1. Open the "**Airline_Modify**" stream created earlier.
2. From the **Output** group, add a **Data Audit** node to the stream and connect it to the Filler node named **Missing value recoding**. Open the node settings.

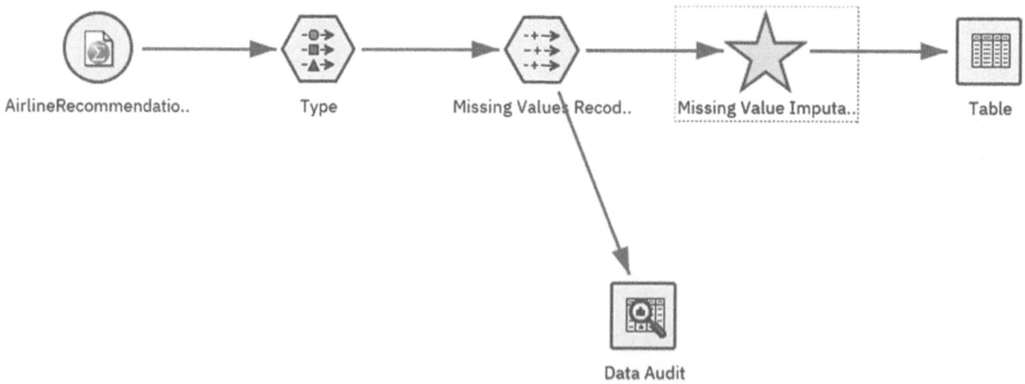

3. Click Apply, and then **Run**.
4. In the **Data Audit** output window, navigate to the **Quality** tab.
5. Make the following changes for four fields with missing values (**Price sensitivity, Arrival delay time, Flight time,** and **Departure delay**):
   - Under **Impute Missing** column, select **Specify**.

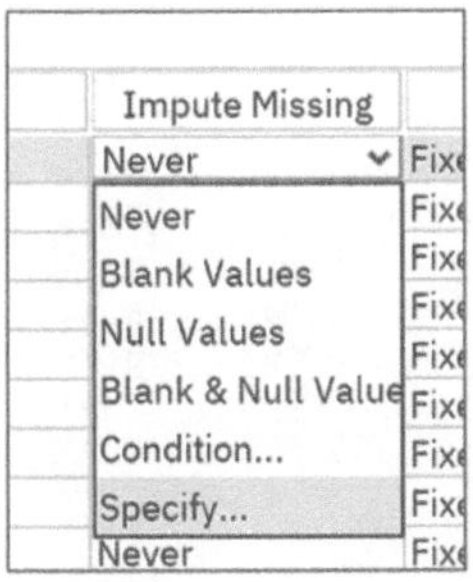

- For **Price sensitivity**, make the following selections:
  - Impute when: **Null Values**
  - Impute Method: **Fixed**
  - Impute Fixed Values: **Mid-Range**

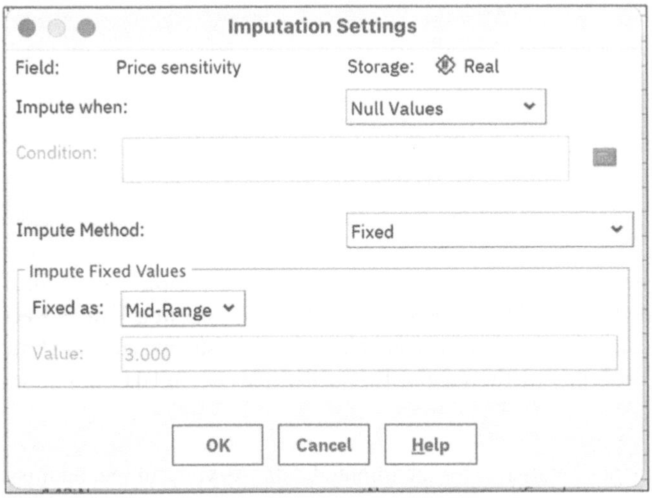

- For the other three fields, follow the same steps, but select **Mean** for the **Impute Fixed Values**.

6. To apply these imputation settings, we need to generate a **SuperNode**. Under the **Generate** menu, select **Missing Values Super Node**, which will create a SuperNode containing all imputation configurations we selected above for the four fields.

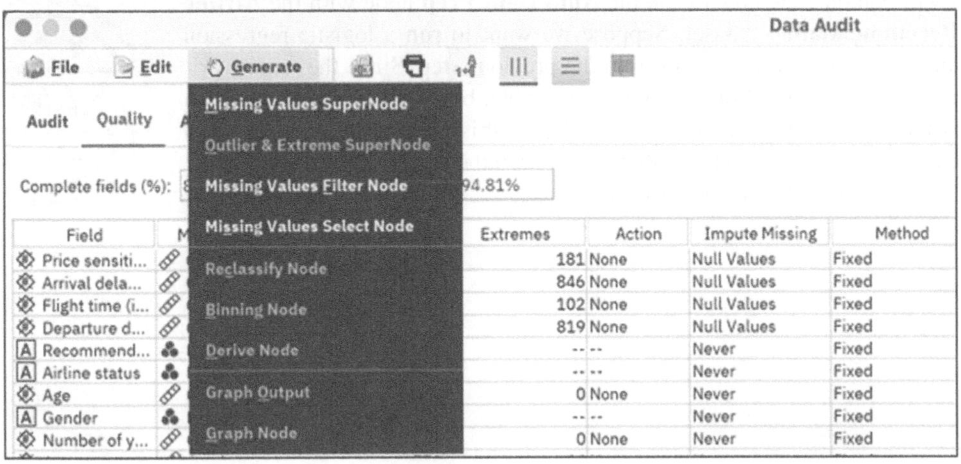

Audit Quality Annotations

Field	Measurement	Min	Max	Mean	Valid
Age	Continuous	18.000	85.000	47.503	124464
Price sensitivity	Continuous	1.000	5.000	1.362	124464
Number of years flying	Continuous	2.000	11.000	6.784	124464
Average number of flights per year	Continuous	0.000	100.000	20.088	124464
Percent of flights with other airlines	Continuous	1.000	110.000	9.438	124464
Number of other loyalty cards	Continuous	0.000	12.000	0.835	124464
Shopping amount at airport	Continuous	0.000	879.000	26.870	124464
Spending for eating and drinking at airport	Continuous	0.000	895.000	68.845	124464
Departure delay (in minutes)	Continuous	0.000	1592.000	15.003	124464
Arrival delay (in minutes)	Continuous	0.000	1584.000	15.380	124464
Flight time (in minutes)	Continuous	8.000	669.000	111.514	124464
Flight distance (miles)	Continuous	31.000	4983.000	793.873	124464
Recommend the airline	Nominal	--	--	--	124464
Airline status	Nominal	--	--	--	124464
Gender	Nominal	--	--	--	124464
Travel purpose	Nominal	--	--	--	124464
Passenger class	Nominal	--	--	--	124464
Origin state	Nominal	--	--	--	124464
Destination state	Nominal	--	--	--	124464
Flight cancellation	Nominal	--	--	--	124464

FIGURE 5.6 Imputation results.

7. Click OK to confirm.
8. SPSS Modeler will generate a SuperNode, named **Missing Value Imputation** node, into the stream. Connect this node to the Filler node we created before (**Missing Values Recoding**).
9. In order to execute the SuperNode, add an executable node to the stream, such as **Data Audit**. In this case, I added a **Table** node to the stream. This will display the full data after imputation.
10. Run the **Table** node.

Figure 5.6 shows the descriptive statistics for the imputed data. As you can see, no missing values remain for the four fields, indicating successful imputation. The output from the SuperNode will be used as input for predictive modeling in the next step. You can view the imputed data in the Table node's output.

Example of Auto Data Prep Node

I will demonstrate how to use the **Auto Data Prep** node with the **Airline Recommendation** dataset. Suppose we want to run a logistic regression analysis on the imputed data from the previous step. Since the target variable, **Recommendation**, is a binary variable, binomial logistic regression is the appropriate method (we will discuss this method in more detail in Chapter 7). However, when I attempted to run the Logistic Regression node, I received the following error message:

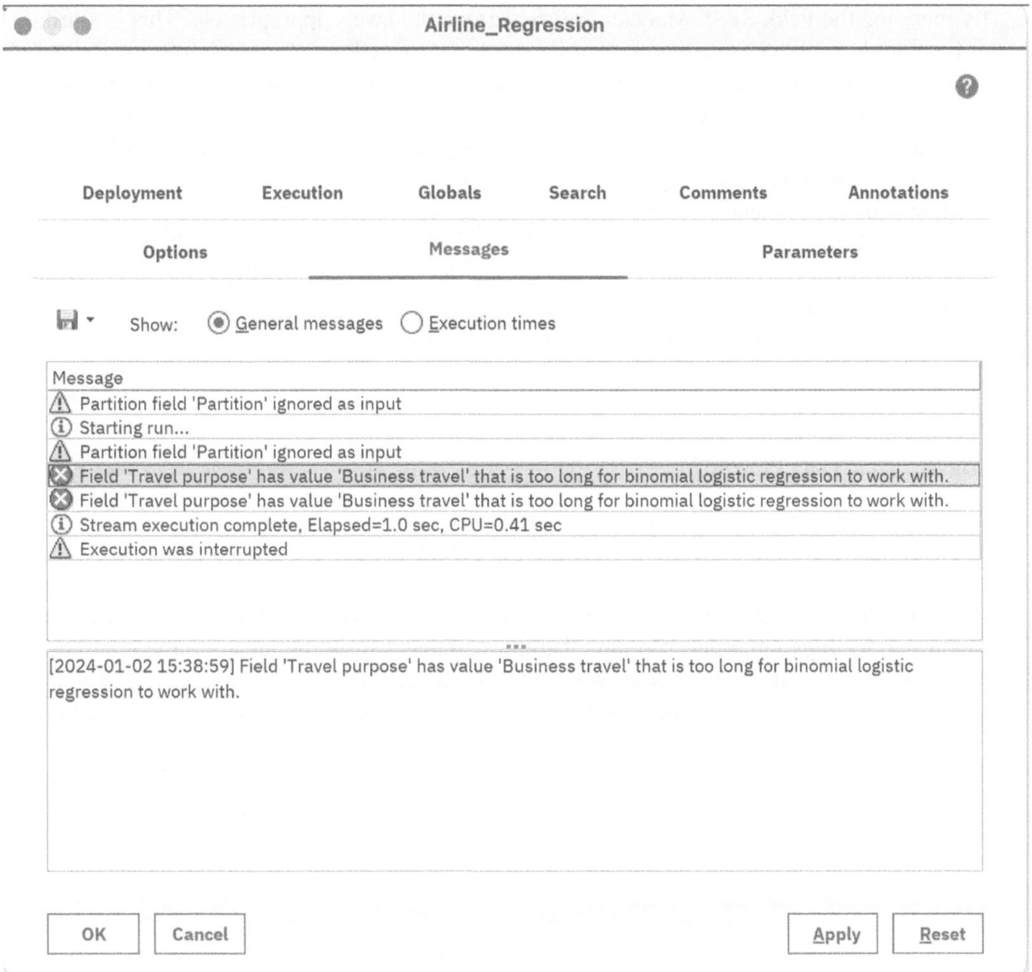

"Field 'Travel purpose' has value 'Business travel' that is too long for binomial logistic regression to work with."

As a result, the execution was interrupted. So, what does this error mean, and how can we resolve it?

If you're familiar with SPSS Statistics or similar software, you may know that error messages can often be vague and confusing. In this case, the error is not suggesting that we shorten the value **Business travel** to something like **Bus_tvl** as you may think because doing that will not resolve the problem. Instead, it points to the fact that SPSS requires numeric values for all input fields, including categorical fields, to run statistical analysis, especially regression.

For example, the **Travel purpose** field is nominal, with three levels: **Business travel, Mileage ticket,** and **Personal travel**. These values are in string format in the dataset, which SPSS Modeler does not accept for regression analysis, even though logistic regression can work with categorical predictors.

To solve this, we need to recode these levels into numeric values as follows:

- 1 = Business travel
- 2 = Mileage ticket
- 3 = Personal travel

By recoding the field, SPSS Modeler will recognize the levels appropriately. This is similar to what I would do in SPSS Statistics, and while the recoding task itself is straightforward, it's only the beginning. Since there are eight categorical fields in this dataset, each needs to be recoded. Now imagine if we were working with hundreds of fields for data mining purposes—it would become an extremely time-consuming task. Manually recoding each field defeats the purpose of using data mining software for large datasets.

So, how can we efficiently build predictive models without spending so much time on recoding? Fortunately, SPSS Modeler provides a solution: the **Auto Data Prep** node, which can handle this issue automatically and efficiently.

> Note that SPSS Modeler only displays this error when running Linear Regression or Logistic Regression. Other methods, such as Decision Trees or Neural Networks, do not encounter this issue. This may be because SPSS Modeler utilizes regression algorithms from SPSS Statistics, which only accept numeric values for statistical analyses.

Follow the steps below to set up and run the Auto Data Prep node:

1. Open the "**Airline_Modify**" stream created earlier.
2. From the **Field Ops** group, add an **Auto Data Prep** node to the stream and connect it to the Type node. Open the node settings.
3. On the **Objectives** tab, select **Custom analysis**, which allows us to customize our selections.
4. On the **Settings** tab, select **Prepare Inputs & Target** from the menu on the left. Then, check the following boxes (as shown in the figure below). These options inform the software to perform: (1) adjust the scales of fields; (2) reorder nominal fields; (3) handle outliers; (4) impute missing values for continuous inputs; and (5) normalize continuous inputs.

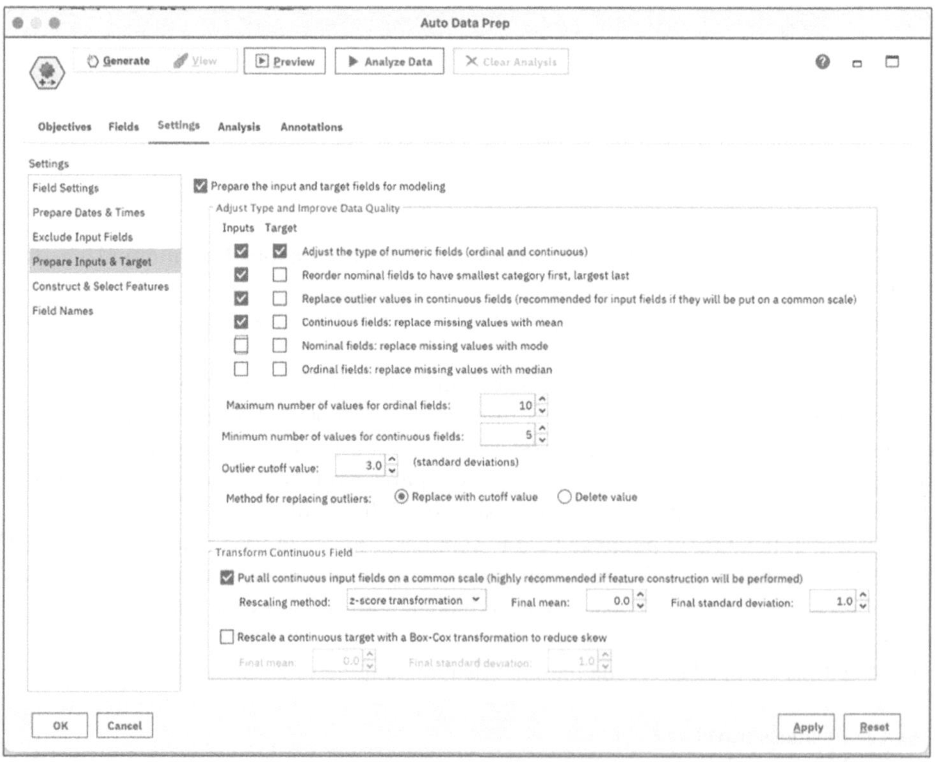

5. Click Apply, and then click Analyze Data (at the top of the menu).
6. Click OK.

TABLE 5.8
Results of the Auto Data Prep Node

Field	Measurement	Min	Max	Mean	Std. Dev	Skewness	Kurtosis	Unique	Valid
Recommend the airline	Flag	–	–	–	–	–	–	2	124464
Age_transformed	Continuous	−1.788	2.273	0	1	0.316	−0.711	–	124464
Price sensitivity_transformed	Continuous	−2.422	2.636	0	1	0.495	−0.075	–	124464
Number of years flying_transformed	Continuous	−1.605	1.414	0	1	−0.072	−1.291	–	124464
Average number of flights per year_transformed	Continuous	−1.427	2.522	0	1	0.679	−0.259	–	124464
Percent of flights with other airlines_transformed	Continuous	−1.072	2.807	0	1	1.502	1.587	–	124464
Number of other loyalty cards_transformed	Continuous	−0.771	2.68	0	1	1.027	−0.105	–	124464
Shopping amount at airport_transformed	Continuous	−0.622	2.502	0	1	1.53	0.958	–	124464
Spending for eating and drinking at airport_transformed	Continuous	−1.473	2.497	0	1	0.76	−0.006	–	124464
Flight distance (miles)_transformed	Continuous	−1.332	2.846	0	1	1.24	1.034	–	124464
Departure delay (in minutes)_transformed	Continuous	−0.588	2.589	−0.034	0.955	1.815	2.016	–	124464
Arrival delay (in minutes)_transformed	Continuous	−0.6	2.536	−0.046	0.938	1.802	1.998	–	124464
Flight time (in minutes)_transformed	Continuous	−1.529	2.808	−0.002	0.998	1.149	0.815	–	124464
Airline status_transformed	Nominal	–	–	–	–	–	–	4	124464
Gender_transformed	Flag	–	–	–	–	–	–	2	124464
Travel purpose_transformed	Nominal	–	–	–	–	–	–	3	124464
Passenger class_transformed	Nominal	–	–	–	–	–	–	3	124464
Origin state_transformed	Nominal	–	–	–	–	–	–	52	124464
Destination state_transformed	Nominal	–	–	–	–	–	–	52	124464
Flight cancellation_transformed	Flag	–	–	–	–	–	–	2	124464

The outcome, shown in Table 5.8, includes several modifications to the data:

- First, all categorical input fields have been transformed into the appropriate numeric format, meeting the requirements for logistic regression analysis.
- Second, missing values have been imputed with mean values, and all cases are valid, indicating no remaining missing values.
- Third, outliers in continuous fields have been addressed, resulting in improved kurtosis and skewness values.
- Fourth, all continuous fields have been normalized to the same scale.

Additionally, Figure 5.7 shows a snapshot of the transformed data for categorical fields. As shown, all levels in those fields have been successfully converted to numeric formats, meeting the requirements for running logistic regression. When I reran the logistic regression with this updated dataset, the error messages were resolved. This approach offers a quick and efficient way to handle this specific issue. Be sure to remember this step, as we will use the Auto Data Prep node again in Chapter 7.

Gender_transformed	Travel purpose_transformed	Passenger class_transformed	Origin state_transformed	Destination state_transformed	Flight cancellation_transformed
1	2	0	22	22	1
0	2	2	22	22	1
1	2	0	22	22	1
1	2	1	22	22	1
1	2	2	22	22	1
1	2	2	22	22	1
0	2	2	22	22	1
1	1	2	22	22	1
1	1	2	22	22	1
0	1	2	22	22	1
0	1	2	22	22	1
1	1	1	22	22	1
1	1	2	22	22	1
0	2	0	51	51	1
1	2	2	51	51	1
0	2	2	51	51	1
1	2	1	51	51	1
0	2	2	51	51	1

FIGURE 5.7 A snapshot of transformed data for categorical fields.

EXAMPLES OF DATA MODIFICATION WITH THE BOSTON HOUSING DATASET

In the second example, we will use the **Boston Housing** dataset to perform imputation and normalization tasks. The data description is in Table 5.9.

- **Dataset**: Boston.sav
- **Target variable**: Imv (Continuous—median home value)

Before we begin, let's explore the dataset to understand its structure. Table 5.10 provides the descriptive statistics for all fields, which are continuous variables. Notably, only one field, the target field (**Imv**), has missing values—specifically, 54 missing entries.

We will address this issue using imputation and follow up with normalization to ensure all fields are on the same scale for further analysis.

TABLE 5.9
Boston Housing Data—Variable Descriptions

Variable Name	Description	Scale	Role
CrimePC	Crime rate per capita	Continuous	Input
ZonePr	Proportion of residential land zoned for lots over 25,000 sq.ft.	Continuous	Input
IndustPr	Proportion of non-retail business acres	Continuous	Input
Charles_River	Charles River dummy variable (1 if tract bounds river; 0 otherwise)	Continuous	Input
NiOxSQ	Nitric oxides concentration (parts per 10 million)	Continuous	Input
RmSQ	Average number of rooms per dwelling	Continuous	Input
AgePre40	Proportion of owner occupied units built prior to 1940	Continuous	Input
LogDisstance	Weighted distances to five Boston employment centers	Continuous	Input
LogAccess	Index of accessibility to radial highways	Continuous	Input
Tax	Full value property tax rate per $10,000	Continuous	Input
PtRatio	Pupil-teacher ratio	Continuous	Input
B	1000(Bk – 0.63)^2 where Bk is the proportion of blacks by town	Continuous	Input
Lstat	Percentage of lower status of the population	Continuous	Input
Imv	Median value of owner-occupied homes in $1,000s	Continuous	Target

TABLE 5.10
Descriptive Statistics Results—Boston Housing

Field	Role	Mean	Standard Deviation	Non Missing	Missing	Min	Median	Max
AgePre40	INPUT	58.74	33.10	506	0	1.14	65.20	100.00
B	INPUT	332.79	125.32	506	0	0.32	390.64	396.90
Charles_River	INPUT	0.14	0.31	506	0	0.00	0.00	1.00
CrimePC	INPUT	1.27	2.40	506	0	0.00	0.14	9.97
IndustPr	INPUT	9.21	7.17	506	0	0.00	6.96	27.74
LogAccess	INPUT	78.06	203.54	506	0	1.00	5.00	666.00
LogDistance	INPUT	6.17	6.48	506	0	1.13	3.92	24.00
Lstat	INPUT	11.54	6.06	506	0	1.73	10.36	34.41
NiOxSQ	INPUT	1.10	1.65	506	0	0.39	0.54	7.31
PtRatio	INPUT	42.61	87.59	506	0	2.60	18.90	396.90
RmSQ	INPUT	15.68	27.22	506	0	3.56	6.32	100.00
Tax	INPUT	339.32	180.67	506	0	20.20	307.00	711.00
ZonePr	INPUT	13.30	23.05	506	0	0.00	0.00	100.00
mv	TARGET	23.75	8.81	452	54	6.30	21.90	50.00

Imputation Example for the Boston Housing

Since the missing values account for about 11% of the data, we will use the mean imputation method. As in the previous example, we will perform the imputation using a combination of the **Data Audit** node and a **SuperNode**.

Follow the steps below to implement data imputation:

1. Open SPSS Modeler.
2. Create a new stream and name it "**Boston_Modify**."
3. From the **Sources** group, add the **Statistics File** node to the stream and import the file **Boston.sav** from your computer.
4. From the **Field Ops** group, add a Type node to the stream and connect it to the data source.
5. Check the measurement scales for all fields. Change the role of the target field **Imv** to **Target.**
6. From the **Output** group, add a Data Audit node to the stream and connect it to the Type node.

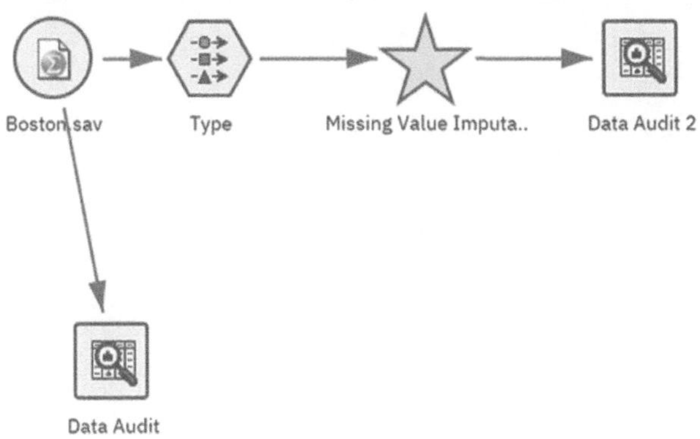

7. Click **Apply** and **Run**.
8. In the **Data Audit** output window, select the **Quality** tab.

9. For the field **Imv**, under the **Impute Missing** column, select **Specify**, and make the following changes:
   - Impute when: **Null Values**
   - Impute method: **Fixed**
   - Impute fixed values: **Mean**

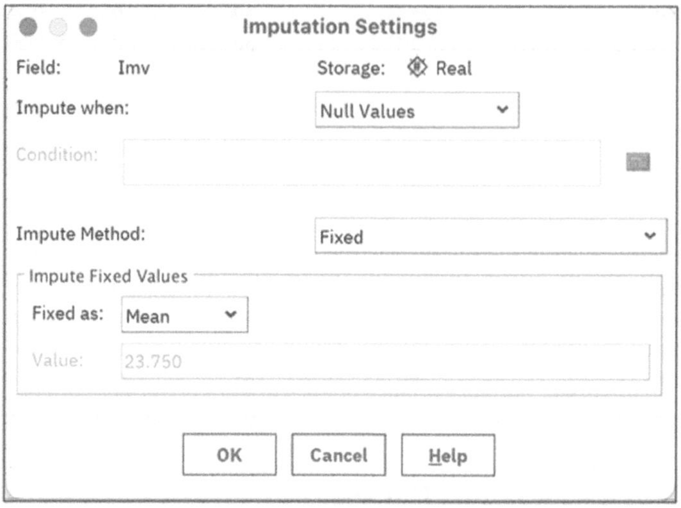

10. To apply these imputations, generate a **SuperNode**. Under the **Generate** menu, select **Missing Values SuperNode**. This will generate a SuperNode containing the imputation configurations for **Imv**.
11. Click OK.
12. SPSS Modeler will create a SuperNode, named **Missing Value Imputation** node, into the stream. Connect this node to the Type node.
13. In order to execute the SuperNode, add another **Data Audit** node and connect it to the SuperNode.
14. Run the **Data Audit** node.

The results displayed in Figure 5.8 show that the field **Imv** now contains all valid records, with no missing values. This confirms that the missing values in this field have been successfully imputed with the mean value.

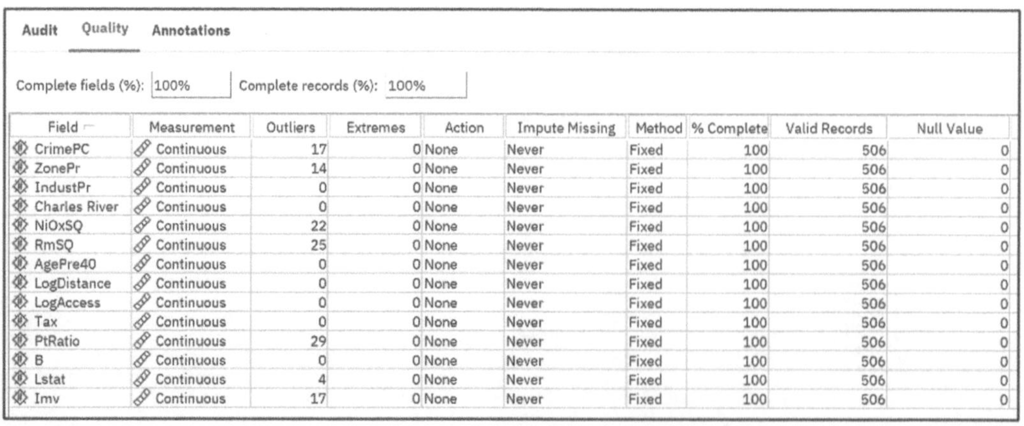
Audit | Quality | Annotations

Complete fields (%): 100% Complete records (%): 100%

Field	Measurement	Outliers	Extremes	Action	Impute Missing	Method	% Complete	Valid Records	Null Value
CrimePC	Continuous	17	0	None	Never	Fixed	100	506	0
ZonePr	Continuous	14	0	None	Never	Fixed	100	506	0
IndustPr	Continuous	0	0	None	Never	Fixed	100	506	0
Charles River	Continuous	0	0	None	Never	Fixed	100	506	0
NiOxSQ	Continuous	22	0	None	Never	Fixed	100	506	0
RmSQ	Continuous	25	0	None	Never	Fixed	100	506	0
AgePre40	Continuous	0	0	None	Never	Fixed	100	506	0
LogDistance	Continuous	0	0	None	Never	Fixed	100	506	0
LogAccess	Continuous	0	0	None	Never	Fixed	100	506	0
Tax	Continuous	0	0	None	Never	Fixed	100	506	0
PtRatio	Continuous	29	0	None	Never	Fixed	100	506	0
B	Continuous	0	0	None	Never	Fixed	100	506	0
Lstat	Continuous	4	0	None	Never	Fixed	100	506	0
Imv	Continuous	17	0	None	Never	Fixed	100	506	0

FIGURE 5.8 Imputation results for the **Boston** dataset.

Normalization Example for the Boston Housing

If you completed the data exploration for the **Boston Housing** dataset in Chapter 3, you may have noticed that the fields are not normally distributed. Since we plan to use the linear regression method to analyze this data, and linear regression works best with normally distributed data, we need to normalize all continuous fields in this dataset.

We will continue using the imputed data from the previous step. To perform normalization in SPSS Modeler, we will use the **Transform** node.

Follow the steps below to perform normalization:

1. Open the "**Boston_Modify**" stream.
2. Ensure the data source, Type node, and the **SuperNode** (**Missing Value Imputation**) created earlier are in the stream.
3. From the **Output** group, add the **Transform** node to the stream and connect it to the **SuperNode**. Then, rename the Transform node **Normalization** and open its settings.

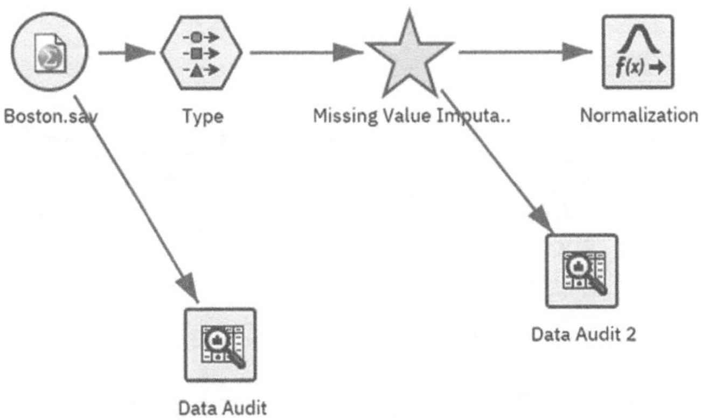

4. Under the **Fields** tab, select all fields by holding the Control button.

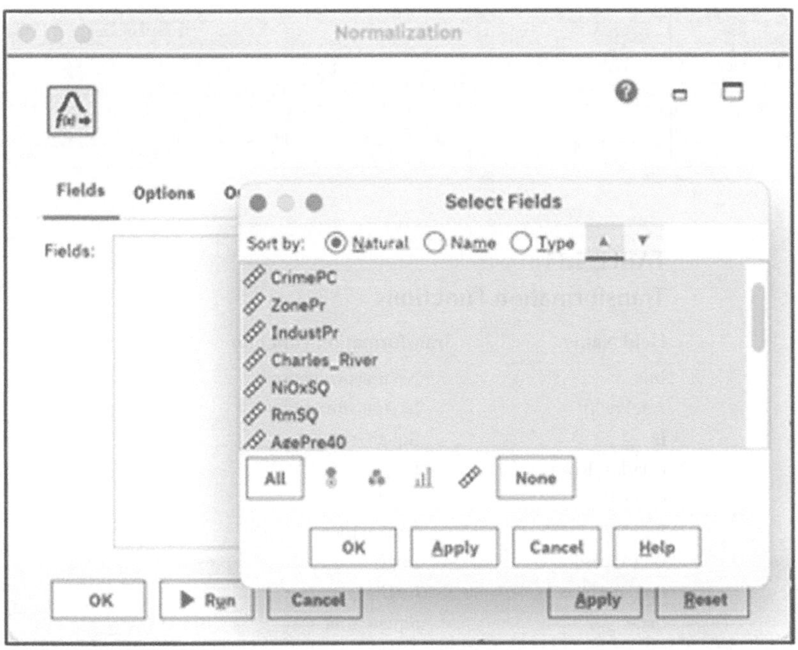

5. In the **Options** tab, select **All Formulas** (this option allows SPSS Modeler to apply all available normalization functions).
6. Click Apply and Run.
7. Review the transformation function applied to each field. Note that normalization is not always guaranteed, even with all functions. Select the transformation that gives a distribution closest to the normal curve for each field. Table 5.11 provides the specific transformation functions for the fields. For fields like **Imv, AgePre40,** and **Tax**, which are already close to normal distributions, I opted to keep their current values (no transformation).

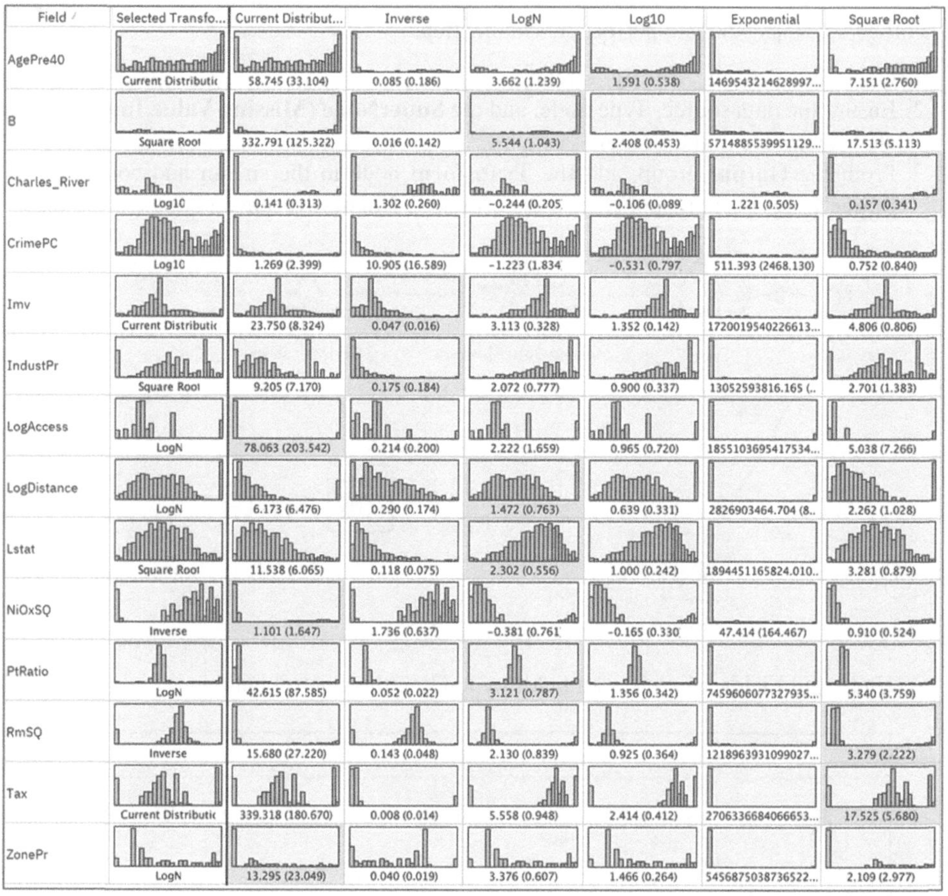

TABLE 5.11
Transformation Functions

Field Name	Transformation Function
Imv	No transformation
AgePre40	No transformation
B	Square root
Charles_River	Log10
CrimePC	Log10
IndustPr	Square root
LogAccess	LogN
LogDistance	LogN
Lstat	Square root

(Continued)

TABLE 5.11 *(Continued)*
Transformation Functions

Field Name	Transformation Function
NiOxSQ	Inverse
PtRatio	LogN
RmSQ	Inverse
Tax	No transformation
ZonePr	LogN

8. Select **Generate/Derive** node, and then choose **non-standardized transformation**. SPSS Modeler will generate a SuperNode named **Supernode_Transform** and add it to the stream.
9. Connect the **Supernode_Transform** node to the **Missing Value Imputation** node created earlier.

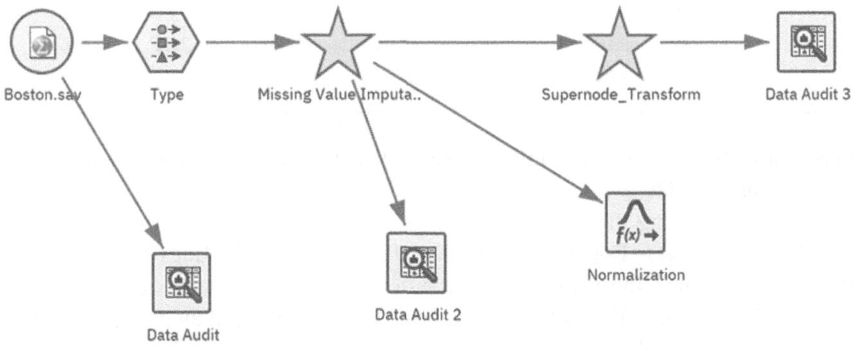

10. Add a third **Data Audit** node to the stream and connect it to the **Supernode_Transform**.
11. Run the **Data Audit** node.

Table 5.12 provides a comparison between the original and normalized data. By reviewing the mean and standard deviation values, we observe that before normalization, the fields had widely varying ranges, with mean values ranging from 0.14 to 339. After normalization, these fields exhibit much more consistent ranges, which aids in the interpretation and comparison of results.

Regarding normal distribution, we can assess the skewness and kurtosis values, which are expected to be closer to zero for a normal distribution. After normalization, we observe improvements in skewness and kurtosis for all fields except for **B**. However, it's worth noting that **PtRatio_LogN** and

TABLE 5.12
Comparison between Original and Normalized Data

Field	Min	Max	Mean	Std. Dev	Skewness	Kurtosis	Valid	Null Value
AgePre40	1.137	100	58.745	33.104	−0.394	−1.215	506	0
B	0.32	396.9	332.791	125.322	−1.953	2.037	506	0
B_Square root	0.566	19.922	17.513	5.113	−2.089	2.61	506	0
Charles_River	0	1	0.141	0.313	1.902	1.947	506	0
Charles_River_Log10	−0.237	0	−0.106	0.089	0.197	−1.663	89	417
CrimePC	0	9.967	1.269	2.399	2.199	3.779	506	0
CrimePC_Log10	−2.199	0.999	−0.531	0.797	0.434	−0.893	452	54
Imv	6.3	50	23.75	8.324	1.37	2.231	506	0
IndustPr	0	27.74	9.205	7.17	0.589	−0.795	506	0
IndustPr_Square Root	0	5.267	2.701	1.383	−0.339	−0.568	506	0

(Continued)

TABLE 5.12 *(Continued)*
Comparison between Original and Normalized Data

Field	Min	Max	Mean	Std. Dev	Skewness	Kurtosis	Valid	Null Value
LogAccess	1	666	78.063	203.542	2.549	4.529	506	0
LogAccess_LogN	0	6.501	2.222	1.659	1.675	1.989	506	0
LogDistance	1.13	24	6.173	6.476	2.143	3.261	506	0
LogDistance_LogN	0.122	3.178	1.472	0.763	0.877	0.229	506	0
Lstat	1.73	34.41	11.538	6.065	0.898	0.65	506	0
Lstat_Square root	1.315	5.866	3.281	0.879	0.288	−0.36	506	0
NiOxSQ	0.385	7.313	1.101	1.647	2.647	5.285	506	0
NiOxSQ_Inverse	0.137	2.597	1.736	0.637	−1.344	1.181	506	0
PtRatio	2.6	396.9	42.615	87.585	3.526	10.818	506	0
PtRatio_LogN	0.956	5.984	3.121	0.787	2.94	7.8	506	0
RmSQ	3.561	100	15.68	27.22	2.604	4.878	506	0
RmSQ_Inverse	0.01	0.281	0.143	0.048	−2.048	3.283	506	0
Tax	20.2	711	339.318	180.67	0.373	−0.074	506	0
ZonePr	0	100	13.295	23.049	2.068	3.625	506	0
ZonePr_LogN	2.526	4.605	3.376	0.607	0.715	−0.883	188	318

RmSQ_Inverse are still not close enough to normal distributions, indicating further adjustments may be needed for these fields.

Additionally, while running the transformation for the complete dataset, we observed data loss in three fields: **Charles_River_Log10** (417 missing values), **CrimePC_Log10** (54 missing values), and **ZonePr_Log10** (318 missing values). All three fields were transformed using the LogN function. Typically, applying a logarithm function to a value of zero results in an undefined value, which is recorded as missing. Since these fields had multiple records with zero values, the transformation introduced many missing values.

To address this issue, I will change the transformation functions for these fields to **Square Root**, which will return a zero for any record with a zero value (as shown in Table 5.13).

We repeat steps 7–11 from the previous process to run the transformation using the new Square Root functions. Table 5.14 shows the updated comparison between the original and normalized data.

TABLE 5.13
Revised Transformation Functions

Field Name	Transformation Function
Imv	No transformation
AgePre40	No transformation
B	Square root
Charles_River	Square root
CrimePC	Square root
IndustPr	Square root
LogAccess	LogN
LogDistance	LogN
Lstat	Square root
NiOxSQ	Inverse
PtRatio	LogN
RmSQ	Inverse
Tax	No transformation
ZonePr	Square root

TABLE 5.14
Updated Comparison between Original and Normalized Data

Field	Min	Max	Mean	Std. Dev	Skewness	Kurtosis	Valid
AgePre40	1.137	100	58.745	33.104	−0.394	−1.215	506
B	0.32	396.9	332.791	125.322	−1.953	2.037	506
B_Square root	0.566	19.922	17.513	5.113	−2.089	2.61	506
Charles_River	0	1	0.141	0.313	1.902	1.947	506
Charles_River_Square root	0	1	0.157	0.341	1.758	1.184	506
CrimePC	0	9.967	1.269	2.399	2.199	3.779	506
CrimePC_Square root	0	3.157	0.752	0.84	1.495	1.056	506
Imv	6.3	50	23.75	8.324	1.37	2.231	506
IndustPr	0	27.74	9.205	7.17	0.589	−0.795	506
IndustPr_Square root	0	5.267	2.701	1.383	−0.339	−0.568	506
LogAccess	1	666	78.063	203.542	2.549	4.529	506
LogAccess_LogN	0	6.501	2.222	1.659	1.675	1.989	506
LogDistance	1.13	24	6.173	6.476	2.143	3.261	506
LogDistance_LogN	0.122	3.178	1.472	0.763	0.877	0.229	506
Lstat	1.73	34.41	11.538	6.065	0.898	0.65	506
Lstat_Square root	1.315	5.866	3.281	0.879	0.288	−0.36	506
NiOxSQ	0.385	7.313	1.101	1.647	2.647	5.285	506
NiOxSQ_Inverse	0.137	2.597	1.736	0.637	−1.344	1.181	506
PtRatio	2.6	396.9	42.615	87.585	3.526	10.818	506
PtRatio_LogN	0.956	5.984	3.121	0.787	2.94	7.8	506
RmSQ	3.561	100	15.68	27.22	2.604	4.878	506
RmSQ_Inverse	0.01	0.281	0.143	0.048	−2.048	3.283	506
Tax	20.2	711	339.318	180.67	0.373	−0.074	506
ZonePr	0	100	13.295	23.049	2.068	3.625	506
ZonePr_Square root	0	10	2.109	2.977	1.048	−0.19	506

The results are similar to those from the previous attempt but without any data loss. Therefore, the normalization process has been successfully completed.

You may ask: Why don't we perform normalization for continuous variables in the Airline Recommendation dataset?

That's a great question. Let me explain.

Normalization is typically important for machine learning algorithms that rely on the ordinal least squares (OLS) estimation method, such as linear regression. These algorithms perform best with normally distributed data, which helps achieve a better model fit. Since the Boston Housing dataset has a continuous target variable, linear regression is an appropriate method, making data normalization beneficial.

However, other algorithms—such as logistic regression, decision trees, or random forests—are based on different estimation methods and do not require normally distributed data. For instance, logistic regression uses maximum likelihood estimation (MLE), which does not assume normality. Therefore, normalizing data may not provide any benefits for these algorithms and can complicate result interpretation due to changes in measurement scales.

For example, if we transform a predictor by taking its square root, the model will be trained on this transformed value. When analyzing the results, such as a regression coefficient, interpreting the predictor's effect on the target variable in terms of its square root can be difficult. To make the interpretation meaningful, we would need to convert the predictor back to its original scale, which adds unnecessary complexity.

In the **Airline Recommendation** dataset, we have a binary target variable, making logistic regression the appropriate choice for predictive modeling. Moreover, advanced machine learning algorithms like neural networks also do not require normally distributed data, though they may benefit from data standardization to improve model fit. Some advanced methods even include built-in data standardization options, allowing us to experiment with different methods dynamically. Therefore, there is no need to normalize continuous variables in this dataset.

Discretization Example for the Boston Housing

To illustrate variable discretization, let's consider a hypothetical scenario. Suppose, we want to convert the continuous target field **Imv** into a categorical field for prediction purposes. To achieve this, we need to perform variable discretization. We will use the **Binning** node and apply the **Quantiles** (equal-frequency binning) method.

Follow the steps below to perform discretization:

1. Open the "**oston_Modify**" stream created earlier.
2. From the **Field Ops** group, add a **Binning** node to the stream and connect it to the Type node. Open the node settings.
3. For the **Bin field**, select **Imv.**
4. For the **Binning method**, select **Tiles (equal count).**
5. Select **Quartile (4)** to divide the data into four equal-frequency bins.

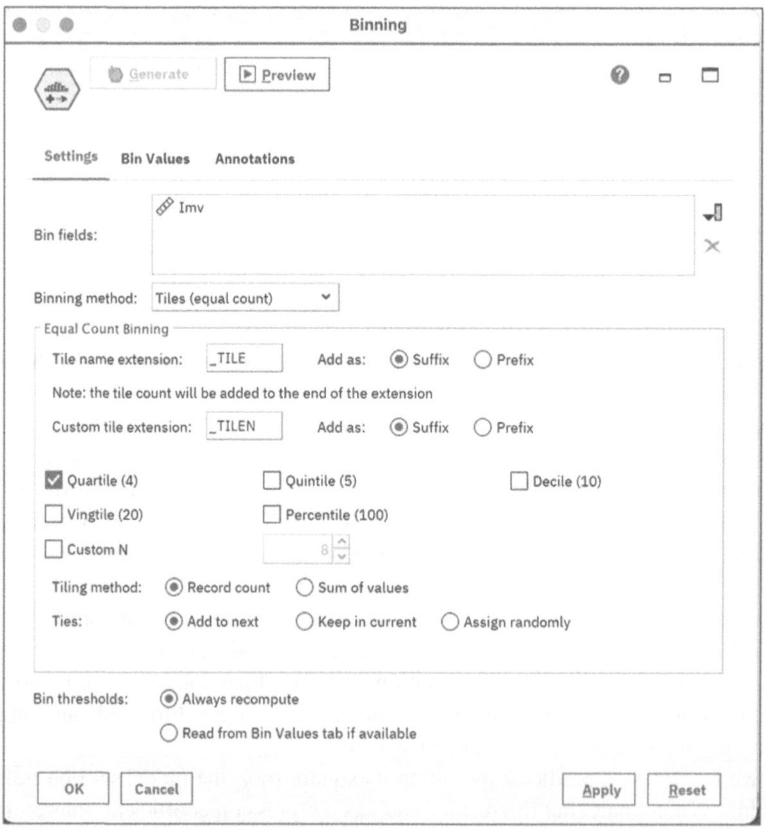

6. Click Apply, and then click OK.

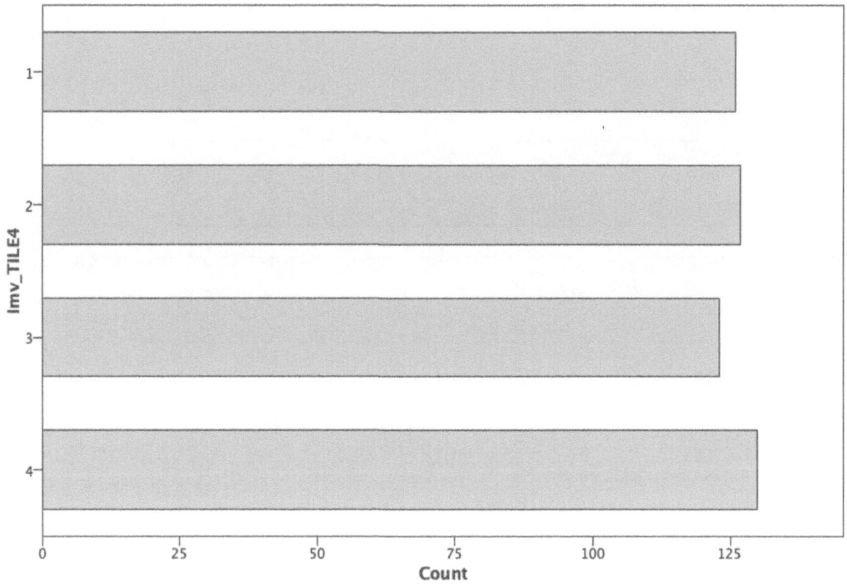

FIGURE 5.9 Discretization results—Quantile/binning method.

By adding a Data Audit node, you can see the result of the discretization. Figure 5.9 shows the new target field, **Imv_TILE4**, which is now a categorical field with four levels, each having an equal frequency count.

Trying a Different Discretization Method

Now, let's try a different discretization method. Repeat the same process, but change the **Binning method** to **Fixed-width/binning (equal-width binning)** and set the **Bin width** to 10. This method divides the data into evenly spaced intervals based on the specified bin width. The result is a new categorical target field, **Imv_BIN**, which also has four levels but with varying frequency counts (see Figure 5.10).

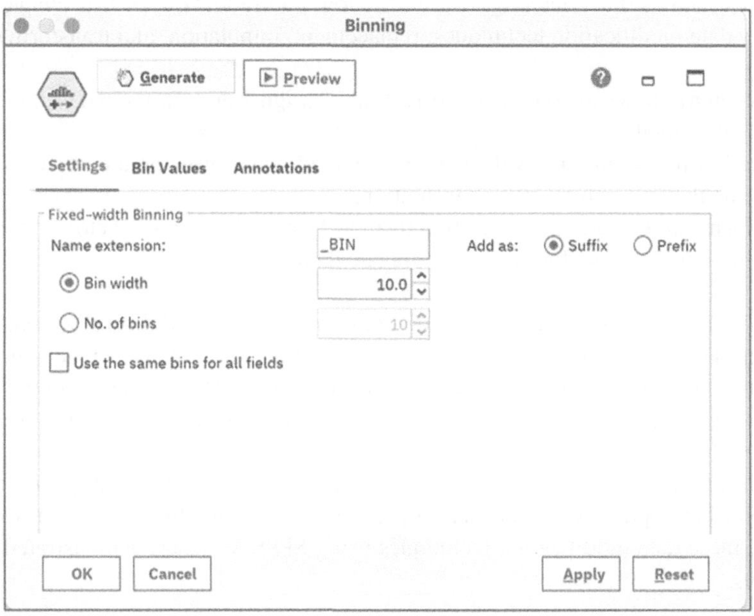

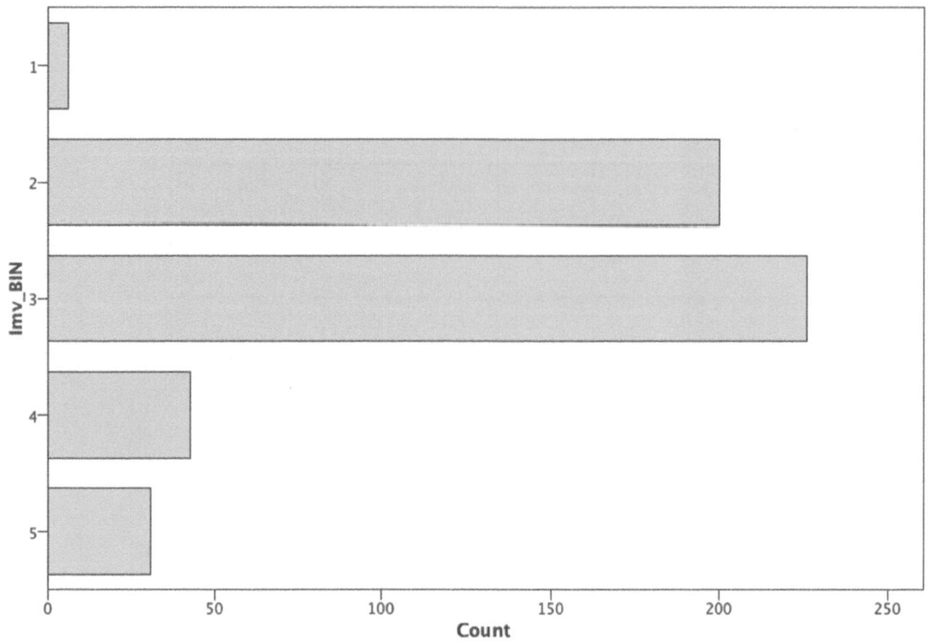

FIGURE 5.10 Discretization results—Fixed width/binning method.

Discretization is a powerful tool in data mining when converting continuous fields into categorical fields for analysis. SPSS Modeler makes this task easy with the Binning node, allowing users to apply different discretization methods without any technical challenges.

CHAPTER SUMMARY

Data modification is a crucial process in data mining that involves altering the original dataset to enhance its quality, completeness, and compatibility for predictive modeling. It encompasses various techniques such as data cleaning, transformation, and reduction. In this chapter, we covered three primary data modification techniques: replacement, imputation, and transformation.

- **Replacement** allows us to recode, correct, and assign values in the dataset, often used to improve data quality.
- **Imputation** fills in missing values with estimated ones, ensuring a complete dataset and preventing the loss of important information.
- **Transformation** creates new variables by modifying existing ones, helping standardize or normalize data, convert formats, or aggregate data.

We introduced several methods for imputation, normalization, and discretization. The choice of method depends on the dataset's characteristics, the machine learning technique being applied, and the goals of the data mining project. It requires careful consideration of the potential impact on the data, such as the introduction of bias during imputation or information loss during normalization or aggregation.

Data modification is a dynamic and iterative process, often requiring multiple rounds of preprocessing to refine and optimize the dataset for predictive modeling. In this chapter, we also learned how to apply these data modification techniques using SPSS Modeler, demonstrated with two different datasets.

CHAPTER DISCUSSION QUESTIONS

1. What is data modification, and why is it important in the data mining process?
2. Provide some practical examples where data modification is necessary.
3. What are the benefits of data modification, and what challenges might arise when using data modification techniques?
4. What are the primary data modification techniques? Provide examples.
5. What are the differences between replacement, imputation, and transformation techniques in data modification? Provide examples.
6. How do we select the appropriate data modification techniques for a given dataset?
7. How can the choice of data modification techniques influence the performance and outcome of a data mining project?
8. Provide examples of situations where replacement techniques can be used to clean data.
9. What are missing values, and how do they affect the data mining process?
10. What are the trade-offs of data imputation? How do we balance improving data completeness with preserving the quality and representativeness of the original data? Provide examples.
11. What are the common imputation methods, and how do we select the right one for our project?
12. What are the main purposes of data normalization, and in which cases should normalization be used?
13. What challenges might we encounter when normalizing data?
14. What are the common normalization methods, and when should each be applied?
15. What is the main purpose of variable discretization? Describe the steps in the discretization process. Provide examples.
16. What are the common discretization methods, and what are their typical applications?

CHAPTER EXERCISES

EXERCISE 1: IMPUTATION FOR THE BANK DATASET

1. View the field description of the **bank** dataset (Appendix 1).
2. Open the **bank** dataset in SPSS Modeler.
3. Create a new stream named **Bank_Modify** and add the data source to the stream.
4. Add a **Type** node to the stream and connect it to the data source.
5. Add a **Data Audit** node to the stream and connect it to the *Type* node.
6. Double-check the measurement scales and roles for all fields.
7. Run the Data Audit node and report the descriptive statistics. Identify fields with missing values.
8. Configure the imputation using the mean imputation method on the **Quality** tab of the **Data Audit** output.
9. Generate a SuperNode called **Missing Value Imputation** for data imputation.
10. Add a second Data Audit node to the stream and connect it to the **SuperNode**.
11. Run the **Data Audit** node to perform data imputation.
12. Report and explain the imputation results. Were all missing values imputed correctly?

EXERCISE 2: NORMALIZATION FOR THE BANK DATASET

1. Continue from Exercise 1.
2. Add a **Transform** node to the stream and connect it to the imputed data (the SuperNode created in Exercise 1).
3. Select all continuous fields and run the **Transform** node using **All Formulas**.

4. Report the results, showing the transformation functions for all fields.
5. Select the appropriate transformation function for each field and generate a SuperNode named **Supernode_Transform**, which incorporates those transformation functions.
6. Add a **Data Audit** node to the stream and connect it to the **SuperNode**.
7. Run the **Data Audit** node to perform data normalization.
8. Compare the mean, standard deviation, skewness, and kurtosis between the original and normalized data. Explain the results.

Exercise 3: Normalization for the Cardiac Dataset

1. Open the **cardiac** dataset in SPSS Modeler.
2. Create a new stream named **Cardiac_Modify** and add the data source to the stream.
3. Add a **Type** node to the stream and connect it to the data source.
4. Add a **Data Audit** node to the stream and connect it to the **Type** node.
5. Double-check the measurement scales and roles for all fields.
6. Run the **Data Audit** node and report the descriptive statistics. Identify the distributions for all continuous fields.
7. Add a **Transform** node to the stream and connect it to the Type node.
8. Select all continuous fields and run the Transform node using **All Formulas**.
9. Report the results, showing the transformation functions for all fields.
10. Select the appropriate transformation function for each field and generate a **SuperNode** called **Supernode_Transform**, which incorporates those transformation functions.
11. Add a **Data Audit** node to the stream and connect it to the SuperNode.
12. Run the Data Audit node to perform data normalization.
13. Compare the mean, standard deviation, skewness, and kurtosis between the original and normalized data. Explain the results.

Exercise 4: Discretization for the Cardiac Dataset

1. Continue from Exercise 3.
2. Add a **Binning** node to the stream and connect it to the **Type** node.
3. Convert the **bmi** field into a categorical field using the **Fixed-width/binning** method. Report the results.
4. Convert the **bmi** field into a categorical field using the **Quantile/binning** method. Report the results.
5. Compare the results from Steps 3 and 4.

Exercise 5: Replacement for the Passengers Dataset

1. View the field description of the **passengers** dataset (Appendix 2).
2. Open the **passengers** dataset in SPSS Modeler.
3. Create a new diagram named **Passenger_Modify** and add the data source to the stream.
4. Add a **Type** node to the stream and connect it to the data source.
5. Add a **Data Audit** node to the stream and connect it to the **Type** node.
6. Double-check the measurement scales and roles for all fields.
7. Run the Data Audit node and report the descriptive statistics. Identify fields with missing values.
8. Identify Likert-scale fields with zero values (these zeros must be re-coded as missing values).
9. From the **Field Ops** group, add the **Filler** node to the stream and connect it to the Type node. Open the node settings.

10. Configure the condition in the Filler node to replace the zero values in those fields with **undef** (blank or null).
11. Report the replacement results, showing the zero values replaced by missing values for each field.

Exercise 6: Imputation for the Passengers Dataset

1. Continue from Exercise 5.
2. Add a **Data Audit** node to the stream and connect it to the **Filler** node.
3. Run the **Data Audit** node and open the **Quality** tab.
4. Under the **Impute Missing** column, select the mean imputation method.
5. Generate a **SuperNode** called **Missing Value Imputation**, incorporating the mean imputation configuration.
6. Add another **Data Audit** node to the stream and connect it to the **SuperNode**.
7. Run the Data Audit node.
8. Report the imputation results. Were all missing values imputed?

Part II

Data Mining Methods

6 Model Evaluation

LEARNING OUTCOMES

Upon the completion of this chapter, you will be able to:

1. Explain the purpose and importance of model evaluation.
2. Compare model fit and predictive power.
3. Describe overfitting problems and identify potential root causes.
4. Discuss strategies to avoid overfitting.
5. Differentiate between model reliability and validity.
6. Evaluate the reliability and validity of predictive models.
7. Define the champion model and explain the significance of model comparison.
8. Explain the purpose and importance of model fit metrics.
9. Select appropriate model fit metrics for association and classification models.
10. Calculate model fit metrics for association models.
11. Define a confusion matrix and describe its components.
12. Calculate model fit metrics for classification models.
13. Explain cost-sensitive learning.

INTRODUCTION

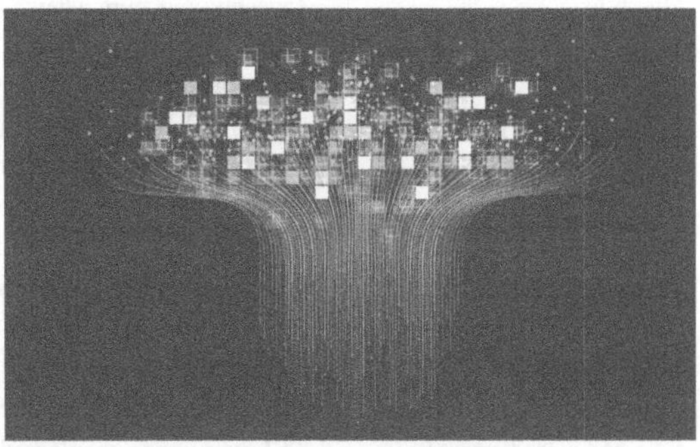

Evaluating the performance of predictive models is a crucial part of the data mining process. As mentioned earlier, there is no such thing as a perfect predictive model and a margin of error is always present. Understanding how well a model can predict the target variable allows us to determine whether it is suitable for actual predictions and what level of prediction error is acceptable. Model performance evaluation is often challenging due to the many concepts, processes, and metrics involved. Additionally, these concepts are not always presented consistently in books or online resources, which can lead to confusion. Using the wrong process or metrics can result in an invalid predictive model, which is why it's essential to address this topic thoroughly before working with machine learning methods.

In this chapter, we will focus on the key concepts of model performance evaluation, the evaluation process, and the critical metrics needed to assess model fit. As discussed in Chapter 2, a common practice in data mining is to build multiple models using different machine learning methods and then compare their performance to select the best, or "champion," model. Before comparing models, we must first decide on the appropriate metrics for evaluating model fit.

There are some interchangeable terms in this area, which can be confusing. Therefore, I will provide clear definitions and explanations of these terms and when to use them. Having this foundational understanding is essential for choosing the right metrics and following the proper evaluation process. It is important to note that the metrics covered in this chapter apply only to supervised learning, where there is a predefined target variable, and the goal is to predict that variable. These metrics do not apply to unsupervised learning, as there is no prediction involved. Furthermore, the metrics used for association models (with a continuous target variable) differ from those used for

DOI: 10.1201/9781003468196-8

classification models (with a categorical target variable). Therefore, attention must be paid to the scale of the target variable to ensure the correct metrics are selected.

BASIC CONCEPTS

Before we go further into the topic of model evaluation, I will introduce some key basic concepts, which are outlined in Table 6.1. These concepts are essential for understanding the terminology used throughout this chapter. Use the table as a reference as we proceed. I will explain each concept in detail, and you can always return to the table to refresh your memory or check the definition of a specific term when needed.

TABLE 6.1
Basic Concepts in Model Performance Evaluation

Terms	Descriptions
Model fit	Model fit measures how well the model fits the data. It assesses the degree to which the model captures the patterns and relationships within the data.
Predictive power	Predictive power refers to a model's ability to accurately capture and represent the underlying patterns, relationships, or trends in the data, and effectively generalize these results to new, unseen datasets.
Overfitting	Overfitting occurs when a model is overly trained to fit the training sample, capturing even the noise and minor fluctuations in the data. As a result, the model fails to generalize to other datasets, making it invalid and unusable for real-world predictions.
Model reliability	Model reliability refers to the internal consistency of a model. It means the model produces consistent results across different runs using different datasets.
Model validity	Model validity refers to the accuracy of the model. It means the model consistently produces an accurate prediction of the target variable.
Champion model	Champion model is the model that demonstrates the best performance among all the models we train and test.
Prediction errors	Errors measure the extent to which the trained model misfits the data. The lower the error, the more accurate the model.
Sum of squared error (SSE)	SSE measures unexplained variance, which is how far the data are from the model's predicted values.
Sum of squared regression (SSR)	SSE measures explained variance, which is how far the mean is from the model's predicted values.
Sum of squared total (SST)	SST measures total variance, which is how far the data are from the mean.
Root mean squared error (RMSE)	RMSE is a common metric for association models. It measures the average difference between the predicted values and the actual values in a dataset, expressed in the same units as the predicted variable.
Mean absolute percentage error (MAPE)	MAPE is a common metric for association models. It measures the percentage difference between the predicted values and the actual values in a dataset.
Relative squared error (RSE)	RSE is a common metric for association models. It measures the relative improvement of the model's predictions compared to a simple baseline model that always predicts the mean of the actual values.
Explanatory power (EP)	EP refers to the ability of a predictive model, especially association model, to explain or predict the variability observed in the data. It measures how well the model captures the underlying relationships between the input variables (predictors) and the target variable.

(Continued)

TABLE 6.1 *(Continued)*
Basic Concepts in Model Performance Evaluation

Terms	Descriptions
Actual by Predicted Chart	The *Actual by Predicted Chart* compares the actual values with the predicted values, showing us how well the trained model fits the data.
Residuals	Residuals measure the difference between actual values and predicted values in a dataset.
Residual by predicted chart	The residual by predicted chart presents not only how close the actual values are to the predicted values but also indicates any patterns in the residuals.
Confusion matrix	Confusion matrix, also called the classification matrix, is the table that summarizes the performance of a classification algorithm.
True positive (TP)	TP is the prediction that correctly indicates the presence of the event.
True negative (TN)	TN is the prediction that correctly indicates the absence of the event.
False positive (FP)	FP is the prediction that wrongly indicates the presence of the event.
False negative (FN)	FN is the prediction that wrongly indicates the absence of the event.
Misclassification rate (MR)	MR is a measure of how often a model misclassifies (makes incorrect predictions) the target variable (including actual positives and actual negatives).
Prediction accuracy (ACC)	ACC is a measure of how accurate a model predicts the target variables (including actual positives and actual negatives).
Sensitivity, recall, or true positive rate (TPR)	Sensitivity is the probability of a positive result (event) or how well a model can predict true positives (events).
Specificity, selectivity, or true negative rate (TNR)	Specificity is the probability of a negative result (non-event) or how well a model can predict true negatives (non-events).
F1 score	F1 score is the harmonic mean of the sensitivity (recall) and precision (positive predicted value); a high F1 score means low false positives (FP) and low false negatives (FN).
Receiver operating characteristic (ROC) chart	The ROC chart displays the true positive rate on the vertical axis and the false positive rate on the horizontal axis, reflecting the balance between sensitivity and specificity.
Lift chart	Lift is an indicator of a predictive model's effectiveness, calculated as the ratio of the results achieved with the model compared to those without it. The Lift chart, or Lift curve, presents the graphic representation between the lift values and response rates.
Gini coefficient	Gini coefficient is a measure of the model's discriminatory power, which is the effectiveness of the model in differentiating between positives and negatives. The Gini coefficient measures how equal the positive rates are across the attributes of a characteristic.
Cost sensitive learning	Cost sensitive learning is a method to evaluate and select a model based on the costs of misclassification.
Decision matrix	Decision matrix is the table that presents the costs of misclassifications, including costs of false positives and costs of false negatives.

MODEL FIT AND PREDICTIVE POWER

First, let's clarify some terminology. **Model fit** and **predictive power** are commonly used terms in data mining, but their distinction is often unclear. In many books and online resources, these terms are used interchangeably, leading to confusion. While they do overlap, they serve different purposes. **Model fit** is more commonly associated with traditional statistical analyses, whereas **predictive power** is used specifically in the context of predictive modeling. Although predictive power is less precise than model fit, it remains a crucial concept in

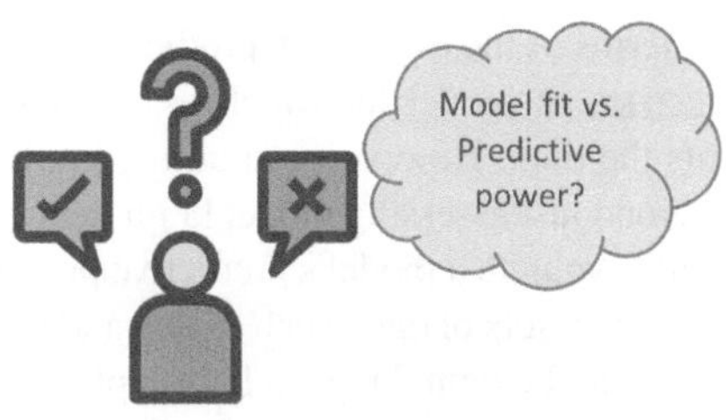

data mining. I will differentiate these terms based on their purposes rather than their measurement since there are overlaps in how they are assessed.

Model Fit

What is model fit? Imagine a model, represented by mathematical expressions, logic, or rules, that describes the relationship between the target variable and its predictors. This model is used to predict the value of the target variable, whether it is developed from a theoretical foundation or derived from a machine learning algorithm. The critical question is how well this model fits the actual data, which we refer to as **model fit**. The difference between the model's predictions and the actual data is called **residuals** in association models or **misclassification** in classification models. The smaller the residuals or misclassifications, the better the model fit.

> **DEFINITION**
>
> Model fit measures how well the model fits the data. It assesses the degree to which the model captures the patterns and relationships within the data.

Model fit is assessed through **goodness-of-fit** metrics, which estimate how well the model predicts the target variable based on the predictors. Some goodness-of-fit metrics are universal and apply to all models, while others are specific to certain machine learning methods. For instance, **prediction accuracy, misclassification rate, sensitivity, specificity, Lift charts,** and **ROC charts** are applicable to all classification models. Similarly, **root mean squared error, mean absolute percentage error,** and **residual by predicted charts** apply to all association models. However, metrics like the **F-test** and **R^2** are specific to linear regression models, while **−2LL difference** and **pseudo R^2** are used for logistic regression models. It's important to recognize both general model fit metrics and specific ones, with the former being more common since not all machine learning methods have their own unique metrics.

It's important to note that the term **goodness-of-fit** is commonly used in other statistics books and materials. I understand this shift in terminology may cause some confusion. Essentially, **goodness-of-fit indices** are specific metrics used to evaluate model fit. To maintain consistency and simplicity throughout this book, I will primarily use the term **model fit** instead of **goodness-of-fit**. Likewise, I will refer to **model fit metrics** instead of **goodness-of-fit indices**. This approach should make it easier for you to follow along.

> We use the term **model fit** consistently throughout the book to maintain clarity. This term is synonymous with **goodness-of-fit** and refers to the same concept of how well a model captures the underlying patterns in the data.

One important thing to note is that model fit is typically evaluated during the model training phase using the training sample. However, achieving a good model fit on the training data does not guarantee that the model will perform well in real-world predictive tasks. This is where predictive power comes into play.

Predictive Power

In terms of measurement, predictive power uses the same metrics as model fit, which assess how well the model fits the data. However, evaluating predictive power goes beyond just assessing model fit on the training sample. To truly evaluate a model's predictive power, we need to test it on new sets of data, such as test and validation samples. This can be confusing, so let's explore an example.

> **DEFINITION**
>
> Predictive power refers to a model's ability to accurately capture and represent the underlying patterns, relationships, or trends in the data, and effectively generalize these results to new, unseen datasets.

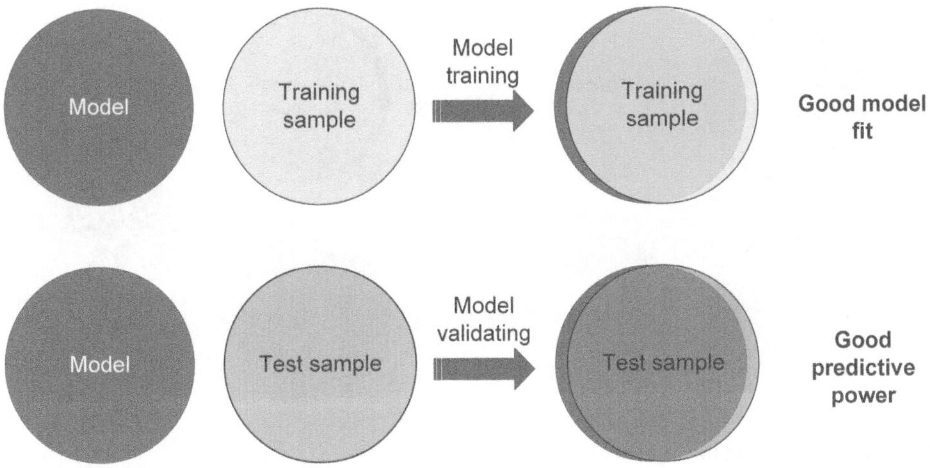

FIGURE 6.1 Model fit and predictive power.

Suppose we build and train a predictive model using a training sample, and we achieve a good model fit, indicating the model fits the training data well. Next, we want to see if this model works with new data. To do this, we apply the model to a test sample and, later, to a validation sample for the final model assessment. Does the model still fit the data well? If yes, then the model is generalizable and can make good predictions with real-world data, meaning it has good predictive power. If not, the model lacks predictive power because it fails to generalize to different datasets, making it unreliable for real-world use. In this case, the model may suffer from **overfitting**, which we will discuss shortly.

In short, while model fit and predictive power use similar metrics, they serve different purposes. Model fit evaluates the model's performance on a specific dataset, such as the training data, while predictive power assesses the model's performance on new data, ensuring it can be generalized to other datasets. In essence, predictive power ensures that the model can make accurate predictions in real-life scenarios. Predictive power is a broader concept than model fit because it requires the model to have a good fit across all datasets, not just the training set. Achieving good predictive power should be our ultimate goal.

> Model fit ensures that the model accurately fits a specific dataset.
>
> Predictive power indicates that the model can generalize and fit multiple datasets, including new or unseen data.

Figure 6.1 illustrates how model fit and predictive power are measured. To keep things simple, the figure only includes training and test samples. In the first case, we have a good model fit with the training sample (the model fits the training data well). However, this does not guarantee good predictive power, as we still need to ensure the model fits the test sample, indicating its ability to generalize to other datasets. Before evaluating predictive power, we must first achieve a good model fit with the training data—if the trained model doesn't fit the training data, it's unlikely to fit other datasets, leading to poor predictive power.

OVERFITTING

Overfitting is a significant concern in both statistical analyses and predictive modeling because it affects a model's predictive power. It is a common issue, especially when working with large datasets in data mining. However, overfitting can be a confusing concept, often inconsistently explained in various materials, leading to reader confusion. This confusion arises from a lack of clear definitions or explanations of the problem.

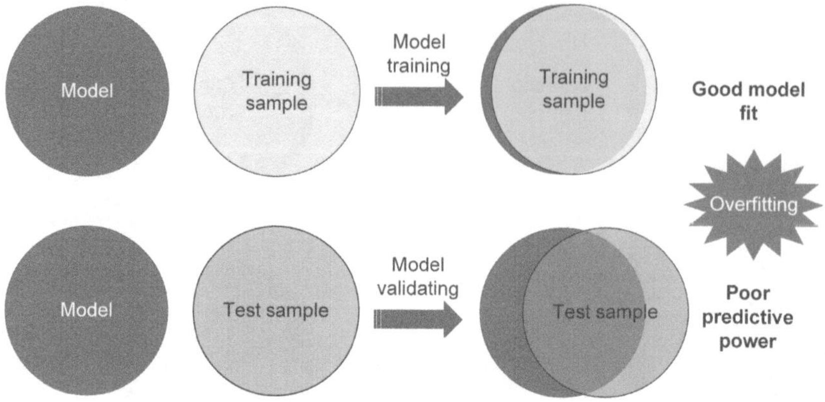

FIGURE 6.2 Overfitting illustration.

What Is Overfitting?

The distinction between **model fit** and **predictive power** helps clarify the concept of overfitting. To break it down step by step, Figure 6.2 illustrates this problem visually. By following these steps, we can identify overfitting in a data mining project:

> **DEFINITION**
>
> Overfitting occurs when a model is overly trained to fit the training sample, capturing even the noise and minor fluctuations in the data. As a result, the model fails to generalize to other datasets, making it invalid and unusable for real-world predictions.

1. We train our model using a training sample and achieve an excellent model fit, meaning the model fits the training data very well.
2. Next, we test the model with a new dataset (the test sample), but the model fit is very poor. What happened?
3. We try again with another dataset (the validation sample), but the result remains unsatisfactory—the model fit is still poor.
4. What does this mean? It means our model is biased toward the training sample and only works well with that particular dataset. It cannot generalize to new datasets and, therefore, lacks predictive power, making it unusable for real-world predictions.
5. What caused this? The model was overtrained on the training sample, capturing even the noise and minor fluctuations in the data, leading to overfitting. This issue could stem from several factors, such as an imbalanced dataset, sampling bias in the training sample, or improper configurations of the machine learning algorithm used.

How to Address Overfitting?

Now that we can identify an overfitting problem, the next step is to address it. When a model suffers from overfitting, it becomes invalid and unusable. So, how can we prevent or resolve this issue? There is no one-size-fits-all solution, but several strategies can help. Table 6.2 outlines common solutions for preventing and addressing overfitting.

By applying these strategies, we can minimize overfitting and ensure that the model generalizes well to new datasets, improving its performance in real-world applications.

TABLE 6.2
Solutions for Overfitting

Solution	Description
Perform data sampling correctly	Using the correct sampling method ensures that we obtain a dataset that is generalizable to the target population. This means that the dataset shares similar trends and characteristics with the population we want to predict. If we use a biased training sample, the model will also be biased. By using a quality and representative dataset to train the model, we can avoid overfitting and ensure the model works well with new data.
Get a large dataset	Training predictive models often requires large datasets, especially since some machine learning algorithms need substantial data to function properly. A small dataset may lead to a poor model fit, as it doesn't provide enough information for the model to learn. Having a sufficiently large sample size increases the model's explanatory power, especially when dealing with many variables. Additionally, a large dataset allows us to split the data into multiple sub-samples for validation. However, it's important to avoid overtraining the model, which could still lead to overfitting.
Clean the data well	A clean dataset is crucial for model training. The saying "garbage in, garbage out" holds true in data mining. A model can only perform as well as the quality of the input data allows. If the data are full of noise or irrelevant information, the machine learning algorithm may attempt to fit these distractions, leading to poor predictive power. We need to carefully examine the data, correct errors, remove excessive noise, irrelevant predictors, and handle missing values or outliers. Data cleaning is one of the most time-consuming tasks in data mining, often accounting for 60–80% of the project's time. Despite its effort, it plays a vital role in avoiding overfitting.
Partition data properly	It is essential to partition the dataset into sub-samples for training, testing, and validation purposes. Proper partitioning ensures that the sub-samples reflect the trends and characteristics of the original data. This minimizes biases in the model training and validation process, reducing the chances of overfitting. Without proper partitioning, the model may perform well on one sample but fail to generalize to others, which is a clear sign of overfitting.
Maintain balance in the data	A highly imbalanced dataset can lead to overfitting because the model may focus only on the majority class. In this case, the model may predict most outcomes as negative cases, leading to high apparent accuracy, but it fails to predict the minority class accurately. This makes the model invalid. An imbalanced dataset is a primary cause of overfitting, especially in classification problems. If the dataset is imbalanced, resampling techniques or other strategies should be used to balance the class distribution, which helps the model avoid overfitting.
Use multiple model fit metrics	Relying on a single model fit metric can be misleading since each metric measures only one aspect of model performance. For example, accuracy alone might not tell the full story if the dataset is imbalanced. Using multiple metrics allows for a more comprehensive evaluation of the model's fit. This multi-metric approach helps us identify weaknesses in the model and provides strategies for improvement, ultimately reducing the risk of overfitting.
Use cost-sensitive learning	In real-world applications, the costs associated with misclassifications are rarely equal. For example, in medical diagnostics, the cost of a false negative (missing a serious condition) is much higher than the cost of a false positive. Therefore, choosing a model purely based on overall accuracy is not always the best approach. By incorporating the costs of misclassification into the model evaluation, we can select the model that minimizes these costs, leading to better real-world performance and less risk of overfitting.
Use an ensemble model	Sometimes, even after trying various machine learning methods, no single model provides the best fit for the data. Different models might even give conflicting predictions. In such cases, using an ensemble model—where predictions from multiple models are combined to create a consensus—can help. Ensemble models are generally more accurate and robust than individual models, as they leverage the strengths of each model while reducing their individual weaknesses, effectively mitigating the risk of overfitting.

MODEL RELIABILITY AND VALIDITY

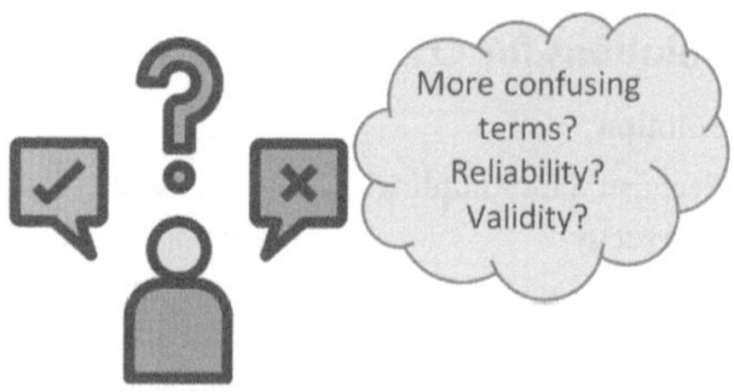

A common question we often receive is about the quality of our predictive model: **Is it reliable and valid?** But what exactly is the difference between model reliability and validity? These concepts may seem simple, and while they are frequently mentioned in statistics and research methodology, they are rarely clearly explained in the context of data mining. Understanding the distinction between **reliability** and **validity** is crucial for evaluating predictive models correctly. I will break down these concepts in a straightforward way to show why both are essential.

In essence, **reliability** refers to the internal consistency of a model, while **validity** indicates the model's accuracy. A **reliable model** should consistently produce similar results when run on different datasets. In contrast, a **valid model** should make accurate predictions of the target variable. Both are important, but reliability alone does not guarantee validity. Therefore, we must evaluate both aspects to ensure the quality of the model.

Still unclear? Let's use a simple analogy: imagine you're using a scale to measure the weight of a 25 lb bag of rice. A reliable scale would give you consistent readings each time you weigh the bag. However, reliability doesn't necessarily mean the scale is accurate. For instance, if the scale consistently shows a weight between 22.1 and 22.8 lbs, it is reliable but not valid because the readings are off by about 2 lbs. In this case, the scale is consistent but inaccurate, meaning it is reliable but not valid. Ideally, we want a reliable scale that also accurately reads 25 lbs.

Now, you may wonder: does having a valid scale imply reliability as well? The answer is **Yes**, but you won't know for sure until you test it multiple times. For example, if you buy a new scale, and it gives an accurate reading of 25 lbs on the first try, it seems promising. However, to confirm its reliability, you would need to weigh the bag several times. If the scale continues to provide accurate readings, then it is both reliable and valid. But if the readings fluctuate, the scale is neither reliable nor valid.

> **DEFINITION**
>
> **Model reliability** refers to the internal consistency of a model. It means the model produces consistent results across different runs using different datasets.
>
> **Model validity** refers to the accuracy of the model. It means the model consistently produces an accurate prediction of the target variable.

In data mining, this is akin to overfitting, where a model predicts well with the training data but fails to generalize to other datasets. This is why evaluating both model reliability and validity is essential. We cannot fully confirm a model's validity until it demonstrates reliability. In other words, for a model to be valid, it must consistently produce accurate predictions of the target variable. Essentially, model validity is the same as predictive power. Figure 6.3 illustrates a situation where we have model reliability but not validity.

With a clear understanding of the difference between reliability and validity, the next step is to learn how to assess these qualities in a predictive model. It's important to note that we cannot assess reliability and validity through separate processes since the same model fit metrics are used for both. Therefore, they must be evaluated within the same process, even though we provide separate evidence for each.

MODEL RELIABILITY AND VALIDITY ASSESSMENT PROCESS

Follow the steps below to evaluate the reliability and validity of a predictive model:

1. **Split the sample** into sub-samples, including training, test, and validation sets. The recommended split ratio is 60–40, with 60% for training, 20% for testing, and 20% for final validation.
2. **Configure a predictive model** based on the chosen machine learning method (details on specific methods will be covered in the next chapters).

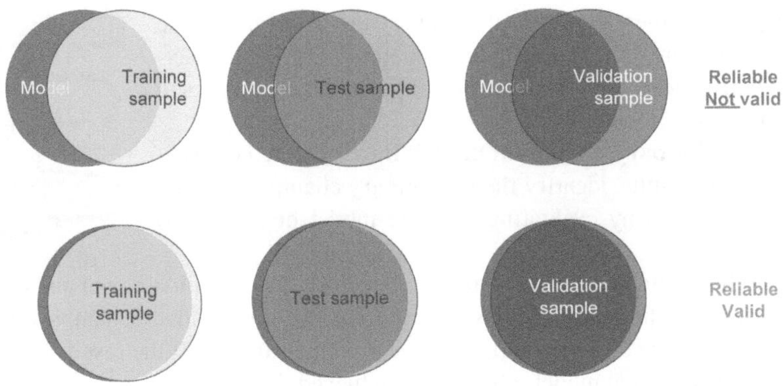

FIGURE 6.3 Model reliability and validity.

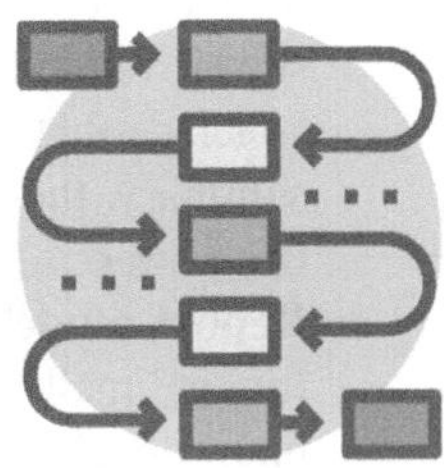

3. **Train the model** using the training sample and then test it with the test sample. Most data mining software automates this step.
4. **Evaluate the model fit** of the training sample using pre-determined metrics (discussed in the following sections). If the model fit is satisfactory, proceed to Step 5. If not, return to Step 2 to reconfigure the model.
5. **Compare the model fit metrics** between the training and test samples. Use the same metrics to assess both.
6. **Assess model reliability** by comparing two samples. If the comparison shows consistency between the training and test samples, you can conclude that the model is reliable. The level of acceptable consistency depends on the context of the study, but if the discrepancy is less than 5%, it can be considered reliable. **Conclusion**: Model reliability is achieved.
7. **If model reliability is not achieved**, adjust the sampling and modeling processes in Steps 1 and 2. Solutions may include resampling data, reconfiguring the model, trying alternative machine learning methods, or using an ensemble model. Repeat Steps 2–5 until reliability is achieved.
8. **Evaluate the overall model fit** for the test sample to confirm it meets the desired criteria (additional metrics may be used). If the model fit is satisfactory, move to Step 9. If not, further adjustments are needed to improve performance.
9. **Evaluate the model fit for the validation sample** as the final assessment. If the model shows a good fit with the validation sample, then model validity is achieved. **Conclusion**: Model validity is achieved.

How to Report Model Reliability and Validity Results

Suppose we have completed the predictive modeling process and achieved both model reliability and validity. How do we report the results effectively? Since the process is iterative, presenting results for every step could make the report overly complex and potentially confusing for readers. To avoid this, I recommend the following steps to present the results clearly and effectively. These steps are particularly useful for projects involving multiple models.

1. **Present model comparison results**: If there are multiple models, compare their performance using key model fit metrics from the test sample. Explain whether the model fit is satisfactory and why. While it is necessary to evaluate the model fit for the training sample

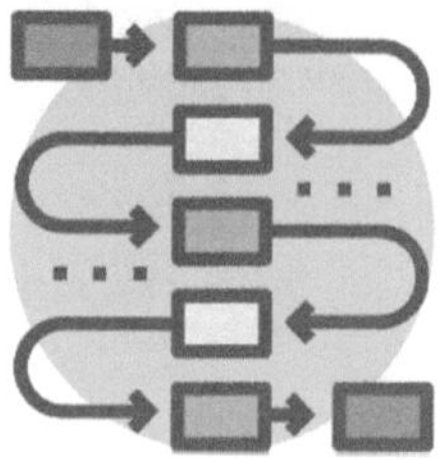

during the process, there is no need to report the training sample results separately. Since the test sample evaluates the model's generalization, reporting test results will be most informative for readers to understand the tested model fit.

2. **Identify the preliminary champion model**: Based on the comparison of test sample results, identify the preliminary champion model. Confirm this selection by evaluating multiple model fit metrics to ensure robustness.
3. **Compare model fit metrics across samples**: Focus on the champion model and present a comparison of model fit metrics across the training, test, and validation samples. Explain whether the results are consistent and reliable across the samples. If the results are consistent, you can conclude that model reliability is achieved.
4. **Evaluate complete model fit**: Present additional model fit metrics, particularly focusing on the validation sample. This step ensures a thorough evaluation of the model fit and confirms whether the champion model fits well with the validation data.
5. **Conclude model validity**: If the model fit metrics for the test sample are satisfactory and the model shows reliability across the samples, you can conclude that the model has achieved validity. In summary, the champion model has good predictive power.
6. **Report model structure and variable importance**: Finally, explain the structure of the champion model, the relationships between variables, and the importance of each variable in the model.

I will use this process to present predictive modeling results in most chapters. By following these steps, we ensure a consistent approach to evaluating predictive models' reliability and validity.

MODEL COMPARISON

In data mining, the goal is to discover unknown patterns from large and often noisy datasets. This process involves mining the data to uncover new insights and relationships among variables that may not be found in existing literature or identifying new predictors. Because of the exploratory nature of data mining, it's difficult to predict in advance which machine learning methods and configurations will work best. The ideal machine learning model depends on the data itself. Our objective is to find the model that best fits the data. Importantly, using a more advanced or complex method does not guarantee that it will be the best model.

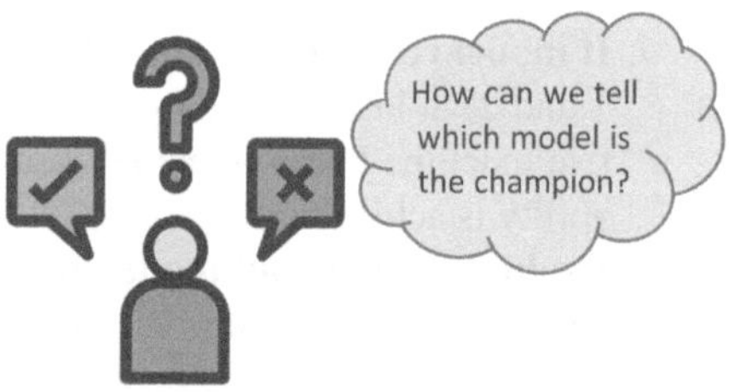

Let's consider a hypothetical scenario illustrated in Figure 6.4. Imagine we build two models for the same dataset. One model uses an advanced method, such as neural networks (on the right), which captures nonlinear relationships between predictors and the target variable and seems to fit the data very well. The other model uses a simpler method, linear regression (on the left), which assumes a linear relationship between the predictors and the target variable.

During the model training process, the neural network model may appear to outperform the linear regression model based on prediction accuracy. However, if we examine the data visually using a scatter plot, we might notice that the data follows a linear pattern. This suggests that, despite being simpler, the linear regression model captures the underlying pattern better than the more complex neural network model.

When we test both models with a new dataset (the test sample), we find that the linear regression model consistently provides a good model fit. In contrast, the neural network model's fit for the

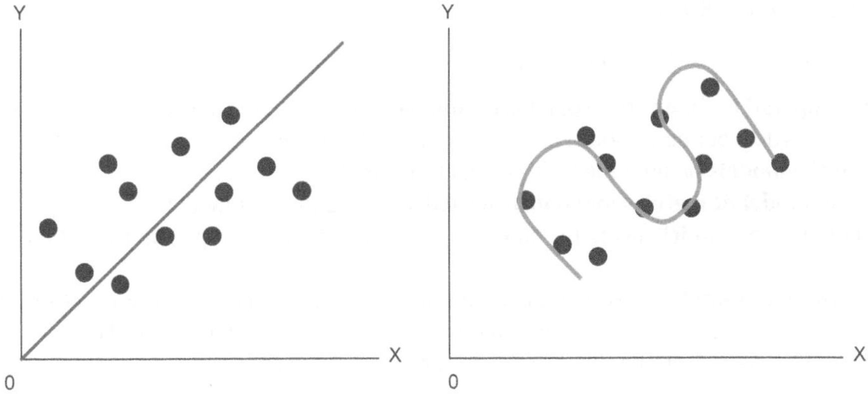

FIGURE 6.4 Simple linear model vs. a complex, nonlinear model—A hypothetical scenario.

test sample is significantly worse than it was during training. What's happening here? The neural network model is likely overfitting the training data, meaning it tries to account for noise in the data and incorporates this noise into the model. While this results in a nearly perfect fit for the training data, it performs poorly when applied to the test sample because the noise is not part of the actual pattern in the data.

In this scenario, the nonlinear relationships derived by the neural network model do not capture the true pattern, and the model is not generalizable—it has poor predictive power. Therefore, despite the complexity of the neural network model, the linear regression model is the better choice for this dataset.

This example demonstrates that it is difficult to pre-determine or speculate which machine learning method will perform best without applying various methods to the actual data. Otherwise, we could always default to the most advanced method and avoid considering simpler approaches. Similarly, even when using the same method, we can create numerous analytical configurations to find a model with strong predictive power. For example, in the scenario above, we could reconfigure the neural network model to capture the linear pattern rather than the nonlinear one. In fact, the simplest form of a neural network closely resembles linear regression.

> There is no such thing as a perfect data mining method. However, there is always a method that works best for a specific dataset, depending on the characteristics of the data and the goals of the analysis.

With that in mind, the best practice for finding a model with good predictive power is to build multiple models using different methods and configurations. Then, we compare the performance of these models using model fit metrics to identify the one with the best overall performance. This is known as the **champion model**, which can be used as the final predictive model for real-life predictions. However, it is important to note that this is only the champion model among the models we trained and tested with the given data. There is still a possibility that the champion model does not have the expected predictive power, particularly if none of the models we trained have a good model fit. In this case, the best-performing model is relatively better but may not be ideal. Therefore, we must still evaluate the champion model and decide whether further adjustments are needed or if additional models should be trained.

> **DEFINITION**
>
> Champion model is the model that demonstrates the best performance among all the models we train and test.

> Build and train multiple models and compare them to find the champion model.

Model Comparison Steps

Follow the steps below to build and compare models:

1. **Develop and train multiple predictive models** using different machine learning methods and analytical configurations appropriate for the given data.
2. **Run the models** on both the training and test samples.
3. **Select model fit metrics** for comparing and selecting the best model.
4. **Compare the models** using the chosen metrics (for the test sample) and select the champion model.
5. **Examine the model fit** of the champion model to determine if it is satisfactory. If not, adjust the configurations of existing models or consider adding new models.
6. **Repeat Steps 1–5** until a satisfactory champion model is achieved.
7. **Evaluate the model fit** of the champion model with the validation sample to confirm its predictive power.

The champion model is simply the best model among those we train and test with the given data. However, there is no guarantee that it has good predictive power. We need to examine it carefully to determine whether additional models should be trained.

MODEL FIT METRICS

We have discussed model fit and predictive power, but how do we evaluate them? In other words, what specific metrics should we use to assess model fit? As mentioned earlier; to avoid confusion, I use the term "model fit" instead of "goodness-of-fit" and "model fit metrics" instead of "goodness-of-fit indices." The same metrics are used to evaluate predictive power as well. These metrics also help in comparing models and identifying the champion model.

It's important to note that different metrics are required for association models and classification models due to the differences in their target variables. As a reminder, association models have continuous target variables, while classification models have binary or nominal target variables. Therefore, I will break down the model fit metrics for each type of model. Table 6.3 summarizes these metrics.

TABLE 6.3
Model Fit Metrics Summary

	Association Models	Classification Model
Target variable	Continuous	Binary or nominal
General model fit metrics	• Root mean squared error (MSE) • Mean absolute percent error (MAPE) • Relative squared error (RSE) • Explanatory power (EP) • Actual by Predicted Chart • Residual by predicted chart	• Misclassification rate • Prediction accuracy • Specificity • Sensitivity • F1 score • ROC chart (only available for models with binary target variables) or ROC index • Lift chart or Lift value • Gini coefficient
Method-specific model fit metrics	Additional model fit metrics are determined based on the specific machine learning method used, which we will cover in the upcoming chapters. However, note that not all machine learning methods require additional metrics.	

In this chapter, we focus on general model fit metrics that apply to all machine learning methods within each category. Each method may have its own specific model fit metrics, which will be described in detail in the relevant chapters (though not all machine learning methods have their own specific metrics).

MODEL FIT METRICS FOR ASSOCIATION MODELS

Model fit metrics are used to assess how well our trained model aligns with the data, typically measured by either the accuracy or the error of its predictions. For association models with continuous target variables, these metrics can be categorized into error measures and charts. Error measures allow for a numerical evaluation of model fit, while charts provide a visual examination of the model's performance.

Prediction Error Measures

Prediction errors reflect how much our trained model deviates from fitting the data. Generally, the lower the error, the more accurate the model. There are numerous error measures in statistics, ranging from simple to more complex. Using all of them can be excessive and unnecessary, as many are quite similar. Here, I will describe a few common error measures and recommend the most practical ones. The following formulas outline these common error measures and demonstrate how to calculate them using actual values, predicted values, and mean values:

Sum of squared error (SSE)

$$SSE = Sum\ of\ (Actual - Predicted)^2$$

Sum of squared regression (SSR)

$$SSR = Sum\ of\ (Predicted - Mean)^2$$

Sum of squared total (SST)

$$SST = Sum\ of\ (Actual - Mean)^2$$

Mean absolute error (MAE)

$$MAE = \frac{Sum\ of\ |Actual - Predicted|}{Sample\ size}$$

Mean squared error (MSE)

$$MSE = \frac{Sum\ of\ (Actual - Predicted)^2}{Sample\ size}$$

Root mean squared error (RMSE)

$$RMSE = \sqrt{\frac{Sum\ of\ (Actual - Predicted)^2}{Sample\ size}}$$

Mean absolute percent error (MAPE)

$$MAPE = \frac{Sum\ of\ \left|\frac{Actual - Predicted}{Actual}\right|}{Sample\ size}$$

Relative squared error (RSE)

$$RSE = \frac{Sum\ of\ (Actual - Predicted)^2}{Sum\ of\ (Actual - Mean)^2} = \frac{SSE}{SST} = \frac{Unexplained\ Variance}{Total\ Variance}$$

Model's explanatory power (EP)

$$EP = 1 - RSE = \frac{Explained\ Variance}{Total\ Variance}$$

We can observe that SSE, SSR, and SST are fundamental components of other error measures, even though they are not typically used as final model fit metrics. Therefore, I will explain what each represents:

- **Sum of squared error (SSE)**—*unexplained variance*: This measures how far the actual data points are from the model's predicted values.
- **Sum of squared regression (SSR)**—*explained variance*: This measures how far the predicted values are from the mean of the actual data.
- **Sum of squared total (SST)**—*total variance*: This measures how far the actual data points are from the mean of the dataset.

It is important to note that in these measures, the mean represents the average of the actual values in the dataset and reflects the expected prediction if no model is used. Figure 6.5 illustrates the relationship between SSE, SSR, and SST using linear regression as an example.

> Note that the relationship among SSE, SSR, and SST shown in Figure 6.5 is accurate only for linear regression. The relationship may differ when dealing with nonlinear patterns. Therefore, in such cases, we should rely on SSE and SST to calculate metrics like RSE and EP.

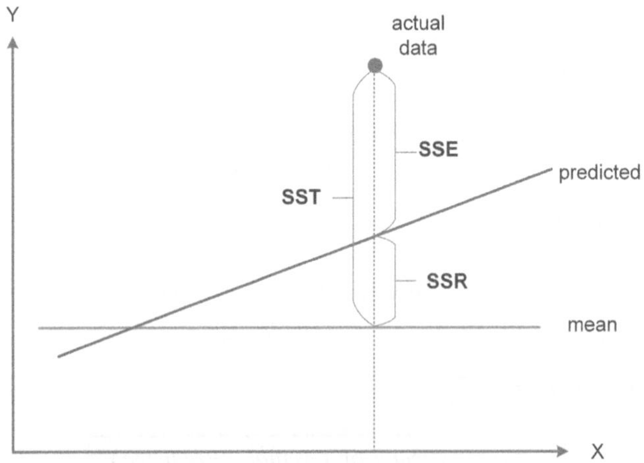

FIGURE 6.5 SSE, SSR, and SST for *linear regression.*

Let's discuss some popular model fit metrics. **Mean absolute error (MAE)** is the average of all absolute errors, representing the average distance between actual and predicted values. MAE is easy to interpret because it provides the average absolute error in the same scale as the target variable. However, the main drawback of MAE is that it treats all errors equally and does not penalize large errors more heavily.

Mean squared error (MSE) is similar to MAE but uses squared errors instead of absolute errors. MSE is the average of all squared errors, and its key difference from MAE is that MSE penalizes large errors more due to squaring the errors. This can be important when prioritizing models that minimize larger errors. Thus, MSE can offer a better measure of model fit in some cases. The drawback of MSE is that, since it squares the errors, the result is not on the same scale as the data.

Root mean squared error (RMSE) addresses this issue by taking the square root of MSE, placing the result on the same scale as the data. For this reason, I recommend using RMSE.

RMSE is **scale-dependent**, meaning it can only be used to compare prediction errors across models with the same target variable and scale. RMSE averages out the effect of sample size and neutralizes the effect of the sign by squaring the differences. However, RMSE does not directly indicate prediction accuracy, as it is not expressed in percentage terms. For instance, if one model has an RMSE of 24.12 and another has an RMSE of 29.36, we can conclude that the first model fits better. However, we cannot interpret 24.12 in terms of prediction accuracy because it represents a root mean squared error, which is less intuitive.

In this context, **mean absolute percent error (MAPE)** is a better measure for comparing prediction accuracy because it is expressed as a percentage. MAPE is scale-independent and tells us the percent error, with lower values indicating higher prediction accuracy. It also accounts for sample size by using the mean and neutralizes the sign by using absolute values. There is no standard cutoff for MAPE, but my rule of thumb is that a MAPE of 0.05 or lower indicates a great model fit, and a MAPE of 0.10 or lower indicates a good model fit.

In addition to MAPE, **relative squared error (RSE)** is another scale-independent metric, also expressed as a percentage. As shown in the formula, RSE is the ratio of SSE to SST, indicating the percentage of unexplained variance in the model. RSE is useful for calculating **Explanatory Power (EP)**, which is the ratio of SSR to SST and indicates the percentage of explained variance in the model. EP is calculated as 1 − RSE.

EP is an important metric because it measures how well the model explains the variance in the target variable based on the variance in the predictors. For example, an EP of 92% means the model explains 92% of the variance in the target variable. I strongly recommend using both RSE and EP to evaluate model fit.

While there is no standard rule for EP, my rule of thumb is that an EP of 90% or higher is considered a great fit, and an EP between 70% and 90% is considered a good fit. In most cases, an EP of 50–70% is acceptable, particularly when the model includes numerous predictors. Table 6.4 presents my recommended model fit metrics and cutoff values.

> While most software programs generate MSE and RMSE values, they typically do not provide RSE and EP directly in the output. In such cases, we need to export the output data to Excel and perform the calculations manually using the appropriate formulas.

Actual by Predicted Chart and Residuals by Predicted Chart

In addition to error measures, two important charts can be used to evaluate the model fit for association models: the **Actual by Predicted Chart** and the **Residuals by Predicted Chart**. These

TABLE 6.4
Recommended Model Fit Metrics for Association Models

Metrics	Descriptions	Recommended Cutoff Values
RMSE	Scale-dependent; a measure of the average difference between the predicted values and the actual values in a dataset.	None; depending on the scale of the data
MAPE	Scale-independent; a measure of the percentage difference between the predicted values and the actual values in a dataset.	MAPE = 0.05 or lower: Great MAPE = 0.05–0.1: Good MAPE = 0.1–0.15: Acceptable
RSE	Scale-independent; a measure of the relative improvement of the model's predictions compared to a simple baseline model that always predicts the mean of the actual values.	RSE = 0.1 or lower: Great RSE = 0.1–0.3: Good RSE = 0.3–0.5: Acceptable
EP	Scale-independent; a measure of how well the model captures the underlying relationships between the input variables (predictors) and the target variable.	EP = 90% or higher: Great EP = 70–90%: Good EP = 50–70%: Acceptable

visualizations provide valuable insights into how well the model fits the data and can help assess the performance of a predictive model.

The **Actual by Predicted Chart** compares actual values with predicted values, helping us understand how well the trained model fits the data. In this chart, the actual values are plotted on the y-axis and the predicted values on the x-axis. The closer the data points are to the diagonal line, the better the model fit, as this indicates small residuals between actual and predicted values (see Figure 6.6 for an example). If the points are scattered far from the diagonal line, it may indicate a poor model fit.

Residuals are calculated as the difference between actual and predicted values. The **Residuals by Predicted Chart** is crucial because it not only shows how close the actual values are to the predicted values but also reveals any patterns in the residuals. In this chart, the residuals are plotted on the y-axis and the predicted values on the x-axis. The distance between each point and the line represents the accuracy of the prediction. Ideally, this distance should not be too large, ensuring the model's predictions are reasonably accurate.

Additionally, it's important to examine this chart for any apparent patterns. A **non-random pattern** indicates a correlation between residuals and predicted values, suggesting that some

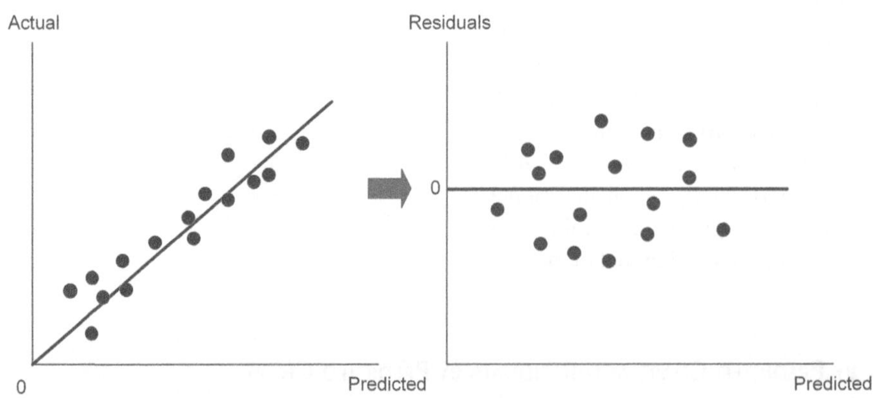

FIGURE 6.6 Actual by Predicted Chart and Residual by Predicted Chart—An example of a good model fit.

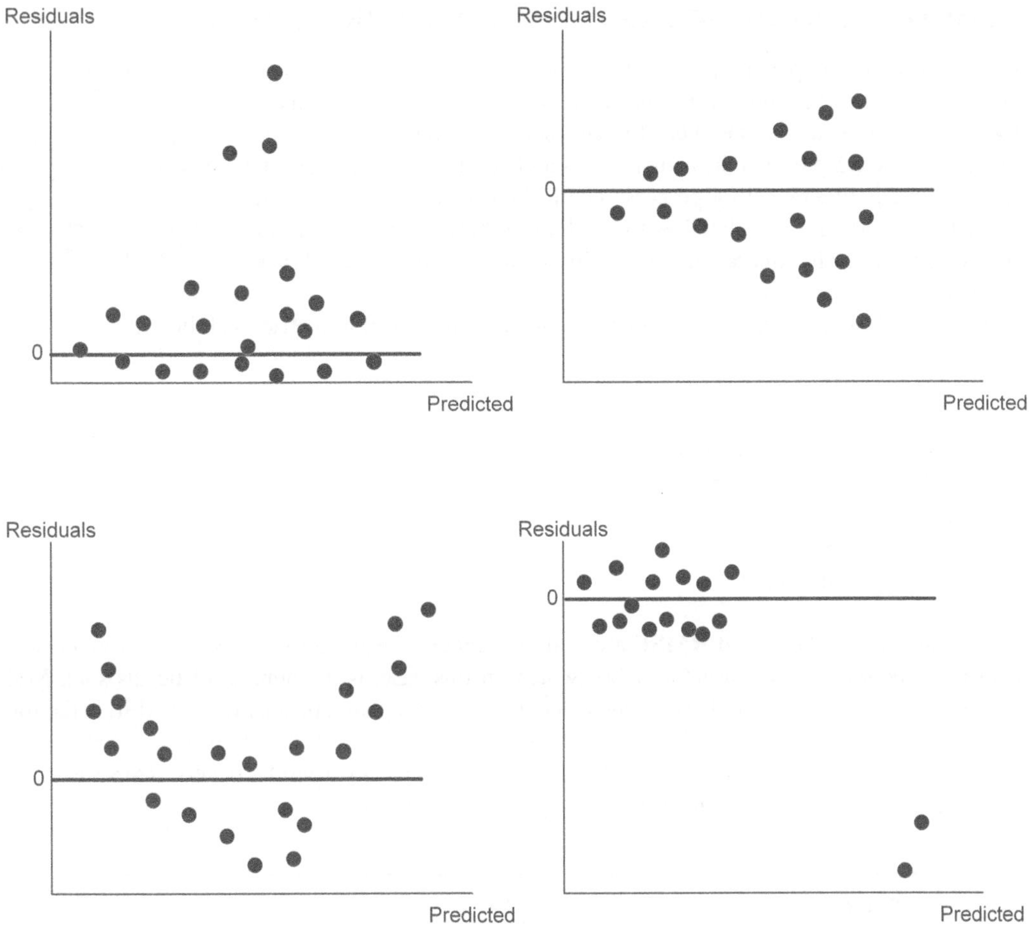

FIGURE 6.7 The residual by predicted charts with non-random patterns.

explanatory information is missing from the model. This implies that the predictors in the model may not be fully capturing certain variables, which are instead reflected in the residuals. Typically, we aim for a chart with **no clear pattern**—a random distribution, also known as "white noise." To meet this requirement, points should be symmetrically distributed and clustered toward the center of the plot (see Figure 6.6).

> Note that not all software programs generate these charts automatically. In such cases, we must manually create the charts using the output data.

Figure 6.7 presents several examples of scatter plots with non-random patterns, such as an uneven distribution of points (Chart A), clear shapes (Charts B and C), and outliers (Chart D). In these cases, the model does not achieve a satisfactory fit and requires improvement. Potential solutions could include reconfiguring the model, using an ensemble model, resampling to add more data points, or incorporating additional predictors.

Overall, I recommend using **RMSE**, **MAPE**, **RSE**, and **EP** to compare models, and using **MAPE**, **RSE**, **EP**, along with the **Actual by Predicted Chart** and the **Residuals by Predicted Chart** to evaluate the model fit of a single model.

EXAMPLE: MODEL FIT METRICS FOR AN ASSOCIATION MODEL

Let's review an example of an association model and see how these metrics are calculated and interpreted. For instance, we built and trained a predictive model to forecast the number of tickets sold for a nighttime boat tour in a city. Table 6.5 presents the actual values and predicted values resulting from the predictive modeling process. The residuals, which represent the prediction errors, are calculated as the difference between the actual and predicted values. The squared errors are simply the squared values of these residuals.

Using the formulas presented earlier, we calculate the model fit metrics as follows:

- SSE = 68,037
- SST = 220,908.95
- MSE = 68,037/20 = 3,401.85
- RMSE = $\sqrt{3,401.85}$ = 58.33
- MAPE = 0.066
- RSE = SSE/SST = 0.308
- EP = 1 − RSE = 0.692

As you can see, **MSE** and **RMSE** are scale-dependent, meaning they must be interpreted with respect to the scale of the target variable, which, in this case, is the number of tickets sold. MSE represents the average squared error, calculated as 3,401.85 squared tickets, while RMSE is the root mean squared error, which is 58.33 tickets. These metrics, particularly RMSE, heavily penalize large errors, as seen in cases 2, 5, 7, 11, and 15, where absolute errors exceed 80 tickets. RMSE is useful for comparing models but doesn't easily convey prediction accuracy.

TABLE 6.5
Actual and Predicted Values

No	Actual Values	Predicted Values	Residuals (Errors)	Squared Errors
1	779	750	29	841
2	958	878	80	6,400
3	860	850	10	100
4	882	821	61	3,721
5	606	508	98	9,604
6	766	813	−47	2,209
7	669	761	−92	8,464
8	856	847	9	81
9	686	652	34	1,156
10	887	829	58	3,364
11	995	907	88	7,744
12	779	724	55	3,025
13	953	995	−42	1,764
14	771	726	45	2,025
15	804	886	−82	6,724
16	992	1034	−42	1,764
17	895	951	−56	3,136
18	843	858	−15	225
19	752	799	−47	2,209
20	738	797	−59	3,481

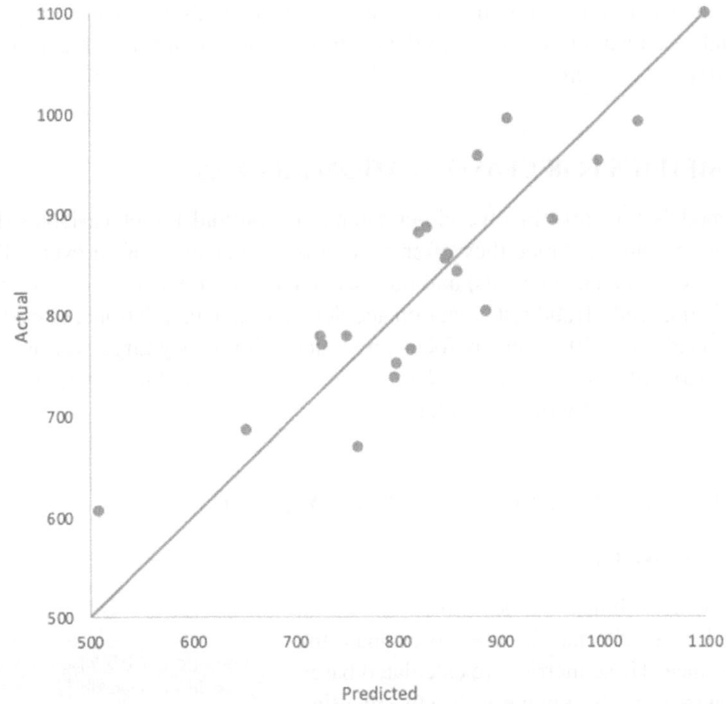

FIGURE 6.8 Actual by Predicted Chart.

Use Excel to construct these charts using the data presented in Table 6.5.

In contrast, **MAPE** is scale-independent and expressed as a percentage. In this example, the MAPE is 0.066 or 6.6%, meaning the model has a prediction error of 6.6%. This implies a prediction accuracy of 1 – 0.066 = 0.934 or 93.4%, which indicates a good model fit.

We can also assess the model's explanatory power using **RSE** and **EP**. Here, RSE is 0.308 and EP is 0.692, meaning the model explains 69.2% of the variance in the target variable, which is considered acceptable.

To further evaluate model fit, we can analyze the **Actual by Predicted Chart** and the **Residuals by Predicted Chart**. As shown in Figure 6.8, the points in the **Actual by Predicted** Chart are relatively close to the diagonal line, with no visible outliers, indicating a good model fit. Additionally, in the **Residuals by Predicted Chart** (Figure 6.9), the residual points are randomly

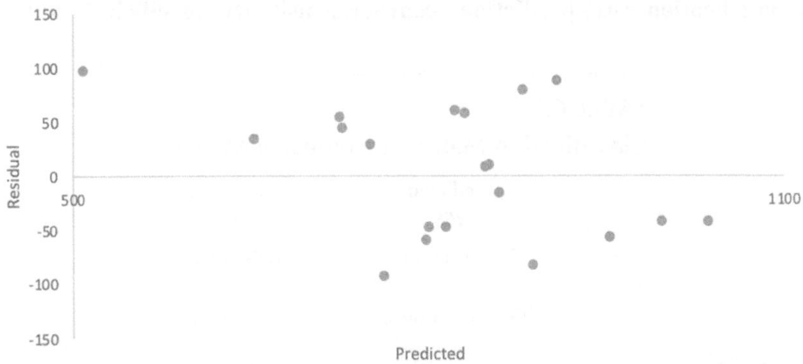

FIGURE 6.9 Residual by Predicted Chart.

and symmetrically distributed on both sides of the centerline, with no discernible patterns or outliers. The residuals are also not too far from the centerline. Based on this, we can conclude that the model demonstrates a good fit.

MODEL FIT METRICS FOR CLASSIFICATION MODELS

Classification models typically involve either binary or nominal target variables. **Binary target variables** are more common since they often predict the occurrence of an event. The target variable has two levels: occurrence (events) and non-occurrence (non-events). For example, this can be applied to predicting credit fraud risk, aviation accidents, flight cancellations, or customer purchase decisions. As a result, we will primarily focus on models with binary target variables. Models with **nominal target variables** (more than two levels) use the same model fit metrics but with some calculation differences, which I will explain later.

MODEL FIT METRICS FOR BINARY TARGET VARIABLES

Confusion Matrix Metrics

When dealing with a binary target variable, several important model fit metrics can be used to evaluate the model's performance. These metrics are calculated based on the **confusion matrix**, also known as the **classification matrix**, which summarizes the performance of a classification algorithm. It presents all the results of the prediction, including both correct and incorrect predictions. Specifically, it shows the number of **True Positives (TP)**, **True Negatives (TN)**, **False Positives (FP)**, and **False Negatives (FN)**. Table 6.6 shows the structure of a confusion matrix. These values (TP, TN, FP, FN) are extracted from the results of the modeling process. Definitions of these terms are provided in the definition box.

In most data mining projects, we aim to predict events and non-events. For this, we usually use a binary target variable with values of 1 (event) or 0 (non-event). Events (1) are considered positives, and non-events (0) are considered negatives. For example, when predicting the risk of flight cancellation, 1 indicates a cancellation (positive) and 0 indicates no cancellation (negative).

Several model fit metrics are derived from the confusion matrix. The most common metrics include **misclassification rate**, **prediction accuracy**, **sensitivity**, **specificity**, and the **F-1 score**.

DEFINITION

Confusion matrix, also known as the classification matrix, is a table that summarizes the performance of a classification algorithm.

True positive (TP): the prediction that correctly indicates the presence of the event.

True negative (TN): the prediction that correctly indicates the absence of the event.

False positive (FP): the prediction that wrongly indicates the presence of the event.

False negative (FN): the prediction that wrongly indicates the absence of the event.

TABLE 6.6
Classification Matrix or Confusion Matrix

	Predicted YES	Predicted NO
Actual YES	TP (true positive)	FN (false negative)
Actual NO	FP (false positive)	TN (true negative)

TABLE 6.7
Model Fit Metrics and Descriptions

Model Fit Metrics	Description
Misclassification rate (MR)	How often a model misclassifies (makes incorrect predictions) the target variable (including actual positives and actual negatives)
Prediction accuracy (ACC)	How accurate a model predicts the target variables (including actual positives and actual negatives) ACC = 1 − MR
Sensitivity, recall, or true positive rate (TPR)	The probability of a positive result (event) or how well a model can predict true positives (events)
Specificity, selectivity, or true negative rate (TNR)	The probability of a negative result (non-event) or how well a model can predict true negatives (non-events)
F1 score	Alternative accuracy metric; it is the harmonic mean of the sensitivity (recall) and precision (positive predicted value); a high F1 score means low false positives (FP) and low false negatives (FN)

The formulas for calculating these metrics based on the confusion matrix are listed below, with Table 6.7 providing descriptions of each metric.

$$Misclassification\ rate\ (MR) = \frac{FP+FN}{Total} = \frac{FP+FN}{TP+TN+FP+FN}$$

$$Accuracy\ (ACC) = \frac{TP+TN}{Total} = \frac{TP+TN}{TP+TN+FP+FN} = 1 - MR$$

$$Sensitivity,\ Recall,\ or\ True\ positive\ rate\ (TPR) = \frac{TP}{TP+FN}$$

$$Specificity,\ Selectivity,\ or\ True\ negative\ rate\ (TNR) = \frac{TN}{TN+FP}$$

$$F1\ score = \frac{2TP}{2TP+FP+FN}$$

It is essential to examine multiple metrics rather than relying on just one. While the **accuracy rate (ACC)** gives an overall sense of how accurate the model is, it doesn't reveal potential issues with **false negatives (FN)** or **false positives (FP)**. For instance, a model with a 90% accuracy rate, which implies a 10% misclassification rate, may seem like a good predictive model at first glance. However, if we further examine the metrics and discover that **sensitivity (recall)** is only 50%, this signals a deeper problem.

Sensitivity, also known as the **true positive rate**, indicates how well the model predicts positive cases (events). A low sensitivity suggests a high number of false negatives. So, why does the model still appear to have good overall accuracy? As discussed in **Chapter 3**, one primary reason could be an **imbalanced dataset** with rare events. For example, if the imbalance ratio is only 5%, as in a flight cancellation scenario, where only 5% of cases are cancellations, the model could classify all cases as non-cancellations and still achieve 95% accuracy. However, looking at the sensitivity reveals that the model is failing to predict the events (cancellations), which was the primary objective of the project. This renders the model invalid, unusable, and not generalizable. In this case, the accuracy and misclassification rates can be misleading, underscoring the importance of evaluating multiple metrics.

There are no standard cutoff values for these metrics. Ultimately, the decision depends on the industry and the tolerance for error, which may vary based on business needs. For instance, as a decision-maker, you must consider: **What margin of error is acceptable for your business?** Is it 1%, 5%, or 10%? My general rule of thumb is similar to the previous section. A **misclassification rate (MR)** of 5% or lower is considered great, and an MR between 5% and 10% is considered good. For **accuracy rate**, **sensitivity**, **specificity**, and the **F1 score**, a value of 95% or higher is considered great, while 90–95% is considered good.

Rule of thumb:
- MR= 5% or lower: Great
- MR = 5-10%: Good
- ACC, TPR, TNR, F1 = 95% or higher: Great
- ACC, TPR, TNR, F1 = 90-95%: Good

EXAMPLE: MODEL FIT METRICS FOR A CLASSIFICATION MODEL

This example demonstrates how to calculate model fit metrics for a classification model. For example, we have a confusion matrix for a house-buying prediction, as presented in Table 6.8.

Using the formulas above, we calculate our model fit metrics as follows:

$$Misclassification\ rate\ (MR) = \frac{53+17}{982+864+53+17} = 0.037\ or\ 3.7\%$$

$$Accuracy\ (ACC) = \frac{982+864}{982+864+53+17} = 0.963\ or\ 96.3\%$$

$$or\ ACC = 1 - MR = 1 - 0.037 = 0.063$$

$$Sensitivity,\ Recall,\ or\ True\ positive\ rate\ (TPR) = \frac{982}{982+17} = 0.983\ or\ 98.3\%$$

$$Specificity,\ Selectivity,\ or\ True\ negative\ rate\ (TNR) = \frac{864}{864+53} = 0.942\ or\ 94.2\%$$

$$F1\ score = \frac{2*982}{2*982+53+17} = 0.966\ or\ 96.6\%$$

Overall, the model's **misclassification rate** is 3.7%, meaning the **accuracy rate** is 96.3%, which is considered great. Further examination reveals a **sensitivity** (true positive rate) of 98.3%, which is also great, and a **specificity** (true negative rate) of 94.2%, which is good. Lastly, the **F1 score** is 96.6%, indicating great performance. In conclusion, the model demonstrates excellent fit, and all model fit metrics show consistent results.

TABLE 6.8
Example of a Confusion Matrix for a House Buying Prediction

	Predicted Buying	Predicted Not Buying
Actual Buying	TP = 982	FN = 17
Actual Not buying	FP = 53	TN = 864

ADDITIONAL MODEL FIT METRICS

In addition to the previously discussed metrics, several other metrics should be evaluated to further confirm model fit. These metrics are not based on the confusion matrix but provide a different perspective by showing how well the model explains the target variable. These include the **ROC chart**, **Lift chart**, and **Gini coefficient**.

Receiver Operating Characteristic (ROC) Chart

The **ROC chart** (Receiver Operating Characteristic chart) displays the **true positive rate (sensitivity)** on the vertical axis and the **false positive rate (1 − specificity)** on the horizontal axis. Essentially, the ROC chart illustrates the trade-off between sensitivity and specificity, metrics we've already discussed.

DEFINITION

The ROC chart displays the true positive rate (sensitivity) on the vertical axis and the false positive rate on the horizontal axis, reflecting the trade-off between sensitivity and specificity.

In the ROC chart, the straight diagonal line represents the baseline (random guessing), while the vertical line on the left and the horizontal line at the top indicate perfect classification. The larger the area between the model's ROC curve and the diagonal line, the better the model fit. Figure 6.10 illustrates the ROC chart and various scenarios for model fit evaluation.

It is important to note that the ROC chart can only be produced with binary target variables because it operates in two dimensions. For nominal target variables with more than two levels, we can only calculate the **ROC index**. An ROC index of 0.9 or higher is generally considered a good fit.

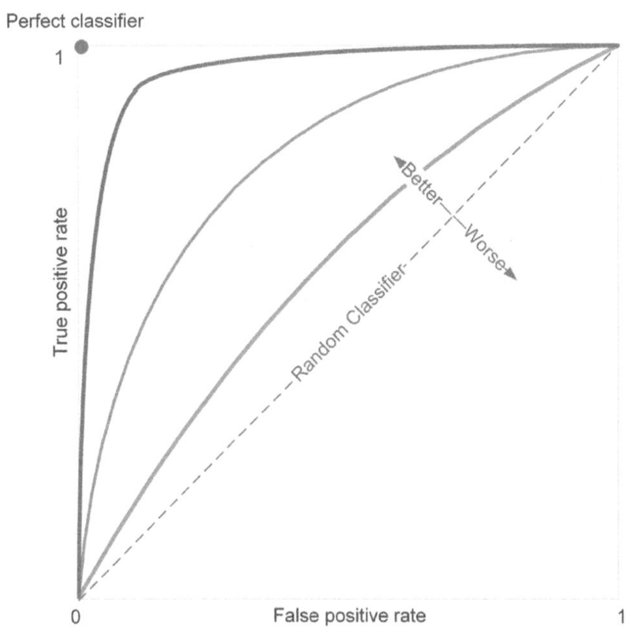

FIGURE 6.10 ROC chart.

Lift Chart

The **Lift chart** is another common metric used to evaluate model performance in classification models. **Lift** measures the effectiveness of a predictive model by calculating the ratio between results obtained with the model and those without it. The Lift chart, or Lift curve, provides a graphical representation of the relationship between lift values and response rates. A model with a higher Lift value at a given response rate performs better.

DEFINITION

Lift measures the effectiveness of a predictive model by calculating the ratio between the results obtained with the model and those without it. The Lift chart provides a graphical representation of the relationship between lift values and response rates.

In the Lift chart, the x-axis represents percentiles, and the y-axis represents lift values. For any given number of cases (X-value), the lift value indicates the expected number of positives that would be predicted. The baseline, or no model, is represented by a horizontal line intersecting the y-axis at 1. For example, selecting 10% of the data without a predictive model would predict 10% of the positives, representing **no lift** (Lift = 1), meaning cases are selected randomly without the model's influence.

The **Lift curve** illustrates how building a predictive model allows us to predict a higher number of positive cases, demonstrating the model's effectiveness. Figure 6.11 provides an example of a Lift chart, showing that if we select 20% of the cases using the model, we can predict 50% of the positive cases (2.5 times more than without a model), as the Lift value is 2.5. The higher the Lift value, the better the model. To determine which model performs best, the Lift chart should be compared with other models.

The ROC chart and Lift chart are two-dimensional and can only be generated for binary target variables. A common workaround for models with nominal target variables is to create these charts separately for each level of the target variable. Alternatively, you can use the ROC index and Lift value to evaluate model performance across multiple levels.

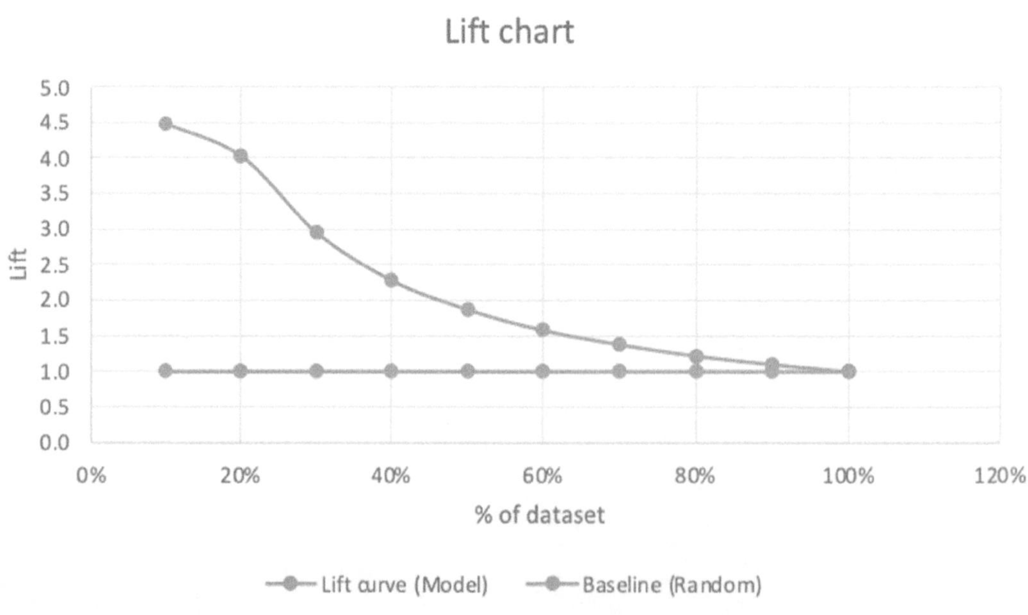

FIGURE 6.11 Lift chart.

Gini Coefficient

The **Gini coefficient** measures a model's discriminatory power or its effectiveness in distinguishing between positives and negatives. In machine learning, the Gini coefficient indicates how evenly positive rates are distributed across the attributes of a characteristic. The Gini coefficient ranges from 0 to 1, with higher values indicating better model performance. Essentially, the Gini coefficient shows how close the model is to being perfect and how far it is from being random.

> **DEFINITION**
>
> The Gini coefficient is a measure of a model's discriminatory power, reflecting how effectively the model differentiates between positives and negatives. It assesses how evenly the positive rates are distributed across a characteristic's attributes.

There are several ways to calculate the Gini coefficient, but a common method is using the **Area Under the ROC Curve (AUROC)**. Most software provides the AUROC value, and once it's determined, the Gini coefficient (also called **Somers' D Gini**) can be calculated as follows. **Somers' D** is named after Robert H. Somers, who proposed the metric in 1962:

$$Gini = 2AUROC - 1$$

While there is no standard cutoff for the Gini coefficient, a **rule of thumb** is that a Gini coefficient of 0.6 or higher indicates a good model fit.

COST SENSITIVE LEARNING (CSL)

As introduced in **Chapter 3**, **Cost-Sensitive Learning (CSL)** is a popular method that allows us to evaluate and select a model based on the costs of misclassification. It is particularly effective in improving a model's performance when dealing with imbalanced datasets. In business decision-making, relying solely on the misclassification rate may not be sufficient for selecting a predictive model. Instead, we must consider the costs of misclassification because, in most cases, the cost of false negatives and false positives is not equal.

> **DEFINITION**
>
> Cost-Sensitive Learning is a method used to evaluate and select models based on the costs associated with misclassification.

For instance, consider a predictive model for detecting **fraudulent credit card transactions**. The cost of correctly classifying a transaction is zero (the model performs as expected). However, the cost of misclassification differs between **false negatives** and **false positives**. Missing a fraudulent transaction (false negative) can result in direct customer loss, further fraudulent uses, and reputational damage to the company. On the contrary, blocking legitimate transactions (false positives) causes customer inconvenience and incurs unnecessary investigation costs. Typically, the cost of false negatives is higher than the cost of false positives.

It's important to note that misclassification costs are not always measured in monetary terms. For example, in **terminal disease prediction**, a false negative may result in serious health consequences for the patient, while a false positive may lead to unnecessary treatment and medical expenses. Therefore, it is crucial for decision-makers to properly define and quantify misclassification costs to build useful predictive models. These costs are represented in a **cost matrix**.

> When predicting a risk:
> Cost of false negatives > Cost of false positives

Once the misclassification costs for all levels are defined in a table (known as a **decision matrix**), we can train and select the model that minimizes the total cost. Table 6.9 explains the decision matrix using misclassification costs.

$$Total\ cost = C(FP) \times FP + C(FN) \times FN$$

TABLE 6.9
Decision Matrix—Costs of Misclassification

	Predicted Event	Predicted Non-event
Actual event	True positive Cost = 0	False negative Cost = C(FN)
Actual non-event	False positive Cost = C(FP)	True negative Cost = 0

Where **C(FP)** is the cost of a false positive and **C(FN)** is the cost of a false negative. Keep in mind that **C(FN)** is often higher than **C(FP)**. Most software includes a feature that allows you to input the decision matrix and set total cost minimization as the model selection metric. In later chapters, I will demonstrate how to perform CSL in SAS Enterprise Miner.

In cases of imbalanced datasets with very low response rates, where the cost of false negatives is much higher than false positives, misclassifications of the minority class (rare cases) will be heavily penalized. When comparing models, the one with the **lowest total cost** will be selected, rather than simply the one with the lowest misclassification rate. **CSL** adjusts the decision boundaries during training to account for these costs, improving generalization on the minority class.

On the contrary, we can also focus on **profits** from accurate predictions. For example, accurately predicting customer purchasing decisions can optimize supply chain, distribution, logistics, and shipment strategies, reducing operational costs and increasing revenue. In this scenario, the profit from true positives (buying) is generally higher than from true negatives (not buying). Of course, there are no profits from misclassification.

If focusing on profits, you can define a decision matrix detailing profits for accurate predictions. When comparing models, the one that maximizes total profit is selected. Table 6.10 presents a decision matrix for profits from accurate classifications.

$$Total\ profit = P(TP) \times TP + P(TN) \times TN$$

Where **P(TP)** is the profit from a true positive, and **P(TN)** is the profit from a true negative. Typically, **P(TP)** is higher than **P(TN)**.

What if we want to include both **profits** and **costs**? For example, in the case of purchasing decisions, misclassifications could result in increased operational costs due to poor planning. In this case, a decision matrix can combine both profits from accurate classifications and costs from misclassifications. A recommended approach is to define the matrix in terms of profits, assigning negative profit margins for misclassification costs. This way, the objective remains to maximize total profit when selecting the best model. Table 6.11 illustrates such a matrix.

$$Total\ profit = P(TP) \times TP + P(TN) \times TN - C(FN) \times FN - C(FP) \times FP$$

TABLE 6.10
Decision Matrix—Profits of Accurate Classification

	Predicted Event	Predicted Non-event
Actual event	True positive Profit = P(TP)	False negative Profit = 0
Actual non-event	False positive Profit = 0	True negative Profit = P(TN)

TABLE 6.11
Decision Matrix with Profits and Costs

	Predicted Event	Predicted Non-event
Actual event	True positive Profit = P(TP)	False negative Profit = −C(FN)
Actual non-event	False positive Profit = −C(FP)	True negative Profit = P(TN)

MODEL FIT METRICS FOR NOMINAL (MULTI-CLASS) TARGET VARIABLES

In most data mining projects, we often work with **binary target variables** to predict the occurrence of events versus non-events. However, in some cases, we encounter **nominal target variables** with multiple levels, also known as **multi-class target variables**.

So, how do we measure model fit in this scenario? Essentially, we use the same metrics as for binary target variables, but with some modifications. The most significant difference is that the **confusion matrix** for multi-class target variables has more levels, making it more complex. Table 6.12 shows a confusion matrix for a target variable with three levels (1, 2, and 3). For example, imagine we want to predict residents' shopping behaviors during the COVID-19 pandemic, where their choices are as follows:

1. Online shopping
2. Curbside pickup
3. In-store shopping

The confusion matrix includes actual and predicted values for all three levels. The challenge is calculating model fit metrics, which typically rely on **True Positives (TP)**, **True Negatives (TN)**, **False Positives (FP)**, and **False Negatives (FN)**. Since there are three classes, we need to determine how to calculate these values for each class.

I explain below a method for calculating **True Positives (TP)**, **True Negatives (TN)**, **False Positives (FP)**, and **False Negatives (FN)** for a multi-class target variable. Essentially, we calculate these metrics separately for each class. For each class, we treat that class as **positive** and the other two classes as **negative**. The following formulas allow us to calculate these values for each class.

For class 1

- TP = a
- TN = b + c + n + z
- FP = x + y
- FN = l + m

TABLE 6.12
Confusion Matrix for a Three-Class Target Variable

	Predicted 1	Predicted 2	Predicted 3
Actual 1	a	l	m
Actual 2	x	b	n
Actual 3	y	z	c

For class 2

- TP = b
- TN = a + c + m + y
- FP = l + z
- FN = x + n

For class 3

- TP = c
- TN = a + b + l + x
- FP = m + n
- FN = y + z

Once we calculate **TP**, **TN**, **FP**, and **FN** for each class, we can apply the same formulas used for binary target variables to compute model fit metrics. However, unlike binary classification, where we calculate the metrics only once, for multi-class target variables, we must calculate these metrics separately for each class. For example, in this case, we need to compute the following metrics for all three classes.

Class 1:

- MR (Class 1)
- ACC (Class 1)
- TPR (Class 1)
- TNR (Class 1)
- F1 score (Class 1)

> 1. Calculate TP, TN. FP, and PN separately for each class.
> 2. Calculate model fit metrics for each class individually.
> 3. Use metrics for all classes to evaluate the overall model performance.

Class 2:

- MR (Class 2)
- ACC (Class 2)
- TPR (Class 2)
- TNR (Class 2)
- F1 score (Class 2)

Class 3:

- MR (Class 3)
- ACC (Class 3)
- TPR (Class 3)
- TNR (Class 3)
- F1 score (Class 3)

When evaluating model fit and comparing between models, we need to use the model fit metrics for all three classes.

As for the **ROC chart** and **Lift chart**, these charts can only be constructed for binary target variables because they are two-dimensional. However, most software provides the following metrics for multi-class models:

- **AUROC (Area Under the ROC Curve)**: Measures the area between the model's ROC curve and the baseline.
- **Lift value**: Measures the Lift of the model compared to the baseline (no model).
- **Gini coefficient**: Evaluates the model's discriminatory power.

CHAPTER SUMMARY

Evaluating predictive model performance is a complex and nuanced task, given the variety of variables, modeling methods, and potential challenges in model development. In this chapter, we explored several key concepts and terminologies related to model fit and predictive power. Model fit refers to how well a model captures the patterns in the data for a specific sample, such as the training sample, while predictive power measures how well the model performs across different datasets, ensuring its generalizability. A model that performs well on the training data but fails to make accurate predictions on new data is said to suffer from overfitting, which compromises its predictive power. We also examined strategies to avoid or minimize overfitting. Additionally, we discussed the differences between model reliability and model validity, highlighting the importance of assessing both.

The second half of this chapter focused on various metrics used to measure model fit for both association models (with continuous target variables) and classification models (with binary or multi-class target variables). We introduced a wide range of metrics, such as RMSE, MAPE, RSE, and EP for association models, and metrics like accuracy, sensitivity, specificity, and the F1 score for classification models. I also provided guidelines on selecting the most appropriate metrics based on the type of model and the scale of the target variable. We emphasized the importance of using multiple metrics to gain a more comprehensive understanding of model performance, as relying on a single metric could lead to a misleading evaluation.

Furthermore, we explored additional metrics, such as the ROC chart, Lift chart, and the Gini coefficient, which provide additional insights into model performance, particularly for classification models. These metrics offer a more detailed view of how well a model differentiates between positive and negative outcomes, helping to fine-tune predictions.

Lastly, we introduced cost-sensitive learning (CSL), an important method for selecting models based on the costs of misclassification. In many real-world scenarios, such as fraud detection or medical diagnosis, the costs of false negatives and false positives are not equal. CSL allows us to account for these differences and choose models that minimize the total cost rather than focusing solely on the accuracy or misclassification rate. This chapter sets the foundation for using CSL in future predictive modeling tasks, which will be demonstrated in Chapter 8.

CHAPTER DISCUSSION QUESTIONS

1. What is the purpose of model evaluation, and why is it important in data mining?
2. What is model fit? Explain the difference between model fit and predictive power. How do they overlap? Provide examples to illustrate.
3. What is overfitting, and how does it affect the validity of a predictive model?
4. What are the primary root causes of overfitting? What strategies can be used to address overfitting?
5. Differentiate between model reliability and model validity. Provide specific examples of each.
6. Why is it important to evaluate both model reliability and validity? List the steps involved in evaluating the reliability and validity of predictive models.
7. What is the purpose of model comparison in data mining, and why is it essential?
8. What is a champion model? Is the champion model always the best model overall? Why do we need a champion model in the context of data mining?
9. What is the recommended model fit metrics for association models, and why are they recommended?
10. Define confusion matrix. Explain its importance in evaluating classification models.
11. Describe the recommended model fit metrics for classification models. Why is it necessary to evaluate multiple metrics rather than relying on just one?
12. Explain Cost-Sensitive Learning (CSL). In which scenarios should CSL be applied?

CHAPTER EXERCISES

Exercise 1: Model Fit Metrics for an Association Model

1. **Download the Excel file** `resp_training.xlsx`, which contains the training results for the **Boston dataset**, including actual and predicted values.
2. **Add new columns** to calculate the **residuals (errors)** for all cases.
3. **Add new columns** to calculate the **squared errors** for all cases.
4. Using the formulas in this chapter, calculate the following **model fit metrics**:
   - **RMSE** (root mean squared error)
   - **MAPE** (mean absolute percent error)
   - **RSE** (relative squared error)
   - **EP** (explanatory power)
5. **Evaluate the model fit** based on these metrics. Does the model exhibit a good fit?
6. **Construct two charts**:
   - **Actual vs. Predicted Chart**
   - **Residuals vs. Predicted Chart**
7. **Evaluate the charts** and confirm the model fit assessment from Step 5.

Exercise 2: Model Fit Metrics for a Classification Model with a Binary Target Variable

You are given the following **confusion matrix** for a binary target variable (1 = accident; 0 = no accident):

TABLE 6.13
Confusion Matrix for a Binary Target

	Predicted Accident	Predicted No Accident
Actual accident	TP = 1685	FN = 158
Actual no accident	FP = 212	TN = 956

Using the formulas from this chapter, calculate the following **model fit metrics**:

- Misclassification rate (MR)
- Accuracy (ACC)
- Sensitivity (Recall)
- Specificity (Selectivity)
- F1 score

Evaluate the model fit based on these metrics.

Exercise 3: Model Fit Metrics for a Classification Model with a Nominal Target Variable

You are given the following **confusion matrix** for a multi-class target variable with three classes: A, B, and C.

TABLE 6.14
Confusion Matrix for a Multi-Class Target

	Predicted A	Predicted B	Predicted C
Actual A	856	125	79
Actual B	111	425	98
Actual C	69	75	327

1. Calculate TP, TN, FP, and FN for all three classes.
2. Using the formulas in this chapter, calculate the following model fit metrics for these three classes (A, B, C) separately:
   - Misclassification rate (MR)
   - Accuracy (ACC)
   - Sensitivity (Recall)
   - Specificity (Selectivity)
   - F1 score
3. Evaluate the model fit based on these metrics.

7 Regression Methods

LEARNING OUTCOMES

Upon completion of this chapter, you will be able to:

1. Describe the purposes and key characteristics of regression methods.
2. Differentiate between linear regression and logistic regression.
3. Explain the advantages and disadvantages of regression methods in data mining.
4. Discuss the primary assumptions for both linear regression and logistic regression.
5. Outline the modeling process for linear and logistic regressions.
6. Compare forward stepwise, backward stepwise, and adaptive stepwise methods.
7. Build and execute predictive models using regression methods in SPSS Modeler.
8. Evaluate the model fit and predictive power of regression models.
9. Identify significant predictors from regression outputs.
10. Interpret the results of regression analysis.

INTRODUCTION

We have covered the first three steps in the data mining process: data exploration, sampling, and modification. All these efforts prepare us for the most critical step in this process—modeling. In this step, researchers select machine learning methods, build and configure multiple models, train the models with data, and validate them. The outcome of these steps is predictive models that generate predictions for the target variable. You can see the importance of this step in data mining and why it is crucial to get it right. Any mistakes made during this stage will result in invalid predictive models. To complete the modeling step effectively, it is essential to understand the underlying foundations of the machine learning methods we select and ensure proper configurations.

We will begin the modeling phase with one of the most popular data mining algorithms: regression. This chapter covers two regression methods—linear regression and logistic regression. Both are widely used across various fields due to their standardized, straightforward analysis process and easily interpretable results. These methods allow us to derive a regression equation that reflects the relationship between predictors and the target variable. For example, regression can be used to predict a customer's likelihood of purchasing a product or the risk of flight delays.

Linear regression is primarily an association method used when the target variable is continuous, with a linear relationship between the target variable and the predictors. In contrast, logistic regression is a classification method used when the target variable is binary (logistic regression) or categorical (multinomial logistic regression). This type of regression reflects a nonlinear relationship between predictors and the target variable, presented as a logit function.

DEFINITION

Regression is a method used to estimate the relationship between a dependent variable (also known as the target variable) and one or more independent variables (predictors).

Dependent variable = Target variable
Independent variable = Predictor

DOI: 10.1201/9781003468196-9

Since regression methods are covered extensively in most statistics textbooks and courses, this chapter provides a brief overview of the statistical foundation, focusing more on how these methods are applied in data mining. I will use practical examples to demonstrate the data mining process and explain how to interpret the results. For those interested in diving deeper into the statistical foundations of these methods, I have included some recommended references for further reading.

WHAT IS THE REGRESSION METHOD?

Regression is a widely used method that estimates the relationship between a dependent variable (target variable) and multiple independent variables (predictors). The goal of regression is to develop a mathematical function or model that can predict the values of the target variable based on the predictors. In many statistics books, the terms "dependent variable" and "independent variable" are commonly used. However, for consistency, I will use the terms "target variable" and "predictors" throughout this book. It's important to remember that the target variable is the same as the dependent variable and predictors are the same as independent variables.

> **Inputs of the regression model:** The target variable and all predictors.
>
> **Output of the regression model:** A regression equation showing the relationship between the target variable and significant predictors.

Let's consider an example: Suppose we want to predict whether customers will choose to dine at a restaurant. Here, the target variable is the dining decision (Yes or No), while predictors may include factors such as gender, race, age, education, income, diet preference, and health condition. The regression model built from this data identifies predictors that significantly influence the dining decision and estimates mathematical relationships between those predictors and the target variable. More specifically, the regression method calculates the coefficients (weights) of the predictors in the model to minimize the difference between predicted values and the actual values of the target variable.

The primary advantage of regression is its simplicity in terms of understanding, usage, and interpretation. Since its mathematical foundation is similar to what is taught in most statistics courses, there is an abundance of resources that explain the algorithm in detail. In fact, most statistics or quantitative analysis books provide comprehensive explanations of this method and how the regression equation is estimated. In this book, I will focus more on the practical application and purpose of regression rather than delving into the detailed mathematical foundations. However, I will explain key terminologies and important aspects of the data analysis process as needed.

LINEAR REGRESSION AND LOGISTIC REGRESSION

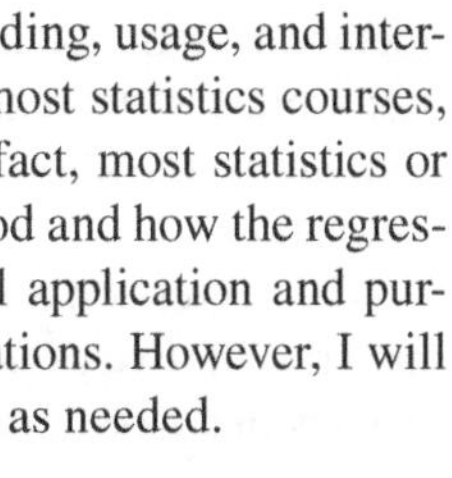

There are two primary types of regression: linear regression and logistic regression. While both are regression methods, they have fundamental differences. Linear regression is used when the target variable is continuous, making it an association model. In some cases, a target variable measured using a Likert scale can be analyzed with linear regression, provided the intervals between levels are equal. As covered in Chapter 3, in such cases, Likert scale variables can be treated as interval variables.

On the other hand, logistic regression is used when the target variable is categorical (binary or nominal), making it a classification model. Binary variables have two levels, whereas nominal variables have more than two levels (multi-class). When the target variable is nominal, the model is often referred to as multinomial logistic regression. However, in data mining, the distinction between binary and nominal target variables is less of a concern as the focus shifts to model performance. To keep things simple, I will use the term "logistic regression" for both

TABLE 7.1
Examples of Linear Regression and Logistic Regression

Regression Methods	Examples
Linear regression example	The target variable is the students' SAT scores, which is continuous, ranging from 0 to 1,600. Predictors include age, gender, race, high school GPA, and family income.
Logistic regression example	The target variable is a student's decision to select a particular major, such as marketing. This is a binary variable (Yes or No). Another example involves the student's choice of one of four majors: marketing, accounting, finance, or management. In this case, the target variable is nominal with four levels: 1 (marketing), 2 (accounting), 3 (finance), or 4 (management). Predictors include age, gender, race, high school GPA, and family income.

binary and nominal target variables, specifying the scale of the target variable for clarification when necessary. Table 7.1 presents examples of both linear and logistic regressions, helping to illustrate the practical differences between these two methods.

DEFINITION

Linear regression: Target variable is continuous
Logistic regression: Target variable is binary
Multinomial logistic regression: Target variable is nominal (multi-class)

Due to differences in the target variable, linear and logistic regressions employ different estimation methods, resulting in variations in their assumptions, outputs, and the interpretation of results. Specifically, linear regression uses the **ordinary least squares (OLS)** method to estimate the regression equation while logistic regression relies on the **maximum likelihood (ML)** method for estimation.

How to choose the right regression method?

If the target variable is continuous or measured on a Likert scale (with equal intervals between levels), then use linear regression.

If the target variable is binary or nominal (categorical), then use logistic regression.

The regression equations for these methods are shown below. To make the equations easier to understand, I've avoided typical mathematical notations and instead named the variables and coefficients in a more straightforward manner rather than using Greek characters.

For linear regression:

$$target = int + coe_1 \times predictor_1 + coe_2 \times predictor_2 + coe_3 \times predictor_3 + \cdots + error$$

For logistic regression:

$$logit(target) = int + coe_1 \times predictor_1 + coe_2 \times predictor_2 + coe_3 \times predictor_3 + \cdots + error$$

where

target is the target variable,
predictor is the predictor variable,
int is the intercept,
coe is the regression coefficient for the corresponding predictor,
error is the error terms.

You may notice that in logistic regression, we do not directly estimate the target variable but rather the **logit value** of that variable. Naturally, the concept of a logit value can be difficult to grasp.

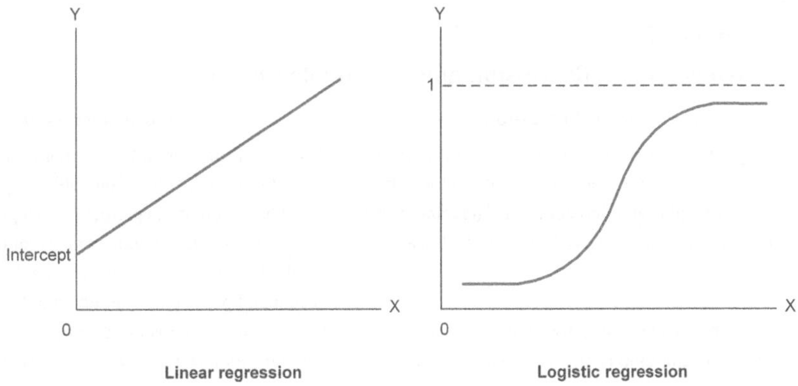

FIGURE 7.1 Linear regression and logistic regression.

Therefore, when interpreting logistic regression results, we use **odds ratios** rather than correlation coefficients. I will explain this in the section "How to Read and Understand Logistic Regression Results" later in this chapter.

The charts in Figure 7.1 illustrate how the linear regression equation differs from the logistic regression equation. As you can see, linear regression detects only the **linear relationship** between the target variable (*Y*) and the predictors (*X*). In contrast, logistic regression captures a **nonlinear relationship** between the target variable (*Y*) and the predictors (*X*) through the logit function. Since the target variable in logistic regression is binary (0 or 1), *Y* can only take on those two values. However, note that the relationship between **logit(*Y*)** and *X* remains linear, as shown in the equation.

Thus, while logistic regression can detect nonlinear relationships via the logit function, its ability to model nonlinearity is limited to this transformation and the linear effect of the correlation coefficients. This is a key difference between logistic regression and other nonlinear methods, such as decision trees, neural networks, or random forests, which can identify more complex forms of nonlinearity.

> **Linear regression:** Detects a **linear relationship** between the target variable and all predictors.
>
> **Logistic regression:** Identifies a **non-linear relationship, in the** form of a logit function, between the target variable and all predictors.

Table 7.2 highlights some key differences between linear regression and logistic regression. In the following sections, we will explore the details of both methods and how to configure models using these approaches.

TABLE 7.2
Differences between Linear Regression and Logistic Regression

Items	Linear Regression	Logistic Regression
Nature of the dependent variable	The target variable in linear regression is continuous and numeric, representing a range of real numbers. The model aims to establish a linear relationship between the predictors and the continuous outcome.	The target variable in logistic regression is binary or categorical, representing two or more discrete outcomes. The model predicts the probability of an event in one category.
Output and range	The output of linear regression is a continuous value, and the predicted values can be any real number. The range of predictions is not constrained.	The output of logistic regression is a probability score between 0 and 1, representing the likelihood of the event occurring.

(Continued)

TABLE 7.2 (Continued)
Differences between Linear Regression and Logistic Regression

Items	Linear Regression	Logistic Regression
Model function	The model function in linear regression is a linear equation that represents the weighted sum of the predictors plus an intercept term. The relationship between the variables is assumed to be linear.	The model function in logistic regression is the logistic (sigmoid) function, which represents the weighted sum of the predictors to a probability score. It models the odds of an event occurring. Note that the relationship between the odds of an event and predictors is assumed to be linear.
Applications	Linear regression is used for predicting and modeling continuous outcomes, such as predicting house prices, temperature, or sales volume based on various factors.	Logistic regression is used for binary or multinomial classification tasks, such as predicting whether an email is spam or not or classifying images into multiple categories.
Evaluation metrics	Linear regression uses common model fit metrics for association models plus F-test and R-squared (coefficient of determination).	Logistic regression uses common model fit metrics for classification models.
Optimization method	The model parameters in linear regression are typically estimated using methods like ordinary least squares (OLS) to minimize the sum of squares errors.	The model parameters in logistic regression are typically estimated using maximum likelihood estimation (MLE) to maximize the likelihood of the observed outcomes.

ADVANTAGES AND DISADVANTAGES

It's important to note that in data mining, no machine learning method is perfect, regardless of its complexity. Every method comes with its own advantages and disadvantages. Therefore, it is crucial to understand these strengths and limitations in order to select the appropriate method and accurately interpret the results. Regression methods, for example, are grounded in statistical foundations, making them relatively easy to implement and understand. However, their linear nature and underlying assumptions can limit their effectiveness in predictive modeling. Table 7.3 summarizes the key advantages and disadvantages of these regression methods.

TABLE 7.3
Advantages and Disadvantages of Regression Methods

Regression Methods	Advantages	Disadvantages
Linear regression	• Simple to implement. • Best for detecting linear patterns. • Outputs are easy to understand and interpret (regression coefficients, direction, p-value). • Works well with a small number of predictors. • Requires less data to work. • Less prone to overfitting. • Short computing time.	• Subject to normality assumption. • Requires the linearity between the target variable and predictors. • Requires continuous variables. • Sensitive to missing values and outliers. • Unable to detect nonlinear relationships. • May not produce a good model fit with a large number of predictors. • Prone to multicollinearity, which limits us from analyzing big data.

(Continued)

TABLE 7.3 *(Continued)*
Advantages and Disadvantages of Regression Methods

Regression Methods	Advantages	Disadvantages
Logistic regression	• Simple to implement and efficient to train. • Provides regression coefficients, direction, and p-value for interpretation. • Provides probabilistic outputs. • Does not require normal distribution. • Works with both continuous and categorical predictors. • Can be extended to multiple classes (multinomial regression). • Able to detect nonlinearity through the logit function. • Less susceptible to overfitting. • Short computing time.	• Requires a categorical target variable. • Assumes linearity between the logit value of the target variable and predictors. • Unable to detect all forms of nonlinear relationships. • Unable to detect complex patterns. • Prone to multicollinearity, which limits us from analyzing big data. • Probabilistic outputs are harder to understand and interpret. • Lack of interpretability for a nominal target variable with multi levels. • Requires more data for a good model fit.

LINEAR REGRESSION

Linear Regression Equation

Let's begin with a linear regression equation to understand the components of this method. As previously mentioned, the regression equation represents the mathematical relationships between the target variable and predictors. Linear regression is applicable only when the target variable is continuous. Here, we will present the linear regression equation in its traditional form, which helps you interpret outputs provided by various software applications. You can always refer to the simplified version shown earlier.

$$Y = \beta_0 + \beta_1 x_1 + \beta_2 x_2 + \cdots + \beta_n x_n + \varepsilon$$

Where

Y is the target variable,
x_1 to x_n are predictor variables ($1 - n$),
β_0 is the intercept,
$\beta_1 - \beta_n$ are the regression coefficients for the corresponding predictors,
ε is the residual or error.

Estimation Approach

An estimation approach refers to the statistical method used to calculate the values of all components in the regression equation, including the intercept, regression coefficients, and error terms. OLS is the standard estimation method for linear regression. It is commonly used to estimate the intercept and regression coefficients by minimizing the sum of squares errors (SSEs). The mathematical foundation of OLS is thoroughly explained in many statistics books, so I won't elaborate here. What is important to emphasize is that using this method imposes strict assumptions on linear regression.

Assumption Requirements

> Linear regression has strict assumptions; however, due to the use of large datasets, data mining projects allow for some flexibility.

In traditional statistical studies, linear regression is subject to strict assumptions to ensure the validity of the results. These include assumptions of normality, homoscedasticity, linearity, non-multicollinearity, continuous variables, and no missing values. If these assumptions are violated, the software can still run the analysis and produce outputs, but the validity of the results—and consequently the conclusions—becomes uncertain. In traditional settings, researchers typically work with a small number of predictors and a small sample size. For instance, with five predictors and an effect size of 0.25, a sample size of just 57 records is required. With ten predictors, the sample size increases only slightly to 75 records.

However, in data mining, we work with very large datasets containing many predictors and a large sample size. As a result, we have more flexibility in terms of meeting these assumptions when using linear regression. The Central Limit Theorem supports this flexibility, as it states that with a large enough sample size, the distribution of the sample can be assumed to be approximately normal. Therefore, when applying linear regression in data mining, we do not need to strictly meet all the assumptions, as is required in traditional studies.

That said, it is still advisable to address key assumptions to improve the performance of the regression model. For instance, handling and imputing missing values is highly recommended, as linear regression performs better with a complete dataset. Normalizing the data can also improve model fit by addressing the normality assumption. Additionally, managing outliers helps reduce noise in the data, leading to a more accurate model fit.

You may ask: Should we be concerned about the validity of the results if we do not meet the test assumptions? It's a good question, as we should always consider the reliability and validity of the results. In data mining studies, we implement additional rigorous steps to ensure the quality of the results. As mentioned in Chapter 3, we typically split the dataset into **training**, **test**, and **validation** samples. The training sample is used to build the model while the test and validation samples ensure the model's performance beyond just the training data, helping prevent overfitting. By comparing the model's results across these three samples, we can assess the model's performance and ensure its reliability and validity. Thus, even if we do not meet all assumptions, these steps allow us to maintain the validity of our predictive models.

Here are a few key assumptions for linear regression, along with the level of flexibility we have in data mining projects:

- **The target variable must be continuous**: This assumption is crucial for establishing an association model. However, a target variable on a Likert scale is acceptable as long as the gaps between levels are equal.
- **Predictors must be continuous**: There is more flexibility with this assumption. Although continuous predictors are preferable for interpretability and model fit, categorical (binary or nominal) predictors can still be used. These variables can be converted into dummy variables, for example, coding an education level variable (1—high school, 2—bachelor, and 3—graduate) into three separate variables: high school, bachelor, and graduate (with two levels, 1 for "yes" and 0 for "no"). While traditional statistical software often requires manual dummy coding, many data mining tools automate this process. Nonetheless, interpreting regression coefficients for dummy variables requires extra caution, as they are not as straightforward as with continuous variables.
- **Normality**: Normality is a common assumption for parametric statistical methods, particularly with the OLS method. In data mining, with larger datasets, fully satisfying the normality

assumption is not always necessary. However, skewed data can still negatively impact model performance. Therefore, reviewing histograms and transforming variables to normalize data are recommended. Additionally, outliers can skew data, so reviewing histograms and box plots to identify and remove extreme outliers is also beneficial. Achieving full normality is not essential, but moving closer to a normal distribution generally improves model fit.

- **Non-multicollinearity**: Linear regression assumes that predictors are independent. In traditional studies, multi-collinearity—inter-correlation among predictors—must be addressed before running the regression analysis. However, research suggests that increasing the sample size can reduce the impact of multi-collinearity. In data mining projects, where large datasets are common, multi-collinearity is typically not a major concern.
- **No missing values**: In traditional statistical studies, regression analysis requires a complete dataset, and missing values must be addressed through imputation or elimination. Some software may even produce an error if missing values are detected. In data mining, however, models can still run with missing values. If the percentage of missing data is small, model performance is generally not significantly affected. Still, if more than 15% of values are missing, imputation is recommended, as a complete dataset generally leads to a better-fitting model.

Linear Regression Model Fit Evaluation

In the OLS process, evaluating the **model fit** is essential. But what exactly is model fit? We discussed this in detail in Chapter 6. Essentially, we have a theoretical regression model, represented by the regression equation, and we also have data for the target variable and all predictors. Our goal is to see how well the data aligns with the theoretical model—this alignment is what we refer to as model fit.

$$Residuals = Actual\ data - Predicted\ data$$

In Chapter 6, I covered the general model fit metrics for association models, including **RMSE**, **MAPRE**, residual sum of squares error **(RSE)**, **explanatory power (EP)**, the **actual vs. predicted** chart, and the **residuals vs. predicted** chart. In this chapter, I will focus specifically on metrics used for linear regression. In particular, when using linear regression, we need to evaluate the **F-test** and **R^2**. Table 7.4 explains these metrics and how to interpret them.

TABLE 7.4
Model Fit Metrics Specific to Linear Regression

Metrics	Description
F-test	The F-test evaluates the significance of the overall model. A significant F-test indicates that the regression model fits the data better than a model with no predictors. A significant F-test is expected for a well-fitting model.
R^2 (coefficient of determination)	R^2 is a common metric for assessing model fit in linear regression. It represents the percentage of variance in the target variable explained by the predictors. There is no absolute cutoff for a good R^2, as it depends on the number of predictors, but a general rule of thumb is as follows: • $R^2 > 0.7$: Good model fit • R^2 between 0.5 and 0.7: Acceptable model fit • $R^2 < 0.5$: Poor model fit Keep in mind that R^2 tends to decrease as the number of predictors increases, which is often the case in data mining. A lower R^2 can still be acceptable with many predictors.
Adjusted R^2	Adjusted R^2 is a modified version of R^2 that accounts only for predictors that significantly impact the model's performance. It is generally recommended to use adjusted R^2 for evaluation. You can apply the same thresholds as suggested for R^2 above.

TABLE 7.5
Key Terminologies and Notations in Linear Regression

Terminologies	Notations	Descriptions
Target variable	y	Actual value of the target variable (dependent variable).
Predictor	x	Actual value of the predictor (independent variable).
Mean of target variable	$\bar{y}$	Mean value of the target variable.
Predicted value	$\hat{y}$	Predicted value of the target variable estimated by the model.
Error	e	Error (residual) is the difference between the actual and predicted values of the target variable.
Sum of squares error	SSE	A measure of how far the data points are from the model's predicted values.
Sum of squares total	SST	A measure of how far the data points are from the mean of the target variable.
Sum of squares regression	SSR	A measure of how far the mean is from the model's predicted values.

In linear regression, when reviewing materials on the F-test and R^2, you may encounter specific terminologies, notations, and formulas. I have included these in Table 7.5 for quick reference. Below are the formulas for calculating sum of squares total **(SST)** and $\mathbf{R^2}$ using sum of squares regression **(SSR)** and **SSE**:

$$SST = SSR + SSE$$

$$R^2 = \frac{SST - SSE}{SST} = \frac{Explained\ variance}{Total\ variance}$$

As explained in Chapter 6, these formulas apply exclusively to linear regression and not to other methods. As you can see, in linear regression, $\mathbf{R^2}$ is essentially the same as the **EP** that we discussed in Chapter 6.

Model Selection Process

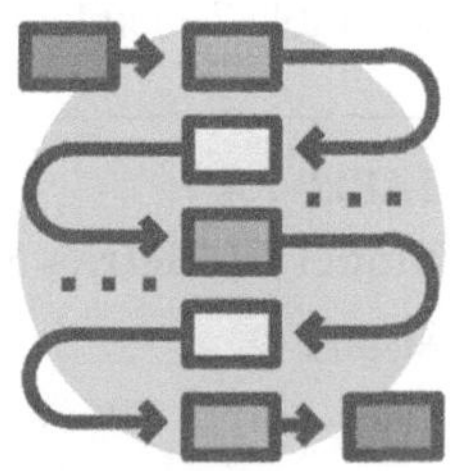

In the regression modeling process, **model selection** is a crucial step in training the model. It determines how predictors are entered into the regression analysis. While it may seem logical to enter all variables at once and let the analysis identify significant predictors, doing so increases computing time and complicates model fitting. Interestingly, there are multiple ways to decide how variables are included in the model, and choosing the right selection method can significantly improve model fit.

Regression literature identifies three primary variable selection methods in linear regression: **backward**, **forward**, and **stepwise**. I will explain these methods simply, accompanied by diagrams for better understanding. Afterward, I will show you the specific methods used in **SPSS Modeler**, as this is the software we use for our analysis.

> Stepwise method is an iterative variable selection technique used in data mining to identify the most relevant subset of variables for a predictive model.
> There are three primary stepwise methods: forward stepwise, backward stepwise, and adaptive stepwise.

One important point to note is that all three methods are forms of the stepwise method, an iterative variable selection process used in data mining to identify the most relevant subset of variables for a predictive model. To avoid confusion, I refer to them as **forward stepwise**, **backward stepwise**, and **adaptive stepwise**. Each method results in multiple models, with the final

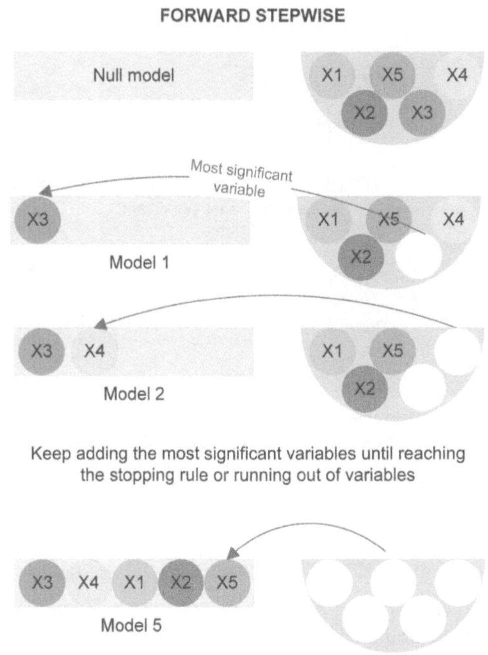

FIGURE 7.2 Forward stepwise method.

one typically being the best. Below, I describe the steps for each method, along with corresponding diagrams (Figures 7.2–7.4).

- **Forward Stepwise** (Figure 7.2)
  i. The process begins with a model without any predictors (null model).
  ii. The most significant variables are added to the model, one after the other. Significant variables are the ones significantly associated with the target variable. Typically, the most significant predictor is the one with the smallest p-value or provides the highest increase in R^2.
  iii. The selection will stop when all variables under consideration are added or a pre-determined stopping rule is met.
- **Backward Stepwise** (Figure 7.3)
  i. The process begins with a model that includes all predictors (full model).
  ii. The least significant variables are removed from the model, one after the other.
  iii. The selection will stop when no variables are left or a pre-determined stopping rule is met.
- **Adaptive Stepwise** (Figure 7.4): This method is similar to forward stepwise but with one key difference: After adding an additional variable to the model, it can remove a variable currently in the model but no longer significant. In other words, it is adaptive. This adaptability allows the model to evolve dynamically, and for this reason, in some cases, **adaptive stepwise** is also known as **bidirectional stepwise**.
  i. The process begins with a model without any predictors (null model).
  ii. The most significant variables are added to the model, one after the other.
  iii. After each addition, any variables already in the model that are no longer significant will be removed.
  iv. The selection will stop when all variables under consideration are added or a pre-determined stopping rule is met.

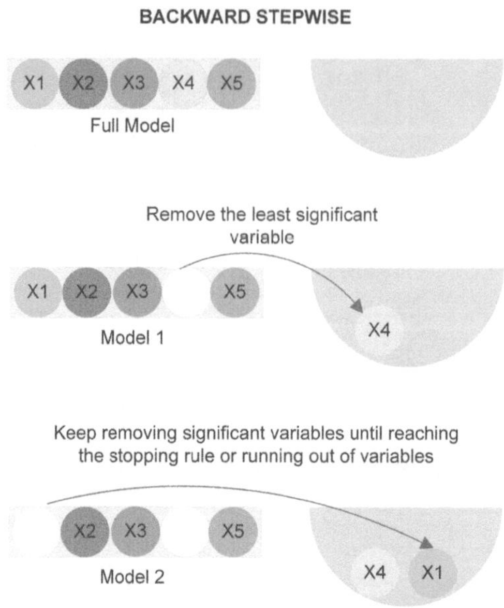

FIGURE 7.3 Backward stepwise method.

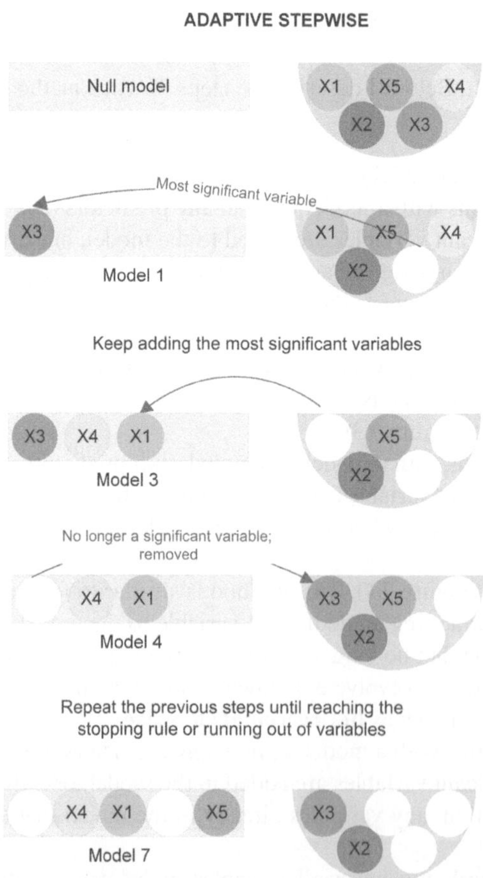

FIGURE 7.4 Adaptive stepwise method.

TABLE 7.6
When to Use Which Stepwise Method?

When to Use Forward Stepwise?	When to Use Backward Stepwise?	When to Use Adaptive Stepwise?
Large number of predictors When dealing with a large number of potential predictors, forward stepwise is computationally more efficient, as it starts with an empty model and incrementally adds variables.	**Multicollinearity concerns** If there are concerns about multicollinearity (high correlation) between predictors, backward stepwise helps identify and remove redundant predictors.	**Exhaustive search** Adaptive stepwise allows for an exhaustive search of significant predictors without becoming overly computationally intensive.
Exploratory analysis Forward stepwise is a good starting point for exploratory analysis, as it allows us to gradually build a model by adding variables that have a significant impact on the target variable.	**Well-founded theoretical basis** When there is a strong theoretical basis or domain knowledge for what predictors are more important than others, backward stepwise is useful in eliminating less significant variables from the model.	**Data exploration and model building** Adaptive stepwise is useful at the initial stages of data exploration and model building as it allows for a comprehensive examination of various predictor subsets and their impact on the model.
No prior knowledge When there is no strong prior knowledge or theory about which predictors are likely to be relevant, forward stepwise helps identify potentially important predictors.	**Complex models** Backward stepwise is particularly useful in complex models, where removing less relevant variables can improve the model's interpretability and generalization.	**Model fit improvement** Since adaptive stepwise considers all possible combinations of predictors, it provides more robust results and improves the model fit.
		Variable interaction analysis When interactions between predictors are essential, adaptive stepwise helps identify relevant interaction terms that might not be captured by individual forward or backward selection.

The important question is: **Which stepwise method should we use?** Table 7.6 outlines scenarios where each method is appropriate. As discussed earlier, **adaptive stepwise** seems to be the most capable of selecting predictors that best predict the target variable. Unlike forward and backward stepwise methods, which follow a linear and sequential process, adaptive stepwise allows for the dynamic addition and removal of variables, resulting in the best model fit. In other words, it can select the best combination of predictors for model-fitting purposes. For this reason, adaptive stepwise is generally recommended for data mining. However, it has a few disadvantages, such as requiring more computing time and, in some cases, leading to overfitting. To mitigate this, cross-validation with sub-samples can be used to evaluate predictive power and avoid overfitting.

> **Rule of thumb**
> Start with the adaptive stepwise method as it typically offers the best model fit by dynamically adding and removing predictors based on their significance.

Since SPSS Modeler uses different terminology for these methods, I have included a comparison in Table 7.7 to avoid confusion. Note that SPSS Modeler provides distinct estimation methods for both linear and logistic regressions. In this book, I refer to the three methods discussed above for both algorithms to ensure consistency.

> SPSS Modeler uses different terminology for these stepwise methods. See Table 7.7 for details.

TABLE 7.7
SPSS Modeler's Terms and Equivalent Terms

Regression Algorithms	SPSS Modeler's Terms	Equivalent Terms
Linear regression	Forwards	Forward stepwise
	Backwards	Backward stepwise
	Stepwise	Adaptive stepwise
Logistic regression	Stepwise	Adaptive stepwise
	Forwards	Forward stepwise
	Backwards	Backward stepwise
	Backward stepwise	Adaptive stepwise (with reverse order)

HOW TO READ AND UNDERSTAND THE REGRESSION EQUATION AND RESULTS?

After running a regression analysis, the first step is to evaluate the model fit using the recommended metrics, such as $\mathbf{R^2}$, **F-test**, **RMSE**, **MAPE**, **RSE**, **EP**, the **actual vs. predicted chart**, and the **residuals vs. predicted chart**. If all criteria are met and indicate a good model fit, the next step is to report and interpret the regression outputs. But how do we make sense of those outputs, and what conclusions can we draw?

These are common questions when conducting any data analysis, and they are even more crucial in data mining, where models often include a large number of predictors. To simplify the interpretation process, let's first review some simple hypothetical regression equations. For ease of understanding, I will limit the number of predictors. Later in this chapter, we will look at actual regression results using SPSS Modeler in a data mining example.

For linear regression, the estimation results typically include the following information:

- Intercept
- Unstandardized regression coefficient (B)
- Standard error for B
- Standardized regression coefficients (beta)
- t-value
- Probability value (p-value)
- 95% confidence interval for B

To interpret the regression results correctly, we focus on the **probability value (p-value)**, **unstandardized coefficients**, and **standardized coefficients**. Table 7.8 explains these terms and how to use them when interpreting regression results.

Probability Value (p-Value)

The **p-value** is a statistical measure used to assess the significance of the estimated coefficients (also known as regression weights or beta values) of the predictors in a regression model. It indicates whether the relationship between each predictor and the target variable is statistically significant or if it could have occurred by chance. The threshold for determining statistical significance is called the **significance level (alpha)**, with the most commonly used significance level being 0.05 (5%).

TABLE 7.8
Probability Value, Unstandardized Regression Coefficient, and Standardized Regression Coefficient

Terms	Description
Probability value (p-value)	The probability value (p-value) is a statistical measure for assessing the significance of the estimated coefficients (also known as the regression weights or beta values) for the predictors in the regression model. It determines whether the relationship between each predictor and the target variable is statistically significant or if it could have occurred by chance.
Significance level (alpha)	The significance level (alpha) is the threshold used in hypothesis testing to determine statistical significance. It represents the maximum probability of rejecting the null hypothesis when it is actually true (commonly set at 0.05).
Unstandardized regression coefficient (B)	The unstandardized regression coefficient (B) quantifies the relationship between a predictor and the target variable in its original units of measurement. It shows the expected change in the target variable for a one-unit change in the predictor, holding all other variables constant.
Standardized regression coefficient (beta)	The standardized regression coefficient (beta) quantifies the relationship between a predictor and the target variable after both are standardized to a common scale (typically in terms of standard deviations). Standardized coefficients range between 0 and 1 and make comparisons between predictors easier since they are on the same scale.

If the p-value is less than or equal to the significance level ($p \leq \alpha$), it suggests that the coefficient is statistically significant. For example, in a linear regression model where the target variable is students' SAT scores, suppose the analysis reveals that **GPA** and **family income** have p-values lower than 0.05. This would indicate that these two predictors are statistically significant and have a meaningful impact on predicting SAT scores.

Unstandardized Regression Coefficients (B)

After identifying significant predictors, the next step is to explain their relationships with the target variable using **unstandardized regression coefficients (B)**. Unstandardized coefficients quantify the relationship between each predictor and the target variable in their original units of measurement.

In our example of predicting SAT scores, suppose the regression results show unstandardized coefficients of 120 for **GPA** and 0.01 for **family income**, with an intercept of 105. The regression equation would be as follows:

$$SAT\ score = 105 + 120 \times GPA + 0.01 \times Family\ income$$

This equation includes the intercept and unstandardized coefficients. It suggests that **GPA** and **family income** are significant predictors of SAT scores, both positively correlated with the SAT score. These predictors represent the student's academic performance (via GPA) and the family's financial resources for education and SAT tutoring (via income).

Using this equation, we can predict a student's SAT score based on known values of these predictors. More importantly, we can interpret how changes in each predictor affect the target variable. For example, with a coefficient of 120 for GPA, a one-point increase in GPA results in a 120-point increase in the SAT score. Similarly, an increase in family income by $1 results in a 0.01-point increase in SAT scores. In practical terms, an increase of $10,000 in family income would raise the SAT score by 100 points ($0.01 \times 10{,}000 = 100$).

Note that the error term is not included in this equation because we have already used it to evaluate model fit. The error represents the difference between predicted and actual SAT scores, and since our model fit is good, the error is considered acceptable. At this stage, the focus is on using the equation to predict changes in SAT scores based on the changes in significant predictors, making the error term less relevant for interpretation.

Standardized Regression Coefficients (Beta)

One thing to notice about unstandardized coefficients is that they depend on the unit of measurement of the predictor, making them scale-dependent. For example, while GPA may have a higher regression coefficient than family income, we can't directly compare their impact because GPA is measured in points and income in USD. To make meaningful comparisons, we use **standardized coefficients.**

> **Unstandardized regression coefficients** and the **intercept** are used in the regression equation to calculate the target variable's value based on the predictors.
>
> **Standardized regression coefficients** are used to compare the effect size or the relative importance of the predictors.

A **standardized regression coefficient** measures the relationship between a predictor and the target variable after both have been standardized to a common scale. Standardized coefficients range from 0 to 1 and are expressed in terms of standard deviations, making them scale-independent and directly comparable across predictors.

For instance, if the standardized coefficients for GPA and family income are 0.68 and 0.34, respectively, we can conclude that GPA has twice the effect size of family income in predicting SAT scores. In this case, GPA is more important than family income by a factor of two.

To identify and rank important predictors, it is recommended to use **standardized coefficients** and rank them from highest to lowest. In data mining, a more generic term used to capture the significance of predictors is **"relative importance."**

EXAMPLE: LINEAR REGRESSION

In this example, we conduct a linear regression analysis using the Boston dataset in SPSS Modeler. In the Chapter 3 exercise, you had the opportunity to explore the data. Now, in this chapter, we will build a predictive model using linear regression.

Data Description

- **Dataset:** Boston.sav
- **Target variable:** Imv (Continuous—Median home value)

The **Boston Housing** dataset contains housing data from the Boston, MA, area collected by the U.S. Census Bureau. The dataset includes 14 fields, as outlined in Table 7.9. All fields are continuous (the **Charles_River** field, originally binary, has been dummy-coded into a continuous variable). The target field is **Imv**, representing the median home values in the area.

The objective of this analysis is to predict the median housing values in Boston, making linear regression the appropriate method for this task.

TABLE 7.9
Boston Housing Data Description

Field	Description	Scale	Role
CrimePC	Crime rate per capita	Continuous	Input
ZonePr	Proportion of residential land zoned for lots over 25,000 sq. ft.	Continuous	Input
IndustPr	Proportion of non-retail business acres.	Continuous	Input
Charles_River	Charles River dummy variable (1 if tract bounds river; 0 otherwise)	Continuous	Input
NiOxSQ	Nitric oxides concentration (parts per 10 million)	Continuous	Input
RmSQ	Average number of rooms per dwelling	Continuous	Input
AgePre40	Proportion of owner-occupied units built prior to 1940	Continuous	Input
LogDistance	Weighted distances to five Boston employment centers	Continuous	Input
LogAccess	Index of accessibility to radial highways	Continuous	Input
Tax	Full value property tax rate per $10,000	Continuous	Input
PtRatio	Pupil-teacher ratio	Continuous	Input
B	1,000 $(Bk - 0.63)^2$ where Bk is the proportion of blacks by town	Continuous	Input
Lstat	% lower status of the population	Continuous	Input
Imv	Median value of owner-occupied homes in $1,000s	Continuous	Target

We have completed three important steps in the data mining process for the Boston dataset: **Data Exploration**, **Transformation**, and **Sampling**. Since these steps were thoroughly covered in Chapters 3, 4, and 5, I won't repeat the procedures or results here. However, I will summarize these tasks and present them in the stream for easier reference.

Data Exploration

In this step, the **Data Audit 1** node is used to generate histograms for all continuous fields and produces descriptive statistics. The data exploration results show that the dataset contains **506** records. The target field (**Imv**) has 54 missing values, but the predictors do not have any missing values. Since linear regression performs better with a complete dataset, it is necessary to impute those missing values.

Additionally, most fields do not follow normal distributions based on their skewness and kurtosis values (these values should ideally be between −0.5 and 0.5 for data to be considered normally distributed). The histograms reveal that some fields, including **Imv**, **Age**, and **Tax**, are slightly skewed, with a few outliers. Other fields exhibit significant skewness and will require normalization (for more details, refer to the Boston dataset example in Chapter 3).

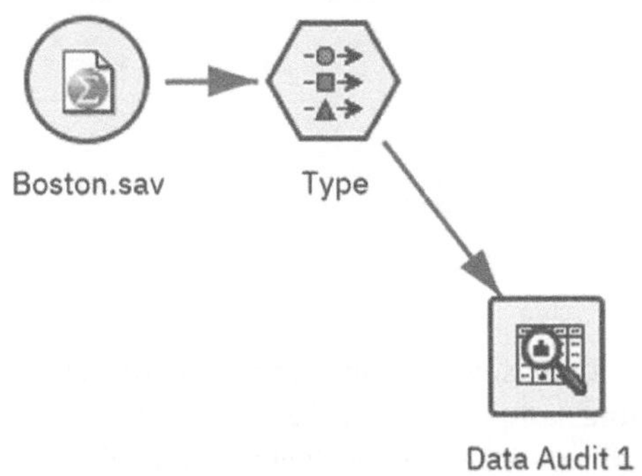

Data Sampling

Typically, the next step is sampling. The dataset contains 506 records, and since this is relatively small, we will use all the records for analysis. Therefore, no data sampling is performed at this stage. The primary task in this step is data partitioning, but we will address that after data modification.

Data Modification

As mentioned earlier, most input fields exhibit non-normal distributions, and the target field (**lmv**) contains missing values. In this step, we perform **data imputation** and **normalization**.

- **Data imputation**: Using the Quality tab from the Data Audit 1 output, we generate a SuperNode labeled "Missing Value Imputation," which replaces missing values in the lmv field with the mean value.
- **Data normalization**: To normalize the data, we apply the Transform node to select appropriate transformation functions. This generates a SuperNode called "Supernode_Transform," which normalizes most input fields except for lmv, Age, and Tax, as they already exhibit near-normal distributions (see the Boston example in Chapter 5 for more details).

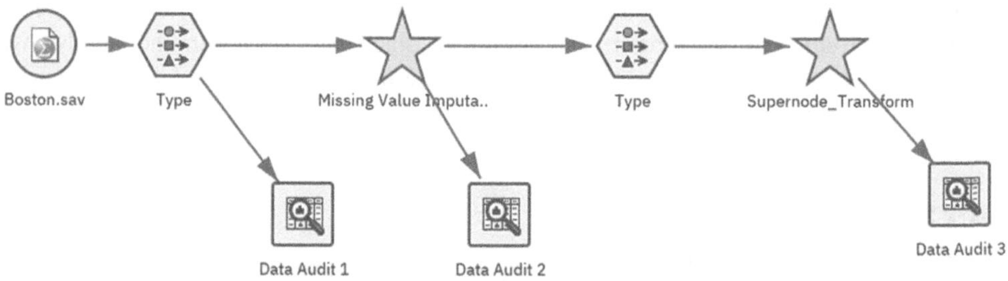

Data Partition

Before building the regression model, it is essential to partition the data. Since the sample size is smaller than 1,000, as recommended in Chapter 4, we partition the data into training and test samples using a 60-40 split ratio. No validation sample is created in this case. The **Partition node** follows the **Super_Transform node** in the workflow.

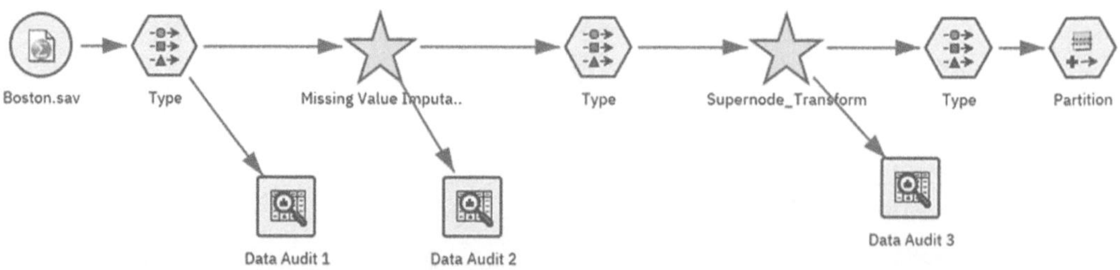

You can set up a data partition using the following selection. This process divides the data into two sub-samples: The **training sample** (60% of the data) will be used to train the model while the **test sample** (40% of the data) will be used to evaluate the model's performance.

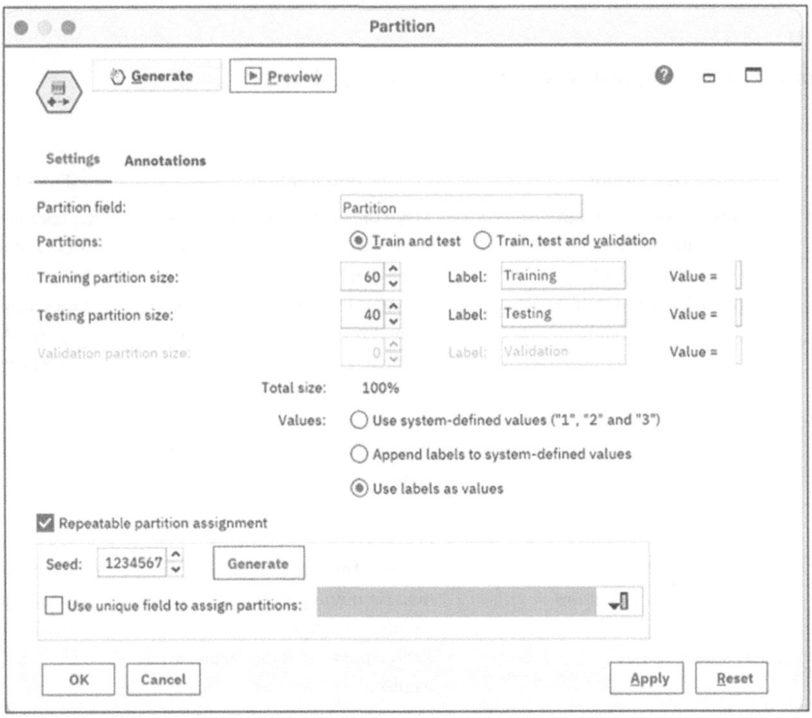

Modeling

Our goal is to build and execute a linear regression model for the target field **lmv**. There are several methods available for model creation, and SPSS Modeler offers a variety of tools in the Modeling group. To explore which method offers the best predictive power, I will build four different linear regression models and compare them. The specifics of these models are outlined in Table 7.10.

With these four regression models, we will have a comprehensive stream that includes all steps of the data mining process. The stream will display how to build these regression models, the flow of nodes, and the connections between steps. I will also provide specific instructions for setting up the configurations for each model.

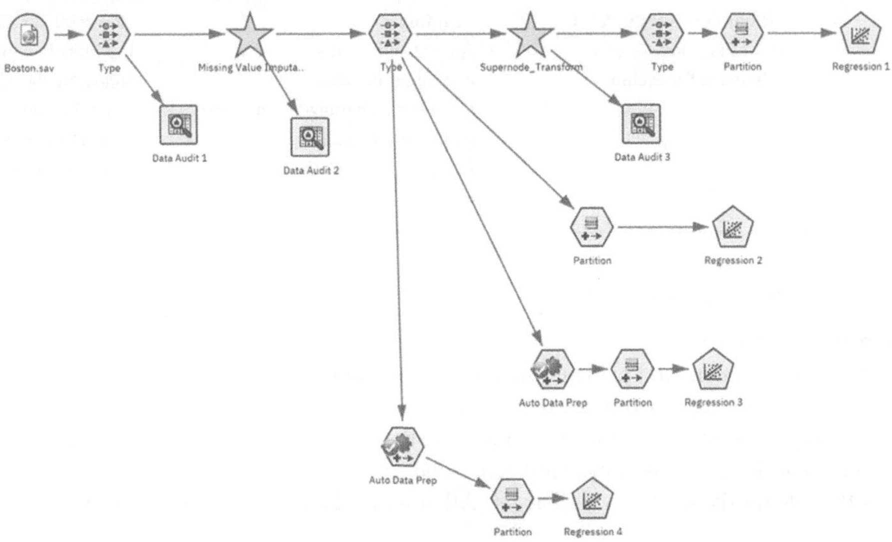

TABLE 7.10
Descriptions and Configurations of the Regression Models

Linear Regression Models	Description	Configurations	SPSS Modeler Tools
Regression 1	Uses the output from the imputation and transformation steps (complete and normalized data used for training)	• Mean imputation method for the field **Imv**. • Specific transformation functions (as indicated in Chapter 5) used to normalize 11 continuous fields (excluding **Imv, Age,** and **Tax**). Transformed input fields are used to train the model. • *Regression model*: the Stepwise method	Regression 1 node (under Modeling group), following the **Missing Value Imputation**, **Supernode_ Transform**, and **Partition nodes**.
Regression 2	Uses the output from the imputation step only (complete data used without normalization)	• Mean imputation method for the field **Imv**. • Imputed **Imv** used as the target field for training the model. • No normalization. • *Regression model*: the Stepwise method.	Regression 2 node (under Modeling group), following the **Missing Value Imputation** and **Partition nodes**.
Regression 3	Balances speed and accuracy—using the **Auto Data Prep** node for data transformation. The complete and normalized data is used to train the model, but normalization is performed automatically by the **Auto Data Prep** node, with the objective set to "**Balance speed and accuracy**."	• Mean imputation method for the field **Imv**. • Auto Data Prep node used to normalize all fields. • *Objectives*: Balance speed and accuracy (no outlier handling). • *Regression model*: the Stepwise method.	Regression 4 node (under Modeling group), following the **Missing Value Imputation**, **Auto Data Prep**, and **Partition nodes**.
Regression 4	Same as Regression 3, but the objective for the **Auto Data Prep** node is set to "Optimize for accuracy"	• Mean imputation method for the field **Imv**. • Auto Data Prep node used to normalize all fields. • *Objectives*: Optimize for accuracy (with outlier handling). • *Regression model*: the Stepwise method.	Regression 4 node (Regression node under the Modeling group), following the Missing Value Imputation node, Auto Data Prep node, and Partition node.

How to Set Up the Regression 1 Model

1. Open SPSS Modeler.
2. Create a new stream and name it "**Boston_Regression**."
3. Add the data source **Boston.sav** to the stream.
4. Add a **Type** node to the stream and connect it to the data source. Open the **Type** node settings.
5. Double-check field names, measurements, and roles.
6. Set the role for the field **Imv** as **Target**. All other fields should be set as **Input**.

7. Perform imputation and transformation using **SuperNodes** as described above. After completing these steps, you should have the **SuperNode Missing Value Imputation** for imputation and the **SuperNode Supernode_Transform** for normalization.
8. Add another **Type** node to the stream and connect it to the **SuperNode Supernode_Transform**. Open the **Type** node settings.
9. Since the transformed data is used for most fields (except for **Imv**, **Age**, and **Tax**) to train the regression model, set the role of **unused fields** as **None**. This step ensures SPSS Modeler does not use those fields in the next analysis. The predefined **Target** and **Input** fields will be passed on to the next node for analysis.

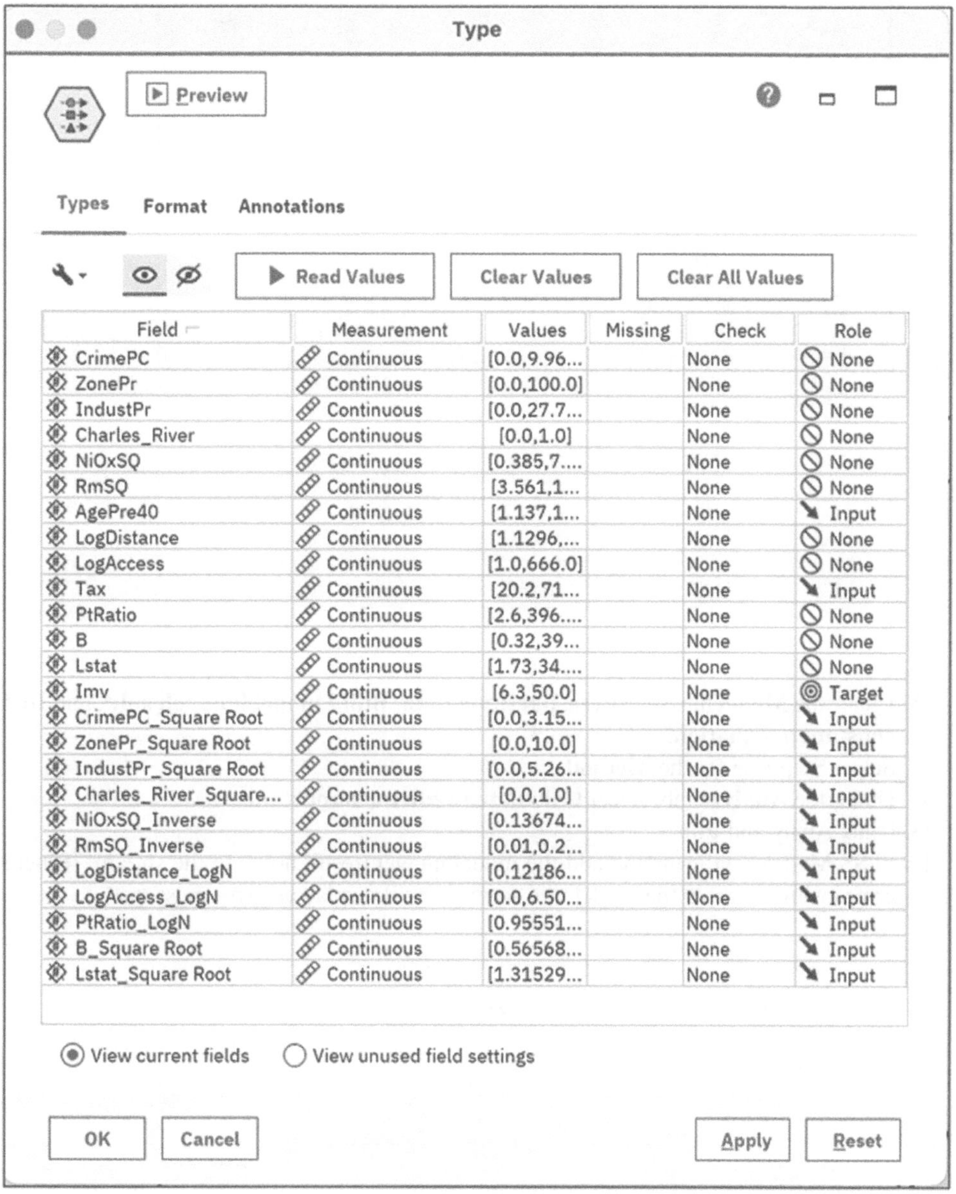

10. Click Apply, and then click OK.

11. Add a **Partition** node to the stream and connect it to this **Type** node. Open the **Partition** node settings.
12. Select **Train and test** and set the **training partition size to 60** and the **testing partition size to 40** (refer to the Data Partition step above for details).
13. Click Apply, and then click OK.
14. From the **Modeling** group, add a **Regression** node to the stream, name it **Regression 1**, and connect it to the **Partition** node. Open the **Regression** node settings.

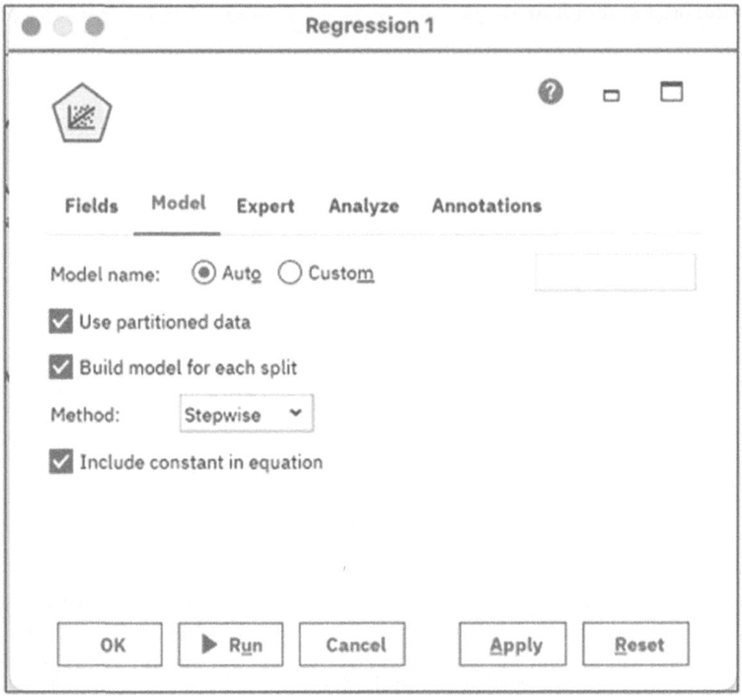

15. Under the **Model** tab, select **Use partition data**, **Build model for each split**, and **Include constant in equation**.
16. Select **Stepwise** for the **Method**.
17. Under the **Analyze** tab, select **Calculate predictor importance**.
18. Click Apply and Run.
19. SPSS Modeler will run and add the model nugget (result node) to the stream. If there are errors while running the model, the software will display an error message.

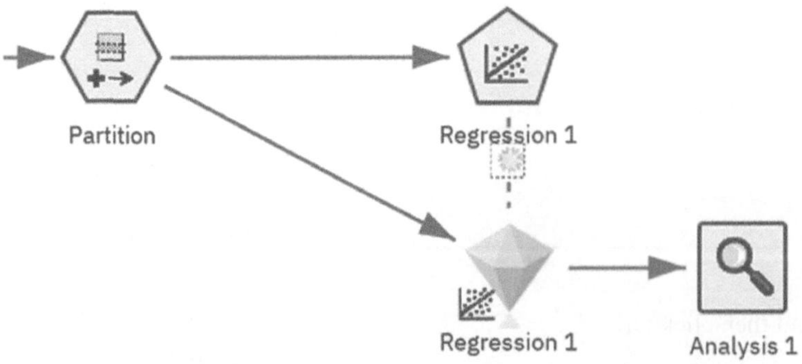

20. From the **Output** group, add an **Analysis** node to the stream and connect it to the model nugget **Regression 1**. This node will compare the model performance between the training and test samples. Open the **Analysis** node settings.
21. Under the **Analysis** tab, select **Performance evaluation**, **Evaluation metric**, **Confidence figures**, and **Separate by partition**.

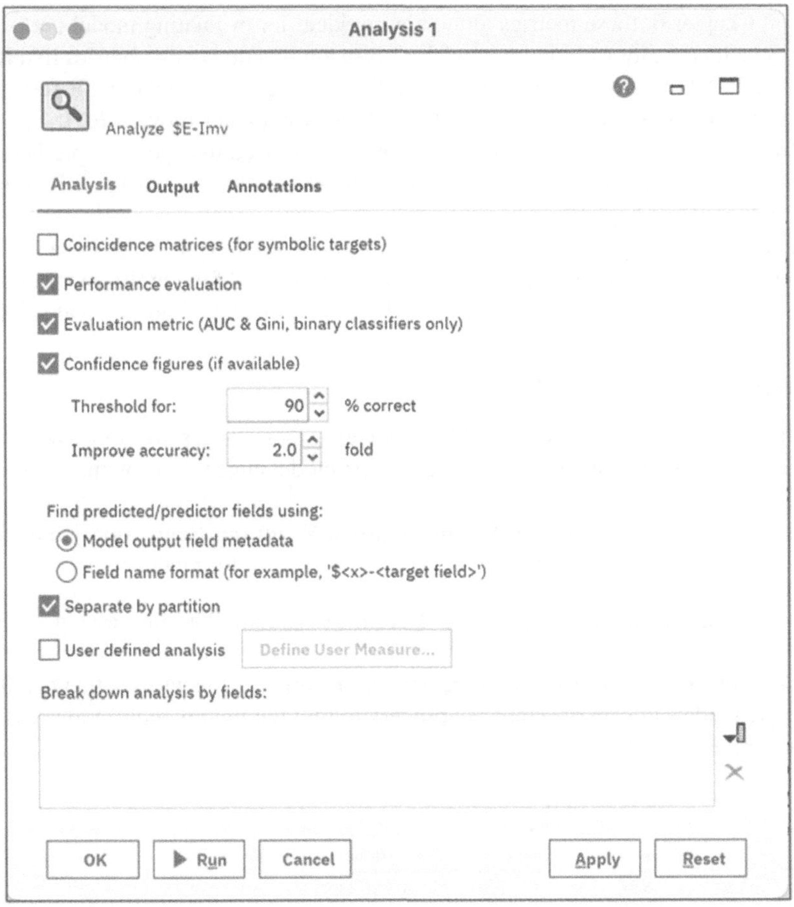

22. Click Apply and Run.

Extracting Model Fit Results

The Analysis node results show the model's performance for both the training and test samples. In this output, the field **$E-Imv** represents the **predicted value** of the target field while **Imv** is the **actual (observed) value** of the target field. Most of the metrics are straightforward. Remember, **error** refers to the difference between observed and predicted values.

Below is a brief description of the key metrics:

- **Minimum error**: The smallest error value.
- **Maximum error**: The largest error value.
- **Mean error**: The average of errors across all records, indicating any potential systematic bias in the model.
- **Mean absolute error (MAE)**: The average of the absolute errors across all records, showing the average magnitude of error.

- **Standard deviation**: The standard deviation of the errors.
- **Linear correlation**: The linear correlation between the predicted and actual values, ranging from –1 to 1.
- **Occurrences**: The sample size. In this case, 304 records are used for training the model and 202 records are used for testing the model.

As discussed in Chapter 6, these metrics alone are not ideal for evaluating model performance and comparison. Specifically, for association models, it is recommended to use metrics like **RMSE**, **MAPE**, **RSE**, and **EP**. Additionally, metrics specific to linear regression—including **F-test**, $\mathbf{R^2}$, and **adjusted** $\mathbf{R^2}$—should be used. However, since SPSS Modeler does not provide **RMSE**, **MAPE**, **RSE**, and **EP** directly through the Analysis node output, we will need to export the prediction outputs to Excel and calculate these metrics separately (see the section "Steps to Export Regression 1 Output to Excel" later in this chapter.).

One important thing to note is that regression-specific metrics, such as the F-test, R^2, and adjusted R^2, are only provided for the training sample. While these can be used for a preliminary evaluation of the regression model, they are not suitable for model comparison and reliability evaluation.

Extracting F-test, R^2, and Adjusted R^2

To extract the F-test, R^2, and adjusted R^2 from SPSS Modeler, these values are stored in the model nugget rather than in the Analysis node. You can access the model nugget to view these regression-specific metrics.

Follow the steps below to extract F-test, R^2, and adjusted R^2 for our Regression 1 model. Note that these results are based on the training sample, which we can use for model comparison.

1. Double-click the model nugget **Regression 1** from the stream. The Regression 1 result window will open.
2. Under the **Advanced** tab, select **Regression/Model Summary**. You will see that the algorithm ran nine models during the stepwise process to find the final model. Model 9 is the final model with $R^2 = 0.735$ and adjusted $R^2 = 0.727$.

Model Summary

Model	R	R Square	Adjusted R Square	Std. Error of the Estimate
1	.733[a]	.537	.536	5.696519
2	.756[b]	.571	.568	5.496741
3	.763[c]	.582	.578	5.434079
4	.787[d]	.620	.615	5.188911
5	.842[e]	.708	.703	4.554442
6	.847[f]	.718	.713	4.483853
7	.851[g]	.724	.717	4.448027
8	.854[h]	.730	.722	4.407756
9	.857[i]	.735	.727	4.369803

3. Select the **ANOVA tab**, where you can view the F-test for Model 9, which is 90.629.

As explained earlier, in linear regression, $\mathbf{R^2}$ is equivalent to **EP**, and **RSE** can be calculated as **1 – EP**. Therefore, we have $\mathbf{EP = R^2 = 0.735}$ and $\mathbf{RSE = 1 - EP = 1 - 0.735 = 0.265}$. In other words, based on the training sample results, this model has an **EP of 73.5%**, meaning it can explain 73.5% of the variance in the target field (**Imv**) based on the variance in the predictors.

That's how you extract these metrics. I will dive deeper into the details of the regression model once we identify the **champion model**.

Regression Models 2, 3, and 4 Setup

Follow the same steps as outlined for **Regression 1** to set up and configure **Regression Models 2, 3, and 4**, using the configurations and nodes described in Table 7.10. Specifically:

Regression 2:

- Use imputed data without normalization.
- Use the **Regression node** with the **Stepwise method**.

Regression 3:

- Use the Auto Data Prep node with the objective set to **Balance speed and accuracy** (no outlier handling).
- Use the Regression node with the **Stepwise method**.

Regression 4:

- Use the Auto Data Prep node with the objective set to **Optimize for accuracy** (with outlier handling).
- Use the Regression node with the **Stepwise method**.

Running the Models

Run each model separately and add the Analysis node, as described for Regression 1. Once the models are run, follow the same steps to extract the model fit metrics from the Analysis node for the testing sample.

Model Comparison in SPSS Modeler

In SPSS Modeler, comparing selected metrics across models is not straightforward. You cannot connect one **Analysis node** to multiple model nuggets to run a comparison, as the **Analysis node** can only be linked to a single preceding node. The process for setting up model comparisons in SPSS Modeler is somewhat unintuitive and requires modifying the stream. Additionally, there are some restrictions on when model comparisons can be performed using the **Analysis node** and when they cannot.

I will demonstrate this approach at the end of this chapter so you can understand how it works.

Champion Model

For the purpose of model comparison, I will use MAE and linear correlation based on the testing samples results. Additionally, I will include the F-test, R^2, and adjusted R^2, but we must remember that these metrics are derived from the training sample only. Table 7.11 presents the model comparison results.

The results indicate that all models have significant F-test statistics, demonstrating the significance of the overall models. When comparing across models, Regression 1 has the lowest errors (MAE and RSE) and the highest explanatory power (EP, R^2, and adjusted R^2) compared to the other three models.

There is no significant difference between Regression 2 and Regression 3, suggesting that the normalization performed by the Auto Data Prep node does not significantly affect the overall model fit. Finally, Regression 4 shows the poorest model fit despite using the Auto Data Prep node with the objective to Optimize for accuracy, which included handling missing values, outliers, and normalizing the data.

TABLE 7.11
Model Comparison

Model Fit Metrics	Sample	Regression 1	Regression 2	Regression 3	Regression 4
MAE	Testing	3.437	3.976	3.976	4.406
Linear correlation	Testing	0.783	0.735	0.735	0.677
RSE (same as R^2)	Training	0.265	0.434	0.434	0.484
EP	Training	0.735	0.566	0.566	0.516
F-test	Training	90.629	64.645	64.645	160.277
R^2	Training	0.735	0.566	0.566	0.516
Adjusted R^2	Training	0.727	0.558	0.558	0.513

Therefore, based on the testing sample's results, Regression 1 has the lowest MAE and the highest linear correlation, a conclusion further supported by the training sample's results. Thus, we can confidently identify Regression 1 as the champion model.

Reliability of the Champion Model

As described in Chapter 6, after identifying the champion model, it is important to further assess its reliability. To do this, we can use the metrics provided by the Analysis node for comparison (see Table 7.12). I have already explained how to extract these metrics earlier.

The results show some discrepancies between the training and testing samples, particularly in the maximum error and minimum error. However, since these metrics can depend on the sample size, they are not the most reliable indicators for evaluating model reliability. Instead, if we focus on metrics that are independent of sample size, such as MAE and linear correlation, the results between the two samples are quite close (with differences of less than 10%).

Therefore, we can conclude that the model's reliability is acceptable. There is no evidence of overfitting, and the model generalizes well across the training and testing samples.

Validity of the Champion Model

The next step is to evaluate the validity of the champion model by using complete model fit metrics. The missing metrics from the earlier model comparison include RMSE, MAPE, the actual vs. predicted chart, and the residual vs. predicted chart. Since SPSS Modeler does not provide these metrics directly, we need to export the prediction results to Excel to calculate RMSE and MAPE, and to construct the necessary charts. As this process is somewhat time-consuming, I will only perform it for the champion model—Regression 1.

TABLE 7.12
Reliability Evaluation of the Champion Model—Regression 1

Metrics	Testing	Training
Minimum error	−11.624	−15.674
Maximum error	29.489	20.466
Mean error	0.353	0
Mean absolute error	3.437	3.16
Standard deviation	5.183	4.304
Linear correlation	0.783	0.857
Occurrences	202	304

Steps to Export Regression 1 Output to Excel:

1. Add a **Type** node to the stream and connect it to the model nugget for **Regression 1**. Open the node settings.
   (Note: This step is necessary to extract the output correctly.)
2. Double-check all fields. You should now see a new field, **$E-lmv**, added to the data. This field represents the predicted value of the target field.
3. From the **Export** group, add an **Excel** node to the stream and connect it to the **Type** node. Open the **Excel** node settings.

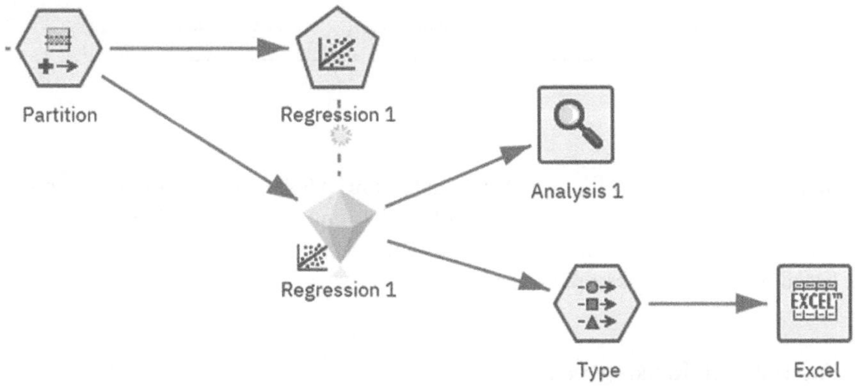

4. For **File name**, enter the name "**Boston_regression1.xlsx**" and select the desired location on your computer. The output will be saved to this file.

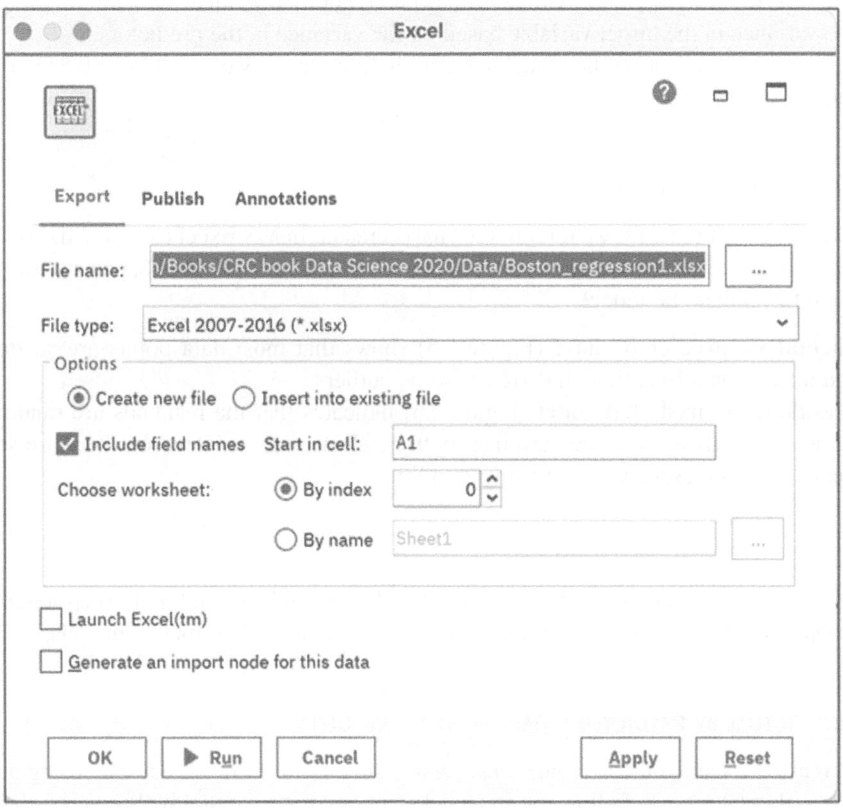

5. Click Apply and Run.

TABLE 7.13
Validity Evaluation of the Champion Model—Regression 1

Metrics	Testing Sample's Results	Evaluation
MAE	3.437	N/A
RMSE	5.182	N/A
MAPE	0.147	Acceptable
RSE	0.394	Acceptable
EP	0.606	Acceptable

You should now find the Excel file "**Boston_regression1.xlsx**" at the location you chose. Then, use the formulas described in Chapter 6 to calculate the required metrics. Specifically, calculate RMSE and MAPE for the testing sample. Table 7.13 will present the model fit metrics for the champion model based on the testing sample's results.

Model Fit Analysis for Regression 1

The RMSE is 5.182, but since it is scale-dependent, it is only useful for model comparison. As mentioned earlier, in linear regression, R^2 is essentially the same as EP. Thus, we have an EP of 0.606 or 60.6%, and a RSE of 0.394 or 39.4%, which is considered good. This means the model can explain 60.6% of the variance in the target variable based on the variance in the predictors.

The MAPE is 0.147, indicating the model has a prediction accuracy of 1 – 0.147 = 0.853 or 85.3%, which is considered acceptable.

Chart Analysis for Testing Sample

To further examine the model fit, we refer to the charts created in MS Excel using the data exported from the Export node. We focus on the testing results, as described in Chapter 6, because the testing sample is used to validate the model.

- The **actual vs. predicted chart** (Figure 7.5) shows that most data points are scattered around the diagonal line, though there are some outliers.
- The **residual vs. predicted chart** (Figure 7.6) indicates that the residuals are randomly distributed on both sides of the zero line, with no clear pattern. However, there are some noticeable outliers, especially above the zero line.

Conclusion

In summary, the model's predictive power is considered acceptable, though not exceptional. In the following chapters, we will explore other methods to build additional predictive models.

Creating the Actual by Predicted Chart in SPSS Modeler

In addition to using Excel, you can create these charts directly in SPSS Modeler by using a scatter plot, as discussed in Chapter 3. Follow the steps below to create the Actual by Predicted chart:

1. Continue with the Regression 1 stream from above.

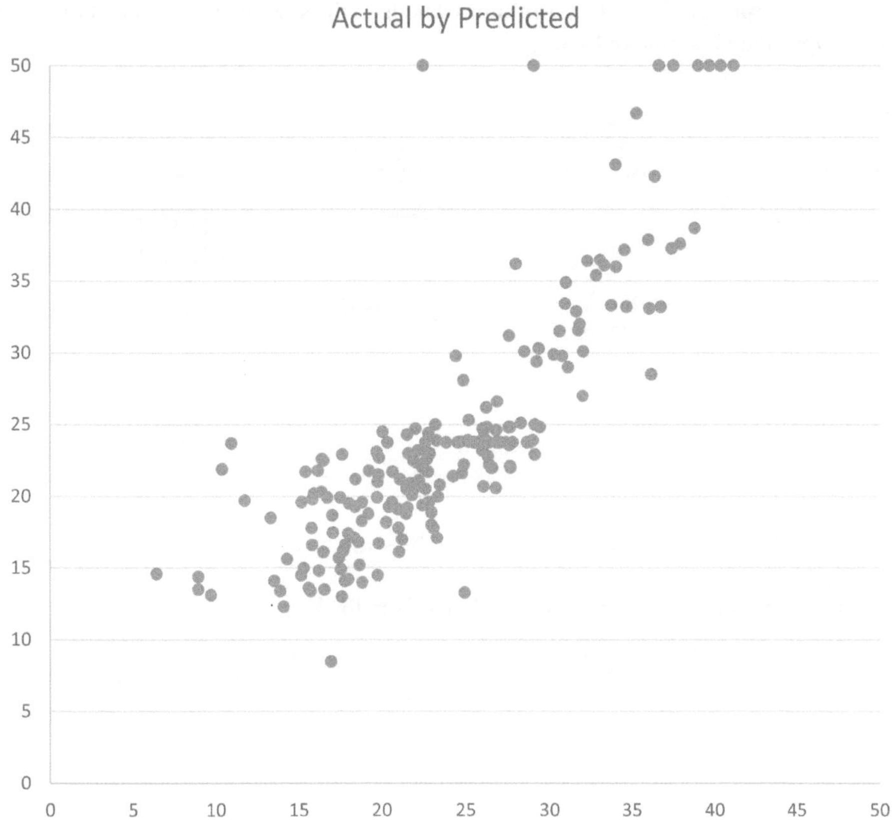

FIGURE 7.5 Actual vs. predicted chart of the champion model—Regression 1.

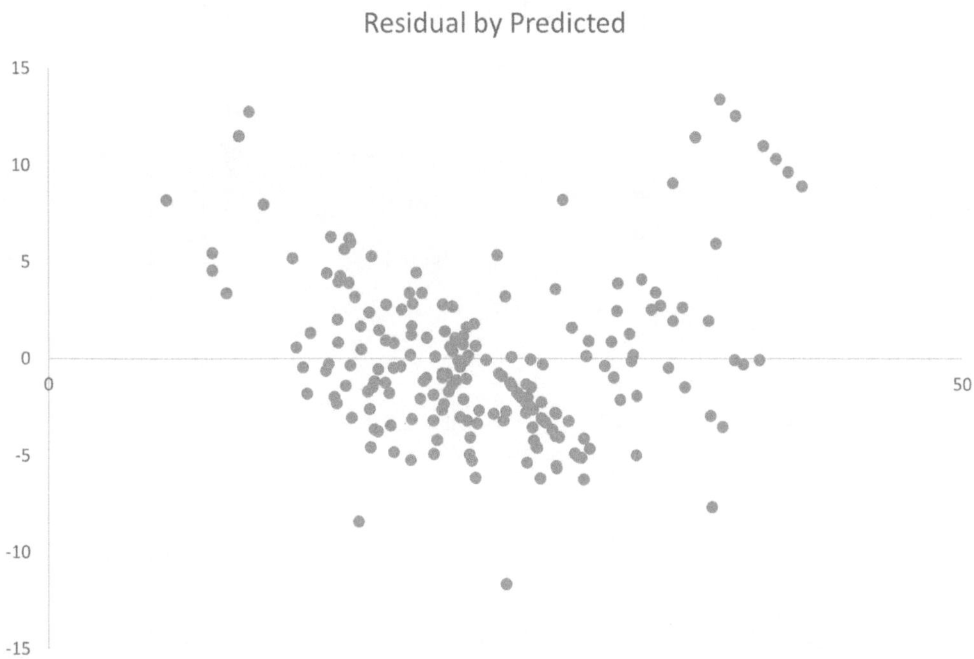

FIGURE 7.6 Residual vs. predicted chart of the champion model—Regression 1.

2. From the **Graphs** group, add a **Graphboard** node to the stream and connect it to the **Type** node. Open the **Graphboard** settings.

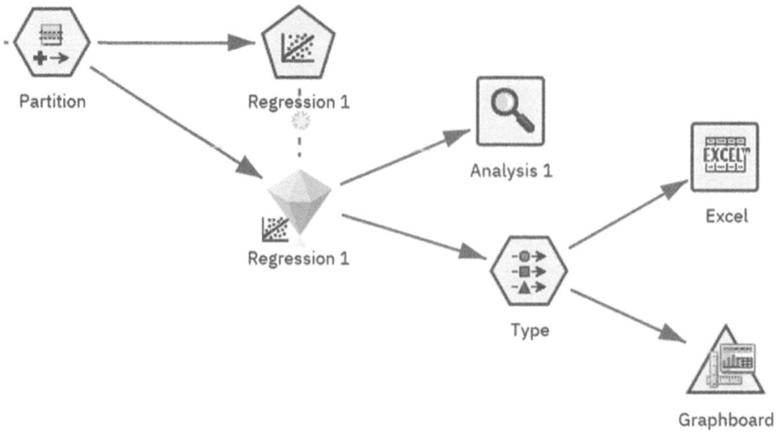

3. Under the **Basic** tab, select the fields **Imv** and **$E-Imv** on the left, and then choose **Scatter** plot on the right.
4. Under the **Detailed** tab, set **$E-Imv** for **X** and **Imv** for **Y**.

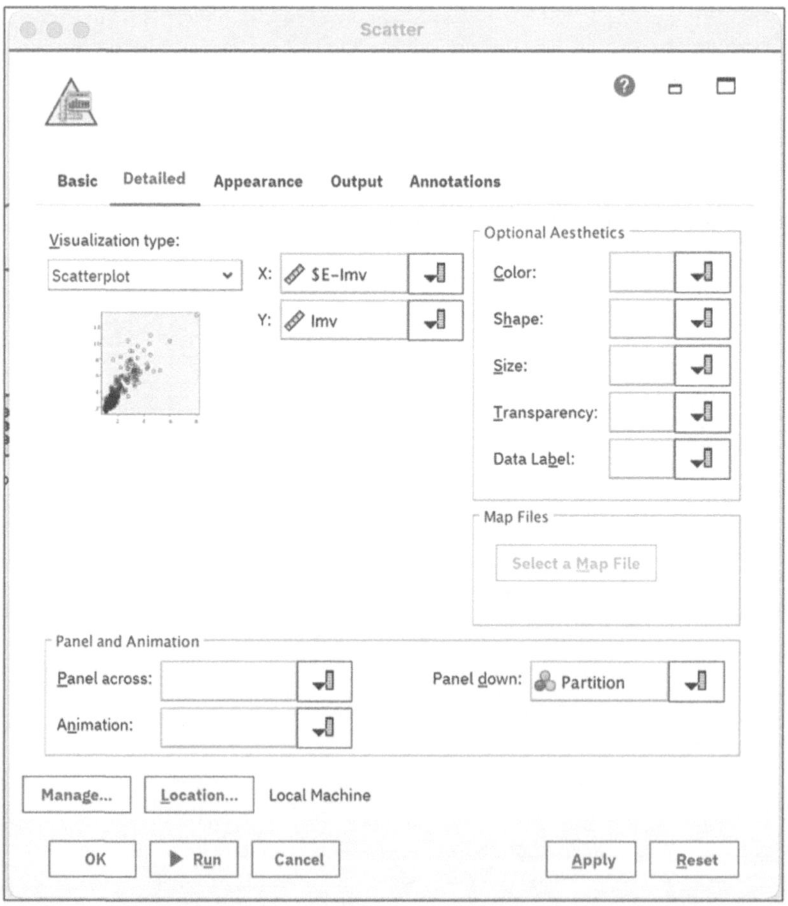

5. Click Apply and Run.

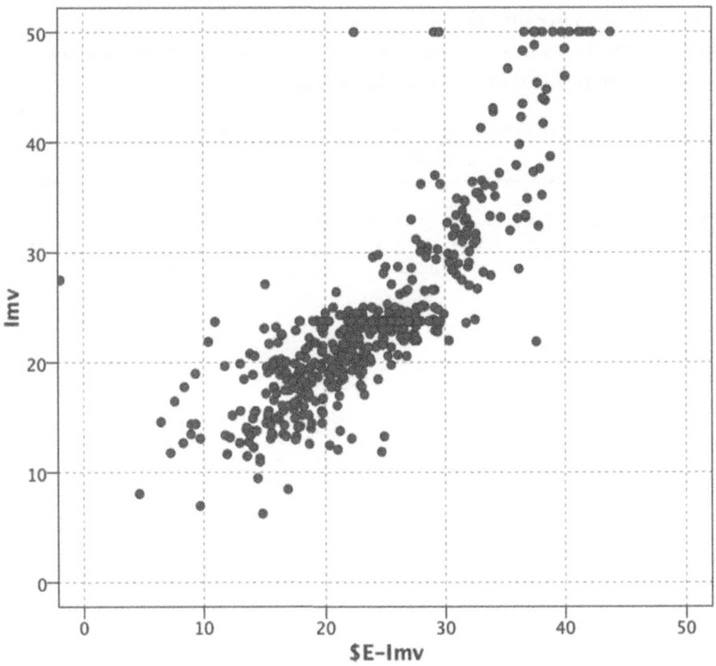

FIGURE 7.7 Actual vs. predicted chart created by SPSS Modeler (Regression 1).

You will generate a scatter plot that shows the Actual by Predicted chart (Figure 7.7). This chart will look the same as the one you created earlier using Excel (Figure 7.5).

Creating the Residual by Predicted Chart in SPSS Modeler

To create the **Residual by Predicted** chart in SPSS Modeler, an extra step is required to calculate the residuals using the **Derive** node. After that, you can use the Graphboard to generate a scatter plot for the residuals against the predicted values (**$E-Imv**).

Follow the steps as given below:

1. Continue with the Regression 1 stream.
2. From the **Field Ops** group, add a **Derive** node to the stream and connect it to the **Type** node. Open the **Derive** settings.

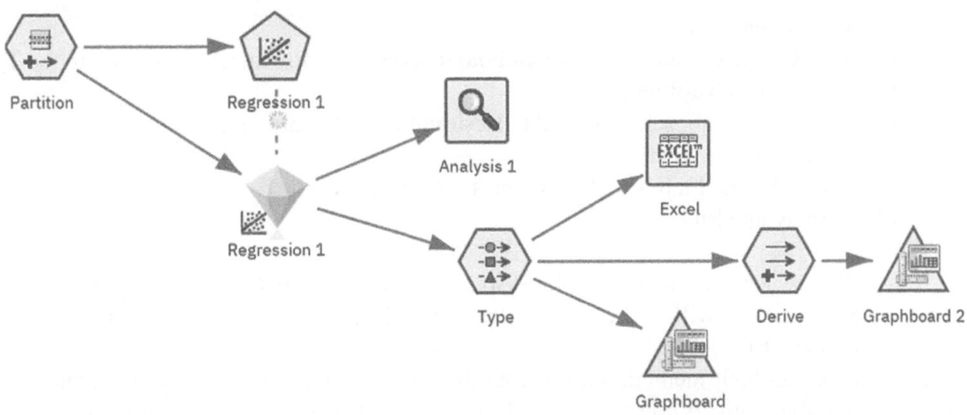

3. For **Derive** field, enter **Residual**.
4. For **Formula**, enter the following formula: `Imv - $E-Imv`. This will calculate the residual values and add them to the dataset in the **Residual** field.

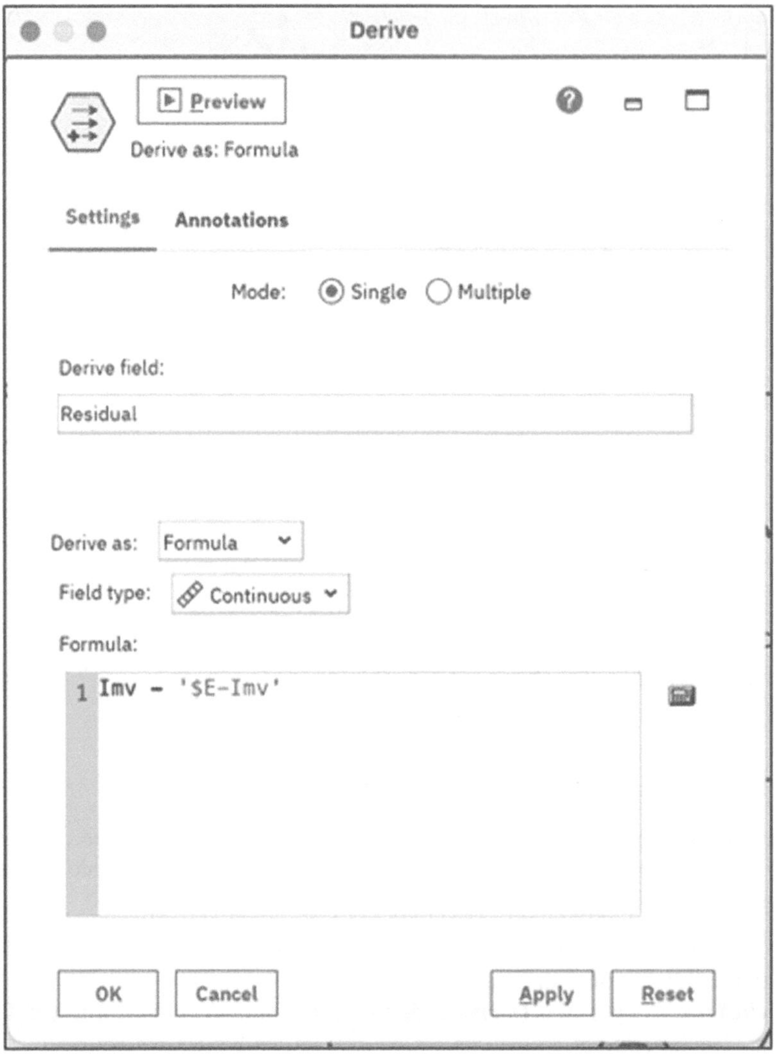

5. Click Apply and OK.
6. From the **Graphs** group, add a **Graphboard** node to the stream and connect it to the **Type** node. Open the **Graphboard** settings.
7. Under the **Basic** tab, select the fields **Residual** and **$E-Imv** on the left, and then choose **Scatter** plot on the right.
8. Under the **Detailed** tab, set **$E-Imv** for *X* and **Residual** for *Y*.
9. Click Apply and Run.

You will now generate a scatter plot displaying the **residual by predicted ($E-Imv)** values (Figure 7.8). This chart will be identical to the **residual by predicted** chart you previously created in Excel (Figure 7.6).

I've shown you both methods for creating these charts, and you can choose whichever you're more comfortable with. While using Excel may be quicker and easier if you're familiar with it,

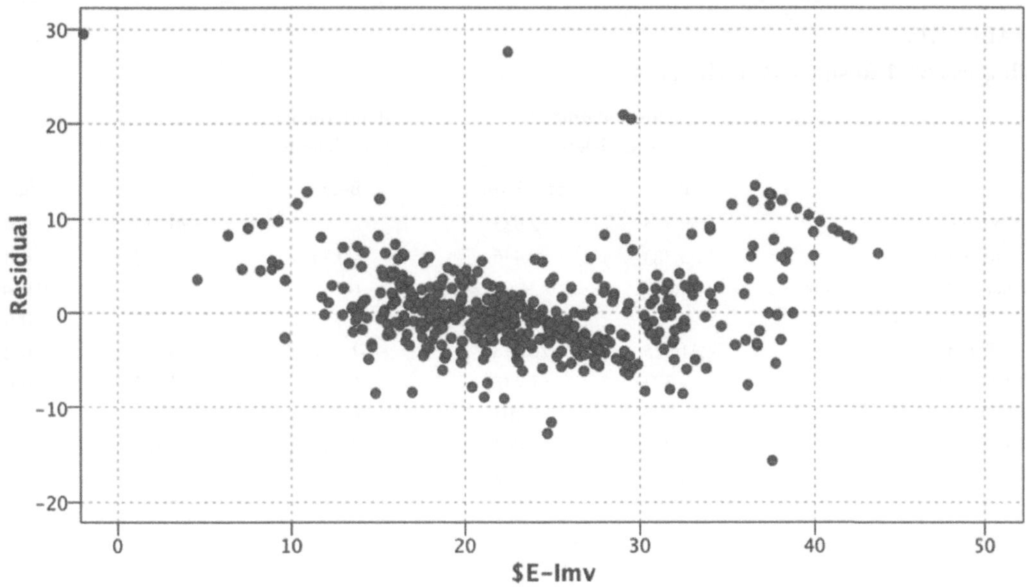

FIGURE 7.8 Residual vs. predicted chart created by SPSS Modeler (Regression 1).

knowing how to create these charts in SPSS Modeler is a valuable skill, as it can be applied to many similar tasks.

Regression Equation Interpretation

We have established that Regression 1 is the champion model, and its overall reliability and validity are acceptable. The next step is to interpret the model structure and regression equation to explain the effects of the significant predictors. For this, we use the regression output from SPSS Modeler. All results are stored in the model nugget for Regression 1.

First, we access the model summary results. Open the model nugget and, under the Advanced tab, extract the model summary results (as shown in Table 7.14). The results indicate that the stepwise process identifies nine steps. By the final step, nine predictors are included in the

TABLE 7.14
Regression 1 Results—Model Summary

Model	R	R^2	Adjusted R^2	Std. Error of the Estimate
1	0.733	0.537	0.536	5.697
2	0.756	0.571	0.568	5.497
3	0.763	0.582	0.578	5.434
4	0.787	0.620	0.615	5.189
5	0.842	0.708	0.703	4.554
6	0.847	0.718	0.713	4.484
7	0.851	0.724	0.717	4.448
8	0.854	0.730	0.722	4.408
9	0.857	0.735	0.727	4.370

TABLE 7.15
Regression 1 Results—Coefficients

	Unstandardized Coefficients		Standardized Coefficients		
	B	Std. Error	Beta	t	Sig.
(Constant)	59.334	2.935		20.218	< .001
Lstat_Square root	−3.583	0.446	−0.39	−8.037	< .001
Charles_River_Square root	3.079	1.06	0.125	2.905	0.004
LogDistance_LogN	−8.383	0.782	−0.778	−10.717	< .001
RmSQ_Inverse	−221.525	20.598	−1.311	−10.755	< .001
NiOxSQ_Inverse	9.387	1.235	0.731	7.602	< .001
B_Square root	0.322	0.103	0.204	3.126	0.002
Tax	−0.012	0.003	−0.264	−3.703	< .001
ZonePr_Square root	0.305	0.12	0.109	2.555	0.011
CrimePC_Square root	1.554	0.627	0.16	2.479	0.014

model, all of which are significant predictors (meaning they have significant correlations with the target variable).

Table 7.14 presents the iterations of the stepwise steps, along with the corresponding **R**, $\mathbf{R^2}$, **adjusted** $\mathbf{R^2}$, and **standard error**. As we can observe, with each additional predictor, the R^2 increases, and the standard error decreases. In Step 9, the model achieves the lowest standard error and the highest R^2. This ninth iteration is considered the final model, as all the relevant fields have been included.

Table 7.15 presents the coefficient results, allowing us to determine the effects of predictors on the target variable (Imv). As shown in Table 7.14, the final model includes nine significant predictors (with p-values less than 0.05). Using the unstandardized coefficients, we can write the regression equation as follows. To keep the equation concise, I've added **_t** at the end of transformed field names to indicate that the field has been transformed.

$$Imv = 59.334 - 3.583 \times Lstat\_t + 3.079 \times Chalers\_River\_t - 8.383 \times LogDistance\_t - 221.525 \times RmSQ\_t + 9.387 NiOxSQ\_t + 0.322 \times B\_t - 0.012 \times Tax + 0.305 \times ZonePR\_t + 1.554 \times CrimePC\_t$$

Using this equation, we can interpret the effects of these predictors on the target variable (**Imv**). Since we used transformed data to train the model, the interpretation must be based on these transformed scales. In some cases, transforming the results back to the original scale may be necessary for meaningful interpretation.

Interpretation of Key Predictors

- Tax: This is a non-transformed variable. For every one-point increase in the full-value property tax rate per $10,000, the median house value decreases by $12 (0.012 × 1,000).
- LogDistance_t: If the logarithm of the weighted distance to Boston employment centers increases by one point, the median house value decreases by $8,383 (8.383 × 1,000).
- Lstat_t: If the square root of the percent lower-status population increases by one point, the median house value decreases by $3,583.
- RmSQ_t: If the inverse of the average number of rooms per dwelling increases by one room, the median house value increases by $221,525.

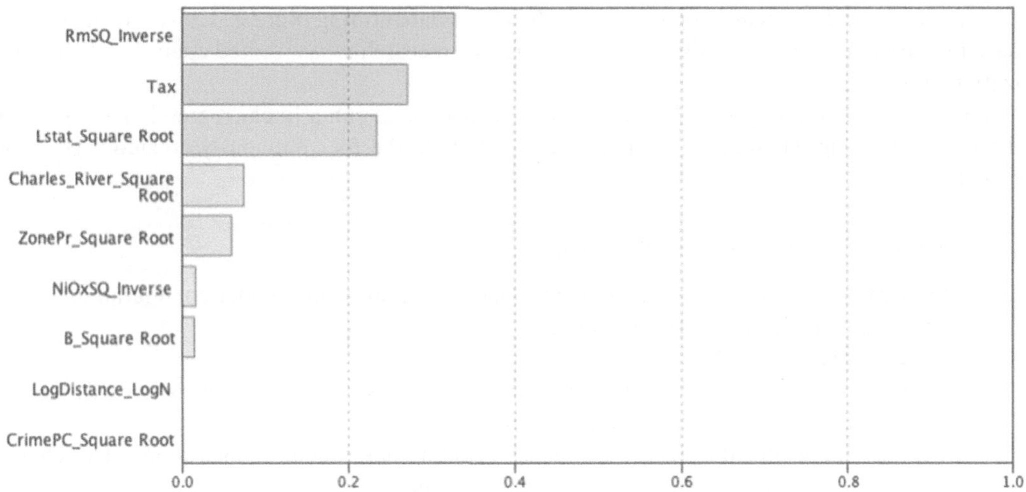

FIGURE 7.9 Variable importance chart (Regression 1).

- ZonePR_t: If the square root of the proportion of residential land zoned for large lots increases by one point, the median house value increases by $305.
- CrimePC_t: If the square root of the crime rate per capita increases by one point, the median house value increases by $1,554.

Conclusion on Data Transformation

While transforming fields to normalize the data can improve model fit, it also introduces challenges in interpreting the results. Therefore, data transformation should only be performed when necessary. If a model with transformed data performs similarly to one without transformation, it is often preferable to use the non-transformed data for ease of interpretation.

Comparing the Effects of Predictors on Imv

To compare the effects of the predictors on the target field **Imv**, we should use **standardized regression coefficients** (also known as **beta coefficients**). These values, presented in Table 7.14, allow for a direct comparison of the relative impact of each predictor, regardless of the units of measurement.

However, SPSS Modeler offers a more intuitive method for comparing predictor importance through the variable importance chart (Figure 7.9). This chart visually ranks the predictors based on their importance in the model.

The chart shows that **RmSQ** is the most important predictor, followed by **Tax**, **Lstat**, **Charles_ River**, and **ZonePR**. On the other hand, **CrimePC** and **LogDistance** are the least important predictors in the model.

What is the Linear Node in SPSS Modeler?

In addition to the Regression node, which is commonly used for performing linear regression in a traditional way, SPSS Modeler offers another node called **Linear**. The Linear node is an automatic linear modeling tool that handles all the steps in the linear regression process automatically. This includes data preparation, transformation, model selection, ensemble construction, and result presentation. Essentially, once the measurement scales and roles for the fields in the data are defined, the Linear node takes over the entire data mining process. However, since most decisions are made automatically by the software, the researcher has limited control and customization options.

For demonstration purposes, I will use the Linear node to run a linear regression analysis on the Boston housing dataset and compare the results with the four previously created regression models.

Description and Configuration of Regression 5

- **Description**: Automatic linear modeling (data preparation and model configuration are handled automatically by the software).
- **Configurations**:
  - **Objective**: Create a standard model (we are not constructing ensembles here, as this topic will be covered later in Chapter 10).
  - **Basis**: Automatically prepare the data, including adjustment of measurement levels, handling of outliers, and treatment of missing values.
  - **Model selection**: Forward stepwise.
  - **Criteria for entry/removal**: Information criterion (AICC).

Steps to Add and Set Up the Linear Node

1. Continue with the stream that contains the four previous regression models.
2. From the **Modeling** group, add a **Linear** node to the stream and connect it to the **Type** node (the one following the data source). Name it **Regression 5**. Open the **Linear** node settings.

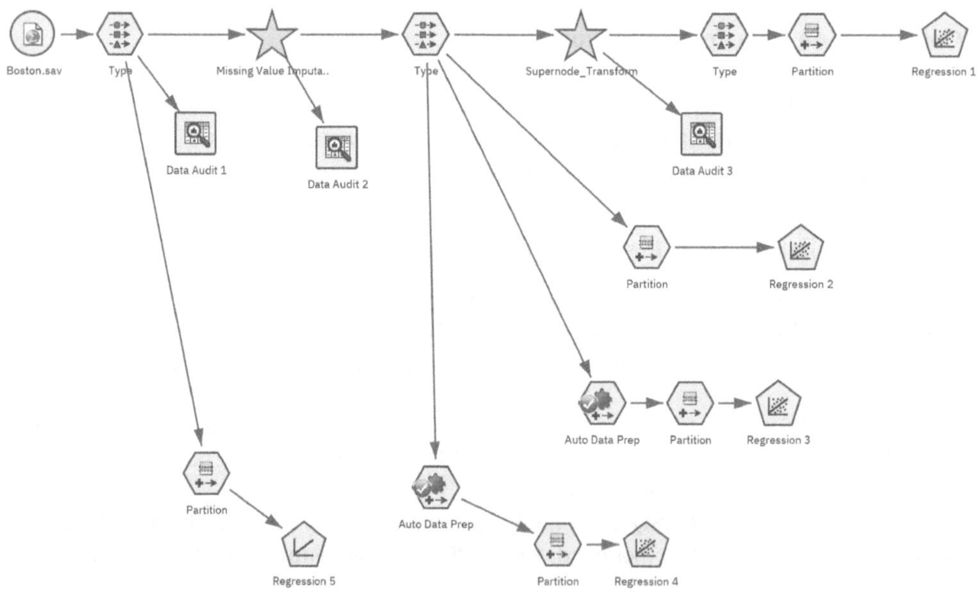

3. Accept all the default settings of the node. Verify that the configurations are as follows:
   - **Objective**: Create a standard model.
   - **Basis**: Automatically prepare the data.
   - **Model selection**: Forward stepwise.
   - **Criteria for entry/removal**: Information criterion (AICC).
4. Click Apply and then Run the node.
5. Once the run is complete, a model nugget labeled **Regression 5** will be added to the stream.

6. Add an **Analysis** node to the stream and connect it to the model nugget **Regression 5**. Open the **Analysis** node settings.
7. Under the **Analysis** tab, select the following options: **Performance evaluation**, **Evaluation metric**, **Confidence figures**, and **Separate by partition**.
8. Click Apply and then Run.

As you can see, the process is simple, easy, quick, and seamless, requiring no prior understanding of the regression algorithm. The output from the Analysis node provides metrics such as the MAE, mean error, standard error, and correlation (as described earlier). To access additional metrics, you need to open the model nugget.

You'll notice that this output differs significantly from the Regression node's output. First, you still receive some overall results, including the adjusted R^2 and the predictor importance chart. Additionally, the output automatically generates the predicted vs. observed chart (similar to the actual vs. predicted chart) without requiring any extra steps.

However, unlike the Regression node, this output does not include an Advanced tab, which would normally provide the model summary, analysis of variance (ANOVA), and coefficients. As a result, we don't have access to metrics such as the $\mathbf{R^2}$ and the **F-test**. To compensate, I added the **MAE** and used the **adjusted** $\mathbf{R^2}$ as the EP to compare this model with the four other models we created earlier (refer to Table 7.16).

Interestingly, the comparison shows that Regression 5 outperforms the other models. Specifically, it has the lowest error (in terms of MAE and RSE) and the highest explanatory power (both EP and adjusted R^2).

To further evaluate the model fit, we will examine the predicted by observed chart (Figure 7.10). The data points appear scattered around the diagonal line, indicating a good model fit. Figure 7.11 shows the predictor importance, with **RmSQ** identified as the most important predictor, followed by **PtRatio**, **Lstat**, and **LogDistance**. In contrast, **Charles River** and **B** are among the least important predictors. While these results differ from those of **Regression 1**, they still show **RmSQ** and **Lstat** as the most significant predictors.

Unfortunately, the Linear node does not provide detailed regression outputs, such as the model summary, ANOVA, or coefficients. As a result, we cannot observe the number of steps in the stepwise process nor the changes in R^2, F-tests, or coefficients for significant predictors. Keep in mind that the Linear node is designed to perform linear regression analysis quickly and automatically, which means it does not guarantee the production of a well-fitting model every time.

For these reasons, I recommend using the Regression node to gain more control and customization over the modeling process. The Regression node allows for adjustments to the model configuration, offering greater flexibility in achieving a good fit. However, the Linear node can be useful for running a preliminary regression analysis, providing quick insights into your model and data. Based on these initial results, you can make informed decisions regarding data modifications and modeling plans to improve your predictive model.

TABLE 7.16
Model Comparison with the Addition of Regression 5

Model Fit Metrics	Sample	Regression 1	Regression 2	Regression 3	Regression 4	Regression 5
MAE	Testing	3.437	3.976	3.976	4.406	3.217
RSE	Testing	0.265	0.434	0.434	0.484	0.214
EP	Training	0.735	0.566	0.566	0.516	0.786
F-test	Training	90.629	64.645	64.645	160.277	
R^2	Training	0.735	0.566	0.566	0.516	0.786
Adjusted R^2	Training	0.727	0.558	0.558	0.513	0.786

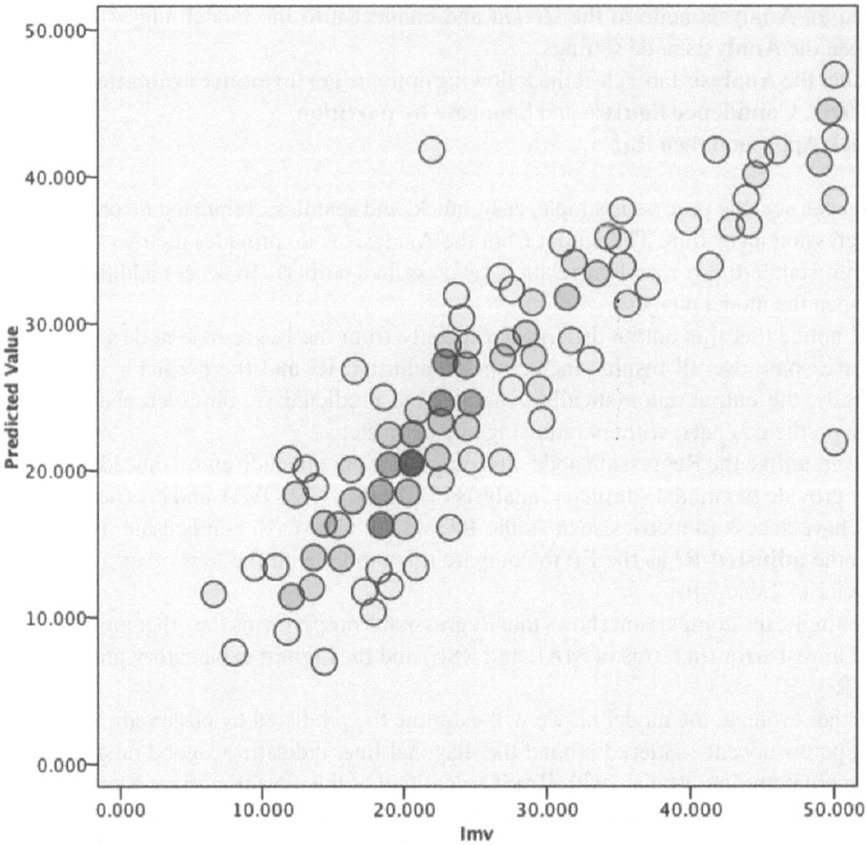

FIGURE 7.10 Predicted by observed chart (Regression 5).

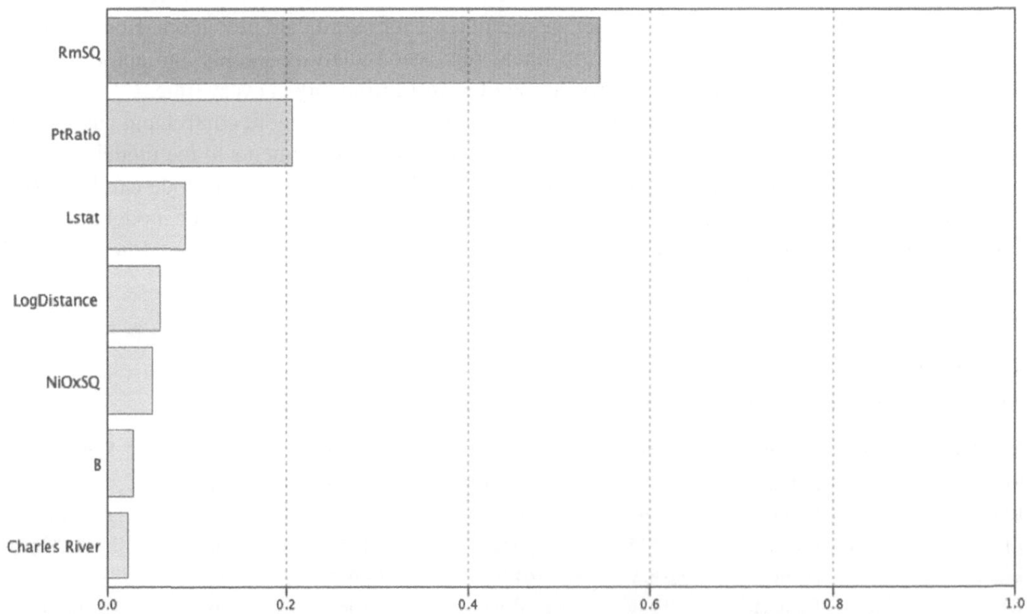

FIGURE 7.11 Variable importance chart (Regression 5).

LOGISTIC REGRESSION

Logistic Regression Equation

Logistic regression is used with a binary or nominal target variable. The traditional logistic regression equation is as follows:

$$Logit(Y) = log\left(\frac{p}{1-p}\right) = \beta_0 + \beta_1 x_1 + \beta_2 x_2 + \cdots + \beta_n x_n + \varepsilon$$

where

Y is the target variable,
p is the probability of the event in the target variable,
x_1 to x_n are predictor variables $1 - n$.
β_0 is the intercept
$\beta_1 - \beta_n$ are regression coefficients for the corresponding predictors,
ε is the residual or error.

Another form of this equation, which is easier to interpret, expresses the odds ratio as follows:

$$Odds(Y) = \left(\frac{p}{1-p}\right) = e^{(\beta_0 + \beta_1 x_1 + \beta_2 x_2 + \cdots + \beta_n x_n + \varepsilon)}$$

Don't be intimidated by these equations! We won't use them directly in the analysis or interpretation of the results. I will explain the results in a simpler way.

Estimation Approach

Logistic regression estimates the target variable and regression coefficients differently from linear regression. The method used is **maximum likelihood estimation (MLE)**. Simply put, we first assume a probability distribution for the target variable. Then, a likelihood function is used to calculate the probability of an outcome based on the predictors. This method is useful for estimating the probability of an event and does not require strict assumptions like **OLS**.

Assumption Requirements

As mentioned, due to the nature of the MLE method, logistic regression does not require strict assumptions like normality, homoscedasticity, or linearity. However, logistic regression typically involves the following assumptions:

- **Binary target variable**: The target variable must be binary, meaning it has two distinct levels (e.g., 1—accident, 0—no accident).
- **Multiple predictors**: Logistic regression requires multiple predictors but does not require them to be continuous. It works well with binary or nominal predictors, offering flexibility in scale. This is advantageous when analyzing large datasets with predictors on different scales.
- **Non-multicollinearity**: Similar to linear regression, predictors must be independent of each other. However, with large sample sizes, the effect of multicollinearity is often tolerable.
- **Linearity**: There must be a linear relationship between the predictors and the logit of the target variable. The equation shows a linear regression relationship between the logit of the target variable and the predictors.

In conclusion, logistic regression offers more flexibility in its assumptions, making it a robust method for data mining across various types of data.

TABLE 7.17
Method-Specific Fit Metrics for Linear Regression and Logistic Regression

Linear Regression	Logistic Regression
Sum of squares total (SST)	–2LL of the base model
Sum of squares error (SSE)	–2LL of the proposed model
Sum of squares regression (SSR)	Difference of –2LL between the base and proposed models
F-test of model fit	Chi-square test of –2LL difference between the base and proposed models
Coefficient of determination (R^2)	Pseudo R^2 measures

Source: Hair et al. (2018).
Note: –2 LL = –2 Log Likelihood.

Model Fit Evaluation

Logistic regression is a classification model, so we evaluate its performance using classification metrics, as described in Chapter 6. These include the following:

- Misclassification rate
- Accuracy rate
- Sensitivity
- Specificity
- F1 score
- Gini coefficient
- Receiver Operating Characteristic (ROC) curve
- Lift chart

Additionally, logistic regression has specific model fit metrics similar to those used in linear regression. Table 7.17 summarizes these metrics, allowing us to compare model fit between these regression methods.

The **Chi-square test** and **pseudo R^2** are specific metrics used to evaluate logistic regression models. To understand the Chi-square test for the –2 Log Likelihood (–2LL) difference, it's important to first grasp the concept of –2LL. For this reason, I have included a description of –2LL in Table 7.18.

TABLE 7.18
Recommended Method-Specific Metrics for Logistic Regression

Logistic Regression Metrics	Description
–2LL (–2 Log Likelihood)	A measure of how well the model fits the likelihood. A good model is characterized by a high likelihood of the observed results or a low –2LL. In the ideal scenario where a model perfectly fits the data, the likelihood attains a value of 1, and the –2LL score becomes 0.
Chi-square test of –2LL difference (Omnibus test of model coefficients)	The difference in –2LL between two models in a stepwise process. A significant Chi-square test indicates improvement in the modeling step.
Pseudo R^2	A measure of how well the model fits the data, compared to a model that always predicts the mean. The two most common pseudo R^2 are • McFadden's R^2 • Cox and Snell R^2 Note that pseudo R^2 does not range from 0 to 1 as it has a much lower upper bound. Hence, the general rule of thumb is that a pseudo R^2 between 0.2 and 0.4 is considered a good fit.

Model Selection Methods

In logistic regression, we also have three stepwise methods, similar to those used in linear regression: **forward stepwise**, **backward stepwise**, and **adaptive stepwise**. Even though logistic regression deals with a binary target variable rather than a continuous one, it utilizes the same types of stepwise methods.

Start with the adaptive stepwise method.

I recommend starting with the **adaptive stepwise** method for variable selection, as it allows us to find the best combination of predictors that leads to a well-fitting model.

How to Read and Understand Logistic Regression Results

Before interpreting logistic regression results, we must ensure we have a good model fit. Let's assume that we have already checked the logistic regression model fit metrics, such as **–2LL**, the **Chi-square test for –2LL**, **pseudo R^2**, **RMSE**, **RSE**, **EP**, **misclassification rate**, **accuracy**, **sensitivity**, **specificity**, **F1 score**, **Gini coefficient**, **ROC chart**, and **Lift chart**, and found the model fit to be satisfactory.

The next step is to interpret the logistic regression results to identify significant predictors and their effects on the target variable.

Typically, the logistic regression estimation results include the following:

- Logistic coefficient (B)
- Standard error
- Wald statistic—a test for the significance of the coefficients
- Degrees of freedom
- Probability value (p-value)
- Exp(B) or odds ratio
- 95% confidence interval for Exp(B)

We focus on the key outputs that help us interpret the results. Logistic regression can be more challenging to read due to the use of the logit function. To simplify the process, we will primarily focus on the **p-value**, **logistic coefficients**, and **odds ratio**.

For example, suppose the regression equation for a model predicting students' decision to choose major A is as follows:

$$Logit(majorA\_choice) = 0.5 - 0.3 \times gender + 0.4 \times age$$

In this equation, 0.5 is the intercept, –0.3 is the logistic coefficient for **gender**, and 0.4 is the coefficient for **age**. The sign of the coefficient indicates the direction of the effect. In this case, **gender** has a negative effect on the choice of major A, while **age** has a positive effect.

While coefficients indicate the size of the effect, interpreting them in relation to the logit value (instead of the actual target variable) can be difficult. To make interpretation easier, we use the **odds ratio**. The equivalent equation in terms of odds is as follows:

$$Odds(majorA\_choice) = e^{(0.5 - 0.3 * gender + 0.4 * age)}$$

Most statistical software provides the odds ratio for each predictor alongside the coefficients. The **sign** of the coefficient indicates the direction of the effect, while the **odds ratio** helps us interpret the magnitude of the predictor's effect on the target variable.

For example, let's say we obtain the following odds ratios for gender and age:

- Odds ratio for **gender** = 0.74
- Odds ratio for **age** = 1.49

How do we interpret these numbers? First, let's consider **gender**, which in this case is a binary variable (e.g., Male and Female). Suppose **Male** is the reference category. This means switching from **Male** to **Female** reduces the odds of choosing major A by 26% (since 0.74 – 1 = –0.26). In other words, male students are 1/0.74 = 1.35 times more likely than female students to choose major A.

For **age**, the odds ratio of 1.49 means that for every one-year increase in age, the odds of a student choosing major A increase by 49% (since 1.49 – 1 = 0.49).

As you can see, interpreting logistic regression results can be more challenging than interpreting linear regression results. However, by following this process, we can effectively understand the results. We'll explore this further with an actual example later.

EXAMPLE: LOGISTIC REGRESSION

To illustrate how logistic regression works in data mining, we will use the Airline Recommendation dataset, which was introduced in previous chapters. For convenience, I have included the variable descriptions in Table 7.19. More details about this dataset can be found in Chapter 3.

- **Dataset:** AirlineRecommendation.sav
- **Target variable:** Recommend (Flag: Yes/No)

We have completed three key steps in the data mining process in Chapters 3, 4, and 5: data exploration, data modification, and data sampling. I will not repeat the detailed steps here, as you can revisit those chapters to see the process in SPSS Modeler and the results. However, I will summarize the stream and primary findings from each step below so you can observe the connections between them and the decisions required to build a predictive model.

TABLE 7.19
Airline Recommendation Data Description

Field Name	Description	Scale	Role
recommend	Recommend the airline to others	Binary/flag (Yes/No)	Target
airline_st	Airline status	Nominal	Input
age	Age	Continuous	Input
gender	Gender	Binary/flag (Male/Female)	Input
price_sensitivity	Price sensitivity	Continuous	Input
no_years_flying	Number of years flying	Continuous	Input
no_flt_pa	Average number of flights per year	Continuous	Input
pct_ flt_other_airline	Percent of flights with other airlines	Continuous	Input
type_of_travel	Travel purpose	Nominal	Input
no_other_lty_cards	Number of other loyalty cards	Continuous	Input
shopping_amt_airpt	Shopping amount at the airport	Continuous	Input
eat_drink_airpt	Spending for eating and drinking at the airport	Continuous	Input
class	Passenger class	Nominal	Input
origin_state	Origin state	Nominal	Input
destin_state	Destination state	Nominal	Input
depart_delay	Departure delay (in minutes)	Continuous	Input
arrival_delay	Arrival delay (in minutes)	Continuous	Input
flt_cancel	Flight cancellation	Binary/flag (Yes/No)	Input
flt_time	Flight time (in minutes)	Continuous	Input
flt_distance	Flight distance (miles)	Continuous	Input

Data Exploration

Data exploration was primarily conducted using the **Data Audit 1** node, which generated graphs and descriptive statistics for the relevant fields. The exploration revealed a total of **124,464 records** in the dataset. Three continuous fields were identified with missing values: **arrival delay**, **departure delay**, and **flight time**. Since the missing values accounted for less than 3% of the data, they will need to be imputed to prevent information loss.

Additionally, the **price sensitivity** field, which is based on a Likert scale, should have values ranging from 1 to 5. However, the presence of zero values indicates missing data that was not properly coded. By examining the histograms for the continuous fields, we found that most of them are not normally distributed. However, since the logistic regression algorithm does not require normality assumptions, normalization is not necessary.

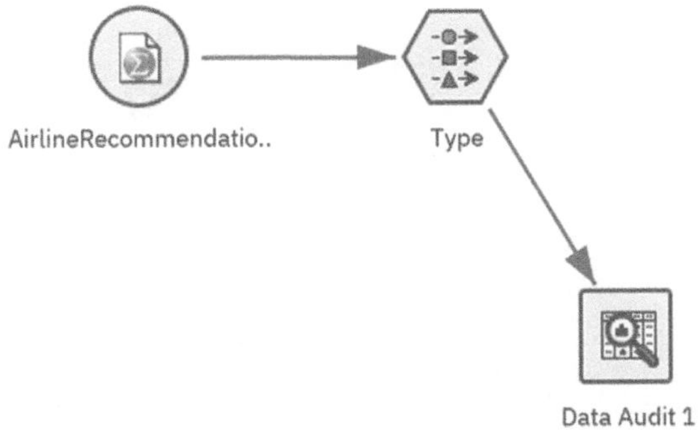

Data Sampling

Typically, the next step is sampling. The dataset contains 124,464 records, and although it's possible to sample a subset for analysis, we will use the entire dataset since this sample size is not particularly large. Some advanced algorithms we'll be using later, such as neural networks or random forests, often require a large dataset to train the model effectively. Therefore, no data sampling will be performed here. The main task in this step is data partitioning, which will be done after data modification.

Data Modification

As identified in the data exploration phase, there are three main issues with the data:

1. Incorrect coding of missing values in the **price sensitivity** field.
2. Missing values in three other fields: **departure delay**, **arrival delay**, and **flight time**.
3. Non-normal distributions in most continuous fields.

Since we are using logistic regression, normalization is unnecessary. Therefore, the two primary data modification techniques we'll use are replacement and imputation.

- **Replacement**: We use the **Filter node** to recode the zero values in the **price sensitivity** field, which are incorrectly coded as missing values.
- **Imputation**: Now, four fields have missing values: **price sensitivity**, **departure delay**, **arrival delay**, and **flight time**. To handle these missing values, we configure imputation methods using the Quality tab from the Data Audit 2 output. Specifically:
  - **Mean imputation** is applied to **departure delay**, **arrival delay**, and **flight time**.
  - **Mid-range imputation** is applied to **price sensitivity** (since it uses a Likert scale).

A **SuperNode** labeled **Missing Value Imputation** is generated to carry out the imputation. The results show the following number of missing values imputed:

- **Price sensitivity**: 3,976 records
- **Departure delay**: 2,201 records
- **Arrival delay**: 2,565 records
- **Flight time**: 2,565 records

Similar to the Boston Housing example, I added the **Type** node after each step, allowing us to check the field names, measurement scales, and roles. This helps ensure the fields are correctly defined for the next step. Additionally, I use the **Data Audit** node to view descriptive statistics and charts after each step, aiding in decision-making for subsequent steps.

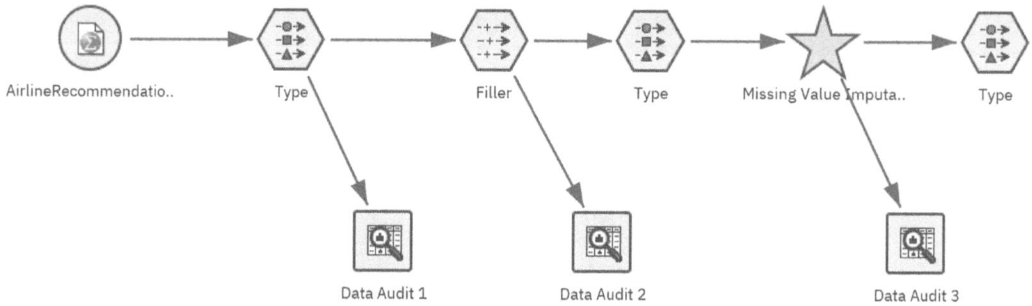

Data Partition

In this step, we partition the data before building the logistic regression model. Since we have enough data, as recommended in Chapter 4, we partition the dataset into three sub-samples using a 60–20–20 split ratio:

- Train (60%): Used to train the model.
- Test (20%): Used to test the model.
- Validation (20%): Used for final model assessment.

The Partition node is applied after the SuperNode Missing Value Imputation.

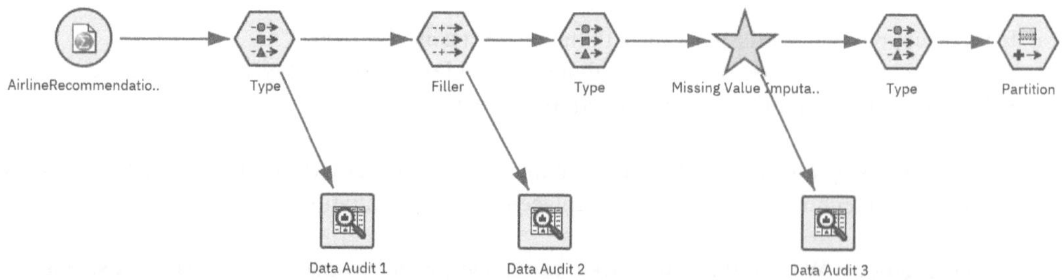

Refer to the screenshot below for setting up the data partition. It divides the dataset into the three sub-samples mentioned above. The training sample (60% of the data) is used to train the model, the test sample (20%) for testing the model, and the validation sample (20%) for final model evaluation.

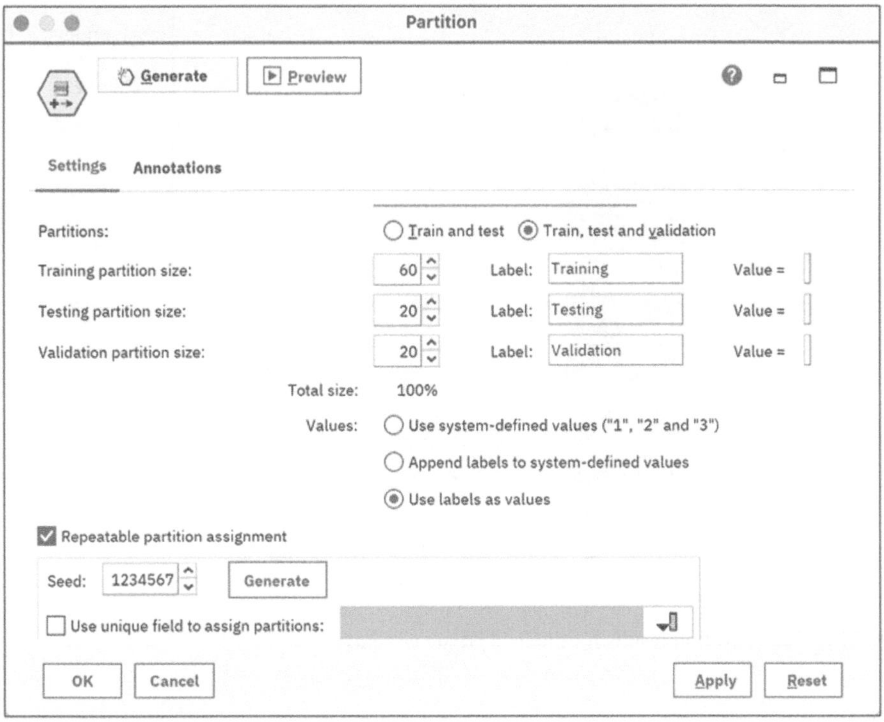

As shown in Figures 7.12 and 7.13, the Partition node splits the data into three sub-samples based on the 60-20-20 ratio. The imbalance ratio between the levels for the two fields—**airline recommendation** and **passenger class**—remains consistent across the sub-samples. You can create additional charts for other fields as well. Therefore, we can conclude that the sub-samples are representative of the full dataset.

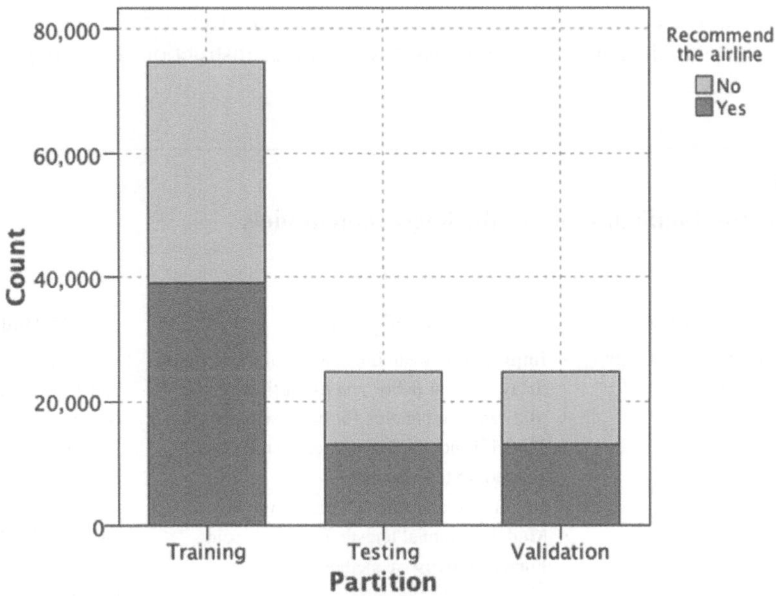

FIGURE 7.12 Data partition results for the field *recommendation the airline*.

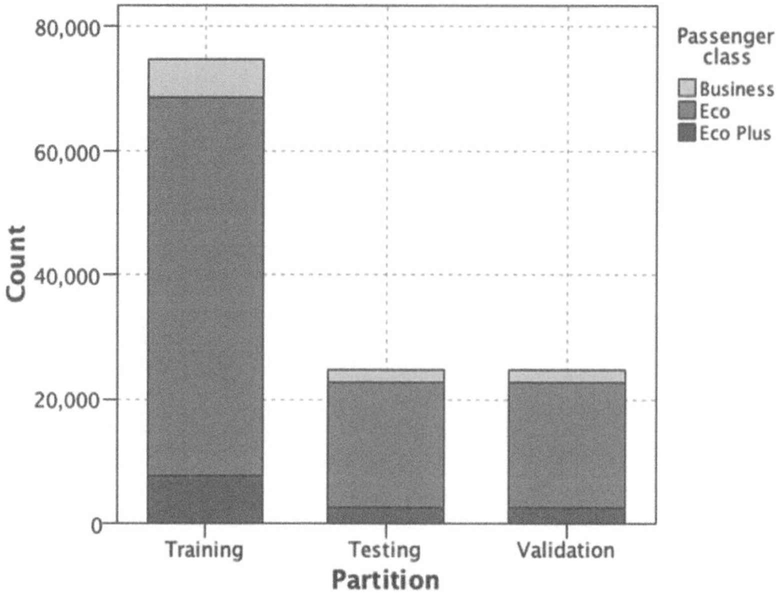

FIGURE 7.13 Data partition results for the field *passenger class.*

Modeling

Our objective is to build and run a logistic regression model with the target field being airline recommendation. Since this is a binary field, we will use binomial logistic regression rather than multinomial regression, which is used for target fields with multiple levels.

There are various approaches to building the model, using different data transformation options and SPSS Modeler tools. To explore the best approach, I will build four different logistic regression models and compare them to determine which one offers the best predictive power. Table 7.20 provides descriptions and configurations for each model.

With these four regression models, we will create a comprehensive stream that shows the flow of nodes and the steps in the data mining process. I will provide instructions on setting up configurations for each model.

TABLE 7.20
Descriptions and Configurations of the Regression Models

Logistic Regression Models	Description	Configurations	SPSS Modeler Tools
Logistic 1	Using the dataset with imputation	• **Imputation**: Mean imputation for **departure delay**, **arrival delay**, and **flight time**. • Mid-range imputation for **price sensitivity**. • **Model**: Binomial logistic regression using **Forward Stepwise** method.	Logistic 1 node (Logistic node under the Modeling group), following the SuperNode "Missing Value Imputation" and Partition node.
Logistic 2	Using the dataset with missing values (no imputation)	• No imputation (four fields have missing values). • **Model**: Binomial logistic regression using **Forward Stepwise** method.	Logistic 2 node (Logistic node under the Modeling group), following the Filter node and Partition node.

(Continued)

TABLE 7.20 *(Continued)*
Descriptions and Configurations of the Regression Models

Logistic Regression Models	Description	Configurations	SPSS Modeler Tools
Logistic 3	Balancing speed and accuracy—using the Auto Data Prep node	• **Auto Data Prep node** for imputing missing values, handling outliers, and standardizing all continuous fields. • **Objective**: Balance speed and accuracy (no outlier handling). • **Model**: Binomial logistic regression using **Forward Stepwise** method.	Logistic 3 node (Logistic node under the Modeling group), following the Filter node, Auto Data Prep node, and Partition node.
Logistic 4	Optimizing for accuracy—using the Auto Data Prep node	• **Auto Data Prep node** for imputing missing values, handling outliers, and standardizing all continuous fields. • **Auto Data Prep node** to normalize all fields. • **Objective**: Optimize for accuracy (with outlier handling). • **Model**: Binomial logistic regression using **Forward Stepwise** method.	Logistic 4 node (Logistic node under the Modeling group), following the Filter node, Auto Data Prep node, and Partition node.

One important note: If we connect the **Partition** node directly to the **Logistic** node when running the analysis, SPSS Modeler will return an error message stating that the categorical input fields have incorrect values (due to string formats). Therefore, we need to use the **Auto Data Prep** node to reformat these fields.

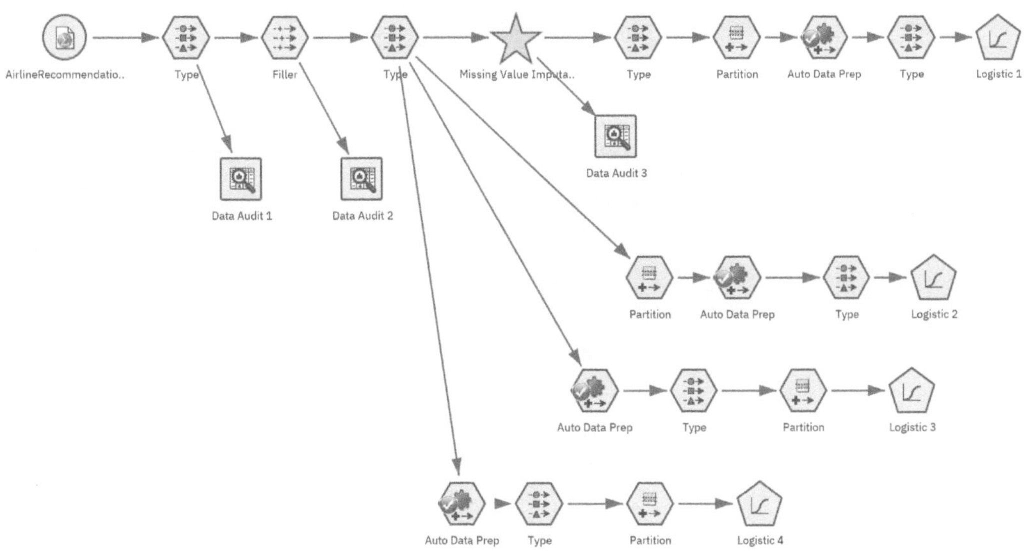

Steps to Build Logistic Model 1

1. Open SPSS Modeler and create a new stream. Name the stream "**Airline_Regression**."
2. Add the data source file "**AirlineRecommendation.sav**" to the stream.
3. Add a **Type** node to the stream and connect it to the data source. Open the **Type** node settings.

4. Double-check the **field names**, **measurement scales**, and **roles** for accuracy.
5. Set the **role** of the field **recommendation** as **Target**. All other fields should be set as **Input**.
6. Perform replacement using the **Filter node** to replace zero values in the **price sensitivity** field with missing values (see Chapter 5 for reference).
7. Perform imputation using **SuperNodes** as described earlier (see Chapter 5 for reference). After this step, you should have a SuperNode named **Missing Value Imputation** for handling missing values.
8. Add another **Type** node to the stream and connect it to the SuperNode **Missing Value Imputation**. Open the **Type** node settings and double-check the fields, measurements, and roles.
9. Add a **Partition** node to the stream and connect it to the **Type** node. Open the **Partition** node settings.
10. Select **Train, test, and validation** and set the **training partition** size to **60%**, the **testing partition** size to **20%**, and the **validation partition** size to **20%** (refer to the Data Partition step for more details).
11. Click **Apply**, then click **OK**.
12. Add an **Auto Data Prep** node to the stream and connect it to the **Partition** node.
13. Under **Settings > Prepare Inputs & Target**, check the box for **Reorder nominal fields to have the smallest category first, largest last** and uncheck all other boxes. This ensures proper numeric formatting for categorical input fields required to run the **Logistic** node.

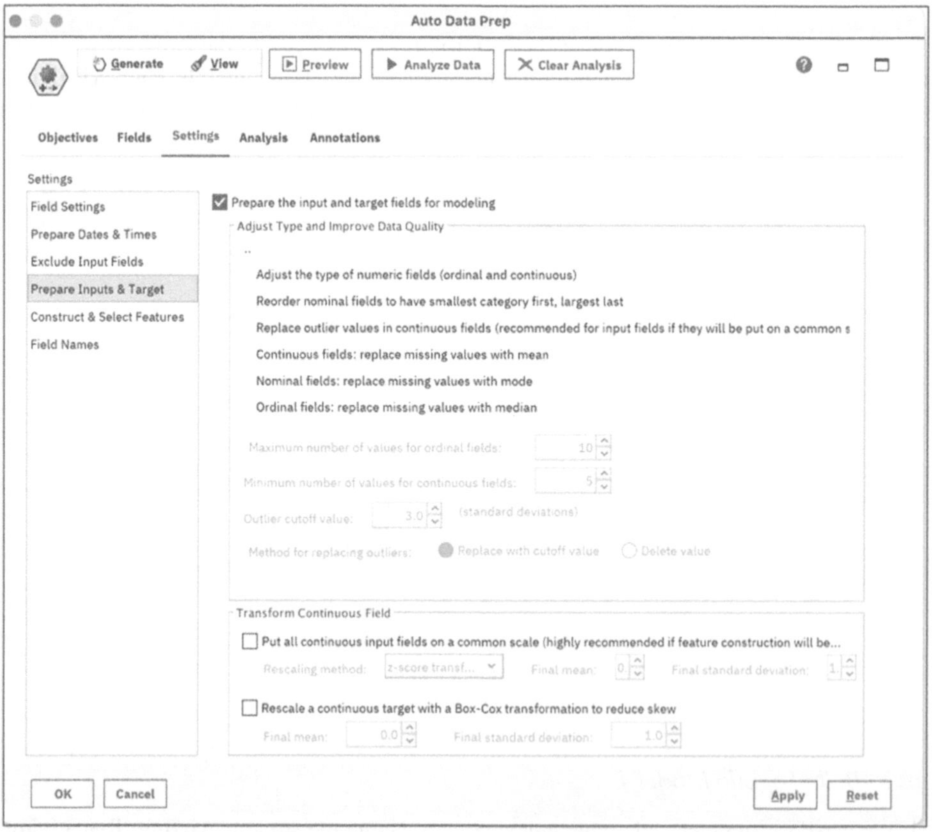

14. Click Apply and then Analyze Data.

15. Add another **Type** node to the stream and connect it to the **Auto Data Prep** node. Open the **Type** node settings.

16. Confirm that all categorical input fields have been transformed (I use a **_t** suffix to denote transformed fields).

17. From the **Modeling** group, add a **Logistic** node to the stream. Name it **Logistic 1** and connect it to the **Type** node. Open the **Logistic** node settings.
18. Under the **Model** tab, select **Use partition data** and **Build model for each split**.
19. Select **Binomial** for the **Procedure**.
20. Choose **Forward Stepwise** for the **Method**.

21. Select and add all **categorical inputs** to the model. The **Base Category** column shows the reference level for these categorical inputs. You can change it to either the **First** or **Last** level. For this example, leave the setting as **First** for all fields.

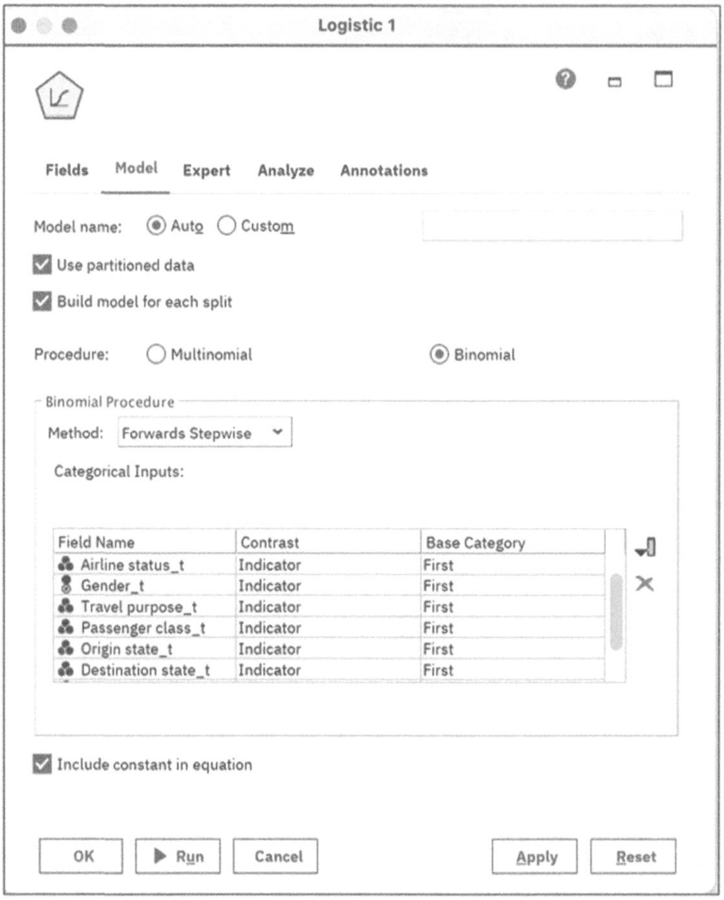

22. Under the **Analyze** tab, select **Calculate predictor importance**.
23. Click Apply and Run.
24. SPSS Modeler will run the analysis and add the **model nugget** (result node) to the stream. If there are errors during the run, the software will display an error message.
25. From the **Output** group, add an **Analysis** node to the stream and connect it to the **model nugget Regression 1**. This node will compare the model's performance between the training and test samples. Open the **Analysis** node settings.

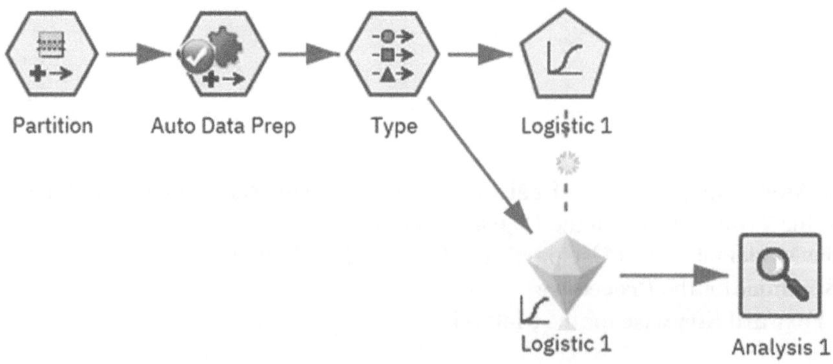

26. Under the **Analysis** tab, select the following options:
    - Coincidence matrices (for symbolic targets)
    - Performance evaluation
    - Evaluation metric
    - Confidence figures
    - Separate by partition

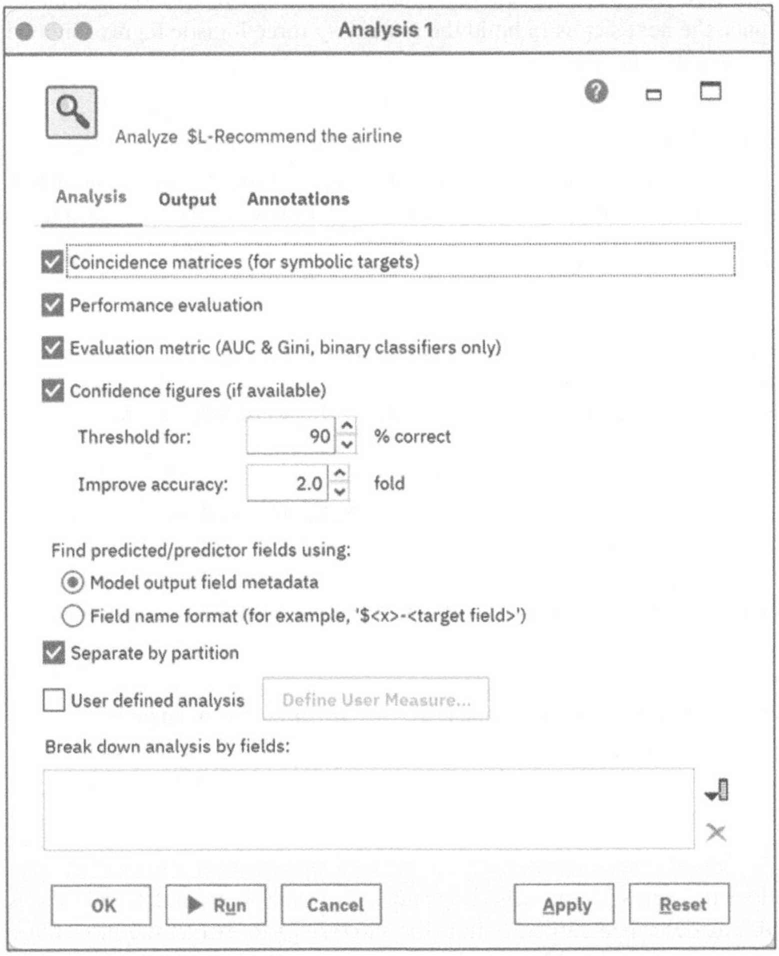

27. Click Apply and Run.

> Since we are using a large dataset to train the model in this example, you may notice that the run time is significantly longer compared to previous examples. Typically, it can take anywhere from 10 to 20 minutes to complete the analysis.

The output from the Analysis 1 node provides us with key metrics as follows:

- **Coincidence matrix**: Equivalent to the confusion matrix, showing true positives (TP), true negatives (TN), false positives (FP), and false negatives (FN). This matrix is crucial because it allows us to calculate important model fit metrics.
- **Correct percent**: Equivalent to prediction accuracy (ACC).

- **Wrong percent**: Equivalent to the misclassification rate (MR).
- **Evaluation metrics**: Includes AUC (area under the curve), also known as the ROC index, and the Gini coefficient.

Using the formulas described in Chapter 6, we can calculate additional metrics, such as sensitivity, specificity, and F1 score, using the information from the confusion matrix.

Now that we know how to build a logistic regression model and extract model fit metrics from SPSS Modeler outputs, the next step is to build the remaining three logistic regression models and compare them to identify the champion model.

Logistic Models 2, 3, and 4

Follow the same steps outlined earlier to set up and configure Logistic Regression models 2, 3, and 4, using the configurations and nodes described in Table 7.20. Below are the specific configurations for each model.

Logistic 2:

- Use the dataset with missing values (no imputation).
- No normalization.
- Configure the Logistic node: Binomial model with the Forward Stepwise method.

Logistic 3:

- Use the Auto Data Prep node with the objective to "**Balance speed and accuracy**."
- Perform automatic data preparation, data imputation, and standardization (without outlier handling).
- Configure the Logistic node: Binomial model with the Forward Stepwise method.

Logistic 4:

- Use the Auto Data Prep node with the objective to "**Optimize for accuracy**."
- Perform automatic data preparation, data imputation, standardization, and outlier handling.
- Configure the Logistic node: Binomial model with the Forward Stepwise method.

Champion Model

After running the logistic regression models (Logistic 1, 2, 3, and 4) separately, add the Analysis and Evaluation nodes as described earlier. Follow the same steps to extract the model fit metrics, including for the testing sample. Table 7.21 provides a comparison of the results for the testing sample across the four models.

TABLE 7.21
Model Comparison—Logistic 1, 2, 3, and 4 (Testing Sample Results)

Metrics	Logistic 1	Logistic 2	Logistic 3	Logistic 4
ACC	76.60%	72.90%	76.60%	76.70%
MR	23.40%	27.10%	23.40%	23.30%
Sensitivity or TPR	0.893	0.851	0.893	0.878
Specificity or TNR	0.624	0.593	0.625	0.642
F1	0.801	0.768	0.801	0.799
Gini	0.658	0.601	0.658	0.665
AUC (ROC index)	0.829	0.801	0.829	0.833

The results indicate that Logistic 2 has the lowest performance, likely due to the number of missing values affecting prediction accuracy. The performance of Logistic 1, Logistic 3, and Logistic 4 is very similar. Upon deeper inspection, we see that Logistic 1 and Logistic 3 have nearly identical metrics. Logistic 4 shows a slight improvement in accuracy (ACC), AUC, and the Gini coefficient compared to Logistic 1 and 3. However, Logistic 4 has a marginally lower F1 score, primarily due to a lower sensitivity (true positive prediction rate), although its specificity is better than the other models.

Overall, these three models—Logistic 1, Logistic 3, and Logistic 4—have very similar model fit metrics, suggesting comparable performance. Technically, any of them could be selected as the champion model. However, it's important to note that Logistic 3 and Logistic 4 use the Auto Data Prep node, which automatically prepares the data, imputes missing values, handles outliers, and normalizes the data. This transformation makes result interpretation more challenging because the fields have been altered.

Typically, at this stage, we would also present **ROC** and **Lift charts** for all models in a single chart to facilitate comparison. However, in this example, we built multiple models using different types of datasets: one with missing values, one with imputed data, and one with transformed data.

As I will explain later in the chapter when using the **Analysis** node for model comparison, all models must have **identical field names**; otherwise, SPSS Modeler will generate an error message. Due to this constraint, we are unable to produce combined charts for these models. As a result, we primarily rely on the metrics shown in **Table 7.21** to compare the models.

Reliability of the Champion Model

Having determined Logistic 1 as the champion model, the next step is to evaluate its reliability. We extract the model fit metrics from the Analysis node for all three samples (training, testing, and validation) and compile the results in Table 7.22 to assess the model's reliability.

Keep in mind that we need to calculate sensitivity, specificity, and F1 score separately using the information from the confusion matrix. As shown in the table, the metric results are highly consistent across the three samples, which suggests that the model is reliable. In other words, there are no signs of overfitting with this model.

We can also compare the ROC and Lift charts across the three samples to further confirm the model's reliability. To produce these charts, we will use the Evaluation Graph node.

Rule of thumb
If the differences among the three sub-samples (training, testing, and validation) are 10% or lower, the model's reliability is considered satisfactory.

TABLE 7.22
Model Reliability—Logistic 1

Metrics	Training	Testing	Validation
ACC	76.70%	76.59%	76.82%
MR	23.30%	23.41%	23.18%
Sensitivity or TPR	0.8946	0.8927	0.8933
Specificity or TNR	0.6269	0.6242	0.6296
F1	0.8007	0.8010	0.8020
Gini	0.666	0.658	0.667
AUC (ROC index)	0.833	0.829	0.833

Steps to Construct the ROC Chart:

1. Continue from the stream "**Airline_Regression**" created earlier.
2. From the **Graphs** group, add an **Evaluation** node to the stream and connect it to the model nugget **Logistic 1**.
3. Open the **Evaluation** node settings.

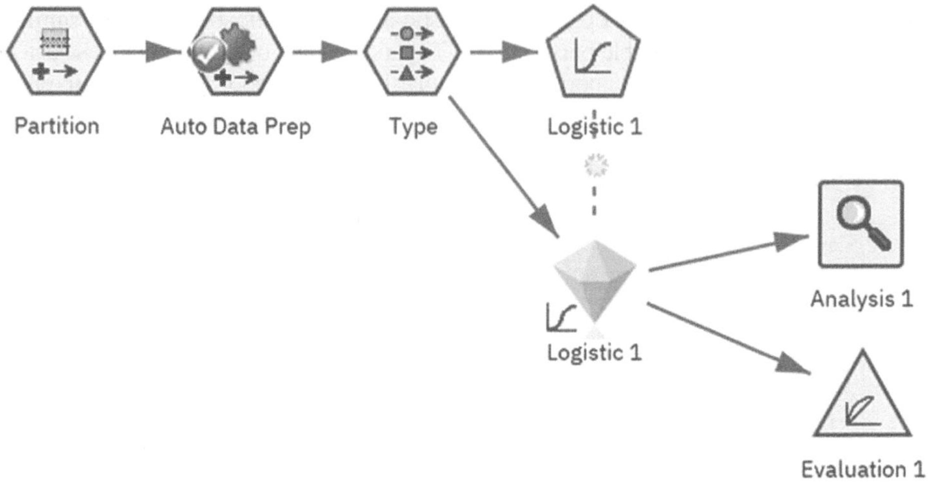

4. Under the **Plot** tab, select **ROC** as the **Chart Type**.
5. Click Apply and then Run.

You will receive three ROC charts corresponding to the training, testing, and validation samples (Figure 7.14). Follow the same steps to produce Lift charts for these samples (Figure 7.15).

As we can see, the ROC and Lift charts are very consistent across the three samples, further confirming the reliability of the model.

Validity of the Champion Model

The next step is to evaluate the validity of the champion model by reviewing its complete model fit metrics. First, let's focus on the metrics specific to the logistic regression algorithm. From the Advanced tab of the model nugget Logistic 1, we can extract key metrics for this model.

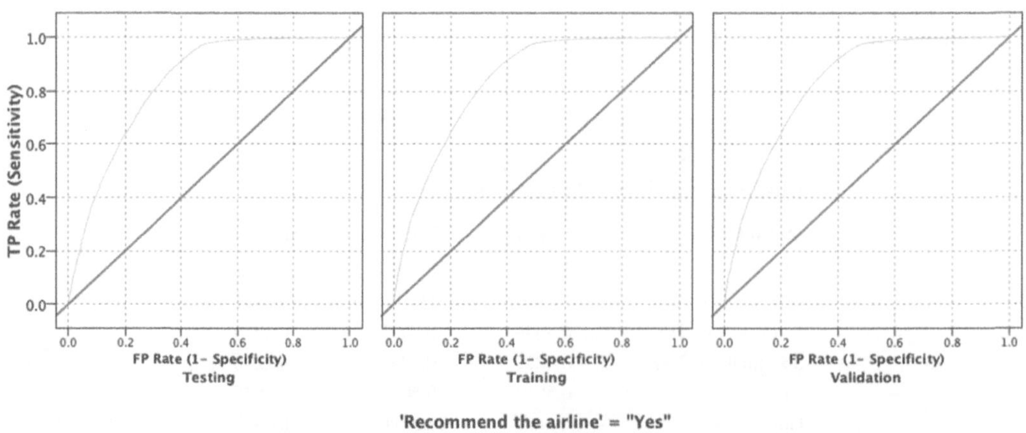

FIGURE 7.14 ROC chart comparison—Logistic 1.

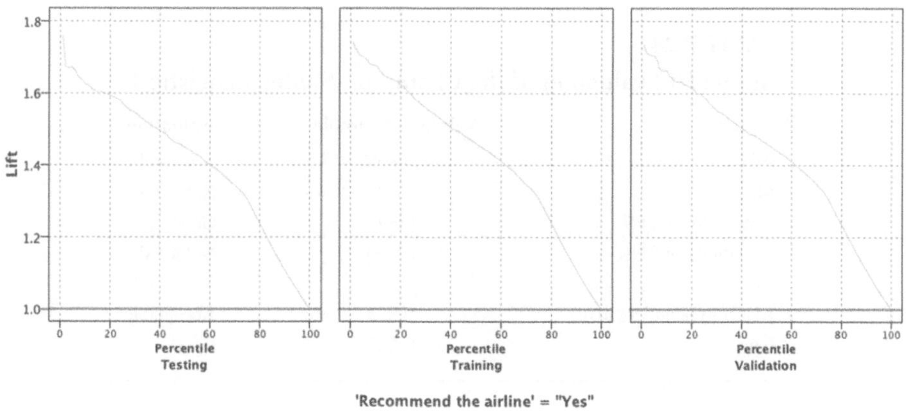

FIGURE 7.15 Lift chart comparison—Logistic 1.

Notably, the stepwise process took 14 steps to reach the final model. Table 7.23 presents the Omnibus test results for all 14 steps, showing whether there is a significant difference between the –2LL of the baseline model and each new model. The results indicate that the Chi-square test is significant ($p < 0.05$) at each step, confirming that the final model has improved significantly over the baseline model.

We can also review the Pseudo R^2 values for the final model as follows:

- **Cox and Snell R^2** = 0.346
- **Nagelkerke R^2** = 0.462

As mentioned earlier, a Pseudo R^2 value between 0.2 and 0.4 is considered indicative of a good model fit. Therefore, in this case, we have achieved a good fit for the logistic regression model.

Next, we will perform a comprehensive assessment of this model using common classification model fit metrics based on the results of the validation sample. Table 7.24 presents the results for the champion model, along with an evaluation based on the suggestions from Chapter 6.

TABLE 7.23
Omnibus Tests of Model Coefficients

Models	Chi-Square	df	Sig.
Step 1	24,762.613	2	< .001
Step 2	4,976.427	3	< .001
Step 3	619.043	1	< .001
Step 4	428.218	1	< .001
Step 5	358.937	1	< .001
Step 6	310.529	1	< .001
Step 7	81.18	2	< .001
Step 8	37.04	1	< .001
Step 9	35.013	1	< .001
Step 10	32.755	1	< .001
Step 11	26.204	1	< .001
Step 12	15.521	1	< .001
Step 13	9.542	1	0.002
Step 14	71.825	51	0.029

TABLE 7.24
Model Fit Evaluation of the Champion Model—Logistic 1

Metrics	Validation Sample	Evaluation
ACC	76.80%	Not good
MR	23.20%	Not good
Sensitivity or TPR	0.893	Good
Specificity or TNR	0.630	Not good
F1	0.802	Acceptable
AUC (Area under curve)	0.833	Acceptable
Gini	0.667	Good

Model Fit Assessment

Based on the validation results, the prediction accuracy seems slightly low at 76.8%, with a misclassification rate (MR) of around 23.2%. However, examining sensitivity and specificity provides more insight. The model demonstrates relatively strong sensitivity at 89.3%, indicating it can effectively predict positive cases (i.e., recommending the airline). On the other hand, specificity is moderate at 63%, suggesting the model struggles to predict negative cases (i.e., not recommending the airline).

The confusion matrix shows that false positives (FP) are higher than false negatives (FN). The F1 score, which balances sensitivity and precision, is 80.2%, which is acceptable. Lastly, the Gini coefficient of 0.667 is considered good.

Overall, while MR and ACC alone may indicate a lower model fit, examining other metrics, such as sensitivity, specificity, F1 score, and Gini coefficient, gives a clearer understanding of the model's performance.

Before making a final conclusion about model fit, we should review the ROC and Lift charts for confirmation. Using the validation sample:

- The ROC chart (Figure 7.14) shows that the model lies between the baseline (diagonal line) and the perfect-fit model (upper left corner). With an AUC index of 0.833, the model's performance is acceptable (an ROC index of 0.9 or higher indicates a strong model fit).
- The Lift chart (Figure 7.15) shows that with 20% of the dataset, the Lift value is 1.6, meaning the model captures 32% of positive cases (1.6 times more than using no model), which is also considered an acceptable fit.

Conclusion

These charts confirm the earlier results. If the objective is to predict positive cases (airline recommendations), this model performs well with a sensitivity of 89.3%. However, it does a poorer job of predicting non-recommendations (negatives).

Ultimately, when faced with mixed results, the decision on whether the model is acceptable depends on the business context. Alternatively, further refinements can be made by reconfiguring the algorithm or selecting a different one.

Logistic Regression Result Interpretation

Assuming the model fit is acceptable, the next step is to interpret the logistic regression results. Table 7.25 presents the detailed regression results (coefficient values) extracted from the Advanced tab of the model nugget for Logistic 1.

First, we examine the p-values for the Wald Chi-square test. Since these p-values are lower than 0.01, it indicates that the corresponding predictors are significant in predicting the target field, which is the recommendation to use the airline.

TABLE 7.25
Coefficient Results of the Champion Model—Logistic 1

Fields	B	S.E.	Wald	Df	Sig.	Exp(B)
Age	–0.013	0.001	370.319	1	< .001	0.987
Price sensitivity	–0.101	0.016	39.135	1	< .001	0.904
Number of years flying	–0.012	0.003	16.12	1	< .001	0.988
Average number of flights per year	–0.012	0.001	303.08	1	< .001	0.988
Number of other loyalty cards	–0.047	0.009	27.092	1	< .001	0.954
Shopping amount at airport	0.001	0	9.199	1	0.002	1.001
Departure delay (in minutes)	0.005	0.001	34.961	1	< .001	1.005
Arrival delay (in minutes)	–0.011	0.001	161.538	1	< .001	0.989
Airline status_t			4,004.021	3	< .001	
Airline status_t(1)	0.448	0.05	81.217	1	< .001	1.566
Airline status_t(2)	0.832	0.034	601.802	1	< .001	2.297
Airline status_t(3)	1.651	0.027	3,759.086	1	< .001	5.213
Gender_t(1)	0.39	0.019	417.081	1	< .001	1.477
Travel purpose_t			11,450.006	2	< .001	
Travel purpose_t(1)	–0.359	0.031	133.121	1	< .001	0.698
Travel purpose_t(2)	–2.865	0.027	11,426.678	1	< .001	0.057
Passenger class_t			75.68	2	< .001	
Passenger class_t(1)	0.299	0.034	75.573	1	< .001	1.349
Passenger class_t(2)	0.037	0.031	1.414	1	0.234*	1.037
Destination state_t			71.538	51	0.03	
Destination state_t(1)	–0.122	1.204	0.01	1	0.919*	0.885
Destination state_t(2)	0.047	0.828	0.003	1	0.955*	1.048
…	0.136	0.385	0.125	1	0.724*	1.146
Flight cancellation_t(1)	–0.435	0.072	36.316	1	< .001	0.647
Constant	1.451	0.058	628.038	1	< .001	4.269

* Not significant.

The B column shows the logistic coefficients, and the S.E. column provides the standard errors.

As mentioned earlier, the sign of the B coefficient indicates the direction of the effect (positive or negative). For example:

- Price sensitivity has a negative effect on the likelihood of recommending the airline.
- Airline status has a positive effect on the target field.

However, to understand the effect size, we must interpret the results using the odds ratio. In the output, Exp(B)represents the odds ratio.

It is also essential to know the reference level when interpreting the effect of nominal or binary predictors. If you recall, we set the first level as the reference level for all categorical input fields. To view these reference levels, simply open the settings for the Type node at the beginning of the stream, which will display the detailed levels for all categorical fields in the dataset (Figure 7.16). The first level listed is the reference level.

There are many predictors with varying results, making it challenging to explain each one in detail. Therefore, I have summarized the interpretation of the significant predictors in a summary table (Table 7.26). Please note that we only interpret predictors that have significant effects on the target field.

Field	Measurement	Values
Airline status	Nominal	Blue,Gold,Platinum,Silver
Travel purpose	Nominal	"Business travel","Mileage tickets","Personal Travel"
Passenger class	Nominal	Business,Eco,"Eco Plus"
Origin state	Nominal	Alabama,Alaska,Arizona,Arkansas,California,Colorado,Connecticut,Delaware,Florida,Geor...
Destination state	Nominal	Alabama,Alaska,Arizona,Arkansas,California,Colorado,Connecticut,Delaware,Florida,Geor...
Recommend the airline	Flag	Yes/No
Gender	Flag	Male/Female
Flight cancellation	Flag	Yes/No

FIGURE 7.16 Levels of categorial fields.

TABLE 7.26
Logistic Regression Result Interpretation

Predictors	Sign of Effect	Level	Odds Ratio	Interpretation
Age	–	Interval	0.987	As age increases by one year, the odds of airline recommendation decrease by 1 – 0.986 = 0.013 or 1.3%
Price sensitivity	–	Interval	0.904	As price sensitivity increases by one unit, the odds of airline recommendation decrease by 1 – 0.904 = 0.096 or 9.6%
Number of years flying	–	Interval	0.988	As the years flying increase by one year, the odds of airline recommendation decrease by 1 – 0.988 = 0.012 or 1.2%
Number of flights per year	–	Interval	0.988	As flights number increases by one flight per year, the odds of airline recommendation decrease by1 – 0.988 = 0.012 or 1.2%
Number of other loyalty cards	–	Interval	0.954	As the loyalty cards increase by one per year, the odds of airline recommendation decrease by 1 – 0.954 = 0.046 or 4.6%
Shopping amount at airport	+	Interval	1.001	As shopping amount increases by one dollar, the odds of airline recommendation increase by 1.001 – 1 = 0.001 or 0.1%
Departure delay	+	Interval	1.005	As departure delay increases by one minute, the odds of airline recommendation increase by 1.005 – 1 = 0.005 or 0.5%
Arrival delay	–	Interval	0.989	As arrival delay increases by one minute, the odds of airline recommendation decrease by 1 – 0.989 = 0.011 or 1.1%
Airline status	+	Gold vs. *Blue*	1.566	The odds of Gold passengers recommending the airline is 1.566 times more likely than Blue passengers.
Airline status	+	Platinum vs. *Blue*	2.297	The odds of Platinum passengers recommending the airline is 12.297 times more likely than Blue passengers.
Airline status	+	Silver vs. *Blue*	5.213	The odds of Silver passengers recommending the airline is 5.213 times more likely than Blue passengers.
Gender	+	Female vs. *Male*	1.477	The odds of female passengers recommending the airline is 1.477 times more likely than male passengers.
Passenger class	+	Eco vs. *Business*	1.349	The odds of Eco class passengers recommending the airline is 1.349 times more likely than Business class passengers.
Travel purpose	–	Mileage tickets vs. *Business travel*	0.698	The odds of business travel passengers recommending the airline is 1/0.698 = 1.432 times more likely than mileage ticket passengers.
Travel purpose	–	Personal travel vs. *Business travel*	0.057	The odds of business passengers recommending the airline is 1/0.057 = 17.54 times more likely than personal travel passengers.
flight_cancelled	–	Yes vs. *No*	0.647	The odds of passengers without flight cancellation experience recommending the airline is 1/0.648 = 1.546 times more likely than the one with cancellation experience.

Note: Italic = Reference level.

For example, while the results indicate that destination state has a significant effect overall, none of the individual states show significant effects on the target field when compared to the reference level (State of Alabama). As a result, we do not interpret the effect of this predictor.

The interpretation method follows the approach previously described for each predictor (we will not interpret the intercept, as it has no practical implications, even though it is necessary for the regression equation). To simplify interpretation, I have grouped interval predictors and nominal/binary predictors separately.

Key points to remember:

- An **odds ratio** greater than 1 indicates a **positive effect**.
- An **odds ratio** less than 1 indicates a **negative effect**.

As you can see, interpreting interval predictors is relatively straightforward, while interpreting nominal and binary predictors requires more care. Additionally, odds ratios greater than 1 are generally easier to interpret than those below 1.

Variable Importance

An important question to consider is: How important are the significant predictors, and which are more influential than others? To answer this, we use the variable importance chart found in the Model tab of the model nugget Logistic 1 (Figure 7.17).

The results indicate that travel purpose is the most important predictor, followed by airline status, age, and arrival delay. Conversely, flight cancellation, price sensitivity, and destination state are among the least important predictors.

As shown in the chart, the importance scores range from 0 to 1 and are scale-independent. Thus, they provide useful insights into the relative importance of each predictor. These findings can assist airline executives in developing marketing and promotion strategies to improve services and attract more passengers.

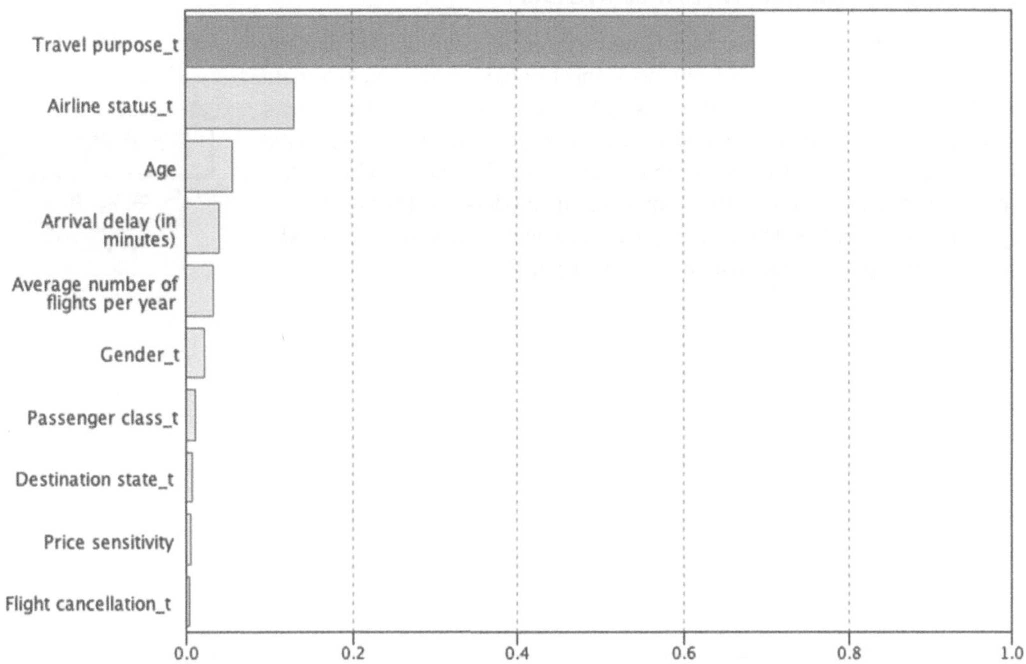

FIGURE 7.17 Variable importance plot for the champion model—Logistic 1.

HOW TO RUN THE ANALYSIS NODE WITH MULTIPLE MODELS?

As mentioned earlier, SPSS Modeler does not have a specific node for comparing selected metrics among all models, like the Model Comparison node in SAS Enterprise Miner. However, there is a way to connect an Analysis node to multiple model nuggets and compare metrics across models. I'll walk you through the process, but keep the following restrictions in mind:

- **Single connection limit**: You cannot simply add an Analysis node and connect it to all model nuggets at once. SPSS Modeler will give an error message because the Analysis node can only connect to one model nugget at a time.
- **Detaching model nuggets**: To connect multiple model nuggets to the Analysis node, you must first detach the model nuggets (except for the first one) from the Partition or Type node. This is because each model nugget can only connect to one preceding node. Next, connect the model nuggets to one another in sequence, and then connect the final nugget in that sequence to the Analysis node before running the analysis.
- **Copying models and nuggets**: If you copy models and nuggets from another stream, ensure the model node and the model nugget remain connected after copying.
- **Identical field names**: Fields in all model nuggets must have identical names. If any model has been transformed in a way that changes the field names, it cannot be compared with other models in the sequence, as this will cause SPSS Modeler to produce an error message. This restriction limits the utility of this method for model comparison purposes.
- **Reconfiguring models**: You can still reconfigure the models and rerun them as needed.

Figure 7.18 provides an example of a stream that compares multiple models using this approach. For better organization, it is recommended to arrange the model nodes and nuggets in sequence for easy tracking. As shown, all model nodes remain connected to their nuggets and the Partition node, but the nuggets must be connected to one another sequentially.

EXAMPLE: COMPARE MULTIPLE MODELS

As mentioned, we built four logistic regression models for this dataset. However, keep in mind that Logistic 1 and Logistic 2 only include data transformation for categorical fields, while Logistic 3 and Logistic 4 involve transformation for all fields. As a result, the field names in Logistic 1 and Logistic 2 differ from those in Logistic 3 and Logistic 4. This difference prevents us from directly comparing all models together using this approach, since SPSS Modeler requires identical field names for model comparison, and it will return an error if they differ.

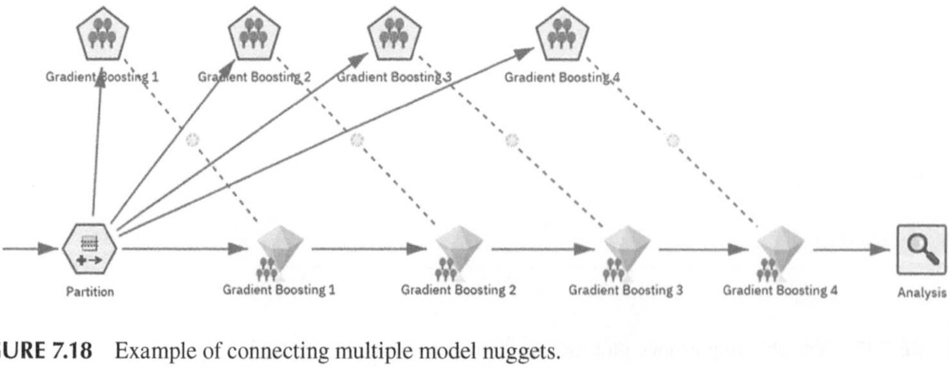

FIGURE 7.18 Example of connecting multiple model nuggets.

In this case, I will guide you on how to compare Logistic 1 and Logistic 2. You can follow the same steps to compare Logistic 3 and Logistic 4 separately.

In the output, you may notice that the models appear under the predicted target names instead of the model names. Specifically:

- Logistic 1 will appear as $L-Recommend the airline
- Logistic 2 will appear as $L1-Recommend the airline

These names can be difficult to remember and track, especially when presenting results. To change the model names in the output, we can edit the field names using the Filter node. I'll show you how to do this in the steps as follows.

Steps to Compare Logistic Models 1 and 2:

1. Start with the stream **Airline_Regression** that we created earlier.
2. Detach the model nugget for **Logistic 2** from the **Type** node. To do this, hover over the connector, right-click, and select **Delete Connection**.
3. Connect the model nugget for **Logistic 1** to the model nugget for **Logistic 2**.
4. Add a **Filter node** to the stream and connect it to the model nugget for **Logistic 2**.

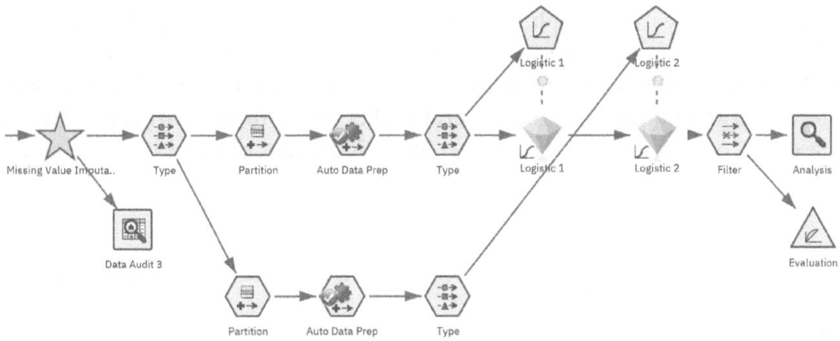

5. Open the **Filter** node settings. (Note that this screen will only appear after you have already run the **Logistic 1** and **Logistic 2** models, which we did earlier in this chapter.)
6. In the settings, change **$L-Recommend the airline** to **Logistic 1**, and change **$L1-Recommend the airline** to **Logistic 2**.

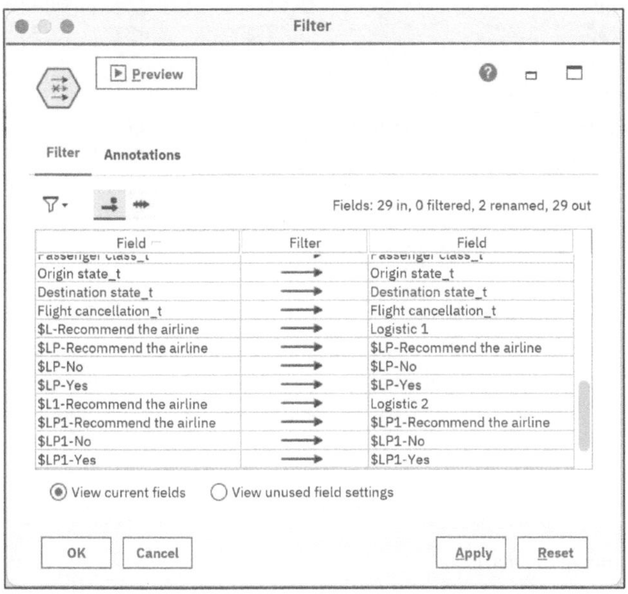

7. Click Apply and OK.
8. Add an **Analysis** node from the **Output** group and connect it to the **Filter node**.
9. Open the **Analysis** node settings and configure it as done previously.
10. Click Apply and Run.

Figure 7.19 shows the results of the Analysis node, which presents three key outputs as follows:

- **Individual models**: Analysis outputs for both models, as shown previously. Since we changed the names, the models are now displayed with the desired labels.
- **Agreement between two models**: The level of similarity in predictions between the two models. In this case, the models show a high level of agreement, with more than 98% similarity.
- **Evaluation metrics**: A comparison of the models based on AUC and the Gini coefficient. As mentioned earlier, Logistic 1 shows slightly better results than Logistic 2.

Please note that we still need to calculate other fit metrics, such as sensitivity and specificity, separately using the confusion matrix.

We can also add an Evaluation Graph node to the model nugget for Logistic 2 and run it to produce an ROC comparison chart or Lift comparison chart for both models. In this case, the ROC and Lift charts for both models are nearly identical, so the difference is not noticeable in the charts (Figure 7.20 and Figure 7.21). However, in other cases, these chart comparisons could be useful. Regardless, we can always review the AUC (ROC index) and Lift score to compare the models.

Results for output field Recommend the airline

Individual Models

Comparing Logistic 1 with Recommend the airline

Comparing Logistic 2 with Recommend the airline

Agreement between Logistic 1 Logistic 2

'Partition'	Testing		Training		Validation	
Agree	24,509	98.58%	73,649	98.53%	24,474	98.47%
Disagree	353	1.42%	1,098	1.47%	381	1.53%
Total	24,862		74,747		24,855	

Comparing Agreement with Recommend the airline

'Partition'	Testing		Training		Validation	
Correct	18,857	76.94%	56,781	77.1%	18,881	77.15%
Wrong	5,652	23.06%	16,868	22.9%	5,593	22.85%
Total	24,509		73,649		24,474	

Coincidence Matrix for Agreement (rows show actuals)

Performance Evaluation

Confidence Values Report for Agreement

Evaluation Metrics

'Partition'	Testing		Training		Validation	
Model	AUC	Gini	AUC	Gini	AUC	Gini
Logistic 1	0.829	0.658	0.833	0.666	0.833	0.667
Logistic 2	0.828	0.656	0.833	0.666	0.833	0.666

FIGURE 7.19 Analysis outputs for model comparison.

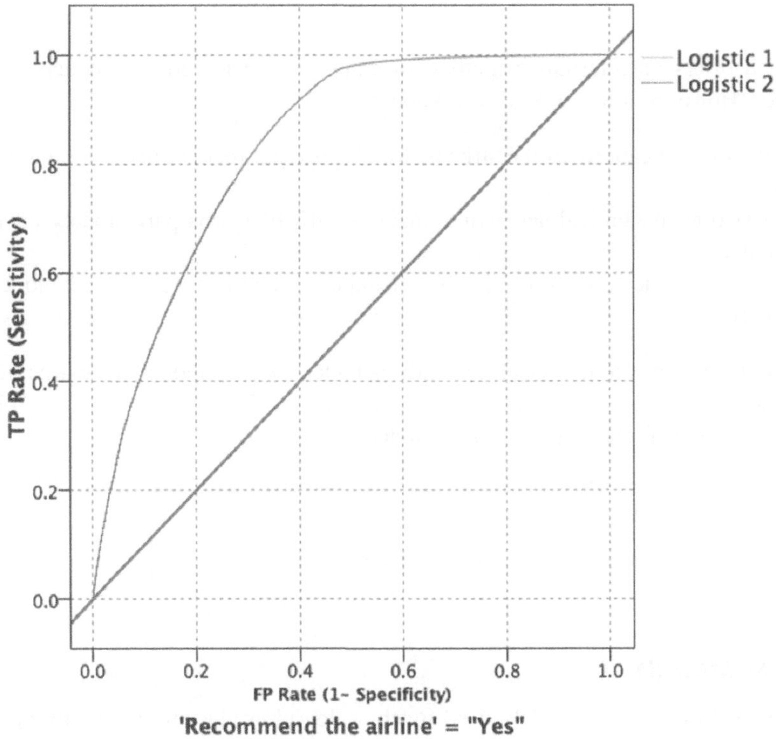

FIGURE 7.20 ROC comparison chart of Logistic 1 and Logistic 2.

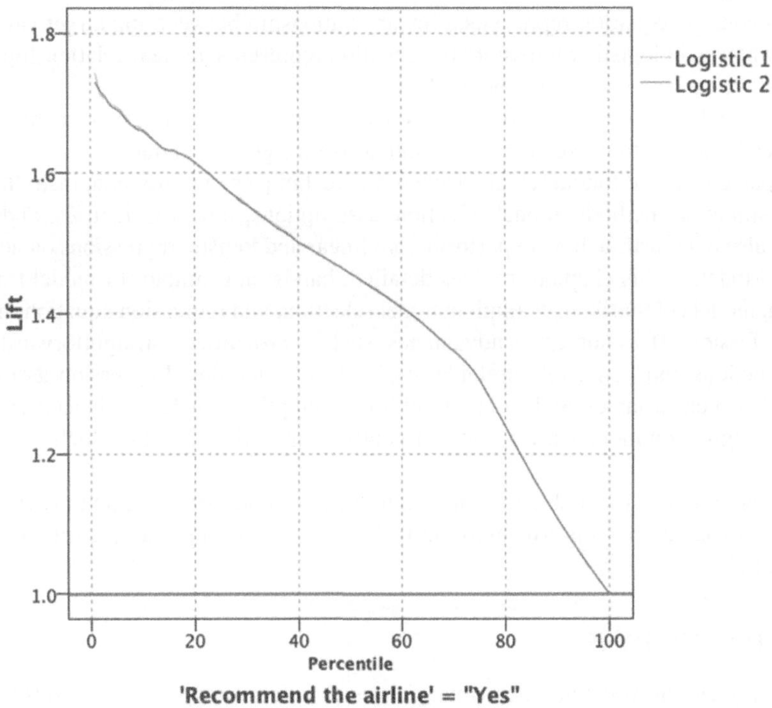

FIGURE 7.21 Lift comparison chart of Logistic 1 and Logistic 2.

Conclusion

This approach offers an alternative method to compare metrics among models using a single Analysis node. However, it has several limitations:

- It requires modifications to the stream, which prevents reconfiguring the models within the stream.
- The restriction on identical field names makes it difficult to compare models that use transformed data.
- Some important metrics still need to be calculated manually based on each model's confusion matrix.

My recommendation is to use separate Analysis nodes for each model and extract the necessary metrics. In cases where field names are identical across models, this method can still be useful for generating ROC and Lift charts for multiple models.

> In **Chapter 10**, we will apply this approach using the **Ensemble** node. You can refer back to this section for the steps to connect multiple models.

CHAPTER SUMMARY

Regression is one of the most widely used machine learning algorithms due to its simplicity in implementation and ease of interpreting results. Regression methods allow us to predict the value of a target variable based on multiple predictors. Linear regression is an association model used for continuous target variables while logistic regression is a classification model used for binary target variables. The key characteristic of regression methods is the assumption of a linear relationship. In linear regression, the equation represents a linear relationship between the target variable and the predictors, whereas in logistic regression, the equation represents a linear relationship between the logit value of the target variable and the predictors.

Regression analysis is widely applied across fields such as economics, finance, healthcare, aviation, and social sciences to make predictions and derive insights from data.

This chapter covers the fundamental concepts and techniques of regression, including regression equations, estimation methods, model selection, assumptions, model fit metrics, and interpreting outputs. We also demonstrate how to perform both linear and logistic regressions on actual datasets using SPSS Modeler. This chapter provides detailed, hands-on examples of model training, result presentation, model comparison, champion model selection, and regression equation interpretation.

While regression offers numerous advantages, such as providing a straightforward approach to making predictions and being interpretable, it also has limitations. Regression assumes a linear relationship between variables, and violations of this assumption may lead to inaccurate predictions. Additionally, regression models are sensitive to outliers and missing data, which can affect model performance.

Overall, regression is a valuable tool in data mining for understanding and predicting relationships between variables, but it is crucial to carefully assess its assumptions and limitations to ensure reliable results.

CHAPTER DISCUSSIONS

1. Define regression. What are the primary characteristics of regression methods?
2. Compare linear regression and logistic regression. Can logistic regression detect nonlinear patterns? Explain.

3. Provide practical examples of using linear regression and logistic regression in real-life scenarios.
4. Explain the advantages and disadvantages of regression methods in data mining.
5. What are the assumptions for linear regression and logistic regression? In data mining projects, must all assumptions be met to use those methods? Why?
6. Which estimation methods are used for linear regression and logistic regression? What is the key difference between these methods?
7. What is the purpose of model selection methods? Compare forward stepwise, backward stepwise, and adaptive stepwise. Which method should we use?
8. What is specific model fit criteria for linear regression and logistic regression?
9. How do we interpret the coefficients of significant predictors in linear regression?
10. How do we interpret the coefficients of significant predictors in logistic regression?

CHAPTER EXERCISES

Exercise 1: Linear Regression Model for the Bank Dataset

1. Open the **bank dataset** in SPSS Modeler.
2. Sample the entire dataset. Refer to Appendix A for variable names and descriptions.
3. Partition the data using a 60-20-20 split ratio (train: 60%, test: 20%, and validation: 20%).
4. Create visualizations for the variables to explore data trends.
5. Explore the data to identify:
   - Missing values
   - Outliers
   - Variables with skewed distributions
6. Perform necessary data modifications:
   - Impute missing values
   - Normalize variables with skewed distributions
7. Set CRSCORE (credit score) as the target variable.
8. Run a linear regression model using the adaptive stepwise method.
9. Evaluate the model fit using both general and regression-specific model fit metrics.
10. Present the results, including significant predictors and their coefficients.
11. Interpret the regression coefficient for each significant predictor and explain how it predicts the target variable.

Exercise 2: Logistic Regression Model for the Passengers Dataset

1. Open the **passengers dataset** in SPSS Modeler.
2. Sample 50% of the dataset. Set return (decision to return to the airline) as the target variable and change its role to binary. Refer to Appendix B for variable names and descriptions.
3. Evaluate the balance of the dataset for the target variable (balance between 1—Yes and 0—No).
4. Partition the data using a 60-20-20 split ratio (train: 60%, test: 20%, and validation: 20%).
5. Explore the data to identify:
   - Missing values
   - Outliers
   - Variables with skewed distributions
6. Perform necessary data modifications:
   - Replace zero values for Likert-scale variables with missing values
   - Impute missing values
   - Normalize variables with skewed distributions

7. Run a logistic regression model using the adaptive stepwise method.
8. Evaluate the model fit using both general and regression-specific model fit metrics.
9. Present the results, including significant predictors, coefficients, and odds ratios.
10. Interpret the odds ratio for each significant predictor and explain how it predicts the target variable.
11. Compare the standardized estimates of the significant predictors and interpret the relative importance of each predictor.

8 Decision Trees

LEARNING OUTCOMES

Upon the completion of this chapter, you will be able to:

1. Explain the purpose of the decision tree method.
2. Describe the tree hierarchy and its structure.
3. Define root nodes, parent nodes, child nodes, and leaf nodes.
4. Discuss the recursive partitioning process in the decision tree method.
5. Evaluate how missing values are handled in decision trees.
6. Discuss the advantages and disadvantages of decision trees.
7. Explain the tree pruning process and its importance.
8. Describe the primary types of tree algorithms.
9. Compare the Classification And Regression Tree (CART), Chi-Squared Automatic Interaction Detection (CHAID) tree, C.50 tree, and Quick, Unbiased, Efficient, Statistical Tree (QUEST).
10. Set configurations for tree algorithms in SPSS Modeler.
11. Interpret the tree plot to analyze the results.
12. Interpret the variable importance from decision tree models.
13. Conduct Cost-Sensitive Learning (CSL) to account for misclassification costs.
14. Present and explain the results of the CSL.

INTRODUCTION

This chapter focuses on the decision tree method, a popular machine learning technique that presents the hierarchical structure of variables in tree form using logical rules. The tree consists of nodes that split into two or more branches based on conditions that best separate individuals into classes. It is one of the most widely used machine learning methods due to its ease of implementation and interpretation. The tree structure visually demonstrates relationships among variables through nodes and branches, making it intuitive to understand. Additionally, decision trees are capable of detecting **nonlinear relationships**.

In this chapter, we will explore the advantages and disadvantages of decision trees, enabling us to compare them with other machine learning methods. The four common types of decision trees—CART (Classification and Regression Tree), CHAID (Chi-squared Automatic Interaction Detector), C5.0, and QUEST (Quick, Unbiased, Efficient, Statistical Tree)—will be discussed. We will examine their characteristics and differences, and I will demonstrate how to build these trees using SPSS Modeler, including the tree pruning process.

DOI: 10.1201/9781003468196-10

Since decision trees are non-parametric and based on logical rules, their theoretical foundation differs significantly from regression methods. Rather than relying on mathematical equations, we follow the branches of the tree to observe how predictors correlate with the target variable. Although mathematical algorithms are still used to build the tree hierarchy, I will focus on explaining the method in a simple and practical way so that you can understand how it works without delving too deeply into the underlying algorithms.

WHAT IS A DECISION TREE? HOW DOES IT WORK?

A **decision tree** is a highly popular machine learning method, frequently used in decision-making processes and data mining projects. It is a **non-parametric** and **nonlinear** method. Decision tree models offer the advantage of **conceptual simplicity**, making them easy to understand while also accommodating nonlinear associations between predictors and the target variable. The tree is represented as a **flowchart**, visually illustrating the relationships among variables, which makes it easy to comprehend and explain—even in complex processes. This is a significant advantage, as many other machine learning algorithms lack this level of interpretability.

> **DEFINITION**
>
> Decision tree is a logical rule-based method that presents a hierarchical structure of variables, including the root node, parent nodes, and child nodes.

Now, let's explore a simple example of a decision tree shown in Figure 8.1. Suppose we want to predict the likelihood of passengers purchasing an airline ticket for a trip, based on factors such as ticket price, family emergency, and hotel price. The flowchart captures the decision-making process or the rules that define the relationship between the passenger's decision (the target variable) and these factors (the predictors). Let's examine the tree structure.

- The primary factor is **ticket price**, which forms the **root node**. The flowchart shows that if the ticket price is high, the next factor to consider is the **family emergency**. If there is a family emergency, the passenger will decide to purchase the ticket, even if the price is

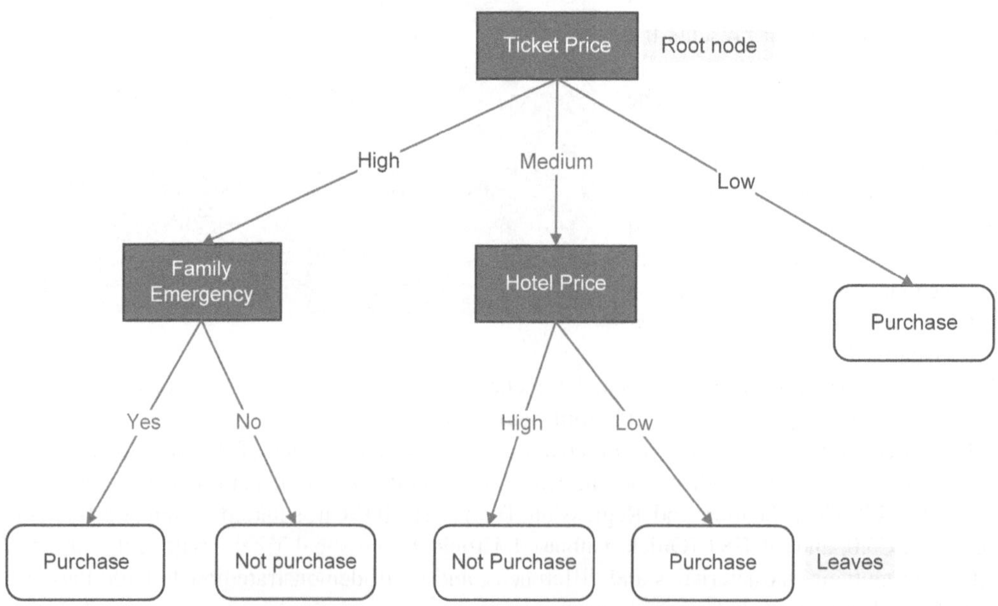

FIGURE 8.1 An example of a decision tree.

high. However, in the absence of an emergency, the ticket will not be purchased since the trip can be postponed.
- If the ticket price is medium, the next factor to consider is the **hotel price**. A high hotel price will likely deter the passengers from purchasing the ticket, as they may prefer to wait for a better opportunity. Conversely, if the hotel price is low, it represents a good deal, and the passenger will proceed to purchase the ticket.
- Finally, if the ticket price is low, it is considered a good deal, and the passenger will purchase the ticket and proceed with the trip without needing to consider other factors.

As you can observe, the decision tree presents the decision-making process in a straightforward and logical manner. It is easy to follow and understand the relationships among variables based on these rules. The tree provides us with a set of rules that can be used to make predictions. Thus, the decision tree is a logical, rule-based method rather than a statistical or mathematical one. The flowchart allows us to create a model that predicts the ticket purchasing decision based on these predictors. When properly trained and validated with enough data, it can predict the purchasing decision (purchase vs. not purchase) with good predictive power.

A decision tree represents a hierarchical structure of variables, consisting of the root node, parent nodes, child nodes, and leaf nodes:

- The **root node** is at the top of the tree and contains the entire dataset.
- A **parent node** is any node that splits into child nodes, forming branches. Whether a node is considered a parent or child depends on the perspective of interpretation.
- Any node that falls under another node is considered a **child node** while the node directly preceding it is the **parent node**. This means a node can be both a child node to one parent and a parent node to other child nodes.
- The final nodes of the decision tree are called **leaf nodes** or **leaves**. For each leaf, a decision is made and applied to all observations within that node. In predictive modeling, the decision is the **predicted value**.

In our example, the tree hierarchy is as follows:

- **Ticket price** serves as the **root node**, and **family emergency** and **hotel price** are child nodes, each of which splits into two branches.
- **Family emergency** is also a **parent node**, with two child nodes: **purchase** and **not purchase**.
- The **rounded rectangle** nodes are the **leaves**, representing the final classification decisions (purchase or not purchase).

BASIC CONCEPTS OF DECISION TREES

Given the number of new terms introduced in this chapter, Table 8.1 provides an overview of the basic concepts in decision tree methods along with their descriptions.

TABLE 8.1
Basic Concepts of Decision Trees

Terms	Descriptions
Decision tree	Decision tree is a logical rule-based method that organizes variables in a hierarchical structure, consisting of a root node, parent nodes, and child nodes.
Root node	A root node is at the beginning of a tree. It represents the initial decision based on a specific feature that has the most significant impact on the target variable. Starting from the root node, the dataset is partitioned based on different features, and subsequently, these subgroups are further divided at each decision node located beneath the root node.
Splitting	Splitting is a process of dividing a node into two or more sub-nodes.
Splitting criterion	Splitting criterion is a measure used to evaluate and select the best predictor to split the data at each node of the tree. It is a measure of impurity or variance reduction.
Decision node	When a sub-node is divided into additional sub-nodes, it is referred to as a decision node.
Leaf node	Nodes that do not split are called leaf nodes. They represent the output or the predicted value/class for the specific subgroup of data that reaches that node.
Pruning	Removing the sub-nodes of a parent node is called pruning. It is a technique used to simplify the decision tree by removing nodes that do not significantly improve its performance on the validation set. A tree is grown through splitting and shrunk through pruning.
Branch or subtree	A branch or a subtree is a sub-section of a decision tree.
Parent node and child node	These terms are relative in nature. Any node situated below another node is typically referred to as a child node or sub-node while the node preceding these child nodes is commonly known as the parent node.
Recursive partitioning	Recursive partitioning denotes an iterative procedure involving the division of data into partitions, followed by further splitting along each branch.
Classification and Regression Tree (CART)	CART is a binary tree algorithm that repeatedly splits a parent node into two child nodes. It can be used for both classification and regression models. CART uses measures like Gini impurity or variance reduction to determine splits.
Chi-squared Automatic Interaction Detection (CHAID)	CHAID is a non-parametric method used to create decision trees by recursively splitting the data into homogeneous subgroups based on categorical predictors. It uses the chi-squared test to assess independence between variables and identify significant splits. CHAID is commonly used for categorical variables and allows for more than two child nodes at each split.
C5.0 tree	C5.0 algorithm is designed for classification tasks and uses the information gain ratio to determine splits. It builds scalable, interpretable decision trees and uses advanced pruning techniques to prevent overfitting. C5.0 is known for being an improvement over its predecessor, C4.5, with faster performance and more flexibility.
Quick, Unbiased, Efficient, Statistical Tree (QUEST)	QUEST is a binary decision tree algorithm used primarily for classification. It uses statistical methods to ensure unbiased splits and is designed to handle large datasets efficiently. It offers quick, unbiased decisions and reduces the bias that can sometimes occur in other decision tree algorithms when dealing with continuous predictors.

HOW DOES NODE SPLITTING WORK?

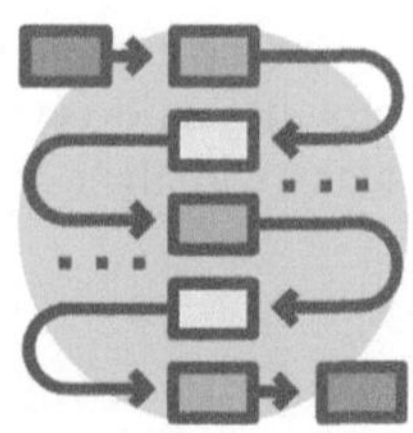

From the example above, we see how a tree is structured based on the hierarchy of decisions made by passengers. The process of node splitting is crucial to constructing that tree. So, how does node splitting work?

Node splitting is a key process in constructing the hierarchical structure of a decision tree. It determines how the data should be partitioned at each level of the tree. During this process, the algorithm selects a variable, and the corresponding node is split into two or more branches based on a condition that best separates individuals into classes.

- If the variable is **categorical**, the branches represent different classes.
- If the variable is **interval-based** (continuous), the branches represent specific ranges.

At each split, a separation condition is applied to determine how the parent node is split into child nodes. This continues until a stopping rule is met. This overall process is called **recursive partitioning**.

Recursive Partitioning Steps

1. **Selecting a splitting criterion**: At each node, the algorithm selects a variable and a corresponding splitting criterion to partition the data into child nodes. The splitting criterion depends on the target variable's scale and may include measures such as the **F-test**, **chi-square test**, **Gini impurity**, or **Entropy**.
2. **Splitting the data**: The selected criterion is used to divide the node into two or more child nodes, with each child node representing a branch from the current node in the tree.
3. **Stopping criteria**: The algorithm checks if any stopping conditions are met, such as follows:
   - Reaching a maximum tree depth
   - Having a minimum number of samples per leaf
   - Achieving a certain level of impurity reduction
4. **Recursive process**: If the stopping criteria are not met, the process continues recursively for each branch. The child nodes become new parent nodes, and the process repeats.
5. **Leaf nodes**: When a stopping criterion is satisfied or no further meaningful splits are possible, the process ends, and the final nodes are called leaf nodes. These leaf nodes represent the final predictions for the target variable.

The recursive partitioning process continues until all stopping criteria are met, or until the tree becomes fully grown. However, a fully grown tree may result in overfitting, where the model captures noise and specific details of the training data, which can lead to poor predictive power on unseen data. To address this issue, **tree pruning** can be applied to simplify the tree by removing branches that do not contribute significantly to the model's performance. Pruning techniques will be covered later in this chapter.

DEFINITION

Recursive partitioning is an iterative process where data are repeatedly split into partitions based on specific conditions, with each partition further divided along the branches of the decision tree. This process continues until the data are adequately segmented, often leading to leaf nodes that represent the final classifications or predictions.

In summary, recursive partitioning in decision trees is a top-down approach that begins with the entire dataset and recursively divides it into smaller, more homogeneous subsets. This process is efficient and enables decision trees to handle both categorical and continuous variables, making them powerful and interpretable models for data mining.

DO MISSING VALUES HAVE EFFECTS ON DECISION TREES?

In traditional statistical studies, missing values are a significant concern because they can represent missing information in the data and negatively affect the results. Regression methods are particularly sensitive to missing values, as they typically require a complete dataset to build models with a good fit. Additionally, missing values can lead to a loss of information and introduce bias into the data.

However, decision trees are a non-parametric method based on logical rules, and they handle missing values much more effectively. Since decision trees are constructed by splitting parent nodes into multiple child nodes, missing values have a less significant impact on the process of building the tree. Decision trees can naturally handle missing values through several mechanisms.

Ways Decision Trees Handle Missing Values:

1. **Splitting criteria with missing values**: Decision trees use different splitting criteria, such as Gini impurity or Entropy, to determine the best variable and threshold for splitting the data at each node. When a variable contains missing values for certain data points, the algorithm can treat these missing values as a separate category or group. Data points with missing values may traverse down both branches of the split, allowing the tree to calculate impurity or error based on the available data. This way, decision trees can leverage the information from non-missing values to make decisions, even when some data points are incomplete.
2. **Surrogate splitting**: Surrogate splitting is another method used by decision trees to handle missing values. When the primary split contains missing values, the decision tree algorithm identifies alternative splits (or surrogates) that closely mimic the primary split based on available data. The surrogate split is then applied when the primary split cannot be used, effectively handling the missing data.
3. **Using imputation**: Some decision tree implementations support the imputation of missing values, where missing data are replaced with estimated values. We have covered imputation techniques in Chapter 5.
4. **Creating multiple branches**: If a data point has missing values for multiple variables, the decision tree may create multiple branches from the current node to account for different combinations of missing and non-missing values. The algorithm evaluates each branch separately, considering the available data for each data point.
5. **Handling missing values during prediction**: When making predictions on new data with missing values, the decision tree can still generate predictions by following the appropriate branches based on the available values. If a data point has missing values for multiple variables, the tree may traverse down multiple paths to reach the leaf nodes. The predictions from these paths are then combined to make the final prediction.

- Decision trees can naturally handle missing values during the modeling process.
- It is recommended that software programs with built-in algorithms, such as SPSS Modeler, be used to effectively handle missing values in decision tree modeling.

It is important to note that some data mining software programs, such as SPSS Modeler, include built-in features that handle missing values more efficiently during decision tree modeling. It is recommended to use software with these built-in algorithms for the best results.

ADVANTAGES AND DISADVANTAGES OF DECISION TREE METHODS

Decision tree is a very popular method in data mining. It is a non-parametric method that can detect nonlinear relationships. It can be used with various measurement scales and is not sensitive to missing values or outliers. The tree structure is logical and easy to understand and interpret. Since we use a strategy that requires developing multiple models using different methods and configurations, it is important to understand the advantages and disadvantages of decision trees, which allows us to compare this method with the regression methods we covered in Chapter 7 and other more advanced methods we will cover later. I summarize the key advantages and disadvantages of this method as follows:

Advantages of Decision Trees

- **Interpretability**: Decision trees provide a transparent and intuitive representation of the decision-making process. The tree's hierarchical structure is easy to understand and interpret, making it a valuable tool for explaining the predictions to non-experts.
- **Handling non-linearity and complex relationships**: By using decision trees, researchers can effectively capture nonlinear relationships and complex interactions between variables through the hierarchical structure.
- **Flexibility with data types**: Decision trees can be used to build either association models or classification models, making them versatile and suitable for a wide range of applications. They can also work with both continuous and categorical predictors.
- **Irrelevant features**: Decision trees are useful to model irrelevant variables in the data since they focus on variables that are more informative for splitting the data.
- **Scalability**: Large datasets are handled more efficiently in decision trees, which do not require high computing power.
- **Handling missing values**: Due to the nature of the recursive partitioning process, decision trees are able to handle missing values effectively through the splitting process.
- **Ensemble methods**: Decision trees can be combined using ensemble methods like Random Forests and Gradient Boosting, which often lead to improved predictive performance and reduced overfitting.
- **Providing variable importance**: A relative variable importance plot is available in a decision tree's output, which makes it easier for us to identify important predictors.

Disadvantages of Decision Trees

- **Overfitting**: Decision trees are sensitive to overfitting, especially when the tree is deep and complex.
- **Instability and variance**: Small changes in the data can lead to significantly different tree structures, which results in high variance in the predictions.
- **Bias toward dominant classes**: In classification models with imbalanced datasets, decision trees may be biased toward the majority class, which leads to suboptimal performance in minority classes.
- **Difficulty handling numeric outliers**: Decision trees are sensitive to the scale of continuous variables, and outliers can have a disproportionate influence on the tree's structure and predictions.
- **Global optimum vs. local optimum**: The greedy nature of the decision tree algorithm may lead to suboptimal splits at individual nodes, even though they contribute to the overall best split.
- **Lack of coefficients**: Unlike regression, the output of a decision tree model does not provide coefficients for evaluating the effects of predictors on the target variable. These effects must be evaluated through the tree structure and the probability values, which could be challenging to understand.

Despite their limitations, decision trees remain a valuable and widely used machine learning method, especially for data mining projects that require interpretability, handling nonlinear relationships, and dealing with both categorical and continuous variables. However, to address some of the drawbacks, techniques like pruning, ensemble methods, and careful hyperparameter tuning are often applied to improve the model performance.

TREE PRUNING

Pruning is a crucial step in ensuring a decision tree model has good predictive power. Pruning involves removing redundant subtrees and replacing them with leaf nodes. Large trees with many branches and layers can lead to overfitting, which occurs when the model performs well on training data but poorly on new data. Additionally, large trees are difficult to interpret and understand. These trees often use the same predictors to construct the hierarchy, repeatedly splitting nodes to create numerous branches in an attempt to improve model fit. However, this process can result in overly complex trees that only fit the training data, leading to overfitting.

Pruning helps to prevent overfitting by simplifying the tree, making it more generalized to new datasets, without sacrificing model accuracy. Furthermore, a pruned tree is easier to understand and interpret. Figure 8.2 illustrates the process of decision tree pruning.

The process of tree pruning consists of six steps:

1. **Growing the full tree**: The first step is to grow the decision tree to its maximum depth or until a stopping criterion is met. During this stage, the algorithm recursively partitions the data and creates branches and nodes based on a chosen splitting criterion.
2. **Evaluating node importance**: After growing the full tree, assess the importance of each node in improving model performance. Nodes that contribute little to predictive accuracy or are not statistically significant are candidates for pruning.
3. **Cost-complexity pruning**: Cost-complexity pruning, also known as minimal cost-complexity pruning, is a popular pruning method. It introduces a complexity parameter (alpha) that balances model accuracy and complexity. The algorithm evaluates the impact of removing each node and calculates a cost-complexity score for the subtree rooted at that node.

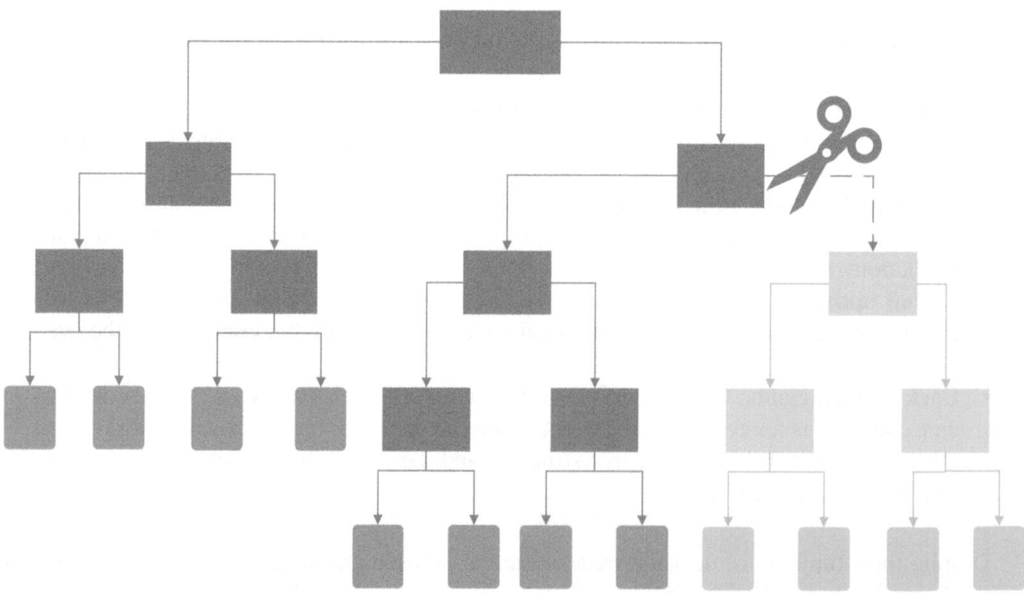

FIGURE 8.2 Decision tree pruning.

4. **Selecting the optimal pruning level**: The algorithm iteratively prunes nodes with the smallest cost-complexity score while evaluating model performance on a validation set. Pruning continues until further reductions result in a significant drop in predictive performance. The optimal pruning level corresponds to the subtree that provides the best balance between accuracy and simplicity.
5. **Pruning the tree**: Once the optimal pruning level is determined, the tree is pruned by removing nodes and branches corresponding to subtrees above the selected level. The resulting pruned tree is a simplified version of the full tree.

Pruning helps control overfitting by reducing the complexity of the decision tree, thereby improving its predictive power. It also enhances model interpretability by providing a more concise representation of the decision-making process. However, pruning is a data-driven process, and the optimal pruning level may vary depending on the specific dataset. Regular validation is essential to determine the best pruning strategy for each model.

Types of Decision Tree Pruning

Decision tree pruning can be classified into two types: pre-pruning and post-pruning.

- **Pre-pruning** (also known as the Early Stopping Rule): Pre-pruning prevents the growth of a tree beyond a specific size or depth based on certain measures like the Gini index or entropy. In this method, the tree-building process is halted when a pruning condition is met at a node. If the condition is satisfied, the subtree is pruned by replacing the decision node with a leaf node. Otherwise, the tree continues to grow. Pre-pruning is faster and more efficient, as it prevents the creation of overly complex subtrees that may overfit the training data. However, pre-pruning can stop the tree's growth prematurely based on the stopping condition.
- **Post-pruning**: In post-pruning, the entire tree is grown first, and then subtrees are pruned from the bottom-up. Starting from the bottom decision nodes, measures such as Gini impurity or information gain are used to decide whether to keep a decision node or replace it with a leaf node.

Manual vs. Automatic Pruning

- **Manual pruning**: Manual pruning is generally not recommended for data mining projects because it relies on the researcher's judgment, which can introduce bias. Manual pruning also requires a clear justification for removing a subtree, which may not always be evident. For example, how do we know which branches are redundant to remove, and how do we justify that decision? Mistakes in manual pruning can lead to poorer model fit or render the tree ineffective. While manual pruning may be feasible for simple trees, it is impractical for large trees with many nodes and branches. It also defeats the purpose of mining large datasets.
- **Automatic pruning**: The recommended approach is automatic pruning. This involves setting specific criteria for tree development and subtree selection, allowing the software to perform pruning based on those configurations. Typically, you define how the software should select and assess subtrees and whether to apply cross-validation. The software then uses search algorithms to find the best tree that meets these criteria and requirements.

DIFFERENT TYPES OF DECISION TREES

As demonstrated in the previous example, **how a node is split** is crucial to developing a good decision tree model. In data mining, decision trees are derived from data, not from common sense or business rules. Therefore, having valid splitting rules is essential to producing a decision tree model with strong predictive power. Splitting algorithms allow us to categorize different types of decision trees.

Over time, many types of decision tree algorithms have been developed. In this book, I introduce four tree algorithms that are widely used in data science and data mining projects:

- **CART**
- **Chi-Squared Automatic Interaction Detection (CHAID)**
- **C5.0**
- **QUEST**

I will provide a general overview of these trees, focusing on their splitting rules, pruning methods, and the appropriate scenarios for using each. As always, I will avoid mathematical formulations and emphasize practical applications.

> SPSS Modeler uses the term C&R Tree instead of CART tree. They are the same.

CLASSIFICATION AND REGRESSION TREE (CART)

CART is a popular decision tree method developed by Breiman (1984). For more detailed information, you can refer to their book, *Classification and Regression Trees*. In this explanation, I will skip the mathematical details and focus on practical applications.

CART is a binary tree, meaning that it repeatedly splits a parent node into two child nodes. This process can be used for both categorical (classification) and continuous (regression) target variables. CART is widely employed in machine learning due to its simplicity, interpretability, and effectiveness.

> **DEFINITION**
>
> CART is a binary tree algorithm that repeatedly splits a parent node into two child nodes. It can be used for both classification and regression models. CART uses measures like Gini impurity or variance reduction to determine splits.

BINARY RECURSIVE SPLITTING

CART uses a binary recursive splitting approach. At each node of the tree, the algorithm selects a feature and a threshold to split the data into two subsets. For categorical variables, CART performs multi-way splits by creating branches for each category, while for continuous variables, it identifies a threshold for splitting.

Splitting Criteria

- For **classification models**, CART typically uses **Gini impurity** as the splitting criterion. Gini impurity measures the likelihood of incorrectly classifying a randomly chosen instance. The algorithm selects splits that minimize impurity.

- For **regression models**, CART uses **variance reduction** as the splitting criterion, with **least squared deviation (LSD)** being the common measure for variance. The goal is to find splits that reduce variance within the resulting subsets.

Most data mining software integrates both classification and regression capabilities in their CART implementation, so you don't need to manually separate the two tree types.

Recursive Tree Building

The process of building the tree is recursive. After the initial split, the algorithm applies the same procedure to each resulting subset, continuing to split nodes until a stopping criterion is reached. Common stopping criteria include the following:

- A minimum number of instances in a leaf node.
- A maximum tree depth.
- No further improvement in impurity or variance reduction.

Figure 8.3 illustrates what a binary tree looks like, where each parent node splits into exactly two child nodes.

Pruning

Once the tree is fully grown, it may be pruned to improve generalization and avoid overfitting. Pruning involves removing branches that do not significantly enhance predictive accuracy. CART typically uses cost-complexity pruning, which balances the complexity of the tree with its accuracy on a validation set.

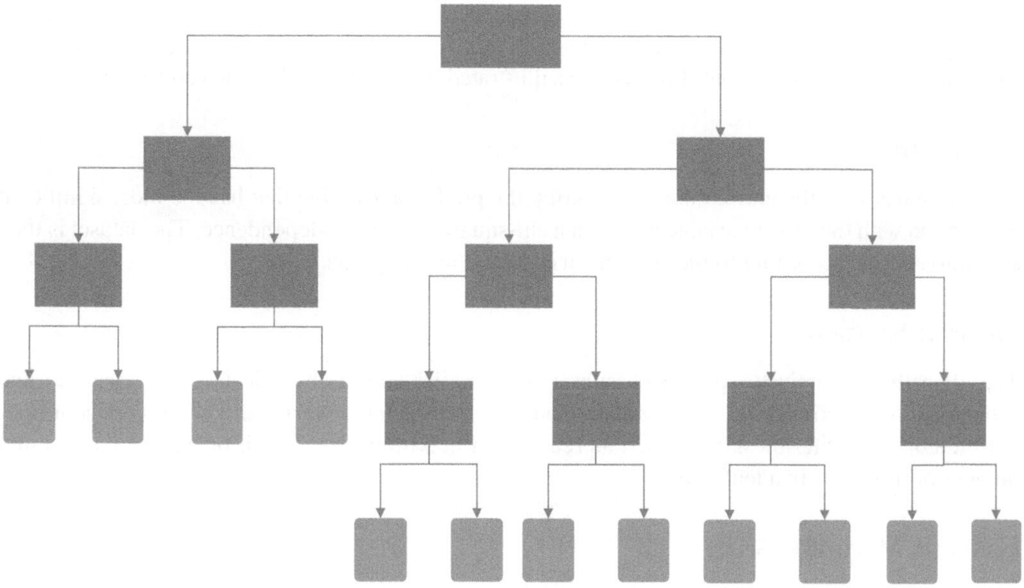

FIGURE 8.3 An illustration of a binary tree.

Handling Missing Values

CART is capable of handling missing data. When a feature contains missing values, the algorithm distributes instances with missing data among child nodes based on the available information from other features.

CHI-SQUARED AUTOMATIC INTERACTION DETECTION (CHAID)

CHAID was developed by Kass in 1980. This method is based on the chi-square statistic, a non-parametric test used to examine differences between categorical variables. CHAID aims to identify significant relationships between variables by applying chi-squared tests for independence. The test produces a probability value (p-value) between 0 and 1. A chi-square value closer to 0 indicates a significant difference between the two variables while a value closer to 1 indicates no significant difference. Unlike CART, CHAID can split a parent node into more than two child nodes, making it useful for creating multi-way splits.

DEFINITION

CHAID is a non-parametric method used to create decision trees by recursively splitting the data into homogeneous subgroups based on categorical predictors. It uses the chi-squared test to assess independence between variables and identify significant splits. CHAID is commonly used for categorical variables and allows for more than two child nodes at each split.

Splitting Rule

At each split, CHAID assesses the statistical significance of the relationships between predictor variables and the target variable using chi-squared tests. The algorithm seeks to create homogeneous groups, where the target variable is similar within each branch of the tree.

Handling Continuous and Categorical Variables

CHAID is primarily designed for categorical variables. While it works for both association and classification models, it requires categorical predictors, either ordinal or nominal. However, many software implementations of CHAID can accept continuous variables by automatically grouping values into ranges before growing the tree. Since Kass' original algorithm, CHAID has evolved to handle continuous target variables by using the p-value from an F-test. A key distinction from CART is that CHAID can split a node into two or more child nodes, as illustrated in Figure 8.4, which depicts a multi-way tree.

Initial Split

CHAID starts with the entire dataset, selecting the predictor variable that has the most significant association with the target variable based on a chi-squared test for independence. The dataset is then split into subsets according to the chosen variable and its categories.

Recursive Splitting

The algorithm recursively splits each subset into smaller subsets by selecting the variables and categories that maximize the statistical association with the target variable. The process continues until a stopping criterion is met, such as reaching a maximum tree depth or having a minimum number of instances in a leaf node.

Handling Multi-Way Splits

CHAID can perform multi-way splits for categorical variables with more than two categories, creating branches for each category. This ability makes it well-suited for datasets with multiple categorical variables.

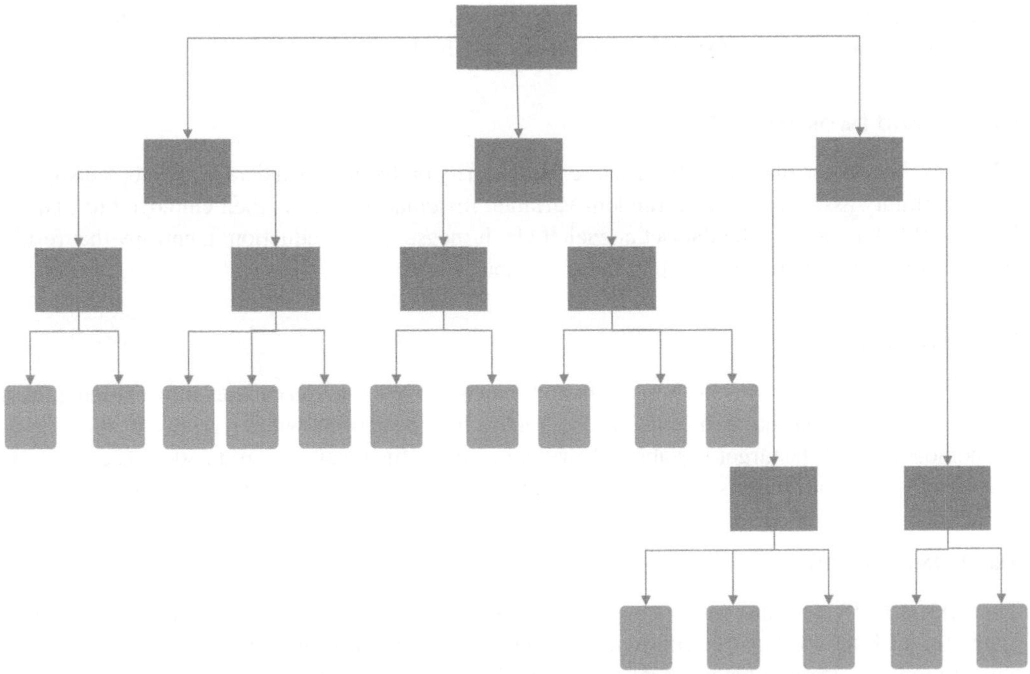

FIGURE 8.4 An illustration of a multi-way tree.

Merge Procedure for Pruning

After the tree is constructed, CHAID applies a merge procedure to prune the tree and simplify its structure. This process merges adjacent categories or nodes that do not significantly differ in terms of the target variable. The merge procedure helps prevent overfitting and produces a more interpretable final tree.

Handling Missing Values

CHAID handles missing values using surrogate splits. Missing values are treated as separate categories, and if a variable has missing data, CHAID identifies surrogate variables—alternative predictors that closely mimic the behavior of the primary variable. These surrogates act as backup rules when making decisions for instances with missing values.

C5.0 TREE

C5.0 is a powerful decision tree algorithm developed by Ross Quinlan in 1993. Primarily designed for classification models, C5.0 can also handle association models. The algorithm uses the information gain or gain ratio criterion to select the best attributes for node splitting, creating a tree structure that optimally classifies or predicts the target variable. C5.0 is known for its efficiency, scalability, and ability to handle both categorical and continuous variables. Additionally, it incorporates advanced pruning techniques to improve model interpretability and prevent overfitting, making it a widely

DEFINITION

C5.0 is a decision tree algorithm that uses the information gain ratio as a splitting rule, allowing it to effectively select variables that contribute the most to class distinctions. Designed for classification models and known for its scalability and interpretability, C5.0 incorporates advanced pruning methods to enhance model generalization.

used and effective tool for predictive modeling. It is considered an improvement over its predecessor, C4.5, offering faster performance and greater flexibility.

Entropy and Information Gain

C5.0 uses the concept of entropy to measure the impurity or disorder in a dataset. Entropy quantifies the uncertainty associated with a random variable. Information gain is then employed to find the best variable for splitting the dataset at each node. It measures the reduction in entropy that results from splitting the data based on a particular variable.

Splitting Criteria

The algorithm evaluates various predictors and selects the one that maximizes information gain or other splitting criteria like gain ratio. These criteria help determine which variable best improves the homogeneity of the target variable's distribution across the resulting child nodes. The goal is to partition the data into subsets that are more homogeneous in terms of the target variable.

Tree Construction

The dataset is divided into subsets based on the selected variable, and the process is recursively applied to each subset. The tree continues to grow until a stopping criterion is met, such as reaching a maximum tree depth or having a minimum number of instances in a leaf node.

Handling Continuous and Categorical Variables

While C5.0 is primarily designed for building classification trees, it can handle both continuous and categorical variables. For continuous variables, the algorithm determines optimal split points based on information gain. It can also handle categorical variables with multiple levels, finding the best way to split the categories for maximum information gain.

Pruning

Once the tree is fully grown, it may be pruned to avoid overfitting. The C5.0 algorithm uses a combination of cost-complexity pruning and reduced-error pruning to enhance the tree's ability to generalize to new data.

Handling Missing Values

C5.0 handles missing values in the dataset using surrogate splits. These are alternative rules that serve as backups when the primary variable for splitting has missing values for certain instances.

> In SPSS Modeler, the C5.0 tree algorithm is specifically designed to work with categorical target variables. As a result, it cannot be used for association models, which typically require continuous target variables.

QUEST (QUICK, UNBIASED, EFFICIENT, STATISTICAL TREE)

QUEST is a decision tree algorithm developed by Loh and Shih in 1997. It is designed for building trees in classification models, with a particular emphasis on handling large datasets efficiently. QUEST is known for its speed and the robust statistical foundation that guides its split decisions.

Overview of QUEST

QUEST operates by using a statistical approach to construct decision trees. The algorithm iteratively selects variables and split points based on statistical criteria, aiming to maximize homogeneity within the branches while minimizing statistical impurity. Unlike traditional tree algorithms, QUEST performs simultaneous tests on multiple variables at each node, allowing for efficient and unbiased selection of the best split. The algorithm focuses on statistical significance and quick tree construction, adapting its structure to the complexity of the data. This makes QUEST particularly useful in situations where interpretability and statistical validity are key.

DEFINITION

QUEST is a binary decision tree algorithm used primarily for classification. It uses statistical methods to ensure unbiased splits and is designed to handle large datasets efficiently. It offers quick, unbiased decisions and reduces the bias that can sometimes occur in other decision tree algorithms when dealing with continuous predictors.

Binary Tree Structure

QUEST generates a **binary tree**, meaning each internal node splits into exactly two child nodes. This binary structure simplifies the decision-making process and enhances the efficiency of data separation.

Splitting Criteria

QUEST uses statistical measures to make splitting decisions. It often applies statistical tests, such as chi-squared tests or other association measures, to quantify the relationship between a potential splitting variable and the target classes. The goal is to maximize the statistical significance of each split, ensuring the resulting child nodes are as distinct as possible in terms of the target variable.

Recursive Binary Splitting

Similar to other decision tree algorithms, QUEST employs recursive binary splitting. The process begins with the entire dataset, selecting the predictor variable and split point that maximizes the statistical difference in the response variable. The data are then split into two subsets, and the process is repeated for each subset.

Stopping Criteria

The tree-building process continues until a predefined stopping criterion is met. Common stopping criteria include reaching a specified tree depth, having a minimum number of instances in a node, or other conditions designed to prevent overfitting. QUEST strives to find an optimal tree size based on statistical considerations.

Unbiased Approach

One of QUEST's main strengths is its unbiased approach to decision-making. The algorithm ensures that decisions are driven by statistical significance rather than biased criteria, providing an accurate and objective representation of the data.

Pruning

To avoid overfitting, QUEST incorporates pruning techniques. The pruning process uses a cost-complexity pruning strategy, where the algorithm evaluates the cost-complexity measure at various levels of tree complexity. It then selects a subtree that strikes an optimal balance between accuracy and simplicity.

Speed and Efficiency

A key feature of QUEST is its speed and efficiency, especially when dealing with large datasets. The algorithm is designed to be fast and unbiased, making it a suitable choice for applications where computational resources are limited or where processing time is critical.

> In SPSS Modeler, the QUEST tree algorithm is specifically designed to work with categorical target variables. As a result, it cannot be used for association models, which typically require continuous target variables.

TREE ALGORITHM COMPARISON

Table 8.2 highlights key differences between CART and CHAID decision trees. Overall, these two types of trees share many similarities, but they differ primarily in how nodes are split. As a reminder, CART uses a binary split algorithm, while CHAID uses a multi-split approach. Both methods can be applied to build either association or classification models. However, CART generally performs better with continuous or ordinal predictors while CHAID is more suited to categorical predictors. That said, most data mining software allows for a combination of predictor types.

Because of these differences in splitting algorithms, CART tends to produce smaller (narrower) but taller trees due to its binary splits, whereas CHAID typically results in larger but shorter trees because of its multi-split structure. Although CHAID may require more data than CART to train a model, it is generally less prone to overfitting. Some researchers suggest that pruning is unnecessary for CHAID, but I recommend pruning for both tree types to prevent overly large trees. We will explore examples of this in later sections.

TABLE 8.2
Differences between CART, CHAID, C5.0, and QUEST

Items	CART	CHAID	C5.0	QUEST
Type of predictive model	Can be used for building both classification and regression trees.	Primarily designed for categorical data and categorical target variables.	Primarily used for building classification trees but can also handle regression trees.	Primarily used for building classification trees.
Tree structure	Binary tree.	Multi-way tree.	Multi-way tree.	Binary tree.
Target variable	Works with both continuous and categorical target variables	Works with both continuous and categorical target variables	Works with both continuous and categorical target variables	Only works with categorical target variables
Predictors	Can handle both categorical and continuous predictors.	Primarily designed for categorical predictors but can handle continuous predictors.	Can handle both categorical and continuous predictors.	Can handle both categorical and continuous predictors.

(Continued)

TABLE 8.2 *(Continued)*
Differences between CART, CHAID, C5.0, and QUEST

Items	CART	CHAID	C5.0	QUEST
Splitting criteria	Binary split only; uses Gini impurity for classification and variance reduction for regression.	Multi-way split; uses chi-squared tests for independence.	Multi-way split; uses information gain or gain ratio.	Binary split only; uses statistical tests such as chi-squared tests or other measures of association.
Handling missing values	Can handle missing values.	Usually does not handle missing values explicitly.	Can handle missing values using surrogate splits.	Can handle missing values.
Pruning	Prunes the tree to avoid overfitting using cost-complexity pruning.	Uses a merge procedure after tree construction to simplify and prune.	Prunes the tree to avoid overfitting using surrogate splits.	Prunes the tree to avoid overfitting using cost-complexity pruning.
Tree size	Smaller, taller tree	Larger, shorter trees	Larger, shorter trees	Smaller, taller tree
Speed and efficiency	Generally efficient and widely used.	Efficient for categorical data.	Known for efficiency and scalability.	Specifically designed for efficiency, particularly with large datasets.
Statistical tests	Does not rely on explicit statistical tests; uses Gini impurity or variance reduction.	Uses chi-squared tests for independence.	Does not rely on explicit statistical tests; uses information gain or variance reduction.	Uses chi-squared test or association measures for splitting.
Interpretability	Produces interpretable trees, especially after pruning.	Provides easily interpretable trees based on categorical variables.	Offers interpretable trees, especially in rule form.	Known for interpretability and simplicity.
Variable importance	Provides a measure of variable importance based on the contribution to tree splits.	Typically does not offer a direct measure of variable importance.	Provides a measure of variable importance based on information gain.	May provide information about variable importance.
Sensitive to overfitting	More	Less	More	Less
Built-in pruning in SPSS Modeler	Yes	No	No	Yes
Handling both classification and regression trees in SPSS Modeler	Yes	Yes	No (only build classification trees)	No (only build classification trees)

Overall, each decision tree algorithm has its strengths and is suitable for different types of models and data. It is hard to speculate in advance which type of tree works better because they all have strengths and weaknesses. The choice of algorithm depends on the characteristics of the data and the specific requirements of the data mining. My recommendation is to build multiple trees using both types with different configurations and compare them to determine which tree works best for the data.

EXAMPLE: CLASSIFICATION MODEL USING DECISION TREES

In this example, we will build classification models for the Airline Recommendation dataset using decision trees. The target variable is whether a passenger recommends the airline, represented as a binary variable (Yes—recommend; No—not recommend). We will construct multiple decision tree models using different algorithms and assess which performs best for this dataset.

- **Dataset**: AirlineRecommendation.sav
- **Target variable**: Recommend (Flag: Yes/No)

To evaluate various algorithms under different conditions, I have developed the following scenarios to guide our decision tree model configurations:

- **Scenario 1**: Compare a pruned model with a non-pruned model. This scenario examines the performance difference between pruning and not pruning a decision tree. It will help us determine whether pruning is a beneficial practice in predictive modeling. To simplify, we will use the CART algorithm for this scenario, as it is the most commonly used decision tree algorithm.
- **Scenario 2**: Compare a model using imputed data, a model with missing values, and a model using the Auto Data Prep node (optimized for two objectives: accuracy and a balance between speed and accuracy). This scenario will assess how well decision tree algorithms handle missing data and whether it is advantageous to use non-transformed data for training. For consistency, we will use the CART algorithm again in this scenario.
- **Scenario 3**: Compare different tree algorithms. In this scenario, we will build decision trees using four algorithms: CART, CHAID, C5.0, and QUEST. Each algorithm may be configured differently, so we will create more than four models. This scenario will help us identify which algorithm performs best for this dataset.

STARTING POINT

Since we have already covered key steps in data exploration, modification, and partitioning for this dataset in Chapter 7, we will not revisit them here. Instead, we will focus primarily on the modeling process. For details on those earlier steps, please refer to the previous chapters. The starting point for this example will begin with the Partition node, and from there, we will proceed with building the models.

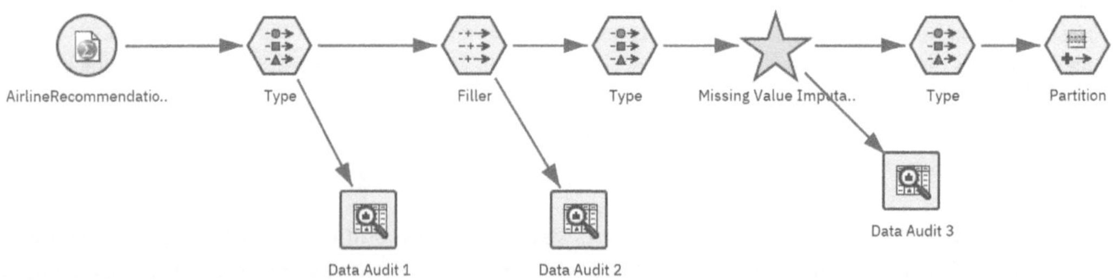

SCENARIO 1: PRUNED VS. NON-PRUNED CART MODELS

In this scenario, we will build two CART models: one with pruning and one without. To ensure a fair comparison, we will use imputed data for both models.

Follow the steps below to build the CART models:

1. Open SPSS Modeler.
2. Create a new stream and name it "Airline_Tree."
3. Copy and paste the stream from "Airline_Regression" (as shown earlier) and double-check the nodes.
4. Run the Data Audit nodes and verify the outputs, ensuring that the replacement and imputation tasks are correctly applied.
5. Continue with the Partition node (recall that we partitioned the data with a 60–20–20 split for training, validation, and testing).
6. From the **Modeling** group, add a C&R Tree node to the stream, name it **C&R Tree 1** and connect it to the Partition node. Open the node settings.
7. Under the **Fields** tab, confirm that the Target and Predictor fields are correct.
8. Under the **Build Options** tab, select **Objective** from the left panel and check the box labeled "Build a single tree" (we are not using ensemble models in this chapter, as that will be covered in Chapter 10).
9. In the **Basics** section, check the option **Prune tree to avoid overfitting**.

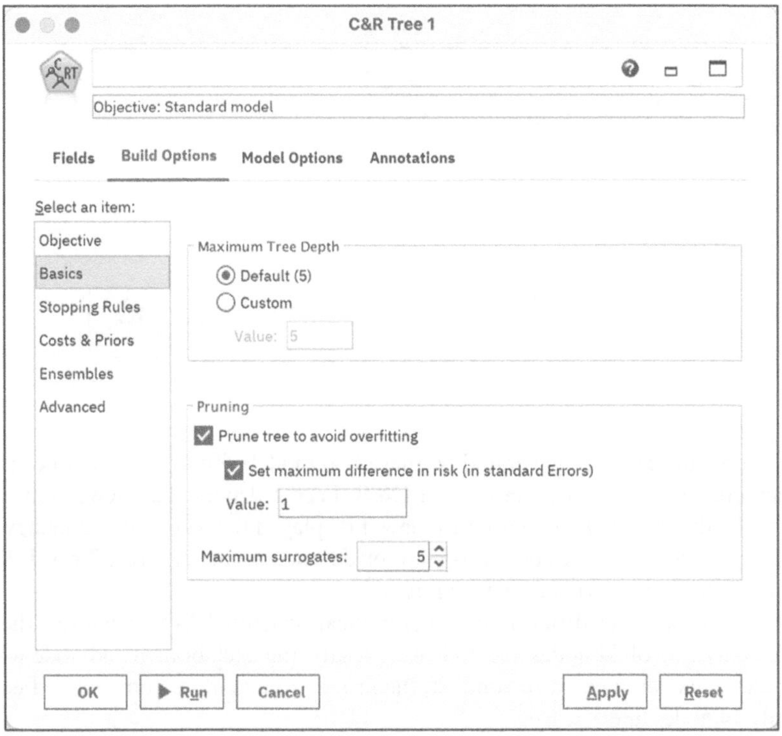

10. Keep the default settings for **Stopping Rules**.
11. Under the **Model Options** section, select "Calculate predictor importance."
12. Click Apply and run the model.
13. SPSS Modeler will generate a model nugget labeled **C&R Tree 1** which contains the results.
14. Add an Analysis node to the stream and connect it to the model nugget **C&R Tree 1**. Open the Analysis node settings.
15. Under the **Analysis** tab, select the following options: **Performance evaluation, Evaluation metric, Confidence figures**, and **Separate by partition**.
16. Click Apply and run the analysis.

This will generate performance metrics for the pruned tree model. We will hold onto these results for comparison. Next, follow the same steps to create and configure a second tree, naming it **C&R Tree 1 Non-Pruned**. The only difference is that in Step 9, uncheck the box labeled **Prune tree to avoid overfitting**, instructing SPSS Modeler to build the full, unpruned tree. Run the analysis for this non-pruned model as well. The final stream configuration should resemble the setup shown above.

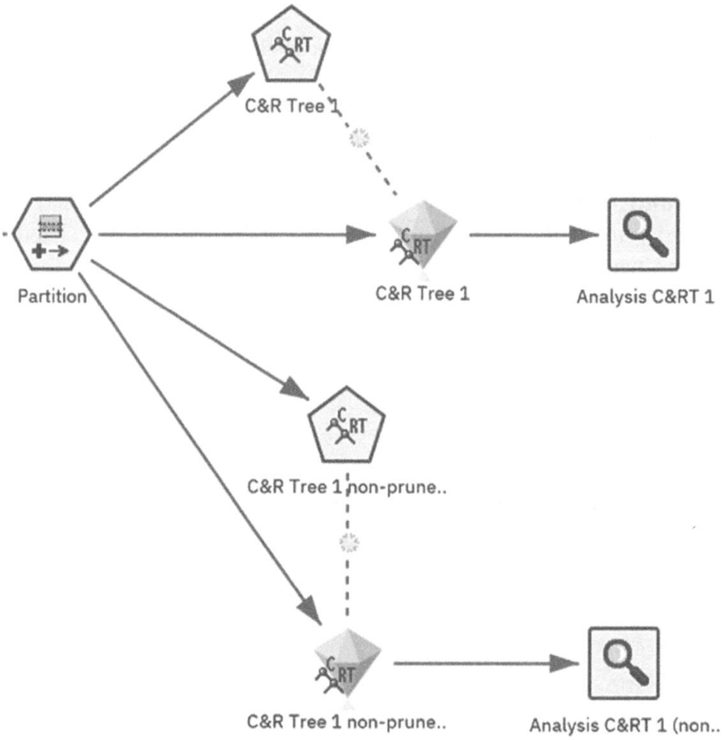

First, let's compare the trees generated by these two models. To extract the tree from the C&R Tree 1 model, open the model nugget labeled **C&R Tree 1**. Under the **Viewer** tab, you will find the tree. To make the tree structure easier to view, I displayed the tree with frequency information shown as a graph (Figure 8.5). Repeat this step for the model nugget **C&R Tree 1 Non-Pruned**. This will give us the full, unpruned tree (Figure 8.6).

So, how do these two trees differ in their hierarchical structure? As you can see, the non-pruned tree is larger, consisting of 34 nodes and 18 leaves. Analyzing each branch and node would be quite challenging due to its complexity. In contrast, the pruned tree is more compact and easier to interpret, with only 14 nodes and 8 leaves.

Note that the node numbers shown on the tree may not accurately reflect the sequential order, as some nodes are skipped during the tree-building process. It is recommended to manually count the nodes on the tree (excluding Node 0).

We compared the model fit metrics between the two trees using the validation results to assess their differences. The ACC, misclassification rate (MR), Gini coefficient, and AUC are directly

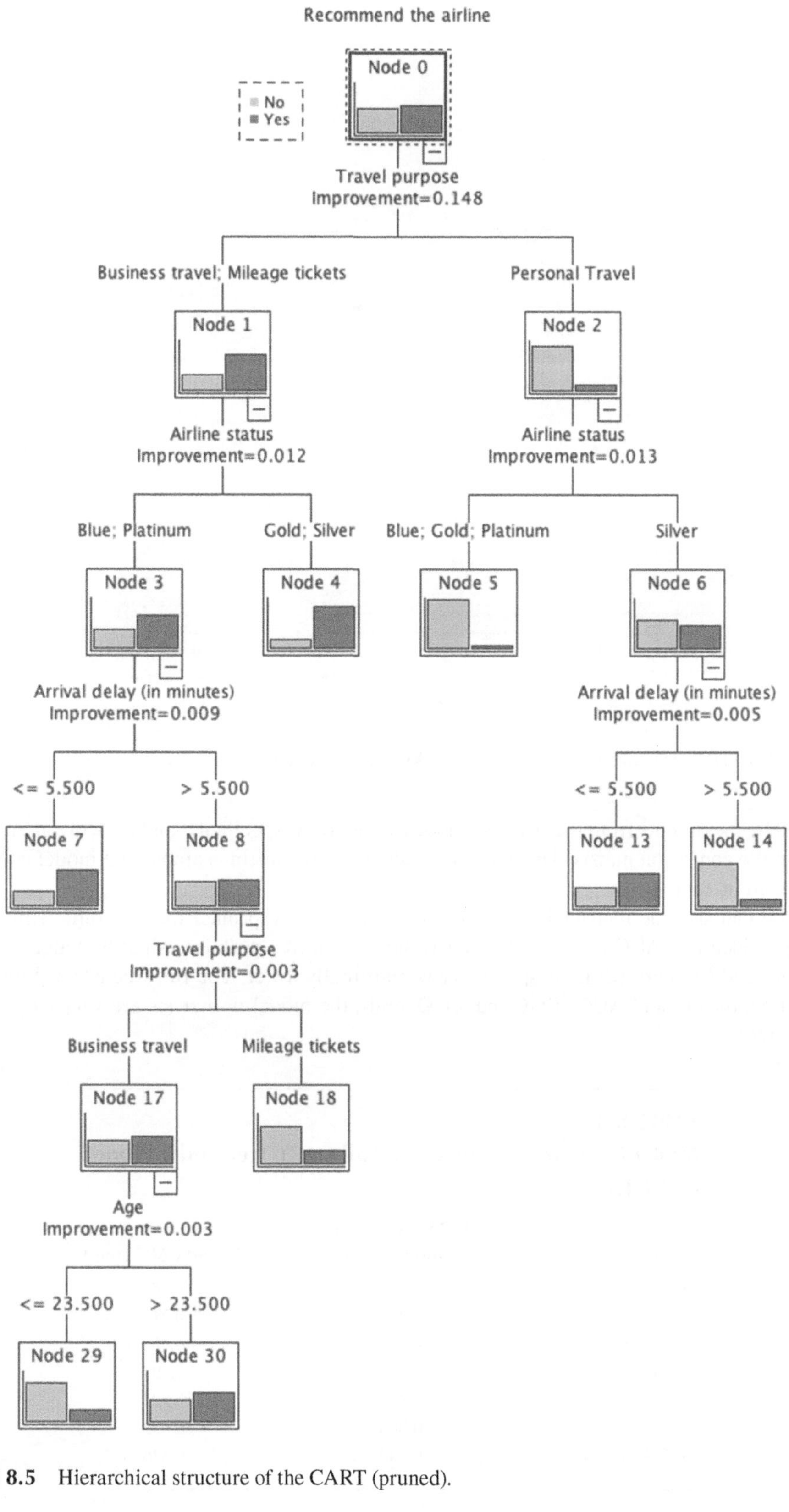

FIGURE 8.5 Hierarchical structure of the CART (pruned).

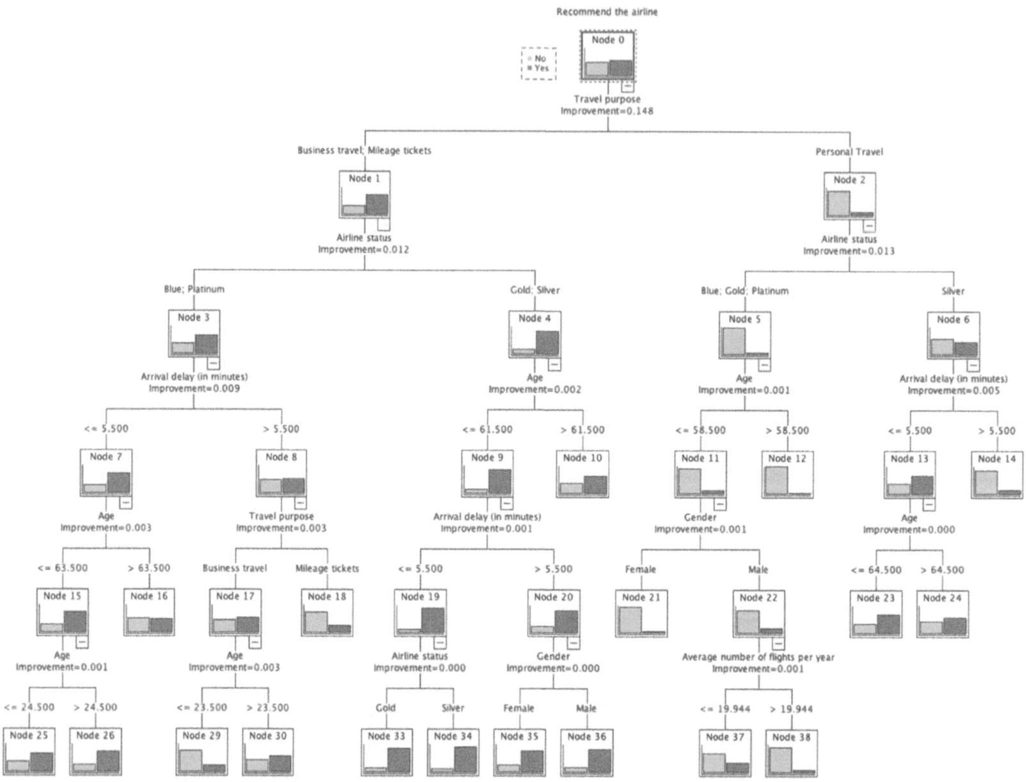

FIGURE 8.6 Hierarchical structure of the full CART (non-pruned).

provided by the outputs from the Analysis nodes. Sensitivity, specificity, and F1 score are calculated based on the confusion matrix. Using these results from the testing sample, the model comparison is presented in Table 8.3.

As you can see, the metrics between the full tree and the pruned tree are quite similar. Both trees show identical ACC and MR. However, the pruned tree performs slightly better in terms of sensitivity and F1 score while its specificity is marginally lower. The full tree has a slight edge in the Gini coefficient and AUC (ROC index). Overall, the model fit metrics are very close between the two trees.

TABLE 8.3
Model Comparison between a Full CART Tree and a Pruned CART Tree

Metrics	Full CART Tree 1 (Non-pruned)	Pruned CART Tree 1
ACC	0.786	0.786
MR	0.214	0.214
Sensitivity or TPR	0.937	0.960
Specificity or TNR	0.617	0.591
F1	0.822	0.826
Gini	0.680	0.652
AUC (ROC index)	0.840	0.826

Thus, building a full tree does not offer a significant advantage over a pruned tree. Additionally, the full tree is much larger, making it harder to interpret, requiring more computation time, and increasing the risk of overfitting. For these reasons, using the pruned tree is the more practical choice.

Scenario 2: CART Models with Missing Values vs. Imputed Data

In this scenario, we will compare a CART model that retains missing values with one that uses imputed data. Decision trees, in general, handle missing values better than regression algorithms. Since decision trees are rule-based rather than statistics-based, they do not typically require a complete dataset to function. Decision trees work by splitting data into distinct bins, so they offer various options for handling missing values without disrupting the model-building process. These options include ignoring missing values, using surrogate splits, treating missing values as a separate category, assigning a default value, or applying probability weighting. As such, using imputation does not necessarily provide any advantage when constructing a decision tree. In fact, imputation can introduce bias or increase prediction errors. Similarly, data normalization and outlier handling may not be necessary for decision tree algorithms.

In this scenario, we will compare three pruned CARTs, each constructed differently, and evaluate their performance:

- **C&R Tree 1**: Uses imputed data without normalization or outlier handling.
- **C&R Tree 2**: Retains missing values without normalization or outlier handling.
- **C&R Tree 3**: Uses the Auto Data Prep node with the objective **Optimize for accuracy**, which imputes missing values, adjusts field scales, normalizes data, and handles outliers.

Follow the same steps we used previously to add **C&R Tree 2** and **C&R Tree 3** to the stream with the specified configurations. Your stream should resemble the one shown in the following diagram.

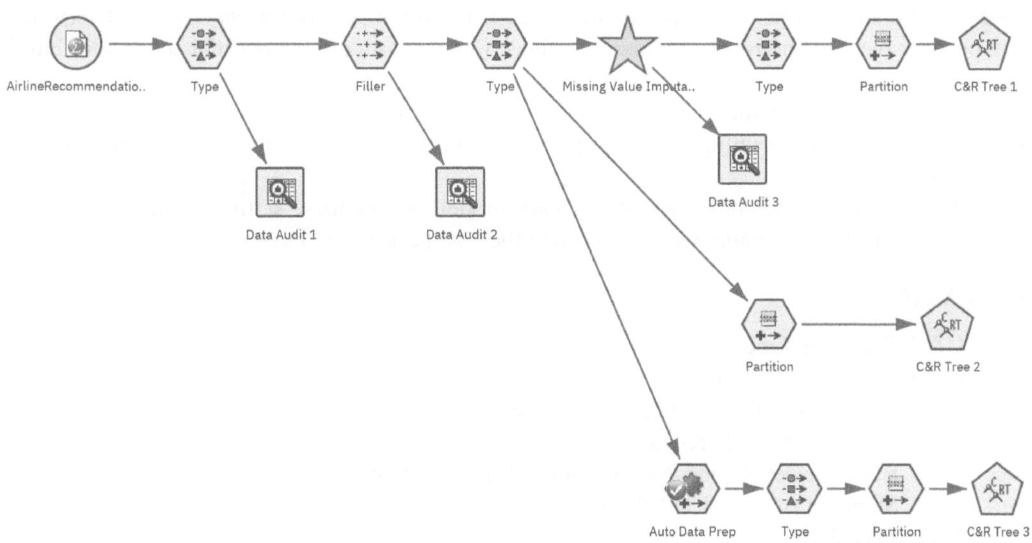

Run all three CART models and use the Analysis node to extract the model fit metrics for each. As before, you can obtain the ACC, MR, Gini coefficient, and AUC directly from the output of the Analysis node. To calculate sensitivity, specificity, and F1 score based on the confusion matrix, refer to the formulas discussed in Chapter 6. Make sure to use the results from the testing sample for model comparison. Table 8.4 presents the model comparison results for Scenario 2.

TABLE 8.4
Model Comparison between C&R Tree 1, C&R Tree 2, and C&R Tree3

Metrics	CART Model Using Imputed Data (C&R Tree 1)	CART Model with Missing Values (C&R Tree 2)	CART Model Using Auto Data Prep (C&R Tree 3)
ACC	0.786	0.787	0.786
MR	0.214	0.213	0.214
Sensitivity or TPR	0.960	0.964	0.960
Specificity or TNR	0.591	0.589	0.591
F1	0.826	0.827	0.826
Gini	0.652	0.652	0.652
AUC (ROC index)	0.826	0.826	0.826

The comparison results show that **C&R Tree 1** and **C&R Tree 3** have identical model fit metrics, indicating that data normalization and outlier handling do not enhance model performance. Furthermore, **C&R Tree 2** (which retains missing values) slightly outperforms the other models, with higher accuracy, better sensitivity, and a higher F1 score. This suggests that decision tree algorithms handle missing values more effectively through the built-in methods of the tree algorithm compared to traditional imputation. Therefore, keeping the missing values appears to be the best approach, and we will adopt it for the third scenario.

Scenario 3: CART vs. CHAID vs. C5.0 vs. QUEST

In this scenario, we will compare decision tree models developed using four different algorithms: CART, CHAID, C5.0, and QUEST. The objective is to identify which algorithm, and configurations work best for the Airline Recommendation dataset. From the previous scenarios, we determined that pruning the tree and retaining missing values in the dataset are the optimal choices for building decision trees. Therefore, in this scenario, we will only build pruned trees using data that contain the original missing values. For each algorithm, we will build two trees: one with the default settings for tree depth, pruning, and stopping rules and the other with modified settings to see if performance improves.

Below are the specific configurations for each model. Follow these settings to set up the trees. The models will then be compared to determine the best-performing model.

CART Models

- **C&R Tree 1 (default)**
  - Under the Build Options tab:
    - Objectives: Select "Build a single tree"
    - Basic: Maximum tree depth: Default (5)
    - Basic: Select "Prune tree to avoid overfitting"; Maximum surrogate = 5
    - Stopping rules: Select "Use percentage"
- **C&R Tree 2**
  - Under the Build Options tab:
    - Objectives: Select "Build a single tree"
    - Basic: Maximum tree depth: Custom = 10
    - Basic: Select "Prune tree to avoid overfitting"; Uncheck "Set maximum difference in risk"; Maximum surrogate = 5
    - Stopping rules: Select "Use absolute value"

CHAID Models

- **CHAID 1 (default)**
  - Under the Build Options tab:
    - Objectives: Select "Build a single tree"
    - Basic: Tree grow algorithm: CHAID
    - Basic: Maximum tree depth: Default (5)
    - Stopping rules: Select "Use percentage"
  - *Note:* CHAID does not have pruning options.
- **CHAID 2**
  - Under the Build Options tab:
    - Objectives: Select "Build a single tree"
    - Basic: Tree grow algorithm: Exhaustive CHAID
    - Basic: Maximum tree depth: Custom = 10
    - Stopping rules: Select "Use absolute value"

C5.0 Models

- **C5.0 1 (default)**
  - Under the Model tab:
    - Select "Use partition data" and "Build model for each split"
    - Output type: Decision tree
    - Mode: Simple
    - Favor: Accuracy
- **C5.0 2**
  - Under the Model tab:
    - Select "Use partition data" and "Build model for each split"
    - Output type: Decision tree
    - Mode: Select Expert
    - Set pruning severity to 100
    - Set minimum records per child branch to 5
    - Select "Use global pruning"

QUEST Models

- **QUEST 1 (default)**
  - Under the Build Options tab:
    - Objectives: Select "Build a single tree"
    - Basic: Maximum tree depth: Default (5)
    - Basic: Select "Prune tree to avoid overfitting"; Maximum surrogate = 5
    - Stopping rules: Select "Use percentage"
- **QUEST 2**
  - Under the Build Options tab:
    - Objectives: Select "Build a single tree"
    - Basic: Maximum tree depth: Custom = 10
    - Basic: Select "Prune tree to avoid overfitting"; Uncheck "Set maximum difference in risk"; Maximum surrogate = 5
    - Stopping rules: Select "Use absolute value"

Model Comparison Setup

Since all models use the same input and identical field names, follow the method described in Chapter 7 to connect all model nuggets for comparison. As shown in the following diagram, use the Filler node to rename the models, the Analysis node to compare model performance, and the Evaluation Graph node to produce ROC charts and Lift charts for all models.

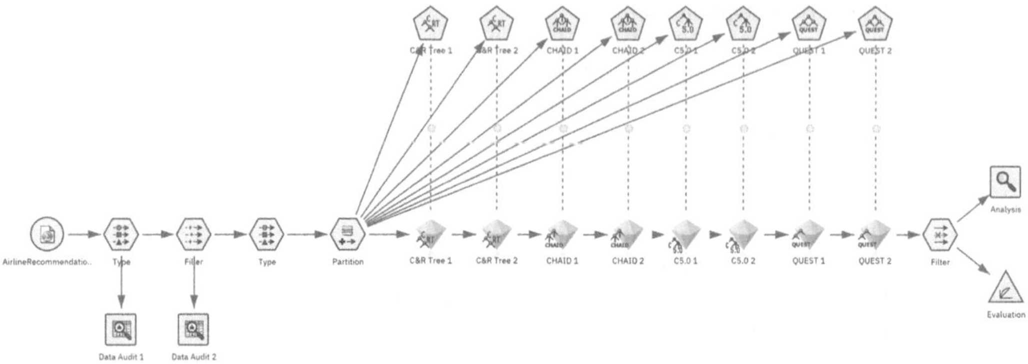

Below are the settings for the Filler node, where we rename the models to reflect their respective configurations. This step ensures that each model is clearly labeled for comparison in the subsequent analysis.

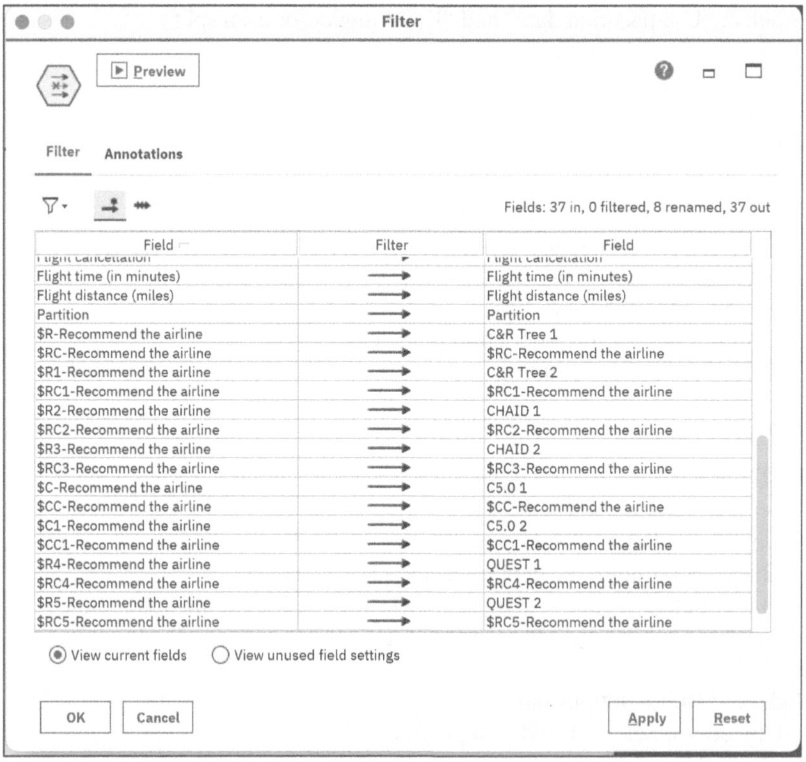

Champion Model

With the eight models built, the next step is to compare their model fit metrics to identify the champion model. Table 8.5 shows the comparison of these models using the testing sample results. As before, extract ACC, MR, Gini coefficient, and AUC (ROC index) from the Analysis node. Sensitivity, specificity, and F1 scores can be calculated from the confusion matrix. Figures 8.7 and 8.8 provide the ROC and Lift comparison charts for all models.

TABLE 8.5
Model Comparison between Decision Tree Algorithms (Using the Testing Sample's Results)

Metrics	C&R Tree 1	C&R Tree 2	CHAID 1	CHAID 2	C5.0 1	C5.0 2	QUEST 1	QUEST 2
ACC	78.70%	79.30%	79.00%	77.70%	79.00%	79.20%	74.50%	78.00%
MR	21.30%	20.70%	21.00%	22.30%	21.00%	20.80%	25.50%	22.00%
Sensitivity or TPR	0.964	0.952	0.924	0.883	0.912	0.949	0.814	0.915
Specificity or TNR	0.589	0.615	0.641	0.659	0.654	0.617	0.668	0.628
F1	0.827	0.829	0.823	0.807	0.821	0.828	0.771	0.814
Gini	0.652	0.665	0.700	0.671	0.680	0.618	0.521	0.617
AUC (ROC index)	0.826	0.832	0.850	0.836	0.840	0.808	0.760	0.808
Lift	1.577	1.577	1.639	1.582	1.587	1.442	1.422	1.471

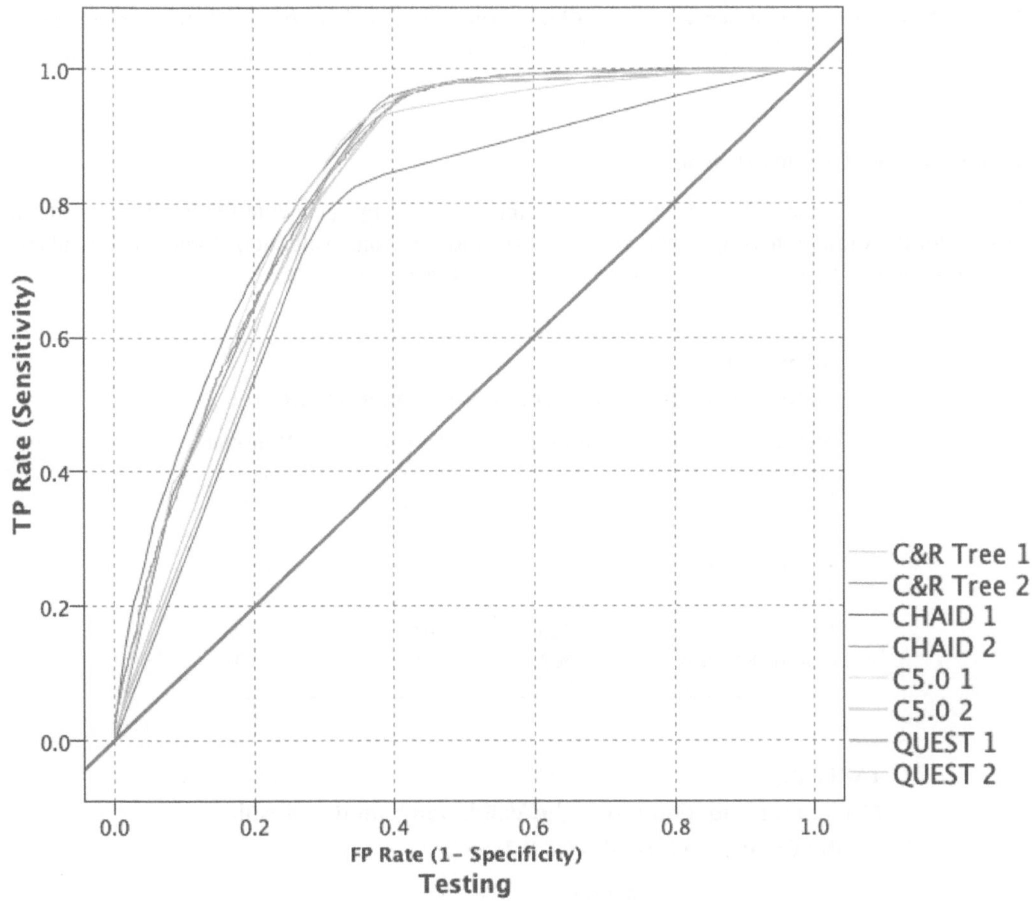

FIGURE 8.7 ROC comparison chart.

Based on the results, CHAID 2, QUEST 1, and QUEST 2 underperform compared to the other models, with the lowest ACC. Among the remaining models, C&R Tree 2 and C5.0 2 outperform the others, having the highest ACC. We can narrow the comparison down to these two models, which show very close performance. C&R Tree 2 has slightly higher prediction ACC, with better ACC, lower MR, and improved sensitivity and F1 score. The specificity between the two models is almost identical. Moreover, C&R Tree 2 has a better Gini coefficient and ROC index than C5.0 2.

Interestingly, CHAID 1 exhibits the highest Gini coefficient and performs best on the ROC and Lift comparison charts, outperforming other models in these metrics. However, CHAID 1's prediction ACC and F1 score are lower than both C&R Tree 2 and C5.0 2. QUEST 1 has the poorest ROC and Lift performance, while C5.0 2 and QUEST 2 also show very poor Lift results.

Ultimately, C&R Tree 2 and CHAID 1 are the top contenders. While CHAID 1 excels in Gini coefficient, ROC, and Lift, C&R Tree 2 outperforms in prediction ACC and F1 score. For demonstration purposes, I have chosen C&R Tree 2 as the champion model due to its higher prediction ACC and F1 score. Additionally, C&R Tree 2 results in a smaller, more streamlined tree compared to CHAID 1, which tends to produce larger trees that are more prone to overfitting and more difficult to interpret. Given the close performance of these models, C&R Tree 2 is the better choice for demonstration.

Reliability of the Champion Model

The detailed model fit metrics for the three sub-samples of C&R Tree 2 are presented in Table 8.6. The results show consistent metrics across all three samples, with less than a 1% difference between them. This consistency demonstrates the reliability of the model and confirms that there are no signs of overfitting.

Validity of the Champion Model

The next step is to assess the validity of the champion model by reviewing the complete model fit metrics for the validation sample, which was reserved for the final evaluation. Table 8.7 presents the full set of metrics for C&R Tree 2 and the evaluation of its performance.

TABLE 8.6
Model Reliability of the Champion Model—C&R Tree 2

Metrics	Testing	Training	Validation
ACC	79.30%	79.20%	79.50%
MR	20.70%	20.80%	20.50%
Sensitivity or TPR	0.952	0.950	0.951
Specificity or TNR	0.615	0.618	0.623
F1	0.829	0.827	0.830
Gini	0.665	0.670	0.673
AUC (ROC index)	0.832	0.835	0.837

TABLE 8.7
Model Fit Evaluation Using the Validation Sample's Results for the Champion Model—C&R Tree 2

Metrics	Validation Sample	Evaluation	Logistic 1 Model
ACC	79.5%	Not good	76.8%
MR	20.5%	Not good	23.2%
Sensitivity or TPR	0.951	Great	89.3%
Specificity or TNR	0.623	Not good	63.0%
F1	0.830	Acceptable	80.2%
Gini	0.673	Good	0.667
AUC (ROC index)	0.837	Acceptable	0.833

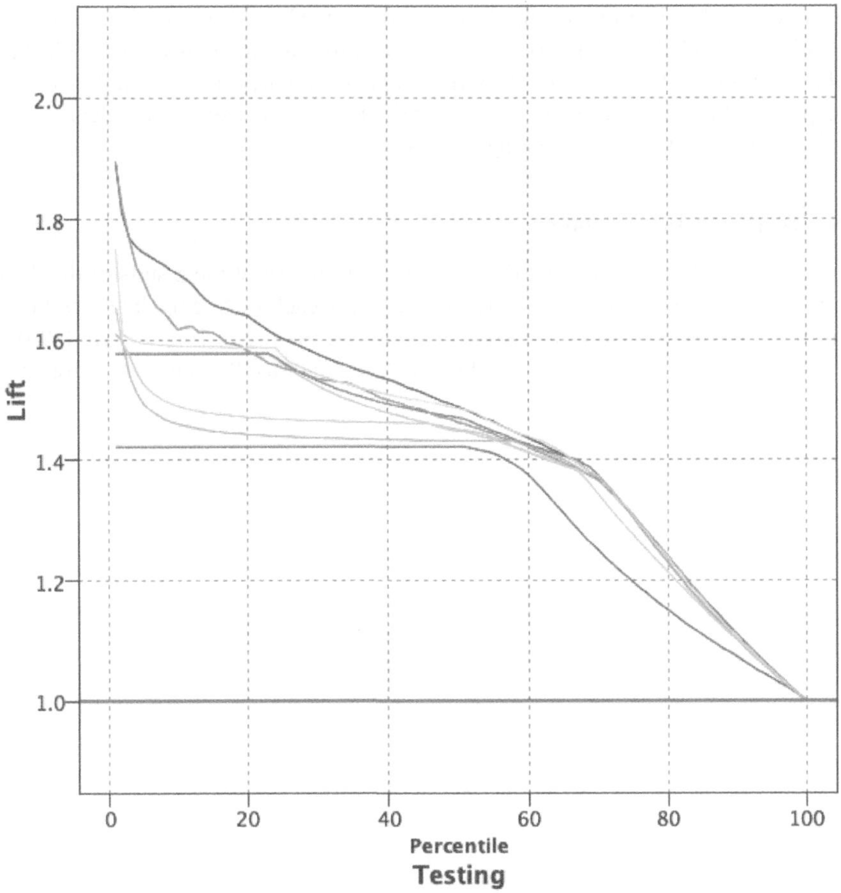

FIGURE 8.8 Lift comparison chart.

As shown, the prediction ACC is below 80%, which is generally considered suboptimal. However, the model demonstrates excellent sensitivity (true positive rate) at 95.1%. The lower ACC can be attributed to poor specificity (true negative rate), meaning the model has a high false positive rate. Despite this, the F1 score—representing the balance between sensitivity and specificity—is considered acceptable. The Gini coefficient also indicates a good fit, and the ROC index shows an acceptable fit, as confirmed by the ROC curve for this model (Figure 8.7).

The Lift curve for the validation sample (Figure 8.8) shows that using 20% of the dataset yields a Lift value of 1.6, meaning the model can capture 1.6 times the positive cases compared to not using a model at all ($1.6 \times 20\% = 32\%$). This is considered an acceptable result.

In summary, while the MR and ACC alone suggest a lower model fit, evaluating additional metrics such as sensitivity, F1 score, Gini coefficient, and ROC demonstrates that the overall fit of C&R Tree 2 is acceptable.

Conclusion

Although some fit metrics for this predictive model are considered acceptable, the model may not fully meet the expectations of airlines for predicting passenger recommendations. While the model performs well in predicting recommendation cases (positives), it struggles with predicting non-recommendation cases (negatives). The overall prediction ACC, particularly in reducing false positives, needs further improvement.

To illustrate how this decision tree model compares to the logistic regression model developed in Chapter 7, I have included the fit metrics for Logistic 1, the champion model from Chapter 7, in the same table. As shown, C&R Tree 2 outperforms Logistic 1, indicating some progress in improving predictive power. We will continue to explore additional models using other algorithms and compare their performance with these existing models.

Decision Tree Result Interpretation

Assuming the model fit is acceptable, the next step is to interpret the results of the decision tree. First, we need to understand the tree hierarchy and how each node contributes to predicting the target variable, which is whether a passenger recommends the airline. Figure 8.9 displays the full tree plot of our champion model. Notably, the tree contains 20 nodes (excluding Node 0) and 12 leaves.

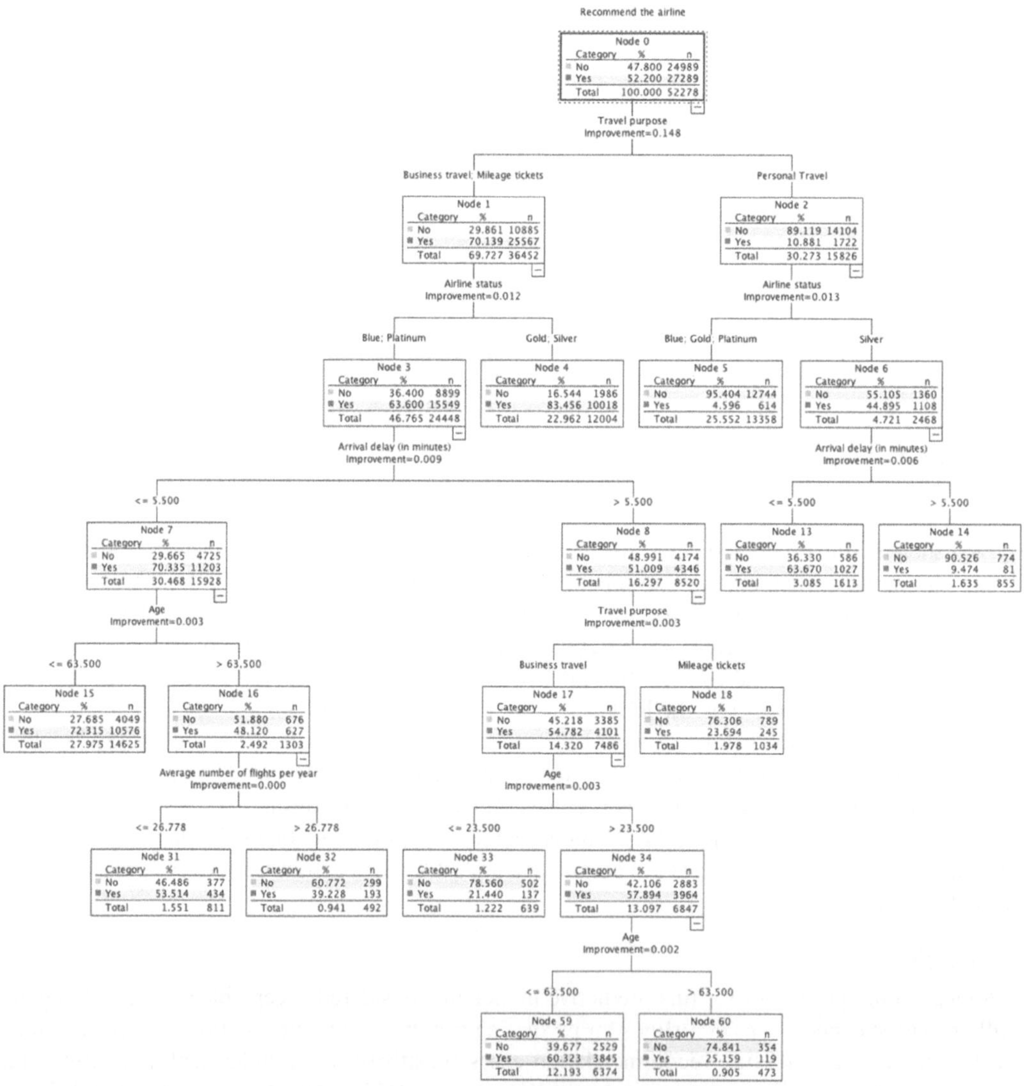

FIGURE 8.9 Full tree plot of the champion model—C&R Tree 1.

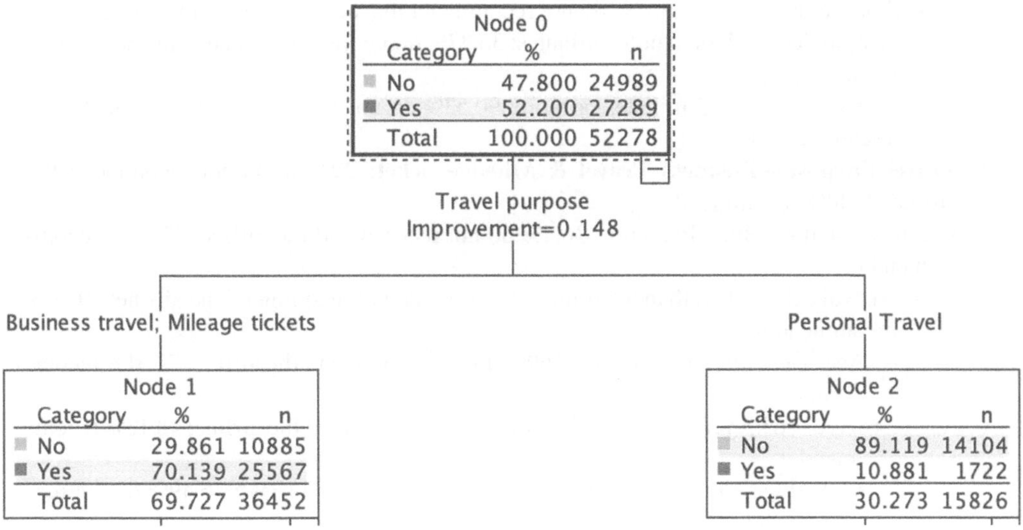

FIGURE 8.10 Detailed node information.

I have extracted the top three nodes to illustrate the detailed information within each node and explain their significance (Figure 8.10). Each node typically consists of the following components:

- **Node ID**
- **Statistics** for all levels of the target variable within the sample, including:
  - The percentage of passengers who do **not** recommend the airline (NO).
  - The percentage of passengers who **do** recommend the airline (YES).
  - The total count of records in the sample for that node.

For example, in **Figure 8.10**, if we look at the node **Personal Travel**, we can interpret the following:

- There are **15,826** records in the sample for passengers who travel for personal purposes.
- **89.119%** of these passengers do **not** recommend the airline.
- Only **10.881%** of these passengers **do** recommend the airline.

How to Interpret a Tree Plot

Tree plots can have many nodes and branches, which may seem confusing at first. Without a clear approach, it's easy to get lost in the structure and misinterpret the plot. The simplest way to explain and interpret a tree plot is to follow the hierarchy from top to bottom, tracing each branch.

In this example, I break the tree plot down by nodes and branches, following the natural hierarchy of the tree. I also create a bullet-point hierarchy that mirrors the tree structure. This method makes it easier to explain the outputs and reduces the chances of errors.

> - Follow the tree hierarchy and create a bullet hierarchy that matches the tree hierarchy.
> - Focus on the levels of the target variable.

Tree Plot Interpretation

- **Travel Purpose—Personal travel:** 89.12% do not recommend the airline; 10.88% recommend it.
  - Airline status—Blue, Gold, Platinum: 95.40% do not recommend the airline; 4.60% recommend it.

    - Airline status—Silver: 55.11% do not recommend the airline; 44.89% recommend it.
        - Arrival delay—less than 5.5 minutes: 36.33% do not recommend the airline; 63.67% recommend it.
        - Arrival delay—5.5 minutes or more: 90.53% do not recommend the airline; 9.47% recommend it.
- **Travel Purpose—Business Travel & Mileage ticket:** 29.86% do not recommend the airline; 70.14% recommend it.
    - Airline status—Blue, Platinum: 36.4% do not recommend the airline; 63.60% recommend it.
        - Arrival delay—less than 5.5 minutes: 29.66% do not recommend the airline; 70.34% recommend it.
            - Age—less than 63.5 years: 27.69% do not recommend the airline; 72.31% recommend it.
            - Age—63.5 years or more: 51.88% do not recommend the airline; 48.12% recommend it.
                - Average number of flights per year—26.78 flights or less: 46.49% do not recommend the airline; 53.51% recommend it.
                - Average number of flights per year—more than 26.78 flights: 60.77% do not recommend the airline; 39.23% recommend it.
        - Arrival delay—5.5. minutes or more: 48.55% do not recommend the airline; 51.45% recommend it.
            - Travel purpose—mileage tickets: 76.31% do not recommend the airline; 23.69% recommend it.
            - Travel purpose—business travel: 45.22% do not recommend the airline; 54.78% recommend it.
                - Age—less than 23.5 years: 78.56% do not recommend the airline; 21.44% recommend it.
                - Age—23. 5 years or more: 42.11% do not recommend the airline; 57.89% recommend it.
                    - Age—less than 63.5 years: 39.68% do not recommend the airline; 60.32% recommend it.
                    - Age—63.5 years or more: 74.84% do not recommend the airline; 25.16% recommend it.
    - Airline status—Gold, Silver: 16.54% do not recommend the airline; 83.46% recommend it.

As you can see, it is not too hard if we make sure the bullet hierarchy matches the tree hierarchy. Typically, in data mining reports, we just need to present the tree plot and explain some key nodes and branches. There is no need to cover all nodes and branches because it could result in many pages of details. Nonetheless, I provide a full interpretation here as a demonstration so you can refer back to this page to learn how to interpret the tree plot.

You can also notice some nodes, such as age, travel purpose, or destination state, are presented multiple times in the tree. That selection is determined by the splitting algorithm to maximize the model fit. In the case of age, this node needs to be split multiple times because CART only uses the binary split.

Variable Importance

Another key output of a decision tree model is the variable importance chart, which displays the relative importance of the predictors in predicting the target variable. This importance is calculated using complex formulas involving the sum of squared errors, but I won't delve into the mathematical

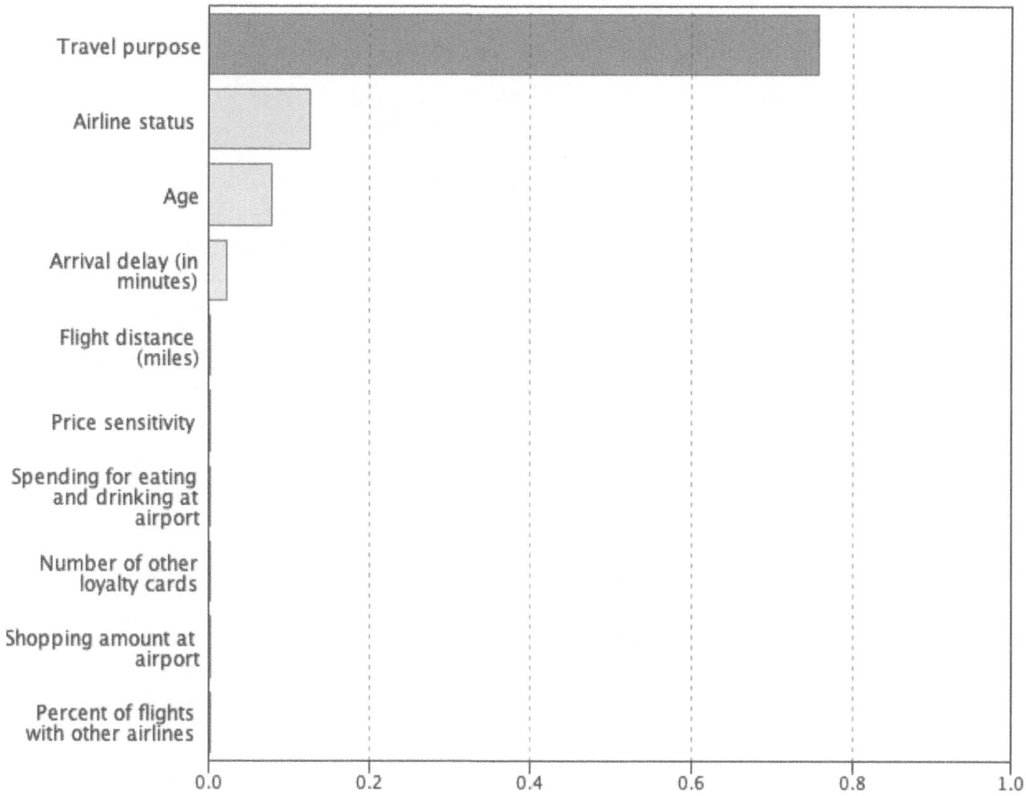

FIGURE 8.11 Variable importance of the champion model—C&R Tree 2.

details here. What's important to understand is that this score reflects how crucial each predictor is in forecasting the target variable, with values ranging from 0 to 1. The closer the score is to 1, the more important the predictor.

This chart is valuable because it highlights which predictors are most significant in predicting the target variable and ranks them accordingly. Figure 8.11 shows the predictor importance chart for the champion model, C&R Tree 2.

The results indicate that **travel purpose** is the most important predictor, followed by **airline status**, **age**, and **arrival delay**. On the other hand, **percent of flights with other airlines**, **shopping amount at the airport**, and **number of loyalty cards** are the least important predictors. Any predictors not included in this list are considered insignificant, meaning they do not contribute to predicting the target variable (whether to recommend the airline).

Tree Size Differences Between Tree Algorithms

Before concluding this example, I want to highlight how tree plots differ based on the tree algorithms used. Below are the example tree plots for all four algorithms:

- **Figure 8.12**: CART plot—skinny, long tree.
- **Figure 8.13**: QUEST plot—skinny, long tree.
- **Figure 8.14**: CHAID tree plot—large, short tree.
- **Figure 8.15**: C5.0 tree plot—large, short tree.

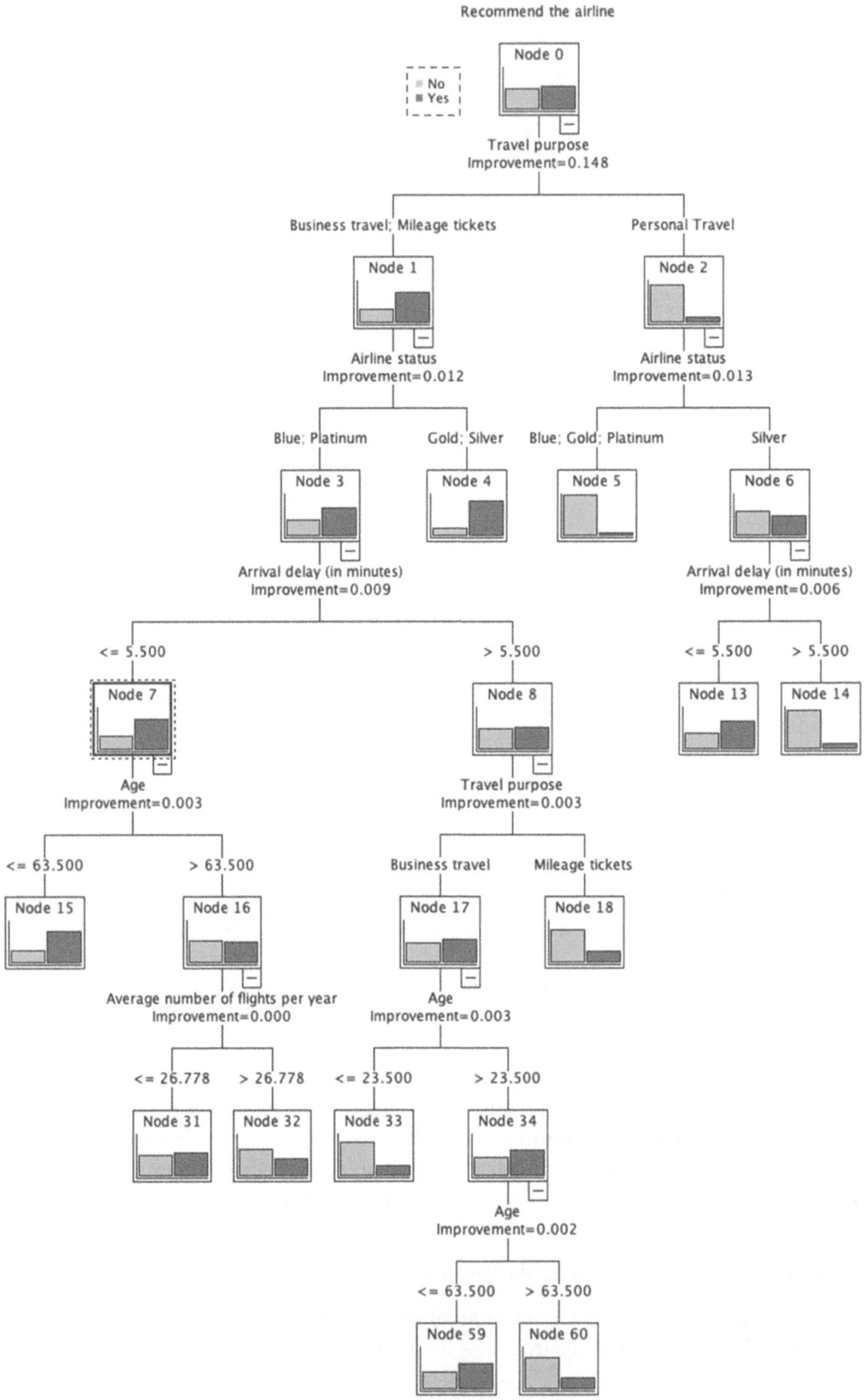

FIGURE 8.12 CART example.

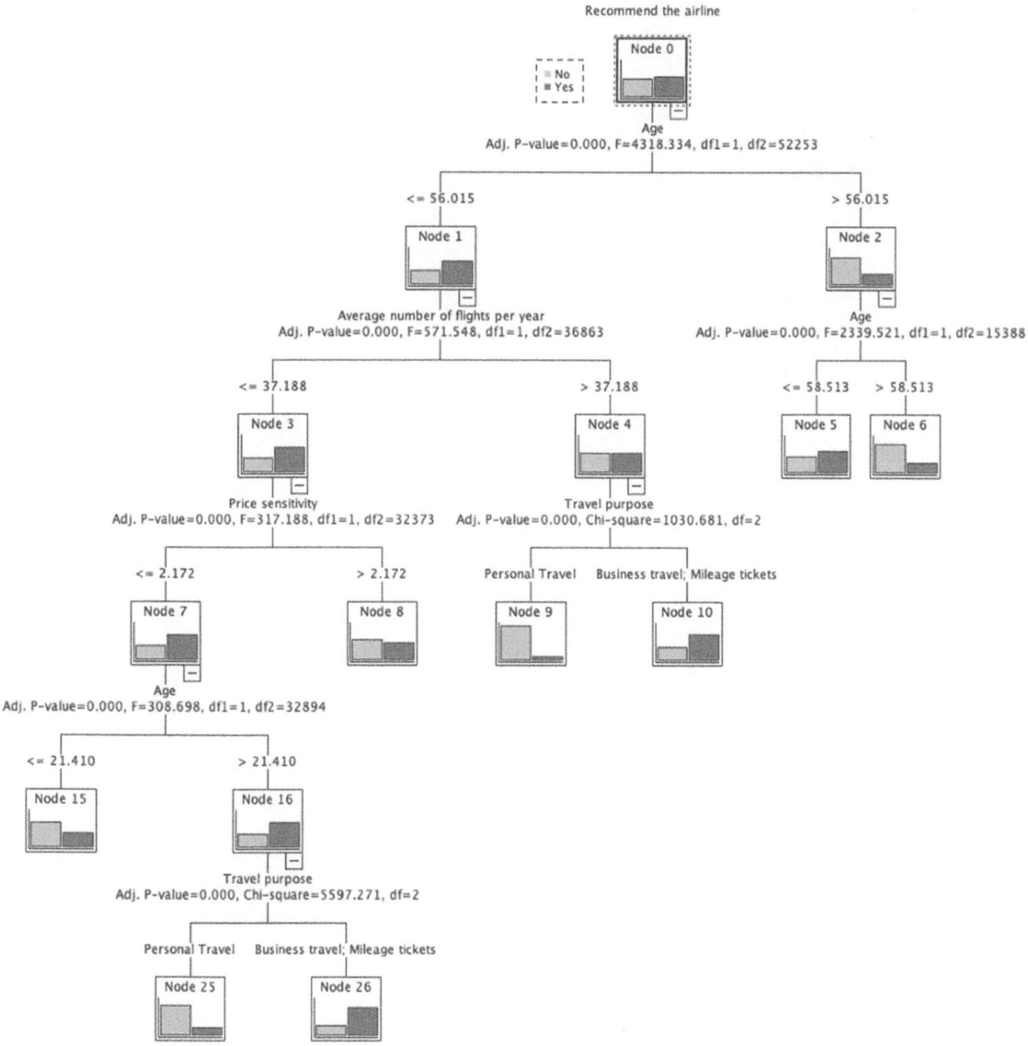

FIGURE 8.13 QUEST example

As you can see, the CART and QUEST algorithms use binary splits, resulting in skinny but long trees. In contrast, the CHAID and C5.0 algorithms use multiple splits, producing short but very large trees.

> Smaller trees are easier to understand and interpret, particularly when we need to present and explain the tree hierarchy.

In my opinion, smaller trees are easier to understand and interpret, especially when presenting and explaining the tree hierarchy. Additionally, large trees tend to raise concerns about overfitting, which is why pruning the trees is important.

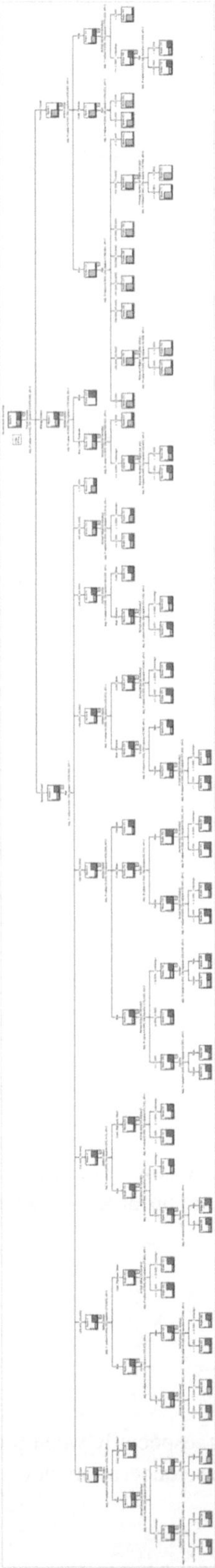

FIGURE 8.14 CHAID tree example.

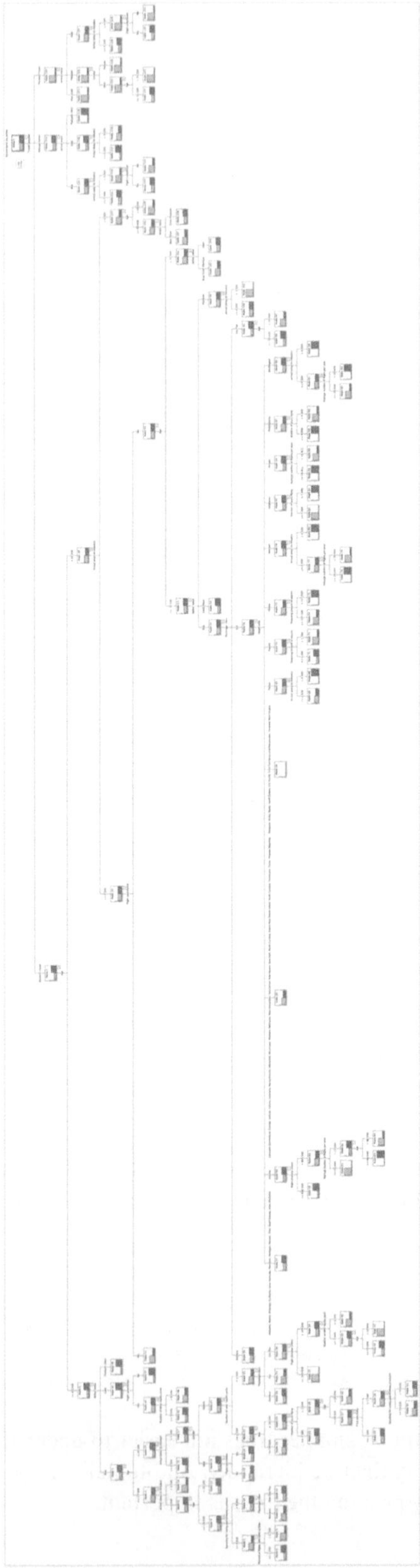

FIGURE 8.15 C5.0 tree example.

EXAMPLE: ASSOCIATION MODEL USING DECISION TREES

In this example, we build decision tree models for the **Boston housing** dataset. The target variable is a continuous variable, **lmv**, representing the median home value in the Boston area. Since in SPSS Modeler only the CART and CHAID tree algorithms can handle continuous target variables, we will focus on building CART and CHAID tree for this dataset. (As noted earlier, the C5.0 and QUEST algorithms in SPSS Modeler are limited to categorical target variables.)

- **Dataset**: Boston.sav
- **Target variable**: lmv (continuous—median home value)

Starting Point

Having completed the key steps of data exploration, modification, and partitioning for this dataset in Chapter 7, we will not repeat them here. Instead, our focus will be on the modeling process. For detailed configurations and insights from those earlier steps, please refer to the previous chapters. As a result, we will start with the Partition node and move forward to build the models, as illustrated in the following diagram.

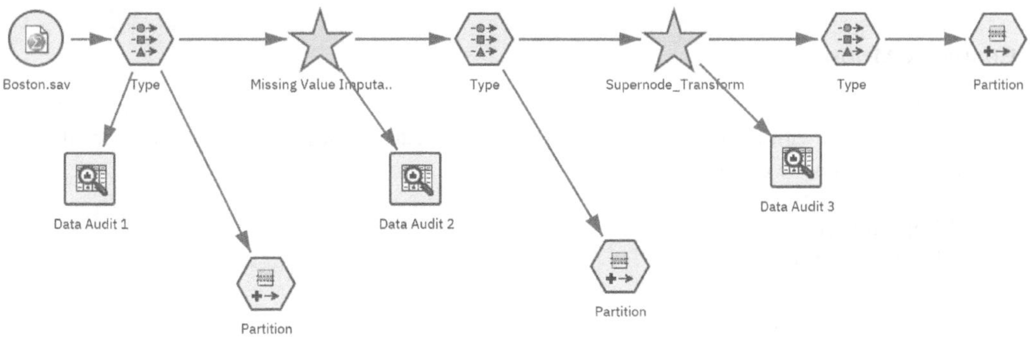

Modeling

You may recall that in the Boston housing project, after data imputation and modification, we created three datasets:

- A dataset with missing values in the target field **lmv**.
- A dataset with imputed data for the field **lmv**.
- A dataset with normalized data.

Since all the missing values are in the target field, our goal is to determine whether the decision tree algorithms can handle these missing values effectively. Additionally, because all predictors are continuous, we aim to evaluate how well these algorithms manage skewed data. For each of the three datasets, we will apply both the CART and CHAID algorithms, resulting in a total of six tree models. I will use the default settings for these algorithms.

- **C&R Tree 1:** Imputed, normalized dataset using the CART algorithm (default settings)
- **CHAID Tree 1**: Imputed, normalized dataset using the **CHAID** algorithm (default settings)
- **C&R Tree 2**: Imputed, non-normalized dataset using the **CART** algorithm (default settings)
- **CHAID Tree 2**: Imputed, non-normalized dataset using the **CHAID** algorithm (default settings)

- **C&R Tree 3**: Original dataset (with missing values, non-normalized) using the **CART** algorithm (default settings)
- **CHAID Tree 3**: Original dataset (with missing values, non-normalized) using the **CHAID** algorithm (default settings)

Follow the same steps as before to add and set up these trees. Your stream should resemble the one shown below.

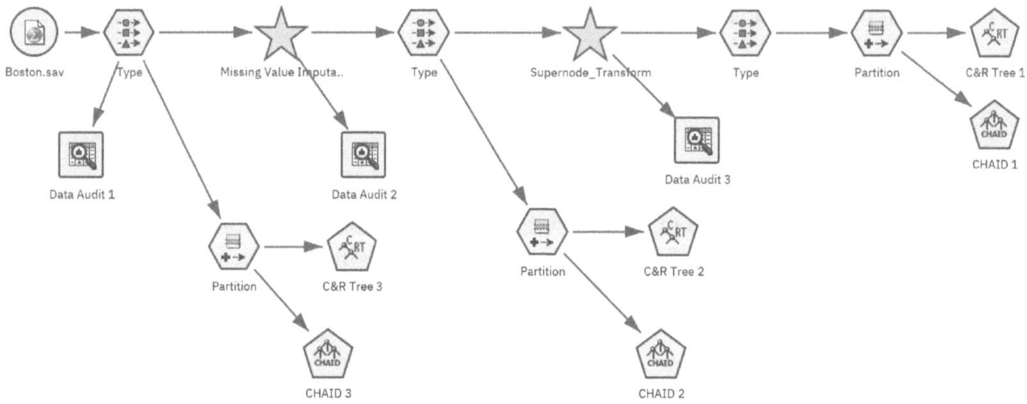

Champion Model

Run all six tree models and use the Analysis node to extract the model fit metrics for each. For association models with continuous target variables, you can extract the following metrics directly from the Analysis node output:

- Minimum error
- Maximum error
- Mean error
- Mean absolute error (MAE)
- Standard deviation
- Linear correlation

Other metrics would need to be calculated separately by exporting the output datasets to Excel, which is a time-consuming process. Unlike other software, SPSS Modeler offers limited fit metrics for association models. For the purpose of model comparison, we will focus on MAE. While it doesn't provide the full picture of model performance and is scale-dependent, it is sufficient for comparing models. As always, we will use the results from the testing sample to compare the models.

Table 8.8 presents the comparison results for the six models. Among the CART models, the one using the original dataset (C&R Tree 3) performs best. For the CHAID models, the one using imputed, normalized data (CHAID 1) performs best while the worst-performing CHAID model is the one using the original dataset.

TABLE 8.8
Model Comparison for the Boston Housing (Using the Testing Sample's Results)

Metrics	C&R Tree 1	C&R Tree 2	C&R Tree 3	CHAID 1	CHAID 2	CHAID 3
MAE	4.014	4.014	3.834	3.168	3.227	4.272
Linear correlation	0.696	0.696	0.741	0.781	0.77	0.694

Overall, CHAID 1 has the lowest error among all models (MAE = 3.168). In Table 8.8, I have also included the linear correlation results, which indicate that CHAID 1 exhibits the strongest correlation between the predictors and the target field, **Imv**. Therefore, we can conclude that **CHAID 1** is the champion model.

Reliability of the Champion Model

In this step, we evaluate the reliability of our champion model. Table 8.9 compares the metrics between the testing sample and the training sample. The results reveal some discrepancies, particularly in the MAE and standard deviation. The difference is notable, with the testing sample showing lower ACC than the training sample, indicating a potential reliability issue.

One reason for this inconsistency could be the relatively low sample size (training sample size: 304 records; testing sample size: 202 records). Typically, a small sample size combined with a large number of variables can result in higher error during the modeling process. To improve the model's reliability, collecting more data and retraining the models would be advisable.

Although we cannot gather more data for this example, it's important to recognize this issue and consider potential resolutions in future modeling efforts.

Validity of the Champion Model

To assess the validity of the champion model, additional model fit metrics need to be calculated. Follow the same steps outlined in Chapter 7 to export the output data of CHAID 1 to Excel for further analysis.

As mentioned in Chapter 7, SPSS Modeler does not automatically produce certain metrics, so we need to calculate them manually using the prediction data. To do this, we will use the **Export** node to export the prediction data to Excel files.

Follow the steps below to export the CHAID 1 output to Excel:

1. Add a **Type** node to the stream and connect it to the model nugget for **CHAID 1**. Open the node settings. (Note: You won't be able to extract the output without completing this step).
2. Double-check all fields. You should see a new field, **$E-Imv**, which represents the predicted value of the target field.
3. From the **Export** group, add an **Excel** node to the stream and connect it to the **Type** node. Open the Excel node settings.
4. For **File name**, enter the file name **Boston_chaid1.xlsx** and select the location on your computer where you want to save it.
5. Click Apply and Run.

TABLE 8.9
Reliability Evaluation of the Champion Model—CHAID 1

Metrics	Testing Sample	Training Sample
Minimum error	–8.983	–10.03
Maximum error	28.07	28.07
Mean error	0.698	0
Mean absolute error	3.168	2.36
Standard deviation	5.175	3.546
Linear correlation	0.781	0.906

TABLE 8.10
Validity Evaluation of the Champion Model—CHAID 1

Metrics	Testing Sample	Evaluation	Regression 1
MAE	3.168	N/A	3.437
RMSE	5.209	N/A	5.182
MAPE	0.125	Acceptable	0.147
RSE	0.398	Acceptable	0.394
EP	0.602	Acceptable	0.606

Once completed, you should find the Excel file **Boston_chaid1.xlsx** at the specified location. Then, use the formulas from Chapter 6 to compute **RMSE, MAPE, residual sum of squares error (RSE)**, and **explanatory power (EP)**. You can also use the same data to construct the **actual vs. predicted chart** and the **residuals vs. predicted chart**. Table 8.10 presents these metrics for the testing sample and their evaluation.

Since RMSE and MAE are scale-dependent, it is difficult to assess whether their values are considered good or not. Therefore, our focus will be on MAPE, RSE, and EP. As recommended in Chapter 6, an MAPE between 0.1 and 0.15 is deemed acceptable. Similarly, an EP between 50% and 70% (or an RSE between 0.3 and 0.5) is considered acceptable.

In addition, a detailed review of the actual vs. predicted chart and the residuals vs. predicted chart is crucial. As shown in Figure 8.16, the actual vs. predicted chart displays data points scattered

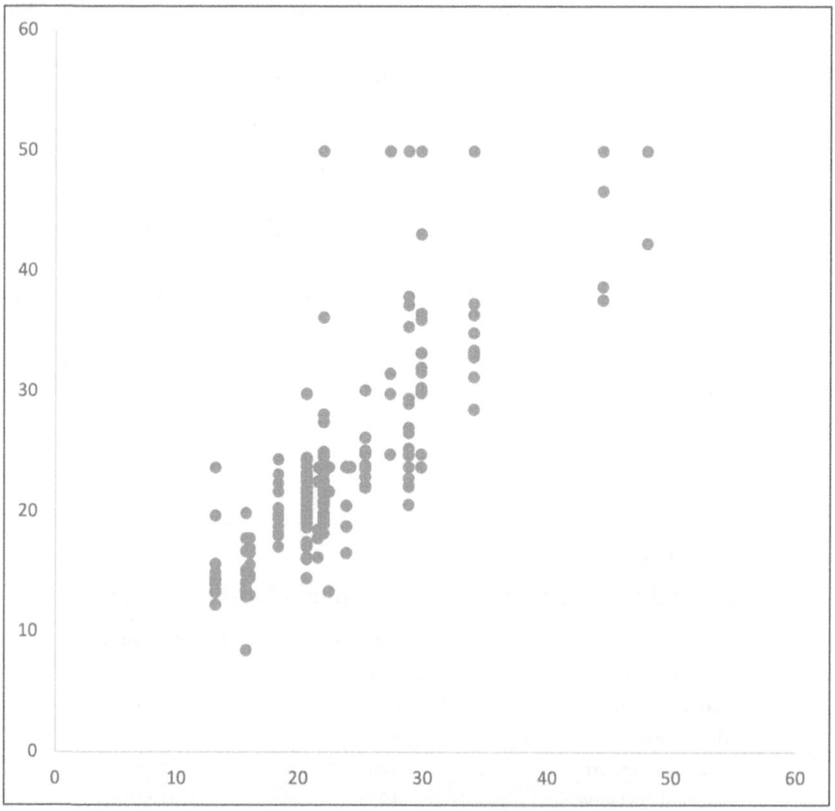

FIGURE 8.16 Actual by predicted chart of the champion model—CHAID 1.

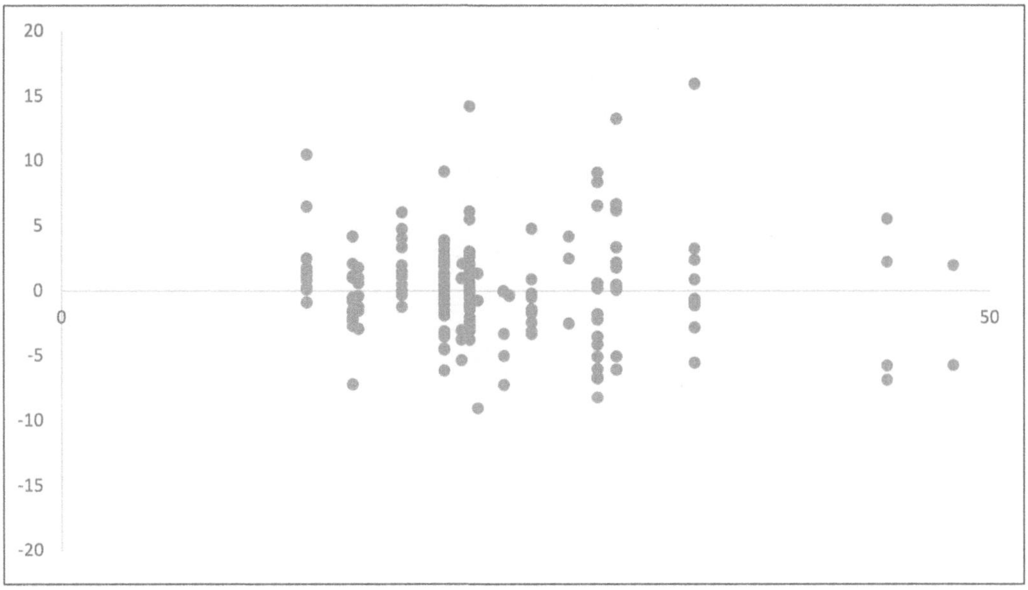

FIGURE 8.17 Residual by predicted chart of the champion model—CHAID 1.

around the diagonal line, with some outliers, and is therefore considered acceptable. The residuals vs. predicted chart (Figure 8.17) shows that the residuals are randomly distributed around the zero line with no clear patterns, indicating a good model fit.

Considering all the fit metrics, we can conclude that the champion model exhibits acceptable predictive power, though it may not be considered excellent. The next step will be to construct predictive models using different algorithms in the pursuit of a potentially better model. For comparison purposes, I have included the fit metrics of the champion model identified in Chapter 7, Regression 1, in Table 8.10. Notably, CHAID 1 produces a model with slightly higher ACC (lower MAE and lower MAPE) but nearly identical EP and RSE compared to Regression 1.

Decision Tree Result Interpretation

Figure 8.18 presents the tree plot for CHAID 1. As you can see, it is a relatively short but larger tree, with 29 nodes and 20 leaves. Based on the hierarchy, the variable **Lstat** serves as the root node, and each subsequent node is split into multiple branches. Using the same approach as discussed earlier, you can explain the tree hierarchy and how each node is split, contributing to the prediction of the target field, **Imv**.

Variable Importance

As expected, the final key result is the variable importance. Figure 8.19 illustrates the relative importance of the predictors, showing that **Lstat** is the most crucial predictor, followed by **Tax**, **RmSQ**, and **IndustPr**. On the other hand, **Charles River** and **CrimePc** rank among the least important predictors in forecasting the target field, **Imv**. This insight allows decision-makers to better assess the housing market in the Boston area and develop appropriate strategies to enhance housing values.

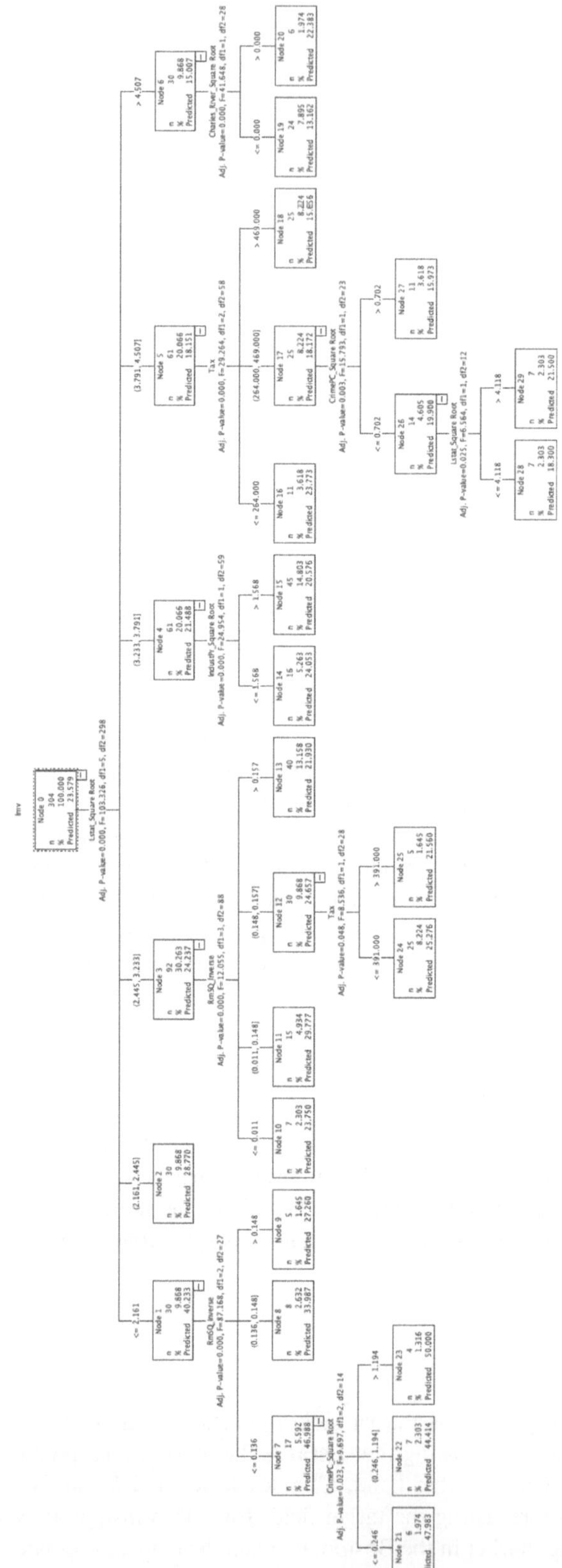

FIGURE 8.18 CHAID 1 tree plot.

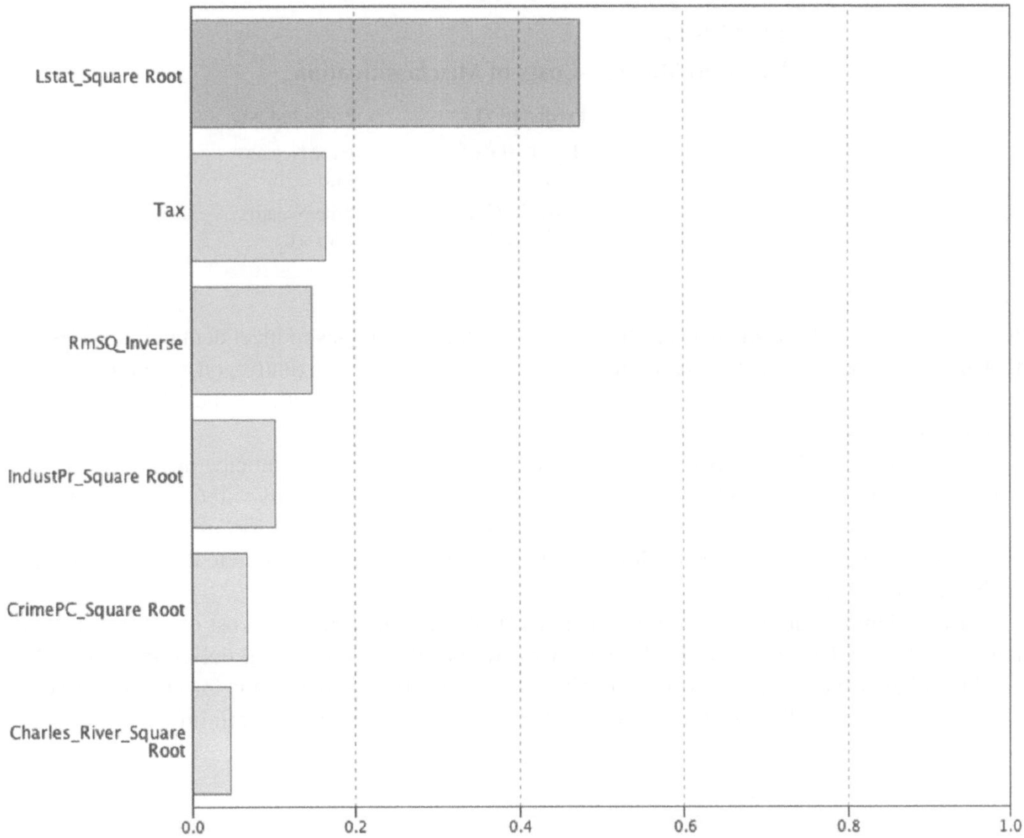

FIGURE 8.19 Variable importance chart of the champion model—CHAID 1.

EXAMPLE: COST-SENSITIVE LEARNING (CSL) USING SPSS MODELER

As explained in Chapters 3 and 6, CSL is a widely used method that allows us to evaluate and select a model based on the costs associated with misclassification rather than simply focusing on the MR. The key assumption is that different types of misclassification may incur varying costs, and our goal is not to minimize the MR itself but to minimize the total costs of misclassification.

I will demonstrate how to apply CSL using the Airline Recommendation dataset. The target variable is a binary variable with two levels: **YES** (recommend the airline) and **NO** (not recommend the airline). The CSL process will consider these two levels in the decision-making. I will illustrate CSL in two different scenarios:

1. **Objective**: Minimize the total costs of misclassification.
2. **Objective**: Maximize the total profits of accurate classification.

Cost-Sensitive Learning (CSL) with the Cost Minimization Objective

Before building our model, it is essential to develop the decision matrix for this scenario. Table 8.11 presents a hypothetical decision matrix. In this case, the goal is to predict passenger recommendations

TABLE 8.11
Decision Matrix—Costs of Misclassification

	Predicted YES	Predicted NO
Actual YES	True Positive Cost = 0	False Negative Cost = 20
Actual NO	False Positive Cost = 200	True Negative Cost = 0

for the airline, which the airline can then use to formulate strategies to meet demand. Let's assume that false positives (predicting recommendations when there are actually non-recommendations) cost the airline more than false negatives (predicting non-recommendations when there are actually recommendations).

The reason is that false positives could lead to the airline incorrectly anticipating higher demand, causing them to increase flights and aircraft, thus raising operating costs without a corresponding increase in revenue. On the other hand, false negatives result in opportunity costs for the airline, as they miss out on potential demand. However, these costs are lower compared to the costs associated with false positives.

For this demonstration, we assume a hypothetical scenario where the cost of false positives is ten times higher than the cost of false negatives. Rather than using actual dollar amounts (which we do not have), we use a unit cost: 200 units for false positives and 20 units for false negatives (as shown in Table 8.11). The objective is to identify the predictive model that minimizes the total cost of misclassification.

Software programs offer various tools for implementing Cost-Sensitive Learning (CSL). For instance, SAS Enterprise Miner uses the Decision node, which allows users to apply CSL across all types of algorithms and specify whether the objective is to minimize costs or maximize profits. The output provides a champion model based on either the lowest cost or the highest profit.

In contrast, SPSS Modeler offers more limited CSL functionality. It only incorporates misclassification costs within decision tree algorithms. Moreover, the results do not display the total cost for each model. Instead, the misclassification costs are factored in as penalties during the model-building process.

Since SPSS Modeler does not provide a built-in tool to handle cost minimization or profit maximization as expected, I will demonstrate how to perform CSL using manual calculations. As outlined in Chapter 6, the formula for calculating the total cost of misclassification is as follows:

$$\textit{Total cost} = FP \times C(FP) + FN \times C(FN)$$

where:

- **FP** = False positives
- **FN** = False negatives
- **C(FP)** = Cost of a false positive
- **C(FN)** = Cost of a false negative

TABLE 8.12
CSL Results—Cost Minimization

Trees	C&R Tree 2	CHAID 1	C5.0 2
FP	4,522	4,211	4,498
FN	634	1,000	669
C(FP)	200	200	200
C(FN)	20	20	20
Total cost	917,080	862,200	912,980
ACC	79.26%	79.04%	79.22%

We will use the results from the decision tree models developed earlier in this chapter. For this example, I selected the top three models: C&R Tree 2, CHAID 1, and C5.0 2. These are the models with the best prediction accuracy. To calculate the total cost of misclassification for each model, we need to extract the **false positives** and **false negatives** from the **Analysis** node for each model and apply the formula above.

Table 8.12 presents the false positives, false negatives, and total cost for each model. I also included the ACC for comparison. Note that the results are based on the testing sample.

As seen in the table, based solely on ACC, C&R Tree 2 has the highest ACC, followed by C5.0 2 and CHAID 1. However, when we factor in the costs of misclassification, CHAID 1 has the lowest total cost, followed by C5.0 2 and C&R Tree 2. In other words, if the objective is to minimize the total cost of misclassification, CHAID **1** emerges as the champion model with the lowest overall cost.

Cost-Sensitive Learning (CSL) with the Profit Maximization Objective

In this scenario, we focus on maximizing the profits from accurate predictions. We assume that the airline is experiencing a decrease in demand, which necessitates making operational decisions regarding the number of flights and investments in new airplanes. Here, the prediction of **non-recommendations** (true negatives) becomes more valuable than the prediction of **recommendations** (true positives).

In other words, accurately predicting how many passengers will not recommend the airline enables the company to adjust flight volume, airplane capacity, and flight paths accordingly. These adjustments can lead to significant cost savings, which in turn increases overall profits.

Table 8.13 presents the decision matrix for this scenario. The unit profit for true negatives is set at 1,000 units while the unit profit for true positives is set at 100 units. The objective is to identify the predictive model that maximizes the total profit from accurate classification.

TABLE 8.13
Decision Matrix—Profits of Accurate Classification

	Predicted YES	Predicted NO
Actual YES	True Positive Profit = 100	False Negative Profit = 0
Actual NO	False Positive Profit = 0	True Negative Profit = 1,000

TABLE 8.14
CSL Results—Profit Maximization

Trees	C&R Tree 2	CHAID 1	C5.0 2
TP	12,484	12,118	12,449
TN	7,222	7,533	7,246
P(TP)	100	100	100
P(TN)	1,000	1,000	1,000
Total cost	8,470,400	8,744,800	8,490,900
ACC	79.26%	79.04%	79.22%

In this example, we will use the following formula to calculate the total profit from accurate predictions (as explained in Chapter 6):

$$Total\ profit = TP \times P(TP) + TN \times P(TN)$$

where:

- **TP** = True positives
- **TN** = True negatives
- **P(TP)** = Profit of a true positive
- **P(TN)** = Profit of a true negative

We can extract the values for true positives and true negatives from the Analysis node (for the testing sample) for the three models. Then, using the formula above, we calculate the total profit of accurate predictions for each model.

Table 8.14 presents the results. As shown, when we factor in the profit from accurate predictions, CHAID 1 achieves the highest total profit, followed by C5.0 2 and C&R Tree 2. Therefore, if the objective is to maximize the total profit from accurate predictions, CHAID 1 is the champion model with the highest profit.

As you can see, we can easily perform CSL by considering the costs of misclassification or the profits of accurate classification in our model comparison process. Using cost minimization or profit maximization, we can select the champion model accordingly. In this book, for consistency, I will continue to use the MR as the primary selection statistic in the following chapters. However, if your project requires using costs or profits as the deciding factor, you can return to this chapter and follow the steps to conduct CSL using the provided formulas.

How to Conduct CSL Using SPSS Modeler?

The approach outlined above requires manual calculation of total misclassification costs or total profits from accurate predictions. While SAS Enterprise Miner provides the Decision node to handle both types of CSL, SPSS Modeler does not offer a similar feature. So, how do we handle CSL in SPSS Modeler?

In SPSS Modeler, misclassification costs are only available for decision tree algorithms, including CART, CHAID, C5.0, and QUEST. Other machine learning methods do not support this setting.

Misclassification costs in SPSS Modeler allow users to assign different weights to specific misclassifications, such as false positives and false negatives. These weights are integrated into the model, influencing predictions to avoid costly errors. When misclassification costs are incorporated into a decision tree algorithm, the algorithm adjusts predictions by imposing higher penalties on certain misclassifications, which may alter the prediction outcomes.

How to Apply Misclassification Costs in SPSS Modeler?

Incorporating misclassification costs into a tree algorithm is straightforward. You follow the same steps used in constructing tree models, with one additional step as follows:

1. In the tree setting window, go to the **Build Options** tab.
2. Select **Costs & Priors**.
3. Check the box labeled **Use misclassification costs**.
4. Enter the misclassification cost information in the matrix.
5. Complete the remaining steps (as described above) and run the tree model.

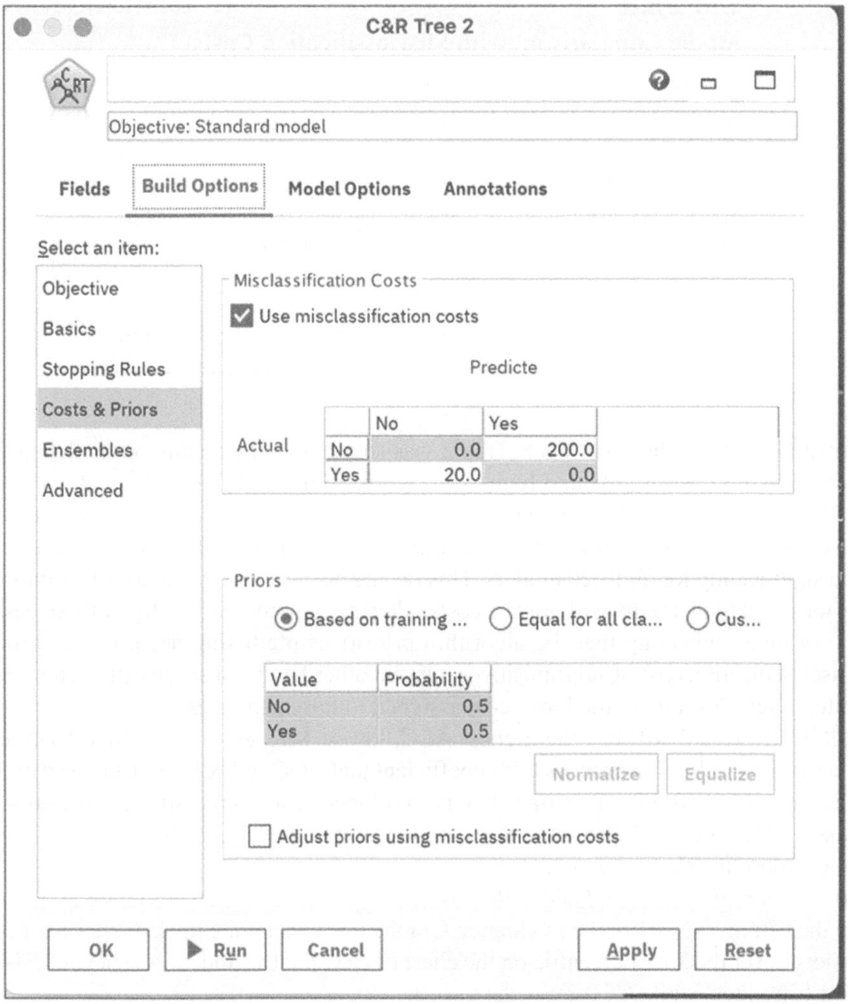

Make sure to apply this step to the other tree models, CHAID 1 and C5.0 2. As usual, we use the results from the testing sample to compare the three models. For easier comparison, I've included the model fit metrics for the models without misclassification costs in Table 8.15 while the results with misclassification costs are shown in Table 8.16.

TABLE 8.15
Model Comparison (without Misclassification Costs)

Metrics	C&R Tree 2	CHAID 1	C5.0 2
ACC	0.793	0.790	0.792
MR	0.207	0.210	0.208
Sensitivity or TPR	0.952	0.924	0.949
Specificity or TNR	0.615	0.641	0.617
F1	0.829	0.823	0.828
Gini	0.665	0.700	0.618
AUC (ROC index)	0.832	0.850	0.808

TABLE 8.16
Model Comparison (with Misclassification Costs)

Metrics	C&R Tree 2	CHAID 1	C5.0 2
ACC	0.511	0.484	0.484
MR	0.489	0.516	0.516
Sensitivity or TPR	0.081	0.023	0.023
Specificity or TNR	0.992	1.000	1.000
F1	0.150	0.044	0.044
Gini	0.590	0.700	0.027
AUC (ROC index)	0.795	0.850	0.514
Lift index	1.63	1.65	1.04

Previously, we established that C&R Tree 2 was selected as the champion model, as it had the highest ACC, highest sensitivity, and highest F1 score. While its Gini coefficient and ROC index were not the best, they were close enough to CHAID 1 to still make C&R Tree 2 the top choice.

However, when we look at the results in Table 8.16, we notice that the prediction ACC has decreased significantly for all three models. This occurs because the tree algorithms impose higher penalties for misclassifications with higher costs, altering the predictions. In all three cases, specificity is very high, indicating that the algorithm prioritizes predicting negatives to minimize the higher misclassification cost of false positives. On the other hand, sensitivity (the true positive prediction rate) is very low due to the lower cost assigned to false negatives.

Overall, C&R Tree 2 still has the highest ACC, lowest MR, and highest sensitivity compared to the other two models. However, its Gini coefficient and ROC index have decreased and are now lower than those of CHAID 1, meaning they are no longer considered strong indicators of model performance. Since the ACCs for all models are low, relying solely on this metric may not be the best way to choose the champion model.

To provide more information, I added the Lift index (for the 20% percentile) for all three models. To obtain the Lift index, we use the Evaluation Graph node to generate the Lift charts for these trees and then hover over the 20% percentile on the chart to view the Lift index. As shown, CHAID 1 has the highest Lift index, while C5.0 2 has the lowest.

Thus, we could justify selecting either C&R Tree 2 or CHAID 1 as the best model. However, considering the very low ACCs across all models, it might be more appropriate to evaluate model performance using metrics other than just prediction ACC. From this perspective, CHAID 1 appears to outperform the other two models. As a result, for this example, I will select CHAID 1 as the champion model.

Undoubtedly, utilizing the misclassification cost approach in SPSS Modeler introduces challenges in interpreting results, compared to the more straightforward approaches of cost minimization or profit maximization discussed earlier. This method also complicates the selection and justification of the champion model. Furthermore, it cannot be applied to other algorithms such as neural networks, regression, or random forests. For these reasons, I recommend using the cost minimization or profit maximization approaches for CSL.

CHAPTER SUMMARY

This chapter introduces **decision trees**, a machine learning method that may have been unfamiliar before reading this book. We cover the basics of decision trees, including how they are constructed and how to interpret the tree hierarchy. The hierarchical structure of a decision tree makes it easier to understand the logic behind predictions, which are made by splitting nodes into branches. This non-parametric method is capable of detecting nonlinear relationships and is not subject to assumptions or restrictions. Additionally, missing values do not impact the validity of the tree, as it has specific techniques for handling them. For these reasons, decision trees are among the most popular machine learning methods.

To prevent the creation of overly large trees, which are difficult to interpret and prone to overfitting, it is necessary to prune the tree by removing redundant branches. Auto-pruning is recommended because it allows the user to set criteria for subtrees, and the algorithm will continue building the tree until those criteria are met. This process is more efficient and reduces the likelihood of human error.

We also explored four different types of tree algorithms: CART, CHAID, C5.0, and QUEST. These algorithms differ in their splitting rules. CART and QUEST use binary splits, where a node is split into only two branches, while CHAID and C5.0 use multi-splits, where a node can split into more than two branches. CHAID and QUEST use statistical measures, such as the chi-squared test, as splitting criteria, whereas CART and C5.0 use methods like Gini impurity or information gain. Each method has its advantages and disadvantages. Importantly, decision trees can be used to build either association or classification models and work with both continuous and categorical predictors, making them very versatile. However, note that in SPSS Modeler, C5.0 and QUEST can only construct classification trees.

The outputs of decision trees include the tree plot, which shows the full hierarchy and explains how predictions are made at each leaf node, and the variable importance chart, which ranks the significant predictors in terms of their relative importance. This allows us to determine which predictors are most crucial for predicting the target variables. We continue to use the Airline Recommendation and Boston housing datasets in examples, so you can compare decision trees with other models we've built.

This chapter also explains and demonstrates the CSL method. CSL is a popular technique for comparing models and selecting the champion model based on the costs of misclassification rather than the misclassification rate. Essentially, the algorithm chooses the model that minimizes the total cost of misclassification. CSL can also be used when the objective is to maximize the total profit of accurate predictions. It is a valuable method for considering profit and loss and can also help address the issue of overfitting.

CHAPTER DISCUSSION QUESTIONS

1. Explain the purpose of the decision tree method. How does the decision tree method work in machine learning?
2. What are the main components of a decision tree? Explain the tree's hierarchical structure.
3. Discuss the process of constructing a decision tree and the criteria used for splitting nodes.
4. What are the advantages and disadvantages of using decision trees compared to other machine learning algorithms?

5. Do missing values affect decision trees? How do decision tree algorithms handle missing values?
6. Decision trees can handle missing values by imputation or using surrogate splits. Compare and contrast these techniques, discussing their advantages, limitations, and potential impact on decision tree accuracy.
7. What are the purposes of tree pruning? Describe and explain different pruning approaches.
8. Compare CART, CHAID, C5.0, and QUEST algorithms. Discuss their similarities and differences.
9. Interpretability is a key advantage of decision trees. How can the decision tree structure be interpreted to gain insights into the important predictors and their impact on the target variable? Discuss methods for visualizing and analyzing decision trees.
10. Decision trees are versatile and applicable to various domains. Give examples of real-world applications where decision trees have been successfully employed and discuss the advantages of using decision trees in those contexts.
11. Discuss the purposes and benefits of CSL. In which cases should CSL be used? Provide practical examples.

CHAPTER EXERCISES

Exercise 1: Decision Tree Model for the Bank Dataset

1. Open the **Bank dataset** in SPSS Modeler.
2. Sample the entire dataset. Refer to the variable names and descriptions in Appendix A.
3. Set **CRSCORE** (credit scores) as the target variable.
4. Perform data modification as follows:
   - Impute missing values.
   - Normalize variables with skewed data distributions.
5. Partition the data using a 60–20–20 split ratio (train: 60%, test: 20%, and validation: 20%).
6. Configure four decision tree models for this dataset as follows (using auto pruning and default settings):
   - Pruned **CART**
   - Pruned **CHAID** tree
   - Pruned **C5.0** tree
   - Pruned **QUEST**
7. Run the models.
8. Compare the model fit metrics between the four models using the testing results.
9. Select the champion model and justify the selection with supporting evidence.
10. Evaluate the reliability and validity of the champion model.
11. Explain the tree plot of the champion model.
12. Present the variable importance information and interpret the results.

Exercise 2: Decision Tree Models for the Passengers Dataset

1. Open the **Passengers dataset** in SPSS Modeler.
2. Sample 50% of the full dataset. Set **return** (decision to return to the airline) as the target variable and change its role to binary. Refer to the variable names and descriptions in Appendix 2.
3. Perform data modification as follows:
   - Replace zero values in Likert-scale variables with missing values.
   - Impute missing values.
4. Partition the data using a 60–20–20 split ratio (train: 60%, test: 20%, and validation: 20%).

5. Configure four decision tree models for this dataset as follows (using Auto Pruning):
   - Pruned **CART** using the built-in missing value algorithm (under Splitting rules, set **Missing Values** to **Largest Branch**).
   - Pruned **CART** using imputation for missing values.
   - Pruned **CHAID** tree using the built-in missing value algorithm (under Splitting rules, set **Missing Values** to **Use in search**).
   - Pruned **CART** using imputation for missing values.
6. Compare the model fit metrics between these four models using the validation results.
7. Select the champion model and justify the selection with supporting evidence.
8. Evaluate the reliability and validity of the champion model.
9. Explain the tree plot of the champion model.
10. Present the variable importance information and interpret the results.

Exercise 3: Cost-Sensitive Learning Model for the Passengers Dataset

1. Repeat Steps 1–5 from **Exercise 2**.
2. Conduct **CSL** with the objective of **minimizing** the total costs of misclassification. Use the decision matrix given in Table 8.17.

TABLE 8.17
Exercise 3.2

	Predicted YES	Predicted NO
Actual YES	True Positive Cost = 0	False Negative Cost = 150
Actual NO	False Positive Cost = 2,000	True Negative Cost = 0

3. Compare the four decision tree models using the total costs of misclassification (based on the testing sample's results).
4. Select the champion model and justify this selection with supporting evidence.
5. Conduct **CSL** with the objective of **maximizing** the total profits of accurate predictions. Use the decision matrix given in Table 8.18.

TABLE 8.18
Exercise 3.5

	Predicted YES	Predicted NO
Actual YES	True Positive Profit = 200	False Negative Profit = 0
Actual NO	False Positive Profit = 0	True Negative Profit = 2,500

6. Compare the four decision tree models using the total profits of accurate predictions (based on the testing sample's results).
7. Select the champion model and justify this selection with supporting evidence.

9 Neural Networks

LEARNING OUTCOMES

Upon the completion of this chapter, you will be able to:

1. Define neural networks.
2. Explain the importance of neural networks in data mining and provide practical examples.
3. Describe common types of neural networks.
4. Explain the foundational concepts of neural networks, including the structure and function of artificial neurons, activation functions, and network architectures.
5. Compare forward propagation and backward propagation algorithms used in training neural networks.
6. Differentiate different activation functions.
7. Define the multi-layer perceptron (MLP) method.
8. Define the Radial Basis Function Network (RBFN) method.
9. Demonstrate how to normalize data for neural network training.
10. Discuss the advantages and disadvantages of neural networks.
11. Explain the neural network training process.
12. Describe the search process to find the optimal network.
13. Differentiate different stopping rules.
14. Apply neural network configurations in the model training using SPSS Modeler.
15. Interpret the neural network results, including the model fit, network structure, and variable importance.

INTRODUCTION

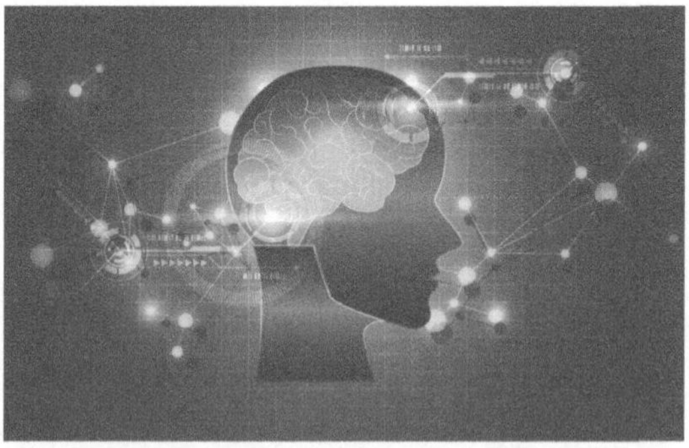

In this chapter, we focus on neural networks, a widely used method in the era of machine learning and AI. Although it is popular, the concept is complex and can be challenging to fully understand and implement. Many resources are available, including videos, websites, blogs, and books, that explain neural networks. However, books often present the material at an advanced level, requiring a solid mathematical background. On the other hand, we now see more engaging videos that simplify neural networks to help audiences grasp the basics.

In this book, my goal is to explain neural networks in a straightforward manner, avoiding mathematical formulas, but not just to make it entertaining. I want to keep it practical, enabling you to conduct meaningful neural network research.

While neural networks are not a new concept, their complexity meant that, in the past, only a few scientists truly understood and applied them. With advancements in data mining and machine learning, we now see more real-world applications of neural networks, making them increasingly popular. Some common examples include Tesla's autopilot, facial recognition on smartphones, voice recognition, and Deep Fake technology, which can convincingly swap one person's face with another's.

DOI: 10.1201/9781003468196-11

However, I won't go deeply into the specifics of how these advanced technologies work, as they require extensive neural network applications, vast resources, and teams of data scientists working over extended periods. Instead, my aim is to provide a simple introduction to neural networks, guiding you in building predictive models.

The key question is: How can we understand neural networks without getting into complex mathematical formulas? Let's get started.

WHAT IS NEURAL NETWORK?

DEFINITION

Neural networks are computing systems composed of interconnected nodes, functioning similarly to neurons in the human brain. These systems utilize algorithms to identify hidden patterns and correlations within raw data, enabling them to cluster and classify information. Over time, they continuously learn and improve their performance (SAS Institute).

Simply put, a neural network (NN) mimics the biological neural network of the human brain. This explains the term "neural network" and also hints at its complexity, as the human brain is the most intricate network known. As we delve deeper into this method, its complexity increases. It's similar to approaching an intersection—more roads appear ahead, and the further we go, the larger and more intricate the network becomes.

Figure 9.1 illustrates some common neural network architectures. Each node, also referred to as a **cell** or **neuron**, represents a variable or unit in the network. These nodes can move forward, backward, travel through time, or interact with other nodes,

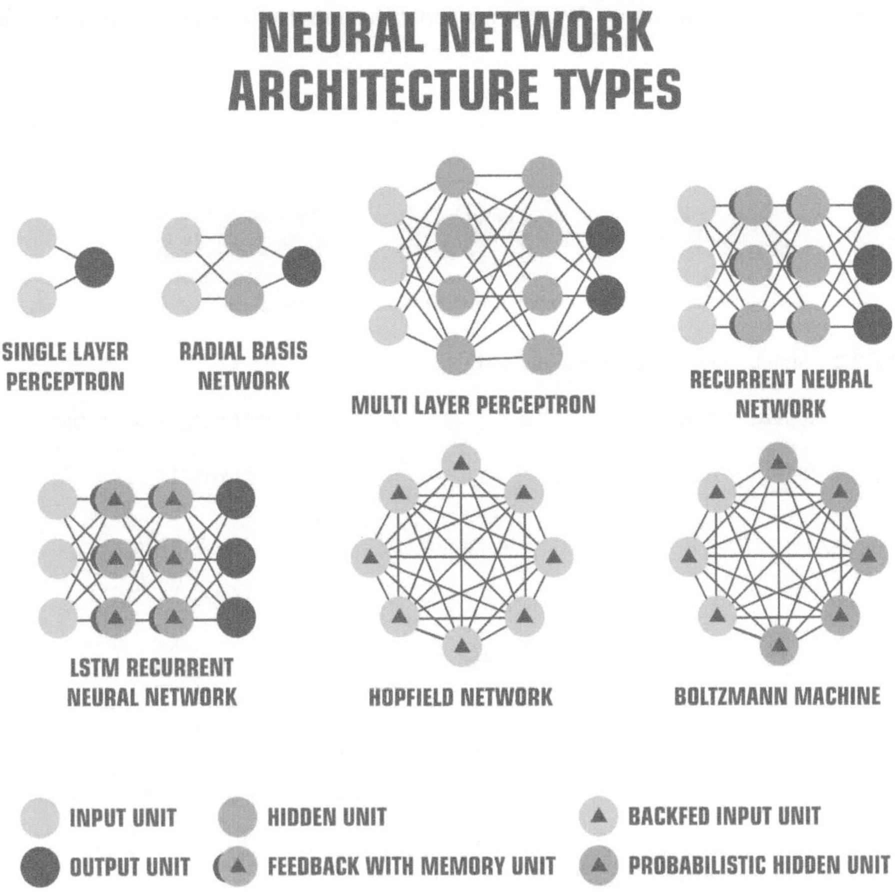

FIGURE 9.1 Various neural network architecture types.

demonstrating just how complex the network can become. Since neural networks are fundamentally math-based, each type comprises various mathematical expressions to capture the relationships between these nodes.

According to the SAS Institute, neural networks are computing systems made up of interconnected nodes that function similarly to neurons in the human brain. Using algorithms, they can detect hidden patterns and correlations in raw data, allowing them to cluster, classify, and continuously learn and improve over time. That's not too hard to grasp, right?

Figure 9.2 illustrates a typical neural network structure. Simply put, a neural network consists of three key components: the input layer, hidden layers, and the output layer. The input layer includes the predictors in the model, while the output layer contains the target variable(s) we aim to predict. The hidden layers, situated between the input and output layers, are made up of multiple hidden nodes that connect these layers in various ways. These hidden layers are crucial for identifying patterns in the data.

DEFINITION

Neurons or cells represent nodes in a neural network architecture.

Neural network model is the same as neural network architecture.

Artificial Neural Network (ANN) is a neural network with more than one hidden layer.

A key aspect of neural network modeling is determining the number of hidden layers required and how many nodes should be in each layer. We'll get to that later.

Let's examine two actual neural network diagrams in Figure 9.3. On the left, we see a typical neural network model with one hidden layer. As mentioned before, the nodes in the network are also referred to as "cells" or "neurons." When a network has more than one hidden layer, it is called an

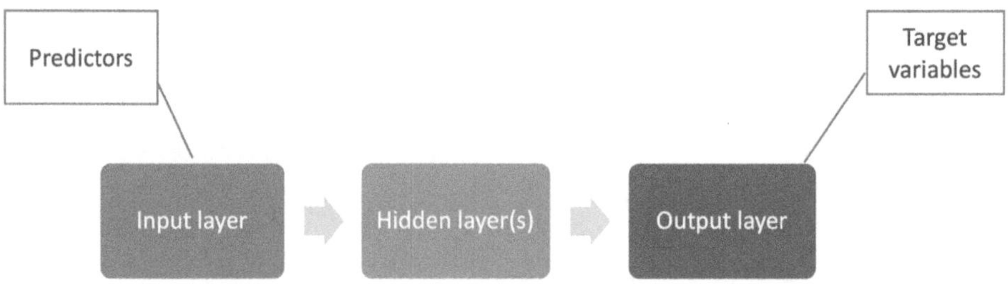

FIGURE 9.2 Simple diagram of a neural network.

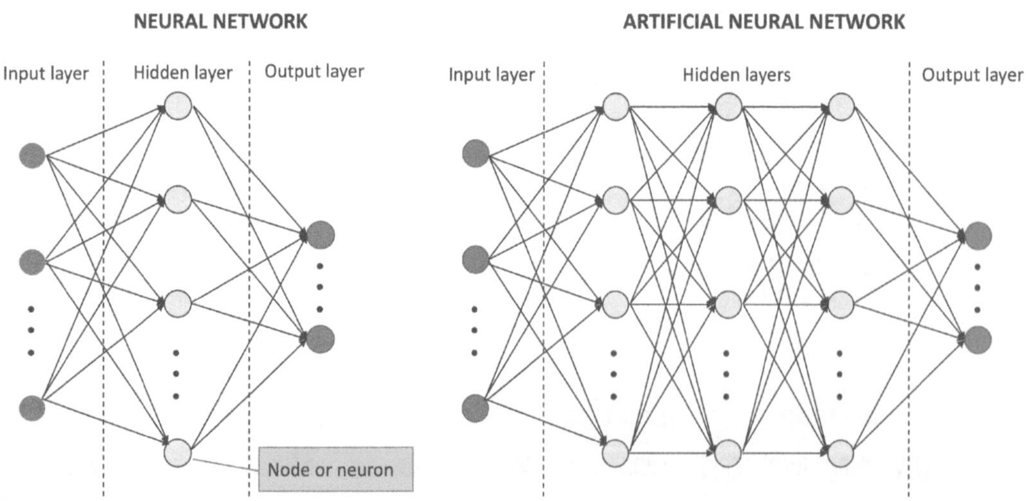

FIGURE 9.3 Neural network and artificial neural network.

"**Artificial Neural Network**" (**ANN**), commonly known as deep learning. In practice, a deep learning network consists of far more than just two or three hidden layers.

In the diagram, the input nodes represent the predictors in our model, and the output nodes represent the target variables. Neural networks are highly versatile, capable of handling both categorical and continuous target variables. The hidden nodes represent connect the input nodes to the output nodes. These connections form the **neural network's architecture** or **model**.

PRACTICAL APPLICATIONS OF NEURAL NETWORKS

Neural networks have a broad range of applications across various domains due to their ability to learn complex patterns from data and make accurate predictions. Below are some practical applications of neural networks:

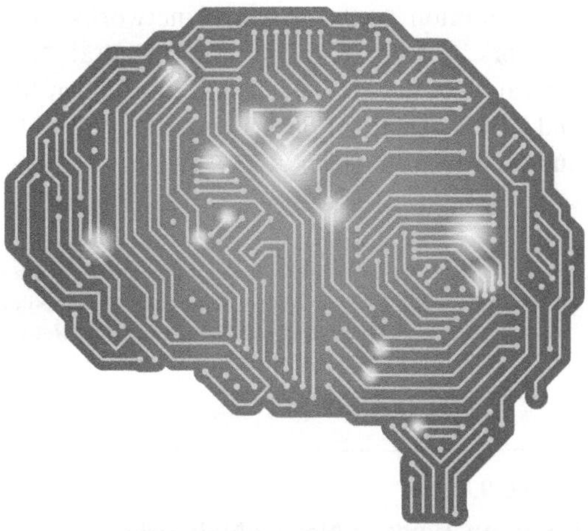

- **Image recognition and computer vision**: Neural networks, particularly convolutional neural networks (CNNs), excel at image recognition tasks. They can identify objects, faces, and scenes in images, enabling applications such as facial recognition, object detection, and self-driving cars.
- **Natural language processing (NLP)**: Recurrent neural networks (RNNs) and transformer networks have revolutionized NLP tasks. Neural networks are used in machine translation, sentiment analysis, text generation, speech recognition, and chatbots.
- **Healthcare**: Neural networks play a key role in medical image analysis, such as detecting abnormalities in X-rays and MRI scans. They are also used to assist in diagnosing diseases and predicting patient outcomes based on medical records and genomic data.
- **Finance**: Neural networks are employed for credit risk assessment, fraud detection, stock market prediction, and algorithmic trading, helping analyze financial data and make informed decisions.
- **Media and product recommendations**: Neural networks power recommendation engines that personalize content and product suggestions on streaming platforms, e-commerce websites, and social media.
- **Flight control systems**: Neural networks optimize aircraft performance and enhance stability in flight control systems. They can model complex flight dynamics, adapt to changing conditions, and assist pilots in handling critical situations.
- **Autonomous vehicles**: Neural networks are essential in self-driving cars, enabling them to interpret sensor data and make real-time decisions for safe navigation.
- **Gaming and virtual environments**: Neural networks create intelligent agents in gaming and virtual environments that learn and adapt to player behavior.
- **Flight path prediction**: Neural networks are used to predict flight paths and optimize flight trajectories for fuel efficiency and reduced emissions, taking into account factors such as weather conditions, wind patterns, and air traffic congestion.
- **Airline passenger behavior analysis**: Airlines use neural networks to analyze passenger data and identify behavior patterns, preferences, and customer satisfaction. This information helps personalize services and improve the passenger experience.

- **Robotics**: Neural networks are used in robotics for perception, object recognition, grasping, and motion control.
- **Music and art generation**: Artists utilize neural networks to generate music, create visual art, and produce content based on learned patterns and styles.

These are just a few examples of the diverse applications of neural networks. Their versatility and ability to process complex data make them a crucial tool in modern machine learning and artificial intelligence, influencing many industries.

COMMON TYPES OF NEURAL NETWORKS

Neural networks come in various architectures and designs, each tailored to handle specific types of data and the purpose of the project. Common types of neural networks include Feedforward Neural Network (FNN), Convolutional Neural Network (CNN), Recurrent Neural Network (RNN), Long Short-Term Memory Network (LSTM), and RBFN. Table 9.1 describes those types of networks, their strengths, and their weaknesses.

> Note that in this book, I do not get into the more complex forms of neural networks. We primarily focus on Multi-Layer Perceptron Feedforward (MLP), a popular Feedforward Neural Network, and Radial Basis Function Network (RBFN).

TABLE 9.1
Common Types of Neural Networks

Type of NN	Description	Strengths	Weaknesses
Feedforward Neural Network (FNN)	The simplest type of neural network. Data flow in a single direction from the input layer to the output layer.	• Easy to understand and implement. • Suitable for a wide range of tasks.	• Less accurate than other types of NNs for complex tasks. • Slower to train.
Convolutional Neural Network (CNN)	A type of feedforward neural network that is specifically designed for image processing. It uses convolution operations to extract features from images.	• Very good at image processing tasks. • High accuracy.	• Complex to understand and implement. • Requires a lot of data to train.
Recurrent Neural Network (RNN)	A neural network that can process sequential data. It has feedback loops, which allow it to remember past inputs and use that information to process future inputs.	• Good at tasks that involve sequential data. • Suitable for long-term dependencies.	• Complex to understand and implement. • Requires a lot of data to train.
Long Short-Term Memory (LSTM) Network	A variant of RNN that is specifically designed to handle long-term dependencies in sequential data.	• Very good at tasks that involve long-term dependencies. • High accuracy.	• Complex to understand and implement. • Requires a lot of data to train.
Radial Basis Function Network (RBFN)	A neural network that uses radial basis functions to compute the output of the network.	• Good at regression tasks. • High accuracy.	• Complex to understand and implement. • Requires a lot of data to train.

NEURAL NETWORK TRAINING

Forward propagation and backward propagation are two fundamental processes in training feedforward neural networks. They work together to enable the network to learn from data and improve its predictions. Forward propagation is the fundamental mechanism used to process data and make predictions for a given input. It passes input data through a neural network to generate predictions or outputs. Backward propagation, also called backpropagation, is a learning algorithm used to train feedforward neural networks. It evaluates the difference between the predicted outputs and the actual target values (errors) and then adjusts the network's weights and biases to minimize the errors during the training process.

> **DEFINITION**
>
> Forward propagation is the process of passing input data through the neural network to generate predictions.
>
> Backward propagation is a learning algorithm that computes the prediction error in terms of the loss function and then adjusts the network's parameters to minimize the errors during the training process.

1. **Forward propagation**
    - Input data are fed into the neural network, and they pass through the layers in a forward direction, from the input layer to the output layer.
    - At each layer, the input is multiplied by the layer's weights and biases, and then an activation function is applied to the result to introduce nonlinearity.
    - The output of each layer becomes the input to the next layer, and this process continues until the final output is generated at the output layer.
    - Forward propagation is essentially the process of making predictions based on the current weights and biases of the neural network, without considering the actual target values.
2. **Backward propagation**
    - Backward propagation is the learning algorithm used to update the weights and biases of the neural network based on the prediction errors during the forward pass.
    - After the forward propagation and predictions, the prediction error is computed using a loss function, which quantifies how far off the predictions are from the true values and serves as a measure of the network's performance.
    - Backward propagation works by computing the gradients of the loss function. The gradients represent the sensitivity of the loss to changes in the network's parameters (weights and biases), indicating which direction and how much each parameter should be adjusted to reduce the error.
    - Using the computed gradients, the weights and biases are adjusted using an optimization algorithm to minimize the loss and improve the network's performance.

In summary, forward propagation passes input data through the network to generate predictions, while backward propagation computes gradients of the loss function and uses them to update the network's parameters during training. Together, these processes form the basis of training feedforward neural networks, enabling the network to learn from data and improve its predictive accuracy.

ADVANTAGES AND DISADVANTAGES OF NEURAL NETWORKS

Neural networks are a very powerful machine learning algorithm. This method has been used in many areas and plays a vital role in the research and development of artificial intelligence. Nonetheless, neural networks are not perfect and still have challenges. It is important for us to understand the advantages and disadvantages of this method so that we can decide when it is the right time to use neural networks for our data mining projects.

Advantages of Neural Networks

- **Nonlinear**: Neural networks can capture well nonlinear relationships in data, which makes this method well-suited for complex problems where traditional linear models might fail.
- **Adaptive learning**: They can learn and adapt to new data in real time. Thus, they are quite effective for modeling dynamic environments where data are constantly changing.
- **Unlimited information**: This method is not limited to input data since they use many hidden layers to create connections between layers. In the training process, more information is created to produce a good model.
- **Parallel processing**: Neural networks can process multiple inputs at the same time. Hence, they can be used for tasks requiring high efficiency, such as image recognition and natural language processing.
- **Fault tolerance**: This method is resilient to noise and errors in data. The network can still perform well even when some data are missing or corrupted because it stores information on the network instead of the database.
- **Variable extraction**: Relevant variables can be automatically extracted from raw data, eliminating the need for manual variable selection.

Disadvantages of Neural Networks

- **Black box**: A neural network is often considered a "black box" because it can be difficult to understand how the model works and how variables are correlated. In other words, we feed the input data to the model, and it predicts the values of output variables. But we do not really understand what is going on in the black box. This lack of interpretability can make the method less suitable for certain applications where transparency is required.
- **Overfitting**: Neural networks can be prone to overfitting, especially when we use a very large amount of data. It means they perform well on the training data but not on new, unseen data. In other words, the model is not generalizable.
- **Data requirements**: This method requires large amounts of labeled data to train effectively, which can be a significant bottleneck in some applications.
- **Computationally intensive**: The training process can be computationally intensive, requiring high computing power and longer training time.
- **Hyperparameter configuration**: Setting hyperparameters for a neural network can be challenging and must be done carefully to achieve optimal performance. Since there are no straightforward rules for determining the right configurations for the network architecture, the modeling process requires experimentation to achieve optimal performance.

Despite those disadvantages, neural networks are still a powerful machine learning method, and their applications are growing rapidly, especially deep learning methods. By understanding the strengths and weaknesses, we can have a good strategy to use this method for our research. Again, the purpose of this book is to provide a simple introduction and guidelines for implementing neural networks, a complex machine learning algorithm. So, we will not go deeply into the complex methods in neural network modeling.

BASIC CONCEPTS OF NEURAL NETWORKS

As mentioned earlier, neural networks are a complex machine learning method. Even when focusing on the simplest forms, it can still be challenging to fully understand how the model operates. The following basic concepts form the foundation of neural networks and will help you grasp the architecture of a neural network and its training process (Table 9.2).

TABLE 9.2
Basic Concepts of Neural Networks

Terms	Description
Neuron (node)	A neuron, also called a node or unit, is a basic building block of a neural network. It receives inputs, processes them using an activation function, and produces outputs.
Connection (edge)	A connection connects neurons. Each connection has an associated weight that determines the level of the influence between connected neurons.
Activation function	An activation function is a function that determines the output of a neuron based on the weighted sum of its inputs. Common activation functions include sigmoid, ReLU (Rectified Linear Unit), and tanh (hyperbolic tangent).
Weight	Each connection between neurons is assigned a weight, which represents the strength or importance of that connection. These weights determine how much influence the input of one neuron has on the output of another.
Bias	A bias term is added to the weighted sum of inputs in each neuron. It allows the network to learn a bias toward certain values and improves the accuracy of the model.
Layers	Neurons are organized into layers in a neural network. The input layer receives the input data, the output layer produces the final output, and there can be one or more hidden layers in between.
Forward propagation	Forward propagation is the process of transmitting input signals through the network, layer by layer, to produce an output. Each neuron in a layer receives input from the previous layer, performs computations using its activation function, and passes the output to the next layer.
Backward propagation (or backpropagation)	Backward propagation is the process of adjusting the weights and biases in the network based on the prediction error. It involves calculating the gradients of the loss function and adjusting the parameters to minimize the error.
Training	Training a neural network involves iteratively adjusting the weights and biases using a training dataset. The goal is to minimize the error or loss function and improve the network's prediction accuracy.
Loss function	A loss function measures the discrepancy between the predicted output of the neural network and the actual target values. It quantifies the error and guides the adjustment of the parameters during training.
Optimization algorithm	The optimization algorithm is used during training to adjust the network's parameters based on the computed gradients. Gradient descent is a common optimization algorithm used for this purpose.
Epoch	An epoch refers to one complete pass through the entire training sample during the training of a neural network.

NEURAL NETWORK MODEL CONNECTIONS

Now that we have a basic understanding of neural networks, let's revisit the structure of neural network models as shown earlier in Figure 9.3. These models consist of connections between input nodes, hidden nodes, and output nodes. But how can we explain these connections without diving into complex mathematics? To illustrate this, I'll use a neural network architecture with a single hidden layer. Figure 9.4 shows the connections between the nodes in a simple way.

Network Connections

Essentially, an output node is a function of the hidden nodes connected to it, along with the corresponding weights. Each connection from a hidden node to an output node carries a weight, much like the regression weights in traditional models. These weights capture the importance of each

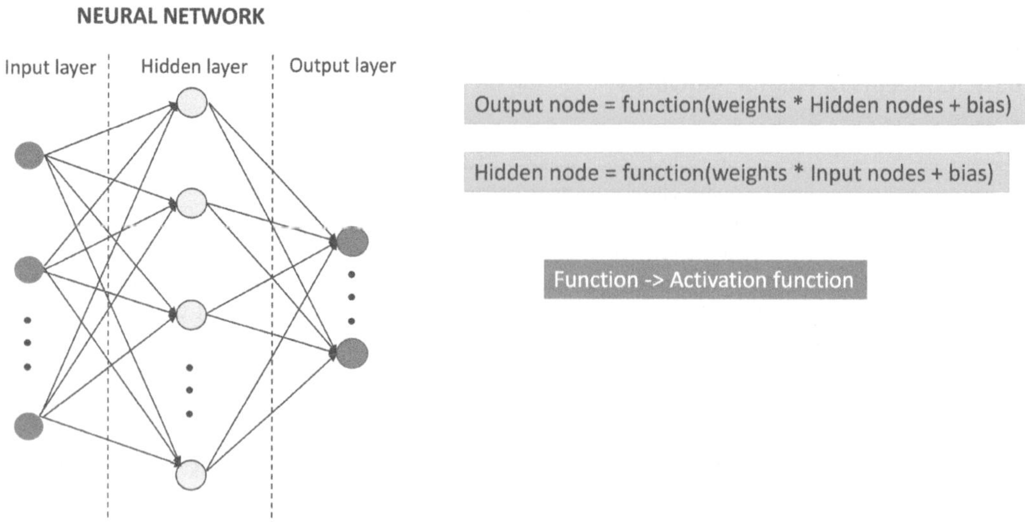

FIGURE 9.4 Connections between nodes in a neural network model.

connection in the model. Similarly, each hidden node is a function of the input nodes connected to it, along with their corresponding weights.

In each of these connections, we also have a bias node. A bias is a learnable parameter associated with each node that allows the model to shift the function left or right, which is crucial during training. The function we're referring to is called the **activation function**. Although the mathematical expressions governing these connections are more complex, at their core, this is what they represent. We will soon explore some common activation functions in more detail.

You may wonder why I don't present these equations like a linear regression equation (with an intercept, weights, input variables, and errors). The reason is that the **activation function** can be either linear or nonlinear. This flexibility is one of the key strengths of neural networks, as it allows them to capture nonlinear relationships between input, hidden, and output nodes within a complex network architecture.

ACTIVATION FUNCTIONS

As mentioned earlier, an **activation function** determines the output of a node based on an input or a set of inputs. It is a critical component of neural networks because it introduces nonlinearity, allowing the network to model complex, nonlinear relationships in data. Selecting the right activation function plays a key role in building an effective neural network.

As explained, the activation function takes the weighted sum of the inputs to a node (neuron), adds a bias term, and applies a nonlinear transformation to generate the node's output. This output is then passed to the next layer as input. In essence, activation functions mathematically represent the connections between nodes, which form the neural network model. Choosing and configuring the activation function is a fundamental task in neural network modeling.

There are several types of activation functions, each with its own strengths and weaknesses. Table 9.3 presents the most common activation functions available in SPSS Modeler. Without delving into complex mathematical formulas, I provide the function name, range, expression in terms

TABLE 9.3
Common Activation Functions (Available in SPSS Modeler)

Function Name	Range	Expression in Terms of x	Chart
Identity	$(-\infty, \infty)$	x Efficient; linear	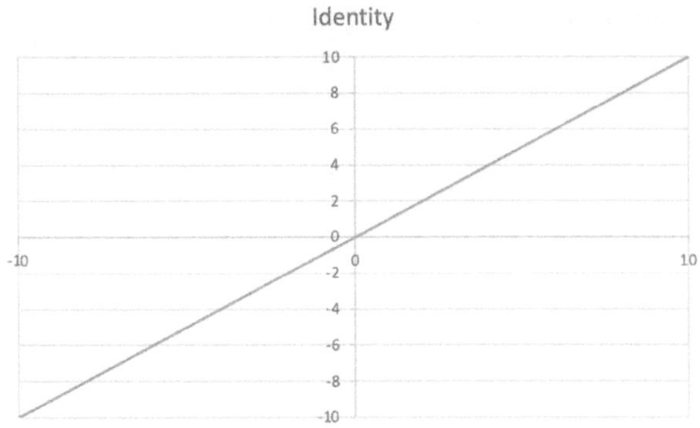
Logistic (or sigmoid)	(0, 1)	1/(1 + exp(−x)) Binary classification model	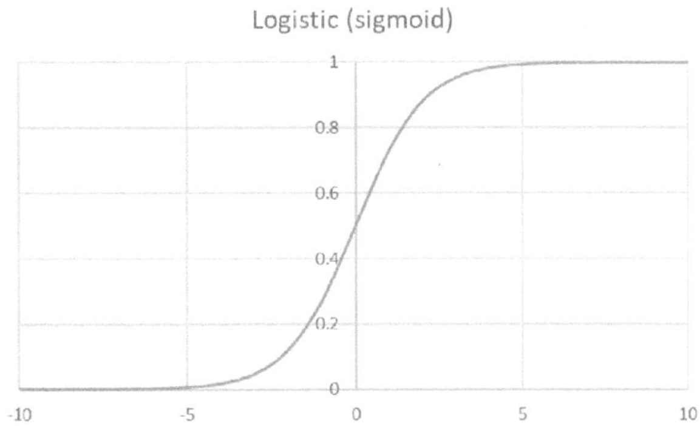
Hyperbolic tangent (or Tanh)	(−1, 1)	tanh(x)=[1 – [2/(1 + exp(2x)] Classification model with negative target values	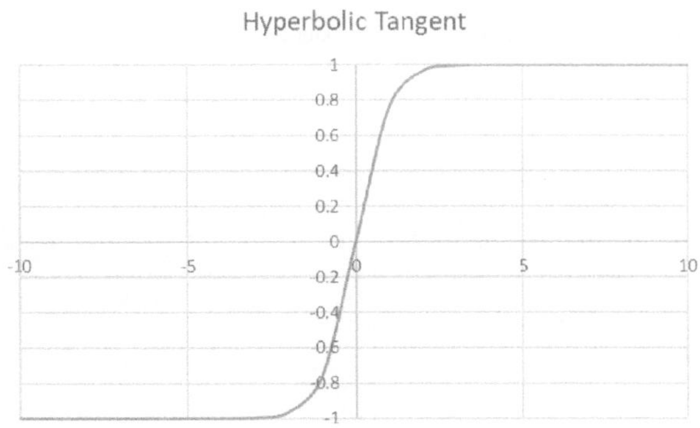
Softmax	(0, 1)	[exp(x)/Sum(exp(x))] Multiclass classification problems	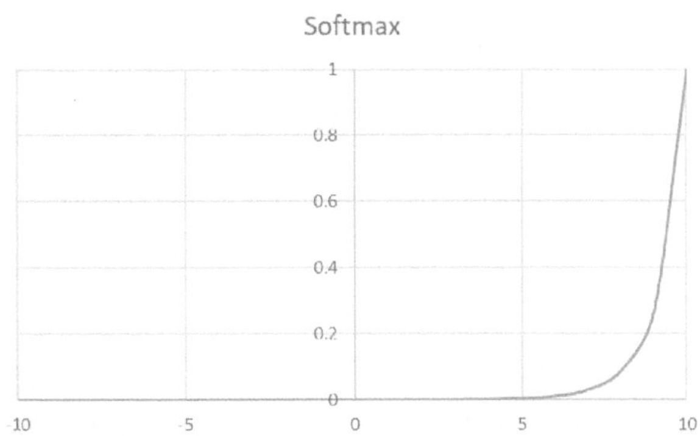

(Continued)

TABLE 9.3 *(Continued)*
Common Activation Functions (Available in SPSS Modeler)

Function Name	Range	Expression in Terms of x	Chart
Exponential	(0, ∞)	exp(x)	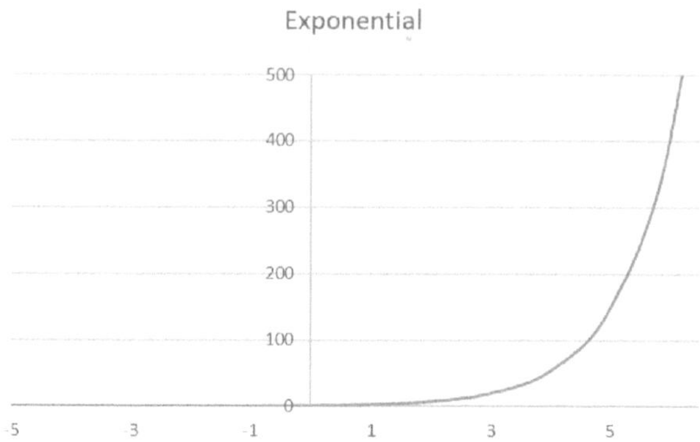
Reciprocal	(0, ∞)	1/x	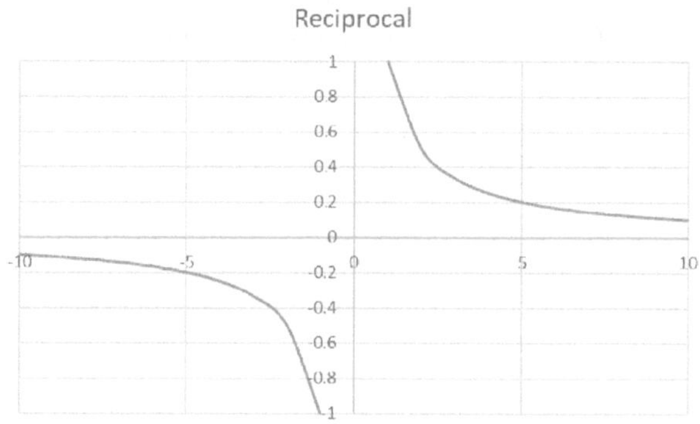
Sine	[−1, 1]	sin(x)	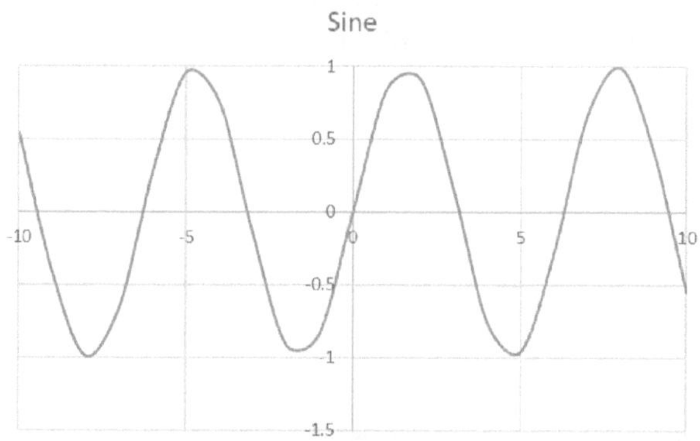
Square	(0, ∞)	x^2	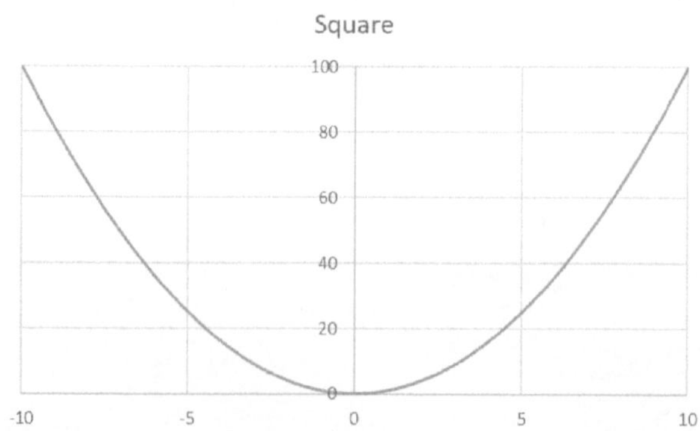

(Continued)

TABLE 9.3 (*Continued*)
Common Activation Functions (Available in SPSS Modeler)

Function Name	Range	Expression in Terms of x	Chart
Normal	(0, ∞)	$(1/\text{sqrt}(2\pi\sigma^2) \exp[-(x-\mu)^2]/2\sigma^2]$ Normal distribution function	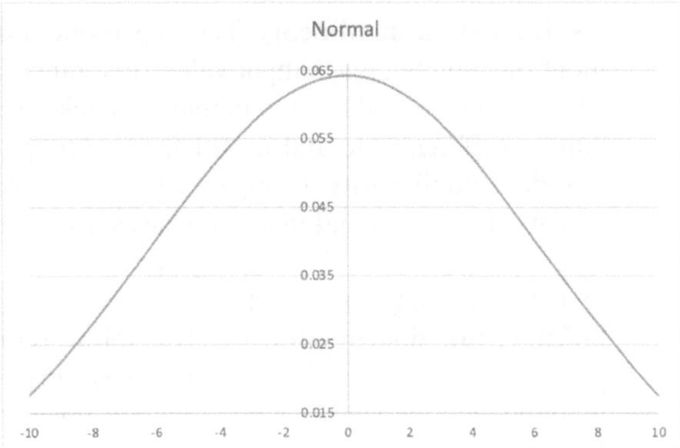
Rectified Linear Unit (ReLU)	(0, ∞)	=0 if x < 0 =x if x > = 0	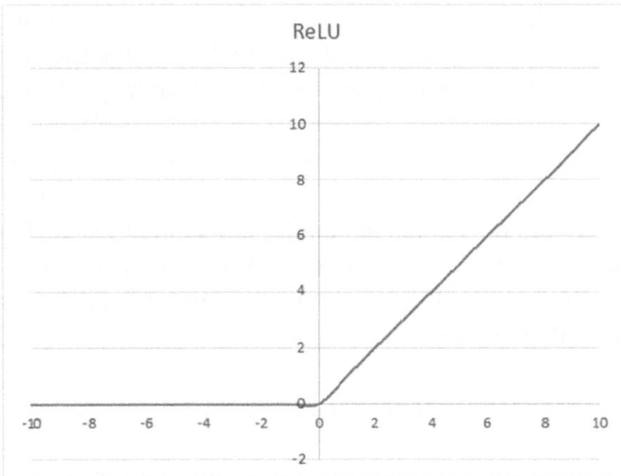

of xx (the input), and a chart showing the graphical representation of each function. Below are some of the most common activation functions:

DEFINITION

The activation function is a function that determines the value of an output node based on input from one or more input nodes.

- **Identity (linear)**: The identity function is a simple linear function that returns the input value unchanged. It is useful in specific neural network architectures where the goal is to reconstruct the input from the output. This function is commonly used as the default activation function in the output layer of regression models, where the aim is to predict continuous values. In such cases, the output of the neural network is a linear combination of the inputs, similar to a regression function. However, because the identity function does not introduce nonlinearity, it limits the network's ability to model complex functions.
- **Sigmoid function (nonlinear)**: The sigmoid function has an S-shaped curve that maps input values to a range between 0 and 1, making it useful for binary classification tasks. Its

differentiability makes it ideal for training neural networks via backpropagation, a popular optimization algorithm. The sigmoid function is often used in the hidden layers of neural networks to introduce nonlinearity, allowing the network to model complex relationships. However, it suffers from the **vanishing gradient problem**, where gradients become very small as input values grow larger or smaller, making it difficult to train deep networks with many layers.

- **Tanh function (nonlinear)**: The hyperbolic tangent (tanh) function is similar to the sigmoid function but maps input values to a range between −1 and 1. This is useful for classification models where the output may take negative values. Like the sigmoid function, tanh is differentiable and useful for backpropagation, often applied in hidden layers to introduce nonlinearity. Compared to the sigmoid, tanh is symmetric around the origin, which can be beneficial in certain cases. However, it too suffers from the vanishing gradient problem, though it is generally considered a better activation function than sigmoid due to its broader range of outputs.
- **Softmax function (nonlinear)**: The softmax function converts real-valued inputs into a probability distribution. It maps input values to positive outputs that sum to 1, making it ideal for multi-class classification tasks, such as image recognition or natural language processing. By using softmax, the neural network can output a probability distribution over all possible classes, with the class having the highest probability being the predicted output. While the softmax function is differentiable and interpretable as probabilities, it can be sensitive to outliers and may suffer from the vanishing gradient problem when inputs are extreme.

The choice of activation function significantly impacts the performance of a neural network and is often a topic of experimentation and research in deep learning. A good activation function should be computationally efficient, easily differentiable, and capable of modeling complex, nonlinear relationships in data. While there are more advanced activation functions used in deep learning, we will focus on those commonly applied in multilayer perceptron feedforward (MLP) neural networks.

MULTI-LAYER PERCEPTRON FEEDFORWARD (MLP) NETWORK ARCHITECTURE

As mentioned at the beginning of this chapter, neural networks come in various architectures that can model complex systems. However, in this book, we focus only on the **Multi-Layer Perceptron (MLP)** network. Even within the MLP network, there are multiple architectures to choose from, highlighting the complexity of this machine learning algorithm and its depth.

Figure 9.5 illustrates four primary architectures that can be used when building MLP neural network models. During the modeling step, you can decide which architecture is best suited for constructing your neural networks.

Below are brief descriptions of each architecture to help you understand the figure:

- **Single hidden layer network**: In this architecture, new hidden nodes are added one by one, selected based on which activation function provides the greatest benefit. Many models can fit successfully with just a single hidden layer.
- **Funnel network**: New hidden nodes are added to form a funnel pattern, with each node selected based on which activation function offers the most benefit.
- **Block network**: New hidden nodes are added as new layers in the network, determined by which activation function provides the most benefit.
- **Cascade network**: In this architecture, new hidden nodes are added and connected to all existing input and hidden neurons. Once trained, the weights and biases from previous layers are frozen, and each additional step further optimizes the network by fitting it to the residuals from earlier steps.

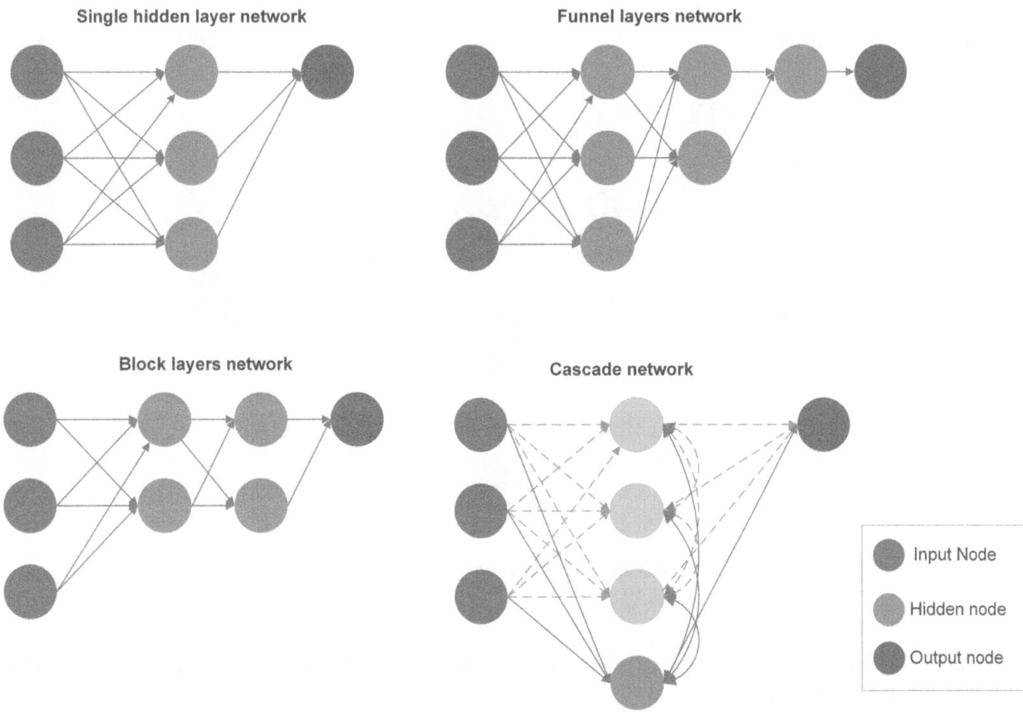

FIGURE 9.5 MLP network architectures.

The training of an MLP network involves **backpropagation**, where weights and biases are iteratively adjusted to minimize the difference between predicted and actual outputs. This process includes initialization, forward and backward passes, gradient descent optimization, and hyperparameter tuning, enabling the model to learn complex nonlinear relationships from the data.

The advantages of MLPs include their ability to serve as **universal approximators**, automatically learning features from data and handling various types of data. MLPs are highly versatile, applicable to both regression and classification tasks, and benefit from parallel processing capabilities. However, challenges such as sensitivity to hyperparameter settings and the risk of overfitting should be carefully considered.

RADIAL BASIS FUNCTION NETWORK (RBFN) ARCHITECTURE

A **Radial Basis Function Network (RBFN)** is a type of neural network that uses radial basis functions as activation functions. RBFNs are particularly well-suited for tasks such as function approximation, classification, and pattern association. Similar to other neural networks, the RBFN architecture consists of three layers: the input layer, hidden layer, and output layer. However, unlike MLPs, RBFNs are limited to having only one hidden layer (as shown in Figure 9.6).

In the hidden layer, the **Radial Basis Functions (RBFs)** are applied. Each node in this layer corresponds to a prototype or center of an RBF. The activation of a node is determined by the similarity between the input data and the prototype, typically measured using a Gaussian function. The activation function in the hidden layer is radial, meaning it depends on the distance between the input data and the prototype.

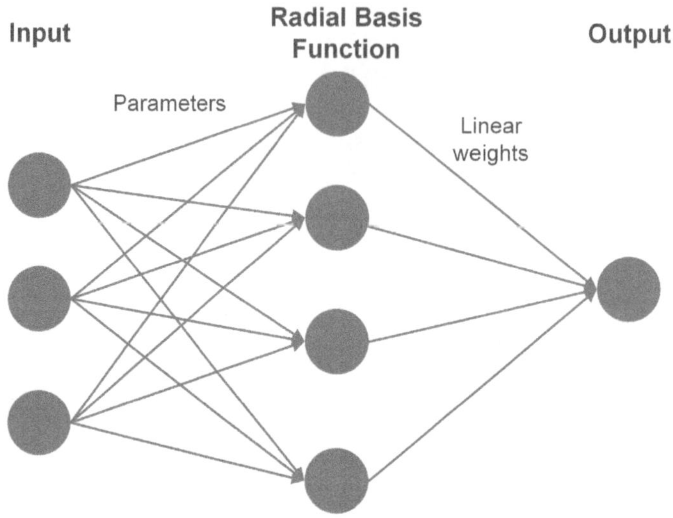

FIGURE 9.6 Radial basis function network architecture.

The radial basis function (RBF), in its simplest form, is often referred to as the Gaussian RBF, which is given by the following formula:

$$\phi(x) = e^{-\frac{\|x-c\|^2}{2\sigma^2}}$$

where:

- $\phi(x)$ is the radial basis function.
- x is the input vector.
- c is the center or prototype vector.
- $\|x - c\|$ represents the Euclidean distance between x and c.
- σ is a parameter that controls the width or spread of the function.

During training, the RBFN adjusts the weights associated with the connections between the hidden and output layers. This adjustment is made to minimize the difference between the predicted output and the actual target values. Training also involves finding the optimal locations for the RBF centers and adjusting the spread parameters to ensure accurate predictions.

RBFNs are widely used in applications such as time series prediction, pattern recognition, and function approximation. Their ability to model complex relationships and handle nonlinear patterns makes them particularly effective for tasks where traditional linear models fall short.

$$u(x)=\Sigma i=1N\lambda i\varphi(\|x-xi\|),\quad x\in Rd$$

SPSS Modeler offers two primary types of neural networks: **Multi-Layer Perceptron (MLP)** and **Radial Basis Function (RBF) Networks.** In MLP models, up to two hidden layers can be included, with the option to specify the number of nodes (neurons) in each layer. However, RBF models are limited to a single hidden layer. Additionally, stopping rules, which control when the training process should stop, are only available for MLP models. In contrast, these options are disabled (grayed out) for RBF models.

TABLE 9.4
Differences between Multi-Layer Perceptron (MLP) and Radial Basis Function Network (RBFN)

Items	Multi-Layer Perceptron (MLP)	Radial Basis Function Network (RBFN)
Network architecture	Consists of an input layer, one or more hidden layers, and an output layer. Neurons in hidden layers use activation functions.	Comprises an input layer, ONE hidden layer with radial basis functions, and an output layer. Each RBF neuron computes its activation based on the distance between the input and a prototype.
Activation functions	Typically employs sigmoid, hyperbolic tangent (tanh), or ReLU activation functions in the hidden layers, and linear activation in the output layer.	Uses radial basis functions, commonly the Gaussian function, as activation functions in the hidden layer. Linear activation is often employed in the output layer.
Training method	Trained using backpropagation and gradient descent algorithms. It adjusts weights and biases iteratively to minimize the error between predicted and actual outputs.	Trained through a two-step process. First, RBF centers and spread parameters are determined using clustering algorithms or heuristics. Second, output weights are adjusted using a linear regression approach.
Data representation	Adaptable for both linear and nonlinear relationships in data. Suitable for a wide range of applications.	Particularly effective for tasks where nonlinear patterns are present. Commonly used for function approximation, classification, and association modeling.
Interpretability	Generally, less interpretable due to the complexity of multiple hidden layers and nonlinear activation functions.	More interpretable, especially in the hidden layer where each neuron corresponds to a radial basis function centered on a prototype.
Robustness to noisy data	May be sensitive to noisy data, as it can learn and memorize noise in the training set.	Can be more robust to noisy data, especially when using localized radial basis functions.
Complexity	Can handle complex relationships and hierarchical features with multiple hidden layers. Suitable for deep learning applications.	Simpler architecture with one hidden layer. Effective for tasks with well-defined radial basis functions.
Applicability	Versatile and widely applicable, used in various domains including image recognition, natural language processing, and regression tasks.	Well-suited for tasks such as function approximation, time series prediction, and pattern recognition, particularly when dealing with nonlinear relationships.

DIFFERENCES BETWEEN MULTI-LAYER PERCEPTRON (MLP) AND RADIAL BASIS FUNCTION NETWORK (RBFN)

Since we focus on two primary types of neural networks in this book, it is important to understand their similarities and differences. Table 9.4 provides a comparison between MLP and RBFN from various perspectives. Although both MLP and RBFN are neural network architectures, they differ in terms of structure, activation functions, training methods, interpretability, and robustness to noisy data. The choice between the two depends on the specific requirements and characteristics of the task at hand.

NEURAL NETWORK SEARCHING

Training a neural network involves finding a network structure (comprising nodes, layers, connections, weights, and activation functions) that best fits the data. Given the vast number of parameters (hyperparameters) and possible network configurations, this process can be time-consuming. It is not feasible to manually determine the best network based on judgment due to the sheer number of potential configurations. Most data mining software applications include an optimization learning feature that iteratively searches for the optimal neural network based on the data.

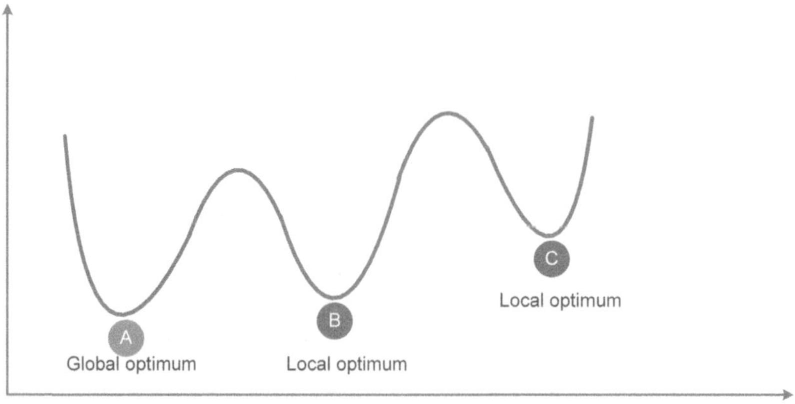

FIGURE 9.7 Global optimum and local optimum.

This iterative training process solves an optimization problem by minimizing error or loss on the training dataset, given the constraints of the parameters and data. As the complexity of the network architecture increases, the learning time tends to increase exponentially due to the growing number of computations required to find the optimal solution.

Local Optimum Situation

During neural network training, we may encounter a situation known as a **local optimum**. A local optimum is a solution that appears optimal (either maximum or minimum) within a specific neighborhood of candidate solutions but is not the best solution overall. This contrasts with the **global optimum**, which is the best solution among all possible solutions.

Figure 9.7 illustrates the difference between a local and global optimum. In the figure, point A represents the global optimum with the lowest error. However, the training process may result in two local optima, represented by points B and C. These local optima are solutions with the lowest error only within a localized set of solutions but do not reflect the true optimal solution. In other words, the training process might incorrectly identify point B or point C as the optimum due to limitations in the learning algorithm.

Of course, we aim for the **global optimum** since a local optimum is not the true optimal solution and may result in a poorly fitting model. Overcoming local optimum problems in neural networks is challenging and requires building a robust optimization model, which is beyond the scope of this book. However, a local optimum is not always a bad outcome. In some cases, particularly when constructing larger neural networks, a local optimum with reasonably low error can be acceptable.

In the example above, point B can be considered acceptable despite not being the global optimum, as it has a reasonably low error. On the other hand, point C is not acceptable due to its high error.

Stopping Rules

Stopping rules for neural networks are criteria or conditions used to determine when the training process should be terminated. These rules are essential to prevent overfitting, optimize computational resources, and improve the generalization performance of the model. The following are common stopping rules for neural networks.

> Note that in SPSS Modeler, stopping rules are only available for the Multilayer Perceptron (MLP) model and are unavailable (grayed out) for Radial Basis Function (RBF) models.

1. **Early stopping**
   - Description: Training is stopped when the performance on the testing sample ceases to improve or starts degrading.
   - Rationale: This rule helps prevent overfitting by monitoring the testing error during training. Once the testing error increases or remains stagnant for a predefined number of epochs, training is halted.
2. **Fixed number of epochs (training cycle)**
   - Description: Training is stopped after a pre-defined number of epochs.
   - Rationale: This is a straightforward rule where the algorithm is trained for a specific number of iterations, regardless of the model's performance. It is useful when computational resources are limited.
3. **Minimum error threshold**
   - Description: Training is terminated when the error (loss) on the training or testing sample falls below a certain threshold.
   - Rationale: Ensures that the model stops when it has achieved an acceptable level of performance, avoiding unnecessary computational costs.
4. **Resource constraints**
   - Description: Training can be stopped if computational resources, such as time or memory, are exhausted.
   - Rationale: Especially important in resource-intensive deep learning models, this rule prevents excessive use of resources and helps manage computational costs.

These stopping rules can be used individually or in combination, depending on the specific requirements of the task and the characteristics of the data. Implementing stopping rules is crucial for achieving an optimal balance between model performance and computational efficiency during the training process.

An epoch (or training cycle) refers to one complete pass through the entire training dataset during the training of a machine learning model, particularly in neural networks.

NORMALIZATION

Before we begin constructing and training a neural network model, there is one more important concept to cover: **normalization**. The primary purpose of normalization is to convert input and output variables to the same scale, which improves model training efficiency and helps with result interpretation. Variables often exist on different scales, and in some cases, the differences can be significant. This can lead to longer training times and less accurate predictions. Additionally, interpreting the outputs becomes more challenging when variables are on varying scales. By normalizing these variables, we can enhance training efficiency and make the results easier to interpret, as all variables will be on a consistent scale.

There are three common normalization approaches. Don't be intimidated by the formulas presented below—they are quite simple. I'll focus more on the ranges of values rather than the technical details of the formulas.

1. **Z-score normalization**: This is a widely used approach in statistics. Though technically, Z scores can take any value, they typically range between −4 and +4, with a more common range being −3 to +3.

2. **Min-max normalization**: This method uses the minimum and maximum values to scale data between 0 and 1. It is commonly used with the **sigmoid activation function**.
3. **Adjusted normalization**: Similar to min-max normalization but with some adjustments, this approach scales values between −1 and +1. It is often used with the **hyperbolic tangent (tanh) activation function**.

Below, I provide the formulas for these three methods in simple terms.

Z-score method

$$new\ value = \frac{actual\ value - mean\ value}{standard\ deviation}$$

Normalization method using max and min values

$$new\ value = \frac{actual\ value - \min\ value}{\max\ value - \min\ value}$$

Adjusted normalization method

$$new\ value = \frac{2 * (actual\ value - \min\ value)}{\max\ value - \min\ value} - 1$$

> SPSS Modeler does not offer built-in normalization features within the Neural Net node. To perform data normalization, you must use either the **Auto Data Prep** node or the **Transform** node.

EXAMPLE: A CLASSIFICATION MODEL USING NEURAL NETWORKS

In this example, we will build neural network models using the **Airline Recommendation dataset**.

- **Dataset:** AirlineRecommendation.sav
- **Target variable:** Recommendation (Flag: Yes/No)

Explore-Sample-Modify-Partition

Refer to previous chapters for more details on these steps. For this example, we will use the original data with corrected codes for missing values, meaning we will not use the previously imputed and transformed data. Instead, we will utilize the data imputation option in the Neural Net node. Additionally, we will experiment with both untransformed and normalized data. The **Auto Data Prep** node will be employed to normalize the continuous fields.

Neural Network Modeling

We will build nine neural network models using both the MLP and RBFN methods, as follows:

1. **Model MLP 1**: Uses the MLP algorithm; allows the software to automatically select the number of units for the hidden layer.
2. **Model MLP 2**: Uses the MLP algorithm with two hidden layers:
   - Layer 1: 10 units
   - Layer 2: 10 units

3. **Model MLP 3**: Uses the MLP algorithm with two hidden layers:
   - Layer 1: 15 units
   - Layer 2: 15 units
4. **Model MLP 4**: Uses the MLP algorithm with two hidden layers:
   - Layer 1: 50 units
   - Layer 2: 50 units
5. **Model MLP 5**: Uses the Auto Data Prep node to normalize the data, with the same configuration as Model MLP 3.
6. **Model RBF 1**: Uses the RBFN algorithm; allows the software to automatically select the number of units for the hidden layer.
7. **Model RBF 2**: Uses the RBFN algorithm with one hidden layer containing 15 units.
8. **Model RBF 3**: Uses the RBFN algorithm with one hidden layer containing 50 units.
9. **Model RBF 4**: Uses the Auto Data Prep node to normalize the data, with the same configuration as Model RBF 3.

Your stream should resemble the one shown below.

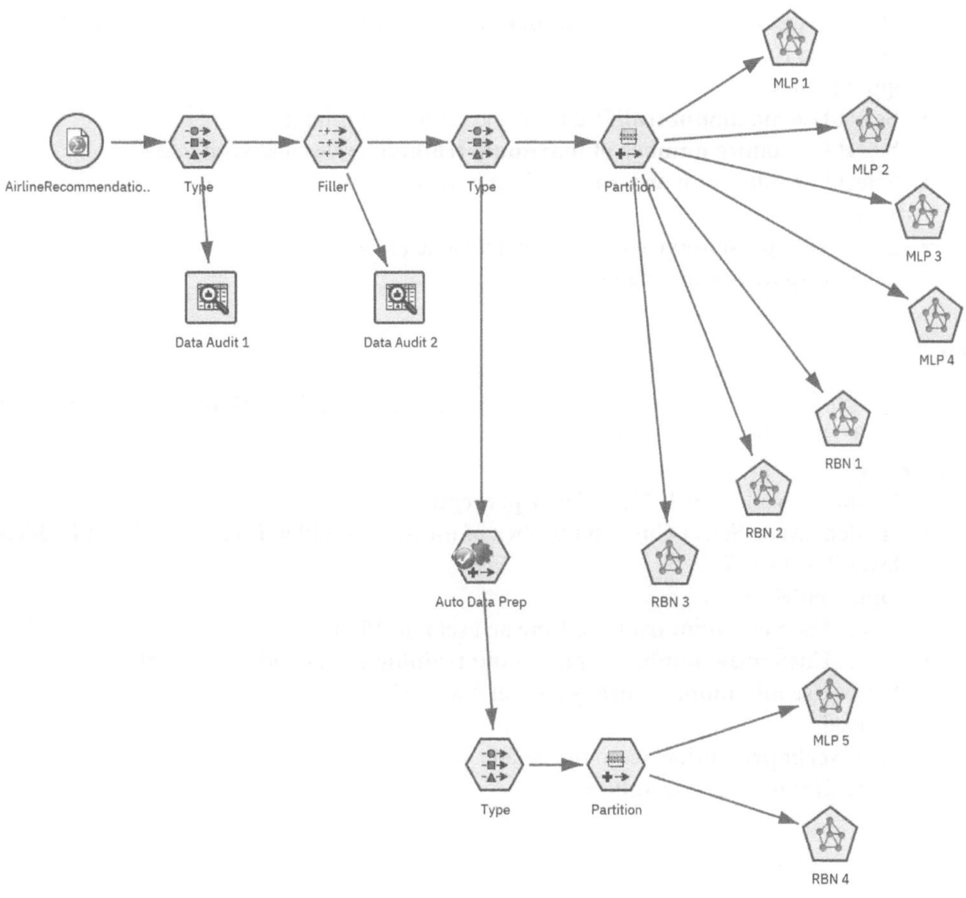

Use the following configurations to set up these nine neural network models.

MLP 1

1. Add a **Neural Net** node to the stream, connect it to the **Partition** node, and name it **MLP 1**. Open the node settings.

2. **Basics**
    - Neural network model: **Multi-layer perceptron (MLP)**.
    - Hidden layers: Select **Automatically compute number of units**.
3. **Stopping rules**
    - Select **Use maximum training time** and set it to 15 minutes.
    - Select **Customize number of maximum training cycles** and set it to 100.
    - Select **Use minimum accuracy** and set it to 90%.
4. **Advanced**
    - Set **Overfit prevention set** to the default value (30%).
    - Select **Impute missing values**.

MLP 2

1. Add a **Neural Net** node to the stream, connect it to the **Partition** node, and name it **MLP 2**. Open the node settings.
2. **Basics**
    - Neural network model: **Multi-layer perceptron (MLP)**.
    - Hidden layers: Select **Custom number of units**; set **Hidden layer 1** to 10 and **Hidden layer 2** to 10.
3. **Stopping rules**
    - Select **Use maximum training time** and set it to 15 minutes.
    - Select **Customize number of maximum training cycles** and set it to 250.
    - Select **Use minimum accuracy** and set it to 90%.
4. **Advanced**
    - Set **Overfit prevention set** to the default value (30%).
    - Select **Impute missing values**.

MLP 3

1. Add a **Neural Net** node to the stream, connect it to the **Partition** node, and name it **MLP 3**. Open the node settings.
2. **Basics**
    - Neural network model: **Multi-layer perceptron (MLP)**.
    - Hidden layers: Select **Custom number of units**; set **Hidden layer 1** to 15 and **Hidden layer 2** to 15.
3. **Stopping rules**
    - Select **Use maximum training time** and set it to 15 minutes.
    - Select **Customize number of maximum training cycles** and set it to 250.
    - Select **Use minimum accuracy** and set it to 90%.
4. **Advanced**
    - Set **Overfit prevention set** to the default value (30%).
    - Select **Impute missing values**.

MLP 4

1. Add a **Neural Net** node to the stream, connect it to the **Partition** node, and name it **MLP 4**. Open the node settings.
2. **Basics**
    - Neural network model: **Multi-layer perceptron (MLP)**.
    - Hidden layers: Select **Custom number of units**; set **Hidden layer 1** to 30 and **Hidden layer 2** to 30.

3. **Stopping rules**
   - Select **Use maximum training time** and set it to 15 minutes.
   - Select **Customize number of maximum training cycles** and set it to 300.
   - Select **Use minimum accuracy** and set it to 90%.
4. **Advanced**
   - Set **Overfit prevention set** to the default value (30%).
   - Select **Impute missing values.**

MLP 5

1. Add an **Auto Data Prep** node to the stream, connect it to the **Type** node after the **Filler** node. Open the node settings.
   - Under **Prepare Inputs & Target**, uncheck all boxes under **Adjust Type and Improve Data Quality**.
   - Under **Transform Continuous Field**, check **Put all continuous input fields on a common scale**.
   - Click **Analyze data** and click **OK**.
2. Add a **Type** node to the stream and connect it to the **Auto Data Prep** node.
3. Add a **Partition** node to the stream and connect it to the **Type** node. Select **Train, test, and validation** and set the partition size to 60–20–20.
4. Add a **Neural Net** node to the stream, connect it to the **Partition** node, and name it **MLP 5**. Use the same configurations as in **MLP 3**.

RBF 1

1. Add a **Neural Net** node to the stream, connect it to the **Partition** node, and name it **RBF 1**. Open the node settings.
2. **Basics**
   - Neural network model: **Radial Basis Function (RBF)**.
   - Hidden layers: Select **Automatically compute number of units**.
3. **Advanced**: Select **Impute missing values**.

RBF 2

1. Add a **Neural Net** node to the stream, connect it to the **Partition** node, and name it **RBF 2**. Open the node settings.
2. **Basics**
   - Neural network model: **Radial Basis Function (RBF)**.
   - Hidden layers: Select **Custom number of units**; set **Hidden layer 1** to 15.
3. **Advanced**: Select **Impute missing values**.

RBF 3

1. Add a **Neural Net** node to the stream, connect it to the **Partition** node, and name it **RBF 3**. Open the node settings.
2. **Basics**
   - Neural network model: **Radial Basis Function (RBF)**.
   - Hidden layers: Select **Custom number of units**; set **Hidden layer 1** to 30.
3. **Advanced**: Select **Impute missing values**.

RBF 4

1. Continue with the **Auto Data Prep** node created for **MLP 5**.
2. Add a **Type** node to the stream and connect it to the **Auto Data Prep** node.

3. Add a **Partition** node to the stream and connect it to the **Type** node. Select **Train, test, and validation** and set the partition size to 60–20–20.
4. Add a **Neural Net** node to the stream, connect it to the **Partition** node, and name it **RBF 4**. Use the same configurations as **RBF 3**.

Run each model separately to generate the corresponding model nugget. Once you have all nine model nuggets, use separate Analysis nodes to obtain the model metrics. You can also attempt model comparison to evaluate the models' performances. However, since MLP 5 and RBN 4 use transformed data, they cannot be compared with the other models. This is because SPSS Modeler requires that all models have the same input with identical field names for model comparison.

In this example, we will compare seven models—MLP 1, MLP 2, MLP 3, MLP 4, RBN 1, RBN 2, and RBN 3—as they all have identical input fields. To perform the comparison, follow the approach described in Chapter 7, where the model nuggets are connected for comparison. As shown in the diagram, the Filler node is used to rename the models to your preferred names. The Analysis node is used to compare the performance of the models, while the Evaluation Graph node is used to produce ROC charts and Lift charts for all the models.

Below are the Filler node settings, where we modify the model names for clarity and consistency during the comparison process.

Champion Model

We follow the standard result presentation process described in Chapter 6, using the testing sample results for comparison. Refer to Chapters 7 and 9 for details on how to extract model fit metrics from the Analysis node and calculate other metrics using the confusion matrix. Table 9.5 compares the model fit metrics for all nine neural network models, while Figures 9.8 and 9.9 show the ROC comparison chart and Lift comparison chart, respectively.

Overall, we observe that MLP 3 and MLP 5 have the highest accuracy (ACC) and lowest misclassification rate (MR). Keep in mind that these two models use the same configurations for the Neural Net node, with the only difference being that MLP 5 uses normalized data for continuous fields (via the Auto Data Prep node), while MLP 3 uses the original data. Similarly, models RBF 3 and RBF 4 share identical model fit metrics, with the only difference being that RBF 4 uses normalized data, while RBF 3 uses non-transformed data.

From this, we can conclude that for this dataset, normalizing data does not improve neural network performance, regardless of whether the MLP or RBFN method is used. Additionally, models

TABLE 9.5
Model Comparison Using the Testing Sample's Results

Metrics	MLP 1	MLP 2	MLP 3	MLP 4	MLP 5	RBF 1	RBF 2	RBF 3	RBF 4
ACC	77.06%	77.33%	78.47%	78.14%	78.47%	70.52%	70.40%	73.78%	73.78%
MR	22.94%	22.67%	21.53%	21.86%	21.53%	29.48%	29.60%	26.22%	26.22%
Sensitivity or TPR	0.913	0.904	0.906	0.905	0.906	0.774	0.756	0.810	0.810
Specificity or TNR	0.612	0.627	0.649	0.643	0.649	0.629	0.646	0.657	0.657
F1	0.808	0.808	0.816	0.814	0.816	0.735	0.729	0.765	0.765
Gini	0.664	0.673	0.695	0.687	0.695	0.534	0.540	0.599	0.599
AUC (ROC index)	0.832	0.837	0.848	0.844	0.848	0.767	0.770	0.799	0.799

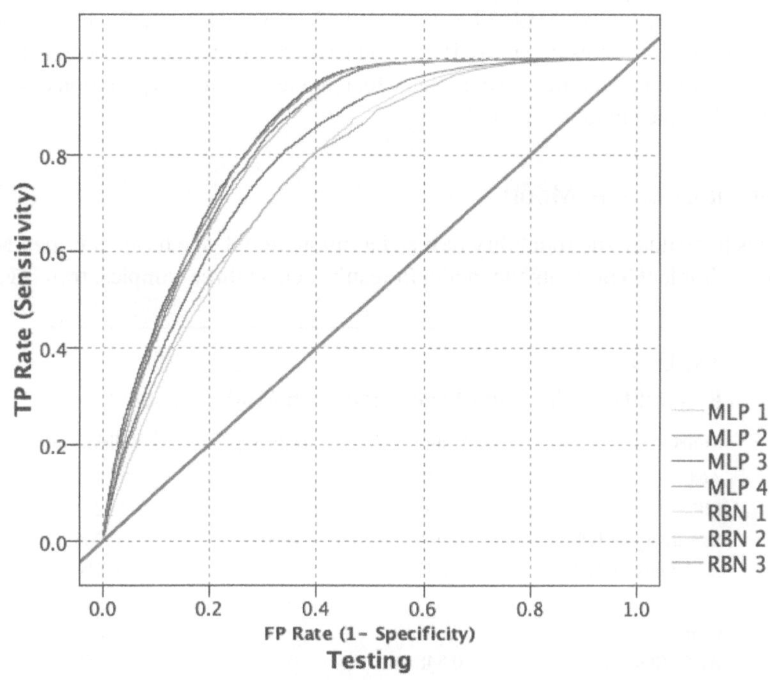

FIGURE 9.8 ROC comparison chart.

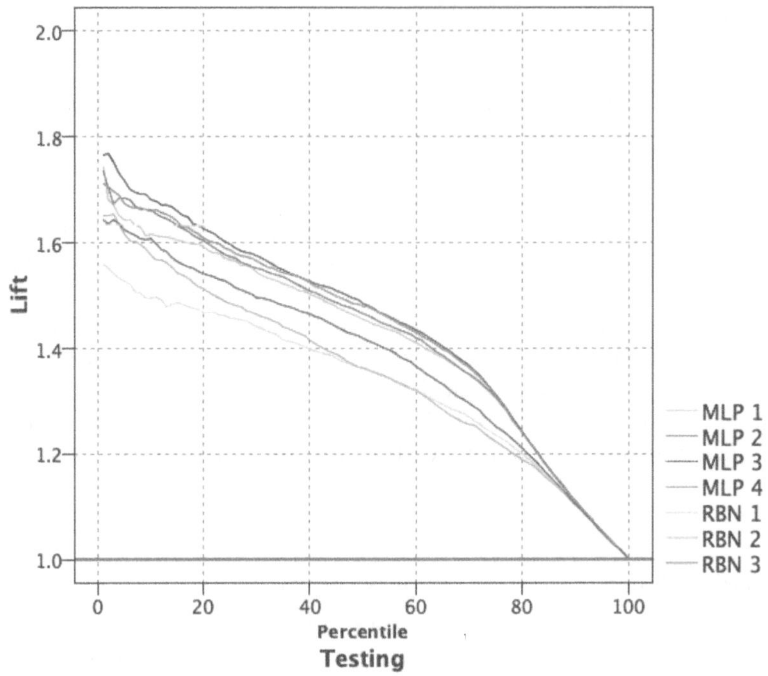

FIGURE 9.9 Lift comparison chart.

using MLP generally perform better than those using RBFN. Based on other model fit criteria, MLP 3 and MLP 5 show the best model fit, with the highest F1 score, Gini coefficient, and ROC index. Although MLP 1 has the highest sensitivity, its low specificity results in a lower F1 score.

Regarding ROC charts, MLP 3 and MLP 4 have better ROC curves compared to other models, while RBF 1 and RBF 2 perform the worst (note that MLP 5 and RBF 4 are excluded from this comparison). As for Lift curves, MLP 3 has the best Lift curve, while RBF 1 has the worst.

Ultimately, the choice comes down to MLP 3 and MLP 5. I would recommend MLP 3, as it uses the original data, avoiding data transformations, which makes result interpretation easier. Therefore, MLP 3 is selected as the champion model.

Reliability of the Champion Model

The next step is to evaluate the reliability of the champion model. We have identified MLP 3 as the champion model. Table 9.6 presents the model fit results across three samples: training, testing, and

TABLE 9.6
Reliability Evaluation of the Champion Model—MLP 3

Metrics	Testing	Training	Validation
ACC	78.47%	78.66%	78.71%
MR	21.53%	21.34%	21.29%
Sensitivity or TPR	0.906	0.907	0.906
Specificity or TNR	0.649	0.654	0.656
F1	0.816	0.817	0.817
Gini	0.695	0.708	0.699
AUC (ROC index)	0.848	0.854	0.850

validation. The results are very consistent between these samples, with a difference of less than 1%. This indicates that MLP 3 demonstrates satisfactory model reliability, with no evidence of overfitting.

Validity of the Champion Model

The next step is to assess the validity of the champion model. For this, we focus on the model fit metrics for the validation sample as the final evaluation, as outlined in Table 9.7. The observed prediction accuracy (ACC) falls below 80%, a threshold typically considered suboptimal. However, the sensitivity (true positive prediction) performs well at 0.906. The lower ACC is primarily due to the subpar specificity (true negative prediction) of 0.656, indicating a higher false positive rate.

The F1 score, which balances sensitivity and specificity, is within acceptable limits. The ROC index, at 0.85, is also considered acceptable, though an index of 0.9 or higher would indicate a better model fit. The Gini coefficient is quite good at 0.699. Additionally, the Lift chart reveals that with 20% of the dataset in use, the Lift value is 1.64, meaning the model captures 1.64 times more positive cases (32.8%) compared to a model-less scenario.

In summary, when we consider all metrics collectively, we can conclude that the model has marginally acceptable predictive power.

In conclusion, while certain fit metrics for this predictive model are deemed acceptable, the model may not meet the airlines expectations for predicting airline recommendations from passengers. Although the model performs well in predicting positive cases of airline recommendations, it falls short in accurately predicting non-recommendations, with a need for improvement in overall accuracy, particularly in reducing false positives. This mirrors the conclusions we reached in Chapters 7 and 9.

For comparison, I also included the best decision tree model we developed in Chapter 8 (C&R Tree 2) in Table 9.7. As shown, the decision tree model has higher accuracy (ACC), lower MR, higher sensitivity, and a better F1 score. However, the MLP 3 neural network model outperforms it in terms of specificity, Gini coefficient, and ROC index. In subsequent chapters, we will continue building more predictive models for this dataset using other algorithms.

Neural Network Interpretation

Suppose the model fit metrics are acceptable to the airline, and the model is selected for making actual predictions; at this point, we would want to examine the neural network itself to understand its architecture and parameters. Figure 9.10 illustrates the network, including input nodes, hidden layers, hidden nodes, and the output node for this model. Specifically, the network consists of two hidden layers, each containing 15 hidden nodes.

TABLE 9.7
Validity Evaluation of the Champion Model—MLP 3

Metrics	Validation	Evaluation	C&R Tree 2
ACC	78.71%	Not good	79.5%
MR	21.29%	Not good	20.5%
Sensitivity or TPR	0.906	Good	0.951
Specificity or TNR	0.656	Not good	0.623
F1	0.817	Acceptable	0.83
Gini	0.699	Good	0.673
AUC (ROC index)	0.85	Acceptable	0.837

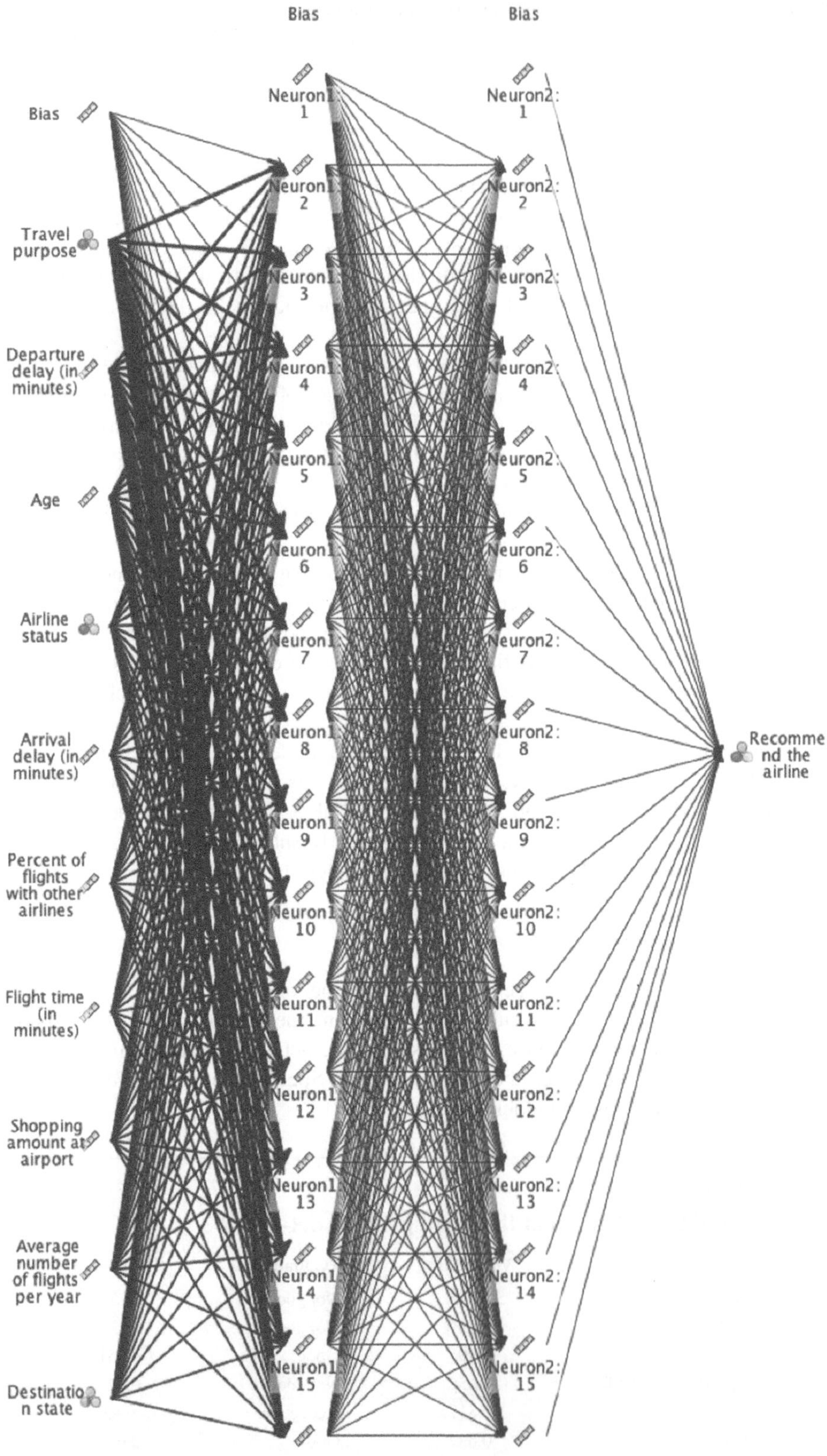

FIGURE 9.10 Network structure of the champion model—MLP 3.

To visualize the synaptic weights and the direction of effects for all paths in the network, we select the Coefficients option for Style, resulting in a network diagram where the weights are color-coded to indicate positive and negative effects (Figure 9.11). However, this visualization does not display the exact values and signs of the weights.

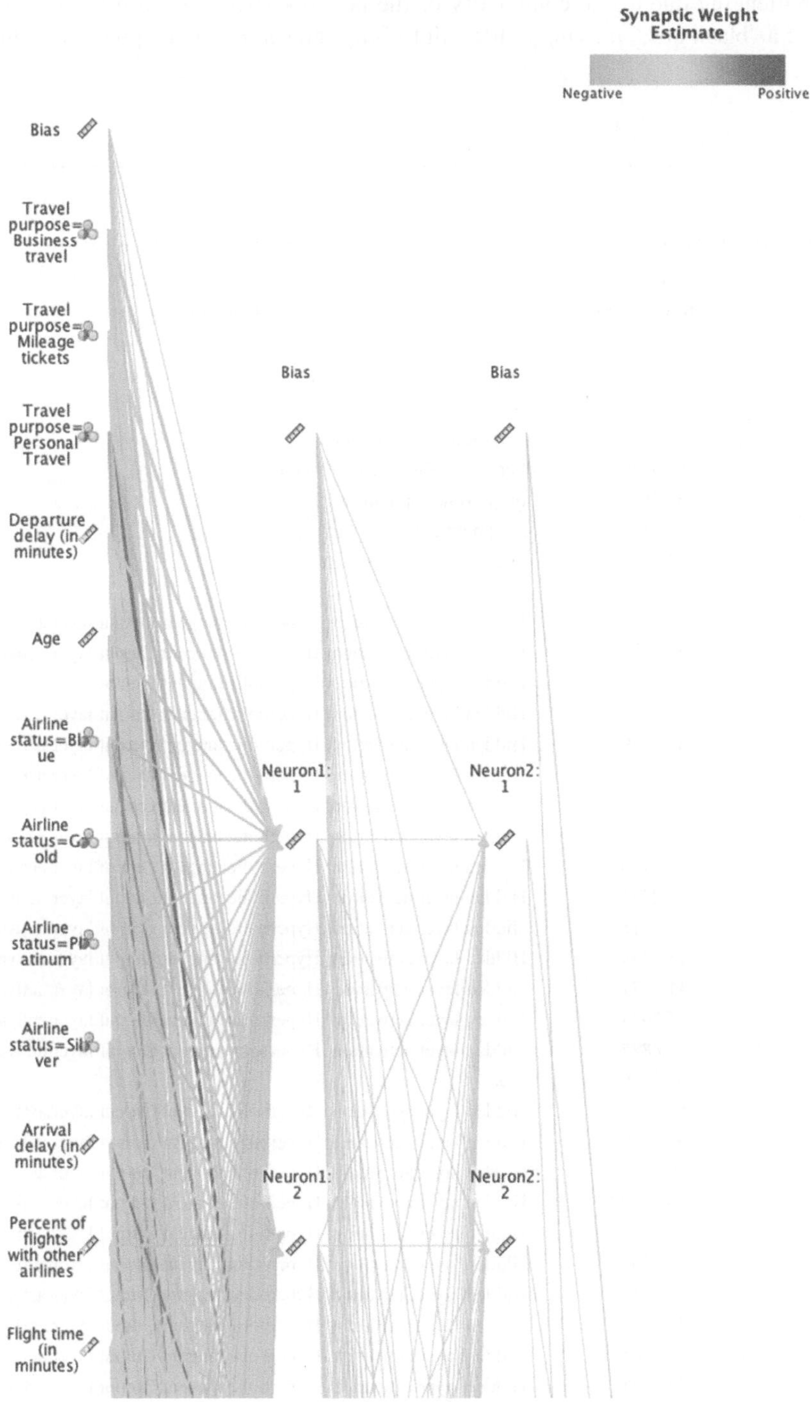

FIGURE 9.11 A snapshot of the network structure of the champion model with synaptic weights.

For more detailed information, we can copy the network into an Excel file, which provides the exact weights, nodes, layers, and activation functions used for each path in the network (Table 9.8). As shown, the hyperbolic tangent function was used for all hidden layers, while the identity function was applied to the output nodes.

Despite this detailed information, interpreting the effect of each predictor on the target field remains challenging due to the complexity of the network structure. In this case, the network appears like a "black box," making it difficult to fully interpret how the predictors influence the outcomes.

TABLE 9.8
Detailed Network Structure with Synaptic Weights, Nodes, and Activation Functions

X (Input Node)	Y (Output Node)	Description
0	65.0294	Bias
0	61.1471	Departure delay (in minutes)
0	60.1765	Age
0	55.3235	Arrival delay (in minutes)
0	54.3529	Percent of flights with other airlines
0	53.3824	Flight time (in minutes)
0	52.4118	Shopping amount at airport
0	51.4412	Average number of flights per year
1	62.1176	Bias
1	58.2353	Hidden layer activation: Hyperbolic tangent Output layer activation: Identity
1	54.3529	Hidden layer activation: Hyperbolic tangent Output layer activation: Identity
1	50.4706	Hidden layer activation: Hyperbolic tangent Output layer activation: Identity
1	46.5882	Hidden layer activation: Hyperbolic tangent Output layer activation: Identity
1	42.7059	Hidden layer activation: Hyperbolic tangent Output layer activation: Identity
1	38.8235	Hidden layer activation: Hyperbolic tangent Output layer activation: Identity
1	34.9412	Hidden layer activation: Hyperbolic tangent Output layer activation: Identity
1	31.0588	Hidden layer activation: Hyperbolic tangent Output layer activation: Identity
1	27.1765	Hidden layer activation: Hyperbolic tangent Output layer activation: Identity
1	23.2941	Hidden layer activation: Hyperbolic tangent Output layer activation: Identity
1	19.4118	Hidden layer activation: Hyperbolic tangent Output layer activation: Identity
1	15.5294	Hidden layer activation: Hyperbolic tangent Output layer activation: Identity
1	11.6471	Hidden layer activation: Hyperbolic tangent Output layer activation: Identity
1	7.7647	Hidden layer activation: Hyperbolic tangent Output layer activation: Identity
1	3.8824	Hidden layer activation: Hyperbolic tangent Output layer activation: Identity
2	62.1176	Bias
2	58.2353	Hidden layer activation: Hyperbolic tangent Output layer activation: Identity
2	54.3529	Hidden layer activation: Hyperbolic tangent Output layer activation: Identity
2	50.4706	Hidden layer activation: Hyperbolic tangent Output layer activation: Identity
2	46.5882	Hidden layer activation: Hyperbolic tangent Output layer activation: Identity
2	42.7059	Hidden layer activation: Hyperbolic tangent Output layer activation: Identity
2	38.8235	Hidden layer activation: Hyperbolic tangent Output layer activation: Identity
2	34.9412	Hidden layer activation: Hyperbolic tangent Output layer activation: Identity
2	31.0588	Hidden layer activation: Hyperbolic tangent Output layer activation: Identity
2	27.1765	Hidden layer activation: Hyperbolic tangent Output layer activation: Identity
2	23.2941	Hidden layer activation: Hyperbolic tangent Output layer activation: Identity
2	19.4118	Hidden layer activation: Hyperbolic tangent Output layer activation: Identity

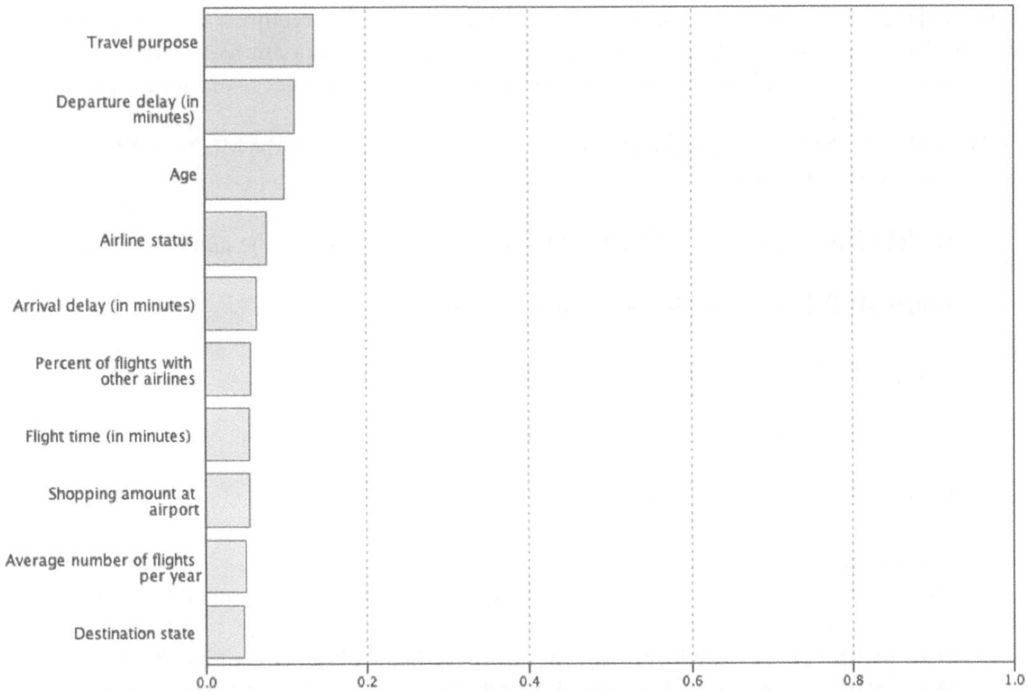

FIGURE 9.12 Variable importance chart of the champion model—MLP 3.

Variable Importance

Since the neural network operates as a "black box" and is difficult to interpret, we can examine the variable importance to understand the relative significance of the predictors. Figure 9.12 displays the variable importance chart for model MLP 3. As shown, **travel purpose** is the most important predictor, followed by **departure delay**, **age**, **airline status**, and **arrival delay**. On the other hand, **destination state** and **average number of flights per year** rank among the least important predictors.

These results are largely consistent with those from the regression and decision tree models discussed in previous chapters.

EXAMPLE: AN ASSOCIATION MODEL USING NEURAL NETWORKS

For this association model, we build several neural network models using the **Boston Housing** dataset.

- **Dataset**: boston.sav
- **Target variable**: lmv (Continuous—Median home value)

Explore-Sample-Modify-Partition

Refer to previous chapters for more details on these steps.

Neural Network Modeling

Given the small sample size of this dataset (506 records) and the fact that all fields are continuous, we will use the imputed, normalized data that was previously completed. Although neural networks can handle missing values and non-normalized data, they typically require a large dataset to train effectively. With a smaller dataset, using complete, normalized data are more likely to result in better-fitting models. Additionally, we partition the data into two samples—training and testing—using a 60–40 split.

We will build the following eight neural network models using either the MLP or RBFN methods with various configurations:

1. **Model MLP 1**: Uses the MLP algorithm; let the software select the number of units for the hidden layer.
2. **Model MLP 2**: Uses the MLP algorithm with two hidden layers:
   - Layer 1: 10 units
   - Layer 2: 10 units
3. **Model MLP 3**: Uses the MLP algorithm with two hidden layers:
   - Layer 1: 20 units
   - Layer 2: 20 units
4. **Model MLP 4**: Uses the MLP algorithm with two hidden layers:
   - Layer 1: 30 units
   - Layer 2: 30 units
5. **Model RBF 1**: Uses the RBFN algorithm; let the software select the number of units for the hidden layer.
6. **Model RBF 2**: Uses the RBFN algorithm with one hidden layer containing 20 units.
7. **Model RBF 3**: Uses the RBFN algorithm with one hidden layer containing 50 units.
8. **Model RBF 4**: Uses the RBFN algorithm with one hidden layer containing 100 units.

Your stream should resemble the one shown below. Note that the neural network models are added to the **Partition** node after the data have been imputed and normalized. This allows us to train the models using the complete, normalized dataset.

You should have a stream looking like the one below. Notice that we added the neural network models to the Partition node after we imputed and normalized the data. Thus, we can use the complete, normalized data to train the models.

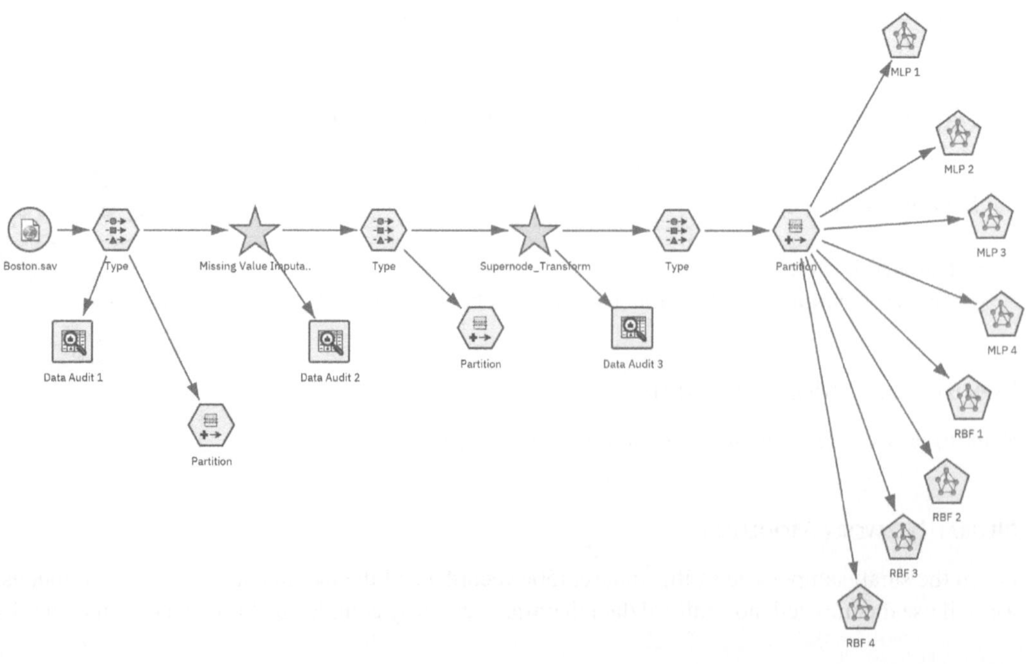

Use the following configurations to set up these nine neural network models.

MLP 1

1. Add a **Neural Net** node to the stream and connect it to the **Partition** node. Name it **MLP 1**. Open the node settings.
2. **Basics**
   - Neural network model: **Multi-layer perceptron (MLP)**
   - Hidden layers: Select **Automatically compute number of units**
3. **Stopping rules**
   - Select **Use maximum training time** and set it to 15 minutes
   - Select **Customize number of maximum training cycles** and set it to 100
   - Select **Use minimum accuracy** and set it to 90%
4. **Advanced**
   - Set **Overfit prevention set** to default (30%)

MLP 2

1. Add a **Neural Net** node to the stream, connect it to the **Partition** node, and name it **MLP 2**. Open the node settings.
2. **Basics**
   - Neural network model: **Multi-layer perceptron (MLP)**
   - Hidden layers: Select **Custom number of units**
   - Set **Hidden layer 1** to 10 and **Hidden layer 2** to 10
3. **Stopping rules**
   - Select **Use maximum training time** and set it to 15 minutes
   - Select **Customize number of maximum training cycles** and set it to 250
   - Select **Use minimum accuracy** and set it to 90%
4. **Advanced**
   - Set **Overfit prevention set** to default (30%)

MLP 3

1. Add a **Neural Net** node to the stream, connect it to the **Partition** node, and name it **MLP 3**. Open the node settings.
2. **Basics**
   - Neural network model: **Multi-layer perceptron (MLP)**
   - Hidden layers: Select **Custom number of units**
   - Set **Hidden layer 1** to 20 and **Hidden layer 2** to 20
3. **Stopping rules**
   - Select **Use maximum training time** and set it to 15 minutes
   - Select **Customize number of maximum training cycles** and set it to 250
   - Select **Use minimum accuracy** and set it to 90%
4. **Advanced**
   - Set **Overfit prevention set** to default (30%)
   - Select **Impute missing values**

MLP 4

1. Add a **Neural Net** node to the stream, connect it to the **Partition** node, and name it **MLP 4**. Open the node settings.
2. **Basics**
   - Neural network model: **Multi-layer perceptron (MLP)**
   - Hidden layers: Select **Custom number of units**
   - Set **Hidden layer 1** to 30 and **Hidden layer 2** to 30

3. **Stopping rules**
   - Select **Use maximum training time** and set it to 15 minutes
   - Select **Customize number of maximum training cycles** and set it to 400
   - Select **Use minimum accuracy** and set it to 90%
4. **Advanced**
   - Set **Overfit prevention set** to default (30%)

RBF 1

1. Add a **Neural Net** node to the stream, connect it to the **Partition** node, and name it **RBF 1**. Open the node settings.
2. **Basics**
   - Neural network model: **Radial Basis Function (RBF)**
   - Hidden layers: Select **Automatically compute number of units**

RBF 2

1. Add a **Neural Net** node to the stream, connect it to the **Partition** node, and name it **RBF 2**. Open the node settings.
2. **Basics**
   - Neural network model: **Radial Basis Function (RBF)**
   - Hidden layers: Select **Custom number of units**
   - Set **Hidden layer 1** to 20

RBF 3

1. Add a **Neural Net** node to the stream, connect it to the **Partition** node, and name it **RBF 3**. Open the node settings.
2. **Basics**
   - Neural network model: **Radial Basis Function (RBF)**
   - Hidden layers: Select **Custom number of units**
   - Set **Hidden layer 1** to 50

RBF 4

1. Add a **Neural Net** node to the stream, connect it to the **Partition** node, and name it **RBF 4**. Open the node settings.
2. **Basics**
   - Neural network model: **Radial Basis Function (RBF)**
   - Hidden layers: Select **Custom number of units**
   - Set **Hidden layer 1** to 100

Champion Model

After running all the models, the Neural Net node provides the accuracy in percentage, which for an association model with a continuous target field is equivalent to explanatory power (EP). To maintain consistency with other chapters and methods, we will refer to this metric as EP. It's important to note that in SPSS Modeler, this metric is only provided for the training sample.

To obtain additional performance metrics, we add an Analysis node to each model and run it. From the output of the Analysis node, we can extract the relevant metrics. However, as discussed in Chapter 8, we will primarily use Mean Absolute Error (MAE) and linear correlation to compare models, as other metrics are less meaningful for comparison.

As shown in Table 9.9, based on the testing sample results, MLP 4 has the highest EP, but it does not have the lowest MAE. MLP 3 shows the lowest MAE and highest linear correlation. This discrepancy arises because EP is calculated from the training sample, while MAE is calculated

TABLE 9.9
Model Comparison Results

Models	EP (%) (Training)	MAE (Testing)	Linear Correlation (Testing)
MLP 1	84.1	2.712	0.864
MLP 2	84.0	2.515	0.889
MLP 3	85.7	2.418	0.89
MLP 4	87.0	2.693	0.877
RBF 1	44.1	4.8	0.564
RBF 2	71.4	3.344	0.788
RBF 3	83.7	3.233	0.777
RBF 4	83.5	3.828	0.632

from the testing sample. In such cases, we should prioritize the results from the testing sample. After recalculating the EP for the testing sample, it was confirmed that MLP 3 has a higher EP than MLP 4. Thus, MLP 3 demonstrates the highest prediction accuracy.

Among the MLP models, MLP 1 (with an automatic number of nodes) has the lowest accuracy. The most accurate model, MLP 3, uses two hidden layers, each with 20 nodes, which is higher than both MLP 1 and MLP 2. However, this does not mean that more nodes always result in better accuracy (e.g., MLP 4, with 30 nodes per hidden layer, did not outperform MLP 3). The key is to find the optimal number of nodes (neurons). SPSS Modeler lacks an automatic optimization feature for network structures, so identifying the best configuration requires trial and error. It's also worth noting that models with more nodes require more computing time to converge.

Among the RBF models, RBF 1 (with an automatic number of nodes) is the least accurate. Interestingly, RBF 3, with 50 nodes, has the highest accuracy among the RBF models. However, overall, the MLP models exhibit higher accuracy than the RBF models for this dataset.

In conclusion, MLP 3 is selected as the champion model due to its superior prediction accuracy.

Reliability of the Champion Model

In this step, we assess the reliability of our champion model. Table 9.10 presents a side-by-side comparison of metrics between the testing and training samples. The findings reveal notable discrepancies between these two samples, particularly in MAE, standard deviation, and linear correlation. Specifically, the differences are 12% for MAE, 16% for standard deviation, and 4% for linear correlation. These differences suggest that the model performs with lower accuracy on the testing sample compared to the training sample, raising potential reliability concerns.

TABLE 9.10
Reliability Evaluation of the Champion Model—MLP 3

Metrics	Testing	Training
Minimum error	−10.665	−24.142
Maximum error	23.489	10.308
Mean error	0.1	0.131
Mean absolute error	2.418	2.126
Standard deviation	3.778	3.154
Linear correlation	0.89	0.926

The relatively small sample size exacerbates this inconsistency (training sample size: 304 records; testing sample size: 202 records). In general, smaller sample sizes, especially when there are numerous variables, can result in higher error during the modeling process. To improve the model's reliability, it is recommended to gather additional data and retrain the models. While such actions are not feasible for this example, it is important to recognize this issue and consider potential solutions for future work.

Validity of the Champion Model

To further assess the performance of the champion model, we need to export the full prediction dataset. As usual, we use the Excel node from the Export group to export the prediction data for this model into an Excel file. Running this node will export both the training and testing prediction data to your computer. Please refer to previous chapters for details on completing this step and downloading the file in Excel. Using the formulas and procedures outlined in Chapter 6, we can calculate three additional fit metrics for MLP 3: MAPE, RSE, and EP, as presented in Table 9.11.

- The MAPE of 0.101 indicates that the model can predict the median house values with an accuracy of 1 − 0.101 = 89.9%, which is considered acceptable.
- The EP of 0.792 shows that the model can explain 79.2% of the variance in the target variable based on the variance in the predictors. As mentioned in Chapter 6, an EP of 0.7 or higher is regarded as a good model fit.

For comparison, I also included the fit metrics for CHAID 1 (the best decision tree model described in Chapter 8) in the table. As you can see, MLP 3 demonstrates higher prediction accuracy (MAPE = 0.101) and much better EP, with an RSE of 0.208 and an EP of 0.792.

We also reviewed two charts—the **Actuals by Predicted** chart and the **Residuals by Predicted** chart—using the validation results. As shown in Figure 9.13, the data points are closely scattered around the diagonal line, with some outliers, indicating a good fit. Additionally, Figure 9.14 demonstrates that the residuals appear random and show no clear patterns. Overall, these charts suggest that MLP 3 provides a good model fit.

Neural Network Interpretation

Having established MLP 3 as the champion model with a good fit, we now dive deeper into its structure. Figure 9.15 illustrates the network's layers and nodes. To generate this chart, open the model nugget for MLP 3 and select the network structure option in the left panel. As shown, the network consists of two hidden layers, each with 20 nodes.

TABLE 9.11
Validity Evaluation of the Champion Model—MLP 3 (The Testing Sample's Results)

Model Fit Metrics	MLP 3	Evaluation	CHAID 1 Tree
MAE	2.418	N/A	3.168
RMSE	3.770	N/A	5.209
MAPE	0.101	Acceptable	0.125
RSE	0.208	Good	0.398
EP	0.792	Good	0.602

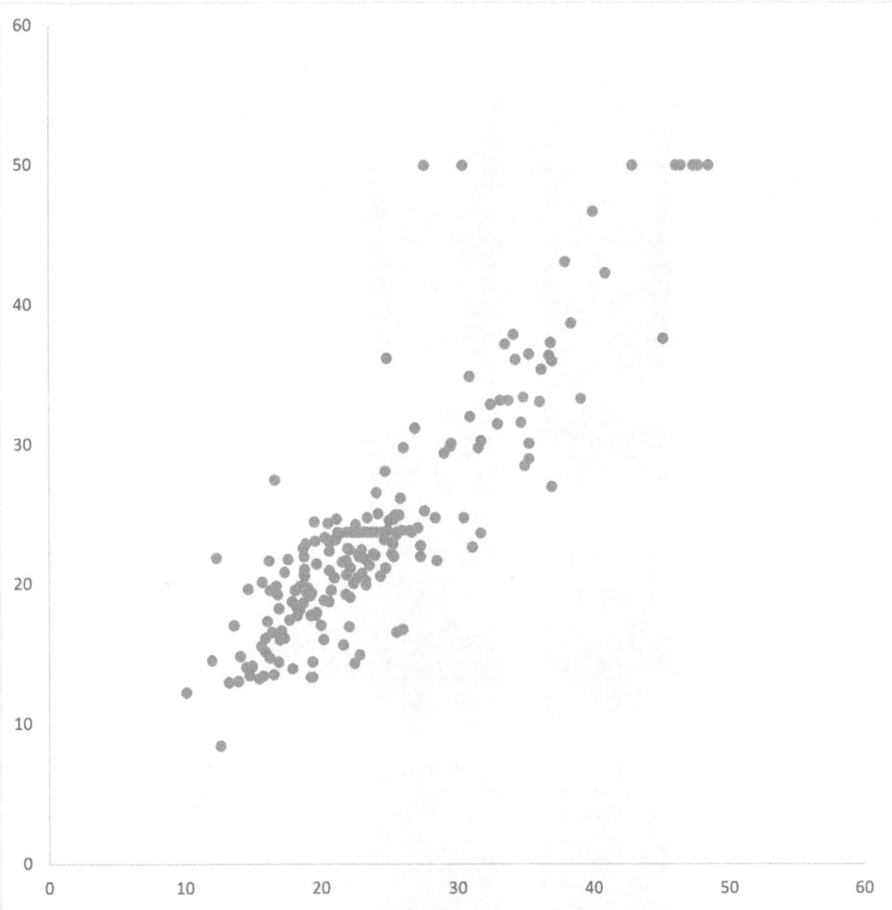

FIGURE 9.13 Actual by Predicted chart of the champion model—MLP 3.

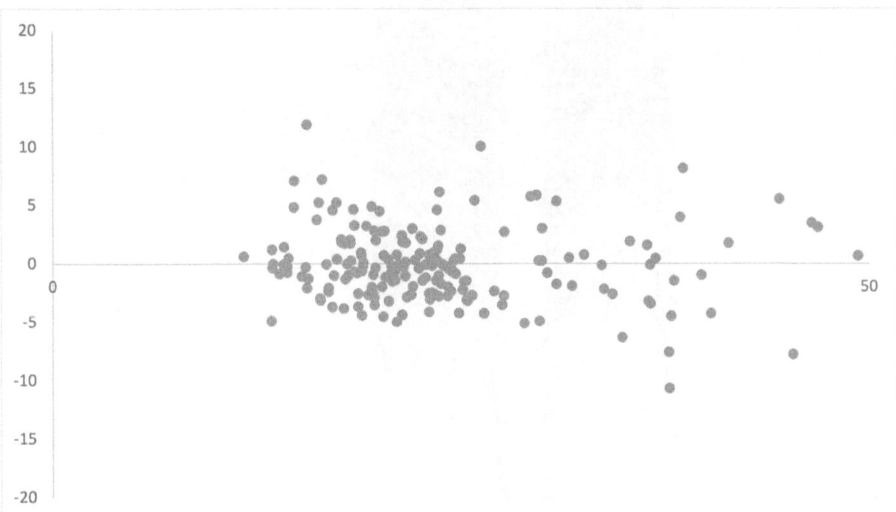

FIGURE 9.14 Residuals by Predicted chart of the champion model—MLP 3.

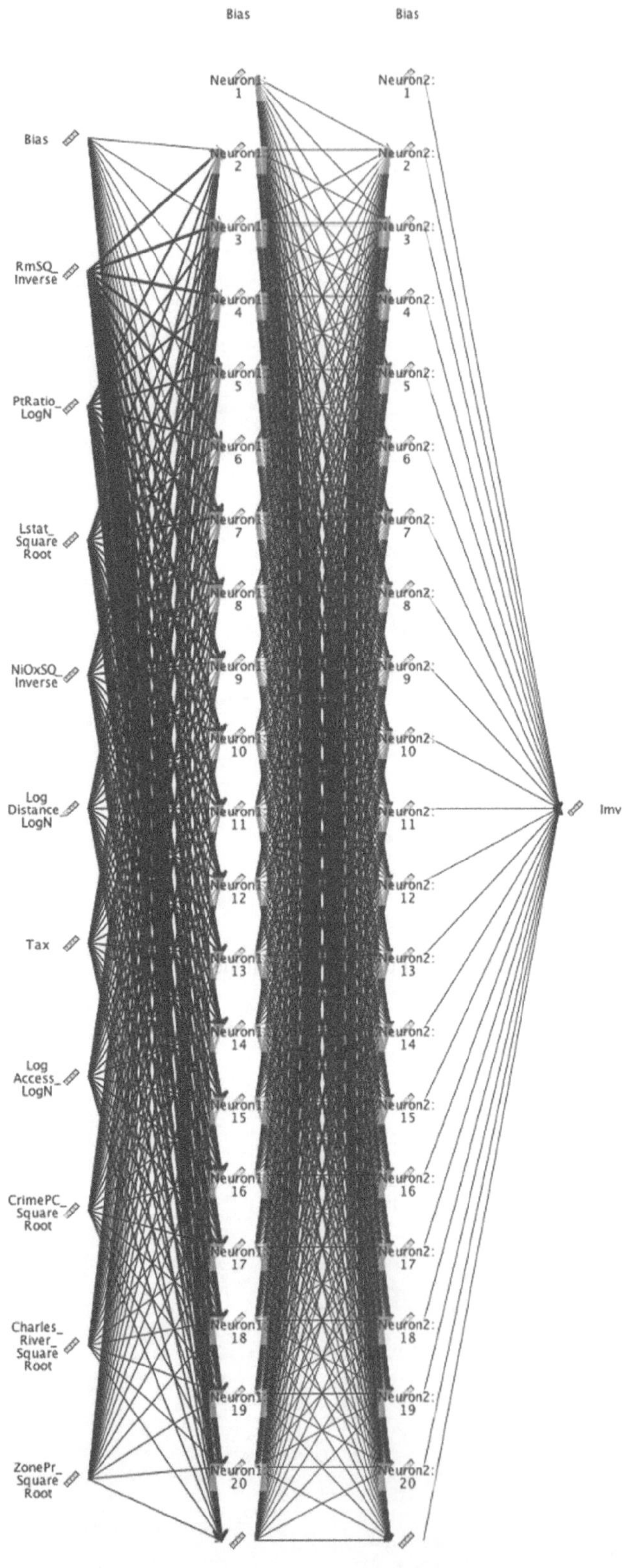

FIGURE 9.15 Neural network structure of the champion model—MLP 3.

To view the synaptic weights for each connection, you can copy the network and paste it into an Excel file, similar to what we did in the **Airline Recommendation** example. However, as previously mentioned, these weights are not easily interpretable due to the complexity of the relationships between the input nodes, hidden nodes, and the output node. In other words, the effect of each predictor on the target variable cannot be clearly understood from this information alone, which is why neural networks are often considered a "black box."

Since the network structure is difficult to interpret, we instead examine the variable importance for a clearer understanding.

Variable Importance

The model nugget of MLP 3 provides a variable importance chart that shows the relative significance of the predictors in predicting the target variable, Imv. As illustrated in Figure 9.16, **RmSQ** is the most influential predictor, followed by **PtRatio**, **Lstat**, and **NiOxSQ**. On the other hand, **ZonePr**, **Charles_River**, and **CrimePC** are considered the least important predictors.

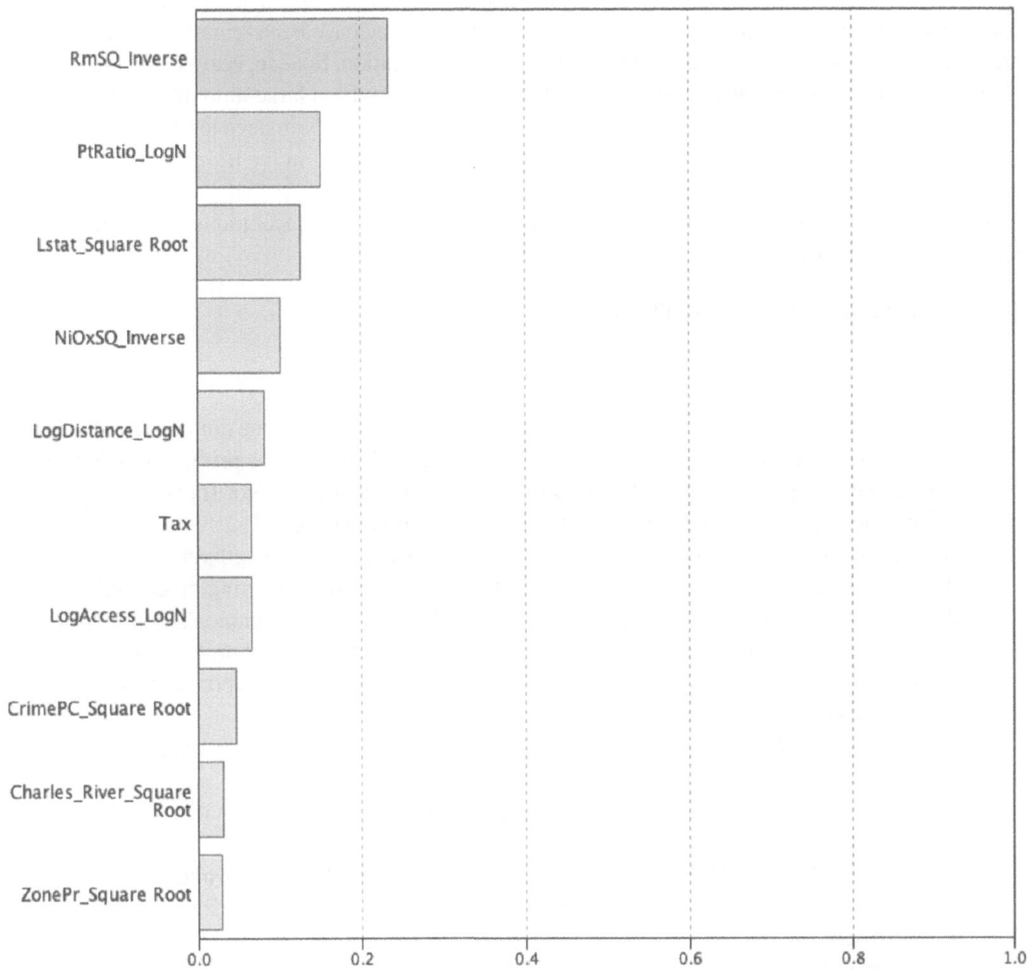

FIGURE 9.16 Variable importance chart.

Note that since we used normalized data, these variables have been transformed. Therefore, it's important to consider the transformation functions when interpreting the results.

CHAPTER SUMMARY

Neural networks, a popular method in data mining, are computational models inspired by the structure and function of biological neural networks. They consist of interconnected artificial neurons organized in layers. These networks process and transmit information from one layer to another to predict outputs based on the inputs. Neural network models can learn complex patterns and relationships in data, making them well-suited for many advanced applications in data mining. There are several types of neural networks, including FNNs, CNNs, RNN, LSTM networks, and RBFNs. In this book, we focus on the MLP and RBFN.

In this chapter, we explored the training process for neural networks. The core concept behind neural networks is to iteratively adjust the weights and biases of the nodes based on input data, enabling the network to learn and make predictions. This learning process is achieved through backpropagation, which computes the gradients of a loss function with respect to the network's parameters. By updating the weights and biases to minimize the loss function, neural networks iteratively enhance their predictive accuracy.

One of the strengths of neural networks is their ability to model complex, nonlinear relationships in data, making them applicable across a wide range of fields. They have been remarkably successful in areas such as computer vision, natural language processing, aviation, finance, healthcare, and robotics. However, neural networks also come with challenges. They require a large amount of labeled training data to perform well and are computationally intensive, particularly in deep learning applications. Overfitting is another concern, as models can become too specialized in the training data and fail to generalize well to new data.

Despite these challenges, neural networks remain a cutting-edge technique in artificial intelligence, driving innovation and transforming industries.

CHAPTER DISCUSSION QUESTIONS

1. What is a neural network, and how does it function?
2. What are the advantages and disadvantages of using neural networks?
3. Neural networks have been successful in various domains. Choose one domain and discuss how neural networks are effectively applied in that field. What are the primary challenges?
4. How do neural networks differ from traditional machine learning algorithms?
5. What is deep learning, and how does it relate to neural networks?
6. What are some common types of neural networks, and what are their strengths and weaknesses?
7. How do you train a neural network, and what are some common training processes?
8. What is a stopping rule in neural network training? What are some common stopping rules?
9. What are the differences between forward propagation and backward propagation?
10. The choice of activation function plays a crucial role in neural network performance. What are some common activation functions?
11. What is the **Multi-Layer Perceptron (MLP)** method? Describe the primary MLP network architectures.
12. What is the **Radial Basis Function Network (RBFN)** method? Describe the primary RBFN network architectures.
13. Compare **MLP** and **RBFN** models. Discuss their advantages and disadvantages.
14. Why is the **Local Optimum Situation** a concern in neural networks? How can this issue be addressed?
15. Interpretability is a critical concern in neural network models. How can the predictions and decision-making processes of a neural network be explained and interpreted?

CHAPTER EXERCISES

Exercise 1: Neural Network Models for the Bank Dataset

1. **Open the Bank Dataset**
   - Load the bank dataset in SPSS Modeler.
2. **Sample the Dataset**
   - Set **CRSCORE** (credit scores) as the target variable.
   - See the variable names and descriptions in Appendix 1.
3. **Partition the Data**
   - Split the data using a 60–20–20 ratio (train: 60%, test: 20%, validation: 20%).
4. **Data Modification**
   - Impute missing values where necessary.
5. **Configure Eight Neural Network Models**

Set up the following eight models for the dataset:

- **MLP 1 model**
  - Method: **MLP**
  - Number of nodes: **Auto selection**
  - Stopping rules: **Default**
- **MLP 2 model** (two hidden layers)
  - Method: **MLP**
  - Hidden layers: **2**
  - Hidden units: **10 per layer**
  - Stopping rules
    - Select **Use maximum training time** (set to 15 minutes)
    - Select **Customize number of maximum training cycles** (set to 100)
    - Select **Use minimum accuracy** (set to 90%)
  - Advanced
    - Set **Overfit prevention** to default (30%)
- **MLP 3 model** (two hidden layers)
  - Method: **MLP**
  - Hidden layers: **2**
  - Hidden units: **20 per layer**
  - Stopping rules
    - Select **Use maximum training time** (set to 15 minutes)
    - Select **Customize number of maximum training cycles** (set to 250)
    - Select **Use minimum accuracy** (set to 90%)
  - Advanced
    - Set **Overfit prevention** to default (30%)
- **MLP 4 model** (two hidden layers)
  - Method: **MLP**
  - Hidden layers: **2**
  - Hidden units: **50 per layer**
  - Stopping rules
    - Select **Use maximum training time** (set to 15 minutes)
    - Select **Customize number of maximum training cycles** (set to 500)
    - Select **Use minimum accuracy** (set to 90%)
  - Advanced
    - Set **Overfit prevention** to default (30%)

- **RBF 1 model** (one hidden layer)
  - Method: **RBF**
  - Number of nodes: **Auto selection**
- **RBF 2 model** (one hidden layer)
  - Method: **RBF**
  - Hidden units: **30**
- **RBF 3 model** (one hidden layer)
  - Method: **RBF**
  - Hidden units: **50**
- **RBF 4 model** (one hidden layer)
  - Method: **RBF**
  - Hidden units: **100**

Note: Refer to the examples in this chapter for additional configuration details.

6. **Compare Model Fit Metrics**
   - Compare the fit metrics between the models using **the** validation results.
7. **Select the Champion Model**
   - Identify the champion model and justify your selection with the appropriate evidence.
8. **Evaluate Reliability and Validity**
   - Assess the reliability and validity of the champion model.
9. **Present the Results of the Champion Model**
   - Summarize and present the performance of the champion model.
10. **Present Variable Importance Information**
    - Provide the variable importance chart and interpret the results.

Exercise 2: Neural Network Models for the Passengers Dataset

1. **Open the Passengers Dataset**
   - Load the passengers dataset in SPSS Modeler.
2. **Sample 50% of the Full Data**
   - Set **return** (decision to return to the airline) as the target variable and change its role to **binary**.
   - Refer to Appendix 2 for the variable names and descriptions.
3. **Partition the Data**
   - Split the data using a 60–20–20 ratio (train: 60%, test: 20%, validation: 20%).
4. **Data Modification**
   - Replace zero values in Likert-scale variables with missing values.
   - Impute missing values.
5. **Configure Eight Neural Network Models**

Set up the following eight models for the dataset:

- **MLP 1 model**
  - Method: **MLP**
  - Number of nodes: **Auto selection**
  - Stopping rules: **Default**
- **MLP 2 model** (two hidden layers)
  - Method: **MLP**
  - Hidden layers: **2**
  - Hidden units: **10 per layer**

- Stopping rules
  - Select **Use maximum training time** (set to 15 minutes)
  - Select **Customize number of maximum training cycles** (set to 100)
  - Select **Use minimum accuracy** (set to 90%)
- Advanced
  - Set **Overfit prevention** to default (30%)

- **MLP 3 model** (two hidden layers)
  - Method: **MLP**
  - Hidden layers: **2**
  - Hidden units: **20 per layer**
  - Stopping rules
    - Select **Use maximum training time** (set to 15 minutes)
    - Select **Customize number of maximum training cycles** (set to 250)
    - Select **Use minimum accuracy** (set to 90%)
  - Advanced
    - Set **Overfit prevention** to default (30%)
- **MLP 4 model** (two hidden layers)
  - Method: **MLP**
  - Hidden layers: **2**
  - Hidden units: **50 per layer**
  - Stopping rules
    - Select **Use maximum training time** (set to 15 minutes)
    - Select **Customize number of maximum training cycles** (set to 500)
    - Select **Use minimum accuracy** (set to 90%)
  - Advanced
    - Set **Overfit prevention** to default (30%)
- **RBF 1 model** (one hidden layer)
  - Method: **RBF**
  - Number of nodes: **Auto selection**
- **RBF 2 model** (one hidden layer)
  - Method: **RBF**
  - Hidden units: **30**
- **RBF 3 model** (one hidden layer)
  - Method: **RBF**
  - Hidden units: **50**
- **RBF 4 model** (one hidden layer)
  - Method: **RBF**
  - Hidden units: **100**

Note: Refer to the examples in this chapter for additional configuration details.

6. **Compare Model Fit Metrics**
   - Compare the fit metrics between the models using the validation results.
7. **Select the Champion Model**
   - Identify the champion model and justify your selection with the appropriate evidence.
8. **Evaluate Reliability and Validity**
   - Assess the reliability and validity of the champion model.
9. **Present the Results of the Champion Model**
   - Summarize and present the performance of the champion model.
10. **Present Variable Importance Information**
    - Provide the variable importance chart and interpret the results.

10 Ensemble Modeling

LEARNING OUTCOMES

Upon the completion of this chapter, you will be able to:

1. Describe the principles and concepts of ensemble methods in machine learning.
2. Explain the concept of diversity in ensemble modeling.
3. Differentiate between an ensemble model and an individual model (base learner).
4. Discuss the advantages and disadvantages of ensemble modeling compared to individual models.
5. Compare different types of ensemble methods, including bagging, boosting, and stacking.
6. Describe aggregation techniques, including averaging, voting, weighted voting, and maximum.
7. Develop and test ensemble models using SPSS Modeler.
8. Evaluate the prediction accuracy of ensemble models.

INTRODUCTION

In this chapter, we explore **ensemble modeling**, a powerful method in machine learning that enhances the accuracy, robustness, and interpretability of models, while also uncovering underlying patterns in the data. The core idea of ensemble modeling is that by combining the predictions of multiple models that each have varying levels of accuracy or error, we can reduce overall error and boost predictive performance. Ensemble modeling is versatile and can be applied to various machine learning models, including classification and association models. It is particularly beneficial for large, complex datasets where individual models may struggle to capture all patterns and relationships within the data.

Ensemble modeling also enhances the interpretability of machine learning models by offering deeper insights into the data's underlying patterns. By combining predictions from multiple models, we can identify the most important predictors for making accurate forecasts and gain a clearer understanding of how the models function. In this chapter, we will practice building ensemble models for data mining projects.

> **DEFINITION**
>
> Ensemble modeling is a machine learning method that involves combining the predictions of multiple models to generate a more accurate and robust overall prediction.

WHAT IS ENSEMBLE MODELING?

Ensemble modeling is based on the idea that combining the predictions of multiple diverse models can lead to better overall performance than relying on any single model alone. In simple terms, ensemble modeling is a

DOI: 10.1201/9781003468196-12

machine learning technique that involves aggregating the predictions of multiple individual models, known as base learners, to achieve a more accurate and robust prediction.

CONCEPT OF DIVERSITY

Diversity is a crucial concept in ensemble modeling, referring to the differences among the individual base learners that form the ensemble. It plays a key role in the effectiveness of the ensemble model. The primary objective is to ensure that the base learners' strengths and weaknesses complement one another, resulting in improved overall performance and robustness. Below are several aspects of diversity in ensemble modeling:

- **Data diversity**: Data diversity refers to using different subsets of the training data to train each base learner. For instance, in bagging-based methods like Random Forest, subsets of the training data are randomly selected with replacement to train individual decision trees. This diversity helps reduce overfitting and allows the ensemble to capture various aspects of the data distribution.
- **Algorithm diversity**: Using different machine learning algorithms as base learners is another way to introduce diversity. Ensemble models often consist of a combination of classifiers or regressors with different underlying principles. For example, a Gradient Boosting Machine (GBM) can be combined with a support vector machine (SVM) and a Neural Network, providing diverse perspectives to the ensemble.
- **Variable diversity**: Ensemble models benefit from variable diversity, which involves using different subsets of variables for training the base learners. This approach is commonly used in ensemble techniques like Random Subspace Method and Random Patches.
- **Configuration diversity**: Each base learner can be trained with different configurations, such as stepwise, number of hidden units, or tree depth. The ensemble model then leverages a range of hyperparameter configurations, which lead to a more comprehensive exploration of the model's parameter space.
- **Model architecture diversity**: In deep learning ensembles, diversity can be achieved by using various architectures or network structures with different depths, widths, or skip connections. Each architecture captures different hierarchical features, enhancing the ensemble's performance.
- **Training diversity**: Diversity can be introduced during the training process by using different random seeds or initialization methods for each base learner. This ensures that each model starts from a different point in the parameter space and explores different regions during optimization.
- **Domain diversity**: In some cases, domain-specific models or models trained on different datasets can be combined in an ensemble. This is particularly useful in transfer learning scenarios, where knowledge from one domain is leveraged to improve performance in another related domain.

By combining diverse base learners, ensemble models can collectively address different aspects of the data mining process and reduce individual model biases. Diversity is crucial for avoiding the problem of "groupthink," where multiple models end up making similar errors due to similar biases or limitations. Instead, a diverse ensemble fosters a more robust and accurate decision-making process, making ensemble modeling a powerful tool for tackling complex real-world challenges.

BASIC CONCEPTS OF ENSEMBLE MODELING

Ensemble modeling is a powerful machine learning technique that integrates the predictions of multiple individual models to enhance overall performance and robustness. Table 10.1 provides a summary of the fundamental concepts in ensemble modeling.

TABLE 10.1
Basic Concepts of Ensemble Modeling

Terms	Description
Ensemble	Ensemble is a machine learning method that involves combining the predictions of multiple models to produce a more accurate and robust prediction.
Diversity	The primary strength of ensemble models lies in the diversity of their individual models. By using different learning algorithms or training on different subsets of the data, the ensemble captures various patterns and reduces the risk of overfitting.
Base learners (individual models)	Base learners are the individual models that constitute the ensemble. Base learners can be any machine learning algorithm, such as decision trees, support vector machines, neural networks, etc.
Aggregation methods	Ensemble models combine the predictions of base learners using aggregation methods. Common aggregation methods include averaging, voting, and weighted voting. The aggregated result is often more accurate and reliable than the predictions of individual models.
Bagging	Bagging is an ensemble method that involves training multiple base learners independently on different random subsets of the training data (sampling with replacement). The final prediction is typically an average or voting over the predictions of all the base learners.
Boosting	Boosting is an iterative ensemble method where base learners are trained sequentially, and each learner focuses on correcting the mistakes made by its predecessors. Boosting assigns higher weights to misclassified instances, effectively giving more attention to challenging samples.
Stacking	Stacking (stacked generalization) is a more complex ensemble technique that combines the predictions of multiple base learners by training a meta-model on their outputs. The meta-model learns to weigh the predictions of individual models effectively.
Ensemble size	The number of base learners in an ensemble is an important consideration. Increasing the ensemble size can lead to better performance up to a point, after which the returns diminish, and the model may become computationally expensive.
Model selection and tuning	Ensemble models require careful selection and tuning of base learners and configurations. Different combinations of models and configurations can significantly impact the model performance.

ENSEMBLE METHODS

Now that we understand the purpose of ensemble modeling, how do we combine the predictions generated by multiple models? There are three primary methods for building ensemble models: **bagging**, **boosting**, and **stacking**.

Bagging

Bagging stands for bootstrap aggregating. Bootstrapping is a sampling technique in which subsets of the dataset are created using a replacement procedure, and aggregation combines the predictions from each model to generate a final result. In bagging, multiple models are created by training each one on a different subset of the training data, which is generated by randomly sampling the data with replacement. The final prediction is the average of all these individual model predictions.

Bagging is especially useful for reducing variance. Since each model in the ensemble is trained on a different subset of the data, they may have different biases and variances. By combining their predictions, the ensemble model produces a prediction with lower variance than any single model. Figure 10.1 illustrates how bagging works.

Boosting

Boosting is an ensemble technique that improves predictions by learning from previous predictor mistakes. In boosting, models are trained sequentially, with each model focusing on correcting

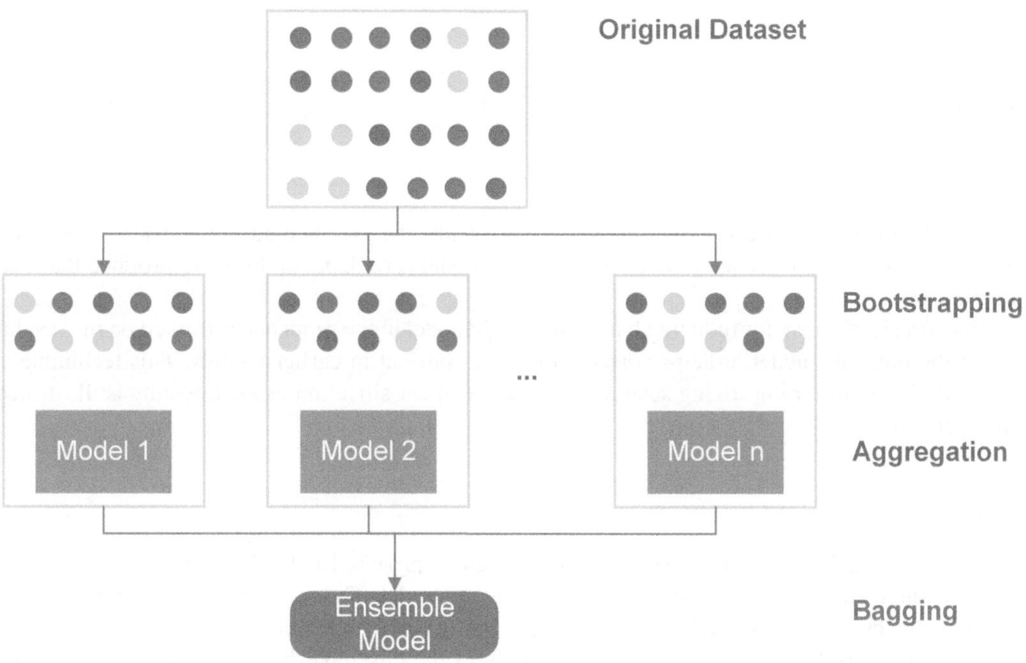

FIGURE 10.1 Bagging method.

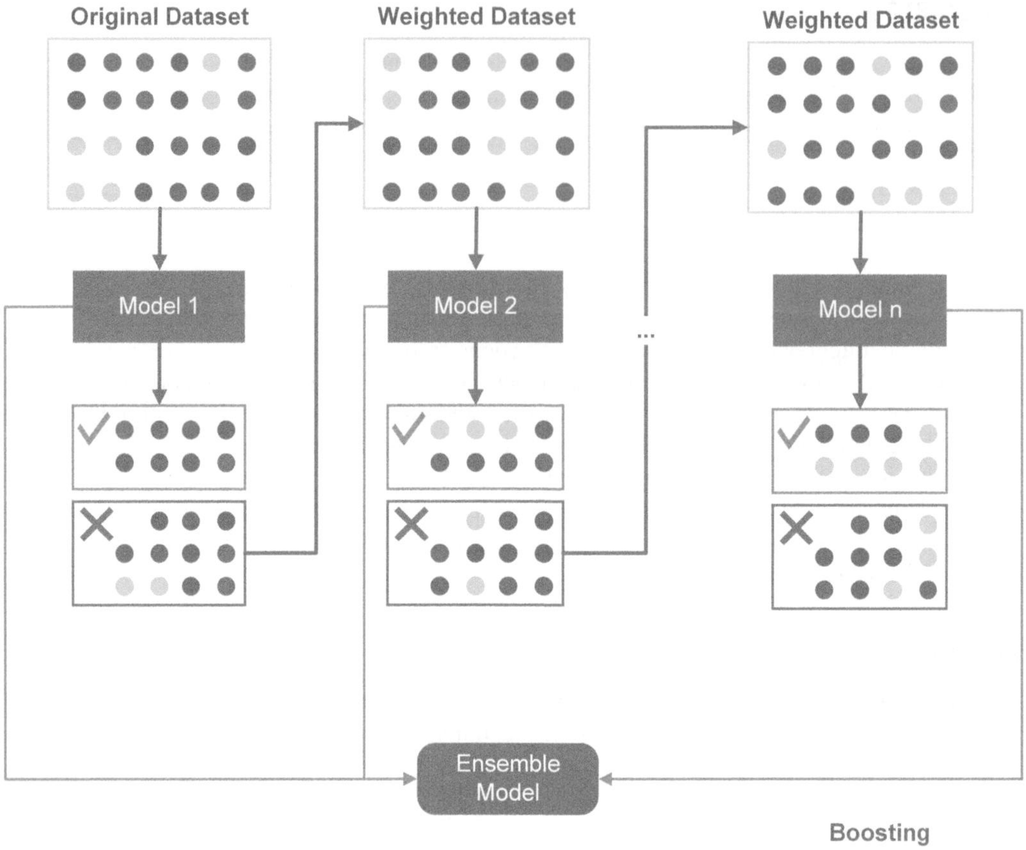

FIGURE 10.2 Boosting method.

the residual errors of the previous model. As a result, the errors made by earlier models are addressed by later ones. The predictions of all the models are then combined to produce the final output.

Boosting is effective for reducing bias. Since each model in the sequence is trained on the residuals of the previous model, it helps correct any biases present in earlier models. This technique is particularly useful for improving accuracy, especially in classification tasks. Boosting is illustrated in Figure 10.2.

Stacking

Stacking, or stacked generalization, takes a different approach. In stacking, multiple models are trained, and their predictions serve as inputs to a meta-model. The meta-model then learns how to combine the predictions of the base models to produce a final, improved prediction.

Stacking is useful when the individual models in the ensemble have different strengths and weaknesses. By combining their predictions and allowing the meta-model to determine the optimal weights for each model's output, the ensemble produces a more accurate final prediction. Figure 10.3 provides a visual explanation of stacking.

In summary, bagging focuses on reducing variance, boosting reduces bias, and stacking leverages the strengths of multiple models to create a more accurate final model.

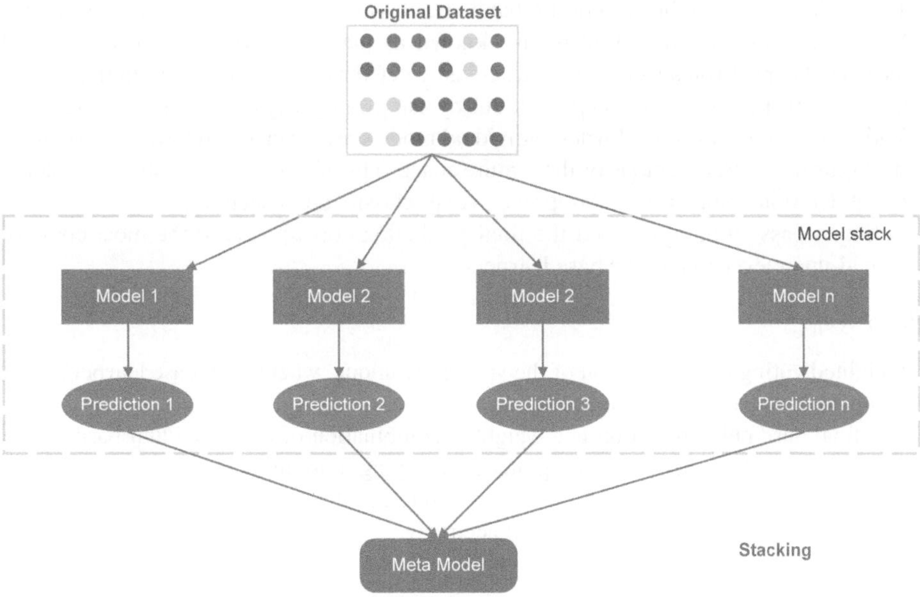

FIGURE 10.3 Stacking method.

AGGREGATION METHODS

Ensemble modeling involves combining the predictions of multiple base learners to obtain a more accurate and robust final prediction. In this process, aggregation methods are used to calculate or determine the final ensemble prediction from individual models. Some common aggregation methods used in ensemble modeling include averaging, voting, weighted voting, and maximum.

Averaging

- In averaging, the predictions of individual base learners are combined by taking their arithmetic mean (for association models) or their weighted mean (for classification models).
- For regression tasks, the ensemble prediction is the average of the individual predictions, resulting in a smoother and more stable final prediction.
- In probabilistic classification, averaging the probabilities of the level (of the categorical target variable) from different base learners helps in obtaining a more confident estimate of the event probabilities.
- It is particularly effective in reducing variance and improving model stability since the aggregated predictions tend to be less sensitive to outliers or noisy predictions from individual models.
- Averaging can be easily implemented and is computationally efficient since it involves simple arithmetic operations.

Voting

- Voting is used for classification tasks, where each base learner predicts the level (of the categorical target variable), and the most frequent prediction is chosen as the final ensemble prediction.

- This technique is suitable for non-probabilistic classifiers, such as decision trees or SVMs.
- In binary classification, voting is also known as majority voting, where the level that receives the most votes from base learners is selected as the ensemble prediction.
- It can handle ties or conflicting predictions by using a majority or weighted voting scheme.
- Voting ensembles work well when individual models are diverse and have complementary strengths in different regions of the feature space. This allows the ensemble to exploit different decision boundaries and improve overall classification accuracy.
- Voting is easy to interpret, and the final prediction corresponds to the most commonly agreed-upon level among the base learners.

Weighted Voting

- Weighted voting is an extension of the voting technique, where each base learner's prediction is assigned a weight based on its performance.
- The final ensemble prediction is a weighted combination of the individual base learners' predictions, with higher-performing models carrying more influence.
- Weighted voting allows the ensemble to emphasize the predictions of stronger models, potentially leading to improved overall accuracy.

Maximum

- In the maximum technique, the final ensemble prediction is obtained by selecting the maximum prediction from the individual models.
- It is commonly used when the ensemble models are non-probabilistic associations, such as decision trees or neural networks.
- For classification models, the maximum of the posterior probabilities (for categorical targets) is used as the final ensemble prediction.
- For association models, the maximum of the predicted values (for interval targets) from different models is used as the final ensemble prediction.
- The maximum technique is suitable for problems where the final prediction should be more conservative and tends to favor the most confident predictions from individual models.

These aggregation methods can be applied to various types of problems and base learners. Each of them has its strengths and applications in ensemble modeling. The choice of the appropriate method depends on the characteristics of the data, the nature of the individual models, and the desired properties of the ensemble model.

ADVANTAGES AND DISADVANTAGES OF ENSEMBLE MODELING

Ensemble modeling is a popular machine learning technique that combines multiple models to produce more accurate predictions. While ensemble modeling has several advantages, there are also some disadvantages that should be considered. In this section, we discuss both the advantages and disadvantages of ensemble modeling.

Advantages of Ensemble Modeling

- **Increased accuracy**: Ensemble modeling can improve the overall predictive accuracy of the model by combining the predictions of multiple models. By leveraging the strengths of each individual model while minimizing their weaknesses, ensemble modeling can produce more accurate predictions than any individual model.

- **Improved generalization**: By using ensemble modeling we can improve the generalization of the model by reducing overfitting. Since the ensemble model is made up of multiple models, it is less likely to be biased toward a particular type of data, making it more suitable for real-world applications where the data may be varied.
- **Robustness**: Ensemble modeling enhances the robustness of the model since it reduces the impact of noisy data. Since the models in the ensemble are trained on different subsets of the data, they may not be affected by the same noise in the data. This can help to produce more accurate predictions even when the data are noisy.
- **Versatility**: Ensemble modeling can be applied to a wide range of machine learning tasks, including association and classification models.

Disadvantages of Ensemble Modeling

- **Increased complexity**: Ensemble modeling can be more complex than training a single model. Since ensemble models involve multiple models, they may require more computing time to train.
- **Overfitting**: While ensemble modeling can reduce overfitting, it can also lead to overfitting if not done correctly. If the individual models in the ensemble are too similar or the ensemble is too complex, it may be prone to overfitting.
- **Interpretability**: Ensemble modeling can be less interpretable than individual models. Since the ensemble model is made up of multiple models, it is more difficult to interpret the contribution of each individual model to the final prediction.
- **Data requirements**: Ensemble modeling may require more data than individual models. Due to the combination of multiple models, we may need more data to train each individual model, which increases the data requirements for the overall model.

Overall, ensemble modeling is a powerful technique that can improve the accuracy and generalization of machine learning models. While there are some disadvantages to ensemble modeling, the benefits often outweigh the drawbacks, making it a popular choice for many machine learning applications.

ENSEMBLE MODELING IN SPSS MODELER

Data mining programs offer different methods for conducting ensemble modeling, with some providing more advanced features than others. In **SPSS Modeler**, there are two main ways to build ensemble models, which we will explore to determine which approach works best:

- **Using the ensemble node (in the Field Ops group)**: This node allows us to combine multiple models to create an aggregated predictive model. For classification tasks, we can choose the aggregation method (such as averaging or voting). However, we cannot select specific ensembling methods like boosting or bagging; we must use the default method set by the program.
- **Using Ensemble Features in Specific Modeling Nodes**: Certain modeling nodes, such as C&R Tree, CHAID, CR5.0, QUEST, and Neural Net, have built-in ensemble features. In these nodes, we can choose specific ensembling methods like boosting or bagging and select the appropriate aggregation method. In Chapters 14 and 15, we will also cover other ensemble algorithms for decision trees, including Random Forest and Gradient Boosting.

By using both approaches, we will compare their performance to identify which method produces better ensemble models.

> SPSS Modeler provides two approaches for building ensemble models:
>
> 1. **Using the Ensemble Node**: This node allows you to create ensemble models by combining the predictions from different base learners.
> 2. **Using the Ensemble Feature in Specific Modeling Nodes**: Certain modeling nodes, such as C&R Tree, CHAID, CR5.0, QUEST, and Neural Net, offer built-in ensemble options to enhance their performance.
>
> In this chapter, we will build ensemble models using both methods and compare their effectiveness to determine which approach works better. It's important to remember that there is no guarantee that the ensemble model will outperform all individual models, but it often provides more robust and accurate predictions.

EXAMPLE: A CLASSIFICATION MODEL USING ENSEMBLE NODE

To build an ensemble model for the **Airline Recommendation** dataset using the **Ensemble node** in SPSS Modeler, follow these steps:

- **Dataset:** AirlineRecommendation.sav
- **Target variable:** Recommendation (Flag: Yes/No)

Explore-Sample-Modify-Partition

Refer to the previous chapters for more details on these steps.

Ensemble Modeling

In this step, we use the **Ensemble node** to combine the best models we previously built using decision trees and neural network algorithms. Regression models were not chosen because the **Auto Data Prep node** transformed data for all regression models, resulting in different field names. For the decision tree and neural network models, only models without data transformation were used to meet SPSS Modeler's requirements for using the Ensemble node. Specifically, we selected the following models for ensembling:

- Four top tree models: **C&R Tree 2**, **CHAID 1**, C5.0 1, and **C5.0 2**.
- Four top neural network models: **MLP 1**, **MLP 2**, **MLP 3**, and **MLP 4**.

Then, we compare this ensemble model with individual models to determine which performs better.

Steps to Create the Ensemble Model:

1. Create a new stream and name it "**Airline_Ensemble1.**"
2. Open the "**Airline_Tree**" stream.
3. Copy the **data source** node, **Type** node**, Filler** node**, and Partition** node from the stream.
4. **Paste** these nodes into the stream "Airline_Ensemble1."
5. Copy the model nodes and nuggets for four models (**C&R Tree 2, CHAID 1, C5.0 1, and C5.0 2**) from the "**Airline_Tree**" stream. Ensure the model node is still connected to the nugget.
6. Paste those model nodes and nuggets into "Airline_Ensemble1."
7. Open the "**Airline_NN**" stream, and copy the model nodes and nuggets for the four neural network models (**MLP 1, MLP 2, MLP 3, and MLP 4**).
8. Paste these model nodes into the stream "Airline Ensemble1."

9. Connect the **Partition** node to the model node and the nugget for **C&R Tree 2**. Ensure the model node remains connected to the nugget.
10. Connect the **Partition** node to the model node **CHAID 1**, and connect the nugget for to the nugget for **C&R Tree 2** to the nugget for **CHAID 1**.
11. Repeat Step 10 to connect all model nodes and nuggets in the sequence: **C&R Tree 2 → CHAID 1 → C5.0 1→ C5.0 2 → MLP 1 → MLP 2 → MLP 3 → MLP 4**.

You should have a diagram with all the models connected in sequence.

12. Add an **Ensemble** node from the **Field Ops** group to the stream and connect it to the last nugget in the sequence, **MLP 4.** Open the node settings.

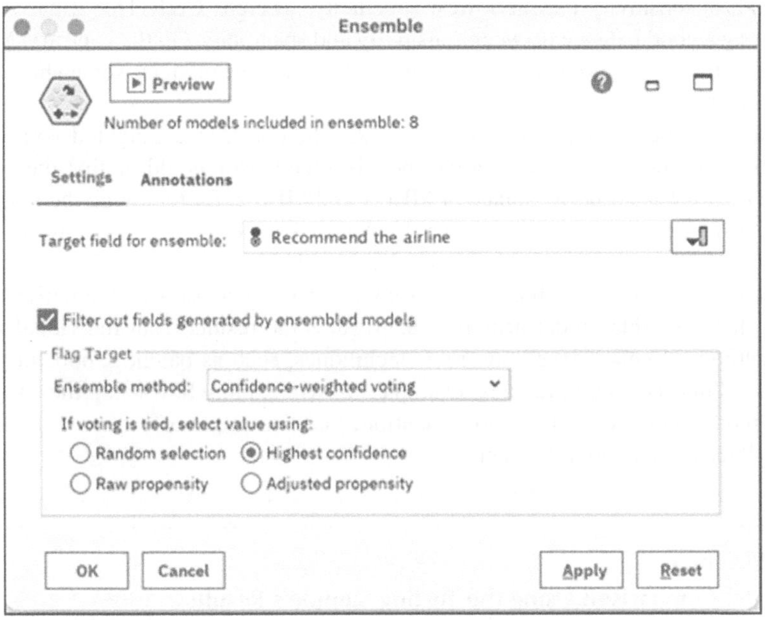

13. For **Target field for ensemble**, select the target field **Recommend the airline**.
14. For **Ensemble method**, select **Confidence-weighted voting**.
15. For **If voting is tied**, select **Highest confidence**.
16. Click Apply and OK.
17. Add an **Analysis** node from the **Output** group to the stream and connect it to the Ensemble node. Open the node settings.
18. Select the following options: **Coincidence matrices**, **Performance evaluation**, **Evaluation metric**, **Confidence figures**, and **Separate by partition**.
19. Click Apply and Run.

TABLE 10.2
Ensemble Model Prediction Accuracy

	Testing		Training		Validation	
Metrics	Number	Percentage	Number	Percentage	Number	Percentage
ACC	19,658	79.07	59,448	79.53	19,770	79.54
MR	5,204	20.93	15,299	20.47	5,085	20.46

This process will allow you to compare the ensemble model with individual models to determine which one performs better.

The results provide the fit metrics for the ensemble model. Table 10.2 presents the prediction accuracy (ACC) and misclassification rate (MR) of this model across the three sub-samples. As shown, the ensemble model does not demonstrate any significant improvement in prediction accuracy, with the ACC remaining slightly below 80% for all three samples.

Let's now compare the ensemble model's performance with two other models: the best decision tree model, C&R Tree 2, and the best neural network model, MLP 3. The comparison results are displayed in Table 10.3, which contains model fit metrics for the testing sample. Note that most of these metrics need to be calculated using the confusion matrix.

From the table, it is clear that C&R Tree 2 has the highest prediction accuracy (ACC), followed by the ensemble model, while MLP 3 ranks last. Looking at other metrics, the ensemble model shows the highest sensitivity but the lowest specificity, whereas C&R Tree 2 has the highest F1 score, indicating a good balance between sensitivity and specificity. On the contrary, MLP 3 has the highest Gini coefficient and ROC index. Therefore, the ensemble model is somewhere in the middle in terms of performance.

In this example, the ensemble model, which uses the confidence-weighted voting aggregation method, does not outperform the individual models. One reason could be that the voting process is influenced by the weaker performance of MLP 1 and MLP 2. By removing the nuggets for MLP 1 and MLP 2 from the stream and rerunning the Analysis node, we obtain a new ensemble model, called Ensemble 2. The fit metrics for this model are added to Table 10.3 for comparison. As we can see, Ensemble 2 shows a significant improvement and outperforms all the individual models.

Note that this ensemble model primarily aggregates the results from individual models using the voting method. More advanced ensemble techniques, such as bagging, boosting, or stacking, were not applied in this case. In the next example, we will explore how using the ensemble feature in a specific modeling node, such as those mentioned earlier, might affect model performance and whether it will lead to any improvement.

TABLE 10.3
Model Comparison Using the Testing Sample's Results

Metrics	Ensemble 1	Ensemble 2	C&R Tree 2	MLP 3
ACC	78.7%	79.41%	79.3%	78.5%
MR	21.3%	20.59%	20.7%	21.5%
Sensitivity or TPR	0.964	0.942	0.952	0.906
Specificity or TNR	0.589	0.629	0.615	0.649
F1	0.827	0.828	0.829	0.816
Gini	0.652	0.707	0.665	0.695
AUC (ROC Index)	0.826	0.853	0.832	0.848

EXAMPLE: A CLASSIFICATION MODEL USING THE ENSEMBLE FEATURE IN A TREE NODE

In this example, we will build several ensemble models for the same dataset using the ensemble feature within a tree node. This feature allows us to run an ensemble using either the boosting or bagging method (note that stacking is not available in SPSS Modeler). For each tree model, we can create two ensemble models: one using boosting and the other using bagging.

The following tree models, developed in Chapter 8, have been selected to create ensemble models based on their algorithms:

- C&R Tree 2 boosting
- C&R Tree 2 bagging
- CHAID 1 boosting
- CHAID 1 bagging
- C5.0 2 boosting (Note: The C5.0 node does not support the bagging method).

Steps:

1. Open the "**Airline_Tree**" stream.
2. Save it as "**Airline_Tree_Ensemble**."
3. Remove all non-selected tree nodes and keep only the nodes listed above.
4. Ensure that all configurations previously set up for these nodes in Chapter 8 **r**emain unchanged.

Once these steps are completed, you should have a stream that resembles the following structure:

- C&R Tree 2 boosting and C&R Tree 2 bagging nodes
- CHAID 1 boosting and CHAID 1 bagging nodes
- C5.0 2 boosting node

This will allow us to evaluate how the boosting and bagging ensemble methods improve the performance of these tree models.

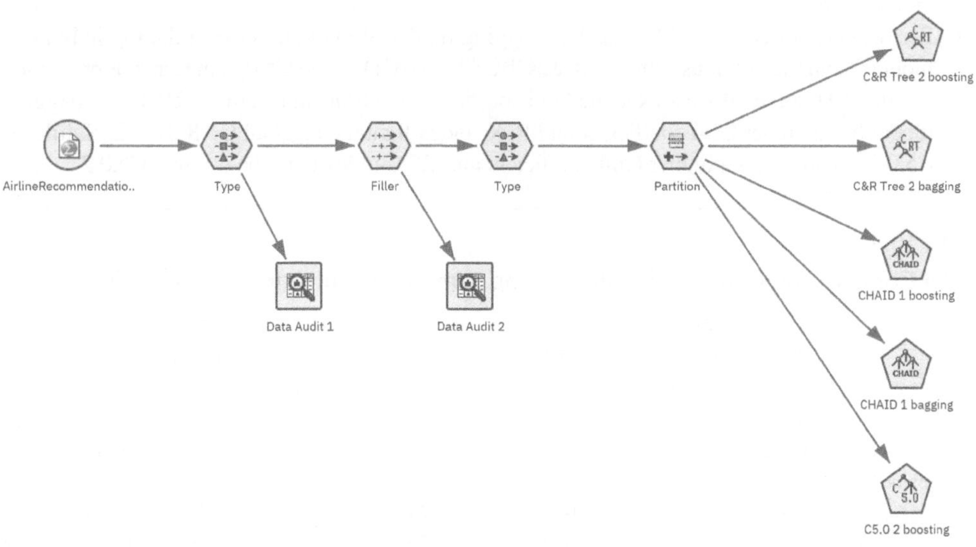

Follow the steps below to build the ensemble models:

- **To build a boosting ensemble model for C&R Tree 2 (boosting):**
  1. Open the node settings.
  2. Under the **Build Options** tab, in the **Objective** menu, select **Enhance model accuracy (boosting)**.
  3. In the **Ensembles** menu, set **Default combining rule for categorical targets** to **Highest probabilities win**. Set the **Number of component models** to **100**.
  4. Click Apply and Run.
- **To build a bagging ensemble model for C&R Tree 2 (bagging):**
  1. Open the node settings.
  2. Under the **Build Options** tab, in the **Objective** menu, select **Enhance model stability (bagging)**.
  3. In the **Ensembles** menu, set **Default combining rule for categorical targets** to **Highest probabilities win**. Set the **Number of component models** to **100**.
  4. Click Apply and Run.
- **To build ensemble models for CHAID 1 (boosting and bagging):** Follow the same steps as above for CHAID 1 boosting and bagging.
- **To build a boosting ensemble model for C5.0 2 (boosting):**
  1. Open the node settings.
  2. Under the **Model** tab, select **Use boosting**.
  3. Set **Number of trials** to **100**.
  4. Click Apply and Run.

Extracting Metrics

After building the models, extract the metrics from the Analysis nodes as done previously. Use the confusion matrix to calculate other required metrics. Table 10.4 shows the model comparison results based on the testing sample's results.

- Bagging vs. Boosting: For the same tree type, bagging models have higher ACC (accuracy) than boosting models. Notably, C&R Tree 2 bagging has the highest ACC.
- Other Metrics: The CHAID 1 bagging model has the highest F-1 score, Gini coefficient, and ROC index.
- Performance of Bagging Models: The bagging models do not outperform the top individual models from the previous example (C&R Tree 2, CHAID 1, and C5.0 2) in terms of prediction accuracy. However, if we look at the Gini coefficient and ROC index, the C&R Tree 2 bagging model has a higher Gini coefficient and ROC index than the original C&R Tree 2. Similarly, C5.0 2 boosting has a higher Gini coefficient and ROC index than the original C5.0 2.

TABLE 10.4
Model Comparison Results Using the Testing Sample's Results (for Tree Models)

Metrics	C&R Tree 2	C&R Tree 2 Boosting	C&R Tree 2 Bagging	CHAID 1	CHAID 1 Boosting	CHAID 1 Bagging	C5.0 2	C5.0 2 Boosting
ACC	79.3%	77.89%	79.22%	79.0%	76.36%	79.08%	79.2%	78.65%
MR	20.7%	22.11%	20.78%	21.0%	23.64%	20.92%	20.8%	21.35%
Sensitivity or TPR	0.952	0.872	0.913	0.924	0.823	0.935	0.949	0.934
Specificity or TNR	0.615	0.675	0.657	0.641	0.697	0.629	0.617	0.619
F1	0.829	0.806	0.823	0.823	0.786	0.825	0.828	0.823
Gini	0.665	0.677	0.683	0.700	0.688	0.700	0.618	0.697
AUC (ROC index)	0.832	0.839	0.842	0.850	0.844	0.850	0.808	0.849

This comparison highlights that bagging can enhance certain model metrics, such as the Gini coefficient and ROC index, even if it does not always improve overall accuracy.

Model Nugget Summary for Ensemble

When we open the model nugget for the ensemble models, we can view a summary that provides a snapshot of the ensemble's quality. For example, Figure 10.4 shows the model summary for the C&R Tree 2 bagging model. The chart presents the accuracy of the final ensemble model compared to two other benchmarks: a reference model and a naive model.

- In **bagging ensembles**, the reference model is the standard model constructed using the entire training partition.
- In **boosted ensembles**, the reference model refers to the first component model in the ensemble.
- The **naive model** indicates the accuracy that would be achieved if no model was created and all records were assigned to the most frequent category. (Note: The naive model is not calculated for continuous targets.)

In this example, the ensemble model has a higher accuracy than the reference model, and it is selected as the best model. It is important to note that this chart shows the accuracy based on the training sample's results, which may explain any discrepancy with the outputs in Table 10.4 (which are based on the testing sample).

Figure 10.5 displays the detailed performance of all models built during the training process. Since we configured 100 component models, this chart shows the performance metrics for each of these models, including:

- Accuracy of each model
- Number of predictors used
- Number of nodes in each model
- Training sample size

The figure presents a snapshot of these results, and you can scroll down on your computer to view the details of all 100 component models used in the ensemble. This comprehensive view helps in assessing the progression and contribution of each model within the ensemble.

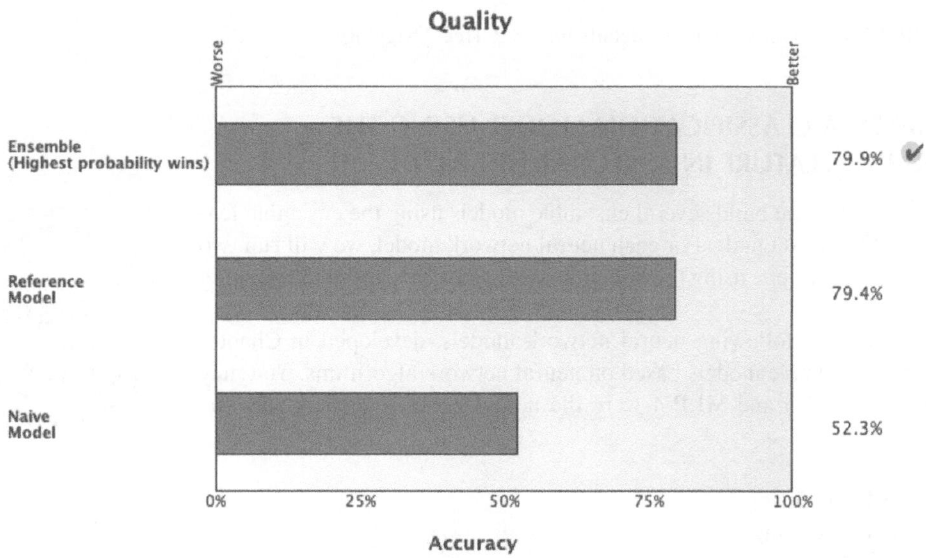

FIGURE 10.4 Model summary with the ensemble results for C&R Tree 2 bagging.

Component Model Details

Model	Accuracy	Method	Predictors	Model Size (Nodes)	Records
1	79.5%		17	41	74,747
2	79.4%		17	43	74,747
3	79.5%		16	35	74,747
4	79.5%		15	37	74,747
5	79.5%		16	37	74,747
6	79.6%		16	41	74,747
7	79.5%		16	39	74,747
8	79.6%		17	53	74,747
9	79.6%		17	51	74,747
10	79.5%		16	45	74,747

FIGURE 10.5 Component model details for C&R Tree 2 bagging.

EXAMPLE: A CLASSIFICATION MODEL USING THE ENSEMBLE FEATURE IN A NEURAL NET NODE

In this example, we build several ensemble models using the ensemble feature in a Neural Net node. For each neural network model, we will run two ensemble models: one using the boosting method and one using the bagging method.

I selected the following neural network models, developed in Chapter 9, to build ensemble models based on neural network algorithms. You may recall that MLP 3 and MLP 4 were the neural network models with the highest accuracy.

- MLP 3 boosting
- MLP 3 bagging
- MLP 4 boosting

- MLP 4 bagging
- MLP 5 boosting
- MLP 5 bagging

Open the "**Airline_NN**" stream and save it as "**Airline_NN_Ensemble**." Then, remove all non-selected tree nodes and retain only the Neural Net nodes listed above. Keep all configurations we previously set up for these neural network nodes in Chapter 9. The resulting stream should look like the one shown below.

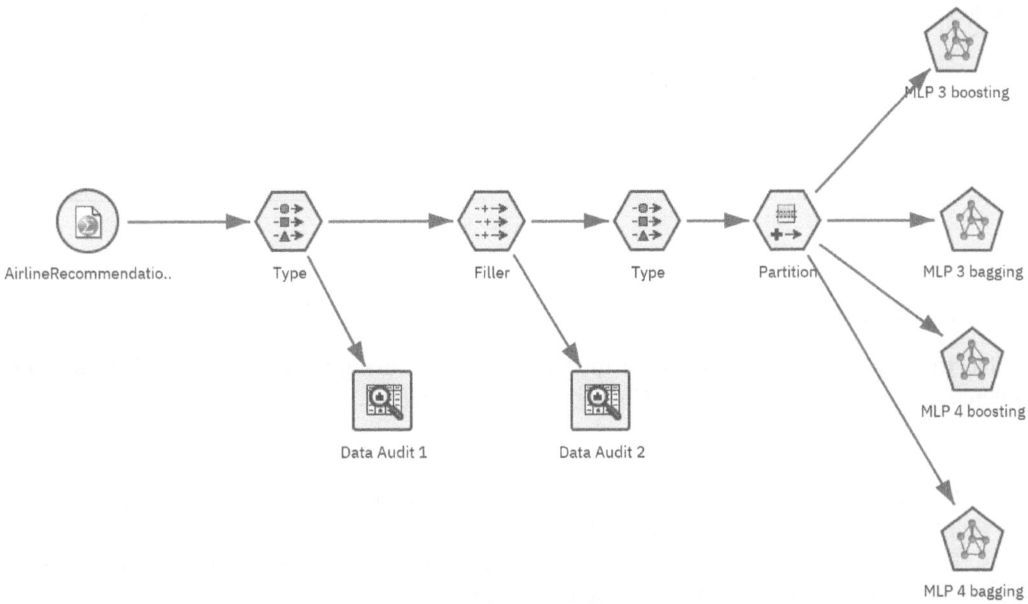

Follow the steps below to build the ensemble models. Be sure to keep the same configurations for the hidden layers, hidden nodes, and interactions as were set in Chapter 9 for these nodes.

- To build a boosting ensemble model for MLP 3 (boosting):
  1. Open the node settings.
  2. Under the **Build Options** tab, in the **Objective** menu, select **Enhance model accuracy (boosting)**.
  3. In the **Ensembles** menu, set **Default combining rule for categorical targets** to **Highest probabilities win**. Set the **Number of component models** to **20**.
  4. Click Apply and Run.
- To build a bagging ensemble model for MLP 3 (bagging):
  1. Open the node settings.
  2. Under the **Build Options** tab, in the **Objective** menu, select **Enhance model stability (bagging)**.
  3. In the **Ensembles** menu, set **Default combining rule for categorical targets** to **Highest probabilities win**. Set the **Number of component models** to **20**.
  4. Click Apply and Run.

Repeat these steps to set up both boosting and bagging models for MLP 4 and MLP 5. After setting up the models, run all the neural network nodes and add separate Analysis nodes to generate model fit results.

TABLE 10.5
Model Comparison Using the Testing Sample's Results (for Neural Network Models)

Metrics	MLP 3	MLP 3 Boosting	MLP 3 Bagging	MLP 4	MLP 4 Boosting	MLP 4 Bagging	MLP 5	MLP 5 Boosting	MLP 5 Bagging
ACC	78.47%	79.01%	78.67%	78.14%	78.35%	76.82%	78.47%	79.01%	78.65%
MR	21.53%	20.99%	21.33%	21.86%	21.65%	23.18%	21.53%	20.99%	21.35%
Sensitivity or TPR	0.906	0.913	0.903	0.905	0.896	0.885	0.906	0.913	0.902
Specificity or TNR	0.649	0.653	0.657	0.643	0.657	0.638	0.649	0.653	0.658
F1	0.816	0.821	0.817	0.814	0.814	0.801	0.816	0.821	0.817
Gini	0.695	0.698	0.692	0.687	0.692	0.661	0.695	0.698	0.693
AUC (ROC index)	0.848	0.849	0.846	0.844	0.846	0.830	0.848	0.849	0.847

Model Comparison

Table 10.5 presents the comparison of model performance using the testing sample's results. For reference, the original models from Chapter 9 are also included.

As you can see:

- For MLP 3 and MLP 5, both the bagging and boosting models show improvements across all metrics compared to the original models.
- For MLP 4, only the boosting model shows some improvement, while the bagging model underperforms compared to the original model.
- Interestingly, in the cases of MLP 3 and MLP 5, the boosting model outperforms the bagging model.
- Notably, MLP 3 boosting and MLP 5 boosting exhibit identical metrics, which aligns with the results we observed in Chapter 9.

This comparison highlights the advantages of using boosting in certain scenarios and emphasizes that ensemble methods can significantly improve model performance, although outcomes may vary depending on the specific model and method used.

Figure 10.6 presents the model summary with ensemble results for MLP 3 boosting. The results indicate that the ensemble model outperforms the reference model and has been selected as the best model. It's important to remember that these results are based on the training sample's performance, which explains the discrepancy when compared to the outputs shown in Table 10.5 (which are based on the testing sample's results).

Variable Importance

Suppose we decide to use MLP 3 boosting as our predictive model. Figure 10.7 shows the variable importance chart, highlighting that travel purpose is the most important predictor, followed by gender, average number of flights, and airline status. On the contrary, flight cancellation and price sensitivity are among the least important predictors. This variable importance ranking provides valuable insights into which factors most significantly influence the model's predictions.

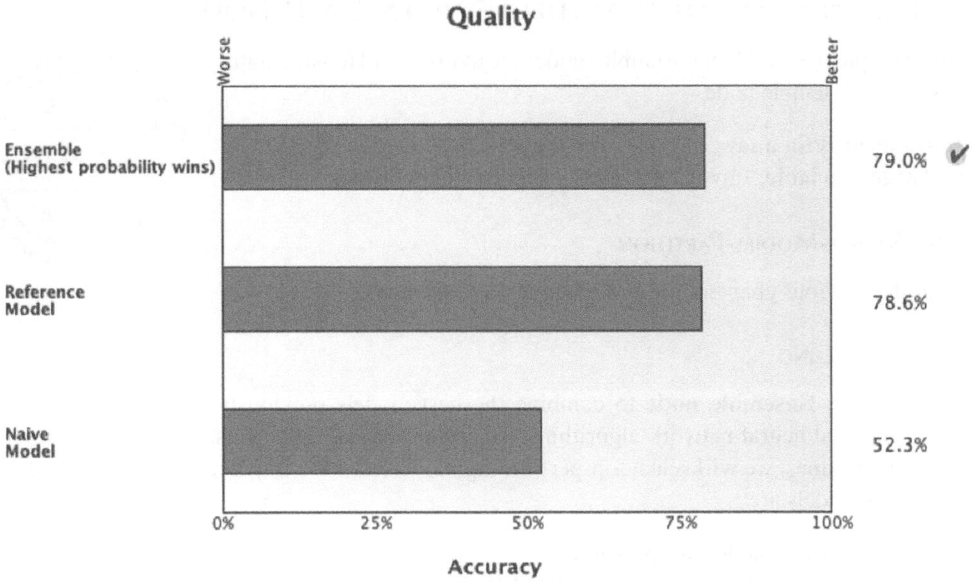

FIGURE 10.6 Model summary with the ensemble results for MLP 3 boosting.

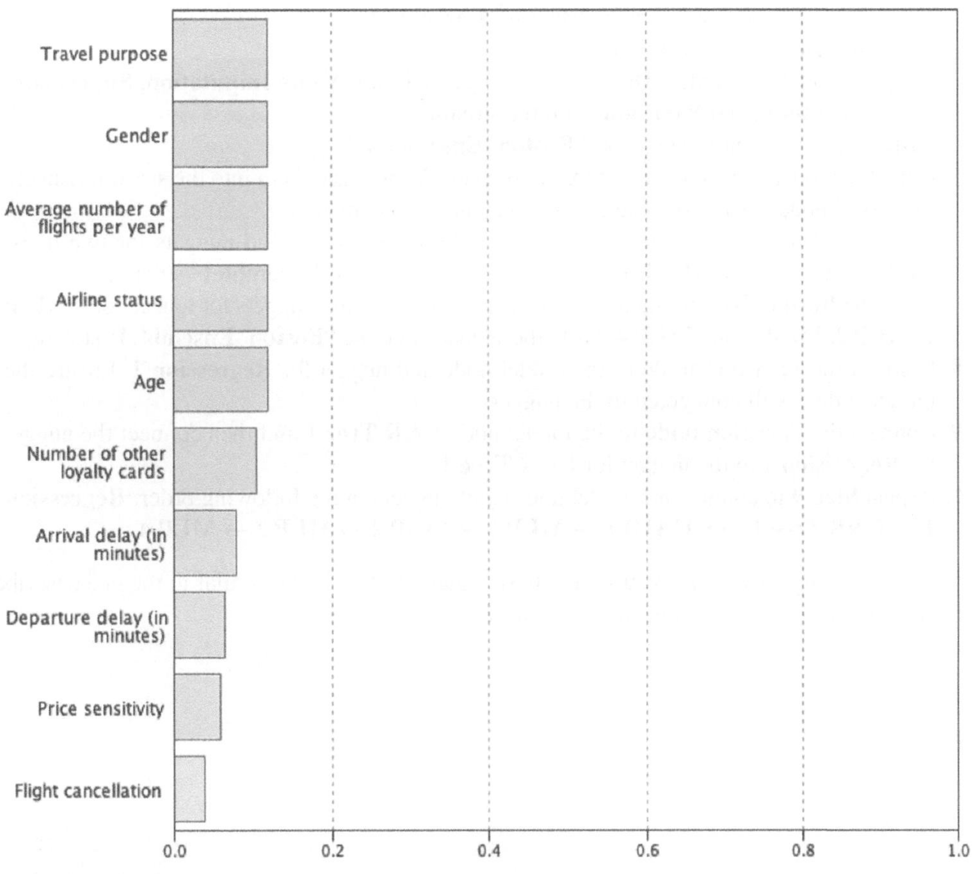

FIGURE 10.7 Variable importance chart—MLP 3 boosting.

EXAMPLE: AN ASSOCIATION MODEL USING ENSEMBLE NODE

In this example, we build an ensemble model for the Boston Housing dataset using the Ensemble node.

- **Dataset:** Boston.sav
- **Target variable:** Imv (Continuous—Median home value)

EXPLORE-SAMPLE-MODIFY-PARTITION

Refer to the previous chapters for more details on these steps.

ENSEMBLE MODELING

We will use the **Ensemble node** to combine the best models previously built using regression, decision tree, and neural network algorithms. To ensure that all models used for ensembling have identical field names, we will select top-performing models that use imputed and transformed data. The selected models are:

- Best regression model: **Regression 1**
- Two tree models: **C&R Tree 1** and **CHAID 1**
- Four neural network models: **MLP 1**, **MLP 2**, **MLP 3**, and **MLP 4**

Steps to Create the Ensemble Model:

1. Create a new stream and name it "**Boston_Ensemble1**."
2. Open the "**Boston_Regression**" stream.
3. Copy the following nodes: **Data source**, **Type**, **Missing Value Imputation**, **Supernode-Transformation**, and **Partition** from the stream.
4. Paste these nodes into the stream "**Boston_Ensemble1**."
5. Copy the model node and nugget for **Regression 1** and paste them into the stream. (Ensure the model node is still connected to the nugget when copied).
6. Open the "**Boston_Tree**" stream, and copy the model nodes and nuggets for two trees: **C&R Tree 1** and **CHAID 1**. Paste them into the "**Boston_Ensemble1**" stream.
7. Open the **Boston_NN** stream and copy the model nodes and nuggets for four models: **MLP 1**, **MLP 2**, **MLP 3**, and **MLP 4**. Paste these nodes into the "**Boston_Ensemble1**" stream.
8. Connect the **Partition node** to the model node and nugget for **Regression 1**. Ensure the model node is still connected to the nugget.
9. Connect the **Partition node** to the model node **C&R Tree 1** and then connect the nugget for **Regression 1** to the nugget for **C&R Tree 1**.
10. Repeat **Step 9** to connect all model nodes and nuggets in the following order: **Regression 1 → C&R Tree 1 → CHAID 1 → MLP 1 → MLP 2 → MLP 3 → MLP 4**.

After completing these steps, you should have a diagram that looks similar to the one described, with all the selected models connected in sequence.

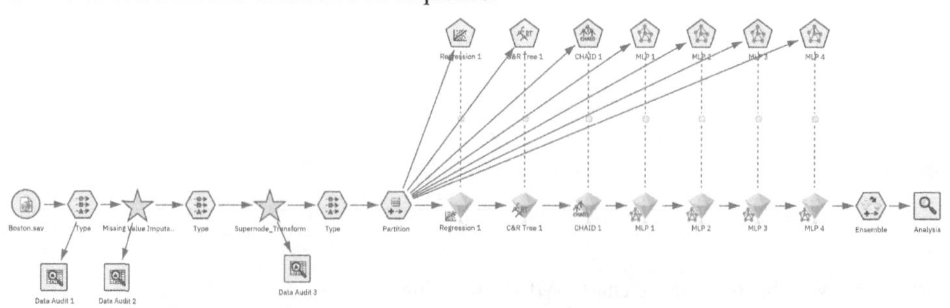

TABLE 10.6
Model Comparison for the Boston Housing (Using The Testing Sample's Results)

Models	MAE (Testing)	Linear Correlation (Testing)
Ensemble 1	2.495	0.882
Ensemble 2	2.331	0.941
Regression 1	3.437	0.783
C&R Tree 1	4.014	0.696
CHAID 1	3.168	0.781
MLP 1	2.712	0.864
MLP 2	2.515	0.889
MLP 3	2.418	0.89
MLP 4	2.693	0.877

11. Add an **Ensemble** node from the **Field Ops** group to the stream and connect it to the last nugget in the sequence **MLP 4**. Open the node setting.
12. For **Target field for ensemble**, select the target field **Imv**.
13. Click Apply and OK.
14. Add an **Analysis** node from the **Output** group to the stream and connect it to the Ensemble node. Open the node setting.
15. Select **Performance evaluation**, **Confidence figures**, and **Separate by partition**.
16. Click Apply and Run.

Model Comparison

The results provide us with the fit metrics for the ensemble model, which we will refer to as Ensemble 1. Table 10.6 shows the fit metrics of the ensemble model, specifically Mean Absolute Error (MAE) and linear correlation, compared to the individual models. As observed, the ensemble model outperforms most of the individual models, except for MLP 3, which remains the champion neural network model.

One possible explanation for this is that the ensemble node combines the models using the average approach, and the outcome may have been influenced by the weaker performance of Regression 1 and C&R Tree 1. If we remove the nuggets for Regression 1 and C&R Tree 1 from the stream and run the process again, we will generate an improved ensemble model, Ensemble 2, with an MAE of 2.331 and a linear correlation of 0.941.

As shown, Ensemble 2 has the best fit metrics compared to all other individual models, confirming its superior performance.

EXAMPLE: AN ASSOCIATION MODEL USING THE ENSEMBLE FEATURE IN DECISION TREE AND NEURAL NET NODES

In this example, we will build several ensemble models using the ensemble feature in a Tree node and a Neural Net node. For each decision tree or neural network model, we will create two ensemble models: one using the boosting method and one using the bagging method.

We selected the following tree and neural network models developed in Chapters 8 and 9 to build ensemble models based on the algorithms configured in those chapters. Since we will create two ensemble models for each node, we selected the best tree model, CHAID 1, and the two best neural network models, MLP 2 and MLP 3.

The selected models are:

- CHAID 1 boosting
- CHAID 1 bagging
- MLP 2 boosting
- MLP 2 bagging
- MLP 3 boosting
- MLP 3 bagging
- MLP 4 boosting
- MLP 4 bagging

Steps:

1. Open the "**Boston_NN**" stream and save it as "**Boston_Ensemble1**."
2. Remove all non-selected nodes and only keep the neural network nodes listed above.
3. Copy the **CHAID 1 node** from the "**Boston_Tree**" stream and paste it into the "**Boston_Ensemble1**" stream.
4. Ensure that all configurations set up for these nodes in Chapters 8 and 9 remain unchanged.

Once completed, you should have a stream that looks like the one described, containing the selected decision tree and neural network nodes ready for ensemble model building using boosting and bagging methods.

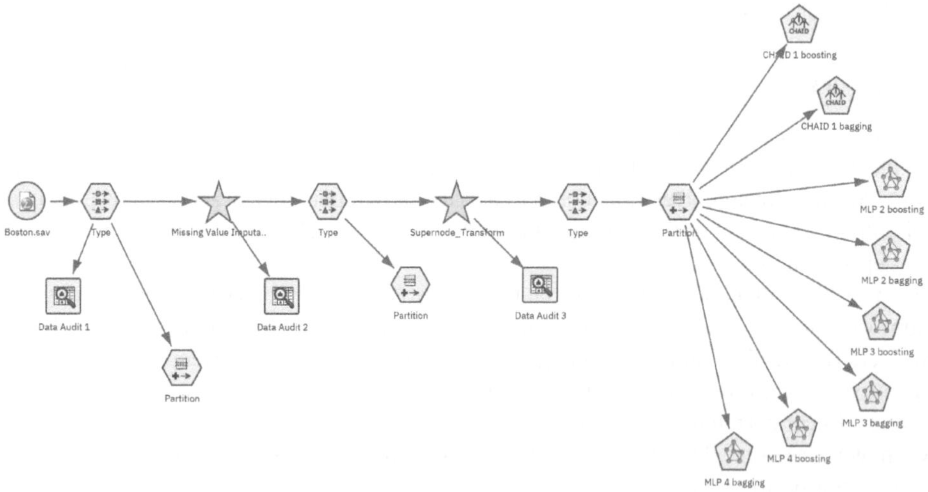

Steps to Build Ensemble Models

Ensure that the hidden layers, hidden nodes, and interactions are kept consistent with the configurations set in Chapter 9 for the following nodes.

- To build a boosting ensemble model for CHAID 1 (boosting):
  1. Open the node settings.
  2. Under the **Build Options** tab, in the **Objective** menu, select **Enhance model accuracy (boosting)**.

3. In the **Ensembles** menu, set **Default combining rule for continuous targets** to **Mean**. Set the **Number of component models** to **100**.
4. Click Apply and Run.

- To build a bagging ensemble model for CHAID 1 (bagging):
  1. Open the node settings.
  2. Under the **Build Options** tab, in the **Objective** menu, select **Enhance model stability (bagging)**.
  3. In the **Ensembles** menu, set **Default combining rule for continuous targets** to **Mean**. Set the **Number of component models** to **100**.
  4. Click Apply and Run.

Follow these same steps to set up boosting and bagging models for MLP 2, MLP 3, and MLP 4. Once all models are configured and run, add separate Analysis nodes to generate the model fit results.

Model Comparison

Table 10.7 presents a comparison of the models. Note that the accuracy, measured as the Explanatory Power (EP) for association models, is reported for the training sample only. The accuracy for the original CHAID 1 is not available since the tree node does not produce this output. The original models from Chapters 8 and 9 are also included for reference.

As shown:

- In all cases, the ensemble models outperform the original models.
- Among the ensemble models, the boosting models consistently show better fit metrics than the bagging models.
- Based on the testing sample's results, the MLP boosting model is the best, exhibiting the lowest MAE and the second highest accuracy based on the training sample's results.
- The quality chart (Figure 10.8) for this model also indicates that it outperforms the reference model, making it the best model overall.

This shows the effectiveness of using boosting and bagging ensemble methods to improve model performance in association tasks.

TABLE 10.7
Model Comparison for the Boston Housing

Models	EP (%) (Training)	MAE (Testing)	Linear Correlation (Testing)
CHAID 1 boosting	93.3	2.558	0.859
CHAID 1 bagging	91.2	2.614	0.851
CHAID 1		3.168	0.781
MLP 2 boosting	96.4	2.006	0.897
MLP 2 bagging	91.5	2.079	0.905
MLP 2	84.0	2.515	0.889
MLP 3 boosting	96.9	2.027	0.892
MLP 3 bagging	91.5	2.069	0.892
MLP 3	85.7	2.418	0.89
MLP 4 boosting	97.0	2.087	0.892
MLP 4 bagging	93.2	2.381	0.873
MLP 4	87.0	2.693	0.877

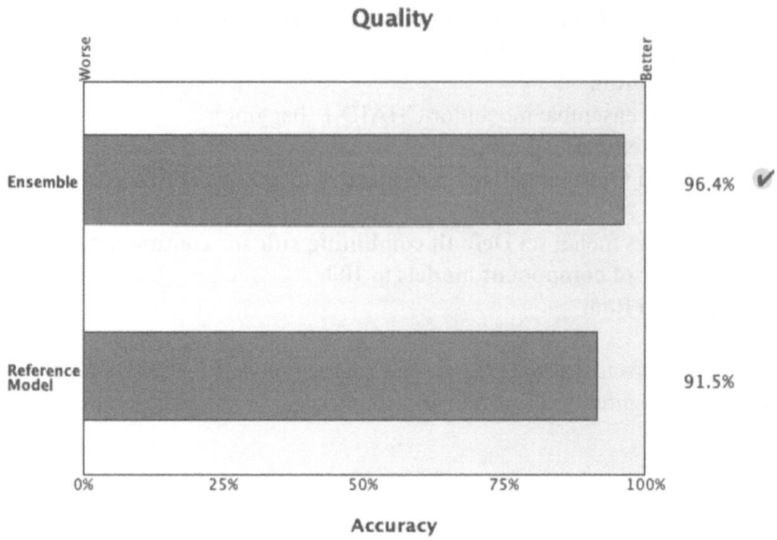

FIGURE 10.8 Model summary with the ensemble results for MLP 2 boosting.

Variable Importance

Figure 10.9 presents the variable importance chart for this model. It indicates that **Lsat (square root)** is the most important predictor, followed by **ZonePr (square root)**, **Tax**, and **CrimePC (square root)**. On the other contrary, **Charles_River (square root)** and **LogDistance (log N)** are among the least important predictors. This ranking of variable importance helps us understand which features contribute the most to the model's predictions, allowing for more informed interpretation of the results.

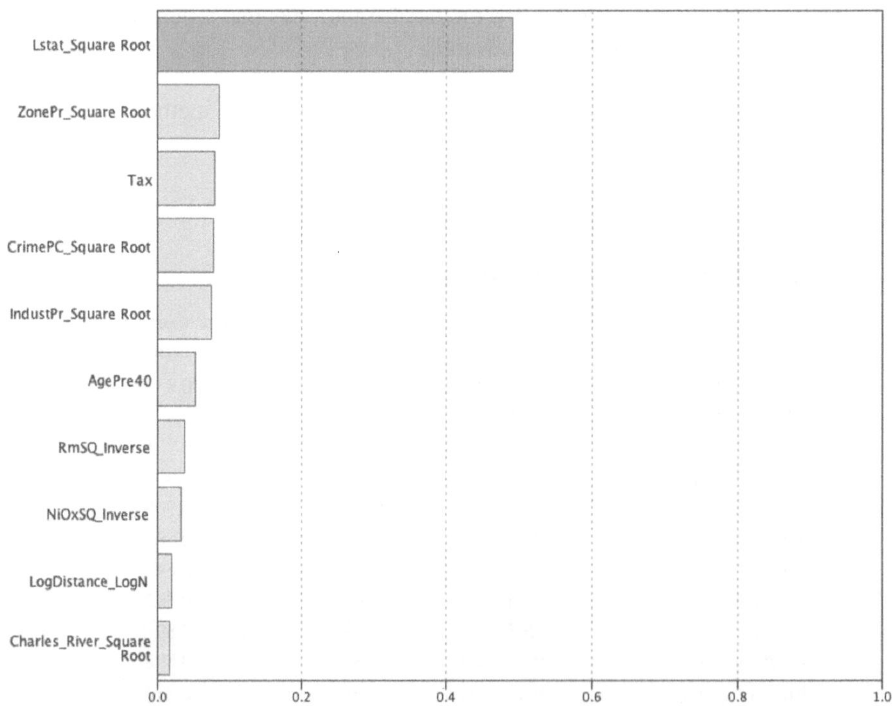

FIGURE 10.9 Variable importance chart—MLP 2 boosting.

CHAPTER SUMMARY

Ensemble methods involve combining multiple models to create a more accurate and robust predictive model. The core idea is that the collective predictions of several models can outperform any individual model. There are various ensemble techniques, including bagging, boosting, and stacking.

- **Bagging methods** create an ensemble by training multiple models independently on different subsets of the training data and then aggregating their predictions.
- **Boosting methods** iteratively train weak learners, focusing on correcting the errors made by the previous models.
- **Stacking** combines the predictions of multiple models as input to a higher-level model, often called a **meta-learner**.

Ensemble methods offer several advantages as follows:

1. **Improved generalization**: By combining diverse models, ensemble methods reduce overfitting and capture a broader range of data patterns, lowering prediction variance.
2. **Robustness**: Ensembles are more resilient against noise and outliers. The aggregated predictions can help mitigate errors caused by noisy data.
3. **Versatility**: They can be applied to various machine learning tasks, such as classification, association, and anomaly detection. Their success spans across domains like finance, healthcare, and natural language processing.

However, it is important to remember that ensemble models do not always outperform individual models. Their success depends on the specific data and the goal of the data mining project. As previously discussed, there is no perfect method; the best model is always relative to the data. Therefore, it is essential to include ensemble modeling in the process to determine if it is beneficial for a given task. Additionally, ensemble modeling requires careful tuning of configurations, especially when combining models from different machine learning methods.

In conclusion, while ensemble modeling is not always guaranteed to be the best approach, it remains a powerful and viable option to enhance prediction accuracy and reliability when building multiple predictive models.

CHAPTER DISCUSSION QUESTIONS

1. How do ensemble methods differ from individual models in terms of predictive performance and generalization capability?
2. What are the advantages and disadvantages of using ensemble methods compared to single models?
3. Can ensemble methods handle imbalanced datasets effectively? If yes, how do they address the issue of class imbalance?
4. Discuss the concept of model diversity in ensemble methods. How does diversity contribute to the overall performance of the ensemble?
5. What are some common methods for generating diversity in ensemble methods, such as bagging, boosting, and stacking? How do these techniques enhance the ensemble's predictive power?
6. In what scenarios would ensemble methods be particularly beneficial and suitable for a data mining project? Are there any specific domains or types of problems where ensemble methods have shown significant success?
7. Discuss the computational complexity of ensemble methods. Are there any techniques or optimizations that can be employed to improve their efficiency for large datasets?

8. Explain the concept of ensemble size and its impact on the performance of ensemble methods. How does the number of models in the ensemble affect prediction accuracy and computational resources?
9. What are some potential challenges or limitations of ensemble methods that researchers and practitioners should be aware of? How can these challenges be addressed or mitigated in practice?

CHAPTER EXERCISES

Exercise 1: Ensemble Models for the Bank Dataset

1. Open the **bank** dataset in SPSS Modeler.
2. Sample the entire dataset. Refer to Appendix A for variable names and descriptions.
3. Partition the data:
   - Split the data using a 60–20–20 ratio (train: 60%, test: 20%, validation: 20%).
4. Perform data modification as follows:
   - Impute missing values.
   - Normalize the data.
5. Set **CRSCORE (credit scores)** as the target variable.
6. Build an ensemble model **using the** Ensemble node for seven individual models. Configure it as follows:
   - **Linear regression** (from Chapter 7 Exercise 1)
   - **Pruned CHAID tree** (from Chapter 8 Exercise 1)
   - **Pruned C5.0 tree** (from Chapter 8 Exercise 1)
   - **MLP 3 model** (from Chapter 9 Exercise 1)
   - **MLP 4 model** (from Chapter 9 Exercise 1)
   - **RBF 3 model** (from Chapter 9 Exercise 1)
   - **RBF 4 model** (from Chapter 9 Exercise 1)
7. Compare the ensemble model **with the seven individual models using the** test results. **Determine which model has the** best performance. Interpret the variable importance results.
8. Build six ensemble models using the Ensemble feature in the Decision Tree and Neural Net nodes as follows:
   - CHAID boosting
   - CHAID bagging
   - MLP 3 boosting
   - MLP 3 bagging
   - MLP 4 boosting
   - MLP 4 bagging
9. Compare the model fit metrics between these ensemble models using the test results. Identify which model has the best performance. Interpret the variable importance results.

Exercise 2: Ensemble Models for the Passengers Dataset

1. Open the **passengers** dataset **in** SPSS Modeler.
2. Sample 20% of the full data. Set return (decision to return to the airline) as the target variable and change its role to binary. Refer to Appendix B for variable names and descriptions.
3. Partition the data:
   - Split the data using a 60–20–20 ratio (train: 60%, test: 20%, validation: 20%).
4. Perform data modification as follows:
   - Replace zero values for Likert-scale variables with missing values.
   - Impute missing values.
   - Normalize data.

5. Build an ensemble model **using the** Ensemble node **for** seven individual models. Configure it as follows:
   - **Logistic regression** (from Chapter 7, Exercise 2)
   - **Pruned CHAID tree** (from Chapter 8, Exercise 2)
   - **Pruned CART tree** (from Chapter 8, Exercise 2)
   - **MLP 1 model** (from Chapter 9, Exercise 2)
   - **MLP 4 model** (from Chapter 9, Exercise 2)
   - **RBF 2 model** (from Chapter 9, Exercise 2)
   - **RBF 4 model** (from Chapter 9, Exercise 2)
6. Compare the ensemble model with the seven individual models using the test results. Determine which model has the best performance. Interpret the variable importance results.
7. Build six ensemble models using the Ensemble feature in the Decision Tree and Neural Net nodes as follows:
   - CART boosting
   - CART bagging
   - MLP 1 boosting
   - MLP 1 bagging
   - MLP 4 boosting
   - MLP 4 bagging
8. Compare the model fit metrics between these ensemble models using the test results. Identify which model has the best performance. Interpret the variable importance results.

11 Presenting Results and Writing Data Mining Reports

LEARNING OUTCOMES

Upon the completion of this chapter, you will be able to:

1. Explain the importance of writing an effective data mining project report.
2. Differentiate between academic research reports, technical reports, and business reports.
3. Identify the target audience for your report and determine the right writing styles.
4. Outline the general structure of a data mining project report.
5. Describe the purpose and content of primary sections in a report.
6. Evaluate approaches for effective writing.
7. Implement various approaches to hook the audience and keep their interest in your report.
8. Write an effective title, abstract, introduction, literature review, methodology, results, discussions, and conclusions for a data mining report.

INTRODUCTION

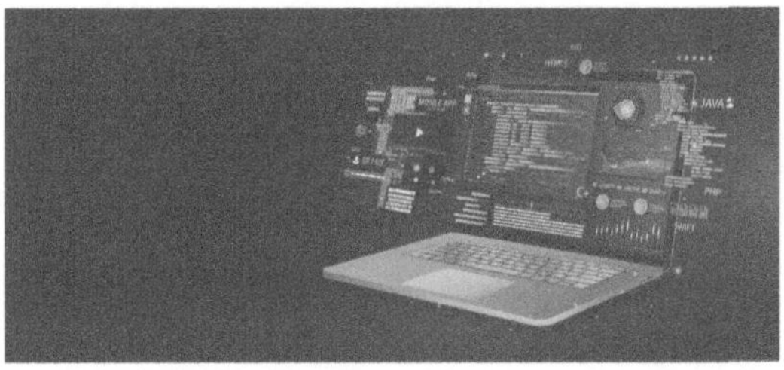

In previous chapters, we have learned the most important parts of data mining: understanding the data, preparing the data, and analyzing the data. We have gone through all the steps in the SEMMA (Sample-Explore-Modify-Model-Assess) process, including **sample, explore, modify, model,** and **assess**. This process led us to the final data mining results, the champion model, and a deep understanding of its structure and patterns. The next step is to write up the report. Sounds simple enough? Not really. Data mining is a complex process with many steps and decisions. It involves machine learning algorithms, configurations, and results, which consist of multiple diagrams, tables, and graphs. Given the abundance of information, it is crucial to select and present the appropriate results and write the report in a well-organized format to avoid confusing the readers. While we may have conducted thorough analyses and achieved useful results, failure to produce a quality report may result in stakeholders misunderstanding our outcomes or making incorrect decisions, leading to business failures or incidents.

In this chapter, I provide specific guidelines for writing a data mining report, with a focus on result presentation. Through examples, tips, and recommendations, I will demonstrate the key components of report writing using a hypothetical study on airline recommendations as an illustrative case. While this isn't an actual research study, I will provide guidelines on how to write critical sections of the report. However, for important sections such as Introduction, Methodology, and Results, I will show specific examples of how to write statements or explanations.

In this chapter, my primary focus lies on enhancing technical writing skills, specifically tailored to the context of data mining reporting. I assume that you already possess proficient general writing skills and are well-versed in standards such as APA formatting. The objective here is not to focus on structuring sentences, forming phrases, or correcting grammar errors. Instead, we direct our attention toward the technical aspects of writing a data mining report, with an emphasis on determining

DOI: 10.1201/9781003468196-13

the appropriate content for each section. Furthermore, I maintain a writing style that strikes a balance between scholarly and business report writing. While scholarly writing fosters objectivity in presenting results and ensures that all statements and conclusions are backed by data and analysis, it tends to be overly formal and relies heavily on passive voice constructions. Recognizing that our target audience may include a diverse range of professionals, such as researchers, executives, business managers, consultants, project managers, and policymakers, I opt for a more active voice and a less academic tone to enhance accessibility and engagement. This approach makes it easier for readers from both academia and industry to grasp complex technical information.

Lastly, writing is a skill that improves with practice. No one becomes an expert in technical writing overnight. The more you write, the more comfortable you will become, and the better you will be at identifying areas for improvement. By writing a little every day, you can continuously refine your technical writing abilities and ensure that your reports are clear, accurate, and impactful.

WHAT ARE THE DIFFERENT TYPES OF REPORTS?

Before we begin writing the report, it is important to understand the different types of reports and the intended audience. Depending on the target audience and the purpose of the study, we can determine the appropriate type of report, structure, and writing style. The saying, "Know your audience," is particularly relevant here. Table 11.1 outlines the three primary types of data mining reports and highlights their key differences.

It's important to note that these characteristics and requirements may vary across disciplines, industries, or organizations. The comparison provides general traits, but it's essential to consult the specific guidelines provided by the institution or organization.

In this chapter, I offer general guidelines and tips without specifying any particular standards. Always review the organization's guidelines to understand the required structure, standards, and style for your report. Remember, they are the ones who will be reading it.

TABLE 11.1
Different Types of Data Mining Reports

Items	Academic Research Reports	Technical Reports	Business Reports
Purpose	The primary goal is to present new knowledge, advance a particular field, and fulfill academic institution requirements. These reports explore research questions, present findings, and contribute scholarly work.	Focused on presenting technical information related to a specific project, often funded by an organization. They communicate technical details, methodologies, and results within a specialized domain.	Prepared to address specific business needs and objectives. These reports provide analysis and recommendations to support business decision-making.
Audience	Typically written for the academic community, including researchers, scholars, and students in a particular discipline.	Targeted toward technical professionals, engineers, project stakeholders, or individuals with expertise in the subject matter.	Tailored for business professionals, stakeholders, executives, and managers involved in strategic decision-making.
Structure	Follows a standard academic structure: abstract, introduction, literature review, methodology, results, discussion, and conclusion, often with an extensive reference list.	Includes sections such as abstract, introduction, problem statement, technical background, methodology, results, discussion, conclusions, and recommendations, often with technical diagrams or charts.	Varies depending on the purpose but often includes executive summary, introduction, business problem, methodology, analysis, findings, recommendations, and conclusion, with financial or market data.

(Continued)

TABLE 11.1 *(Continued)*
Different Types of Data Mining Reports

Items	Academic Research Reports	Technical Reports	Business Reports
Writing style	Uses formal, objective language with thorough analysis. Clarity and concise writing are essential. All statements must be supported by proper citations and references, adhering to academic standards.	Features a precise, technical style, emphasizing accuracy and the use of relevant terminology. Statements and conclusions must be backed by facts and analysis results.	Writing style can vary—ranging from formal to business-oriented. The tone may be persuasive, informative, or analytical, depending on the objective of the report.

GENERAL STRUCTURE OF A PROJECT PAPER USING DATA MINING

A good data mining research report should be well-organized, clearly written, and logically structured. It must provide sufficient details about the research problem, purpose of the study, background information, research process, data, methodology, and results to ensure reproducibility. Additionally, the report should effectively communicate the significance of the findings and their implications within the field of study.

Figure 11.1 illustrates the general structure of a data mining report. Typically, it consists of a project title, abstract, introduction, literature review, methodology, results, discussion and conclusions, references, and appendices. The Literature Review is generally a required section for academic reports. However, in technical or business reports, this section can be replaced with background information describing existing technologies or systems, serving as the foundation for the project framework or model. Additionally, appendices are optional and often used to include detailed information, instruments, design, or supplemental analyses and results that do not need to appear in the main report but are useful for readers who seek further insights or need the details to replicate the study.

Table 11.2 provides a detailed explanation of each section. Each main section usually consists of sub-sections, which help break down the content into specific categories, making it easier for readers to follow along.

> I use the structure and outline of academic research reports due to their standardized organization. This structure is logical for a data mining project and easy to follow. It can also be easily adapted for other types of reports, such as technical or business reports, by making modifications in the Literature Review section and the Discussion and Conclusions section.

HOW TO FOLLOW MY WRITING GUIDELINES?

In order to help you follow my guidelines and examples of report writing, I have developed specific color codes and formats as follows. Keep in mind that I use a hypothetical study of airline recommendations instead of an actual research study since we have been using the airline recommendation dataset throughout this book. Therefore, for certain sections, I provide tips and recommendations rather than actual writing, as conducting a full research study would be required for that.

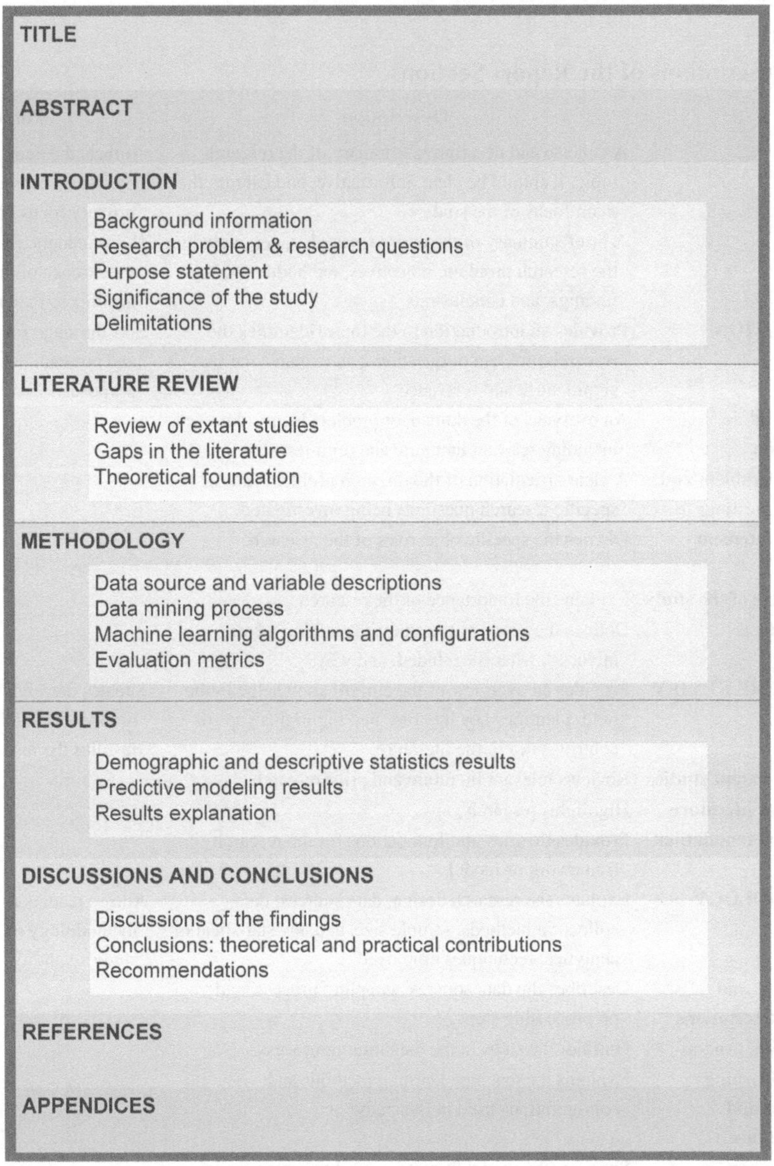

FIGURE 11.1 Structure of a data mining report.

Hypothetical research project: Prediction of passengers' airline recommendations

Figure 11.2 shows the format of my writing guidelines. Below are the instructions for using this format:

- **Bold and capitalized text** indicates the section of interest.
- **Text on a white background** provides the purpose of the section along with general guidelines and tips. These guidelines are typically broken down into key points with specific recommendations.
- **Text on a grey background** contains actual examples of writing with complete sentences and details. These examples provide demonstrations of how to phrase certain statements

TABLE 11.2
Detailed Descriptions of the Report Sections

Sections	Description	Purpose
TITLE	A concise and descriptive summary of the research topic. It should be clear, informative, and capture the main focus of the study.	Attracts the attention of readers by clearly conveying the primary focus of the study.
ABSTRACT	A brief summary of the entire research paper, including the research problem, objectives, methodology, key findings, and conclusions.	Helps readers quickly understand the essence of the study without needing to read the entire paper.
I. INTRODUCTION	Provides an introduction to the topic, identifies the research problem or question, and explains the study's significance and relevance.	Sets the stage for the research and communicates the study's purpose.
• **Background information**	An overview of the domain or problem being addressed, including relevant literature and prior research.	
• **Research problem and research questions**	A clear articulation of the research problem and the specific research questions being investigated.	
• **Purpose statement**	Defines the specific objectives of the research.	
• **Significance of the study**	Explains the importance of the research.	
• **Delimitations**	Defines the scope of the study, explaining what is included, what is excluded, and why.	
II. LITERATURE REVIEW	Provides an overview of the current knowledge in the field, identifies key theories, and highlights gaps or controversies in the literature.	Situates the study within the broader academic context and justifies the research objectives.
• **Review of extant studies**	Reviews relevant literature and prior research.	
• **Gaps in the literature**	Highlights research gaps.	
• **Theoretical foundation**	Provides theories and foundations for the research framework or model.	
III. METHODOLOGY	Explains the research design, data sources, data collection methods, sample size, and any statistical or analytical techniques employed.	Allows readers to replicate the methodology or assess the study's validity.
• **Data source and variable descriptions**	Describes the data sources, sampling process, and preprocessing steps.	
• **Data mining process**	Outlines the steps in the data mining process.	
• **Machine learning algorithms and configurations**	Explains the machine learning methods and configurations used in the study.	
• **Evaluation metrics**	Specifies the model fit metrics used to assess model performance and compare models.	
IV. RESULTS	Presents the findings of the study in a clear and organized manner.	Objectively presents the research outcomes without interpretation or discussion.
• **Demographic and descriptive statistics results**	Presents key demographic and descriptive statistics using charts and tables to justify the generalizability of the study.	
• **Predictive modeling results**	Presents predictive modeling results clearly, with key results including model evaluation, champion model selection, and predictor rankings.	
• **Result explanation**	Provides an in-depth analysis and explanation of the findings, highlighting key patterns, trends, or relationships discovered.	

(Continued)

TABLE 11.2 *(Continued)*
Detailed Descriptions of the Report Sections

Sections	Description	Purpose
V. DISCUSSIONS AND CONCLUSIONS	Discusses the implications, significance, and limitations of the findings, compares them to previous research, and provides recommendations.	Interprets the results in light of the research objectives and broader context, explaining the study's contributions.
• **Discussion of findings**	Discusses key findings and their implications relative to the research objectives, compares with existing literature, and highlights new discoveries.	
• **Conclusions**	Discusses the theoretical and practical contributions of the research and acknowledges any limitations.	
• **Recommendations**	Offers suggestions for future research or potential areas of investigation based on the current findings.	
REFERENCES	Properly cites all references to acknowledge prior work and support claims made in the report (check with the publisher or funding organizations for the reference standard).	Provides readers with reliable and valid sources used to support the study.
APPENDICES	Contains additional information, such as detailed descriptions of algorithms, data samples, supplementary analysis, and full results.	Provides supplementary details that support the main findings without cluttering the main report.

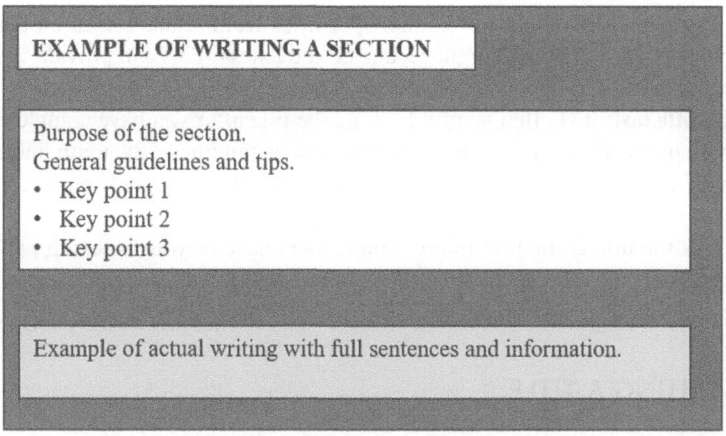

FIGURE 11.2 Format for the writing guidelines.

specific to the airline recommendation project and the data mining context. These will guide you in writing an effective title, abstract, problem statement, purpose statement, research questions, and delimitations. More importantly, you will also learn how to write the methodology and results sections for a data mining study.

> In these examples, I do not strictly adhere to scholarly writing styles typical of academic research papers. Instead, I blend them with business report writing styles, incorporating more active voice in the sentences.

HOW TO WRITE THE TITLE?

When writing a title for a research paper, it is important to create a concise and informative statement that captures the essence of your study. Following are some useful guidelines for writing a good title:

1. **Be clear and specific**: Your title should clearly convey the main focus of your research. Avoid using vague or general terms. Instead, use specific and descriptive language that accurately represents the content of your paper.
2. **Keep it concise**: Aim for a title that is brief and to the point. Typically, research paper titles are around 8–15 words long. Avoid lengthy and convoluted titles that may confuse or lose the reader's attention.
3. **Use keywords**: Include relevant keywords that reflect the main concepts, method, or variables of your research. These keywords will help readers quickly identify the topic and content of your paper. It can also improve the discoverability of your research in online databases or search engines.
4. **Be informative**: Your title should provide a clear idea of what the research paper is about. It should hint at the purpose, scope, or findings of your study, giving readers a preview of what they can expect from the paper.
5. **Consider the tone**: The tone of your title should align with the overall style and purpose of your research. It can be descriptive, objective, provocative, or even creative, depending on the nature of your study and the target audience.
6. **Spell the acronyms**: If you use an acronym in the title, make sure to spell it out since not all readers are familiar with the acronym. Typically, acronyms should only be used if they are absolutely necessary.
7. **Review existing titles**: Look for examples of titles from related research papers in your field. Analyze their structure, language, and effectiveness. This can provide you with inspiration and insights into how to craft your own title.
8. **Write the title last**: It is often helpful to write the title after you have completed the rest of the research paper. This way, you have a clearer understanding of the main findings and contributions of your study, making it easier to create a title that accurately represents your work.

Remember that the title is the first impression readers have of your research paper. It should be informative, engaging, and relevant to attract the attention of potential readers and accurately reflect the content of your study.

EXAMPLE: WRITING A TITLE

Below are examples of four titles we can use for the airline recommendation project. The titles should clearly communicate the study's purpose, which is to predict passengers' airline recommendations. Additionally, the use of data mining methods should be explicitly mentioned in the title.

Predicting Passengers' Airline Recommendations Using
Data Mining Methods

Using Machine Learning Algorithms to Predict Passengers' Airline Recommendations

Predictive Modeling for Passengers' Airline Recommendations

How Do We Predict Passengers' Recommendations of the Airline Using Machine Learning?

HOW TO WRITE AN ABSTRACT?

An abstract is the first thing readers encounter in a paper, making it a crucial part of capturing their attention and interest. Most readers don't immediately dive into the entire paper. Typically, they read the abstract carefully first. Once they understand the study's purpose and findings, they will decide whether to read the full paper. If the abstract highlights the study's relevance, novelty, and usefulness, readers are more likely to explore the details. A well-written abstract draws readers in and motivates them to read further, while a poorly written one can confuse readers and lose their interest right away, causing them to skip the rest.

Writing an abstract for a research paper involves summarizing the key elements of your study in a concise and engaging manner. Here are some steps to help you write an effective abstract:

1. **Understand the requirements**: Familiarize yourself with any specific guidelines or requirements provided by the target journal, conference, or publishing organization. Pay attention to word limits, formatting, and any specific sections or elements they expect to be included in the abstract.
2. **Identify the purpose**: Typically, abstracts aim to provide a concise overview of the research problem, objectives, methodology, key findings, and conclusions. Determine what information is essential to convey and what aspects of your study are most important to highlight.
3. **Structure the abstract**: While the specific structure may vary depending on the requirements, most abstracts include the following elements:
   - Context: Start by providing a brief context of your research. Explain the research problem or question you aimed to address.
   - Purpose: Describe the objectives of the study that allow you to find the answer to the research questions.
   - Methods: Summarize the research design, methodology, and data collection procedures employed in your study. Include briefly any relevant details about the sample size, data analysis techniques, or experimental setup.
   - Results: Present the key findings or outcomes of your research. Highlight the most significant and noteworthy results that directly address the research objectives.
   - Conclusions: State the main conclusions or implications of your study. Explain the significance and relevance of your findings and how they contribute to the broader field of research.
4. **Be clear and concise**: Abstracts are typically limited in word count, so it's crucial to convey your information succinctly. Use clear and concise language and avoid unnecessary jargon or technical terms. Make every word count to effectively communicate the essence of your study.
5. **Add keywords**: Incorporate relevant keywords or phrases that reflect the main concepts, variables, method, or design of your research. It is a common practice to add keywords to the end of the abstract. These keywords can help index your abstract in databases and make it more discoverable to readers searching for related studies.

6. **Write and revise**: Start by drafting your abstract and ensuring that you cover all the necessary elements. Once you have a draft, review and revise it multiple times. Make sure each sentence contributes to the overall clarity and coherence of the abstract. Remove any unnecessary details or repetitions.
7. **Check the style and formatting**: Verify that your abstract adheres to any required style or formatting guidelines, such as font size, margins, or referencing style. Pay attention to any specific instructions provided by the target journal or conference.
8. **Proofread and edit**: Finally, proofread your abstract for grammar, typos, spelling, and punctuation errors. Edit for clarity and readability. Make sure that the abstract is well-written, engaging, and accurately represents the content and contributions of your research.

Remember, the abstract serves as a concise summary of your research, allowing readers to quickly grasp the key aspects and significance of your study. It should be engaging, informative, and well-crafted to capture the attention and interest of potential readers.

EXAMPLE OF WRITING AN ABSTRACT

Following are key points for writing an abstract with demonstrations for the airline recommendation paper.

EXAMPLE

Brief introduction of the research problem.

The airline industry has encountered significant operational and business challenges, including pilot shortages, insufficient ground staff, frequent flight delays, cancellations, and overall service instability. These disruptions have led to some airlines losing passengers to competitors, prompting the need for strategic actions to enhance their competitive edge. A critical challenge is to predict passengers' airline recommendations and identify the factors that influence those recommendations.

Describe the purpose of the study.

The purpose of this study is to develop predictive models that predict passengers' airline recommendations and identify key predictors influencing those recommendations.

Describe the data and methodology.

The data mining method was selected, and various machine learning algorithms were used to construct predictive models. The data obtained from an airline passenger survey, provided by *[add the data source]*, were used to train and validate the models with a sample size of 100,000 responses. Primary machine learning algorithms applied included logistic regression, decision trees, and neural networks.

Present the primary findings.

> The results indicated that the boosted neural network model with two layers was the champion model with the best predictive power. The output of this model showed travel purpose as the most important predictor, followed by gender, average number of flights per year, airline status, and age.

Explain the contributions of this study.

> These findings provide the airlines with useful information and a decision-supporting tool that they can use to predict the demand and establish appropriate strategies to enhance their competitive advantage in the competing market.

Keywords: Select keywords that not only capture the essence of the study but also help improve the index of the paper.

> Keywords: airline industry; airline recommendation; data mining; machine learning; predictive modeling; neural network; decision tree; regression; predictive power; model fit

HOW TO WRITE AN INTRODUCTION SECTION?

The introduction section of a research paper serves as the opening statement, setting the stage for the study and capturing the reader's interest. Readers tend to focus on this section carefully to understand the study's purpose, its significance to both academia and industry, the scope, and any novel approaches or methods employed. Once they grasp these critical points, they can more easily follow the rest of the paper. However, if the introduction is unclear or inconsistent, readers may conclude that the paper lacks value and may stop reading altogether.

It's no exaggeration to say that the introduction is often the most challenging section to write. Following these key steps can help ensure you craft a strong, effective introduction:

1. **Hook the reader**: Begin your introduction with a captivating opening sentence or a compelling fact that grabs the reader's attention. You can use an anecdote, a relevant quote, a startling statistic, or a thought-provoking statement. The goal is to intrigue the reader with curiosity and make them want to continue reading.
2. **Provide context**: Give a brief overview of the study area and provide the necessary context information to help readers understand the relevance of your study. Explain why the topic is important and highlight any real-world implications or applications.
3. **State the research problem**: Clearly state the specific research problem or question that your study aims to address. Be concise and precise in formulating your research problem to avoid ambiguity or confusion. This statement should highlight the gap or limitation in the existing knowledge that your study seeks to fill.
4. **Review previous research**: Summarize the key findings and contributions of previous studies related to your research topic. This literature review should focus on the most

relevant and recent research. Identify any gaps, controversies, or unanswered questions in the existing literature that your study aims to address.

5. **Outline objectives and scope**: Clearly state the research objectives or hypotheses that your study intends to achieve. These objectives should be specific and align with your research problem.
6. **Scope and limitations**: State the scope (delimitations) and limitations of your study to set appropriate expectations for the reader.
7. **Justify the research**: Explain why your study is important and necessary from both theoretical and practical perspectives. Discuss the potential benefits, advancements, or contributions that may arise from your research findings. Demonstrate the significance of your study in advancing the field of research or addressing a practical issue.
8. **Preview the methodology**: Provide a brief overview of the research design and methodology employed in your study. This includes mentioning the data collection methods, target population, or analytical techniques. However, save the detailed description of the methodology for the methodology section of your paper.
9. **Structure and organization**: Briefly outline the organization of your research paper by mentioning the main sections and their order. This helps readers understand the flow of your paper and locate specific information they may be interested in.
10. **Maintain clarity and conciseness**: Ensure that your introduction is clear, concise, and focused. Avoid unnecessary technical jargon, excessive details, or length sentences. Use simple and straightforward language to make it accessible to a wide range of readers.
11. **Include citations**: Provide appropriate citations for the sources you reference in the background section. This helps support the credibility of your research and allows readers to explore the referenced studies for more in-depth information.
12. **Write and revise**: Start by drafting the introduction section and incorporating the points discussed above. Once you have a draft, review and revise it multiple times. Make sure that the information flows logically, the tone is engaging, and the introduction effectively sets the stage for your research.

By following these steps, you can write an introduction that engages your readers and prepares them to fully appreciate the significance and insights of your research.

EXAMPLE OF WRITING AN INTRODUCTION SECTION

Before I break down the section into subsections for the airline recommendation paper, I want to present a complete introduction section from a research paper I published in 2021 in the *Journal of Air Transport Management*. This serves as an example of an actual introduction in a peer-reviewed academic paper published in a top-tier journal. It adheres to the format and guidelines I outlined previously. You can access the full paper via the reference provided below.

EXAMPLE

Truong, D. (2021). Estimating the impact of COVID-19 on air travel in the medium and long term using neural network and Monte Carlo simulation. *Journal of Air Transport Management*, 96. https://doi.org/10.1016/j.jairtraman.2021.102126

Figure 11.3 illustrates the structure of the introduction section, breaking down each paragraph to demonstrate how it addresses key elements in an effective introduction.

- **Paragraph 1**: Captures the reader's attention by discussing the global context of the COVID-19 pandemic, focusing on how travel restrictions and the virus's spread affected individuals' lives, personal trips, and transportation decisions.

1. Introduction

The COVID-19 pandemic has significantly affected the airline industry. In order to reduce the spread of coronavirus, many countries have issued travel restrictions, especially air travel, to limit the travel volume. In the United States, many states limit air travel by mandating self-quarantine for people arriving from other states or countries by air. This requirement has affected many business and personal trips as it causes disruption in business operations or personal activities. Many organizations have moved to working from home and using Zoom meetings to avoid any interruption [1]. People chose to travel by car whenever possible. As a result, air travel has reduced significantly. Figs. 1 and 2 show the comparison of commercial flights between 2019 and 2020. It appears that the number of commercial flights, both domestic and international, has declined significantly amid the pandemic. Before the pandemic, there was an average of 25,000 domestic flights and 4300 international flights per day in 2019. Since early April 2020, when most states issued stay-home orders, the number of flights reduced significantly to less than 10,000 domestic flights and about 500 international flights per day. While these numbers started to rise again in early July 2020, they are still far below the number of flights in the same month in 2019. From July to December 2020, the number of daily domestic flights was about 15,000, whereas daily international flights ranged from 1000 to 1500 (BTS, 2020a).

In this study, air travel is measured by the number of domestic and international flights. While flights during the pandemic may not be full, this number is not too far off the actual travel demand. Fig. 3 shows the actual number of people checked in daily at U.S. airports for both domestic and international flights. We can note the similar trends of these numbers compared to the trend [2] of domestic and international flights. There was a sharp fall in April 2020, followed by a slow gain from June to August 2020. Then, the demand gradually increased toward the end of December 2020. Thus, the number of flights does capture well the patterns of travel demand by U.S. residents. One advantage of using the number of flights is that it allows airlines to easily estimate their revenues and expenses in the future, which helps determine the timeline for the airline industry recovery.

The decreased air travel has had negative impacts on the airline industry. According to BTS (2020b), U.S. airlines reported substantial losses in three consecutive quarters in 2020, with a loss of $5.2 billion in the first quarter, $11 billion in the second quarter, and $11.8 billion in the third quarter in both domestic and international operations. Compared to high profits accumulated in 2019, with $4.5 billion profit in the third quarter and $3.4 billion profit in the fourth quarter, these losses pose a significant risk to [3] the airline industry and its future. In order to cope with these losses, many airlines have furloughed or laid off many employees, which results in a bigger challenge for the economy in the near future. According to Mr. Calio, the CEO of Airlines for America, the airline industry will not fully rebound until 2024 (Blum, 2020). Given the uncertainty in the trend of the pandemic and the vaccine rollout timeline, the time it takes for the airline industry to come back to normal may even be more uncertain.

While it may take some time for the airline industry to fully recover, it is vital to understand what the industry will look like in the meantime. More specifically, the overarching question that needs answers is what will be air travel during and post COVID-19. Air travel will certainly continue to be on the low side during [4] the pandemic and may start growing after the pandemic. But to what extent? What will air travel be in different COVID-19 remediation scenarios? What will air travel be after the pandemic? The answers to these questions can provide useful information about the status of the airline industry in the medium and long term.

The traditional literature in air travel prediction tends to focus on macro factors, such as population demographics, location, economic, consumption expenditure, and deregulation, and micro factors, including airlines' service, pricing, and quality (Wang and Song, 2010; Valdes, 2015). While these factors still play an important role in determining air travel demand, they do not reflect how the demand changes amid a pandemic, especially an unprecedented one like COVID-19. The airline industry has never experienced [5] anything like this before, including the September 11 event. The impact of COVID-19 on society and transport behavior is dynamic and unpredictable with a high uncertainty level. Thus, a good prediction needs a capability to capture the dynamism of the pandemic and the change in people's travel behavior. The literature on COVID-19's impact on air transportation is rather limited and focused primarily on the impact differences between regions and how airlines reacted to travel restrictions. A specific model that can predict air travel during and post COVID-19 is still missing.

The purpose of this paper is to predict air travel in the U.S. during and post COVID-19 using novel variables. These novel variables are U.S. residents' daily trips by distance since their daily travel decisions change dynamically according to their risk perception of COVID-19. Research shows that the more trips people make amid the pandemic, the safer they feel in their travels (Beck and Hensher, 2020). In addition, the trip distance does play an important role in their travel behavior in responding to changes in COVID-19 cases and deaths. More specifically, the residents tend to make short trips right after they notice the decrease in COVID-19 cases and deaths, while it takes them more time to decide to make medium and long trips (Truong and Truong, 2021). Thus, it is possible that residents' daily travel behavior can capture the dynamism of the transport pattern during and post pandemic. Finding whether these novel variables can predict [6] air travel is the purpose of this study. Other input variables include the economic index, COVID-19 variables, and travel restrictions in order to capture the spread and severity of the pandemic and the economic condition. Neural network and Monte Carlo simulation are appropriate methods to find a pattern of relationship between input variables and air travel and to perform what-if-scenarios to predict the air travel volume during and post COVID-19. This paper's findings add value to the air transportation literature by examining the effects of novel variables, daily trips by distance, and modeling the air travel under uncertainty. The sensitivity analysis results provide a prediction of air travel in various scenarios, which can help the government and airlines make informed decisions to formulate policies and strategies to survive this crisis. It is important to note that this paper does not aim at performing time series forecast, as this method assumes temporal dependency of the air travel. Given the rapidly changing situation of COVID-19 and unclear patterns in the past, time series forecast models may need to be updated frequently to be meaningful (Truong and Truong, 2021). Accordingly, the simulation results in this paper can provide more useful information to run several what-if scenarios to examine how air travel will be during and post pandemic.

The paper is organized as follows. Section 2 reviews the existing literature in air travel prediction and COVID-19's impacts on air transportation. Research gaps are discussed, along with the explanation for the contributions of this study. Section 3 describes the data sources, variables, and research methodologies [7], including neural network and Monte Carlo simulation. Section 4 provides detailed results of the neural network models, simulation results, and sensitivity analysis results. Finally, Sections 5 and 6 discuss the research findings, implications of the study, and recommendations for future research.

FIGURE 11.3 Example of an Introduction section (Truong, 2021).

- **Paragraph 2**: Provides the background of the study by emphasizing the drop in air travel volume and explains why the study measures air travel demand by domestic and international flight numbers instead of passenger counts.
- **Paragraph 3**: Clearly articulates the research problem, highlighting the challenges the airline industry is facing, especially the significant financial losses due to the pandemic.
- **Paragraph 4**: Poses the primary research question: What will the air travel demand be during and post-pandemic?
- **Paragraph 5**: Reviews prior studies and points out their limitations, setting up the need for this research.
- **Paragraph 6**: Explains the study's objectives, which are to predict air travel during and after the pandemic using novel variables, with a focus on daily trips by distance. It also references prior research to support the selection of these variables.
- **Paragraph 6 (continued)**: Discusses the study's methodology, including the use of machine learning and Monte Carlo simulations.
- **Paragraph 6 (continued)**: Highlights the significance of the research and how the results could help airlines and governments plan for future demand.
- **Paragraph 6 (continued)**: Outlines the study's scope and limitations, focusing on simulations based on variable relationships instead of time series forecasts.
- **Paragraph 7**: Briefly outlines the structure of the rest of the paper, explaining the content and purpose of each section.

This breakdown helps you understand how to structure an introduction that effectively engages readers, introduces the research problem, and sets up the objectives, methodology, and significance of your study.

Writing an introduction section indeed requires significant effort, as it must convey essential information in a concise and engaging way, capturing the reader's interest from the outset. Now, let me guide you through the process of writing the introduction for the airline recommendation paper.

INTRODUCTION FOR THE AIRLINE RECOMMENDATION PAPER

Since this is not an actual research study and the dataset is used for educational purposes throughout this book, a full research paper cannot be presented. However, I provide specific guidelines, tips, and writing demonstrations on how to craft the introduction section, particularly focusing on the problem statement, research questions, and purpose statement. Let's break down each component step by step:

How to Hook the Reader?

Start by discussing recent incidents in the airline industry that have impacted passengers, such as major flight delays, cancellations, or poor customer service, leading to dissatisfaction. Provide real examples and statistics that show how these events have caused airlines to lose passengers or suffer business losses. Highlight the consequences of these incidents, making it clear why understanding passenger recommendations is crucial for airlines.

How to Provide the Context of This Paper?

Next, explain the significance of airline recommendations in the industry. Define what an airline recommendation entails and why it matters to the success or failure of an airline. Cite studies or statistics to support your argument. Show that passengers' likelihood to recommend an airline is linked to business performance, market share, and customer loyalty.

How to State the Research Problem and Research Questions?

Clearly describe the new challenges the airline industry is facing and explain why it is crucial to understand how passengers decide to recommend an airline to others. It is important to emphasize that if many studies have already addressed this issue, it may no longer be a strong research problem, even if the problem persists. The goal is to focus on a new problem that has not yet been adequately explored in the literature, allowing you to highlight the originality of your study. If the problem has been tackled by other researchers, the topic loses its novelty.

Below is an example of a concise problem statement. Remember, you need to build up to this statement through the context information mentioned earlier.

> The recurring flight delays and cancellations in the past few years have caused serious concerns among passengers regarding airlines' operational efficiency and service quality. With multiple venues for communication and feedback, passengers increasingly share their recommendations of an airline with others, such as family, friends, colleagues, or fellow passengers, via social networks. Positive recommendations can lead to thriving travel businesses, while negative recommendations result in losing passengers and escalating business losses. Understanding airline recommendations is critical for airlines to establish appropriate strategies and make informed decisions to improve flight plans, operations, and service quality ….

Add a research question to this paragraph to highlight the important unanswered question.

> … The primary question is: How can airlines find the right tool to accurately predict passengers' airline recommendations? Misprediction can lead to poor business decisions that result in airline failures. Additionally, airlines need to understand what they can do to increase positive recommendations and reduce negative ones. Hence, the second question is: What are the important factors associated with airline recommendations?

How to Review Previous Studies?

To convince readers that the research problem has not been adequately addressed in the literature, it is essential to conduct a comprehensive literature review. In this paragraph, briefly summarize the previous research, their findings, and any gaps or shortcomings. For instance, you can provide evidence that machine learning methods have not been applied to predict passengers' airline recommendations in earlier studies. Most existing research tends to rely on traditional statistical methods, such as surveys, often with small sample sizes. This limits the generalizability of their findings and excludes many relevant variables. Save the in-depth discussion of these studies for the Literature Review section. If written effectively, this paragraph will justify the importance of conducting your research.

How to Write a Purpose Statement?

For data mining papers, particularly those involving supervised learning, there are typically two main objectives: (1) to find the best-performing model (the "champion model") using machine learning algorithms and (2) to identify and rank the most important predictors. These objectives should align with the research questions stated earlier. Therefore, a purpose statement for such a study could be written as follows:

> The purpose of this study is twofold. First, we aim to develop predictive models and identify a champion model that accurately predicts passengers' airline recommendations. Second, we will use the champion model to identify key predictors and rank them in order of importance.

How to Briefly State the Methodology?

This statement should inform readers of the methodology used, particularly if a novel analytic method is deployed, and explain why it is appropriate. For example, you can write a statement as follows:

> To achieve these objectives, we selected the data mining method. Given the large number of variables and the dataset size, data mining is the most suitable approach to uncover hidden patterns using machine learning algorithms. The predictive models were trained and validated using the airline survey dataset provided by *[add datasource]* in 2021.

How to Justify the Significance of the Study?

Clearly explain how this study adds value to the airline management literature (theoretical implications) and how the findings benefit the airline industry (practical implications). For example, you could state.

> This study contributes to the airline management literature by developing predictive models using novel variables that include a combination of flight operations and efficiency metrics, such as flight delays, cancellations, flight durations, travel purposes, and flight paths. These factors provide fresh insights into airline decision models in a rapidly changing environment *[theoretical]*. By incorporating demographic variables, this study captures a comprehensive view of passenger behavior. Practically, the findings enable airlines to estimate patterns in passengers' airline recommendations, allowing executives to make data-driven decisions that improve competitive advantage and enhance operational strategies *[practical]*.

How to Describe the Delimitations and Limitations?

You should describe the scope (delimitations) of this study. For example:

> This study focuses on U.S. airlines and domestic flights, targeting U.S. adults who have flown at least once in the past year. The analysis primarily examines variables related to flight operations, such as delays, cancellations, travel purposes, and passenger demographics.

Then, clarify the limitations of the study:

> ... It *[this study]* does not account for factors like the impact of the pandemic, competition from international airlines, or security concerns. However, these limitations are not expected to compromise the validity of the results, as the focus remains on domestic airline operations and passengers' recommendations, which align with the research objectives.

How to Describe the Organization of the Paper?

Typically, this part is relatively easy since you only need to explain the remaining sections of the paper, such as the literature review, methodology, results, and discussions and conclusions. It is fairly straightforward. My only suggestion is to avoid repeating phrases to make the writing more professional.

> The paper is organized as follows. Section 2 reviews existing studies in airline management and identifies research gaps related to passengers' airline recommendations. It also explains the foundational framework for variable selection and predictive modeling. Section 3 outlines the reasons for selecting the data mining method, details the data mining process, data collection, sampling, and the machine learning algorithms used. Detailed results of the data mining are presented and explained in Section 4, showing the model comparison, the champion model, and variable importance. Finally, Section 5 provides discussions of the findings, implications of the study, limitations, and recommendations for future research.

HOW TO WRITE A LITERATURE REVIEW SECTION?

When writing a literature review section for a research paper that focuses on data mining, you need to review and summarize existing research and scholarly works specifically related to the topic, applications of data mining techniques in this area, and any advancements. Following are some tips to help you write an effective literature review section.

1. **Define the scope**: Determine the scope and boundaries of your literature review. Use the research questions and study objectives to guide your review within the field of the focused topic and data mining. This helps focus your search and analysis on relevant literature.
2. **Identify relevant sources**: Conduct a comprehensive search to identify relevant sources specifically related to the scope identified above. Use library databases, academic journals, conference proceedings, and reputable organizations that cover relevant topics. Include recent and seminal works to ensure the review is up-to-date and comprehensive.
3. **Categorize the literature**: Organize the literature based on different categories or themes related to the topic of interest. If your paper involves advancements in data mining methods, include topics such as data preprocessing techniques, algorithm development, or relevant subfields within data mining.
4. **Summarize key findings**: Summarize the key findings and contributions of each source within the relevant categories or themes. Provide concise summaries of each work, including the research objectives, methodologies, datasets, sample size, and main results or conclusions. Highlight any unique or innovative approaches, algorithms, or applications discussed. If you have a large number of studies to review, using a table to summarize key findings can make it easier to follow and save space.
5. **Identify gaps and limitations**: Identify any gaps, limitations, or unresolved issues in the existing literature related to your topic. Discuss areas where further research is needed or where conflicting results or opinions exist. This highlights the need and rationale for your own research.
6. **Compare and contrast**: For data mining literature, compare and contrast different approaches, algorithms, or techniques discussed. Analyze the strengths and weaknesses of each, considering factors such as accuracy, scalability, interpretability, or applicability. Identify trends, similarities, and differences in findings across reviewed studies.

7. **Synthesize the literature**: Synthesize the information by integrating findings and conclusions across the reviewed sources. Look for common themes, patterns, or emerging trends. Provide a coherent narrative that connects the literature to your research objectives. This synthesis helps you identify the theoretical foundation for your study.
8. **Describe the theoretical foundation for the study**: For theory-driven research, present the theoretical foundation or framework used to select variables for analysis. Even for data-driven, exploratory research like data mining studies, providing a theoretical basis for selecting variables is still important.
9. **Discuss the novelty of this study**: Explain why your study is original and how it fills research gaps. Identify novel variables or approaches added to the study and describe their significance. This answers the "so what" question for the readers.
10. **Write and revise**: Draft the literature review and incorporate the above points. Review and revise multiple times to ensure a logical flow, proper citation of sources, and a comprehensive overview of the literature.

The literature review section should provide an overview of current knowledge and advancements in your field. It should demonstrate your understanding of existing literature and how it informs and supports your research objectives. By critically analyzing and synthesizing the literature, you establish the context and rationale for your research within the field of data mining.

Since this section requires a thorough review of existing studies, I don't provide detailed examples specific to the airline recommendation study as it's hypothetical. However, by following the steps outlined, you should be able to write an effective literature review that informs readers of current knowledge in airline management, gaps in the literature, and the value this study adds.

Note that for technical and business reports, this section is usually much lighter and less focused on peer-reviewed literature. Instead, it emphasizes case studies, technical reports, or managerial reports published by organizations in the same field. These types of reports typically do not require extensive justification of the novelty of the study or its theoretical contributions. Instead, they focus more on the practical contributions and actionable insights derived from the findings. Therefore, depending on the type of report you are writing, you can adjust the literature review section accordingly to highlight relevant practical applications and insights.

HOW TO WRITE A METHODOLOGY SECTION

The methodology section is crucial as it provides readers with details on the research method and process used in the study. This information allows them to evaluate the study's validity and helps them replicate the methodology in future research. When writing a methodology section for a data mining research paper, it's important to outline the specific steps and techniques used to extract knowledge from large datasets. Here are some key tips for writing an effective methodology section:

1. **Research design**: Explain the overall research design and objectives of the study. Justify why data mining is the appropriate method for addressing the research question(s).
2. **Data collection**: Describe the datasets used in the study, specifying their sources—whether from public repositories, proprietary databases, or data gathered through surveys or experiments. Include details about the sample size, variables, and the scales of those variables.
3. **Data preprocessing**: Detail the steps taken to prepare the data before applying data mining techniques. This may include data cleaning, imputing missing values, handling outliers, transforming or normalizing data, selecting variables, or reducing dimensionality. Explain any specific methods and tools used in this stage.
4. **Machine learning method selection**: Specify the machine learning method used, such as classification or association models. Mention the specific algorithms used (e.g., logistic

regression, decision trees, neural networks, or random forests) and justify your choices based on the research objectives and the nature of the dataset.
5. **Configurations**: Provide information on the parameter settings and configurations used for each machine learning method. If multiple models were built with varying configurations, explain how you arrived at the optimal parameter values. Also, mention any software and tools used for implementing the algorithms.
6. **Evaluation metrics**: Discuss the evaluation metrics used to assess model performance and compare the models. For instance, common metrics for classification models include accuracy, misclassification rate, sensitivity, specificity, F1 score, Gini coefficient, ROC curve, and Lift chart. Justify why these metrics were selected based on the type of models used and the dataset characteristics.
7. **Ethical considerations**: Address any ethical concerns related to data mining, such as data privacy, confidentiality, or bias. Describe how you addressed these issues, whether by de-identifying sensitive information, obtaining informed consent, or adhering to relevant regulations and guidelines.
8. **Write and revise**: Begin drafting the methodology section, incorporating the points discussed above. Review and revise the draft multiple times to ensure that the information is clearly and logically presented. Explain any technical terms or processes to ensure that the methodology is accessible to a broad audience.

By providing detailed explanations of your research design, data, and the techniques employed, readers will be able to better understand and replicate your data mining approach. Justify your choices and demonstrate the suitability of your methodology, enhancing the credibility and reproducibility of your research.

EXAMPLE OF WRITING A METHODOLOGY SECTION

As you can see, I have already covered these details in previous chapters where I used the **airline recommendation** dataset as part of the examples. To avoid unnecessary repetition, I will not go over every detail again here. Instead, I will summarize the key points you should cover in each part of the methodology, provide some writing examples, and indicate which chapter to revisit for more details. This section should be well-organized and concise, avoiding excessive length.

Key Points for the Methodology Section with Examples

Justify the Use of the Data Mining Method

The primary objective of this study is to identify the best predictive model for airline recommendations and to determine the key factors that influence these recommendations. To achieve this, data mining was chosen as the appropriate method. Data mining is a data-driven, exploratory technique that allows researchers to analyze large datasets and uncover hidden patterns, which can then inform decision-making in a business context.

Data Source and Variable Descriptions

In this study, we used data from the 2021 airline passenger survey. The dataset was obtained from *[insert data source]*, which conducted a survey targeting adults in the United States who had been flying for at least one year. Respondents shared their experiences with domestic flights and indicated whether they would recommend the airline. The data were collected, validated, and coded by the data provider. Unusable data were removed to maintain the quality of the dataset. Additionally, all personal and identifying information was de-identified by the data provider to protect the participants' privacy, ensuring no ethical concerns were present. Consequently, Institutional Review Board (IRB) approval was not required for this study.

[Please note that these are hypothetical examples since this is not an actual research study. In your paper, ensure to use the correct data source and describe how the data were actually collected.]

The dataset consists of one target variable and 19 predictors. Table 11.3 presents the variable names, labels, descriptions, scales, and roles. The target variable is **recommendation** (whether the passenger recommends the airline), which is a binary variable (Yes—recommend; No—not recommend). The predictors vary in scale, including interval, binary, and nominal. Notably, price sensitivity is a 5-point Likert-scale variable but is treated as an interval variable because the gaps between levels are considered equal.

[Refer to Chapter 4 for the details of the dataset.]

TABLE 11.3
Data Description

Variable Name	Label/Description	Scale	Role
recommend	Recommend the airline to others	Flag (binary) (yes/no)	Target
airline_st	Airline status	Nominal	Input
age	Age	Continuous	Input
gender	Gender	Flag (binary) (male/female)	Input
price_sensitivity[a]	Price sensitivity	Continuous	Input
no_years_flying	Number of years flying	Continuous	Input
no_flt_pa	Average number of flights per year	Continuous	Input
pct_ flt_other_airline	Percent of flights with other airlines	Continuous	Input
type_of_travel	Travel purpose	Nominal	Input
no_other_lty_cards	Number of other loyalty cards	Continuous	Input
shopping_amt_airpt	Shopping amount at the airport	Continuous	Input
eat_drink_airpt	Spending for eating and drinking at the airport	Continuous	Input
class	Passenger class	Nominal	Input
origin_state	Origin state	Nominal	Input
destin_state	Destination state	Nominal	Input
depart_delay_in_minutes	Departure delay (in minutes)	Continuous	Input
arrival_delay_in_minutes	Arrival delay (in minutes)	Continuous	Input
flt_cancel	Flight cancellation	Flag (Binary) (yes/no)	Input
flt_time	Flight time (in minutes)	Continuous	Input
flt_distance	Flight distance (miles)	Continuous	Input

[a] 5-point Likert scale, treated as continuous scale.

Data Mining Process

In this section, you need to specify and justify the use of the SEMMA process. Then, describe briefly what was done in each step of SEMMA. Remember that you should not just present the general process but must explain what was done in each step with enough specifics so the readers can follow your decision-making process. Below is an example.

In this study, we follow the Sample-Explore-Modify-Model-Assess (SEMMA) process. This process is ideal for the objectives of this research as it guides us through all the necessary steps to build predictive models, identify the champion model, and pinpoint significant predictors. Since it is not the purpose of this study to deploy the model in real-life business operations, the model deployment step is not included. SPSS Modeler was used for the entire data mining process.

In the Sample step, given that the dataset contains 124,464 records, we decided to use the entire dataset for analysis, providing a robust sample size for training models, especially when using advanced machine learning algorithms like neural networks that typically require large datasets for effective model training. The dataset includes 65,297 cases of "recommend" (52.46%) and 59,167 cases of "not recommend" (47.54%). Since the dataset has a relatively balanced distribution between the two levels, re-sampling was not necessary.

The data was partitioned into three samples using a 60–20–20 split ratio: training sample (60%), testing sample (20%), and validation sample (20%). This partition provides us with enough data to train the models, and we still have enough data to validate the models and perform the final test.

[Refer to Chapter 3 for more details on the Sample step.]

In the Explore step, both visualization and descriptive statistics were used to explore the data. We identified three continuous variables with missing data: **arrival delay in minutes, departure delay in minutes**, and **flight time in minutes**. Additionally, while **price sensitivity** is a Likert-scale variable, descriptive statistics revealed zero values, which does not make sense since Likert-scale values do not include zero. These zero values were determined to be missing values that needed to be addressed.

[Refer to Chapter 4 for more details on the Explore step.]

In the Modify step, several steps were taken to correct data errors and handle missing values. **Price sensitivity** was re-coded using the Filler node to replace zero values with missing values, resulting in four variables with missing values. All missing values in these four variables were imputed using the mean value in SPSS Modeler. The mean method was chosen because the percentage of missing values was low (about 2%), and mean imputation is a widely accepted approach. The imputed values were then used for the modeling step.

[Refer to Chapter 5 for more details on the Modify step.]

In the step, multiple classification models were developed since the target variable is binary. Three primary machine learning methods were selected: logistic regression, decision trees, and neural networks. These methods are appropriate for this study because they are known for building robust and valid classification models. They handle large datasets effectively and do not require strict assumptions. Each method captures different types of relationships within the data. Logistic regression, being statistics-based, models linear relationships between the odds ratio of the target variable and predictors. Decision trees, a rule-based method, capture non-linear relationships effectively. Neural networks can model complex, non-linear relationships using multiple hidden layers. Moreover, decision trees and neural networks are less affected by missing values. Detailed configurations for these models will be presented later.

[Refer to Chapters 7–9 for more details on those methods.]

In the last step, Assess, all models were evaluated for model fit and predictive power, and were compared to determine the champion model. The results of the champion model were used to identify important predictors and rank them by their impact. Evaluation metrics will be provided in the next section.

[Refer to Chapter 6 for more details on model evaluation and comparison.]

Machine Learning Algorithms and Configurations

In this part, you need to provide detailed configurations for all models you build. The easiest way to do this is to summarize the configurations in a table. I have extracted the key model configurations presented in Chapter 7 (Regression: Logistic 1), Chapter 8 (Decision Tree: C&R Tree 1 and CHAID 2), Chapter 9 (Neural Networks: MLP 4), and Chapter 10 (Ensemble: MLP 3 boosting), and summarized them in Table 11.4 as an example. Since we developed many models in those chapters, I have selected only five for this example to avoid overwhelming the reader with too much information. Note that some detailed configurations are default values, which can be extracted directly from the modeling nodes.

In this study, we built multiple predictive models using three primary methods: logistic regression, decision trees, and neural networks. Table 11.4 presents the detailed configurations for each model.

TABLE 11.4
Predictive Model Configurations

Predictive Models	Configurations
Model 1: Logistic regression	• Model selection method: forward stepwise • Selection criterion: validation misclassification
Model 2: CART tree	• Binary tree. • Splitting rule: Gini impurity measure. • Maximum tree depth: 5. • Prune tree to avoid overfitting. • Maximum surrogates: 5. • Stopping rule: minimum records in parent branch: 2%. • Stopping rule: minimum records in child branch: 1%. • Minimum change in impurity: 0.0001. • Overfit prevention set: 30%.

(Continued)

TABLE 11.4 *(Continued)*
Predictive Model Configurations

Predictive Models	Configurations
Model 3: CHAID tree	• Multi-way tree. • Splitting rule: Chi-square test—Pearson ratio. • Splitting rule: significance level for splitting: 0.05. • Adjust significance values using Bonferroni method. • Maximum tree depth: 10. • Maximum iterations for convergence: 100. • Stopping rule: minimum records in parent branch: 100. • Stopping rule: minimum records in child branch: 50.
Model 4: Neural Network model with two hidden layers	• Network architecture: multi-layer perceptron. • Number of hidden layers: 2. • Number of hidden neurons: 30. • Stopping rules: maximum training time: 15 minutes. • Stopping rules: maximum training cycles: 300. • Stopping rules: minimum accuracy: 90%. • Overfit prevention set: 30%.
Model 5: Boosted Neural Network model with two hidden layers	• Network architecture: multi-layer perceptron. • Number of hidden layers: 2. • Number of hidden neurons: 15. • Stopping rules: maximum training time: 15 minutes. • Stopping rules: maximum training cycles: 250. • Stopping rules: minimum accuracy: 90%. • Overfit prevention set: 30%. • Ensembling method: boosting. • Aggregation method: highest probability wins. • Number of component models for boosting: 20.

[Refer to Chapters 7–10 regarding the explanation of those configurations.]

Evaluation Metrics

Describe the key metrics used to evaluate model fit and compare the models. These metrics were covered in detail in Chapter 6. Depending on your audience, you can decide how much detail to provide. For an academic audience, you can briefly describe the metrics without too much explanation, as they are already familiar with them. For professional or business audiences, focus more on the purpose of the metrics rather than the scientific background. Below is an example statement.

Several metrics were used to evaluate the model fit and compare the models. Specifically, the model fit metrics include the misclassification rate, accuracy, sensitivity, specificity, F1 score, Gini coefficient, and ROC index. These are common metrics for evaluating classification models. First, we compared all models using the misclassification rate for the validation sample to identify the champion model, which is the one with the lowest misclassification rate. Then, we used sensitivity, specificity, F1 score, Gini coefficient, and ROC index to confirm the champion model's performance. Once identified, we further evaluated the reliability and validity of the champion model to ensure it met the expected predictive power. Finally, we reviewed the structure of the champion model to examine the relationships between the predictors and the target variable. The variable importance results were also analyzed to identify and rank the most important predictors.

[Refer to Chapter 6 for more details on model fit metrics.]

HOW TO WRITE A RESULTS SECTION?

The results section is crucial because readers often focus on it to understand what your study has uncovered. In this section, you need to present and explain the findings derived from your data mining analysis. Although this section is typically straightforward, it requires careful writing to ensure accuracy and avoid interpreting or discussing the findings. Your goal is to present the results clearly and objectively without influencing the reader's interpretation. Before writing, ensure you thoroughly review and understand the analysis results. Below are some tips for writing an effective results section:

1. **Organize the presentation**: Start by organizing the presentation of your results in a logical and coherent manner. Consider the research objectives and the structure of your analysis to determine the most appropriate way to present the findings.
2. **Provide demographic information**: Provide key demographic information on the data. These results help the reader understand the characteristics of individuals, cases, or organizations in the dataset. Demographic information allows you to justify the representativeness of your sample, which ensures the generalizability of the study.
3. **Provide descriptive statistics**: Provide descriptive statistics results of the data. It includes measures such as frequency, mean, median, mode, standard deviation, minimum, maximum, or any other relevant statistical measures that help describe the characteristics of the dataset.
4. **Present data mining results**: Present the main results of your data mining analysis. This part should include the outcomes of all predictive models that you trained and validated. Report the specific model fit metrics for each model, model comparison, and the champion model. Present and explain the predictive power of the champion model, its detailed structure, and data patterns detected by this model. Finally, present the variable importance results to show which predictors are important and rank them by order.
5. **Visualize the results**: Use visualizations such as tables, charts, graphs, or diagrams to enhance the understanding of your results. Choose visual representations that are appropriate for the type of data and the research objectives. Include captions or labels to provide clear explanations of the visualizations. Visualization is usually easier to understand and follow than just texts.
6. **Explain the findings**: Explain the findings of your data mining analysis objectively in the context of your research objectives. It is to help the readers follow and understand the findings. Don't assume that they can understand everything just by looking at tables or graphs. Explain the patterns, relationships, or insights that emerge from the analysis and avoid interpreting and discussing the findings. The objective of this section is to present the results objectively without influencing the readers.
7. **Highlight new discoveries**: Compare your results with the findings from existing literature in the field of data mining. If your results show any new discoveries, such as new significant predictors, new relationships, or new patterns, highlight them.

8. **Provide supplementary analysis**: Consider providing additional supplementary analysis if relevant to support or clarify your main findings. This could include sensitivity analyses, subgroup analyses, or additional exploratory data mining techniques that help deepen the understanding of the results.
9. **Write and revise**: Start by drafting the results section and incorporating the points discussed above. Once you have a draft, review and revise it multiple times. Make sure that the information is presented clearly, concisely, and in a well-structured manner. Use appropriate terminology and provide sufficient context for readers to understand and follow the results.

The results section should provide a comprehensive and clear presentation of the findings derived from your data mining analysis. It should be supported by appropriate statistical measures, visualizations, and interpretation. By effectively communicating the results, you contribute to the overall credibility and impact of your research.

EXAMPLE OF WRITING A RESULTS SECTION

Below are specific examples demonstrating how to write the results section for the airline recommendation paper. These results are extracted from the findings in previous chapters and presented in a clear, objective manner.

Demographic Information

This section provides readers with essential information about the demographic characteristics of individuals, cases, or organizations in the dataset. Presenting demographic information helps the audience understand the sample's makeup and how it connects to the data mining outcomes. It also justifies the representativeness of the sample, supporting the generalizability of the study's findings.

Depending on data availability, you should select relevant demographic variables to present. In Chapter 4, we explored the airline recommendation dataset and used multiple charts to visualize the data. Since the dataset is based on passenger survey responses, you should focus on passengers' primary demographic characteristics. Bar charts and pie charts are effective methods for visualizing this information. Drawing from the data description (Table 4.1 in Chapter 4), you can present key demographic variables such as age, gender, airline status, and travel purposes. In this section, I highlight the most relevant visualizations from Chapter 4 for demonstration purposes.

It's important to remember that the graphs provide detailed demographic information, so repeating the same details in the report isn't necessary. Instead, focus on explaining and highlighting significant patterns or insights for readers to notice.

Here is an example of the demographic information results.

Figures 11.4–11.7 present passengers' age, gender, travel purposes, and airline class, respectively. The results show that most passengers are between the ages of 31 and 49, followed by those in the 49–58 age range. There are fewer older passengers aged 72–85. Additionally, the normal curve indicates that the data points are relatively close to a normal distribution. Overall, the sample covers a wide range of ages.

Regarding gender, there are slightly more female passengers than male (56.33% vs. 43.68%), but the ratio is relatively balanced and similar to the national demographics. The travel purposes reveal interesting insights, with the majority of passengers traveling for business purposes, followed by personal trips. Only a small percentage of passengers are traveling on mileage tickets.

Finally, most passengers hold blue status (around 65%), which mirrors the actual market distribution. The remaining 35% are divided among silver, gold, and platinum statuses. Overall, the sample reflects a broad range of passenger demographics, aligning closely with national airline passenger data. This suggests that the findings of this study can be generalized to the broader U.S. passenger population.

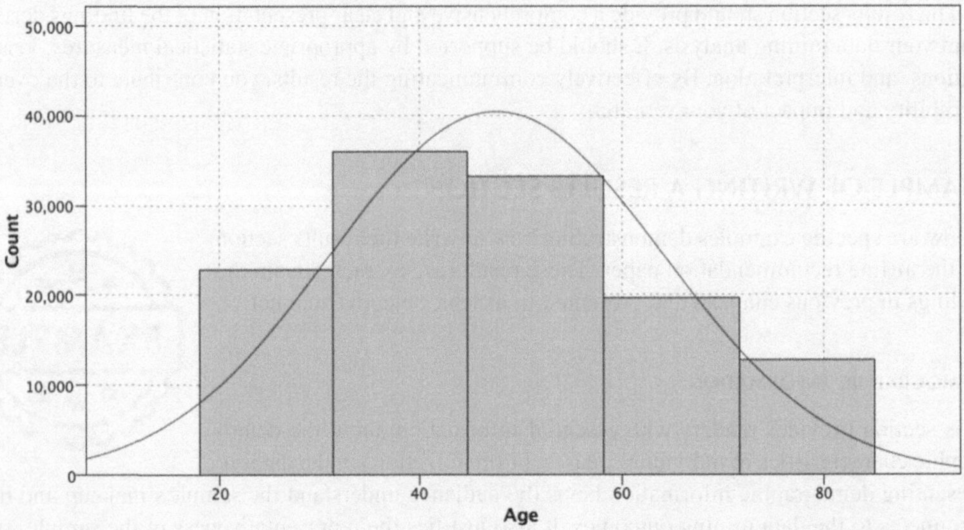

FIGURE 11.4 Age.

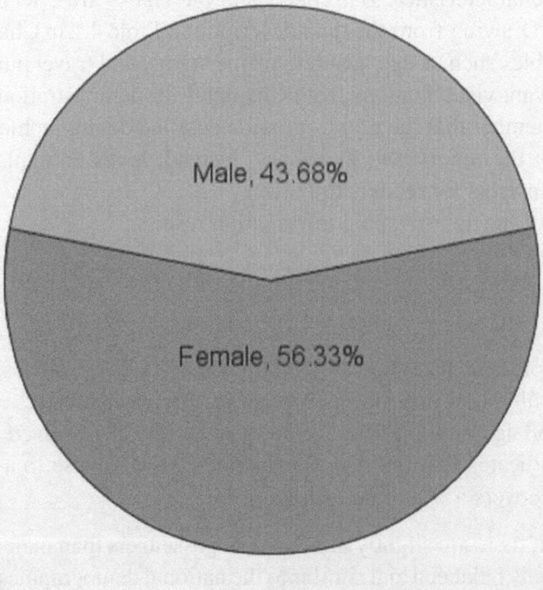

FIGURE 11.5 Gender.

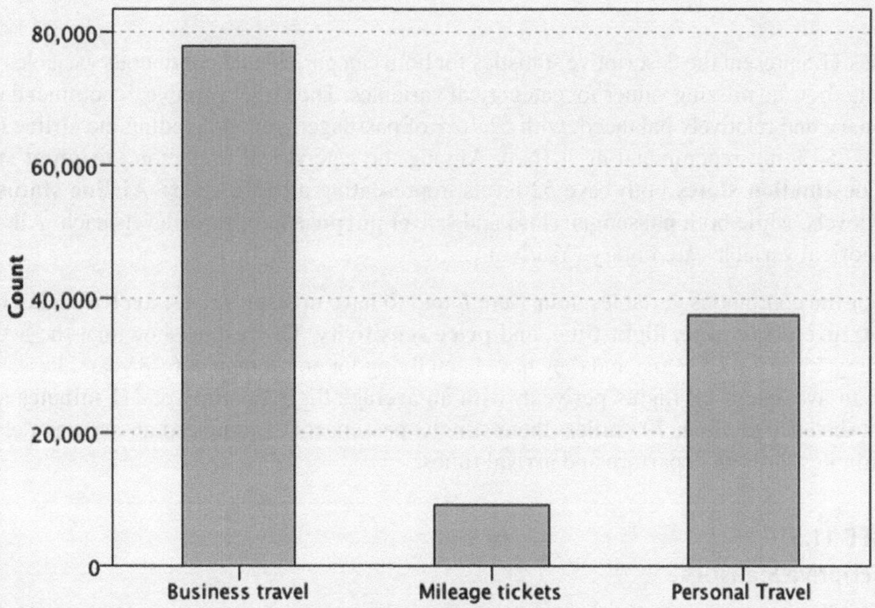

FIGURE 11.6 Travel purposes.

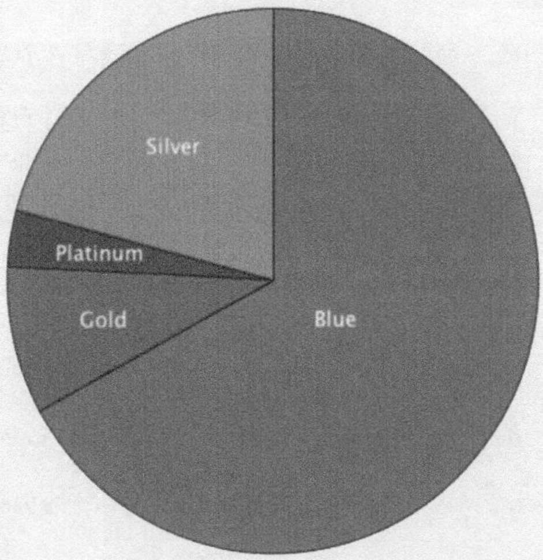

FIGURE 11.7 Airline status.

[Refer to Chapter 4 for detailed results.]

Descriptive Statistics Results

Descriptive statistics provide readers with an overview of the statistical measures summarizing the characteristics of all variables in the dataset. It is helpful to present detailed descriptive statistics for both categorical and interval variables using tables, as this allows for a clear and comprehensive view of the data. However, rather than explaining every detail, the focus should be on key observations or any notable trends that stand out.

Below is an example of how to write this part.

Tables 11.5 present the descriptive statistics for both categorical and continuous variables. The results show no missing values for categorical variables. The target variable, recommendation, is binary and relatively balanced, with 52.46% of passengers recommending the airline (Yes) and 47.54% not recommending it (No). Among the categorical predictors, **original states** and **destination states** both have 52 levels, representing all U.S. states. **Airline status** has four levels, while both **passenger class** and **travel purpose** have three levels each. All other categorical variables are binary (Yes/No).

For the continuous variables, four were found to have missing values: **arrival delay time, departure delay time, flight time,** and **price sensitivity**. The results show that the average passenger age is 47.5 years, and they have been flying for an average of 6.78 years. Passengers take an average of 20 flights per year, with an average flight duration of 111 minutes and a flight distance of about 793 miles. Interestingly, passengers experienced an average delay of 15 minutes for both departure and arrival times.

TABLE 11.5
Descriptive Statistics

Field	Min	Max	Mean	Standard Deviation	Skewness	Kurtosis	Unique	Valid	Missing
Recommend the airline	–	–	–	–	–	–	2	124,464	0
Airline status	–	–	–	–	–	–	4	124,464	0
Age	18	85	47.503	16.498	0.316	–0.711	–	124,464	0
Gender	–	–	–	–	–	–	2	124,464	0
Price sensitivity	1	5	1.308	0.498	1.327	1.173	–	120,488	3976
Number of years flying	2	11	6.784	2.981	–0.072	–1.291	–	124,464	0
Average number of flights per year	0	100	20.088	14.4	0.851	0.468	–	124,464	0
Percent of flights with other airlines	1	110	9.438	8.865	2.346	7.625	–	124,464	0
Travel purpose	–	–	–	–	–	–	3	124,464	0
Number of other loyalty cards	0	12	0.835	1.119	1.357	1.867	–	124,464	0
Shopping amount at the airport	0	879	26.87	53.156	3.309	15.935	–	124,464	0
Spending for eating and drinking at the airport	0	895	68.845	52.596	1.997	9.948	–	124,464	0
Passenger class	–	–	–	–	–	–	3	124,464	0
Origin state	–	–	–	–	–	–	52	124,464	0
Destination state	–	–	–	–	–	–	52	124,464	0
Departure delay (in minutes)	0	1592	15.003	38.448	6.817	100.767	–	122,263	2201
Arrival delay (in minutes)	0	1584	15.38	38.838	6.663	95.221	–	121,899	2565
Flight cancellation	–	–	–	–	–	–	2	124,464	0
Flight time (in minutes)	8	669	111.514	71.768	1.484	2.688	–	121,899	2565
Flight distance (miles)	31	4983	793.873	592.098	1.51	2.778	–	124,464	0

[Refer to Chapters 4 and 5 for detailed results. The only update here is the revised descriptive statistics for the price sensitivity variable, as we re-coded zero values to missing values during the data modification step. This replacement, which was completed in Chapter 5, was done using the

Filler node to replace zero values with missing values. After applying the Filler node, running the Data Audit node again on the entire dataset provided the updated descriptive statistics for the price sensitivity variable.]

Data Mining Results

This part is crucial as it presents the data mining results. Typically, you start with the model comparison to identify the champion model. Then, you provide details about the champion model, including its predictive power, structure, and variable importance. It is important not to repeat the details from tables and charts in the text but rather explain what they mean and highlight what the readers should focus on. You also need to explain the decisions you made throughout the process and provide justification for those choices.

Let's begin by presenting the model comparison results. Usually, both tables and charts are useful to illustrate the comparison between models using various model fit metrics. While ROC charts and Lift charts are commonly used metrics, in this case, it is not feasible to present these charts for all models in a single view because we extracted model fit metrics from different streams. Furthermore, the curves of the models are often very close to each other, making it difficult to distinguish the superior model from the charts alone. Therefore, we chose to present the ROC index, which is easily extracted from the Analysis node output. Additionally, we could use the Lift value (for 20% of responses) from the Lift chart, but since both metrics give similar outcomes, the ROC index was selected for its simplicity. Presenting the results in a table format allows for a clearer comparison.

Below is an example of how to write the model comparison results in this hypothetical scenario. [Refer to Chapters 7–9 for the detailed results.]

To determine the champion model, five predictive models were compared using four model fit metrics: accuracy, misclassification rate, sensitivity, specificity, F-1 score, Gini coefficient, and ROC index. Table 11.6 presents the model comparison using the testing sample's results. The results show that the performance of the five models is quite close. Based on the misclassification rate, Model 5 (boosted neural network) has the lowest misclassification rate of 20.99%, corresponding to the highest overall prediction accuracy of 79.01%, followed by Model 2 (CART tree) and Model 4 (Neural Network). Models 1 and 3 have relatively higher misclassification rates (above 21%) and were not considered further. Since the misclassification rate alone is not sufficient to determine model fit, additional metrics were examined to identify the best model. Model 5 also exhibits the highest Gini coefficient and ROC index. Based on the comparison of these four metrics, Model 5 is identified as the champion model.

TABLE 11.6
Model Comparison—Results for the Testing Sample

Metrics	Model 1: Logistic Regression	Model 2: CART Tree	Model 3: CHAID Tree	Model 4: Neural Network	Model 5: Boosted Neural Network
Accuracy	76.60%	78.70%	77.70%	78.14%	79.01%
Misclassification rate	23.40%	21.30%	22.30%	21.86%	20.99%
Sensitivity	0.893	0.964	0.883	0.905	0.913
Specificity	0.624	0.589	0.659	0.643	0.653
F-1 score	0.801	0.827	0.807	0.814	0.821
Gini coefficient	0.658	0.652	0.671	0.687	0.698
ROC index	0.829	0.826	0.836	0.844	0.849

Now that the champion model has been identified, it is important to further examine its performance to ensure that the results are reliable and valid and that the model possesses strong predictive power.

Below is an example of how to write this part.

The reliability evaluation of the champion model is shown in Table 11.7, comparing the results between the testing sample, training sample, and validation sample. The results are highly consistent, with discrepancies lower than 2%, which suggests that the model is reliable. In other words, the model generates consistent performance across the sub-samples, and there is no indication of overfitting.

TABLE 11.7
Reliability of the Champion Model (Boosted Neural Network Model)

Model Fit Metrics	Testing	Training	Validation
Accuracy	79.01%	79.27%	79.44%
Misclassification rate	20.99%	20.73%	20.56%
Sensitivity	0.913	0.915	0.913
Specificity	0.653	0.658	0.663
F1 score	0.821	0.822	0.824
Gini	0.698	0.712	0.707
ROC index	0.849	0.856	0.854

The validity of the champion model is evaluated based on the validation sample's results (Table 11.8). While the model's accuracy is below 80%, it is important to consider other metrics. Sensitivity is strong at over 90%, indicating that the model effectively identifies positive cases. However, specificity is relatively low at 66.3%, suggesting a higher rate of false positives. The F-1 score, which balances sensitivity and specificity, stands at 0.824, an acceptable value. Furthermore, the Gini coefficient of 0.707 reflects a good model fit, and the ROC index of 0.854 is considered acceptable. Overall, the predictive power of the model is deemed acceptable.

TABLE 11.8
Validity of the Champion Model (Boosted Neural Network Model)

Model Fit Metrics	Validation	Evaluation
Accuracy	79.44%	Marginal
Misclassification rate	20.56%	Marginal
Sensitivity	0.913	Good
Specificity	0.663	Not good
F1 score	0.824	Acceptable
Gini	0.707	Good
ROC index	0.854	Acceptable

[Refer to Chapters 9 and 10 for the detailed results.]

After evaluating the model performance, it is important to present the detailed model structure to inform the readers of the relationships between predictors and the target variable. In this case, since the champion model is an ensemble model, the structure cannot be presented in the same way as traditional models like neural networks. Unlike standard neural networks that display a clear network architecture, ensemble models are composed of multiple individual models with varying architectures, making it difficult to visualize the structure. Nevertheless, we can still describe the configuration, such as the number of hidden layers and neurons used in the boosted neural network model.

Here is an example.

The champion model is a boosted neural network with two hidden layers and 15 neurons in each layer. Since this model is an ensemble model, it is not possible to directly visualize or quantify the effects of the predictors on the target variable (airline recommendation). However, these effects can be analyzed through the variable importance chart. Figure 11.8 presents the variable importance chart, which ranks the significant predictors by order of importance. According to the results, **travel purpose** is the most important predictor, followed by **gender, average number of flights per year, airline status,** and **age**. These predictors are highly significant in predicting passengers' airline recommendations. Other less important predictors include the **number of other loyalty cards, arrival delay,** and **departure delay,** while **flight cancellation** and **price sensitivity** are among the least important predictors.

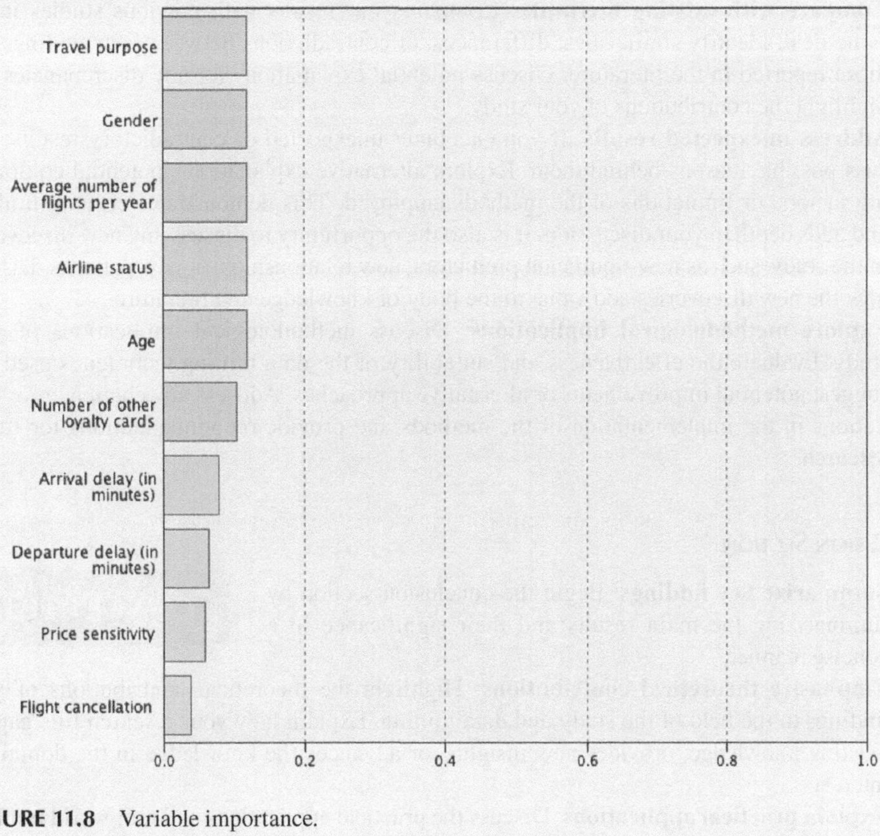

FIGURE 11.8 Variable importance.

[Refer to Chapters 8–10 for the detailed results.]

HOW TO WRITE DISCUSSION AND CONCLUSION SECTIONS?

The discussion and conclusion sections of a research paper are crucial for interpreting the findings, discussing their implications, and summarizing the key takeaways from your study. In some cases, the discussion and conclusion are written as two separate sections. And in other cases, they are included in the same section. It depends on the structure requirements set by the publisher, businesses, or funding organizations. Nonetheless, the strategies for writing the discussion and conclusion sections are the same regardless of the format. Below are some guidelines to help you write these sections effectively.

Discussion Section

1. **Recapitulate the findings**: Begin the discussion section by summarizing the main findings of your data mining analysis. Provide a concise overview of the significant results and patterns that emerged from your study.
2. **Revisit research objectives**: Restate the research objectives or questions that guided your study. Evaluate whether the objectives were met and to what extent the research questions were answered based on your findings.
3. **Interpret the findings:** Interpret the results in the context of your research objectives and the existing literature. Discuss the implications of the findings and explain their relevance to the research questions or problem you investigated.
4. **Compare with existing literature**: Compare your results with previous studies in the same field. Identify similarities, differences, or contradictions between your findings and those reported in the literature. Discuss potential explanations for any discrepancies and highlight the contributions of your study.
5. **Address unexpected results**: If you encounter unexpected or contradictory results, discuss possible reasons behind them. Explore alternative explanations, potential confounding factors, or limitations of the methods employed. This demonstrates critical thinking and adds depth to your discussion. It is also the opportunity to discuss any new discoveries in the study, such as new significant predictors, new relationships, or new patterns. Discuss how the new discoveries add value to the body of knowledge and literature.
6. **Explore methodological implications**: Discuss methodological implications of your study. Evaluate the effectiveness and suitability of the data mining techniques used and suggest potential improvements or alternative approaches. Address any challenges or limitations in the implementation of the methods and provide recommendations for future research.

Conclusion Section

1. **Summarize key findings**: Begin the conclusion section by summarizing the main results and their significance in a concise manner.
2. **Emphasize theoretical contributions**: Highlight the theoretical contributions of your findings to the field of the study and data mining. Explain how your research fills gaps in existing knowledge, provides new insights, or advances the knowledge in the domain of interest.
3. **Explain practical applications**: Discuss the practical applications and real-world implications of your findings. Explore how the insights gained from your data mining analysis can be applied in relevant domains or industries. Highlight potential benefits, challenges, and considerations for implementing the findings in practice.

4. **Provide recommendations to the target audience**: Offer recommendations to the target audience, especially the government, industry, or private organizations, on how to use the results of this study to enhance their operations or performance. All recommendations should be substantiated by the analysis results to avoid bias.
5. **Discuss limitations**: Acknowledge and discuss the limitations of your data mining analysis. This may include constraints in the dataset, assumptions made during the analysis, or potential biases. Explain how these limitations may affect the generalizability or validity of your findings.
6. **Discuss implications and future directions**: Discuss the broader implications of your findings and their potential impact. Highlight any new research avenues that have emerged from your study and suggest future directions for further investigation.
7. **End on a strong note**: Conclude your research paper with a strong and concise statement that summarizes the key takeaways from your study. Emphasize the significance of your findings and their potential contributions to the field of data mining.

Remember to provide a clear and logical flow of information in both the discussion and conclusion sections. Maintain a balance between summarizing the findings and providing critical analysis. By effectively discussing the implications and summarizing the key findings, you ensure that your research paper concludes on a strong note, leaving a lasting impression on the reader.

EXAMPLES OF WRITING DISCUSSION AND CONCLUSION SECTIONS

As mentioned earlier, the airline recommendation study is hypothetical, so I cannot provide an actual writing of the discussion and conclusion sections, which would typically require an in-depth analysis of the theoretical and practical implications of the results. However, I will present the complete **discussion** and **conclusion** sections from my published article (Truong, 2021). This breakdown by paragraph will show you how the content flows, followed by recommendations for writing these sections for the hypothetical airline recommendation case.

DISCUSSION SECTION

Figures 11.9 and 11.10 illustrate the **discussion** section from my article, structured in four paragraphs as follows:

- **Paragraph 1**: Restates the purpose of the study and summarizes the primary findings. The focus is on domestic air travel, identifying the key factors predicting domestic air travel volume during the pandemic. Any new discoveries are mentioned here.
- **Paragraph 2**: Explains the findings related to international travel, discussing the factors influencing international air travel volume. The discussion connects these findings to broader travel behavior patterns.
- **Paragraph 3**: Addresses the simulation results that predict medium- and long-term travel volumes. Detailed explanations of these results are provided, with emphasis on any new insights discovered through the simulations.
- **Paragraph 4**: Concludes the discussion by addressing the broader practical implications of the study's findings, especially for industry professionals and policymakers.

5. Discussions

The paper focuses on predicting air travel in the U.S. during and post COVID-19. The results of this study indicate some interesting and new findings. As for domestic air travel, the results show that WEI is the most important predictor, followed by trips between 50 and 500 miles. Population not staying home and [illegible] COVID-19 variables (new cases and new hospitalized) can be con[illegible]d moderately important along with domestic travel restrictions. Thus, when people decide to increase daily trips in medium or long distances for work, business, or personal visits, they will more likely consider traveling by air because they feel safe traveling. However, this decision is usually made in conjunction with the current U.S. government's travel restrictions and the spread and severity of COVID-19.

The results are somewhat similar for international air travel. The number of international flights is most affected by WEI, followed by COVID-19 new cases, new tests, and international restrictions. The only important daily travel variable is the trips between 100 and 250 miles. Hence, when people make more long distance trips domestically, they feel more comfortable flying [illegible] in the U.S. but not as comfortable flying internationally. The de[illegible] to travel to other countries is highly impacted by the economic condition, COVID-19 new tests, new cases, and current international travel restrictions by the U.S. government and governments of other countries. This finding is explained by the negative COVID-19 test result requirement, quarantine, and safety protocols issued by international travel restrictions policies.

FIGURE 11.9 Example of a discussion section (part 1) (Truong, 2021).

Conclusion Section

Figures 11.11 and 11.12 present the actual **conclusion** section of my article, structured into six different paragraphs, as follows:

- **Paragraph 1**: Restates the importance of the study and explains specific theoretical contributions of the findings. It highlights how the study addresses research gaps and adds value to the body of knowledge, particularly in the literature concerning the impacts of COVID-19 on air travel. The discovery of novel predictors is emphasized here.
- **Paragraph 2**: Discusses the implications of the simulation results, focusing on how they enable predictions of air travel volume in both the medium and long term.
- **Paragraph 3**: Explores the practical implications of the study, particularly how airlines can use the model and prediction results to develop effective business strategies. It also discusses how government bodies can utilize the findings to assess the impact of policies on the economy.

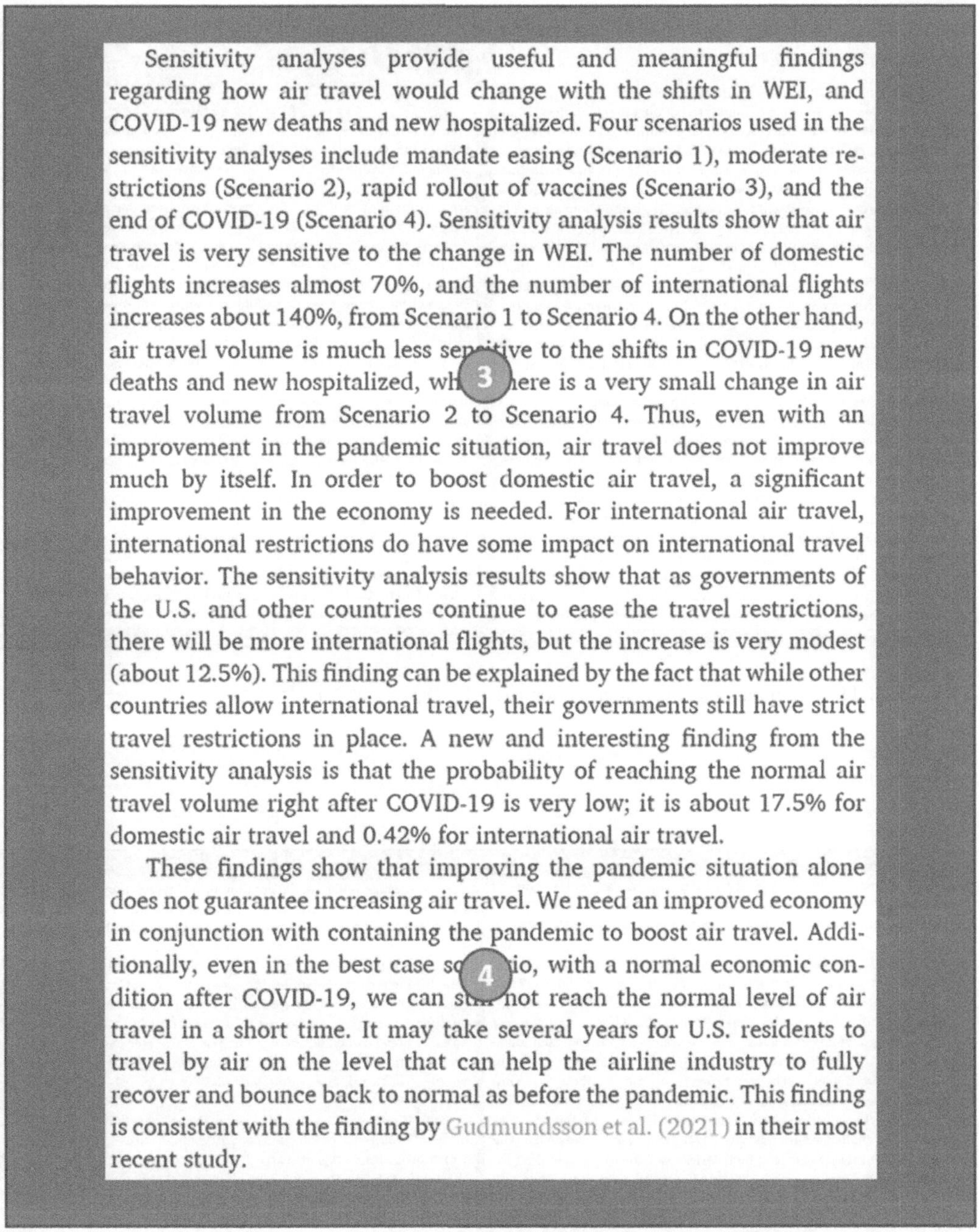

Sensitivity analyses provide useful and meaningful findings regarding how air travel would change with the shifts in WEI, and COVID-19 new deaths and new hospitalized. Four scenarios used in the sensitivity analyses include mandate easing (Scenario 1), moderate restrictions (Scenario 2), rapid rollout of vaccines (Scenario 3), and the end of COVID-19 (Scenario 4). Sensitivity analysis results show that air travel is very sensitive to the change in WEI. The number of domestic flights increases almost 70%, and the number of international flights increases about 140%, from Scenario 1 to Scenario 4. On the other hand, air travel volume is much less sensitive to the shifts in COVID-19 new deaths and new hospitalized, while there is a very small change in air travel volume from Scenario 2 to Scenario 4. Thus, even with an improvement in the pandemic situation, air travel does not improve much by itself. In order to boost domestic air travel, a significant improvement in the economy is needed. For international air travel, international restrictions do have some impact on international travel behavior. The sensitivity analysis results show that as governments of the U.S. and other countries continue to ease the travel restrictions, there will be more international flights, but the increase is very modest (about 12.5%). This finding can be explained by the fact that while other countries allow international travel, their governments still have strict travel restrictions in place. A new and interesting finding from the sensitivity analysis is that the probability of reaching the normal air travel volume right after COVID-19 is very low; it is about 17.5% for domestic air travel and 0.42% for international air travel.

3

These findings show that improving the pandemic situation alone does not guarantee increasing air travel. We need an improved economy in conjunction with containing the pandemic to boost air travel. Additionally, even in the best case scenario, with a normal economic condition after COVID-19, we can still not reach the normal level of air travel in a short time. It may take several years for U.S. residents to travel by air on the level that can help the airline industry to fully recover and bounce back to normal as before the pandemic. This finding is consistent with the finding by Gudmundsson et al. (2021) in their most recent study.

4

FIGURE 11.10 Example of a discussion section (part 2) (Truong, 2021).

- **Paragraph 4**: Addresses the limitations of the study, which stem from the availability and constraints of the data.
- **Paragraph 5**: Explains the reasoning behind the method not used in the study, emphasizing that the purpose was not to explore causal relationships.
- **Paragraph 6**: Provides recommendations for future research, suggesting areas where further investigation is needed or improvements can be made.

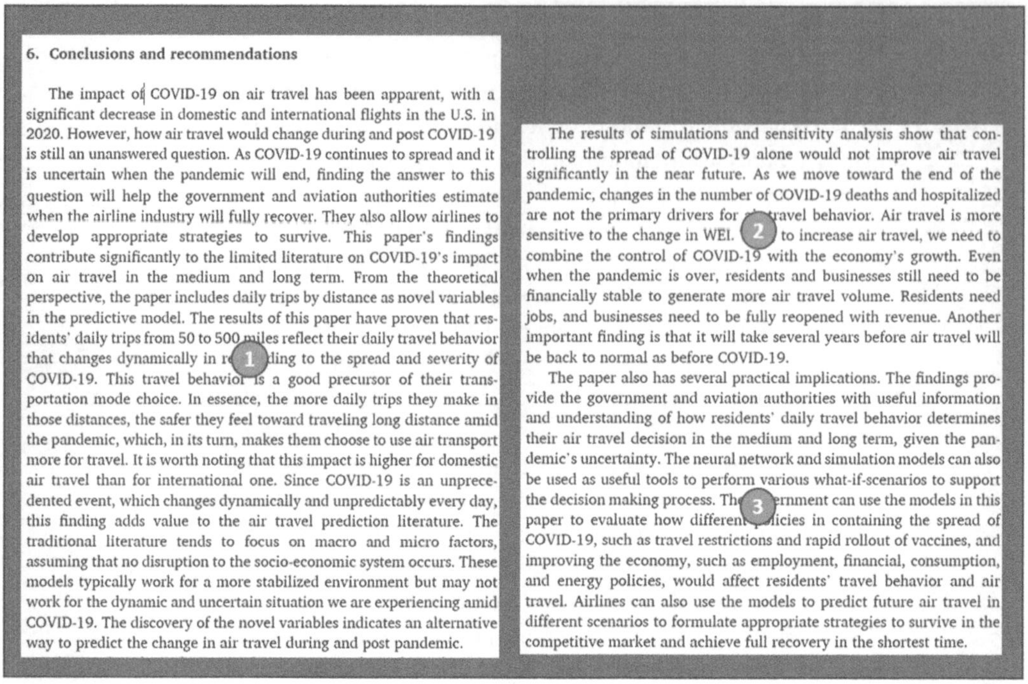

6. Conclusions and recommendations

The impact of COVID-19 on air travel has been apparent, with a significant decrease in domestic and international flights in the U.S. in 2020. However, how air travel would change during and post COVID-19 is still an unanswered question. As COVID-19 continues to spread and it is uncertain when the pandemic will end, finding the answer to this question will help the government and aviation authorities estimate when the airline industry will fully recover. They also allow airlines to develop appropriate strategies to survive. This paper's findings contribute significantly to the limited literature on COVID-19's impact on air travel in the medium and long term. From the theoretical perspective, the paper includes daily trips by distance as novel variables in the predictive model. The results of this paper have proven that residents' daily trips from 50 to 500 miles reflect their daily travel behavior that changes dynamically in r 1 ding to the spread and severity of COVID-19. This travel behavior is a good precursor of their transportation mode choice. In essence, the more daily trips they make in those distances, the safer they feel toward traveling long distance amid the pandemic, which, in its turn, makes them choose to use air transport more for travel. It is worth noting that this impact is higher for domestic air travel than for international one. Since COVID-19 is an unprecedented event, which changes dynamically and unpredictably every day, this finding adds value to the air travel prediction literature. The traditional literature tends to focus on macro and micro factors, assuming that no disruption to the socio-economic system occurs. These models typically work for a more stabilized environment but may not work for the dynamic and uncertain situation we are experiencing amid COVID-19. The discovery of the novel variables indicates an alternative way to predict the change in air travel during and post pandemic.

The results of simulations and sensitivity analysis show that controlling the spread of COVID-19 alone would not improve air travel significantly in the near future. As we move toward the end of the pandemic, changes in the number of COVID-19 deaths and hospitalized are not the primary drivers for travel behavior. Air travel is more sensitive to the change in WEI. 2 to increase air travel, we need to combine the control of COVID-19 with the economy's growth. Even when the pandemic is over, residents and businesses still need to be financially stable to generate more air travel volume. Residents need jobs, and businesses need to be fully reopened with revenue. Another important finding is that it will take several years before air travel will be back to normal as before COVID-19.

The paper also has several practical implications. The findings provide the government and aviation authorities with useful information and understanding of how residents' daily travel behavior determines their air travel decision in the medium and long term, given the pandemic's uncertainty. The neural network and simulation models can also be used as useful tools to perform various what-if-scenarios to support the decision making process. Th 3 ernment can use the models in this paper to evaluate how differen licies in containing the spread of COVID-19, such as travel restrictions and rapid rollout of vaccines, and improving the economy, such as employment, financial, consumption, and energy policies, would affect residents' travel behavior and air travel. Airlines can also use the models to predict future air travel in different scenarios to formulate appropriate strategies to survive in the competitive market and achieve full recovery in the shortest time.

FIGURE 11.11 Example of a conclusion section (part 1) (Truong, 2021).

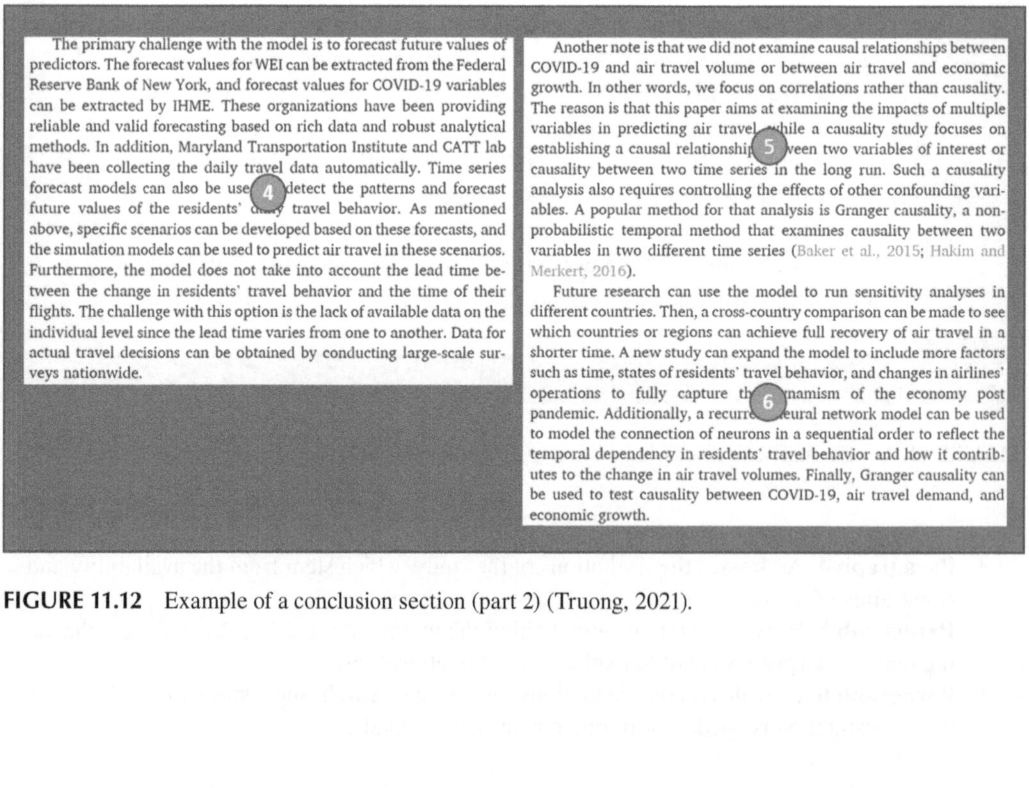

The primary challenge with the model is to forecast future values of predictors. The forecast values for WEI can be extracted from the Federal Reserve Bank of New York, and forecast values for COVID-19 variables can be extracted by IHME. These organizations have been providing reliable and valid forecasting based on rich data and robust analytical methods. In addition, Maryland Transportation Institute and CATT lab have been collecting the daily travel data automatically. Time series forecast models can also be use 4 detect the patterns and forecast future values of the residents' daily travel behavior. As mentioned above, specific scenarios can be developed based on these forecasts, and the simulation models can be used to predict air travel in these scenarios. Furthermore, the model does not take into account the lead time between the change in residents' travel behavior and the time of their flights. The challenge with this option is the lack of available data on the individual level since the lead time varies from one to another. Data for actual travel decisions can be obtained by conducting large-scale surveys nationwide.

Another note is that we did not examine causal relationships between COVID-19 and air travel volume or between air travel and economic growth. In other words, we focus on correlations rather than causality. The reason is that this paper aims at examining the impacts of multiple variables in predicting air travel while a causality study focuses on establishing a causal relationshi 5 veen two variables of interest or causality between two time series in the long run. Such a causality analysis also requires controlling the effects of other confounding variables. A popular method for that analysis is Granger causality, a non-probabilistic temporal method that examines causality between two variables in two different time series (Baker et al., 2015; Hakim and Merkert, 2016).

Future research can use the model to run sensitivity analyses in different countries. Then, a cross-country comparison can be made to see which countries or regions can achieve full recovery of air travel in a shorter time. A new study can expand the model to include more factors such as time, states of residents' travel behavior, and changes in airlines' operations to fully capture th 6 namism of the economy post pandemic. Additionally, a recurr eural network model can be used to model the connection of neurons in a sequential order to reflect the temporal dependency in residents' travel behavior and how it contributes to the change in air travel volumes. Finally, Granger causality can be used to test causality between COVID-19, air travel demand, and economic growth.

FIGURE 11.12 Example of a conclusion section (part 2) (Truong, 2021).

DISCUSSIONS AND CONCLUSIONS FOR THE AIRLINE RECOMMENDATION PAPER

Below are key points that you should cover in the discussion and conclusion sections for the airline recommendation paper:

Discussions

- **Restate the purpose**: Start by restating the main objectives of the study, which were to build an accurate predictive model for passengers' airline recommendations and to identify the most important predictors influencing those recommendations.
- **Discuss findings**: Highlight the main findings in relation to these objectives. In this case, five predictive models were developed and validated. Among them, the neural network model with one layer emerged as the champion model, demonstrating acceptable predictive power. Additionally, the model output identified travel purpose as the most important predictor, followed by gender, average number of flights per year, and airline status.
- **Interpret the findings**: Offer insights into what these findings mean in the context of the airline industry. For instance, the high importance of travel purpose may suggest that airlines should tailor their services based on passengers' reasons for traveling, while gender differences could indicate specific marketing strategies for male and female passengers.
- **Compare with existing literature**: Compare your findings with previous studies, noting similarities or discrepancies. Highlight any new discoveries, such as new predictors or relationships that were not identified in previous research.
- **Implications for practice**: Discuss how these findings can be practically applied by airlines to improve operations, services, or marketing efforts. For example, knowing which factors most influence recommendations can help airlines enhance customer experience and satisfaction, ultimately leading to better business outcomes.

Conclusions

- **Restate the importance**: Summarize the significance of the study and its main findings. Keep this brief since the findings are discussed in more detail in the previous section.
- **Theoretical contributions**: Explain how the findings contribute to the academic literature. Highlight how the study helps fill existing research gaps, particularly in the context of airline recommendation prediction using machine learning.
- **Practical contributions**: Discuss how the results can be applied in the airline industry. Airlines can use the predictive model to forecast passenger recommendations and establish effective business strategies. Additionally, the identification of key predictors like travel purpose and gender can inform operational and marketing improvements to enhance customer satisfaction.
- **Limitations**: Address any limitations in the study. For instance, the exclusion of variables related to customer service or onboard services due to data constraints, and the focus on U.S.-based passengers, may limit the generalizability of the findings.
- **Recommendations for future research**: Suggest areas for future research, such as including additional variables like customer service ratings, expanding the study to include international flights, or using more advanced machine learning techniques to further improve the accuracy of the model.

This structure ensures that both the discussion and conclusion sections provide valuable insights and wrap up the research in a meaningful and comprehensive way.

CHAPTER SUMMARY

Writing an effective data mining report is a crucial step in any data mining project, where the results and insights obtained from analyzing large datasets are documented and communicated clearly. The first step involves identifying the target audience, understanding the report's requirements, and selecting the appropriate writing style. The report should begin by defining the research problem, followed by the scope and objectives of the project, and providing justification for the study's importance. The background section is used to inform the readers about existing research in the field, highlighting gaps that justify the study's originality. Additionally, the literature review provides the theoretical foundation for the models used in the study.

The methodology section is key to enabling readers to assess the validity of the findings and replicate the study. This section should include a description of the dataset, the variables analyzed, the specific machine learning methods used, their configurations, and the evaluation metrics employed.

The next stage in the report is the presentation of results and analysis. This section should present the patterns, associations, or correlations found in the data, often using visualizations such as charts and graphs to make the information more accessible. The report should not only showcase the outcomes of the data mining process but also provide clear explanations and insights into the significance of the results within the context of the research objectives.

The final part of the data mining report is the discussion and conclusion section. Here, the key findings are summarized along with their implications. The practical applications of the results may also be discussed, and the section can suggest areas for future research. If applicable, the report may offer actionable recommendations based on the data analysis.

In summary, writing an effective data mining report requires thorough planning, diligent data collection and analysis, and effective communication of findings. This process demands perseverance, critical thinking, and adherence to ethical standards. A well-structured, concise, and audience-specific report empowers stakeholders to make informed decisions based on the knowledge extracted from the data.

CHAPTER DISCUSSION QUESTIONS

1. What is the main purpose of a data mining report?
2. Why is it important to an effective report for a data mining project?
3. What are the primary differences between academic research reports, technical reports, and business reports? Discuss in detail those differences.
4. Select one data mining project and identify the target audience. What are the expectations of that audience for your data mining report?
5. What is the effective writing style for that audience? Why? Describe the pros and cons of that writing style.
6. What are the primary sections of a data mining report? Describe the purpose and content of each section.
7. What are different ways to hook the audience and keep their interest in your report?
8. How do you tell if a report title is effective? What are strategies to write a good title?
9. What are the key strategies for writing each section of the report?

CHAPTER EXERCISES

Exercise 1: Writing the Report for the Bank Dataset

1. Prepare an outline for a business report for the bank customer project. The purpose is to build a predictive model that accurately predicts a customer's credit score based on their demographic variables and banking information.
2. Write a project title.

3. Write an abstract.
4. Write a brief introduction section, including the problem statement, research questions, purpose statement, significance of the study, and delimitations.
5. Write a methodology section, including data source, variable description, data mining process, machine learning methods and configurations, and model evaluation metrics.
6. Choose the results from the following models you completed for the exercises in Chapters 7 and 8 (choose the best one for each type):
   - Linear regression using stepwise.
   - CART tree.
   - CHAID tree.
   - C5.0 tree.
   - MLP model.
7. Write a results section to present the findings, including:
   - Demographic information.
   - Descriptive statistics results.
   - Model comparison results.
   - Champion model identification.
   - Evaluation of the reliability and validity of the champion model.
   - Detailed structure and relationships in the champion model.
   - Explanation of variable importance.
8. Write a brief discussion section, including discussions of the findings, their implications, and limitations of the analysis.

Exercise 2: Writing the Report for the Passengers Dataset

1. Prepare an outline for a business report for the airline passengers project. The purpose is to build a predictive model that accurately predicts a passenger's decision to return to the same airline based on their demographic variables, flight information, and their evaluation of airline services.
2. Write a project title.
3. Write an abstract.
4. Write a brief introduction section, including the problem statement, research questions, purpose statement, significance of the study, and delimitations.
5. Write a methodology section, including data source, variable description, data mining process, machine learning methods and configurations, and model evaluation metrics.
6. Choose the results of the models you completed for the exercises in Chapters 8 and 9 (choose the best one for each type):
   - CART tree
   - CHAID tree.
   - MLP model
   - RBF model.
7. Write a results section to present the findings, including:
   - Demographic information.
   - Descriptive statistics results.
   - Model comparison results.
   - Champion model identification.
   - Evaluation of the reliability and validity of the champion model.
   - Detailed structure and relationships in the champion model.
   - Explanation of variable importance.
8. Write a brief discussion section, including discussions of the findings, their implications, and limitations of the analysis.

12 Principal Component Analysis

LEARNING OUTCOMES

Upon the completion of this chapter, you will be able to:

1. Describe the concepts of principal component analysis (PCA) as a dimensionality reduction technique in data mining.
2. Explain the principles and algorithms used in PCA for extracting the principal components from a dataset.
3. Discuss the advantages and disadvantages of PCA in the context of data mining.
4. Explain how PCA captures the variance and correlation structure of data through eigenvalues and eigenvectors.
5. Select the proper number of principal components to retain using the scree plot, cumulative eigenvalue, and explained variance.
6. Explain component rotation methods used in PCA to align principal components with the original variables in order to improve interpretability.
7. Conduct PCA to reduce the number of variables using SPSS Modeler.

INTRODUCTION

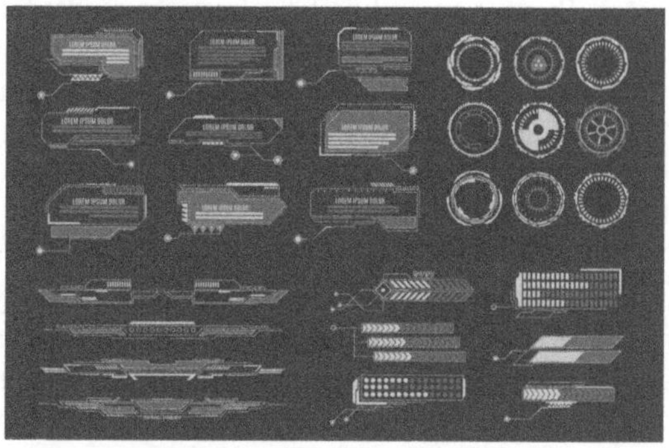

When dealing with big data, we often encounter datasets containing a large number of variables, sometimes numbering in the hundreds. These are referred to as high-dimensional datasets. In an effort to improve the accuracy of predictive models, researchers frequently attempt to include all of these variables in the analysis. While this can sometimes be effective, it often leads to issues. As discussed in previous chapters, the number of variables included in the modeling process can significantly impact the model's explanatory power. The more variables involved, the lower the explanatory power might become because the algorithm must account for the variance in all variables. This can result in a model that fails to achieve satisfactory predictive performance. Additionally, a large number of variables may significantly increase computing time, making the modeling process inefficient.

So, how can we reduce the number of variables in a dataset without compromising the model's validity? The solution largely depends on the characteristics of the data. For instance, during the exploration step, we might discover that some variables are irrelevant to the target variable, meaning they do not significantly contribute to the prediction. In such cases, we can use a variable selection tool to filter out the most relevant predictors, reducing the number of variables included in the model. However, in some cases, variables might be relevant but highly correlated with each other, creating subsets of correlated variables within the dataset. Simply discarding some of these variables could result in the loss of valuable information. A more effective solution is to combine these correlated variables into a smaller set of new composite variables. In simpler terms, we replace the original

DOI: 10.1201/9781003468196-14

variables with a set of new variables, each representing a combination of the correlated variables. This results in a dataset with fewer dimensions. This approach is known as dimensionality reduction.

The challenge with dimensionality reduction is that it alters the original data. Therefore, the key question becomes: How do we reduce the dimensionality of the data while preserving as much of the original information as possible?

Principal component analysis (PCA) is one of the most commonly used techniques for dimensionality reduction in machine learning. PCA is a statistical method that helps summarize the variation in a high-dimensional dataset by identifying the most important variables or combinations of variables, known as principal components. PCA is widely applied in fields such as computer vision, image processing, data mining, and bioinformatics, among others. In this chapter, we will explore the basic concepts of PCA, its practical applications, and the steps involved in implementing PCA in machine learning.

CURSE OF DIMENSIONALITY

Before diving into the details of dimensionality reduction and PCA, it's essential to understand a common issue in data mining known as the **curse of dimensionality**. This term, introduced by Richard Bellman in the context of optimization and dynamic programming, refers to problems that arise when working with high-dimensional data, or datasets with a large number of variables. As the number of dimensions (variables) in a dataset increases, the amount of data needed to ensure that the data are representative and informative grows exponentially. Simply put, when the number of variables increases, we require a very large sample size to accurately model or analyze the data. Without this, many machine learning techniques that perform well in low-dimensional settings become less effective or even infeasible in high-dimensional spaces.

One key issue with high-dimensional data is that it tends to become sparse. As the number of dimensions grows, the available data becomes less representative of the overall distribution, making it harder to model or make accurate predictions. Moreover, high-dimensional data often contains redundant or irrelevant features, further complicating the modeling process and potentially reducing the model's performance.

The curse of dimensionality also significantly affects computational efficiency. Many machine learning algorithms require substantial computation, and as the dimensionality of the data increases, so too does the computational burden. This can result in extended training times, higher memory usage, and decreased scalability, making it difficult to apply these algorithms to very large datasets.

To mitigate the curse of dimensionality, several techniques can be used, such as dimensionality reduction, variable selection, and regularization. These techniques aim to reduce the number of dimensions while preserving the most important information in the dataset, thereby simplifying the modeling process and improving performance. In this chapter, we focus on one of the most widely used dimensionality reduction techniques: PCA. PCA helps reduce the dimensionality of data while retaining as much of the original information as possible.

WHAT IS PRINCIPAL COMPONENT ANALYSIS (PCA)?

DEFINITION

PCA (Principal Component Analysis) is a dimensionality reduction technique that transforms high-dimensional data into a smaller set of uncorrelated variables, called principal components, which capture the most important information and variance in the data.

Principal component analysis (PCA) is a method used to identify patterns and structure in high-dimensional data by reducing its dimensionality while preserving the most important information. In other words, PCA allows us to condense a large set of variables into a smaller set without losing key information. This is achieved by

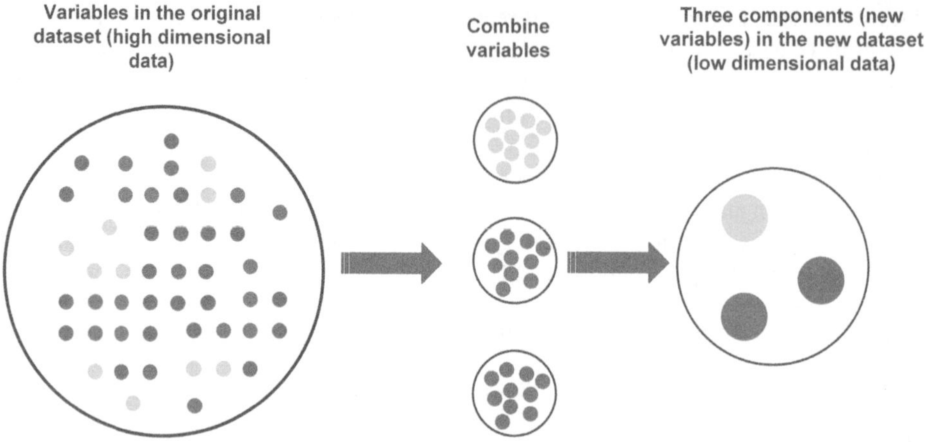

FIGURE 12.1 Illustration of the dimensionality reduction.

transforming the original variables into a new set of variables, called principal components, which are linearly uncorrelated and capture the maximum variance in the data. These principal components are unique and effectively explain the variance in the dataset. Figure 12.1 illustrates this dimensionality reduction process.

Principal Components

The principal components are linear combinations of the original variables, and each principal component represents a new dimension in the reduced dimensional space. The first principal component captures the largest amount of variance in the data, and each subsequent component captures the next largest amount of variance, subject to the constraint that it is uncorrelated with the previous components.

> **DEFINITION**
>
> Principal components are linear combinations of the original variables, with each principal component representing a new dimension in the reduced dimensional space.

Eigenvalue

The importance of each principal component can be measured by its eigenvalue, which represents the amount of variance in the data captured by that component. The eigenvalue of each principal component is equal to the squared length of its corresponding eigenvector, which is a vector that defines the direction of the component. I won't get into the mathematical details of eigenvalue calculation. The key point is that the eigenvalue allows us to determine the right number of components in our dataset.

BASIC CONCEPTS OF PRINCIPAL COMPONENT ANALYSIS

There are several fundamental concepts that are essential to understanding PCA. These concepts form the basis of how PCA works and why it is used. Table 12.1 presents these key concepts along with detailed descriptions, offering a comprehensive overview to help you grasp the important principles and processes behind PCA. This table will guide you through each concept, ensuring a solid understanding of how PCA can be applied to data analysis.

TABLE 12.1
Basic Concepts of PCA and Descriptions

Concepts	Description
Variance	Variance of a variable measures how much the values of that variable vary from their mean. In PCA, the goal is to find the directions in the data with the highest variance as these directions contain the most information about the data.
Covariance	Covariance between two variables measures how much they vary together. If two variables have a high covariance, it means they are positively correlated, while a low covariance indicates no correlation. In PCA, we choose the principal components with the expectation that they are uncorrelated.
Eigenvectors and eigenvalues	PCA seeks to identify the directions of maximum variability in the data. Eigenvectors are the directions in the data that are not affected by a linear transformation, while eigenvalues indicate how much variance is captured by each eigenvector. In PCA, the principal components are the eigenvectors of the covariance matrix sorted by their corresponding eigenvalues.
Principal components	Principal components are the eigenvectors that correspond to the highest eigenvalues of the covariance matrix. They are ordered in terms of their importance, where the first principal component explains the maximum variance in the dataset, followed by the second principal component, and so on.
Explained variance	The proportion of variance explained by each principal component is crucial for understanding the importance of each component in the data. It helps us determine how many principal components to retain for dimensionality reduction while preserving significant information.
Scree plot	A scree plot is a graphical tool used in PCA to visualize the eigenvalues associated with each principal component. It is a simple line segment plot that shows the eigenvalues for each individual principal component. The scree plot helps determine the number of principal components based on the amount of variance they explain.
Component rotation	Component rotation is used to transform the principal components to align those components with the original variables, which achieves a more interpretable and meaningful representation of the data.
Orthogonal rotation	Orthogonal rotation is the rotation method, in which the rotated components are constrained to be orthogonal to each other, meaning they are independent and uncorrelated.
Oblique rotation	Oblique rotation is the rotation method in which the rotated components are allowed to be correlated with each other.
Varimax	Varimax is an orthogonal rotation method that aims to maximize the variance of the squared loadings on each component, leading to a simple and sparse structure where each variable has a high loading on one component and close to zero loadings on other components.
Oblimin	Oblimin is an oblique rotation method that allows for correlation or obliqueness between the rotated components. Thc main objective of Oblimin rotation is to achieve a simpler structure by minimizing the number of variables with high loadings on a component.
Promax	Promax is an oblique rotation method that extends the advantages of Varimax rotation while also accounting for possible correlations between the rotated components. It is considered a compromise between the simplicity of orthogonal rotation and the flexibility of oblique rotation.
Dimensionality reduction	Once the principal components have been identified, PCA can be used to transfer the high-dimensional data into a lower-dimensional data, while still retaining as much of the original variation as possible. The number of principal components retained determines the dimensionality of the new data.
Reconstruction	PCA allows the inverse transformation of the lower-dimensional data back into the original high-dimensional space. Although some information is lost during dimensionality reduction, the reconstruction can still be useful for interpretation or other downstream tasks.

PRACTICAL APPLICATIONS OF PCA IN THE INDUSTRY

PCA has a wide range of practical applications across various industries. Some key examples include:

- **Image compression**: PCA can be used to compress images by reducing the dimensionality of the image data while preserving the essential features. For example, in facial recognition systems, PCA can be used to reduce the number of dimensions in the image data to improve recognition accuracy.
- **Finance**: In finance, PCA can be used to identify the underlying factors that drive the correlation structure of asset returns. By decomposing the covariance matrix of asset returns into principal components, portfolio managers can better understand the sources of risk and return in their portfolios.
- **Genetics**: In genetics, PCA can be used to identify the genetic variations that contribute to the differences between individuals. By analyzing the principal components of genetic data, researchers can identify groups of individuals with similar genetic profiles and gain insights into the genetic basis of diseases.
- **Natural Language Processing**: In natural language processing, PCA can be used to reduce the dimensionality of text data by extracting the most informative features. For example, in sentiment analysis, PCA can be used to identify the keywords and phrases that are most strongly associated with positive or negative sentiment.
- **Marketing**: In marketing, PCA can be used to identify the key factors that drive consumer preferences and behavior. By analyzing the principal components of customer data, marketers can identify the segments of customers that are most similar and tailor their marketing strategies accordingly.

These are just a few examples of the many applications of PCA. PCA is a versatile and widely used technique that can help simplify complex datasets and reveal the underlying structure and patterns in the data.

APPLICATIONS OF PCA IN DATA MINING

In the context of data mining and predictive modeling, PCA serves two main purposes, as follows.

- **In the Modify step of supervised learning**: PCA is used to transform a large number of predictors into a smaller number of predictors (principal components). These components are then used as new predictors in the Model step to develop predictive models for the target variable.
- **In unsupervised learning**: PCA is employed to explore and detect underlying theoretical factors within the data. The principal components represent these factors, which can then be used to build a theoretical research model to examine the relationships among them.

ADVANTAGES AND DISADVANTAGES

PCA is a powerful technique for reducing the dimensionality of datasets in machine learning. It has several advantages, but there are also some challenges associated with its use. Understanding these can help data scientists make informed decisions about whether PCA is suitable for their specific data mining projects.

Advantages of PCA:

- **Reduces dimensionality**: PCA effectively reduces the number of variables in a dataset, making the data easier to analyze, visualize, and interpret.
- **Retains maximum information**: PCA minimizes information loss by retaining the most important variance in the data, ensuring that the reduced dataset still captures most of the original dataset's essential information.
- **Eliminates correlated features**: By transforming correlated variables into uncorrelated principal components, PCA eliminates redundancy and simplifies the structure of the data.
- **Improves model accuracy**: By removing noise and irrelevant variables, PCA can improve the accuracy of machine learning models, as the most relevant data are highlighted.
- **Speeds up computation**: Reducing the number of dimensions can significantly decrease the computational burden of machine learning algorithms, leading to faster processing and model building.

Disadvantages of PCA:

- **Interpretability**: The new set of principal components may not have a clear or intuitive interpretation, making it difficult to understand the meaning of the transformed variables.
- **Information loss**: While PCA aims to retain as much information as possible, reducing dimensions inevitably leads to some loss of detail. The more dimensions you remove, the more information may be lost.
- **Sensitivity to outliers**: Outliers can distort the calculation of principal components, leading to skewed results and potentially undermining the analysis.
- **Assumes linearity**: PCA works best when the relationships between variables are linear. It may not perform well on datasets with complex, non-linear relationships.
- **Scalability**: PCA can struggle with very large datasets, as the computational and memory requirements increase with the size of the data, potentially limiting its scalability in some cases.

Balancing these advantages and disadvantages allows data scientists to decide whether PCA is the best dimensionality reduction technique for their specific needs.

STEPS OF PRINCIPAL COMPONENT ANALYSIS

The general steps involved in performing PCA are outlined as follows:

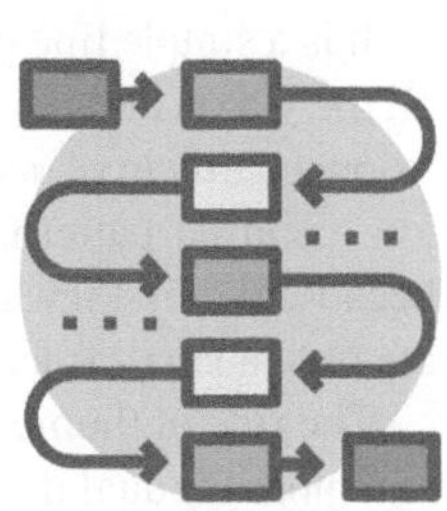

1. **Data standardization**: The first step in PCA is to standardize the dataset by centering the variables around zero and scaling them to have unit variance. This ensures that variables with larger variances do not dominate the analysis.
2. **Compute the covariance matrix**: After standardizing the data, the covariance matrix is calculated to capture the relationships between the variables in the dataset.
3. **Compute the eigenvectors and eigenvalues**: The next step is to compute the eigenvectors and eigenvalues of the covariance matrix. The eigenvectors represent the directions of maximum variability in the data, while the eigenvalues quantify the amount of variability each eigenvector explains.

4. **Rank eigenvectors by eigenvalues**: The eigenvectors are ranked according to their eigenvalues. The eigenvector with the highest eigenvalue represents the first principal component, explaining the most variance.
5. **Select the principal components**: In this step, the number of principal components to retain is determined. Typically, this is based on a threshold for the variance explained (e.g., 70% or higher) or using methods like the Scree Plot to help identify the optimal number of components.
6. **Transform the data**: The selected principal components are then used to transform the data into a lower-dimensional space, resulting in a new set of variables that capture most of the original dataset's variability.
7. **Interpret the results**: Finally, the transformed data can be analyzed to uncover patterns and structures in the reduced-dimensional space.

In SPSS Modeler, most of the PCA steps are automated, and the software manages the technical calculations. However, one crucial decision that requires a solid understanding is selecting the number of principal components to retain. The goal is to capture a high percentage of the total variance in the data without retaining too many components, as this would undermine the purpose of dimensionality reduction.

A general rule of thumb is to stop adding components when an additional component does not significantly reduce the eigenvalue, meaning it no longer improves the explanation of the data's variance. One widely used method for determining this point is the Scree Plot. This graphical tool helps identify when the eigenvalue levels off, indicating that adding more components provides diminishing returns in terms of explained variance. I will discuss the Scree Plot method in the following section.

SCREE PLOT AND NUMBER OF COMPONENTS

A **Scree Plot** is a graphical tool used in PCA to visualize the eigenvalues associated with each principal component. It plots the eigenvalues in descending order, helping to identify the "elbow" point where the eigenvalues start to level off. This point indicates the number of principal components to retain, as it shows where adding more components no longer significantly increases the amount of variance explained by the model. The scree plot is a valuable tool for selecting the optimal number of principal components in a dataset.

> **DEFINITION**
>
> A scree plot is a graphical tool used in PCA to visualize the eigenvalues associated with each principal component. The scree plot helps determine the number of principal components based on the amount of variance they explain.

It is a simple line plot that shows the eigenvalues for each principal component, with eigenvalues on the y-axis and the number of components on the x-axis (Figure 12.2). The plot typically displays a downward curve, as adding additional components decreases the eigenvalue. Most scree plots follow a similar pattern: they start high on the left, fall quickly, and then level off. This shape occurs because the first few components explain most of the variance, while the latter components contribute only a small fraction of the total variability.

The rule of thumb is to identify the "elbow" in the plot, where the line shifts from a steep descent to a more gradual decline. The components before the elbow are typically retained. For example, if we see an elbow at three components, we select these three components for the PCA.

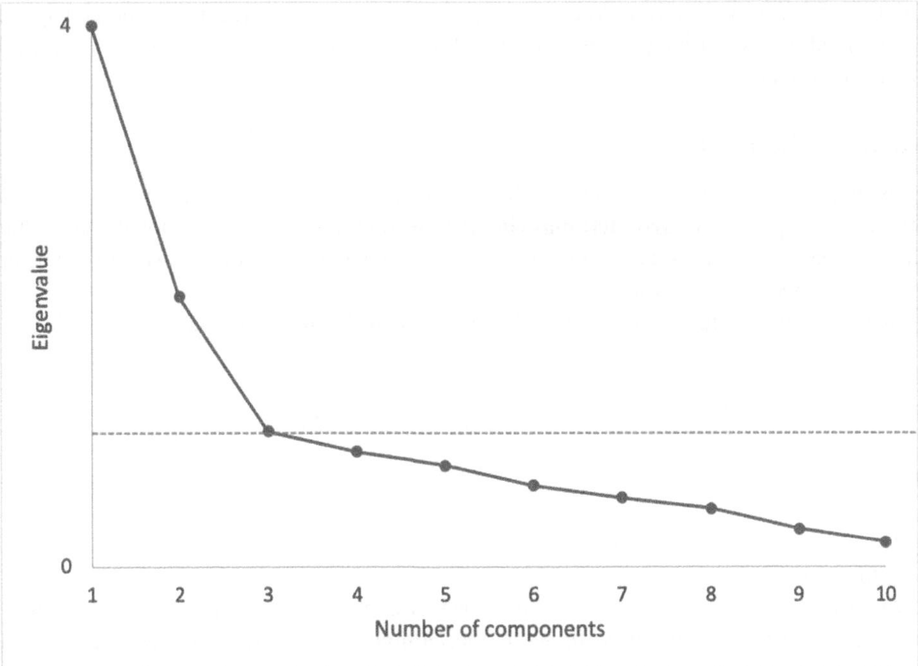

FIGURE 12.2 Example of a scree plot.

RELATIONSHIP BETWEEN EIGENVALUES AND PERCENTAGE OF VARIANCE EXPLAINED

The relationship between **eigenvalues** and the **percentage of variance explained** in PCA is key to understanding the significance of each principal component in capturing the variability within a dataset.

In PCA, eigenvalues represent the amount of variance explained by each principal component. Larger eigenvalues indicate that the corresponding principal component captures more variance in the data. To calculate the percentage of variance explained by each principal component, you divide the eigenvalue of that component by the total sum of eigenvalues (which represents the total variance in the data) and multiply by 100 to express it as a percentage.

Here's the formula to calculate the percentage of variance explained by each principal component:

$$Percentage\ of\ Variance\ Explained_i = \frac{Eigenvalue_i}{\sum_j^n Eigenvalue_j} \times 100$$

where

- i is the index of the principal component.
- n is the total number of principal components.
- $Eigenvalue_i$ is the eigenvalue of the ith principal component.

By calculating the percentage of variance explained for each component, you can assess the relative importance of each in capturing the variability of the original dataset. Most software programs that perform PCA provide a Total Variance Explained table, which includes the percentage of variance explained by each principal component and the cumulative percentage of variance explained up to that component.

Proportional Eigenvalues

In cases where a scree plot using eigenvalues does not show a clear elbow, we can alternatively construct a scree plot using **proportional eigenvalues** on the y-axis. Proportional eigenvalues help emphasize changes in variance explained across components, making it easier to determine the number of components to retain.

Proportional eigenvalues can be calculated using the following formula:

$$Proportional\ Eigenvalue_i = \frac{Percentage\ of\ Variance\ Explained_i}{Cumulative\ Percentage\ of\ Variance_i}$$

where:

- *Proportional eigenvalue*$_i$ represents the proportional contribution of each principal component.
- *Percentage of variance explained*$_i$ is the variance explained by the iith principal component.
- *Cumulative percentage of variance*$_i$ is the cumulative variance explained up to the iith component.

By plotting proportional eigenvalues against the principal components, the scree plot can show more distinct drops in variance, potentially making it easier to identify the elbow and decide how many components to retain.

In case the scree plot using eigenvalues does not show a clear elbow, we can also construct a scree plot using the proportional eigenvalues for the y-axis.

COMPONENT ROTATION METHODS IN PCA

The principal components extracted by PCA are linear combinations of the original variables. They are arranged in descending order of variance and are orthogonal to each other, meaning they are uncorrelated. However, the principal components obtained from PCA are not always aligned with the original variables, and their interpretation may not be straightforward. PCA rotation is used to transform the principal components to achieve a more interpretable and meaningful representation of the data. The goal of rotation is to obtain a simpler and more structured representation of the relationships between the variables.

DEFINITION

Orthogonal rotation is a rotation method in which the rotated components are constrained to be orthogonal to each other, meaning they remain independent and uncorrelated.

Oblique rotation is a rotation method in which the rotated components are allowed to be correlated with each other, meaning the components can share some level of correlation.

The two common rotation methods in PCA are orthogonal and oblique rotations. These methods aim to achieve different goals in terms of maximizing variance, minimizing the number of variables with high loadings on a component, or allowing for correlation between components. Figure 12.3 illustrates the differences between these two methods.

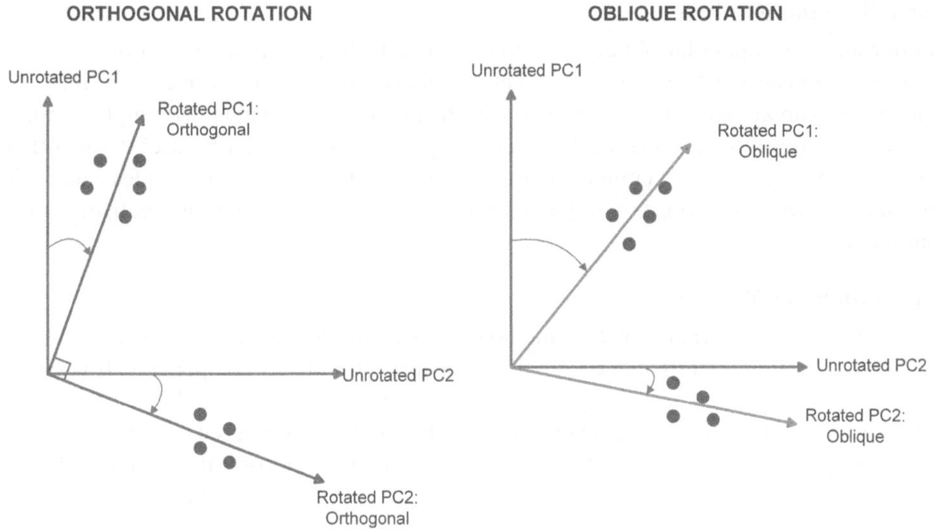

FIGURE 12.3 Orthogonal and oblique rotations.

Orthogonal rotation is the rotation method in which the rotated components are constrained to be orthogonal to each other, meaning they are independent and uncorrelated. Orthogonal rotation is particularly useful when the components are assumed to be uncorrelated, as it provides simpler and more interpretable solutions.

On the contrary, oblique rotation is the rotation method in which the rotated components are allowed to be correlated with each other. This means that the components can have shared variance, and they are no longer independent. Oblique rotation is suitable when the factors are assumed to be correlated or when there are theoretical reasons to allow for such correlation among the factors.

By applying a rotation method, the principal components are transformed to have a more interpretable structure, which allows us to identify the underlying factors or patterns in the data. The rotated components may align more closely with the original variables, making it easier to understand the relationships between the variables and interpret the results.

It is important to note that PCA rotation is an optional step in the PCA analysis and is not always necessary. The decision to perform rotation depends on the specific goals of the analysis and the complexity of the data. If the principal components obtained from PCA are already interpretable and aligned with the original variables, rotation may not be required. However, if the components are complex and difficult to interpret, rotation can be useful to simplify and clarify the results.

Common Rotation Methods

Varimax, Oblimin, and Promax are popular rotation methods used in PCA to achieve a more interpretable and meaningful representation of the extracted components. Below are descriptions of those rotation methods.

Varimax Rotation (Orthogonal)

Varimax rotation is a widely used orthogonal rotation method. It aims to maximize the variance of the squared loadings on each component, leading to a simple and sparse structure where each variable has a high loading on one component and close to zero loadings on other components. This rotation method helps in achieving component interpretability by reducing cross-loadings and emphasizing clear relationships between variables and components.

Oblimin Rotation (Oblique)

Oblimin rotation is a popular oblique rotation method. Unlike Varimax, Oblimin allows for correlation or obliqueness between the rotated components. The main objective of Oblimin rotation is to achieve a simpler structure by minimizing the number of variables with high loadings on a component. It allows for correlation between components, which can be useful when there are underlying factors that are not completely independent. Oblimin rotation provides a more flexible approach by allowing components to be correlated and can result in a more realistic representation of the data.

Promax Rotation (Oblique)

Promax rotation is an oblique rotation method that extends the advantages of Varimax rotation while also accounting for possible correlations between the rotated components. It is considered a compromise between the simplicity of orthogonal rotation and the flexibility of oblique rotation. Promax rotation includes a parameter (kappa) that controls the degree of orthogonality, which allows varying levels of correlation between components. By adjusting the kappa value, Promax rotation can be tuned to emphasize orthogonality (similar to Varimax) or allow for more correlation (similar to Oblimin).

These rotation methods provide different approaches to achieve a more interpretable and structured representation of the components obtained from PCA. The choice of rotation method depends on the specific goals of the analysis, the underlying assumptions about the relationships between variables, and the desired interpretability of the components. It's worth noting that there is no universally "best" rotation method, and the selection depends on the specific characteristics of the data.

PRINCIPAL COMPONENT ANALYSIS IN SPSS MODELER

In SPSS Modeler, PCA/Factor is listed under the **Modeling** group. Keep in mind that PCA is often used to reduce the data dimensionality for predictive modeling purposes. In other words, when we have a very large number of variables, we can use PCA to combine those variables into a smaller set of components and then use these components as predictors for predictive modeling.

SPSS Modeler preconfigures the PCA steps, so we do not have to perform each step in the PCA process as we would when using other statistical software, such as SPSS. This makes the process of running PCA easier but also reduces the level of control we have over the analysis process. However, we should keep in mind that in data mining, we need to automate the process to reduce analysis time, especially since we are dealing with large datasets. SPSS Modeler still allows us to make certain changes in the PCA configurations, enabling us to produce good principal components for dimensionality reduction. I will show you some key configurations in SPSS

EXAMPLE: PRINCIPAL COMPONENT ANALYSIS

In this example, we will build a PCA model for the airline recommendation dataset.

- **Dataset:** AirlineRecommendation.sav
- **Target variable:** Recommendation (Flag: Yes/No)

TABLE 12.2
Airline Recommendation Data Description—Input Variables for PCA

Variable Name	Label/Description	Scale	Role
Recommend	Recommend the airline to others	Flag/binary (Yes/No)	Target
airline_st	Airline status	Nominal	None
Age	Age	Continuous	Input
Gender	Gender	Flag/binary (Male/Female)	None
price_sensitivity[a]	Price sensitivity	Continuous	Input
no_years_flying	Number of years flying	Continuous	Input
no_flt_pa	Average number of flights per year	Continuous	Input
pct_ flt_other_airline	Percent of flights with other airlines	Continuous	Input
type_of_travel	Travel purpose	Nominal	None
no_other_lty_cards	Number of other loyalty cards	Continuous	Input
shopping_amt_airpt	Shopping amount at the airport	Continuous	Input
eat_drink_airpt	Spending for eating and drinking at the airport	Continuous	Input
Class	Passenger class	Nominal	None
origin_state	Origin state	Nominal	None
destin_state	Destination state	Nominal	None
depart_delay	Departure delay (in minutes)	Continuous	Input
arrival_delay	Arrival delay (in minutes)	Continuous	Input
flt_cancel	Flight cancellation	Flag/binary (Yes/No)	None
flt_time	Flight time (in minutes)	Continuous	Input
flt_distance	Flight distance (miles)	Continuous	Input

[a] 5-point Likert scale, treated as continuous scale

Variables for PCA

Table 12.2 lists the variables used for PCA. Since PCA is an unsupervised learning technique, only the input variables are used for the analysis. The target variable, *recommend*, will be excluded from the PCA process. Additionally, we will not include input variables with binary or nominal scales, as these variables are less likely to form a subset of correlated variables.

By focusing on variables that are continuous or interval-scaled, we can more effectively explore the relationships between variables and reduce the dimensionality of the dataset, identifying the principal components that capture the most variance.

> SPSS Modeler automatically detects the input fields with continuous scales and uses them as input fields for PCA. Therefore, we do not need to manually reject the nominal or binary fields. This automation simplifies the process by ensuring that only relevant continuous variables are selected for dimensionality reduction, making the PCA setup more efficient.

Sample-Explore-Modify

Refer to the previous chapters for data preparation. Since there are missing values in the continuous variables, we will use the imputed data from previous steps. Additionally, because PCA is an unsupervised method, we are not making any predictions and have no need for the target variable in this analysis. Therefore, the target variable will be excluded. Moreover, partitioning the data is unnecessary in this case, as we will use the entire dataset for the PCA analysis.

PCA Modeling

Follow the steps below to configure the PCA/Factor node in SPSS Modeler.

1. Create a new stream and name it "Airline_PCA."
2. Copy nodes from an existing stream: Open the stream "Airline_Tree" and copy the nodes from the data source to the Missing Value Imputation node and the following Type node. Paste these nodes into the new "Airline_PCA" stream.
3. Add a PCA/Factor node: From the Modeling group, add a PCA/Factor node to the stream and connect it to the Type node. Open the PCA/Factor node settings.

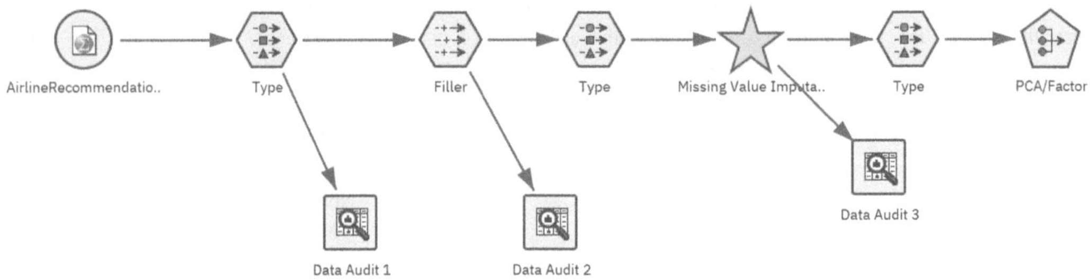

4. Configure the Model tab:
   - Select **Principal Components** for the Extraction Method.
   - Uncheck the box labeled **Use partition data**.
5. Go to the Expert tab:
   - Choose **Expert**.
   - Select **Correlation matrix**.
   - For **Extract Factor**, set **Eigenvalue over** to 1.
   - Check the box for **Sort values**.
   - Check **Hide values below** and set the threshold to **0.4**.

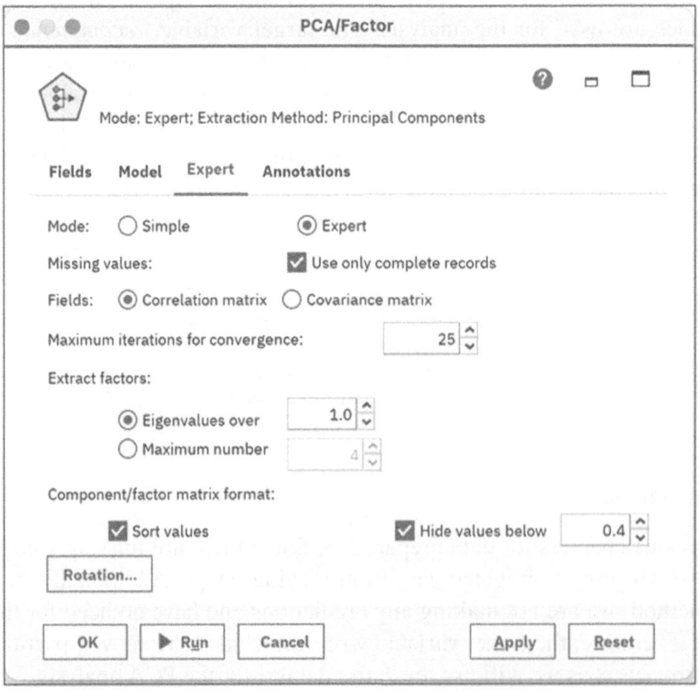

6. Apply rotation:
   - Click the **Rotation** button.
   - Select **Varimax** for the rotation method.
7. Finalize and run the model:
   - Click Apply and Run.

This configuration will extract principal components using the Varimax rotation, focusing on factors with eigenvalues over 1, ensuring you capture the most significant components for dimensionality reduction.

PCA Result Interpretation

To determine the number of principal components, we first examine the results of the total variance explained (Table 12.3). As mentioned, we configured the software to automatically decide the number of components (factors) based on the condition of eigenvalues greater than 1. According to Table 12.3, five components have eigenvalues greater than 1, resulting in a total cumulative variance explained to 67.21%. This means that by using these five components, our model can explain 67.21% of the variability in the data. Therefore, SPSS Modeler extracted five components (factors) in the PCA.

Now, we need to review the component results to determine if any adjustments are needed. Table 12.4 presents the rotated component matrix, which shows the number of components and the loadings of the original variables on each rotated component. Each cell in the matrix represents the correlation between a specific variable and a particular component. Higher absolute values in the matrix indicate stronger relationships between variables and components. As configured in the PCA/Factor node, the results only display loadings of 0.4 or higher.

Interpreting the rotated component matrix involves identifying which variables have high loadings on each component. Variables with high loadings are strongly associated with that component and help us interpret the underlying dimension or factor represented by it. In this case, the results indicate that flight distance and flight time have high loadings on Component 1, suggesting that these fields collectively measure an underlying factor related to flight duration. Similarly, arrival delay and departure delay have high loadings on Component 2, representing an underlying factor related to flight delays. These two components are logical, as the variables within each component are highly correlated and measure the same underlying construct.

TABLE 12.3
Total Variance Explained

	Initial Eigenvalues			Rotation Sums of Squared Loadings		
Component	Total	Percentage of Variance	Cumulative Percentage	Total	Percentage of Variance	Cumulative Percentage
1	1.982	16.518	16.518	1.968	16.404	16.404
2	1.955	16.29	32.807	1.96	16.337	32.741
3	1.946	16.217	49.024	1.947	16.226	48.967
4	1.163	9.691	58.715	1.138	9.487	58.455
5	1.019	8.495	67.21	1.051	8.755	67.21
6	0.986	8.218	75.428			
7	0.918	7.652	83.08			
8	0.822	6.848	89.928			
9	0.644	5.368	95.295			
10	0.493	4.112	99.407			
11	0.045	0.378	99.785			
12	0.026	0.215	100			

TABLE 12.4
Rotated Component Matrix with Five Components

Components	1	2	3	4	5
Flight distance (miles)	0.992				
Flight time (in minutes)	0.992				
Arrival delay (in minutes)		0.99			
Departure delay (in minutes)		0.99			
Number of other loyalty cards			0.801		
Percent of flights with other airlines			0.722		
Age			–0.712		
Average number of flights per year			–0.514		
Spending for eating and drinking at airport				0.719	
Price sensitivity				–0.659	
Shopping amount at airport					0.685
Number of years flying					0.67

We can apply the same interpretation to Component 3, which includes age, number of flights, percentage of flights with other airlines, and number of other loyalty cards. These variables all relate to flying experience, forming a coherent component.

However, issues arise when reviewing Components 4 and 5. For example, spending on eating and drinking at the airport is loaded onto Component 4, while shopping amount at the airport is loaded onto Component 5, despite the expectation that they would be highly correlated and grouped in the same component. Additionally, the number of years of flying is associated with Component 5 but does not seem to fit well with shopping amount at the airport. These inconsistencies suggest that further adjustments may be needed to improve the component structure and interpretability.

Typically, when interpreting the rotated component matrix, we focus on variables with loadings of 0.5 or higher as they indicate a strong association with a particular component. However, in some cases, variables with loadings close to 0.5 can still be considered if they provide valuable insights or contribute meaningfully to the interpretation of the component. These near-threshold loadings can help capture subtle but important relationships between variables and components, depending on the context of the analysis and the specific goals of the PCA.

The concerns regarding the use of five components raise an important question: Is this the best decision for PCA? Since SPSS Modeler extracted five components based on the condition that eigenvalues are greater than 1, it would be prudent to explore an alternative method for determining the optimal number of components. As described earlier, a scree plot is commonly used for this purpose, based on the "elbow" of the plot.

Since SPSS Modeler does not provide a scree plot in the PCA/Factor node output, I generated one in Excel using the eigenvalues and number of components. Figure 12.4 presents this scree plot. However, identifying the exact elbow from this plot is somewhat challenging.

As you may recall, the elbow point on a scree plot indicates where the curve starts to flatten, suggesting that adding more components beyond this point does not significantly enhance the explanation of the variance in the data. In this case, it appears the elbow may occur around four components, as the slope becomes less steep at this point. However, the plot's shape introduces some ambiguity in identifying the elbow.

To identify the elbow more clearly, I reconstructed the scree plot using **proportional eigenvalues**. As mentioned earlier, proportional eigenvalues are calculated by dividing the percentage of

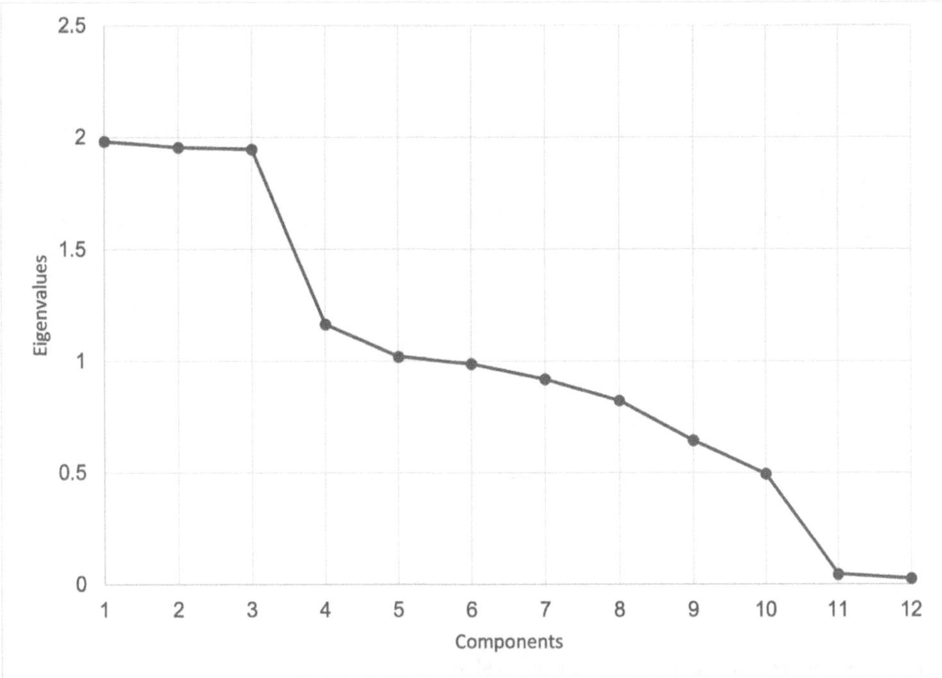

FIGURE 12.4 Scree plot with eigenvalues.

variance explained by the cumulative percentage of variance explained for each component. Table 12.5 displays the results of this calculation.

Figure 12.5 presents the scree plot using proportional eigenvalues on the y-axis. This plot clearly shows the elbow at the point of four components, which confirms our earlier findings. This result indicates that, in this example, using four components is a more optimal choice than five, as it better explains the variability in the dataset. Additionally, it addresses the concerns we previously had with the variables loaded onto Components 4 and 5, as the reduced number of components aligns better with the data structure and the relationships between variables.

TABLE 12.5
Calculations of Proportional Eigenvalues

Component	Eigenvalues	Percentage of Variance	Cumulative Percentage of Variance	Proportional Eigenvalues
1	1.982	16.518	16.493	0.12017
2	1.955	16.29	32.781	0.05964
3	1.946	16.217	49.02	0.03970
4	1.163	9.691	58.696	0.01981
5	1.019	8.495	67.199	0.01516
6	0.986	8.218	75.419	0.01307
7	0.918	7.652	83.092	0.01105
8	0.822	6.848	89.951	0.00914
9	0.644	5.368	95.308	0.00676
10	0.493	4.112	99.397	0.00496
11	0.045	0.378	99.783	0.00045
12	0.026	0.215	100	0.00026

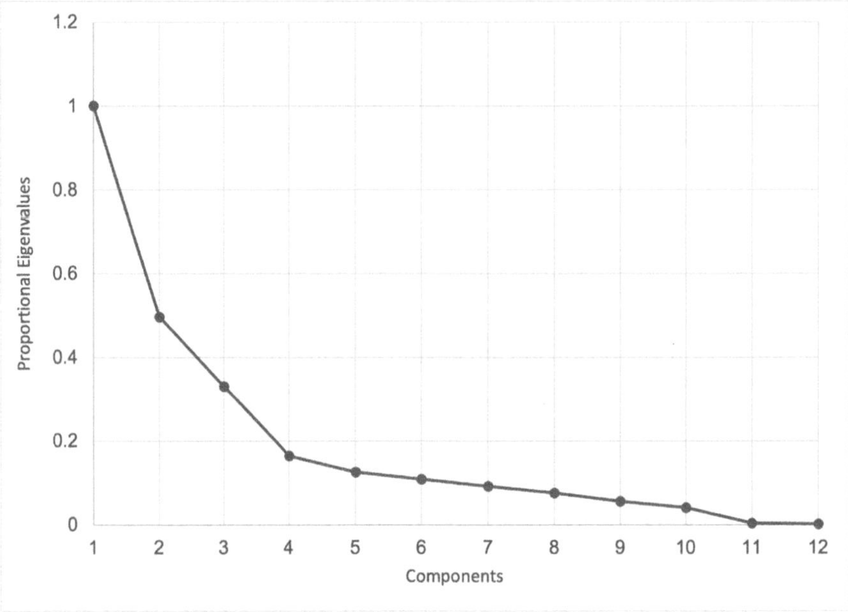

FIGURE 12.5 Scree plot with proportional eigenvalues.

We will now run the PCA/Factor node again using the updated configuration, setting the number of components to 4. Follow the steps below to make this adjustment:

1. In the "Airline-PCA" stream, open the node settings for the PCA/Factor node.
2. Under the **Expert** tab, select **Expert** mode.
3. For **Extract Factor**, choose **Maximum number** and set it to 4.

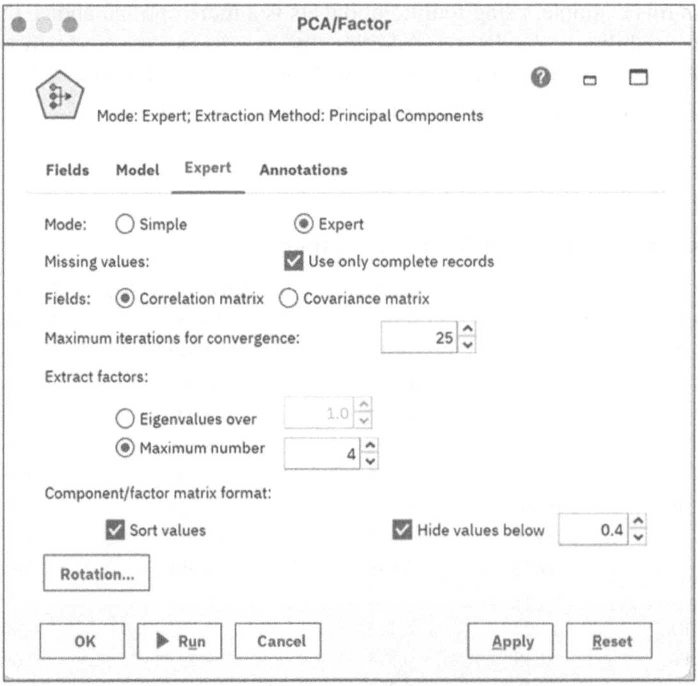

4. Click Apply and Run.

TABLE 12.6
Rotated Component Matrix with Four Components

Component	1	2	3	4
Flight distance (miles)	0.992			
Flight time (in minutes)	0.992			
Arrival delay (in minutes)		0.99		
Departure delay (in minutes)		0.99		
Number of other loyalty cards			0.797	
Age			–0.736	
Percent of flights with other airlines			0.705	
Average number of flights per year			–0.492	
Spending for eating and drinking at airport				0.707
Price sensitivity				–0.543
Shopping amount at airport				0.417
Number of years flying				

The updated PCA/Factor results present the rotated component matrix with four components, shown in Table 12.6. As observed, the variables and loadings for Components 1, 2, and 3 remain the same as before, and these components make sense.

With this new configuration, Component 4 now includes three variables: **spending for eating and drinking at the airport**, **price sensitivity**, and **shopping amount at the airport**. This component structure is logical as it reflects passengers' spending behavior at the airport. Notably, **price sensitivity** is negatively correlated, meaning that passengers with higher price sensitivity tend to spend less at the airport.

Additionally, the variable **number of years flying** does not load significantly onto any component, indicating it is not strongly associated with any of the identified factors. Therefore, for this dataset, we successfully identified four underlying factors from the continuous variables: **flight duration, flight delays, flying experience,** and **spending behavior**.

Dimensionality Reduction

What does dimensionality reduction with these four principal components mean for the dataset? Essentially, the findings tell us that the original continuous variables can be grouped into four underlying factors: **flight duration**, **flight delays**, **flying experience**, and **spending behavior**. These factors replace the original variables and can be used as new predictors in the modeling step.

The variable **number of years flying** remains in the model as a potential predictor. Its low coefficient does not suggest removing it but indicates that it is distinct and cannot be combined with the other principal components. Figure 12.6 shows the input variables that we could use as predictors for the predictive modeling step. The variables labeled **$F-Factor-1**, **$F-Factor-2**, **$F-Factor-3**, and **$F-Factor-4** are the new variables derived from PCA.

Initially, we used 19 input variables for predictive modeling. After applying PCA, we reduced the number of predictors to 12. Although this is a modest reduction in this example, if working with a larger dataset containing many correlated variables, PCA could further reduce the data's dimensionality significantly.

Field	Measurement	Values	Missing	Check	Role
Recommend the ...	Flag	Yes/No		None	Target
Airline status	Nominal	Blue,Gold,Plat...		None	Input
Age	Continuous	[18.0,85.0]		None	Input
Gender	Flag	Male/Female		None	Input
Price sensitivity	Continuous	[0.0,5.0]		None	Input
Number of years ...	Continuous	[2.0,11.0]		None	Input
Average number ...	Continuous	[0.0,100.0]		None	Input
Percent of flights...	Continuous	[1.0,110.0]		None	Input
Travel purpose	Nominal	"Business trav...		None	Input
Number of other ...	Continuous	[0.0,12.0]		None	Input
Shopping amount...	Continuous	[0.0,879.0]		None	Input
Spending for eati...	Continuous	[0.0,895.0]		None	Input
Passenger class	Nominal	Business,Eco,"...		None	Input
Origin state	Nominal	Alabama,Alas...		None	Input
Destination state	Nominal	Alabama,Alas...		None	Input
Departure delay (...	Continuous	[0.0,1592.0]		None	Input
Arrival delay (in ...	Continuous	[0.0,1584.0]		None	Input
Flight cancellation	Flag	Yes/No		None	Input
Flight time (in mi...	Continuous	[8.0,669.0]		None	Input
Flight distance (...	Continuous	[31.0,4983.0]		None	Input
$F-Factor-1	Continuous	[-1.393666131...		None	Input
$F-Factor-2	Continuous	[-0.448247386...		None	Input
$F-Factor-3	Continuous	[-2.794541724...		None	Input
$F-Factor-4	Continuous	[-4.525411218...		None	Input

FIGURE 12.6 Airline recommendation data—new input variables for predictive modeling.

EXAMPLE: PRINCIPAL COMPONENT ANALYSIS

In this example, we will conduct PCA for the Boston Housing dataset with the purpose of identifying underlying factors from the data.

- **Dataset:** Boston.sav
- **Target variable:** Imv (Continuous—Median home value)

Variables for PCA

Since all the input variables in the Boston Housing dataset are continuous, they will all be used for PCA. Note that the target variable **Imv** will not be included in this analysis, as PCA is an unsupervised learning method and target variables are not used.

Sample-Explore-Modify

As discussed in previous chapters, we address missing values by using imputed data for the continuous variables. Since PCA is an unsupervised method and does not require the target variable, there is no need to partition the data into training or testing sets. For this analysis, the entire dataset will be used.

PCA Modeling

Follow the steps below to configure the PCA/Factor node in SPSS Modeler:

1. Create a new stream and name it "Boston_PCA."
2. Copy nodes from an existing stream:
   - Open the "**Boston_Tree**" stream.
   - Copy the nodes from the data source to the **Supernode_transform** node and the subsequent **Type** node.
   - Paste these nodes into the "**Boston_PCA**" stream.

3. Add the PCA/Factor node:
   - From the **Modeling** group, add a PCA/Factor node to the stream.
   - Connect the PCA/Factor node to the **Type** node.
   - Open the PCA/Factor node settings.

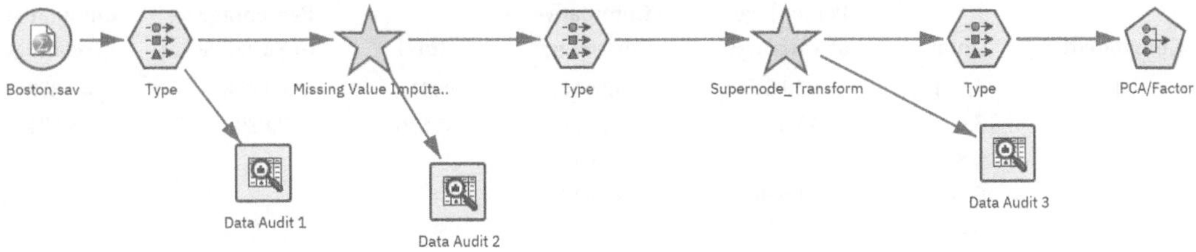

4. Configure the Model settings:
   - Under the **Model** tab, select **Principal Components** for the Extraction Method.
   - Uncheck the box for **Use partition data** since PCA is unsupervised.
5. Configure the Expert settings:
   - Under the **Expert** tab, select **Expert** mode.
   - Select **Correlation matrix**.
   - Set **Eigenvalue over to 1** for extracting factors with an eigenvalue greater than 1.
6. Sort and hide values:
   - Select **Sort values** to organize component loadings.
   - Choose **Hide values below** and set it to 0.4 to display only significant loadings.

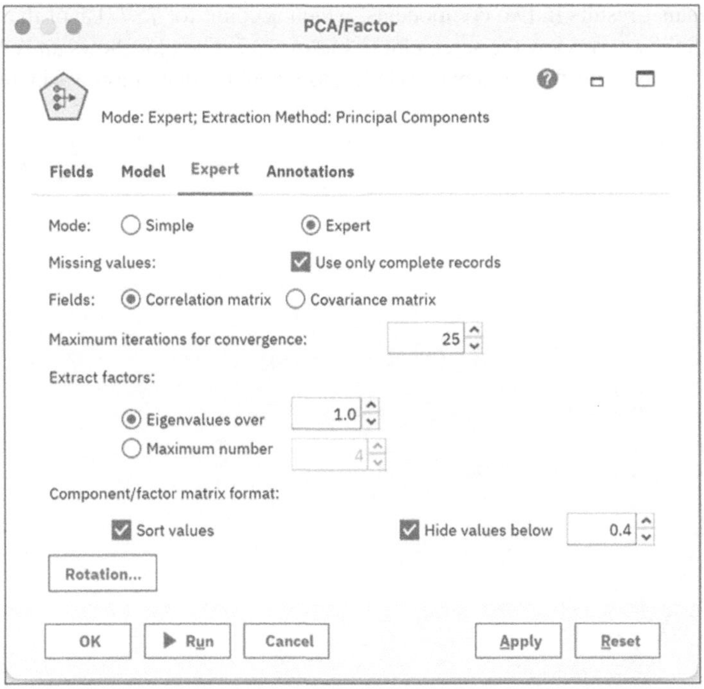

7. Rotation:
   - Click **Rotation** and select **Varimax** for orthogonal rotation, which simplifies component interpretation.
8. Finalize the configuration:
   - Click Apply and Run.

TABLE 12.7
Total Variance Explained

	Initial Eigenvalues			Rotation Sums of Squared Loadings		
Component	Total	Percentage of Variance	Cumulative Percentage	Total	Percentage of Variance	Cumulative Percentage
1	6.781	52.161	52.161	5.314	40.878	40.878
2	3.069	23.61	75.771	4.536	34.893	75.771
3	0.808	6.218	81.989			
4	0.645	4.958	86.947			
5	0.485	3.727	90.674			
6	0.321	2.466	93.14			
7	0.28	2.154	95.294			
8	0.183	1.407	96.701			
9	0.145	1.117	97.817			
10	0.109	0.835	98.652			
11	0.083	0.639	99.291			
12	0.052	0.4	99.692			
13	0.04	0.308	100			

The PCA results show that SPSS Modeler extracted two components from the data based on the total variance explained results (Table 12.7). The table indicates that applying the condition eigenvalues greater than 1 results in two components, which account for 75.771% of the variability in the data. To confirm this, we review the scree plot in Figure 12.7. The plot shows an "elbow," suggesting that using three components may be a better choice, as it could explain more variability in the dataset.

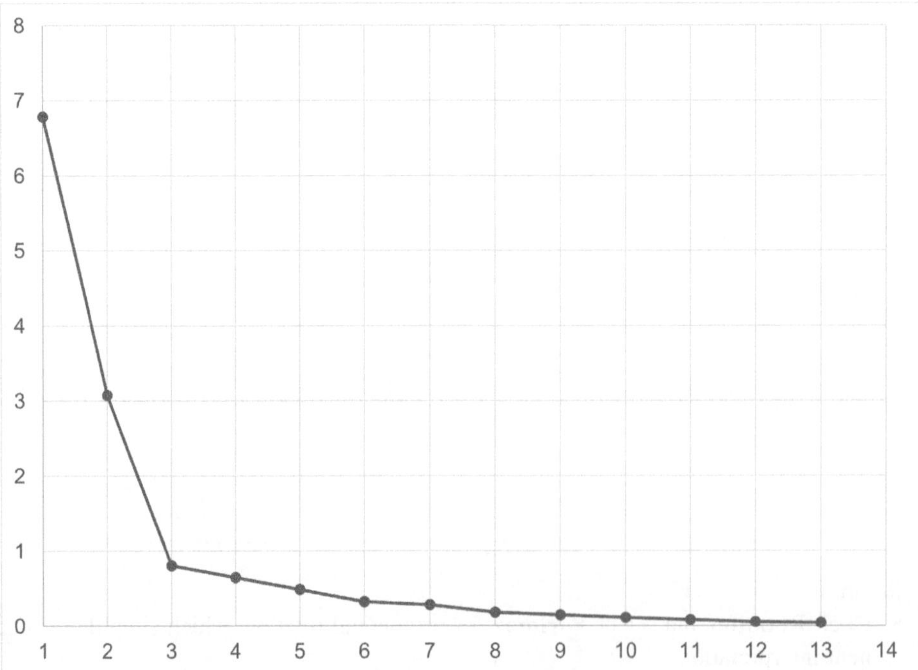

FIGURE 12.7 Scree plot based on Eigenvalues.

TABLE 12.8
Rotated Component Matrix with Two Components

Component	1	2
LogAccess_LogN	–0.949	
NiOxSQ_Inverse	0.943	
B_Square root	0.902	
RmSQ_Inverse	0.84	
PtRatio_LogN	–0.83	
Charles_River_Square root	–0.708	
CrimePC_Square root		0.854
IndustPr_Square root		0.832
AgePre40		0.802
Tax		0.781
LogDistance_LogN		–0.757
ZonePr_Square root		–0.738
Lstat_Square root		0.645

To ensure we choose the best structure, we will compare the results from both the two-component and three-component PCA models to see which one provides a more interpretable and meaningful representation of the data.

Tables 12.8 and 12.9 present the rotated component matrices for two-component and three-component structures, respectively. As observed, Component 1 is similarly structured in both cases, with the only difference being that, in the three-component structure, **LogDistance** is also loaded onto this component. The variables associated with this component capture the primary geographic characteristics of the area.

The major distinction lies in the remaining variables. In the two-component structure, we see a mixture of population characteristics and socio-economic characteristics grouped under Component 2. Additionally, the appearance of **LogDistance** in this component does not seem logical in terms of interpretation.

TABLE 12.9
Rotated Component Matrix with Three Components

Component	1	2	3
LogAccess_LogN	–0.965		
NiOxSQ_Inverse	0.932		
B_Square root	0.906		
RmSQ_Inverse	0.843		
PtRatio_LogN	–0.819		
Charles_River_Square root	–0.698		
LogDistance_LogN	–0.555		
CrimePC_Square root		0.915	
Tax		0.886	
IndustPr_Square root		0.614	
ZonePr_Square root			–0.836
Lstat_Square root			0.724
AgePre40			0.674

On the contrary, the three-component structure cleanly separates socio-economic characteristics (e.g., crime rate, tax rate, and business-housing ratio) in Component 2 from population characteristics (e.g., age, residential land ratio, and lower-status population) in Component 3. This clearer distinction makes the three-component structure more interpretable and logical. Thus, we select the three-component PCA structure and use these factors as new predictors for the modeling step.

CHAPTER SUMMARY

PCA is a widely used dimensionality reduction technique in data mining, aimed at transforming high-dimensional data into a lower-dimensional format while preserving essential information. PCA works by identifying principal components—new variables that capture the maximum variance in the data. The first principal component accounts for the largest portion of the variability, with each subsequent component capturing a smaller portion. By selecting a subset of principal components that represent a significant portion of the total variance, PCA reduces the dimensionality of data without sacrificing meaningful insights.

PCA offers several key advantages in data mining. It simplifies data representation by reducing the number of variables, making it easier to interpret and visualize. It also eliminates redundant or irrelevant variables, which enhances computational efficiency and reduces the risk of overfitting. Additionally, PCA can reveal hidden patterns and relationships in the data, as the principal components are linear combinations of the original variables. PCA is often used as a preprocessing step in predictive modeling to reduce the number of variables, using principal components that retain the majority of the variance.

However, PCA assumes linearity in the relationships among variables and is sensitive to outliers, which can affect the accuracy of the results. Moreover, interpreting principal components can be difficult because they are formed from combinations of multiple original variables. Despite these limitations, PCA remains an essential tool in data mining for dimensionality reduction, variable selection, and exploratory data analysis.

CHAPTER DISCUSSION QUESTIONS

1. What is Principal Component Analysis (PCA) and how does it enable dimensionality reduction?
2. Discuss the advantages and potential limitations of using PCA to reduce the number of variables in a dataset.
3. How does PCA capture and quantify the variability in the data through the concept of principal components?
4. Explain the role of eigenvalues and eigenvectors in determining the importance and direction of each principal component.
5. In what scenarios would PCA be particularly useful in data mining tasks? Discuss specific applications where PCA has been successfully applied.
6. What techniques can be used to determine the optimal number of principal components to retain in a given dataset?
7. Discuss the interpretation of the principal components obtained from PCA. How can the relationships between the original variables and the principal components be understood and analyzed?
8. What is the purpose of the component rotation in PCA? Discuss primary rotation methods.
9. Discuss the trade-off between dimensionality reduction achieved by PCA and the potential loss of information. How can the impact of dimensionality reduction on model performance or data analysis outcomes be assessed and evaluated?

CHAPTER EXERCISES

Exercise 1: Principal Component Analysis for the Bank Dataset

1. Open the **bank** dataset in SPSS Modeler.
2. Sample the entire dataset. Refer to Appendix A for variable names and descriptions.
3. Perform data modification as follows:
   - Impute missing values.
   - Normalize interval variables.
4. Conduct a principal component analysis (PCA) for all continuous input variables. Make sure to NOT include **CRSCORE** (the target variable)
   - **Configure the Model settings:**
     1. Under the **Model** tab, select **Principal Components** for the Extraction Method.
     2. Uncheck the box for **Use partition data.**
   - **Configure the Expert settings:**
     1. Under the **Expert** tab, select **Expert** mode.
     2. Select **Correlation matrix**.
   - **Rotation:**
     1. Click **Rotation** and select ***Varimax*** for orthogonal rotation, which simplifies component interpretation.
5. Identify the number of principal components using the scree plot and the Total variance explained table.
6. Adjust the number of components as needed.
7. Present the PCA results: Identify the principal components from the results and the significant variables composing each component.
8. Interpret the dimensional reduction with those principal components, and give them meaningful names and descriptions.

Exercise 2: Principal Component Analysis for the Passengers Dataset

1. Open the **passengers** dataset in SPSS Modeler.
2. Sample 20% of the full dataset. Set **return** (decision to return to the airline) as the target variable and change its role to binary. See Appendix B for variable names and descriptions.
3. Perform data modification as follows:
   - Replace zero values for Likert-scale variables with missing values.
   - Impute missing values.
   - Normalize interval variables.
4. Conduct a principal component analysis (PCA) for all continuous input variables. Make sure NOT to include **recommend** (the target variable)
   - **Configure the Model settings:**
     1. Under the **Model** tab, select **Principal Components** for the Extraction Method.
     2. Uncheck the box for **Use partition data.**
   - **Configure the Expert settings:**
     1. Under the **Expert** tab, select **Expert** mode.
     2. Select **Correlation matrix**.
   - **Rotation:**
     1. Click **Rotation** and select ***Varimax*** for orthogonal rotation, which simplifies component interpretation.

5. Identify the number of principal components using the scree plot and the Total variance explained table.
6. Adjust the number of components as needed.
7. Present the PCA results: Identify the principal components from the results and the significant variables composing each component.
8. Interpret the dimensional reduction with those principal components, and give them meaningful names and descriptions.

13 Cluster Analysis

LEARNING OUTCOMES

Upon the completion of this chapter, you will be able to:

1. Describe the concept and principle of cluster analysis.
2. Differentiate between cluster analysis and principal component analysis.
3. Discuss the advantages and disadvantages of cluster analysis
4. Explain how cluster analysis groups objects based on similarity.
5. Compare hierarchical and partitioning cluster algorithms.
6. Describe the key steps involved in the cluster analysis process.
7. Explain the role of distance measures in cluster analysis and describe common measures.
8. Describe the two-step cluster analysis. Discuss its advantages and limitations.
9. Conduct cluster analysis using SPSS Modeler.

INTRODUCTION

In this book, we cover two unsupervised learning methods. In Chapter 12, we explored principal component analysis (PCA). Another unsupervised learning method that I want to introduce is cluster analysis. Although both PCA and cluster analysis are forms of unsupervised learning, they serve very different purposes.

Imagine you have a large dataset where each record represents a specific case or individual. If the dataset contains millions of records, there may be instances where you want to determine if certain cases or individuals share similar characteristics, and if so, how to group them into distinct categories. These groups should represent clear segments of cases or individuals relevant to the sector of interest. With this segmentation, businesses can choose the right strategies to make informed decisions, such as addressing safety concerns for certain case segments or targeting products and services toward specific customer segments. These groups, or segments, are referred to as clusters, giving rise to the term cluster analysis.

Cluster analysis is a powerful machine learning technique used to identify groups of data points that exhibit similar characteristics. As an unsupervised learning method, it does not require labeled data to detect patterns. Instead, it employs clustering algorithms to group similar data points based on their distance from one another. This chapter introduces cluster analysis, including its definition, applications, and various clustering algorithms.

DEFINITION OF CLUSTER ANALYSIS

DEFINITION

Cluster analysis is an unsupervised learning method that groups similar data points together into clusters based on their similarity.

Cluster analysis is an unsupervised learning technique that groups data points into clusters based on their similarities, typically measured using distance metrics. The primary objective of cluster analysis is to reveal hidden

DOI: 10.1201/9781003468196-15

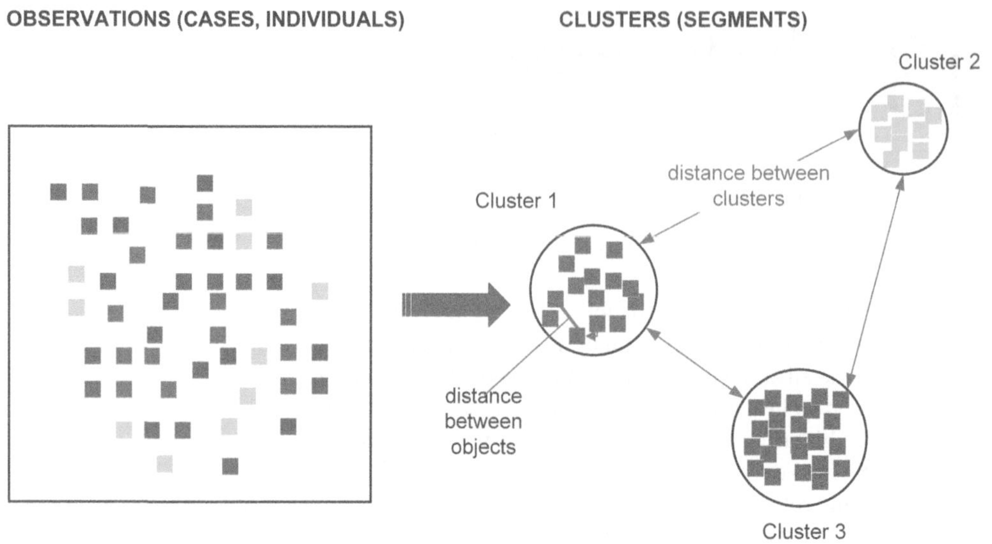

FIGURE 13.1 Illustration of cluster analysis.

patterns or structures within the data by organizing similar data points into distinct clusters or segments. In essence, it allows us to identify groups of data points that share similar characteristics, where the similarity is assessed by their proximity in the feature space. Each cluster should be distinct, meaning the data points within a cluster are more similar to each other than to those in other clusters. The dissimilarity between clusters is measured by the distance between them. Figure 13.1 illustrates the process of cluster analysis.

These clusters represent exclusive and meaningful groups of individuals, objects, or cases. By examining the variables and attributes associated with each cluster, we can differentiate the groups and understand their unique profiles—a process called cluster profiling. This step is crucial because it allows decision-makers to interpret the nature of each cluster and develop tailored strategies based on the cluster characteristics. For example, businesses might use cluster profiles to segment customers and offer targeted marketing, while researchers may use clusters to group organisms based on genetic similarities.

Cluster analysis has wide-ranging applications in various fields, including biology (to classify species based on genetic traits), finance (to group customers by risk profiles), marketing (to segment consumers by purchasing behavior), and social sciences (to classify communities or individuals based on socioeconomic factors). Its versatility makes it a powerful tool for exploratory data analysis and strategic decision-making across multiple domains.

PRACTICAL APPLICATIONS OF CLUSTER ANALYSIS

Cluster analysis is a useful technique for grouping similar objects or observations together based on their attributes or characteristics. It is applied in many domains. Some practical applications of cluster analysis are as follows:

- **Customer segmentation**: In marketing, companies use cluster analysis to group customers based on their demographic, geographic, and behavioral characteristics. This clustering helps them identify customer segments that have similar needs and preferences and allows marketers to tailor their marketing strategies accordingly.

- **Image segmentation**: Cluster analysis is used to segment images into regions with similar color or texture characteristics. This application can be useful for object detection, image recognition, and other image processing tasks.
- **Fraud detection**: Financial companies use this method to detect fraudulent activity by identifying groups of transactions that are similar in nature. Then, they can identify potential fraudsters who are attempting to avoid detection by spreading their activity across multiple transactions.
- **Medical diagnosis**: In healthcare, cluster analysis is used to group patients based on their symptoms and medical history. Doctors can use it to diagnose diseases and develop personalized treatment plans for each patient.
- **Natural language processing**: In this field, researchers use cluster analysis to group similar documents or words together. This tool can be useful for information retrieval, topic modeling, and other text processing tasks.
- **Crime analysis**: By using cluster analysis, law enforcement can identify areas with high crime rates by grouping together crimes that are similar in nature. Police departments use the cluster profiles to allocate resources more efficiently and reduce crime rates in high-risk areas.

These examples demonstrate the versatility of cluster analysis. It is applicable in any field where objects or data points need to be grouped based on shared attributes, making it a vital tool for decision-making and strategic analysis across domains.

DIFFERENCES BETWEEN PCA AND CLUSTER ANALYSIS

PCA (principal component analysis) and cluster analysis are both unsupervised learning methods; however, they serve different purposes and utilize different techniques. While both methods involve identifying patterns in the data by grouping variables or objects, their goals and methodologies differ significantly. It is easier to get confused between them, and misunderstanding these distinctions can lead to the incorrect application of the techniques. Table 13.1 provides a comparison between PCA and cluster analysis.

TABLE 13.1
Comparison between PCA and Cluster Analysis

Items	Principal Component Analysis (PCA)	Cluster Analysis
Purpose	To identify the most important variables or combinations of variables that explain the variation in the data and to reduce the dimensionality of the data.	To group objects or observations together based on their similarity or dissimilarity of specific characteristics and attributes.
Goal	To simplify the data and identify the underlying structure in the data, rather than to group or cluster similar objects together.	To identify natural groupings in the data and to understand the characteristics of each group.
Objects of analysis	Variables	Observations (cases or individuals)
Output	A set of principal components that summarize the information in the original data.	A set of clusters, where each cluster is a group of similar objects.

(Continued)

TABLE 13.1 *(Continued)*
Comparison between PCA and Cluster Analysis

Items	Principal Component Analysis (PCA)	Cluster Analysis
Methodology	PCA involves creating a new set of variables, called principal components, which are linear combinations of the original variables. The principal components are ordered based on their importance, with the first principal component explaining the largest amount of variation in the data.	Cluster analysis involves partitioning the data into a set of groups, called clusters, such that objects within a cluster are as similar as possible to each other, while objects in different clusters are as dissimilar as possible from each other.
Application	PCA is often used in data visualization, variable selection, and dimensionality reduction.	Cluster analysis is used in market segmentation, customer profiling, and anomaly detection.
Similarity and dissimilarity measures	PCA does not involve any similarity or dissimilarity measures.	Cluster analysis requires a similarity or dissimilarity measure to group similar objects together.

BASIC CONCEPTS OF CLUSTER ANALYSIS

Cluster analysis is a data exploration and unsupervised machine learning technique used to group similar objects or observations based on their characteristics. There are several fundamental concepts in cluster analysis that are essential for effectively applying clustering techniques and interpreting the results. Table 13.2 presents these key concepts and their descriptions.

TABLE 13.2
Basic Concepts of Cluster Analysis and Descriptions

Terms	Descriptions
Clustering	A process of grouping objects or observations into clusters (segments) based on their similarity or proximity. The goal is to create clusters that are internally homogeneous (objects within the same cluster are similar) and externally heterogeneous (objects from different clusters are dissimilar).
Cluster or segment	A cluster, also known as a segment, refers to a group of data points or objects that are similar to each other based on certain characteristics or attributes.
Distance or similarity measures	A distance, also known as a similarity measure, quantifies the similarity or dissimilarity between pairs of objects. These measures are used to determine the proximity between objects and form the basis for clustering algorithms. Common distance measures include Euclidean distance, Manhattan distance, and cosine similarity.
Centroids	Centroids are the mean values of the attributes for the objects in a cluster. They are used to represent the cluster and are often used as the basis for assigning new objects to a cluster.
Dendrograms	A dendrogram is a tree-like diagram that shows the hierarchical relationship between clusters in hierarchical clustering. The height of each branch represents the distance between clusters at that level.

(Continued)

TABLE 13.2 *(Continued)*
Basic Concepts of Cluster Analysis and Descriptions

Terms	Descriptions
Clustering algorithms	A clustering algorithm determines the number of clusters and assigns observations to clusters. Cluster algorithms can be categorized into hierarchical clustering and partitioning clustering.
Hierarchical clustering	Hierarchical clustering builds a hierarchy of clusters, forming a tree-like structure called a dendrogram.
Partitioning clustering	Partitioning clustering divides the data into non-overlapping partitions or clusters. It directly assigns each data point to a specific cluster.
K-means clustering	K-means is a partitioning clustering algorithm that partitions the data into k distinct clusters by iteratively updating the cluster centroids and assigning data points to the nearest centroid. The objective is to minimize the sum of squared distances between data points and their respective cluster centroids.
Cluster assignment	Cluster assignment is a process of assigning each object to a specific cluster based on its similarity to the cluster centroids. The assignment is usually done by comparing the distance or similarity of the object to the centroids.
Cluster validation	Cluster validation is a process of assessing the quality and reliability of the clusters obtained. It involves evaluating the internal cohesion and separation of the clusters, as well as their stability and meaningfulness.
Cluster profiling	Cluster profiling involves analyzing the characteristics of each cluster, such as the mean value of the attributes or the distribution of the objects within the cluster.

ADVANTAGES AND DISADVANTAGES OF CLUSTER ANALYSIS

Cluster analysis is a powerful unsupervised technique for identifying patterns in datasets, but it also has its limitations. We should carefully consider this method's strengths and weaknesses when using it in machine learning.

Advantages of Cluster Analysis

- **Pattern recognition**: Cluster analysis helps identify patterns in large datasets that may not be apparent to humans.
- **Unsupervised learning**: It is an unsupervised learning method, which means that it does not require a target variable and labeled data to identify patterns. This makes it useful in situations where labeled data are scarce or expensive to obtain.
- **Data reduction**: Cluster analysis reduces the size of large datasets by grouping similar data points together. It makes it easier to analyze and visualize data. Instead of dealing with individual data points, we have the data represented by the clusters, reducing the complexity and dimensionality of the dataset.
- **Decision making**: It helps with decision making by identifying groups of similar objects or data points that can be used to inform decisions.
- **Anomaly detection**: Clustering identifies outliers or anomalies that deviate significantly from the typical patterns observed in clusters. These anomalies could represent errors, unusual behaviors, or novel observations that necessitate further investigation.

Disadvantages of Cluster Analysis

- **Difficulty with the number of clusters**: One of the main challenges of cluster analysis is determining the optimal number of clusters. If the number of clusters is too small, important patterns may be overlooked. If it is too large, the results may be difficult to interpret.
- **Sensitivity to initial conditions**: This method is sensitive to the initial conditions such as cluster centers. Different starting points can lead to different results. Multiple runs with different initializations might be necessary to achieve more robust results.
- **Outliers**: Cluster analysis is also sensitive to outliers, and outliers can skew the results of the clustering analysis.
- **Difficulty with high-dimensional data**: As the dimensionality of the data increases, the distance measures may become less meaningful, and the curse of dimensionality can impact the clustering results.
- **Interpretability**: While clustering can reveal clusters or segments in the data, the meaning and interpretation of the clusters often require further analysis and domain knowledge. The clusters themselves may not have clear labels or explanations, limiting their direct utility in some applications.
- **Scalability**: This method can be computationally intensive and may not be scalable to very large datasets. This can make it difficult to use in situations where real-time analysis is required.

STEPS OF CLUSTER ANALYSIS PROCESS

In order to conduct cluster analysis, there are typically eight steps as follows:

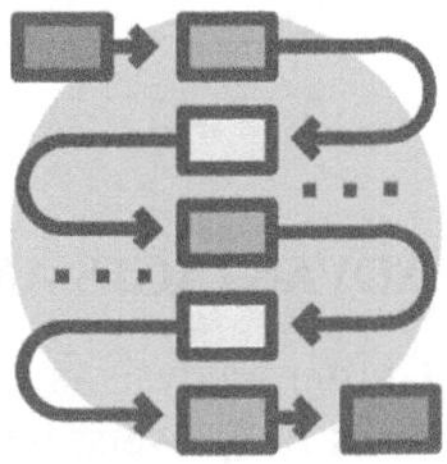

1. **Data preprocessing**: This step involves data cleaning, data transformation, and variable selection. The goal of data preprocessing is to ensure that the data are in a suitable format for clustering.
2. **Variable selection for clustering**: Clusters are formed based on similarity. In this step, we select the variables that are used to determine the similarity between objects.
3. **Distance measure selection**: The next step is to select an appropriate distance measure to quantify the similarity or dissimilarity between pairs of objects. This is an important step, as the choice of distance measure can affect the clustering results. In this book, we cover some most common distance measures in cluster analysis.
4. **Clustering algorithm selection**: There are two primary types of clustering algorithms: hierarchical clustering and partitioning clustering. It is important to select an appropriate clustering algorithm based on the characteristics of the data and the purpose of the clustering analysis.
5. **Selection of the number of clusters**: The next step is to determine the optimal number of clusters for the data. This can be done using visualization techniques such as the elbow method or using statistical validation measures.
6. **Cluster assignment**: In this step, each object is assigned to a cluster. This step is done by applying the chosen clustering algorithm to the data. In other words, objects are assigned to the appropriate clusters based on their similarity.
7. **Cluster profiling**: Once the objects have been assigned to clusters, the next step is to profile the clusters. Cluster profiling involves analyzing the characteristics of each cluster, such as the mean value of the attributes or the distribution of the objects within the cluster.
8. **Cluster validation**: The final step is to validate the clustering results to ensure that the clustering results are meaningful, reliable, and appropriate for the data.

Those are the general steps involved in cluster analysis. The actual implementation of cluster analysis may involve variations and adjustments to those steps depending on the nature of the data and the purpose of the clustering analysis. Many data mining software apps have built-in tools for several steps, which reduce our efforts in the analysis process. We will perform cluster analysis with the airline recommendation dataset using features provided by SPSS Modeler. I won't present all of the above steps explicitly because some of them have been covered in other chapters or incorporated into the software.

CLUSTER DISTANCE MEASURE

Distance measures play a crucial role in the cluster analysis process and in clustering algorithms in particular, as they determine the similarity or dissimilarity between data points. In this section, we cover some common distance measures. Figure 13.2 illustrates these measures and how they differ.

EUCLIDEAN DISTANCE

The most common distance measure, Euclidean distance, is defined as the straight-line distance between two data points in Euclidean space. It is measured as the square root of the sum of the squared differences between the corresponding coordinates of the two points.

The formula for the distance is as follows:

$$d(x,y)=\sqrt{(x_1-y_1)^2+(x_2-y_2)^2+\cdots+(x_n-y_n)^2}$$

where

$d(x, y)$ is the distance between two data points, x and y.
$(x_1, x_2, \ldots, x_n)$ and $(y_1, y_2, \ldots, y_n)$ are the coordinates of the two data points.

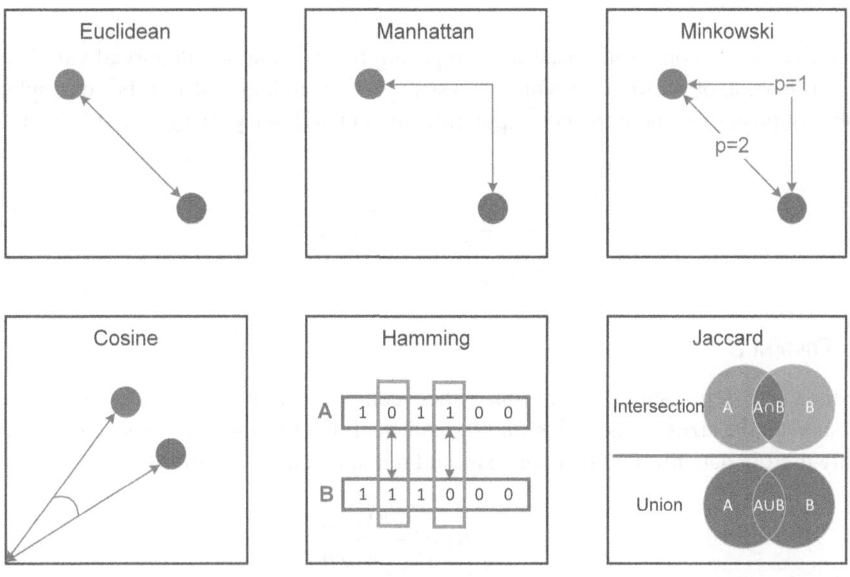

FIGURE 13.2 Different measures of cluster distance.

MANHATTAN DISTANCE

Manhattan distance is also known as city block distance or L_1 distance. It measures the distance between two data points by summing the absolute differences between their coordinates. It is named after the city block grid layout, where the shortest path between two points follows a grid-like pattern.

$$d(x,y) = |x_1 - y_1| + |x_2 - y_2| + \cdots |x_n - y_n|$$

MINKOWSKI DISTANCE

Minkowski distance is a generalization of both Euclidean and Manhattan distances. It is defined as the *n*th root of the sum of the *n*th powers of the absolute differences between corresponding coordinates of two points.

$$d(x,y) = \left(|x_1 - y_1|^p + |x_2 - y_2|^p + \cdots |x_n - y_n|^p\right)^{1/p}$$

Where p is a parameter. When $p = 1$, it becomes Manhattan distance, and when $p = 2$, it becomes Euclidean distance.

COSINE DISTANCE

Cosine distance measures the cosine of the angle between two vectors, representing data points. It is often used when dealing with high-dimensional data or when the magnitude of the vectors is not important, but the orientation is crucial.

$$d(x,y) = \frac{1-(x*y)}{\|\mathrm{x}\| * \|\mathrm{y}\|}$$

Where $x*y$ represents the product of x and y, and $\|x\|$ and $\|y\|$ denote the Euclidean norm of vectors x and y, respectively.

HAMMING DISTANCE

Hamming distance is commonly used for comparing binary data or categorical variables. It measures the proportion of positions at which two strings of equal length differ. For example, it counts the number of positions where the corresponding bits in two binary strings are different.

$$d(x,y) = |x_1 - y_1| + |x_2 - y_2| + \cdots |x_n - y_n|$$

$$If\ x_i = y_i\ then\ x_i - y_i = 0$$

$$If\ x_i \neq y_i\ then\ x_i - y_i = 1$$

JACCARD DISTANCE

Jaccard distance is a measure of dissimilarity between two sets. It is defined as the ratio of the difference between the sizes of the union and the intersection of the sets to the size of the union. It is commonly used in data mining, text analysis, and recommendation systems.

$$d(A,B) = 1 - \frac{|A \cap B|}{|A \cup B|}$$

Where A = set 1 and B = set 2.

These are some of the common distance measures used in cluster analysis. The choice of distance measure depends on the type of data and the specific requirements of the analysis. Each distance measure may be more appropriate for different scenarios, so understanding the characteristics of the data and the goals of the analysis is crucial when selecting a distance measure.

TYPES OF CLUSTERING ALGORITHMS

Two common types of clustering algorithms are **hierarchical clustering** and **partitioning clustering**. In this section, we discuss these algorithms at a high level, avoiding complex mathematical expressions. Most data mining software applications incorporate these algorithms into their interfaces, allowing users to perform clustering analysis without setting up intricate configurations.

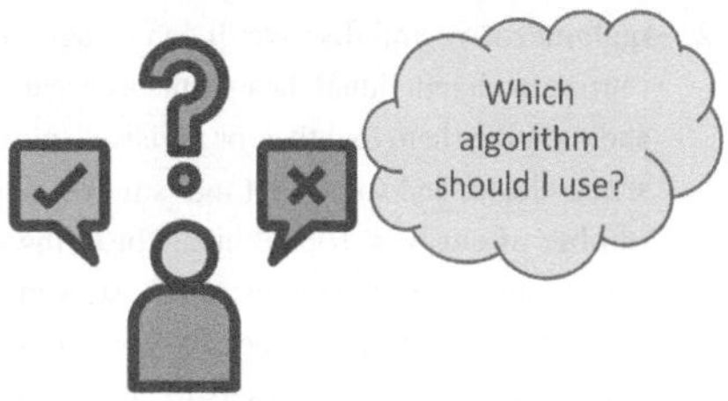

HIERARCHICAL CLUSTERING

Hierarchical clustering is a clustering algorithm that organizes data points into a tree-like structure known as a dendrogram. The process begins by assigning each data point to its own cluster. Then, the algorithm successively merges the closest clusters based on their similarity. This merging continues until all data points are combined into a single cluster or until a specified number of clusters is reached. Hierarchical clustering can be performed in two ways:

- **Agglomerative** (bottom-up approach): Each data point starts as its own cluster, and clusters are merged step by step.
- **Divisive** (top-down approach): All data points start in one cluster, and the cluster is split recursively.

PARTITIONING CLUSTERING

Partitioning clustering divides data points into a predefined number of clusters. The primary objective is to minimize the distance between data points within each cluster while maximizing the distance between different clusters. The most commonly used partitioning clustering algorithm is k-means, which assigns data points to clusters by minimizing the within-cluster variance.

Figure 13.3 presents an illustration of hierarchical and partitioning clustering. Table 13.3 summarizes the key differences between the two methods.

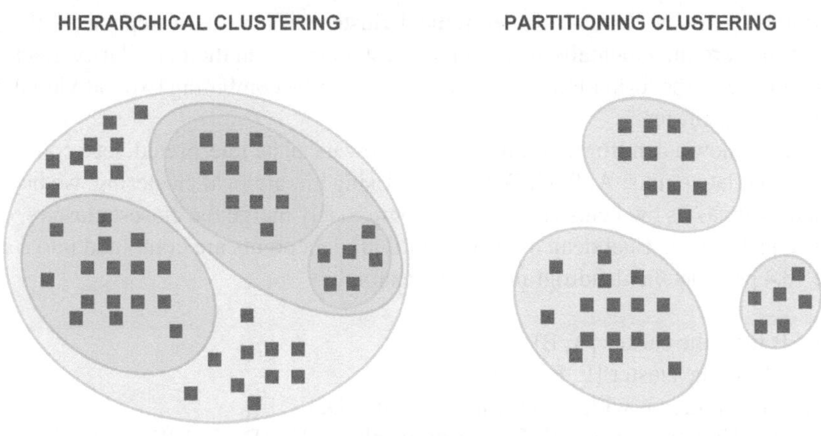

FIGURE 13.3 Hierarchical clustering and partitioning clustering.

TABLE 13.3
Differences between Hierarchical Clustering and Partitioning Clustering

Hierarchical Clustering	Partitioning Clustering
1. **Hierarchy**: Hierarchical clustering creates a hierarchy of clusters by iteratively merging or splitting clusters based on similarity or distance measures.	1. **Fixed number of clusters**: Partitioning clustering methods, such as k-means or k-medoids, require specifying the number of clusters in advance.
2. **Agglomerative and divisive**: It can be agglomerative, starting with individual data points as separate clusters and merging them together, or divisive, starting with a single cluster and splitting it into smaller clusters.	2. **Non-hierarchical**: Partitioning clustering does not create a hierarchy of clusters. It assigns each data point to one and only one cluster based on a predefined criterion.
3. **Number of clusters**: Hierarchical clustering does not require specifying the number of clusters in advance. It produces a hierarchical structure with different levels of granularity, which allows users to choose the number of clusters based on their requirements.	3. **Objective function optimization**: Partitioning methods aim to optimize an objective function, such as minimizing the within-cluster variance or maximizing inter-cluster dissimilarity.
4. **Cluster representation**: The results of hierarchical clustering can be visualized using dendrograms, which show the hierarchical structure of clusters and the distance at which they merge.	4. **Speed**: Partitioning algorithms are generally faster than hierarchical clustering, especially for large datasets, as they do not require pairwise distance calculations for all data points.
5. **Computationally expensive**: Hierarchical clustering can be computationally expensive, especially for large datasets, as it requires computing pairwise distances or similarities between all data points.	5. **Initialization sensitivity**: The initial selection of cluster centroids in partitioning methods can significantly affect the final clustering results, making them sensitive to initialization.
6. **Noisy data handling**: This algorithm is robust to noise and outliers in the data because it does not rely on a global optimization criterion but rather on local similarities.	6. **Difficulty with outliers**: Partitioning methods can be sensitive to outliers because they try to optimize a global criterion. Outliers can influence the cluster centers and distort the resulting clusters.

HIERARCHICAL CLUSTERING

Hierarchical clustering organizes data points into a hierarchy of clusters, producing a tree-like structure called a dendrogram. This method is particularly flexible because it does not require a predetermined number of clusters, unlike other clustering methods.

DENDROGRAM

The dendrogram is a key element of hierarchical clustering. It visually represents the process in which data points are incrementally merged into clusters based on their similarity. Each step in the dendrogram indicates the formation of a new cluster, either by combining two individual data points or merging existing clusters.

To help clarify how a dendrogram functions and how it can be interpreted, let's consider a simple example with six data points: A, B, C, D, E, and F. Using hierarchical clustering, we begin by treating each data point as its own cluster. We then progressively merge the closest clusters, determined by a distance metric (e.g., Euclidean distance), until all data points are combined into a single cluster. Figure 13.4 presents the dendrogram for this example.

- A and B form the cluster [A, B].
- E and F form the cluster [E, F].
- Cluster [E, F] merges with D to form the cluster [D, [E, F]].
- Cluster [D, [E, F]] merges with C to form the cluster [C, [D, [E, F]]].
- All data points now belong to a single cluster.

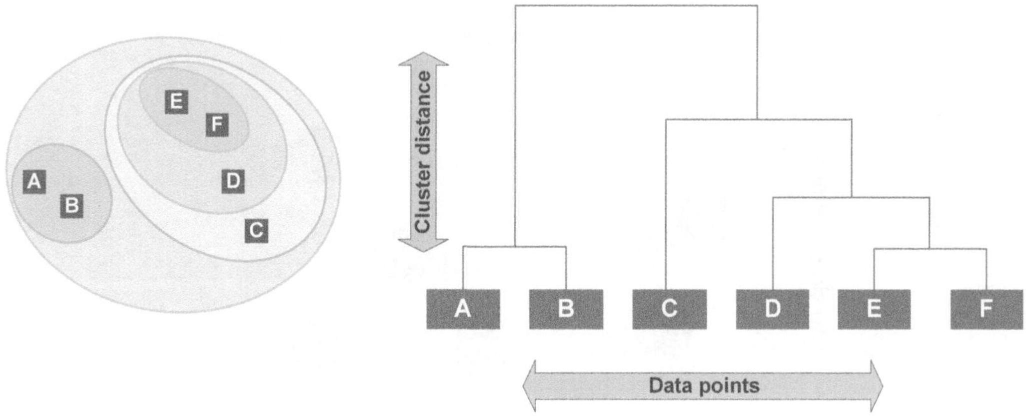

FIGURE 13.4 Example of a dendrogram.

As shown, dendrograms visually depict how data points are gradually grouped into clusters, creating a hierarchy. Each level of the hierarchy shows how clusters are nested and organized into a tree structure.

There are two primary types of hierarchical clustering:

Agglomerative Clustering (Bottom-Up)

Agglomerative clustering begins by treating each data point as its own cluster. It then iteratively merges the most similar clusters until all data points are grouped into a single cluster. At each step, the algorithm selects the closest pair of clusters based on a chosen distance metric and merges them. This process continues until every data point belongs to one final cluster (see Figure 13.5).

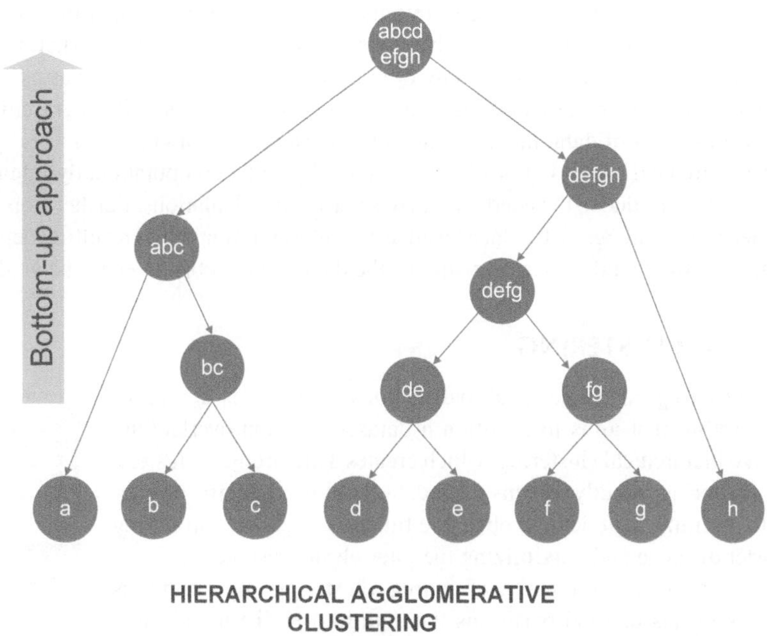

FIGURE 13.5 Agglomerative clustering (bottom-up).

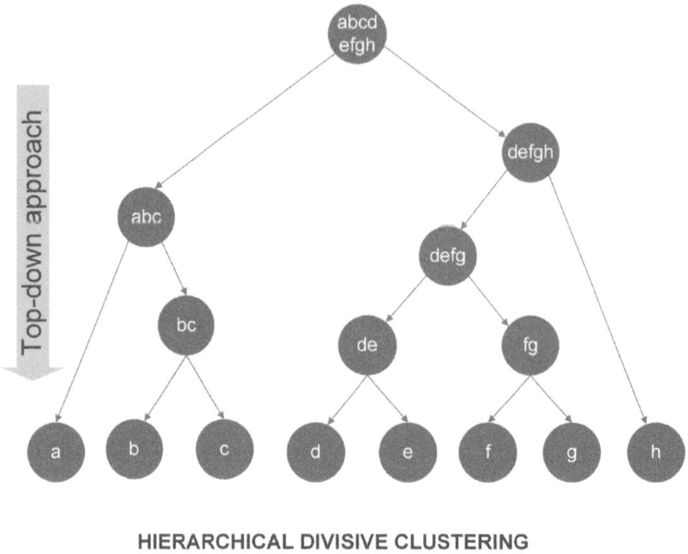

FIGURE 13.6 Divisive clustering (top-down).

Divisive Clustering (Top-Down)

In contrast, divisive clustering starts with all data points in one cluster and then splits them recursively into smaller clusters. At each step, the algorithm identifies the most dissimilar cluster and divides it into two smaller clusters, again based on a selected distance metric. This process is repeated until each data point is isolated in its own cluster (see Figure 13.6).

The choice of distance measure is critical in hierarchical clustering, as it directly influences how clusters are formed. The linkage criterion determines how the distance between clusters is calculated during the merging or splitting process. Common linkage methods include single-linkage, complete-linkage, and average-linkage. We will delve into these distance measures later in this chapter.

Hierarchical clustering offers several advantages. It reveals the inherent structure and relationships within the data by creating a hierarchy of clusters. The dendrogram serves as a visual representation of the clustering process, making it easier to interpret. Additionally, hierarchical clustering can handle various types of data, including numerical, categorical, and binary.

However, hierarchical clustering has its limitations. It can be computationally intensive, particularly with large datasets, due to the need for pairwise distance calculations. Furthermore, the selection of distance metric and linkage criterion, as well as the interpretation of the resulting dendrogram, can vary depending on the specific characteristics of the data and the objectives of the analysis.

PARTITIONING CLUSTERING

Partitioning clustering, also known as partitional clustering, is a type of clustering algorithm that aims to partition a dataset into non-overlapping clusters. Unlike hierarchical clustering, which creates a hierarchy of clusters, partitioning clustering directly assigns each data point to a specific cluster. The goal is to optimize a predefined objective function, such as minimizing the intra-cluster distance and maximizing the inter-cluster distance.

One of the most well-known partitioning clustering algorithms is k-means. The k-means method partitions the data into K distinct clusters by iteratively updating the cluster centroids and assigning data points to the nearest centroid. The algorithm works iteratively to assign data points to clusters and adjust the cluster centroids until convergence. The objective is to minimize the sum of squared distances between data points and their respective cluster centroids.

Steps in K-Means Clustering:

1. **Initialization**: Choose the number of clusters, k, and randomly initialize k centroids within the data range.
2. **Assignment**: Assign each data point to the nearest centroid based on a distance measure, commonly Euclidean distance. This step creates k clusters.
3. **Update**: Recalculate the centroid for each cluster by taking the mean of all data points assigned to that cluster.
4. **Iteration**: Repeat steps 2 and 3 until convergence. Convergence occurs when the centroids no longer change significantly or when a predetermined maximum number of iterations is reached.
5. **Result**: Once the algorithm converges, each data point belongs to a particular cluster, and the cluster centroids represent the center of each cluster.

K-means is an efficient and straightforward algorithm, but its success depends on the initial selection of centroids and the number of clusters (k). The number of clusters is often determined using domain knowledge or techniques like the elbow method or silhouette analysis to select the optimal number of clusters. Multiple runs of the algorithm with different initializations may be required, and the result with the lowest error is typically chosen.

Advantages of K-Means:

- K-means is relatively efficient and can scale well with large datasets.
- It works well when clusters have a spherical shape and similar sizes.
- The algorithm is easy to implement and interpret.

Limitations of K-Means:

- It assumes clusters are spherical and mutually exclusive, which may not always be the case in real-world data.
- It is sensitive to the initial placement of centroids, which can lead to different clustering results.
- The choice of **K**, the number of clusters, can be arbitrary if not properly determined using a suitable method.

TWO-STEP CLUSTER ANALYSIS

Two-step cluster analysis is a hybrid approach that combines elements of hierarchical clustering and k-means clustering. This method is particularly useful for analyzing large datasets or datasets that contain both continuous and categorical variables, making it versatile for real-world applications. The algorithm works in two main stages, as described below:

Process of Two-Step Cluster Analysis:

- **Pre-clustering:**
  In the first step, a hierarchical clustering method (often agglomerative clustering) is used to generate preliminary clusters. This step helps in identifying potential clusters by reducing the computational load, especially when dealing with large datasets. By creating pre-clusters, the process becomes more efficient.

- **Two-step clustering:**
  In the second step, k-means clustering is applied to the pre-clustered data. This step assigns data points to clusters based on their proximity to the centroids. As in standard k-means clustering, the centroids are updated iteratively until the clusters stabilize, ensuring convergence.

Advantages of Two-Step Cluster Analysis:

- **Suitable for large datasets:**
  This method is well-suited for handling large datasets, making it efficient and scalable compared to traditional clustering methods.
- **Automatic determination of the number of clusters:**
  One of the main benefits of two-step cluster analysis is its ability to automatically determine the optimal number of clusters based on statistical criteria, saving the user from having to manually specify the number of clusters.
- **Handles mixed data types:**
  Two-step cluster analysis can accommodate both categorical and continuous variables, providing flexibility for datasets with mixed data types.

Limitations of Two-Step Cluster Analysis:

- **Sensitivity to outliers:**
  Like many clustering algorithms, two-step cluster analysis can be sensitive to outliers, which may distort the results and lead to incorrect cluster assignments.
- **Relies on predefined distance measures:**
  The algorithm assumes that the similarity between data points can be adequately measured using predefined distance measures (e.g., Euclidean or Manhattan), which may not always capture the true relationships in the data.
- **Lack of interpretability:**
  The clusters generated may not always have clear or intuitive meanings, requiring further analysis and validation to understand the significance of each cluster.

In this book, we use both the K-means clustering and the two-step method to perform cluster analysis using SPSS Modeler.

In conclusion, two-step cluster analysis is a valuable tool in data mining, especially when dealing with large datasets or mixed data types. Its ability to automatically determine the optimal number of clusters enhances its efficiency, and its flexibility in handling various variable types makes it applicable across diverse domains. However, like all clustering methods, careful attention should be given to outliers, and further interpretation is often necessary to fully understand the clusters.

CLUSTERING CRITERIA

Clustering criteria are statistical measures used to evaluate the performance and quality of clustering models. Two widely recognized criteria for determining the optimal number of clusters in a dataset are the **Bayesian Information Criterion (BIC)** and the **Akaike Information Criterion (AIC)**. These measures help balance the trade-off between model complexity and goodness of fit, ensuring the clusters are meaningful without overfitting.

Bayesian Information Criterion (BIC)

BIC is a criterion for model selection among a finite set of models. In the context of clustering, BIC is often used to evaluate the goodness of fit of different clustering solutions with varying numbers of clusters. BIC balances the trade-off between the complexity of the model (number of clusters) and the goodness of fit to the data. It is calculated using the formula:

$$BIC = -2 \times log - likelihood + k \times log(n)$$

where:

- Log-likelihood: A measure of how well the model fits the data.
- k: The number of parameters in the model.
- n: The number of data points.

In clustering, a lower BIC value indicates a better model fit with an appropriate level of complexity. The optimal number of clusters corresponds to the model with the minimum BIC value, providing the best trade-off between model complexity and data fit.

Akaike Information Criterion (AIC)

Similar to BIC, AIC is a measure of the relative quality of a statistical model. It penalizes models for their complexity to prevent overfitting. AIC is calculated using the formula:

$$AIC = -2 \times log - likelihood + 2 \times k$$

where the terms have the same meaning as in BIC.

In the context of clustering, AIC serves a similar purpose as BIC in selecting the optimal number of clusters. However, AIC tends to be less stringent in penalizing model complexity compared to BIC. Therefore, AIC may favor more complex models with higher numbers of clusters compared to BIC.

Application of BIC and AIC in Clustering

To determine the optimal number of clusters, analysts typically run several clustering models with varying numbers of clusters and compute the BIC and AIC values for each model. The model with the minimum BIC or AIC value is selected as having the optimal number of clusters for the dataset.

- **BIC** often selects fewer clusters as it penalizes more complex models more heavily.
- **AIC** might suggest more clusters, being less conservative with complexity.

By applying these criteria, analysts can find a clustering solution that achieves a balance between simplicity and model fit, avoiding overfitting while ensuring the model captures the essential patterns in the data.

CLUSTER QUALITY ASSESSMENT

Evaluating the quality of clusters is crucial to understanding how effectively clustering algorithms have grouped data points. One commonly used metric for assessing cluster quality is the silhouette measure, which quantifies how well a data point fits within its assigned cluster compared to other clusters. The silhouette measure considers both cohesion (how closely points are grouped within a cluster) and separation (how distinct the cluster is from other clusters).

Silhouette measure calculation: The silhouette measure for a data point i is calculated as follows:

1. **Mean intra-cluster distance (a)**: This is the average distance between data point ii and all other points within the same cluster.
2. **Mean nearest-cluster distance (b)**: This is the average distance between data point ii and all points in the nearest cluster (i.e., the cluster to which *i* does not belong).
3. **Silhouette coefficient (s)**: The silhouette coefficient for data point *i* is given by:

$$s_i = \frac{b_i - a_i}{\max(a_i, b_i)}$$

The silhouette coefficient ranges from −1 to 1, where:

- $\geq s_i = 1$ indicates that the data point is well-clustered,
- $s_i = 0$ indicates that the data point is on the boundary between two clusters, and
- $s_i = -1$ indicates that the data point may have been assigned to the wrong cluster.

By calculating the silhouette coefficient for each data point and averaging the values across all points, we can obtain an overall measure of cluster quality. A higher average silhouette coefficient implies better-quality clusters, with values closer to 1 indicating more compact and well-separated clusters.

Application of Silhouette Measure

The silhouette measure is especially useful for determining the optimal number of clusters. By comparing the average silhouette coefficient across different numbers of clusters, analysts can find the cluster configuration that best balances cohesion and separation. Additionally, a silhouette plot can provide a visual representation of how well individual points are clustered, which aids in interpreting the clustering results.

Comparison between Silhouette, BIC, and AIC

- BIC and AIC are model selection criteria that focus on finding the best-fit model while penalizing for complexity, whereas silhouette score evaluates the quality of clusters directly.
- BIC and AIC are typically used to compare different clustering solutions based on their overall fit to the data, while silhouette score is used to assess the quality of individual clusters within a solution.
- Lower values of BIC and AIC indicate better model fit, whereas higher silhouette scores indicate better cluster quality.

CLUSTER ANALYSIS IN SPSS MODELER

SPSS Modeler provides several nodes to perform cluster analysis. In this book, we will focus on three primary clustering nodes: k-means node, TwoStep Cluster node, and Auto Cluster node.

K-Means Node

The K-means node performs cluster analysis using the k-means clustering method. K-means operates by initially establishing a predetermined set of cluster centers derived from the dataset. Subsequently, it assigns each data record to the cluster with the highest similarity, as determined by the input field values of the record. Once all records are assigned, the cluster centers are adjusted to accommodate the newly assigned records. Then, the records are reevaluated to determine if they should be reassigned to a different cluster. This iterative process of record assignment and cluster adjustment continues until either the maximum number of iterations is achieved or the change between consecutive iterations falls below a specified threshold.

TwoStep Cluster Node

The TwoStep Cluster node performs cluster analysis using the two-step clustering method. In the initial step, the node makes a single pass through the data, compressing the raw input data into a more manageable set of subclusters. Subsequently, in the second step, a hierarchical clustering technique is utilized to progressively merge these subclusters into larger clusters, all without necessitating another pass through the data. Notably, hierarchical clustering eliminates the need for preselecting the number of clusters in advance. Many hierarchical clustering algorithms commence with individual records as initial clusters, merging them recursively to form increasingly larger clusters. While such methods may struggle with extensive datasets, the initial preclustering performed by TwoStep facilitates swift hierarchical clustering, even with large volumes of data.

Auto Cluster Node

The Auto Cluster node estimates and evaluates clustering models designed to identify groups of records sharing similar characteristics. Operating akin to other automated modeling nodes, it facilitates experimentation with various combinations of options within a single modeling iteration. The node allows for the comparison of models using fundamental measures, aiding in the filtering and ranking of the usefulness of cluster models.

These three nodes—K-means, TwoStep, and Auto Cluster—offer distinct approaches to clustering and are suited to different types of data and analytical needs. Each provides flexibility and ease of use in the SPSS Modeler environment.

EXAMPLE: CLUSTER ANALYSIS

In this example, we perform a cluster analysis for the Airline Recommendation dataset.

- **Dataset:** AirlineRecommendation.sav
- **Target Variable:** recommendation (Flag: Yes/No)

Variables for Cluster Analysis

Table 13.4 presents the variables used in this cluster analysis. Since cluster analysis is an unsupervised learning technique, only input variables are considered. The target variable, **recommend**, is excluded from the analysis.

The goal of this example is to group similar individuals based on shared characteristics in terms of their demographics and travel experience. For this purpose, we have selected six key variables for the cluster analysis, as highlighted in Table 13.4:

- Age
- Gender
- Airline status (e.g., Blue, Silver, Gold, Platinum)
- Average number of flights per year
- Travel purpose (e.g., Business, Personal, Mileage)
- Shopping amount at the airport

TABLE 13.4
Airline Recommendation Data for Cluster Analysis

Variable Name	Label/Description	Scale	Role
recommend	Recommend the airline to others	Binary (Yes/No)	None
airline_st	Airline status	Nominal	Input
Age	Age	Interval	Input
Gender	Gender	Binary (Male/Female)	Input
price_sensitivity	Price sensitivity	Interval	None
no_years_flying	Number of years flying	Interval	None
no_flt_pa	Average number of flights per year	Interval	Input
pct_ flt_other_airline	Percent of flights with other airlines	Interval	None
type_of_travel	Travel purpose	Nominal	Input
no_other_lty_cards	Number of other loyalty cards	Interval	None
shopping_amt_airpt	Shopping amount at the airport	Interval	Input
eat_drink_airpt	Spending for eating and drinking at the airport	Interval	None
Class	Passenger class	Nominal	None
origin_state	Origin state	Nominal	None
destin_state	Destination state	Nominal	None
depart_delay	Departure delay (in minutes)	Interval	None
arrival_delay	Arrival delay (in minutes)	Interval	None
flt_cancel	Flight cancellation	Binary (Yes/No)	None
flt_time	Flight time (in minutes)	Interval	None
flt_distance	Flight distance (miles)	Interval	None

These selected variables enable the identification of distinct market segments among the airline's passengers. The resulting clusters will allow the airline to better understand its customers and develop strategies to effectively target each market segment based on their travel behavior and preferences.

Other variables in the dataset, such as binary or irrelevant categorical fields, are rejected from this analysis to focus on the key factors that contribute to meaningful segmentation.

Sample-Explore-Modify

Refer to the previous chapters for the details. As mentioned earlier, since cluster analysis is an unsupervised learning method, partitioning the data is not necessary. For this example, we will use the entire dataset for clustering.

Cluster Modeling

To begin the analysis, we will use the AutoCluster node to quickly generate initial clustering results. The AutoCluster node is a useful starting point because it provides a fast and efficient means of exploring multiple clustering methods. We can use it to run both K-means and Two-Step cluster models and then compare the results. However, since this node is pre-configured, it does not allow for manual adjustments to the clustering configurations.

After reviewing these initial results, we will construct and run the K-means and Two Step Cluster nodes separately with custom configurations. This approach allows us to fine-tune the clustering process to better meet the specific needs of the airline recommendation dataset.

Auto Cluster Node

Follow these steps to set up and run the AutoCluster node for the airline recommendation dataset:

1. Create a new stream: Name it "**Airline_Cluster**."
2. Copy the necessary nodes: Open the existing stream "**Airline_Tree**." and copy the nodes from the data source up to the **Missing Value Imputation** node, including the **Type** node. Paste these into the new "**Airline_Cluster**" stream.
3. Update variable roles:
   - Open the settings for the **Type** node.
   - Modify the roles of the variables as outlined in Table 13.4. Make sure the fields for clustering are correctly identified.
   - Click Apply to save the changes.

Field	Measurement	Values	Missing	Check	Role
Recommend t...	Flag	Yes/No		None	None
Airline status	Nominal	Blue,Gold,...		None	Input
Age	Continuous	[18.0,85.0]		None	Input
Gender	Flag	Male/Female		None	Input
Price sensitivity	Continuous	[0.0,5.0]		None	None
Number of ye...	Continuous	[2.0,11.0]		None	None
Average num...	Continuous	[0.0,100.0]		None	Input
Percent of fli...	Continuous	[1.0,110.0]		None	None
Travel purpose	Nominal	"Business t...		None	Input
Number of ot...	Continuous	[0.0,12.0]		None	None
Shopping am...	Continuous	[0.0,879.0]		None	Input
Spending for ...	Continuous	[0.0,895.0]		None	None
Passenger class	Nominal	Business,E...		None	None
Origin state	Nominal	Alabama,A...		None	None
Destination st...	Nominal	Alabama,A...		None	None
Departure del...	Continuous	[0.0,1592.0]		None	None
Arrival delay ...	Continuous	[0.0,1584.0]		None	None
Flight cancell...	Flag	Yes/No		None	None
Flight time (in...	Continuous	[8.0,669.0]		None	None
Flight distanc...	Continuous	[31.0,4983.0]		None	None

4. Add the AutoCluster node:
   - From the **Modeling** group, add an **Auto Cluster** node to the stream and connect it to the **Type** node.
   - Open the Auto Cluster node settings.

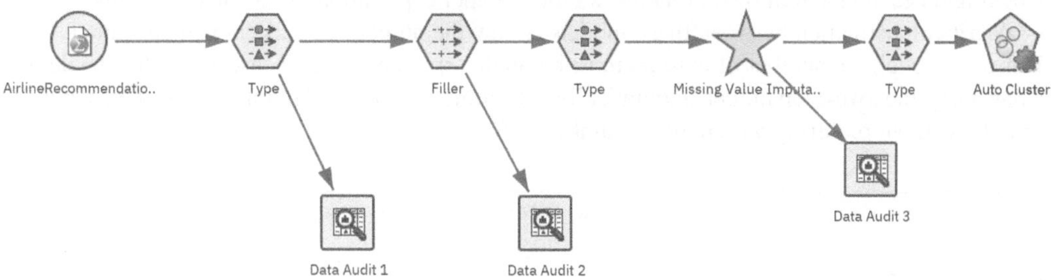

5. Configure the Model tab:
   - Uncheck the box labeled "**Use partitioned data**."
   - For **Rank models by**, select **Silhouette** to rank the models by the silhouette score.

6. Configure the Expert tab:
   • Select two models for comparison: "TwoStep" and "K-means."

7. Click Apply and then Run.

Results of the Auto Cluster Node

Figure 13.7 presents the results of the Auto Cluster node, showing the comparison between the K-means model and the Two-Step model. According to the output, the K-means model utilizes five clusters, while the Two-Step model groups the data into two clusters. The models are ranked based on their Silhouette scores, with the K-means model achieving a score of 0.359, indicating a better fit than the Two-Step model, which has a Silhouette score of 0.309.

Although the Auto Cluster node suggests using the K-means model due to its higher Silhouette score, there are considerations to note. One key observation is that the smallest cluster in the K-means model contains only 6% of the data, which may indicate an uneven distribution of clusters. Clicking on each model in the output allows for a deeper exploration of the cluster details.

A critical distinction between these models is how the number of clusters is determined. In the K-means model, the number of clusters must be predetermined—in this case, set to five by default. Conversely, the Two-Step model automatically determines the optimal number of clusters using BIC or AIC, without requiring a predefined number.

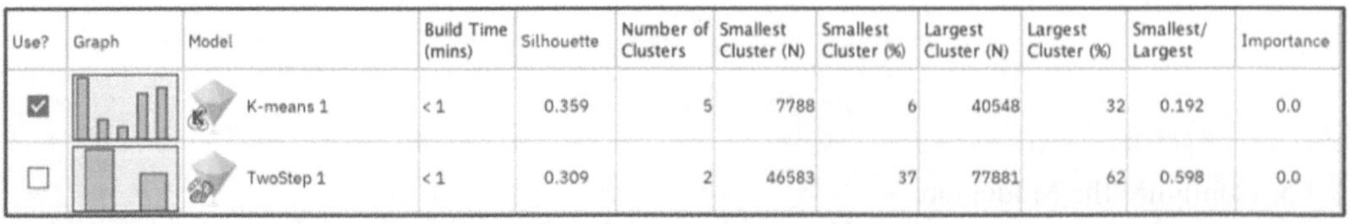

Use?	Graph	Model	Build Time (mins)	Silhouette	Number of Clusters	Smallest Cluster (N)	Smallest Cluster (%)	Largest Cluster (N)	Largest Cluster (%)	Smallest/ Largest	Importance
☑		K-means 1	< 1	0.359	5	7788	6	40548	32	0.192	0.0
☐		TwoStep 1	< 1	0.309	2	46583	37	77881	62	0.598	0.0

FIGURE 13.7 Results of the Auto Cluster node.

When choosing the number of clusters, it's important to balance capturing the complexity of the data with maintaining interpretability and generalizability. Having too many clusters can lead to overfitting, decreased model simplicity, and challenges in understanding and applying the results. Given this, reducing the number of clusters in the K-means model may yield more interpretable results without compromising the quality of the clustering.

Since the Auto Cluster node does not allow for reconfiguring models, we will now conduct the individual K-means and Two-Step cluster models separately. This will enable us to adjust the number of clusters and compare the results more effectively.

Two-Step Cluster Node

Follow the steps below to set up the TwoStep Cluster node:

1. From the stream above, add a **TwoStep Cluster** node from the **Modeling** group and connect it to the **Type** node.
2. Open the **TwoStep Cluster** node settings.
3. Under the **Model** tab, uncheck the box labeled "**Use partitioned data**."
4. Check the box labeled "**Standardize numeric data**" to ensure that all numeric variables are on the same scale.
5. For **Distance measure**, select "**Euclidean**" to measure similarity between the data points.
6. For **Clustering criterion**, select "**Schwarz's Bayesian Criterion (BIC)**" to determine the optimal number of clusters.
7. Select "**Automatically calculate number of clusters**" to let the algorithm determine the most appropriate number of clusters based on the data.

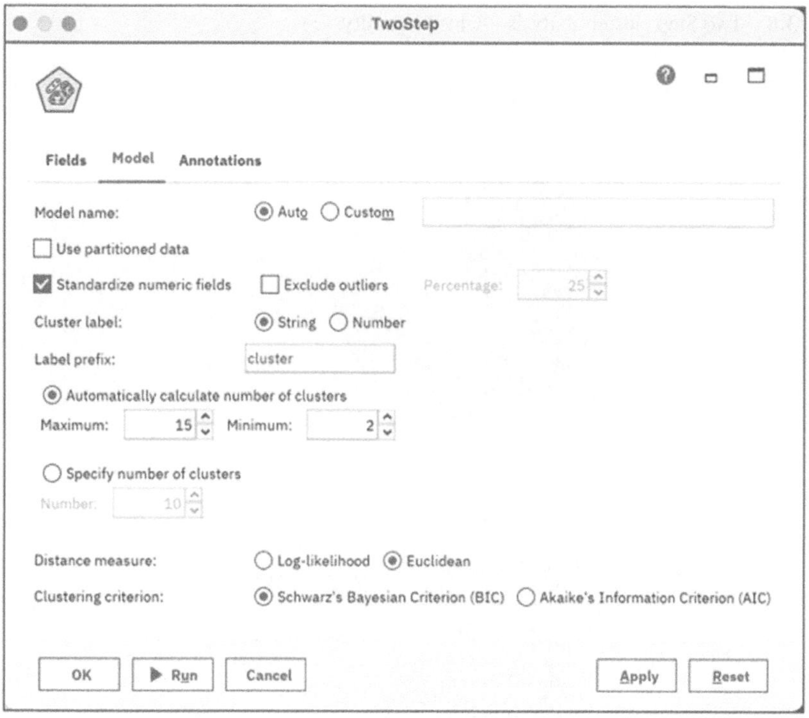

8. Click Apply and then Run.

Results of the Two-Step Cluster Node

The clustering results for the Two-Step Cluster node confirm the same findings as in the Auto Cluster node, showing the selection of two clusters. Figures 13.8 and 13.9 illustrate the cluster quality and cluster sizes, respectively. As observed, the model with two clusters yields a Silhouette score of 0.3.

- **Cluster 1** contains **62.6%** of the data.
- **Cluster 2** contains **37.4%** of the data.

Since the cluster quality is consistent with the previous findings from the Auto Cluster node, we will now proceed to test the K-means model to determine whether it can produce better cluster quality or offer more meaningful segmentation.

Model Summary

Algorithm	TwoStep
Inputs	6
Clusters	2

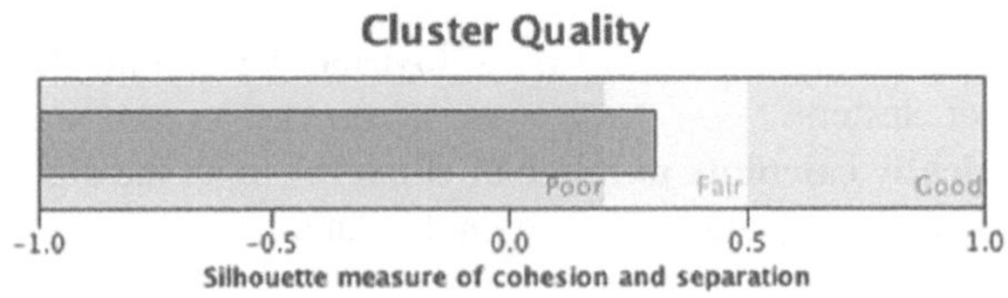

FIGURE 13.8 Two Step cluster analysis—Cluster quality.

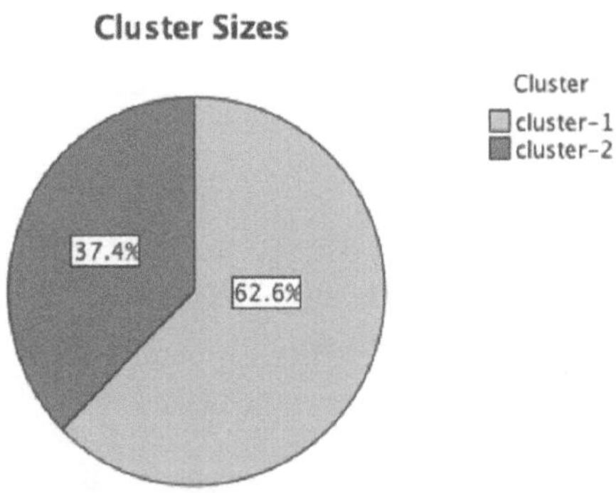

Size of Smallest Cluster	46583 (37.4%)
Size of Largest Cluster	77881 (62.6%)
Ratio of Sizes: Largest Cluster to Smallest Cluster	1.67

FIGURE 13.9 Two Step cluster analysis—Cluster sizes.

K-Means Cluster Node

The K-means cluster node will allow us to perform the K-means clustering with three clusters, a reduction from the five-cluster model that had been automatically determined earlier. Our goal is to improve the interpretability and generalizability of the cluster model while maintaining or improving the model fit.

Follow the steps below to configure the K-means cluster node:

1. Add a K-Means Node:
   - In the existing stream, add a **K-Means** node from the **Modeling** group.
   - Connect it to the **Type** node.
2. Open Node Settings:
   - Open the **K-Means** node settings.
3. Model Tab:
   - Uncheck the box labeled "**Use partitioned data**."
   - Set the **Number of clusters** to **3**.
   - Select the box "**Generate distance field**."

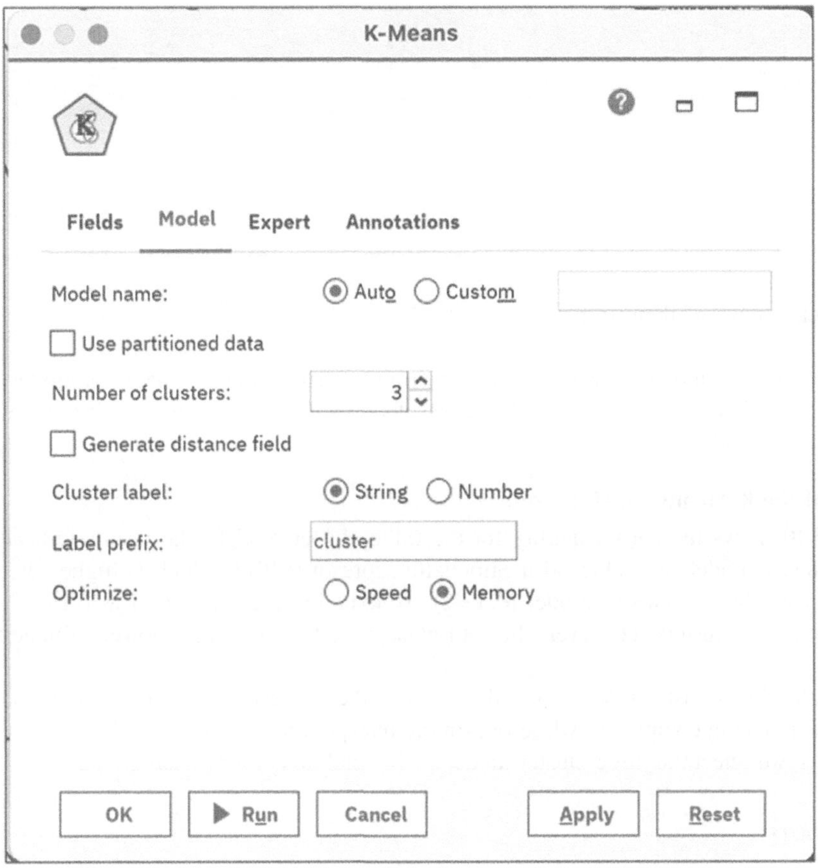

4. Expert Tab:
   - Go to the **Expert** tab.
   - Select "**Expert**" and "**Custom**" options.
   - Set **Maximum iterations** to **100**.

5. Click Apply and then Run.

By adjusting the model to use 3 clusters, we aim to strike a balance between model complexity and interpretability, and see if the clustering results provide more meaningful insights while maintaining a good model fit.

Results of the K-Means Cluster Node

Figure 13.10 shows the cluster quality for the three-cluster model. The results indicate that with a three-cluster model, we achieved a Silhouette score of 0.4014, which is higher than the score obtained with the five-cluster model (0.3542). To explore other options, I also ran the K-means model with four clusters. However, the four-cluster model produced a lower Silhouette score of 0.3086.

Thus, the three-cluster model not only provides the highest cluster quality but also simplifies the model, reducing complexity while improving interpretability and generalizability. Given these advantages, we select the three-cluster model as the final model for this analysis.

Cluster Sizes

Figure 13.11 presents the cluster analysis results, indicating the distribution of cases across three clusters, as determined by the clustering algorithm. The clusters are as follows:

- **Cluster 1** consists of **45,510 cases**, representing **36.6%** of the dataset.
- **Cluster 2** includes **37,533 cases**, accounting for **30.2%**.
- **Cluster 3** contains **41,421 cases**, making up **33.3%** of the dataset.

Model Summary

Algorithm	K-Means
Inputs	6
Clusters	3

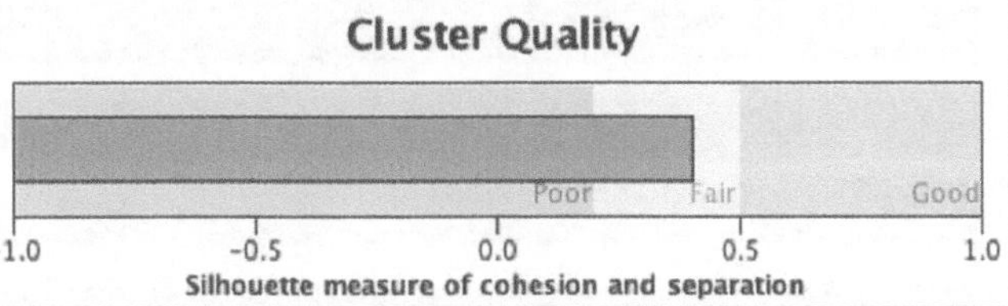

FIGURE 13.10 K-means cluster analysis—Cluster quality.

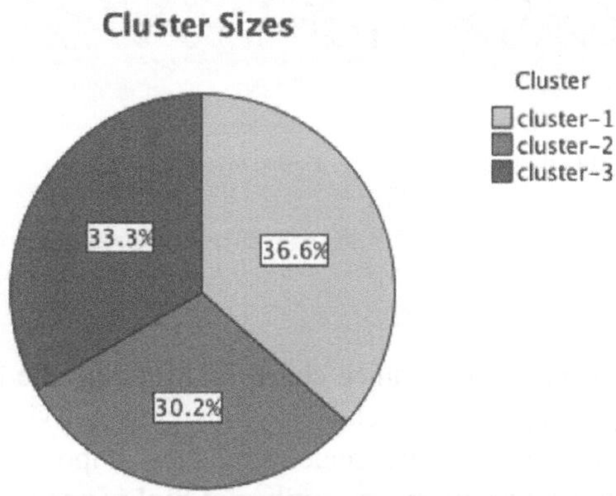

Size of Smallest Cluster	37533 (30.2%)
Size of Largest Cluster	45510 (36.6%)
Ratio of Sizes: Largest Cluster to Smallest Cluster	1.21

FIGURE 13.11 K-means cluster analysis—Cluster sizes.

This distribution suggests relatively balanced cluster sizes, with no single cluster dominating the dataset. Cluster 1 is the largest, followed by Cluster 3, and then Cluster 2. We will explore the characteristics of each cluster in more detail shortly. Additionally, the K-means node has created a new categorical field, $KM-K-Means, in the dataset. This field consists of three levels—cluster-1, cluster-2, and cluster-3—corresponding to the three clusters identified in the analysis. Records are assigned to their respective clusters based on these labels, as depicted in Figure 13.11.

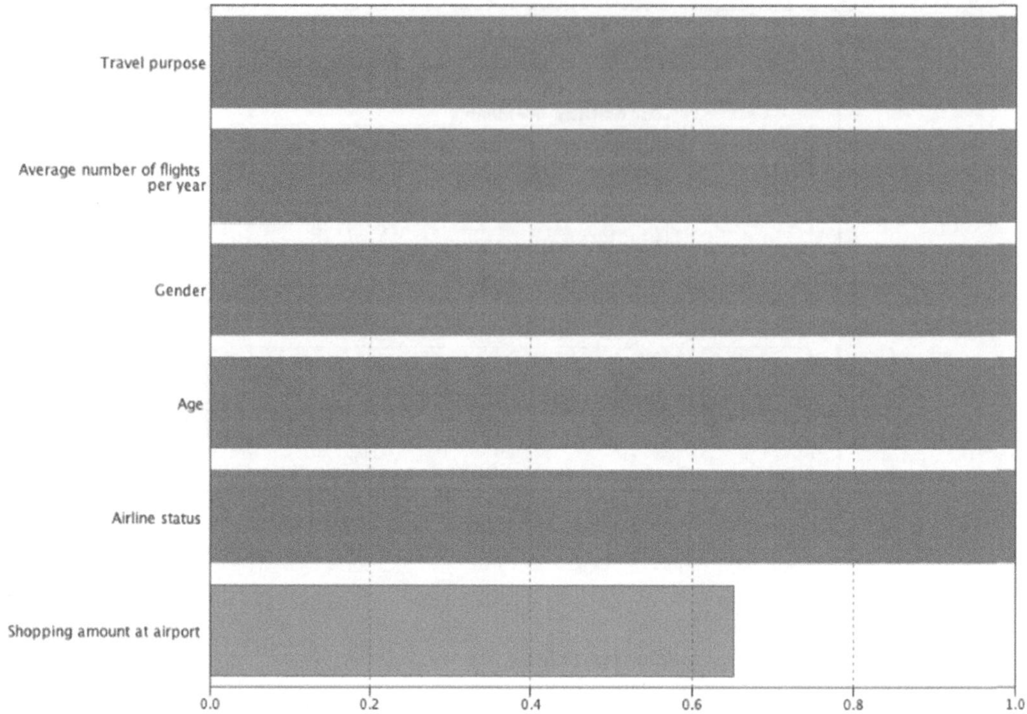

FIGURE 13.12 Variable importance.

Variable Importance

Figure 13.12 displays the variable importance chart, highlighting the key variables that contributed significantly to the formation of the clusters. This chart provides insight into the relative importance of each field in estimating the model. The most important variables in determining the clusters are **travel purpose**, **age**, **average number of flights per year**, and **flight distance**. On the contrary, the **shopping amount at the airport** is identified as the least important variable in cluster formation.

Cluster Distance

Typically, reporting the distance between clusters helps show how distinct clusters are from one another, providing an indication of the quality of the clusters. Unfortunately, SPSS Modeler does not provide cluster distances directly in its results. However, we can still present the cluster distances in a different way. When we configured the K-means node, we selected the option to generate the distance field, resulting in the addition of the new field, **$KMD-K-Means**, to the dataset. This field is a continuous variable that represents the distance of each record from the center of its assigned cluster.

Using this distance field, we can create a histogram to display the distributions of distances for all three clusters (Figure 13.3). The results show that the distance ranges are consistent among the three clusters, falling between 0.35 and 1.4. Furthermore, Cluster 1 and Cluster 3 have similar distributions, while Cluster 2 exhibits a notably different distribution. The mean distances for Cluster 1 and Cluster 3 are 0.605 and 0.602, respectively, which are very close. In contrast, Cluster 2 has a slightly higher mean distance of 0.688.

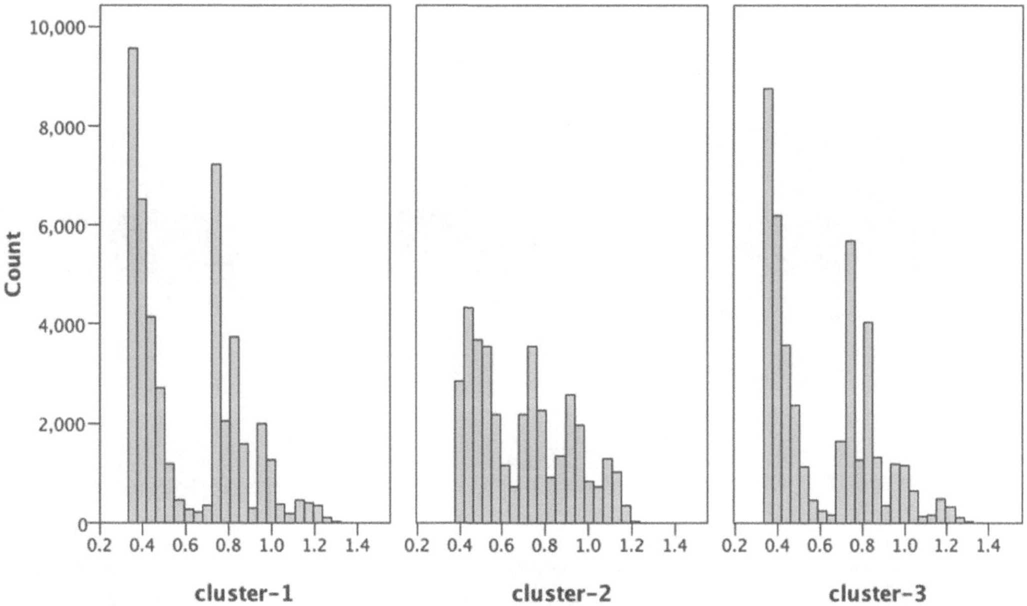

FIGURE 13.13 Distances between data points and the center of each cluster.

Although this histogram does not directly show how distinct one cluster is from another, it does illustrate the differences in the distance distributions among clusters. We will further explore these differences through cluster profiling in the next step.

> SPSS Modeler currently does not provide the distance between each pair of clusters in any clustering node.
>
> The K-Means node, however, has a feature that generates a new field containing the distance of each record from the center of its assigned cluster. We can analyze this data to get some preliminary differences among clusters. Cluster profiling should give us a more in-depth comparison.

Cluster Profiling

The next step is cluster profiling, which involves reviewing the profile of each cluster (segment), examining the differences between those clusters, and identifying their unique characteristics. To do this, we must review the results for all seven variables. Figures 13.14–13.15 present the cluster profiles for these variables. By reviewing each chart, we can summarize the cluster profile for each variable as follows:

- **Age**: Cluster 2 has older passengers, with an average age of 58 years. Clusters 1 and 3 have younger passengers, with the same average age of 43 years.
- **Gender**: Cluster 1 has mostly females, and Cluster 3 has mostly males. Cluster 2 has a more balanced gender split, with more females than males (67%).
- **Airline status**: Cluster 2 has 76% Blue status, while Cluster 1 and Cluster 3 have about 63% Blue.
- **Travel purpose**: Clusters 1 and 3 consist mostly of passengers traveling for business purposes (almost 90% of passengers). On the contrary, Cluster 2 consists mostly of passengers traveling for personal purposes (100%).

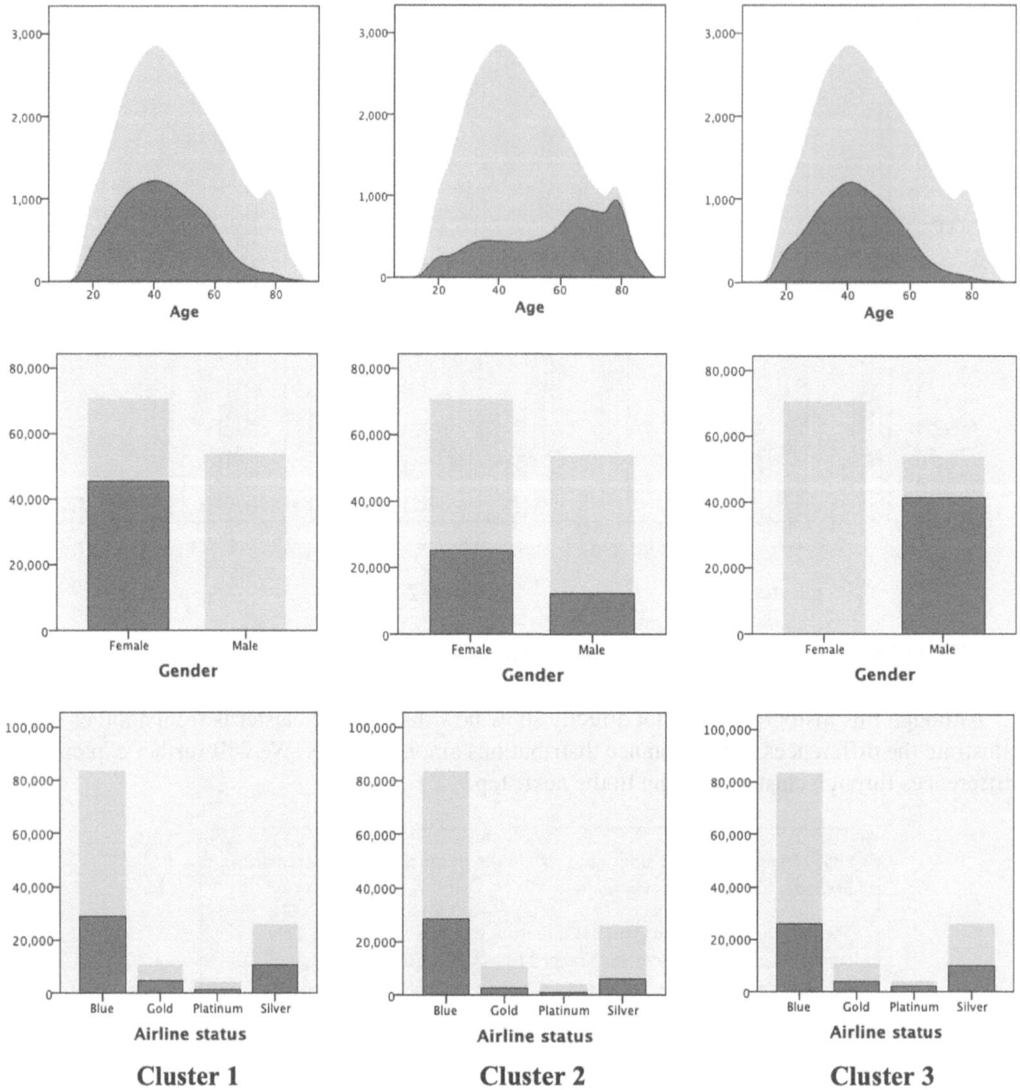

FIGURE 13.14 Cluster profiling with **age, gender,** and **airline status**.

- **Number of flights per year**: Passengers in Cluster 2 fly more often, with an average of 27 flights per year. Clusters 1 and 3 include passengers with fewer flights per year (18 flights per year for Cluster 1 and 18 for Cluster 3).
- **Shopping amount at the airport**: On average, passengers in Cluster 1 spend the most at the airport, with 32 shopping units. Cluster 2's passengers spend an average of 27 units, and Cluster 3's passengers spend an average of 21 units.

Based on this cluster profiling, we can describe the unique characteristics of the three clusters as follows (this process can be somewhat subjective, as it depends on the researcher's interpretation of the results):

- **Cluster 1**: Younger passengers; mostly females; Blue status; business travel; moderate frequent travel; moderate spending at the airport.

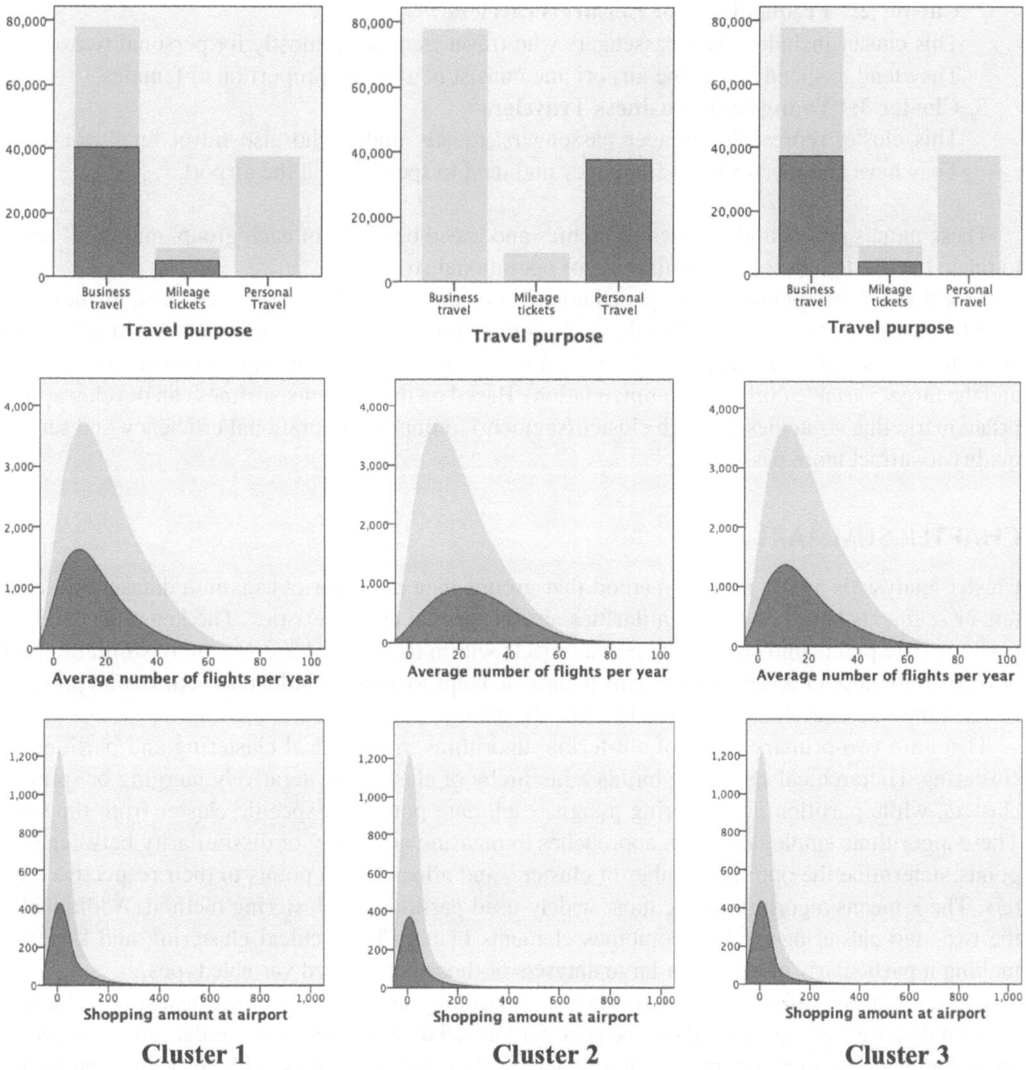

FIGURE 13.15 Cluster profiling with **travel purpose, average number of flights per year,** and **shopping amount at airport**.

- **Cluster 2**: Older passengers; more females than males; Blue status; personal travel; very frequent travel; not spending much at the airport.
- **Cluster 3**: Younger passengers; mostly males; Blue status; business travel; moderate frequent travel; not spending much at the airport.

Based on the cluster profiles described, here are some suggestions for naming the three clusters:

1. **Cluster 1: "Young Female Business Travelers"**
   This cluster represents younger passengers, predominantly female, who travel for business purposes. They have moderate flight frequency and spend a moderate amount at the airport.

2. **Cluster 2: "Frequent Senior Leisure Travelers"**
   This cluster includes older passengers who travel frequently, mostly for personal reasons. They tend to spend less at the airport and consist of a higher proportion of females.
3. **Cluster 3: "Young Male Business Travelers"**
   This cluster represents younger passengers, mostly male, who also travel for business. They have a moderate flight frequency and tend to spend less at the airport.

These names reflect both the demographics and travel behavior of each group, making it easier to relate the clusters to specific marketing or operational strategies.

Now that we understand the unique characteristics (profiles) of these three clusters, the next step is to build a predictive model using these clusters as predictors (they can be combined with other predictors in the dataset). The predictive model will show us the relationships between these clusters and the target variable (airline recommendation). Based on these results, airlines can develop appropriate marketing strategies for each cluster (segment) to improve operational efficiency and service quality to attract more passengers.

CHAPTER SUMMARY

Cluster analysis is a data mining method that groups data points or objects in a dataset into clusters or segments based on their similarities across specific characteristics. The key objective is to partition data points into clusters such that objects within the same cluster are more similar to each other than to those in other clusters. This technique helps to uncover hidden structures and provides insights into the underlying characteristics of the data.

There are two primary types of clustering algorithms: hierarchical clustering and partitioning clustering. Hierarchical clustering builds a hierarchy of clusters by iteratively merging or splitting clusters, while partitioning clustering assigns each data point to a specific cluster from the start. These algorithms employ different approaches to measure similarity or dissimilarity between data points, determine the optimal number of clusters, and allocate data points to their respective clusters. The k-means algorithm is the most widely used partitioning clustering method. Additionally, the two-step clustering method combines elements of both hierarchical clustering and k-means, making it particularly effective for large datasets or those with mixed variable types.

Cluster analysis offers several advantages, particularly as an unsupervised learning technique, where it discovers groupings without the need for labeled data. It is valuable for data summarization, reduction, and visualization, making it a widely applied tool in various fields. It helps identify distinct customer segments, discover meaningful patterns in large datasets, detect anomalies or outliers, and make personalized recommendations. However, cluster analysis comes with challenges, such as sensitivity to parameter settings, determining the optimal number of clusters, and interpreting and evaluating clustering results. Despite these challenges, cluster analysis remains a powerful tool in data mining, effectively organizing data into meaningful groups and uncovering hidden patterns.

CHAPTER DISCUSSION QUESTIONS

1. What is cluster analysis, and how does it group objects into clusters based on their similarities?
2. What are the key similarities and differences between cluster analysis and principal component analysis (PCA)?
3. Provide examples of cluster analysis applications in various domains. What are the benefits and challenges of applying cluster analysis in these areas?
4. How does the choice of distance measure affect the results of cluster analysis? Discuss different distance measures and their suitability for various types of data and clustering algorithms.

5. What are the advantages and disadvantages of cluster analysis?
6. Compare hierarchical clustering and partitioning clustering algorithms. What are their key differences in terms of assumptions, clustering methods, strengths, and limitations?
7. What is k-means clustering? Explain how it works and why it is a popular clustering method.
8. What challenges arise when determining the optimal number of clusters in cluster analysis?
9. What is two-step clustering? Discuss its strengths and weaknesses.
10. What is the role of cluster profiling? How can cluster profiles be used to support the decision-making process?
11. What are the different ways to combine cluster analysis results with predictive modeling to create an effective predictive model?

CHAPTER EXERCISES

Exercise 1: Cluster Analysis for the Bank Dataset

1. Open the **bank** dataset in SPSS Modeler.
2. Sample the entire dataset. Refer to the variable names and descriptions in Appendix A.
3. Do **not** partition the data.
4. Perform data modification as follows:
   - Impute missing values.
5. Select the following variables for cluster analysis:
   - Age
   - Income
   - Length of residency
   - Teller visit
   - Saving balance
   - CD balance
   - Money market balance
   - Mortgage balance
   - Home value
   - Investment balance
   - Loan balance
6. Conduct the cluster analysis using three different approaches:
   - Auto Cluster Node
   - TwoStep Cluster Node
   - K-Means Cluster Node
7. Compare the results across the three models, focusing on:
   - Number of clusters
   - Cluster quality
   - Cluster size
8. Determine the appropriate number of clusters and rerun the analysis using the K-Means Cluster Node.
9. Present the results of the cluster analysis, including:
   - Number of clusters
   - Cluster size
   - Cluster distance
   - Variable importance
10. Perform cluster profiling using the selected variables.
11. Describe the unique characteristics of each cluster based on the cluster profiling results.
12. Discuss how the cluster analysis results can be used to predict individuals' credit scores and suggest strategies that banks and financial organizations can use to target each cluster.

Exercise 2: Cluster Analysis for the Passengers Dataset

1. Open the **passengers** dataset in SPSS Modeler.
2. Sample 20% of the full dataset.
3. Do **not** partition the data.
4. Perform data modification as follows:
   - Replace zero values for Likert-scale variables with missing values.
   - Impute missing values.
5. Select the following variables for cluster analysis:
   - Gender
   - Age
   - Type of travel
   - Customer type
   - Ticket class
   - Flight distance
6. Conduct the cluster analysis using three different approaches:
   - Auto Cluster Node
   - TwoStep Cluster Node
   - K-Means Cluster Node
7. Compare the results across the three models, focusing on:
   - Number of clusters
   - Cluster quality
   - Cluster size
8. Determine the appropriate number of clusters and rerun the analysis using the K-Means Cluster Node.
9. Present the results of the cluster analysis, including:
   - Number of clusters
   - Cluster size
   - Cluster distance
   - Variable importance
10. Perform cluster profiling using the selected variables.
11. Describe the unique characteristics of each cluster based on the cluster profiling results.
12. Discuss how the cluster analysis results can be used to predict passengers' decisions to return to the same airline and suggest strategies that airlines can implement to target each cluster.

Part III

Advanced Data Mining Methods

14 Random Forest

LEARNING OUTCOMES

Upon the completion of this chapter, you will be able to:

1. Describe the principles and concepts of random forest as an ensemble machine learning method.
2. Discuss the advantages and disadvantages of random forest compared to other machine learning methods.
3. Explain how random forest constructs decision trees using random subsets of training data and input variables.
4. Describe how random forest combines predictions from multiple decision trees through voting or averaging.
5. Explore the hyperparameters of random forest and their impact on model performance.
6. Explain the scalability and efficiency of random forest, including its parallelization capabilities.
7. Define out-of-bag evaluation and explain how it affects the model performance.
8. Construct random forest modeling using SPSS Modeler.

INTRODUCTION

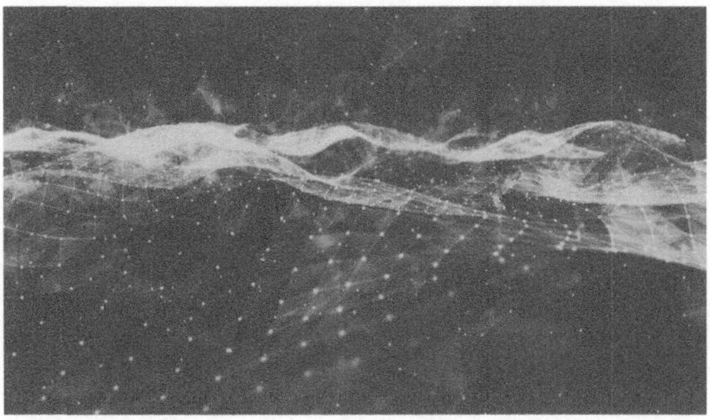

Random forest is one of the most popular and powerful machine learning methods. It is an ensemble learning method that constructs numerous decision trees. We discussed ensemble modeling in Chapter 10, and the concept here is similar, except random forest focuses on combining multiple decision trees. Imagine building many decision tree models (as described in Chapter 8) and then combining those trees into an ensemble model to create a more robust and accurate predictive model. Visually, this forms a "forest" of trees, which is why it is called **random forest**. Essentially, it is a tree-based method.

For classification models, the output of the random forest is the mode (most frequent class) of the individual trees' predictions. For regression models, the output is the mean prediction from all individual trees.

Random forest is one of the most accurate learning algorithms and can handle a large number of input variables without needing to delete any. It is simple to use, faster to compute, and more resistant to noise in the data. Moreover, random forests are effective at preventing overfitting. In this chapter, we will introduce and explain the random forest algorithm, as well as how to perform random forest analysis and interpret the results.

DOI: 10.1201/9781003468196-17

WHAT IS RANDOM FOREST?

Random forest is a machine learning algorithm that combines the predictions of multiple decision trees to make more accurate and robust predictions. It is classified as an ensemble learning method, which combines the results of multiple models to create a stronger predictive model.

Specifically, random forest is a bagging algorithm (refer to Chapter 10 for details on bagging). To recap, bagging is an ensemble technique where multiple individual models are trained independently on different random subsets of the training data (sampling with replacement). The final prediction is determined by averaging all individual model predictions for regression tasks or by taking the majority vote for classification tasks.

A random forest operates under the principle that an ensemble of uncorrelated trees will outperform any of the individual trees. Those individual trees are trained like standard trees with separate training data sampled from the dataset. Then, the final prediction of the random forest model is determined based on a majority vote or average of individual trees' prediction outputs. Figure 14.1 illustrates a random forest model. The idea behind this randomness is to introduce diversity among the trees, reducing the risk of overfitting and improving the predictive power.

DEFINITION

Random forest is a machine learning algorithm that combines the predictions of multiple decision trees to make more accurate and robust predictions. It falls under the category of ensemble learning, where multiple models are combined to form a more powerful predictive model.

Random forest can be applied to both classification and regression models:

- **For classification**, the random forest output is determined by the **mode** (majority vote) of the predictions made by individual trees in the forest.
- **For regression**, the random forest output is the **average** of the predictions from the individual trees.

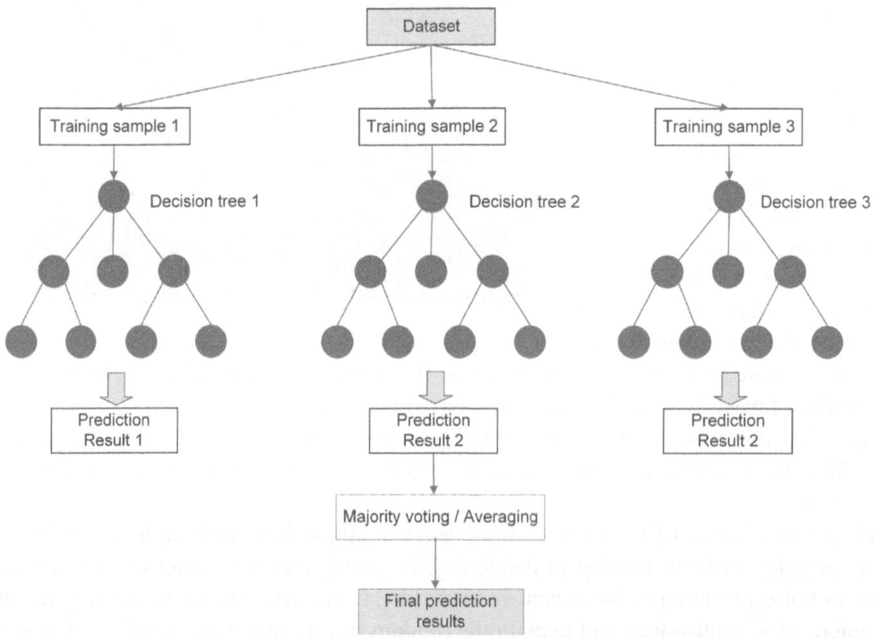

FIGURE 14.1 Illustration of a random forest model.

PRACTICAL APPLICATIONS OF RANDOM FOREST

Random forest has numerous practical applications in many domains. Below are some of the common applications of random forest:

1. **Anomaly detection**: This method can be applied to identify anomalies or outliers in datasets. It has been used for network intrusion detection, fraud detection in financial transactions, and detecting anomalies in sensor data.
2. **Recommendation systems**: Organizations can use random forest in recommendation systems to provide personalized recommendations. It can analyze user behavior and preferences to suggest movies, products, or content tailored to individual users.
3. **Aircraft maintenance and fault detection**: Random forest can be applied to predict potential maintenance issues and detect faults in aircraft systems. By analyzing sensor data and historical maintenance records, it can help identify patterns indicative of impending failures or anomalies.
4. **Flight delay prediction**: Airlines can use this method to predict flight delays by analyzing historical data on flight schedules, weather conditions, airport congestion, and other relevant factors. This information can aid airlines in managing their operations and reducing passenger inconvenience.
5. **Passenger sentiment analysis**: Airlines can also use random forest models to analyze passenger feedback, reviews, and sentiments from various sources, such as social media and surveys. This analysis allows airlines to understand customer preferences and improve their services.
6. **Social media analysis**: Sentiment analysis and predicting user behavior in social media data are other applications of random forest. The model helps analyze trends, identify influential users, and detect spam or fake accounts.
7. **Customer relationship management**: Businesses can use this method in customer lifetime value prediction and customer behavior analysis. They can use these results to understand customer preferences, target marketing campaigns, and optimize customer engagement strategies.
8. **Risk assessment and safety analysis**: Random forest can be used to assess risk factors and analyze safety data in aviation. Airlines use the method to identify potential safety hazards and develop risk management strategies.

Those are just a few examples, and the random forest's versatility allows it to be applied in many other fields where accurate predictions or anomaly detection are required.

DIFFERENCES BETWEEN DECISION TREE AND RANDOM FOREST

While random forest is essentially a collection of decision trees, there are key differences between these two algorithms. Understanding these differences can help in selecting the appropriate method based on the problem at hand. Table 14.1 summarizes some major differences between random forests and decision trees.

TABLE 14.1
Differences between Random Forest and Decision tree

Criteria	Decision Trees	Random Forest
Number of trees	Decision trees consist of a single tree structure.	Random forest is an ensemble method that combines multiple decision trees.
Training process	Decision trees are trained by recursively partitioning the data based on the selected variables and splitting criteria.	Random forest trains each decision tree on a random subset of the training data using bootstrapping.
Variable selection	Decision trees consider all available variables at each split to determine the best predictor and threshold.	Random forest randomly selects a subset of variables at each split, providing randomness and reducing variable correlations.
Prediction	Decision trees make predictions based on the majority class in each leaf node (classification models) or the average of target values (association models).	Random forest combines predictions from multiple trees by voting (classification models) or averaging (association models) to make the final prediction.
Overfitting	Decision trees are prone to overfitting, meaning they can create complex models that fit the training data too closely and generalize poorly to new data.	Random forest mitigates overfitting by aggregating predictions from multiple trees, reducing the variance and improving the model's generalizability.
Bias-variance tradeoff	Decision trees tend to have low bias but high variance as they can be sensitive to small changes in the training data.	Random forest strikes a balance between bias and variance by reducing the variance through averaging predictions.
Handling outliers and noisy data	Decision trees can be sensitive to outliers and noisy data, as they can create biased splits.	Random forest is more robust to outliers and noise due to the ensemble of trees, which reduces the impact of individual trees' mistakes.
Interpretability	Decision trees are often more interpretable because they provide a clear decision path.	Since random forest is an ensemble of trees, it can be less interpretable as it combines multiple decision paths. It is hard to visually follow the decision path with the random forest.
Performance	Decision trees are computationally efficient to train and make predictions.	As an ensemble model, random forest can be more computationally expensive during training and prediction due to the combination of multiple trees.
Variable importance	Decision trees can provide variable importance measures based on the variable's contribution to the splits.	Random forest can also provide variable importance measures, which are aggregated across multiple trees and can be more reliable.

BASIC CONCEPTS OF RANDOM FOREST

Random forest is a bagging ensemble method that combines multiple decision trees to make more accurate predictions. Table 14.2 outlines the key concepts in random forest and their descriptions. Understanding these concepts is crucial for grasping how random forest works and why it is so effective and versatile in various machine learning tasks.

TABLE 14.2
Basic Concepts of Random Forest and Descriptions

Terms	Descriptions
Random forest	Random forest is a machine learning algorithm that combines the predictions of multiple decision trees to make more accurate and robust predictions. It falls under the category of ensemble learning, where multiple models are combined to form a more powerful predictive model.
Ensemble learning	Random forest uses ensemble learning that creates a collection of decision trees and combines their predictions to make a final prediction. The principle of ensemble learning is that the collective decision of multiple models can yield better results than a single model.
Random subset selection	Each decision tree in the random forest is built using a random subset of the training data. This process, known as bootstrap aggregating or "bagging," involves randomly sampling data points with replacement to create diverse subsets for training each tree.
Random variable selection	The random forest algorithm selects a random subset of input variables for each decision tree. This technique, known as variable bagging or variable subsampling, introduces additional randomness and helps to reduce the dominance of any single variable in the ensemble.
Sampling with replacement	Sampling with replacement occurs when a randomly chosen unit from the population is put back into the population, and subsequently, another element is chosen at random. In this process, the population consistently retains all its original units, allowing for the possibility of selecting the same unit multiple times.
Voting/averaging	During prediction, each decision tree independently generates a prediction, and the final prediction of the random forest is determined through voting (for classification models) or averaging (for association models) the individual tree predictions. This ensemble approach helps reduce bias and variance in the modeling and improves the overall accuracy and robustness of the model.
Out-of-bag evaluation	Random forest utilizes an out-of-bag (OOB) evaluation technique. Since each decision tree is trained on a different subset of the training data, the samples not included in a tree's training subset can be used for evaluation. This provides an unbiased estimate of the model performance without the need for a separate validation set.
Variable importance	Random forest provides the variable importance plot, which indicates the relative contribution of each predictor in making predictions. This information can be used to assess the relevance of variables and assist in variable selection.
Robustness	The algorithm is known for its robustness against overfitting, noisy data, and outliers. By aggregating multiple decision trees, it can capture complex relationships and handle noisy datasets more effectively.
Scalability	Random forest can be parallelized, as each decision tree in the ensemble can be built independently. This allows for efficient computation and scalability to large datasets.

ADVANTAGES AND DISADVANTAGES

Random forest is a machine learning algorithm that can be used for both classification and association models. It is a powerful and widely used algorithm, but like all algorithms, it has both strengths and limitations. In this section, we discuss the main advantages and disadvantages of the random forest method, so we can decide whether to use it for a particular task.

Advantages of Random Forest

1. **Robustness**: Random forest is a highly robust machine learning algorithm that can handle missing data and outliers. It is also less sensitive to noise than other methods.

2. **Accuracy**: This method often performs well in both classification and association models, and it is known to be one of the most accurate machine learning algorithms. It has a low bias and moderate variance, which helps reduce overfitting.
3. **Scalability**: It can handle large datasets with many variables and can handle high-dimensional data without the need for variable selection. It can also be easily parallelized, which makes it a good choice for big data applications.
4. **Interpretability**: Random forest can provide variable importance rankings, which are useful for understanding the relative importance of different variables in the model.
5. **Preventing overfitting**: Random forest can be used to prevent overfitting by randomly selecting subsets of the data and variables to build multiple decision trees, which can then be averaged to obtain a final prediction. This process of bootstrapping helps reduce the variance of the model and can improve the predictive power.

Disadvantages of Random Forest

1. **Complexity**: The algorithm can be complex, and it can be difficult to interpret the results of the model.
2. **Computationally expensive**: When dealing with large datasets, random forest can be computationally expensive. This can make it impractical for real time or online applications.
3. **Memory consumption**: This method requires a large amount of memory, especially when dealing with large datasets with many variables.
4. **Non-deterministic**: It is a non-deterministic algorithm, which means results obtained from a random forest model may vary when the algorithm is trained on the same dataset multiple times. This can make it difficult to reproduce results and can lead to the model's unreliability.
5. **Biased toward categorical variables**: Random forest can be biased toward categorical variables that have many levels, which can lead to the overfitting of these variables.

RANDOM FOREST PROCESS

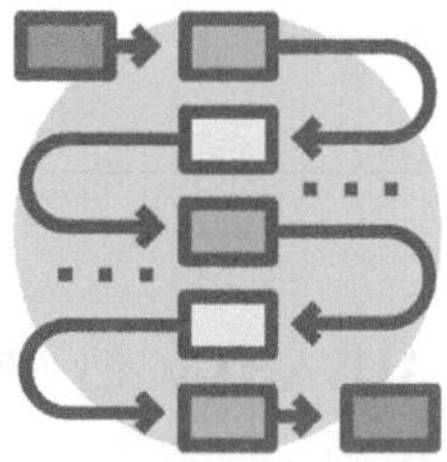

Random forest is one of the most accurate learning algorithms and can handle a very large number of input variables without variable deletion. It is simple to use, faster to calculate, and more resistant to noise in the data. Additionally, random forest can also prevent overfitting.

The implementation of the random forest consists of the following steps:

1. **Data preprocessing**: The first step is to prepare the data for modeling. This includes data cleaning, handling missing values, and transforming data into a suitable format for random forest modeling.
2. **Splitting the data**: The next step is to split the data into training and validation samples. The training sample is used to train the model, while the validation sample is used to evaluate the model performance.
3. **Random forest configuration**: This step involves configuring the hyperparameters of the random forest model, such as the number of trees, the maximum depth of the trees, and other parameters like minimum samples per leaf or the number of variables considered for splitting.
4. **Aggregating the trees**: Once the decision trees have been built, they are combined using a technique called ensemble learning. This involves aggregating the predictions of the individual trees to make a final prediction.

5. **Model training**: In this step, the random forest model is trained with the training sample using the chosen hyperparameters.
6. **Model evaluation**: The final step is to evaluate the performance of the random forest model using the validation sample. This involves evaluating model metrics. If the performance is not satisfactory, the model can be further optimized by adjusting hyperparameters such as the number of trees, maximum depth of the trees, and minimum samples per leaf.

Remember that random forest is an ensemble method, and its strength lies in building multiple decision trees and combining their predictions. The process of conducting random forest modeling may require experimentation with hyperparameters and data preprocessing to achieve the best possible results.

OUT OF BAG

In random forest, the out-of-bag (OOB) error estimation is a technique used to estimate the model performance without the need for a separate validation set. The OOB samples are the data points that were not included in the bootstrap sample used to train each individual decision tree in the random forest.

The OOB error estimation works as follows.

- **Bootstrap sampling**: Random forest generates multiple bootstrap samples by randomly selecting data points, with replacement, from the original training dataset. Each of these bootstrap samples is then used to train an individual decision tree within the random forest model.
- **OOB samples**: Since the bootstrap sampling involves selecting data points with replacement, some data points are left out in each bootstrap sample. These omitted data points are called OOB samples.
- **OOB predictions**: Each decision tree in the random forest is trained on a different bootstrap sample and can be evaluated using the OOB samples that were not included in its training. These OOB samples are passed through the decision tree, and their predictions are recorded.
- **OOB error calculation**: The OOB predictions from all the decision trees in the random forest are aggregated, and the OOB error is calculated by comparing the aggregated predictions to the true target values of the OOB samples. This error estimate provides an approximation of the model performance on unseen data.

> **DEFINITION**
>
> What is sampling with replacement?
>
> Sampling is called with replacement when a unit selected at random from the population is returned to the population, and then a second element is selected at random. Whenever a unit is selected, the population contains all the same units, so a unit may be selected more than once.

The OOB error estimation serves as an internal validation mechanism within the random forest algorithm. It helps assess the model's predictive power without the need for a separate validation sample, saving computational resources and simplifying the training process. The OOB error can be used as a model fit metric to compare different random forest models or to guide hyperparameter tuning decisions, such as determining the optimal number of trees or other parameters. Overall, the OOB estimation is a valuable feature of random forest that provides a reliable estimate of the model performance on unseen data while training the ensemble of decision trees.

RANDOM FOREST AND OVERFITTING PREVENTION

Random forest is known for its ability to mitigate overfitting, which is a common problem in predictive modeling. Below are some ways in which random forest helps prevent overfitting.

1. **Random subsampling**: Random forest applies bootstrapping to create multiple subsets of the original training data. Each decision tree in the random forest is trained on a different bootstrap sample. This random subsampling introduces variability and reduces the chances of overfitting to specific patterns in the training data.
2. **Variable randomness**: At each split in the decision tree construction, the random forest randomly selects a subset of variables to consider. This random variable selection decorrelates the trees and prevents overfitting by reducing the impact of individual variables or subsets of variables that may dominate the decision-making process.
3. **Ensemble of trees**: The final prediction in the random forest is obtained by aggregating the predictions from all the individual trees. By combining the predictions of multiple trees, the random forest reduces the variance and tends to generalize better to unseen data. The ensemble nature of the random forest smooths out the individual tree's idiosyncrasies and reduces overfitting.
4. **Out-of-Bag evaluation**: During the training process of each decision tree in the random forest, some data points are left out due to bootstrapping. These OOB samples can be used for evaluation. OOB evaluation provides an estimate of the model performance on unseen data without the need for a separate validation sample. It serves as an additional measure to assess the model's predictive power and helps detect overfitting.
5. **Hyperparameter tuning**: Random forest has several hyperparameters that can be tuned to control the model's complexity and prevent overfitting. Parameters such as the number of trees, maximum depth of trees, minimum number of samples required for a split, and maximum number of variables considered at each split can be adjusted to find the right balance between underfitting and overfitting.
6. **Early stopping**: Random forest training can be stopped early if there is no significant improvement in the model performance. Monitoring the performance on a validation sample or using techniques like cross-validation can help determine the optimal number of trees to include in the random forest and prevent overfitting.

It's important to note that while random forest helps in reducing overfitting, it may still be possible for individual decision trees within the random forest to overfit the data. However, the aggregation of multiple trees in the ensemble typically compensates for this issue and results in a more robust and generalized model.

RANDOM FOREST HYPERPARAMETER TUNING

Hyperparameter tuning is an essential step in optimizing the performance of a random forest model. I describe below some key hyperparameters that can be tuned in random forest modeling.

1. **Number of trees**: It determines the number of decision trees in the random forest. Increasing the number of trees can improve performance, but it comes with a tradeoff of increased computational time. It is important to find the right balance to avoid overfitting or unnecessary computational burden.

2. **Maximum depth**: It sets the maximum depth allowed for each decision tree in the random forest. A deeper tree can capture more complex relationships but increases the risk of overfitting. Tuning this parameter can help control the complexity of individual trees and prevent overfitting.
3. **Minimum samples for split**: It defines the minimum number of samples required to perform a split at a node. Increasing this value can prevent overfitting by enforcing a higher threshold for node splitting.
4. **Minimum samples per leaf**: It specifies the minimum number of samples required to be in a leaf node. Similar to the minimum samples for split, increasing this value can control overfitting by ensuring a minimum number of samples in each leaf.
5. **Maximum variables**: It determines the number of variables to consider when looking for the best split at each node. Reducing the number of variables can introduce randomness and decorrelation between trees, preventing overfitting.
6. **Variable subset sampling**: It controls whether random variable selection is performed for each tree. Enabling this option introduces additional randomness and diversity among trees, reducing the risk of overfitting.
7. **Out-of-Bag (OOB) evaluation**: It enables the use of OOB samples to estimate the model performance during training. The OOB error can provide a measure of model performance without requiring a separate validation sample. It can be used to compare different hyperparameter settings.

Those are just a few examples of hyperparameters that can be tuned in configuring the random forest model. Hyperparameter tuning can be performed using techniques like grid search, random search, or Bayesian optimization. It involves searching through different combinations of hyperparameter values and evaluating their impact on the model performance using appropriate model fit metrics such as accuracy, precision, recall, or mean squared error. The goal is to find the set of hyperparameter values that optimize the model performance on the specific task at hand.

> In SPSS Modeler, the Random Forest node includes the Hyper-Parameter Optimization feature, which automatically searches for the optimal combination of parameters to minimize the error rate on the samples. This feature simplifies the process by tuning the model to perform effectively without the need for manual adjustments.
>
> Nonetheless, it still allows us to change certain configurations as needed. I will provide some examples of how to configure the Random Forest node to produce better results.

EXAMPLE: RANDOM FOREST

In this example, we build predictive models for the **Airline Recommendation** dataset using the random forest algorithm with various configurations.

- **Dataset:** AirlineRecommendation.sav
- **Target variable:** Recommendation (Flag: Yes/No)

Sample-Explore-Modify

Refer to the previous chapters for more details on these steps.

Random Forest Modeling

In this example, we train four random forest models using the dataset with missing values.

Follow the steps below to set up the Random Forest models:

1. Create a New Stream:
   - Open SPSS Modeler and create a new stream named "Airline_RandomForest."
2. Copy the Existing Stream:
   - Open the stream "Airline_Tree."
   - Copy the section of the stream from the **data source**, **Filler node**, to the **Partition node**.
   - Paste these nodes into your new "Airline_RandomForest" stream.
3. Add Four Random Forest Nodes:
   - From the **Modeling** group, add four **Random Forest nodes** to the stream.
   - Connect them to the **Partition node**.
   - Label the four Random Forest nodes as:
     i. Random Forest 1
     ii. Random Forest 2
     iii. Random Forest 3
     iv. Random Forest 4

At this point, your stream diagram should look like the following:

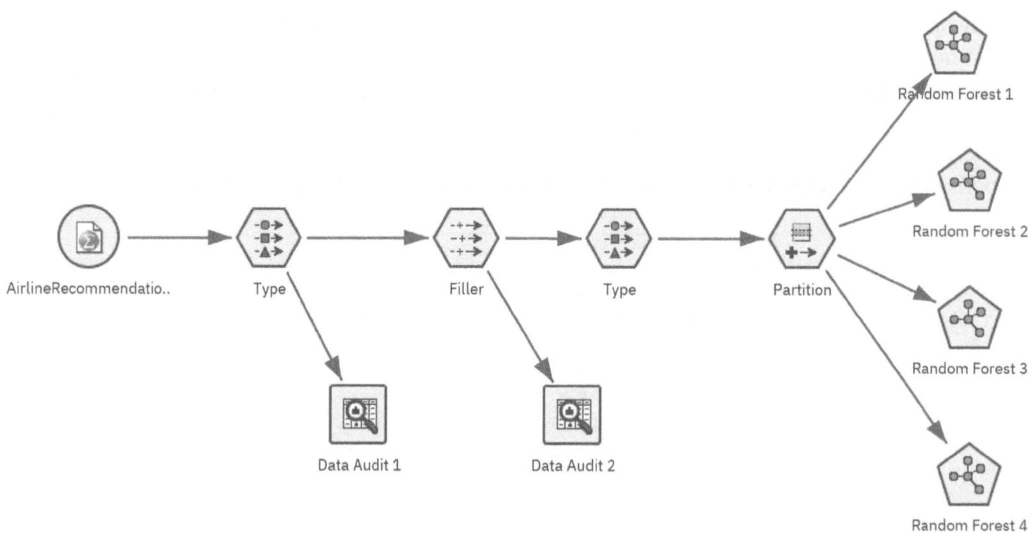

Then, use the details below to set the configurations for each random forest model. We focus on a few key configurations that are available under the Basic option and Advanced option in the Random Forest node settings.

Random Forest 1 Configuration

Basic Settings:

- Number of trees: **10**
- Maximum tree depth: **10**
- Minimum leaf node size: **1**
- Number of features to use for splitting: **auto**

Advanced Settings:

- Use bootstrap samples when building trees: **Checked**
- Use out-of-bag samples to estimate the generalization accuracy: **Checked**
- Hyper-parameter optimization: **Unchecked**

Random Forest 2 Configuration

Basic Settings:

- Number of trees: **20**
- Maximum tree depth: **20**
- Minimum leaf node size: **1**
- Number of features to use for splitting: **auto**

Advanced Settings:

- Use bootstrap samples when building trees: **Checked**
- Use out-of-bag samples to estimate the generalization accuracy: **Checked**
- Enable parallel model building: **Checked**
- Hyper-parameter optimization:
  - Target: **0.01**
  - Max iterations: **1000**
  - Max evaluations: **300**

Random Forest 3 Configuration

Basic Settings:

- Number of trees: **100**
- Maximum tree depth: **20**
- Minimum leaf node size: **1**
- Number of features to use for splitting: **auto**

Advanced Settings:

- Use bootstrap samples when building trees: **Checked**
- Use out-of-bag samples to estimate the generalization accuracy: **Checked**
- Use extremely randomized trees: **Checked**
- Enable parallel model building: **Checked**
- Hyper-parameter optimization:
  - Target: **0.05**
  - Max iterations: **5000**
  - Max evaluations: **500**

Random Forest 4 Configuration

Basic Settings:

- Number of trees: **200**
- Maximum tree depth: **100**
- Minimum leaf node size: **1**
- Number of features to use for splitting: **auto**

Advanced Settings:

- Use bootstrap samples when building trees: **Checked**
- Use out-of-bag samples to estimate the generalization accuracy: **Checked**

- Use extremely randomized trees: **Checked**
- Enable parallel model building: **Checked**
- Replicate results: **Checked**
- Hyper-parameter optimization:
  - Target: **0.05**
  - Max iterations: **10,000**
  - Max evaluations: **500**

To run each of the random forest nodes separately and produce model nuggets, follow the same steps described in Chapter 7:

1. **Run each node separately**: After configuring the four random forest models (Random Forest 1, Random Forest 2, Random Forest 3, and Random Forest 4) with the specified settings, execute each of the nodes separately. This will generate the corresponding model nuggets for each.
2. **Connect model nuggets**: Since all models use the same input data with identical field names, you can use the approach described in Chapter 7 to connect all the model nuggets for comparison. Follow the same steps outlined in Chapter 7 under the section "How to run the analysis node with multiple models?"
3. **Add filter node**: Add a filter node to the stream and connect it to the model nugget for Random Forest 4. Rename the model nuggets to the selected names (Random Forest 1, Random Forest 2, etc.) (see the detailed instructions below).
4. **Add analysis node**: After renaming the model nuggets, use an Analysis node to compare the performance of the models. The Analysis node will compare the models based on key performance metrics.
5. **Add evaluation graph node**: Use an Evaluation Graph node to produce ROC charts and Lift charts for all the models. This will provide a graphical representation of model performance.
6. **Run the diagram**: Run the entire diagram to generate the comparison results for each random forest model.

By following these steps, you'll have a diagram as below, including model comparison through the Analysis node and graphical evaluation with the ROC and Lift charts.

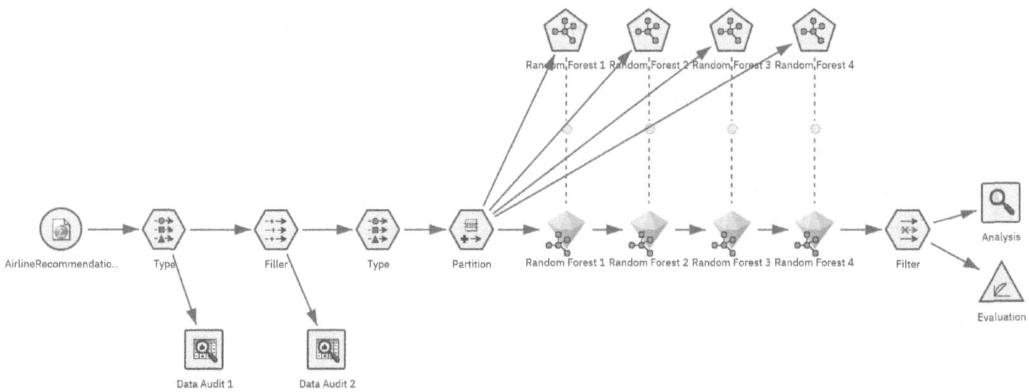

To configure the **Filter node** settings to change the model names for each of the random forest models, follow these steps:

1. **Open the filter node**: After connecting the model nuggets (Random Forest 1, Random Forest 2, Random Forest 3, and Random Forest 4) to the Filter node, open the settings of the Filter node.

2. **Rename models**: Select the field to rename each model as follows:
   - Rename the field **$R-Recommend the airline** to **Random Forest 1**.
   - Rename the field **$R1-Recommend the airline** to **Random Forest 2**.
   - Rename the field **$R2-Recommend the airline** to **Random Forest 3**.
   - Rename the field **$R3-Recommend the airline** to **Random Forest 4**.
3. **Apply the changes**: After setting the model names, click **Apply** to save the changes. The models should now appear with their new names in the stream.

This process ensures that each random forest model is clearly labeled in the final analysis, making it easier to differentiate between the results when comparing performance.

Champion Model

The model comparison results are presented in Table 14.3, which summarizes the model fit metrics for the testing samples. Additionally, Figures 14.2 and 14.3 show the ROC comparison chart and

TABLE 14.3
Model Comparison Using the Testing Sample's Results

Metrics	Random Forest 1	Random Forest 2	Random Forest 3	Random Forest 4
ACC	76.21%	79.30%	77.13%	77.13%
MR	23.79%	20.70%	22.87%	22.87%
Sensitivity or TPR	0.812	0.935	0.922	0.922
Specificity or TNR	0.706	0.633	0.602	0.602
F1	0.783	0.827	0.810	0.810
Gini	0.615	0.647	0.627	0.627
AUC (ROC Index)	0.807	0.824	0.813	0.813

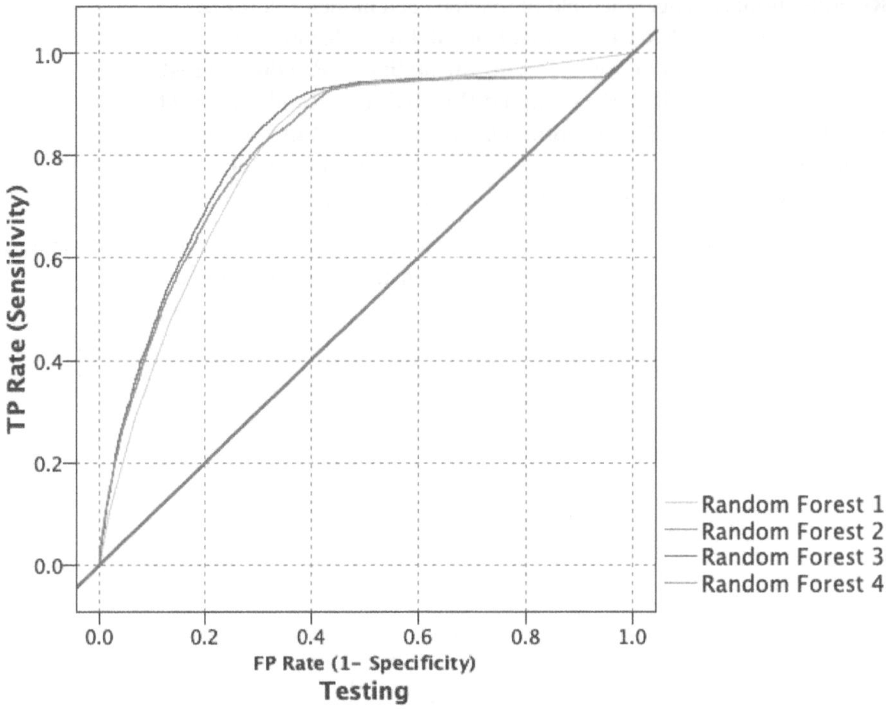

FIGURE 14.2 ROC comparison chart.

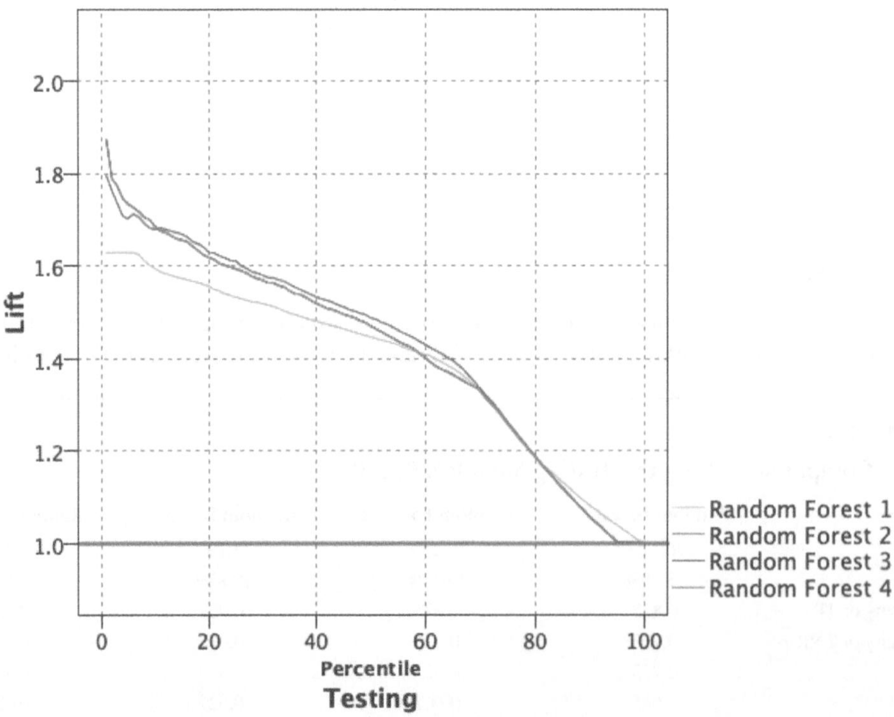

FIGURE 14.3 Lift comparison chart.

Lift comparison chart, respectively. The performance metrics for the models are quite close, but distinct differences emerge in the comparison.

- Random Forest 3 and Random Forest 4 have nearly identical model fit metrics.
- Random Forest 1 performs the worst among all the models.
- Random Forest 2 stands out with the highest accuracy (ACC) and the highest F-1 score. It also has the highest Gini coefficient and ROC index.

These results are further confirmed by the ROC comparison chart, where the ROC curve for Random Forest 2 is closer to the top left corner, indicating better model performance. The Lift comparison chart reveals that Random Forest 1 and Random Forest 4 have the best Lift curves overall. However, at a 20% response rate, Random Forest 2 demonstrates a higher Lift than Random Forest 4.

Based on these results, Random Forest 2 is identified as the champion model due to its superior performance across multiple metrics.

Reliability of the Champion Model

To assess the reliability of the champion model, we compared the fit metrics across the testing, training, and validation samples. The results indicate a high degree of consistency with differences in the model performance metrics across the three samples being less than 10% (see Table 14.4). This consistency suggests that the model generalizes well to unseen data and is not overly tuned to the training sample.

As the discrepancies between the sample metrics are minimal, we conclude that the model's reliability is acceptable, and there is no evidence of overfitting.

Validity of the Champion Model

To assess the validity of the champion model, we evaluated the model fit metrics specifically for the validation sample, as presented in Table 14.5. The prediction accuracy (ACC), at just below 80%, falls short of the generally acceptable threshold, suggesting room for improvement. However, the sensitivity (true positive prediction rate) is notably strong at 0.936, indicating the model's high ability to correctly identify positive cases.

The lower overall accuracy is primarily due to the poor specificity (true negative prediction rate) of 0.644, which reflects a higher false positive rate. Despite this, the F-1 score, which balances both sensitivity and specificity, remains acceptable at 0.83. Additionally, the Gini coefficient shows a good model fit at 0.649, and the ROC index is at 0.825, which is considered an acceptable level of performance.

TABLE 14.4
Reliability Evaluation of the Champion Model—Random Forest 2

Metrics	Testing	Training	Validation
ACC	79.30%	82.96%	79.82%
MR	20.70%	17.04%	20.18%
Sensitivity or TPR	0.935	0.963	0.936
Specificity or TNR	0.633	0.681	0.644
F1	0.827	0.856	0.830
Gini	0.647	0.753	0.649
AUC (ROC index)	0.824	0.877	0.825

TABLE 14.5
Validity Evaluation of the Champion Model—Random Forest 2

Metrics	Validation	Evaluation
ACC	79.82%	Not good
MR	20.18%	Not good
Sensitivity or TPR	0.936	Good
Specificity or TNR	0.644	Not good
F1	0.830	Acceptable
Gini	0.649	Good
AUC (ROC index)	0.825	Acceptable

An analysis of the Lift chart shows that at a 20% response rate, the model achieves a Lift value of 1.65, meaning the model can capture 33% of the positive cases—1.65 times better than random guessing.

In conclusion, considering all these metrics together, we determine that the model demonstrates marginally acceptable predictive power. While the high sensitivity is a strong positive, the relatively low specificity and overall accuracy suggest some limitations in the model's performance.

Model Interpretation

In this section, we would typically delve into the architecture and hierarchy of the model. However, because Random Forest is an ensemble method made up of many decision trees, it does not produce a singular tree structure to analyze. Instead of focusing on the architecture of one tree, we focus on understanding the collective output by examining the variable importance chart.

Variable Importance

To interpret the model, we rely on the variable importance plot generated by the champion model. Figure 14.4 presents this chart, created using the Python algorithm. The plot identifies the most significant predictors contributing to the model's predictions. The most critical variable in the model is

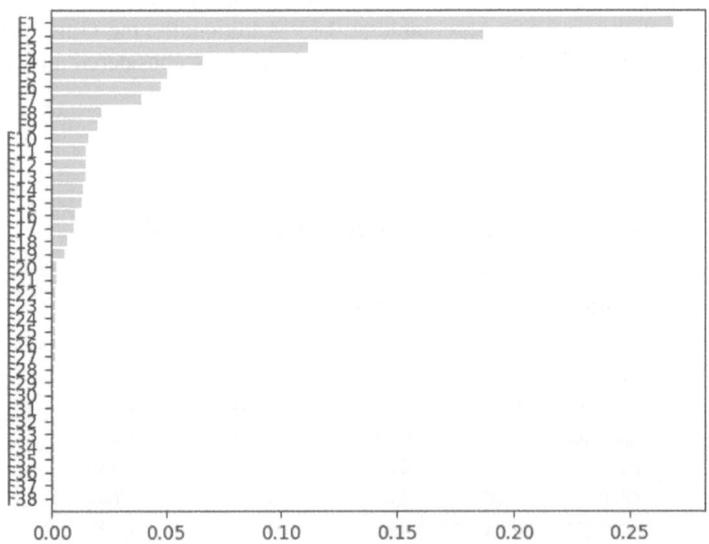

FIGURE 14.4 Variable importance plot.

travel purpose (distinguishing between personal and business travel), followed by age, airline status (with "Blue" and "Silver" statuses being prominent), and the average number of flights per year. These variables play a crucial role in predicting whether a passenger would recommend the airline.

> The variable importance chart for the Random Forest node is generated by the Python algorithm, not directly by SPSS Modeler. One limitation is that Python uses coded variables such as F1, F2, F3, and so on rather than displaying the actual field names or labels from the dataset. Additionally, because of this setup, it is not possible to download or reformat the chart directly. This limitation can make it more challenging to interpret the results clearly, as users need to map the variable codes back to the original field names manually.

CHAPTER SUMMARY

Random forest is a popular ensemble machine learning method that combines multiple decision trees to make more robust and accurate predictions. It is a bagging method that constructs each decision tree using a random subset of the training data and a random subset of input variables. This randomness reduces overfitting and enhances the model's ability to generalize to new data. During prediction, the random forest aggregates the predictions of individual trees through either voting (for classification models) or averaging (for association models), resulting in a final prediction that benefits from the collective predictions of the individual trees. It also uses OOB evaluation for unbiased performance estimation. The random forest also provides a measure of variable importance, allowing users to understand the relative contribution of different predictors in making predictions. This information can be useful for variable selection and identifying the most relevant factors influencing the output.

The random forest offers several advantages. First, it is known for its robustness against noise and outliers in the data. By combining multiple decision trees, each trained on different subsets of the data, the random forest can effectively handle noisy data. Additionally, random forest is versatile and applicable to various domains and data types, including categorical, numerical, and text data. It is capable of handling high-dimensional datasets and is less sensitive to the curse of dimensionality compared to some other algorithms. Random forest is also parallelizable, which ensures efficient computations and scalability to large datasets. Overall, random forest is a powerful and widely used machine learning method that combines the strengths of individual decision trees to deliver accurate and robust predictions.

CHAPTER DISCUSSION QUESTIONS

1. What are the advantages of using random forest over other machine learning algorithms?
2. What are the primary limitations of random forest?
3. How does the random subset selection of training data and input variables contribute to the performance of random forest?
4. Can random forest handle imbalanced datasets effectively? If yes, how does it address the issue of class imbalance?
5. What are the main hyperparameters that can be tuned in a random forest model? How do these hyperparameters affect the model performance?
6. Random forest is known for its ability to handle high-dimensional data. What techniques does it employ to handle the curse of dimensionality?
7. How does the parallelization of random forest impact its scalability and efficiency for large datasets?
8. In what scenarios would random forest be a suitable choice for a machine learning task, and when might it be less appropriate?
9. What is out-of-bag evaluation? How does it help address overfitting?

CHAPTER EXERCISES

Exercise 1: Random Forest Models for the Bank Dataset

1. Open the **bank** dataset in SPSS Modeler.
2. Sample the entire dataset. See the variable names and descriptions in Appendix A.
3. Partition the data using a split ratio of 60–40 (train: 60%; test: 20%; validation: 20%)
4. Perform data modification as follows:
   - Impute missing values.
   - Normalize the data
5. Set **CRSCORE** (credit scores) as the target variable.
6. Configure four random forest models for this dataset as follows:

Random Forest 1

Basic Settings:

- Number of trees: **10**
- Maximum tree depth: **10**
- Minimum leaf node size: **1**
- Number of features to use for splitting: **auto**

Advanced Settings:

- Use bootstrap samples when building trees: **Checked**
- Use out-of-bag samples to estimate the generalization accuracy: **Checked**
- Hyper-parameter optimization: **Unchecked**

Random Forest 2

Basic Settings:

- Number of trees: **30**
- Maximum tree depth: **30**
- Minimum leaf node size: **1**
- Number of features to use for splitting: **auto**

Advanced Settings:

- Use bootstrap samples when building trees: **Checked**
- Use out-of-bag samples to estimate the generalization accuracy: **Checked**
- Enable parallel model building: **Checked**
- Hyper-parameter optimization:
  - Target: **0.01**
  - Max iterations: **1000**
  - Max Evaluations: **300**

Random Forest 3

Basic Settings:

- Number of trees: **100**
- Maximum tree depth: **50**
- Minimum leaf node size: **1**
- Number of features to use for splitting: **auto**

Advanced Settings:

- Use bootstrap samples when building trees: **Checked**
- Use out-of-bag samples to estimate the generalization accuracy: **Checked**
- Use extremely randomized trees: **Checked**
- Enable parallel model building: **Checked**
- Hyper-parameter optimization:
    - Target: **0.05**
    - Max iterations: **5000**
    - Max Evaluations: **500**

Random Forest 4

Basic Settings:

- Number of Trees: **300**
- Maximum tree depth: **200**
- Minimum leaf node size: **1**
- Number of features to use for splitting: **auto**

Advanced Settings:

- Use bootstrap samples when building trees: **Checked**
- Use out-of-bag samples to estimate the generalization accuracy: **Checked**
- Use extremely randomized trees: **Checked**
- Enable parallel model building: **Checked**
- Replicate results: **Checked**
- Hyper-parameter optimization:
    - Target: **0.05**
    - Max iterations: **10,000**
    - Max Evaluations: **500**

7. Compare the model fit metrics between these four models using the testing results.
8. Select the champion model and justify this selection with the necessary evidence.
9. Present the results of the champion model.
10. Evaluate the reliability and validity of the champion model.
11. Present the variable importance chart and interpret the results.

Exercise 2: Random Forest Models for the Passengers dataset

1. Open the **passengers** dataset in SPSS Modeler.
2. Sample 20% of the full data. Set **return** (decision to return to the airline) as the target variable and change its role to binary. See the variable names and descriptions in Appendix B.
3. Partition the data using a split ratio of 60–40 (train: 60%; test: 20%; validation: 20%)
4. Perform data modification as follows:
    - Replace zero values for Likert-scale variables with missing values.
    - Impute missing values.
5. Configure five random forest models for this dataset as follows.

Random Forest 1

Basic Settings:

- Number of trees: **10**
- Maximum tree depth: **10**
- Minimum leaf node size: **1**
- Number of features to use for splitting: **auto**

Advanced Settings:

- Use bootstrap samples when building trees: **Checked**
- Use out-of-bag samples to estimate the generalization accuracy: **Checked**
- Hyper-parameter optimization: **Unchecked**

Random Forest 2

Basic Settings:

- Number of trees: **30**
- Maximum tree depth: **30**
- Minimum leaf node size: **1**
- Number of features to use for splitting: **auto**

Advanced Settings:

- Use bootstrap samples when building trees: **Checked**
- Use out-of-bag samples to estimate the generalization accuracy: **Checked**
- Enable parallel model building: **Checked**
- Hyper-parameter optimization:
  - Target: **0.01**
  - Max iterations: **1000**
  - Max evaluations: **300**

Random Forest 3

Basic Settings:

- Number of trees: **100**
- Maximum tree depth: **50**
- Minimum leaf node size: **1**
- Number of features to use for splitting: **auto**

Advanced Settings:

- Use bootstrap samples when building trees: **Checked**
- Use out-of-bag samples to estimate the generalization accuracy: **Checked**
- Use extremely randomized trees: **Checked**
- Enable parallel model building: **Checked**
- Hyper-parameter optimization:
  - Target: **0.05**
  - Max iterations: **5000**
  - Max evaluations: **500**

Random Forest 4

Basic Settings:

- Number of trees: **300**
- Maximum tree depth: **200**
- Minimum leaf node size: **1**
- Number of features to use for splitting: **auto**

Advanced Settings:

- Use bootstrap samples when building trees: **Checked**
- Use out-of-bag samples to estimate the generalization accuracy: **Checked**

- Use extremely randomized trees: **Checked**
- Enable parallel model building: **Checked**
- Replicate results: **Checked**
- Hyper-parameter optimization:
  - Target: **0.05**
  - Max iterations: **10,000**
  - Max evaluations: **500**

6. Compare the model fit metrics between these four models using the testing results.
7. Select the champion model and justify this selection with the necessary evidence.
8. Present the results of the champion model.
9. Evaluate the reliability and validity of the champion model.
10. Present the variable importance chart and interpret the results.

15 Gradient Boosting

LEARNING OUTCOMES

Upon the completion of this chapter, you will be able to:

1. Describe the principles and concepts of gradient boosting as a powerful machine learning algorithm.
2. Discuss the advantages and disadvantages of using gradient boosting compared to other machine learning methods.
3. Explain the iterative process of building an ensemble of weak learners in gradient boosting.
4. Categorize different loss functions in gradient boosting.
5. Define gradient descent and explain its important role in gradient boosting.
6. Compare traditional gradient boosting, Gradient Boosting Additive Models (GBAM), and Stochastic Gradient Boosting (SGB).
7. Describe the importance of hyperparameter tuning in gradient boosting and its impact on model performance.
8. Build and train gradient boosting models using SPSS Modeler.

INTRODUCTION

This chapter introduces gradient boosting, an ensemble method that also utilizes decision trees. Gradient boosting is a tree-boosting technique that constructs a series of decision trees, which together form a single predictive model. It is a highly effective method for making predictions and selecting important variables. Does it sound similar to random forest? Yes, both methods involve combining multiple decision trees to produce more robust and accurate predictions. However, they operate differently, and we will compare these two methods later in this chapter.

Gradient boosting begins with an initial decision tree model and improves it by using the residuals from the previous model as the target for the next step. Imagine multiple trees connected in sequence, with each tree building on the errors of the previous one—this forms the structure of a gradient boosting model. If each tree in the sequence is created from a random sample of the training data, the method is referred to as stochastic gradient boosting. This chapter covers the principles and concepts of gradient boosting, how it differs from random forest, the gradient boosting process, and how to implement gradient boosting using SPSS Modeler.

WHAT IS GRADIENT BOOSTING?

> **DEFINITION**
>
> Gradient boosting is a machine learning ensemble method that sequentially combines weak base learners (individual decision trees) to build a powerful predictive model by minimizing errors using gradient descent.

Gradient boosting is a widely used machine learning method for both classification and association models. It is an ensemble learning technique that combines multiple weak models to form a stronger, more accurate model. The key idea behind gradient boosting is

DOI: 10.1201/9781003468196-18

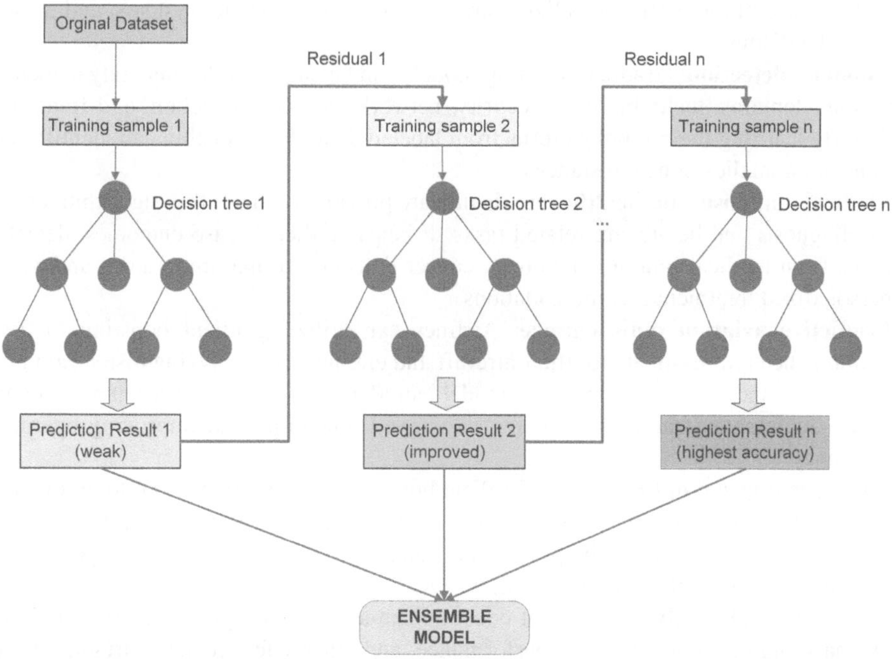

FIGURE 15.1 Gradient boosting algorithm.

to iteratively add new models to the ensemble, each one designed to correct the errors made by the previous models.

Here's how gradient boosting works: the algorithm uses a loss function to measure the difference between predicted and actual values (prediction errors) and then trains new models iteratively to reduce those errors. At each iteration, a new model is trained to predict the difference between the current predictions and the actual values, also known as residuals. The new model is trained to predict these residuals using the variables in the data. Once trained, the new model is added to the ensemble, and the ensemble's predictions are updated by adding the predictions of this new model. This process continues iteratively, with each new model correcting the errors of the previous models, until the overall error is minimized or a predefined number of iterations is reached.

The term "gradient" in gradient boosting refers to the use of the gradient of the loss function to optimize the model. The gradient provides information about the direction and magnitude of the error, helping update the model parameters effectively.

Unlike random forest, which is a bagging method, gradient boosting is a boosting method. We discussed boosting methods in Chapter 10, where weak learners are combined sequentially to form strong learners, ultimately leading to the highest accuracy in the final model. Figure 15.1 illustrates the working of the gradient boosting algorithm using multiple decision tree models. It is important to note that gradient boosting is commonly used with decision trees.

PRACTICAL APPLICATIONS OF GRADIENT BOOSTING

Gradient boosting is a powerful machine learning algorithm that has been applied in various domains. I provide some practical applications of gradient boosting as follows.

- **Ranking and recommendation systems**: Gradient boosting can be used in ranking and recommendation systems. It learns to rank items or recommend personalized content by optimizing ranking metrics such as the pairwise or listwise loss functions. This capability makes it useful for applications like search engines, recommender systems, and personalized advertising.

- **Anomaly detection**: Gradient boosting models can be employed for anomaly detection in various domains, including cybersecurity, network intrusion detection, and fraud detection. By learning the normal patterns from labeled data, the algorithm can identify deviations or anomalies in new instances.
- **Medical diagnosis and healthcare**: Healthcare providers can use this algorithm for medical diagnosis and healthcare-related tasks. It helps predict disease outcomes, detect diseases from medical imaging data (e.g., cancer detection in mammograms), and assist in personalized treatment recommendations.
- **Predictive aviation maintenance**: Airlines can utilize gradient boosting to predict maintenance requirements for their aircraft and engines. By analyzing historical maintenance data and sensor readings, the model can identify potential issues and recommend timely maintenance actions, reducing unplanned downtime and improving operational efficiency.
- **Fuel consumption optimization**: Gradient boosting models can be used to optimize fuel consumption by analyzing various factors, including aircraft type, route, weather conditions, and flight altitude. Such optimization can lead to significant cost savings and reduce the environmental impact of air travel.
- **Safety and risk analysis**: Gradient boosting can aid in identifying potential safety risks by analyzing incident data, pilot performance, and other safety-related variables. By proactively addressing risks, airlines can enhance safety standards and maintain a positive reputation.
- **Customer sentiment analysis**: By analyzing customer feedback, reviews, and social media interactions, service providers can use gradient boosting to perform sentiment analysis and understand customer perceptions. This insight helps improve services and address customer concerns promptly.
- **Natural Language Processing (NLP)**: NLP tasks such as text classification, sentiment analysis, named entity recognition, and document categorization can benefit from gradient boosting. It allows us to capture complex language patterns to make accurate predictions on textual data.
- **Time series forecasting**: Gradient boosting can be applied to time series forecasting problems, such as predicting stock prices, electricity demand, or weather forecasting. By incorporating lagged values and other time-dependent variables, it can capture temporal patterns and make accurate predictions.

As you can see, the applications of gradient boosting are tremendous. The algorithm's flexibility, accuracy, and ability to handle complex relationships make it a valuable tool in various fields where accurate predictions or pattern recognition are required.

DIFFERENCES BETWEEN RANDOM FOREST AND GRADIENT BOOSTING

Random forest and gradient boosting are popular machine learning algorithms used for supervised learning tasks. In addition, they both are ensembles of multiple decision trees. While they share some similarities, there are significant differences between the two. Understanding these differences will help us select the right method for our data mining projects. The choice between the two algorithms depends on the specific problem, the characteristics of the data, and the tradeoff between interpretability and performance. Table 15.1 summarizes the key differences between them.

TABLE 15.1
Differences between Random Forest and Gradient Boosting

Items	Random Forest	Gradient Boosting
Ensemble method	Random forest is an ensemble bagging method that combines multiple decision trees parallelly to make predictions.	Gradient boosting is an ensemble boosting method that combines weak decision trees in a sequential manner.
Training process	Random forest builds each decision tree independently, where each tree is trained on a random subset of the training data (bootstrap aggregating or bagging).	Gradient boosting trains the decision trees sequentially. Each new tree is trained to correct the mistakes made by the previous tree.
Bias-variance tradeoff	Random forest reduces variance by averaging predictions from multiple trees. It is less prone to overfitting and performs well with noisy data.	Gradient boosting aims to reduce bias and variance by minimizing the errors made by the previous trees. It tends to achieve lower bias but can be more prone to overfitting.
Variable selection	Random forest randomly selects a subset of variables for each tree, which leads to a relatively lower correlation among trees. It can handle a large number of variables and can provide estimates of variable importance.	In gradient boosting, variable selection is based on the information gain or gradient importance, and variables are typically used in a deterministic order.
Parallelism	Random forest can train trees in parallel since each tree is independent. This makes it computationally efficient and suitable for parallel and distributed computing.	Gradient boosting trains trees sequentially, limiting parallelism. However, there are techniques like parallelized gradient boosting implementations that can be used to overcome this limitation.
Hyperparameters	Random forest has hyperparameters such as the number of trees, the maximum depth of trees, and the number of variables considered for each split.	Gradient boosting has hyperparameters such as the learning rate, number of trees, and maximum depth of trees.
Prediction	Random forest predictions are made by averaging or voting the predictions of all the trees.	In gradient boosting, predictions are made by summing the predictions of all the trees, where the contribution of each tree depends on its learning rate.
Interpretability	Random forest is relatively easier to interpret since it provides variable importance measures and can show how much each variable contributes to the overall prediction.	Gradient boosting, especially with a large number of trees, can be more complex to interpret due to the sequential nature of the training process.

BASIC CONCEPTS OF GRADIENT BOOSTING

Gradient boosting is a powerful ensemble learning method that iteratively combines weak learners to create a strong predictive model. There are several important concepts that need to be understood in order to perform this method correctly. Table 15.2 presents these key concepts along with their descriptions, helping to clarify how gradient boosting works and how to effectively apply it in machine learning tasks.

TABLE 15.2
Basic Concepts of Gradient Boosting

Terms	Description
Gradient boosting	Gradient boosting is a machine learning ensemble method that sequentially combines weak learners to create a powerful predictive model by minimizing errors using gradient descent.
Ensemble learning	Gradient boosting is an ensemble learning method, which combines multiple weak learners (individual models with limited predictive power) to create a robust and accurate predictive model. The final prediction is obtained by aggregating the predictions of all individual weak learners.
Weak learners	In the context of gradient boosting, weak learners are individual models that perform slightly better than random guessing but are not highly accurate on their own. Decision trees are commonly used as weak learners in gradient boosting, but other models can also be used.
Gradient descent	Gradient boosting minimizes a specified loss function by using gradient descent optimization. It computes the gradients of the loss function and adjusts the subsequent weak learner to correct the errors made by previous learners. The learning process aims to iteratively reduce the loss and improve the overall model performance.
Boosting	Gradient boosting employs a boosting algorithm, which builds the ensemble sequentially, with each weak learner focusing on the mistakes of its predecessors. Subsequent learners give more weight to the misclassified examples in the training data to correct the errors made by previous models.
Residuals	In gradient boosting, the subsequent weak learners are trained to predict the residuals or errors made by the previous learners. The goal is to progressively reduce the remaining errors by refining the model's predictions.
Loss function	The loss function measures the model performance and quantifies the difference between predicted and actual values. In gradient boosting, the algorithm minimizes the loss function during the training process to iteratively improve the model's predictive power.
Learning rate	The learning rate is a hyperparameter that controls the step size in gradient descent. It determines how much each weak learner's contribution affects the final ensemble. A lower learning rate makes the algorithm converge slowly but can improve generalizability, while a higher learning rate leads to faster convergence but may increase the risk of overfitting.
Tree depth and number of trees	These are hyperparameters that control the complexity and size of the individual decision trees. The tree depth affects the model's capacity to capture complex relationships in the data, while the number of trees determines the boosting iterations and overall model complexity.
Stochastic Gradient Boosting	Stochastic Gradient Boosting (SGB) is a variant of the traditional gradient boosting algorithm that introduces randomness during the training process. It combines the concepts of gradient boosting and stochastic learning to enhance the model's generalizability and reduce the risk of overfitting.
Gradient Boosting Additive Model	Gradient Boosting Additive Models (GBAM) is a variant of gradient boosting that extends the traditional algorithm to handle additive models. An additive model is a model that consists of a sum of functions of individual input variables, where each function depends only on a single input variable.
Variable importance	Gradient boosting provides a measure of variable importance, indicating the relative contribution of each input variable in making predictions. It allows users to identify the most influential predictors and gain insights into the underlying data patterns.

ADVANTAGES AND DISADVANTAGES OF GRADIENT BOOSTING METHOD

Gradient boosting is a powerful and widely used machine learning algorithm that has several advantages and disadvantages. Understanding those strengths and weaknesses allows us to decide when to use this method and choose the right configurations for the predictive modeling step.

Advantages of Gradient Boosting

1. **High predictive accuracy**: Gradient boosting is known for its exceptional predictive accuracy. By combining multiple weak learners, it creates a strong ensemble model that can capture complex relationships in the data and produce highly accurate predictions.
2. **Capturing nonlinear relationships**: Nonlinear relationships between predictors and the target variable can be captured effectively by a gradient boosting model, making it suitable for a wide range of real-world applications where the data exhibit intricate patterns.
3. **Variable importance**: The method provides insights into variable importance, which allows us to identify which predictors are most influential in making predictions. This information aids in variable selection and provides a better understanding of the underlying problem.
4. **Flexibility**: Gradient boosting can handle various data types, including numerical, categorical, and textual data. It can be used for both association and classification models, making it a versatile method.
5. **Robustness to outliers**: Due to its ensemble nature, gradient boosting is relatively robust to outliers. The impact of outliers is minimized as the model learns from the collective behavior of multiple weak learners.
6. **Handling missing data**: The method can handle missing data without requiring explicit data imputation techniques. Decision trees in the gradient boosting can handle missing values internally during the training process.
7. **No assumptions about data distribution**: Gradient boosting does not assume any specific distribution of the data, making it applicable in situations where the underlying data distribution is unknown.

Disadvantages of Gradient Boosting

1. **Potential overfitting**: Gradient boosting can be susceptible to overfitting, especially when the model is complex or when the dataset is relatively small. Proper hyperparameter tuning and configurations are essential to mitigate this risk.
2. **Computationally intensive**: Training a gradient boosting model can be computationally intensive and time-consuming, especially when dealing with large datasets or deep trees. Parallelization and optimized implementations can help address this issue.
3. **Hyperparameter tuning**: Finding the optimal hyperparameters for gradient boosting can be a challenging task. The performance of the model is sensitive to hyperparameter settings, and extensive experimentation is often required to find the right configurations.
4. **Black box model**: The ensemble nature of gradient boosting makes it a black box model. In other words, it is challenging to interpret the complete structure of the model and explain its predictions fully.
5. **Data imbalance**: Gradient boosting can struggle with imbalanced datasets, where one class significantly outnumbers the other. Techniques such as class weighting and data resampling may be necessary to address this issue.

GRADIENT BOOSTING STEPS

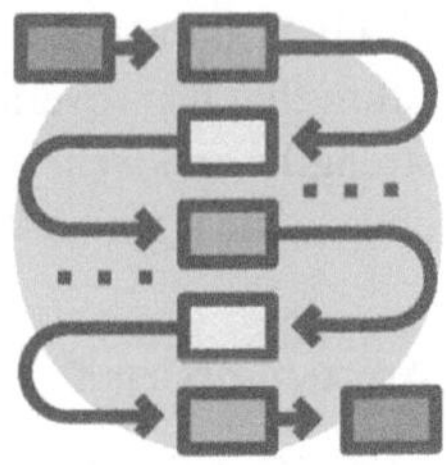

Gradient boosting is an iterative algorithm that combines weak learners (individual decision trees) in a sequential manner to create a strong predictive model. So, how does it work? Below are the general steps involved in training a gradient boosting model:

1. **Initialize the model**: Initialize the model by defining the initial prediction, which is often set to the average value of the target variable in association models or the logarithm of the odds ratio in classification models.
2. **Calculate residuals**: Calculate the residuals or errors between the actual target values and the initial predictions. These residuals represent the information that the model needs to learn and improve upon.
3. **Train weak learners**: Train a weak learner (e.g., a decision tree) on the variables and the calculated residuals. The weak learner is typically shallow, meaning it has a small depth and few splits. The goal is to find the best split points that minimize the loss function (e.g., mean squared error [MSE] for association models or log loss for classification models) based on the residuals.
4. **Compute the learning rate**: Decide on a learning rate, which controls the contribution of each weak learner to the final prediction. The learning rate helps prevent overfitting and control the impact of each weak learner on the ensemble.
5. **Update the model**: Update the model by adding the weak learner's prediction to the previous predictions. The predictions are weighted by the learning rate, which determines how much each weak learner contributes to the overall model.
6. **Update residuals**: Update the residuals by subtracting the predicted values from the actual target values. These updated residuals represent the remaining information that the next weak learner needs to focus on.
7. **Repeat Steps 3–6**: Repeat Steps 3–6 for a specified number of iterations or until a stopping criterion is met. In each iteration, train a new weak learner on the updated residuals and update the model and residuals accordingly.
8. **Finalize the model**: Once all the iterations are completed, the final model is obtained by combining the predictions of all the weak learners. The learning rate determines the contribution of each weak learner to the final prediction.
9. **Prediction**: To make predictions for new data, pass the data through each weak learner and combine their predictions using the learning rate. The final prediction is the sum of the predictions from all the weak learners.

It's important to note that there are variations of gradient boosting algorithms, which include additional optimizations and enhancements to improve training speed and model performance. These steps outline the general process of training a gradient boosting model. In practice, the actual implementation may involve additional hyperparameter tuning and handling specific nuances based on the chosen gradient boosting algorithm.

GRADIENT BOOSTING LOSS FUNCTION

In gradient boosting, the choice of the loss function is crucial as it determines how the algorithm updates the model at each iteration. The loss function measures the difference between the predicted values and the actual values, and the algorithm aims to minimize this difference by iteratively adding new models to the ensemble.

The most commonly used loss functions in gradient boosting are MSE for association models and cross-entropy loss (also called log loss) for classification models. MSE measures the average squared difference between the predicted values and the actual values, while cross-entropy loss measures the difference between the predicted probability distribution and the actual probability distribution. A list of common loss functions will be provided in this section.

Regardless of the specific loss function used, the algorithm updates the model parameters by calculating the negative gradient of the loss function with respect to the predicted values. This gradient provides information about the direction and magnitude of the prediction error and is used to update the model in the direction that minimizes the loss.

The learning rate is also used to control the size of the updates at each iteration. A lower learning rate results in smaller updates and a slower convergence, while a higher learning rate results in larger updates and a faster convergence. The choice of learning rate is important as it determines the balance between accuracy and convergence speed.

> The loss function measures the quality of a model's coefficients and how well it fits the underlying data. The residual is defined as the derivative of the loss function, representing the difference between the predicted and actual values. To minimize prediction errors, the loss function must be optimized during the model training process.

Below are the most common loss functions for association models and classification models and when they can be used.

Loss functions for association models with a continuous target variable

1. **Mean squared error (MSE)**: The most widely used loss function for association problems, MSE calculates the average squared difference between predicted and actual values.
2. **Mean absolute error (MAE)**: MAE measures the average absolute difference between predicted and actual values, making it less sensitive to outliers than MSE.
3. **Huber loss**: Huber loss is a combination of MSE and MAE, acting like MSE when the error is small and like MAE when the error is large. It provides a balance between the two and is robust to outliers.
4. **Quantile loss**: Quantile loss is useful for quantile regression, where the goal is to predict specific quantiles of the target distribution. It optimizes the difference between predicted quantiles and actual quantiles.
5. **Poisson loss**: This loss function is used when the target variable follows a Poisson distribution, such as in count data regression problems.

Classification loss functions for classification models with a categorical target variable

1. **Log loss (binary cross-entropy)**: The most commonly used function for binary classification models. Log loss measures the performance of a classifier by penalizing false classifications with logarithmic terms.
2. **Multi-class log loss (multi-class cross-entropy)**: For multi-level (multinomial) classification models, this loss function extends binary cross-entropy to handle multiple classes, measuring the dissimilarity between predicted class probabilities and the actual class values.

3. **Exponential loss (AdaBoost)**: Primarily used for binary classification models in the AdaBoost algorithm. This loss function assigns higher weights to misclassified samples to give them more emphasis during boosting.

These loss functions play a critical role in guiding the gradient boosting algorithm to optimize the model performance. Each loss function has its strengths and weaknesses, and the choice of the appropriate loss function depends on the characteristics of the data, the type of target variable, and the specific objectives of the modeling task.

GRADIENT BOOSTING ADDITIVE MODELS AND STOCHASTIC GRADIENT BOOSTING

In addition to the traditional gradient boosting algorithm we have discussed so far, there are two popular variants that extend the traditional approach: **Gradient Boosting Additive Models (GBAM)** and **Stochastic Gradient Boosting (SGB)**. We will discuss the key differences among these algorithms, providing a deeper understanding of this powerful machine learning method. Table 15.3 summarizes the similarities and differences between these three types of gradient boosting algorithms.

TABLE 15.3
Differences among Traditional Gradient Boosting, GBAM, and SGB

Items	Traditional Gradient Boosting	Gradient Boosting Additive Models	Stochastic Gradient Boosting
Training approach	Traditional gradient boosting trains weak learners, usually decision trees, sequentially, with each new learner focused on minimizing the errors made by the previous learners.	GBAM extends traditional gradient boosting by allowing the inclusion of different types of weak learners, not limited to decision trees. It allows the combination of diverse models, such as linear models, decision trees, or neural networks.	SGB introduces randomness into the training process by using random subsets of the training data for training each weak learner.
Learning rate	Traditional gradient boosting uses a fixed learning rate that determines the contribution of each weak learner to the overall ensemble. It is typically manually tuned.	GBAM uses a learning rate that controls the contribution of each model component, similar to traditional gradient boosting.	SGB uses a fixed learning rate similar to traditional gradient boosting.
Split selection	Deterministic split selection: Traditional gradient boosting selects the best split points deterministically, aiming to minimize the loss function based on the residuals.	Flexibility in weak learner selection: GBAM provides flexibility in selecting the weak learners. This allows for the use of different models suitable for different types of variables and data characteristics.	Random subsampling: SGB randomly samples a subset of the training data for each iteration, leading to diversity in the weak learners and reducing overfitting.
Model	Traditional gradient boosting tends to create complex models with high variance, especially when the number of iterations is large.	GBAM can offer better model interpretability compared to traditional gradient boosting, especially when linear models or other inherently interpretable models are included.	SGB introduces a variance-reduction tradeoff due to the random subsampling. It can reduce overfitting but may increase bias.

(Continued)

TABLE 15.3 *(Continued)*
Differences among Traditional Gradient Boosting, GBAM, and SGB

Items	Traditional Gradient Boosting	Gradient Boosting Additive Models	Stochastic Gradient Boosting
Training efficiency	Traditional gradient boosting can be slower compared to other boosting algorithms due to its sequential nature and the need to train each weak learner in order.	Training GBAM can be slower than traditional gradient boosting, especially if it involves computationally intensive models like neural networks.	SGB can be faster compared to traditional gradient boosting since it trains each weak learner on a smaller subset of data. This makes it more scalable for large datasets.

Gradient Boosting Additive Models (GBAM)

Essentially, GBAM is a variant of Gradient boosting that extends the traditional algorithm to handle additive models. An additive model is a model that consists of a sum of functions of individual predictors, where each function depends only on a single predictor. Unlike traditional gradient boosting, in which each new model in the ensemble is trained on the residuals of the previous model, in GBAM, the new models are not trained on the residuals, but on the original input variables.

The GBAM algorithm works by constructing a set of basis functions for each predictor, and then fitting a linear combination of these functions to the target variable. The basis functions can be any type of function, and they can be chosen based on the specific problem and the complexity of the predictors. At each iteration, the algorithm adds a new basis function to the linear combination by fitting a model to the gradient of the loss function with respect to the current prediction. This allows the algorithm to gradually refine the model and improve its performance on the training data.

GBAM has several advantages over traditional gradient boosting. First, it handles well nonlinear relationships between the predictors and the target variable, which is important for many real-world problems. Second, it incorporates domain knowledge into the model by selecting appropriate basis functions for each predictor. Finally, it provides a more interpretable model by decomposing the prediction into a sum of functions of individual predictors.

However, GBAM also has some limitations. First, it requires the choice of appropriate basis functions for each predictor, which can be time-consuming and require expert knowledge. Second, it may not be able to capture complex interactions between the predictors, which may limit its performance on certain problems. Finally, it may be slower and require more memory than traditional gradient boosting due to the need to store the basis functions for each predictor.

Stochastic Gradient Boosting (SGB)

SGB is a variant of gradient boosting that introduces randomness into the training process. In traditional gradient boosting, each new model in the ensemble is trained on the same set of samples and variables, which can lead to overfitting and poor model generalizability. SGB addresses this issue by randomly sampling a subset of the data and variables at each iteration, leading to a more diverse set of models and better predictive power.

The key difference between SGB and traditional gradient boosting is that SGB only uses a subset of the training data at each iteration, which is randomly sampled from the original dataset. This sampling process is typically done with replacement (refer to Chapter 15 regarding sampling with replacement), and the size of the sample is a hyperparameter that can be tuned based on the size of the original dataset and the complexity of the problem.

In addition to sampling the data, SGB also randomly samples a subset of the variables at each iteration. This helps reduce the correlation between the models in the ensemble and makes the algorithm more robust to noisy data. The sampling process introduces randomness into the training process, which can improve the performance of the algorithm and reduce overfitting. However, it also increases the variance of the model, which can lead to a higher training error and slower convergence. To address this issue, SGB introduces a new hyperparameter called the subsample rate, which controls the fraction of the training data that is used at each iteration. A lower subsample rate results in a more diverse set of models, but it also increases the variance of the model and requires more iterations to converge. A higher subsample rate results in a more stable set of models, but it may also increase the risk of overfitting.

In summary, traditional gradient boosting trains weak learners sequentially, uses a fixed learning rate, and selects splits deterministically. GBAM extends this approach to include different types of weak learners and offers flexibility and interpretability. Finally, SGB introduces random subsampling to improve training efficiency and reduce overfitting but introduces a variance-bias tradeoff. The choice between these algorithms depends on factors such as the dataset size, interpretability requirements, and tradeoff preferences between accuracy and training efficiency.

GRADIENT BOOSTING HYPERPARAMETER TUNING GUIDE

As mentioned before, gradient boosting involves hyperparameter tuning, which is a crucial step in finding the optimal configurations and addressing overfitting. There are many hyperparameters in gradient boosting, given its complexity and numerous variants. When tuning hyperparameters, it's important to consider the computational resources, time constraints, and the size of the dataset. It is recommended to use cross-validation to evaluate the model performance for different hyperparameter settings. By iteratively tuning the hyperparameters and evaluating the model performance, you can find the optimal combination that maximizes the model's accuracy and generalizability.

Below are some important hyperparameters to consider and techniques for tuning them.

1. **Number of trees**: This hyperparameter determines the number of weak learners (trees) in the ensemble. Increasing the number of trees can improve the model performance, but it also increases training time. It is essential to find the right balance to avoid overfitting. You can try different values and use cross-validation to select the optimal number.
2. **Learning rate**: The learning rate controls the contribution of each weak learner to the overall model. A lower learning rate makes the model more robust but increases the number of iterations needed for convergence. A higher learning rate may lead to faster convergence but could result in overfitting. Experiment with different learning rates to find the optimal value.
3. **Maximum depth of trees**: This hyperparameter determines the depth of the individual decision trees. A deeper tree can capture more complex relationships in the data but is prone to overfitting. Shallower trees reduce overfitting but may result in underfitting. It's important to tune the maximum depth based on the complexity of the problem and the available data.
4. **Subsampling**: Subsampling refers to the fraction of training samples to be used for training each weak learner. Using a value less than 1.0 introduces randomness and can reduce overfitting. However, setting it too low may result in underfitting. Cross-validation can help identify the optimal subsampling value.
5. **Variable subsampling**: This hyperparameter determines the fraction of variables to be used for training each weak learner. It introduces randomness and helps reduce overfitting. You can try different values (e.g., 0.5, 0.8) to find the optimal variable subsampling ratio.

6. **Automated hyperparameter optimization**: There are also automated hyperparameter optimization techniques available, such as Bayesian optimization, genetic algorithms, or gradient-based optimization. These methods can efficiently search the hyperparameter space and find optimal values, often using a guided exploratory process.

> In SPSS Modeler, we use the XGBoost Tree node to build gradient boosting models. This node includes the Hyper-Parameter Optimization feature, which automatically searches for the optimal combination of parameters to minimize the error rate on the samples. This feature simplifies the process by tuning the model to perform effectively without requiring manual adjustments.
>
> Nonetheless, it still allows us to change specific configurations as needed. I will provide some examples of how to configure the XGBoost Tree node to produce better results, both with and without the Hyper-Parameter Optimization feature.

EXAMPLE: GRADIENT BOOSTING

In this example, we build predictive models for the **Airline Recommendation** dataset using the gradient boosting algorithm with various configurations.

- **Dataset:** AirlineRecommendation.sav
- **Target variable:** Recommendation (Flag: Yes/No)

Sample-Explore-Modify

Refer to the previous chapters for more details on these steps.

Gradient Boosting Modeling

In this example, we train four gradient boosting models using the dataset with missing values. Follow the steps below to set up the Gradient Boosting models:

1. Create a New Stream:
   - Open SPSS Modeler and create a new stream named "Airline_GradientBoosting."
2. Copy the Existing Stream:
   - Open the stream "Airline_Tree."
   - Copy the section of the stream from the **data source**, **Filler node**, to the **Partition node**.
   - Paste these nodes into your new "Airline_GradientBoosting" stream.
3. Add Five **XGBoost Tree** Nodes:
   - From the **Modeling** group, add four **XGBoost Tree Nodes** to the stream.
   - Connect them to the **Partition node**.
   - Label the four XGBoost Tree nodes as:
     i. Gradient Boosting 1
     ii. Gradient Boosting 2
     iii. Gradient Boosting 3
     iv. Gradient Boosting 4

At this point, your stream diagram should look like the following:

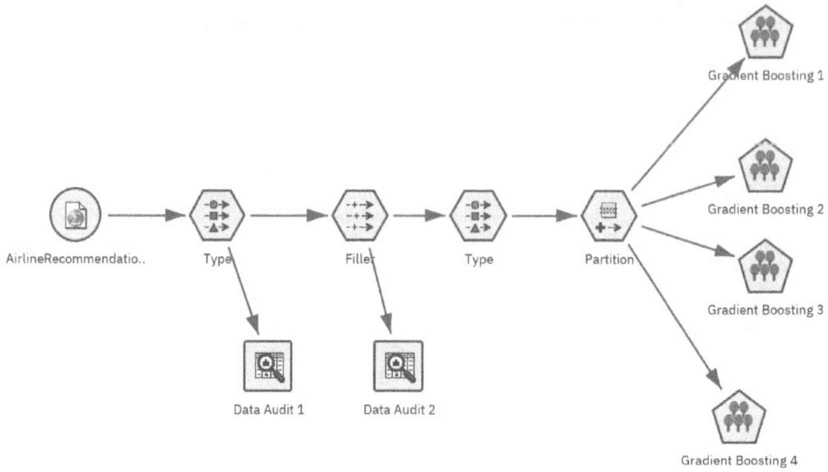

Then, use the details below to set the configurations for each gradient boosting model. We focus on a few key configurations that are available under the Basic options, Learning task, and Advanced options in the XGBoost Tree node settings.

Gradient Boosting 1: Default

- Basic options
  - Hyper-parameter optimization: **unchecked**
  - Tree method: **auto**
  - Number boost round: **10**
  - Max depth: **6**
  - Min child weight: **1**
- Learning task
  - Objective: **binary:logistic**
  - Early stopping: **unchecked**
- Advanced options
  - Sub sample: **1.0**
  - Gamma: **0.0**

Gradient Boosting 2

- Basic options
  - Hyper-parameter optimization: **unchecked**
  - Tree method: **auto**
  - Number boost round: **50**
  - Max depth: **20**
  - Min child weight: **1**
- Learning task
  - Objective: **binary:logistic**
  - Early stopping: **unchecked**
- Advanced options
  - Sub sample: **1.0**
  - Gamma: **0.0**

Gradient Boosting 3

- Basic options
  - Hyper-parameter optimization: **unchecked**
  - Tree method: **auto**
  - Number boost round: **100**
  - Max depth: **50**
  - Min child weight: **1**
- Learning task
  - Objective: **binary:logistic**
  - Early stopping: **unchecked**
- Advanced options
  - Sub sample: **1.0**
  - Gamma: **0.0**

Gradient Boosting 4 (hyper-parameter optimization)

- Basic options
  - Hyper-parameter optimization: **checked** (note: change the configurations below before checking this box).
  - Tree method: **auto**
  - Number boost round: **100**
  - Max depth: **50**
  - Min child weight: **1**
- Learning task
  - Objective: **binary:logistic**
  - Early stopping: **unchecked**
- Advanced options
  - Sub sample: **1.0**
  - Gamma: **0.0**

To run each of the XGBoost Tree nodes separately and produce model nuggets, follow the same steps described in Chapter 7:

1. **Run each node separately**: After configuring the four gradient boosting models (Gradient Boosting 1, Gradient Boosting 2, Gradient Boosting 3, and Gradient Boosting 4) with the specified settings, execute each of the nodes separately. This will generate the corresponding model nuggets for each.
2. **Connect model nuggets**: Since all models use the same input data with identical field names, you can use the approach described in Chapter 7 to connect all the model nuggets for comparison. Follow the same steps outlined in Chapter 7 under the section "How to Run the Analysis Node with Multiple Models?"
3. **Add filter node**: Add a Filter node to the stream and connect it to the model nugget for Gradient Boosting 4. Rename the model nuggets to the selected names (Gradient Boosting 1, Gradient Boosting 2, etc.) (see the detailed instructions below).
4. **Add analysis node**: After renaming the model nuggets, use an Analysis node to compare the performance of the models. The Analysis node will compare the models based on key performance metrics.
5. **Add evaluation graph node**: Use an Evaluation Graph node to produce ROC charts and Lift charts for all the models. This will provide a graphical representation of model performance.
6. **Run the diagram**: Run the entire diagram to generate the comparison results for each gradient boosting model.

By following these steps, you'll have a diagram as below, including model comparison through the Analysis node and graphical evaluation with the ROC and Lift charts.

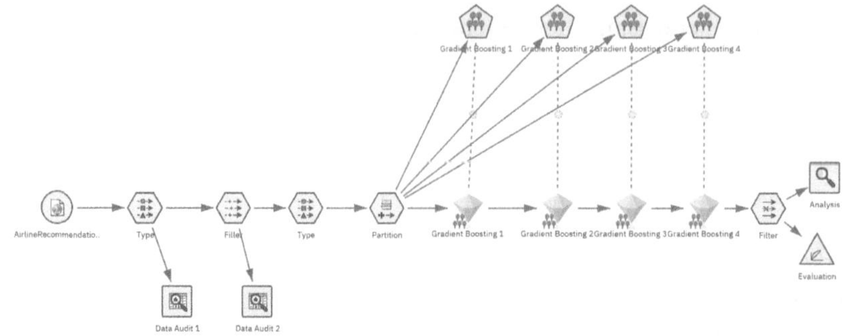

To configure the Filter node settings to change the model names for each of the gradient boosting models, follow these steps:

1. **Open the Filter node**: After connecting the model nuggets (Gradient Boosting 1, Gradient Boosting 2, Gradient Boosting 3, and Gradient Boosting 4) to the Filter node, open the settings of the Filter node.
2. **Rename models**: Select the field to rename each model as follows:
   - Rename the field **$XGT-Recommend the airline** to **Gradient Boosting 1**.
   - Rename the field **$XGT1-Recommend the airline** to **Gradient Boosting 2**.
   - Rename the field **$XGT2-Recommend the airline** to **Gradient Boosting 3**.
   - Rename the field **$XGT3-Recommend the airline** to **Gradient Boosting 4**.
3. **Apply the changes**: After setting the model names, click Apply to save the changes. The models should now appear with their new names in the stream.

This process ensures that each gradient boosting model is clearly labeled in the final analysis, making it easier to differentiate between the results when comparing performance.

Champion Model

The model comparison results are presented in Table 15.4, which summarizes the model fit metrics for the testing samples. Additionally, Figures 15.2 and 15.3 show the ROC comparison chart and Lift comparison chart, respectively. Gradient Boosting 1 and Gradient Boosting 4 perform the best overall, with the following comparisons highlighting key differences:

- Gradient Boosting 1 shows a slightly higher accuracy (ACC) than Gradient Boosting 4, though the difference was minimal.
- Sensitivity (true positive prediction) for Gradient Boosting 1 is higher than for Gradient Boosting 4, but Gradient Boosting 1 has lower specificity (true negative prediction).
- Gradient Boosting 1 has a marginally better F-1 score, indicating a better balance between sensitivity and specificity.

TABLE 15.4
Model Comparison Using the Testing Sample's Results

Metrics	Gradient Boosting 1	Gradient Boosting 2	Gradient Boosting 3	Gradient Boosting 4
ACC	79.61%	77.87%	78.16%	79.56%
MR	20.39%	22.13%	21.84%	20.44%
Sensitivity or TPR	0.937	0.866	0.868	0.921
Specificity or TNR	0.638	0.682	0.685	0.655
F1	0.829	0.805	0.808	0.826
Gini	0.713	0.681	0.682	0.719
AUC (ROC Index)	0.856	0.841	0.841	0.859

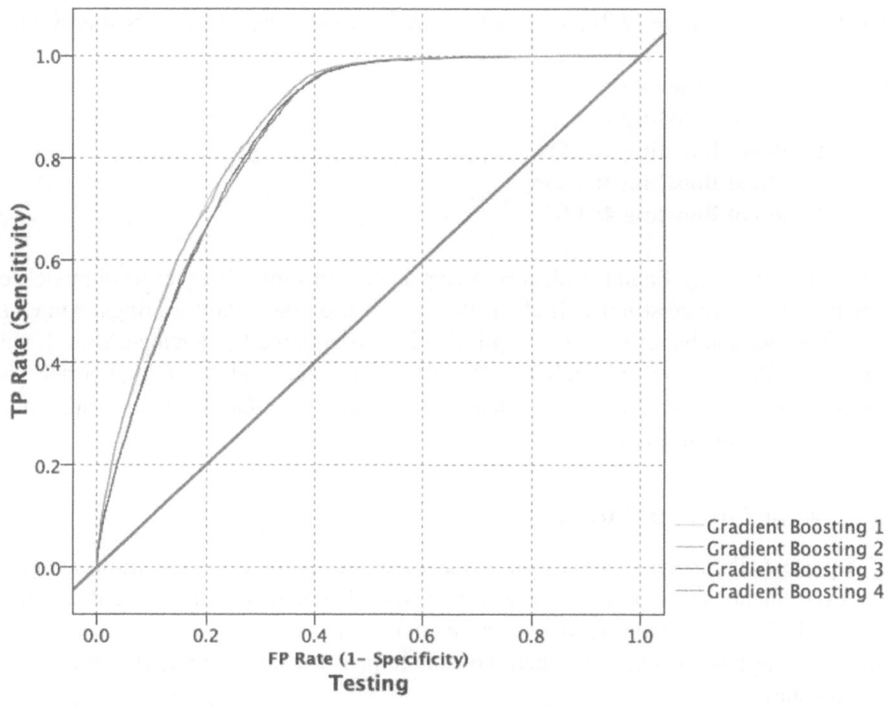

FIGURE 15.2 ROC comparison chart.

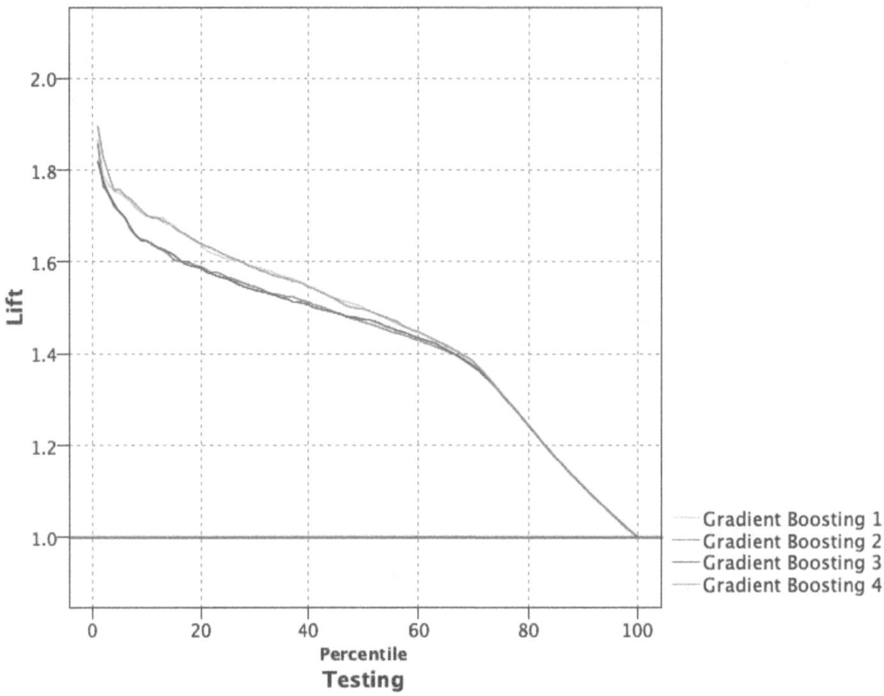

FIGURE 15.3 Lift comparison chart.

- Gradient Boosting 4 outperforms Gradient Boosting 1 in:
  - Gini coefficient
  - ROC index

  The ROC chart (Figure 15.2) confirms that Gradient Boosting 4 has the best ROC curve.
- Lift Scores (derived from Figure 15.3):
  - At a 20% response rate, the Lift scores for the models are as follows:
    - **Gradient Boosting 1:** 1.635
    - **Gradient Boosting 2:** 1.588
    - **Gradient Boosting 3:** 1.586
    - **Gradient Boosting 4:** 1.651

Both Gradient Boosting 1 and Gradient Boosting 4 perform similarly in terms of model accuracy and predictive ability. Interestingly, Gradient Boosting 1 used the default settings, while Gradient Boosting 4 incorporated hyper-parameter optimization through the Python algorithm. If we had to choose, Gradient Boosting 4 would be selected as the champion model due to its superior Gini coefficient, ROC index, and Lift score. These metrics demonstrate that Gradient Boosting 4 provides a slightly better fit and predictive power overall.

Reliability of the Champion Model

Table 15.5 presents the metric comparison across the three samples: testing, training, and validation. The results demonstrate very consistent values, with differences being less than 2%, indicating the model's reliability. Essentially, there is no evidence of overfitting based on this comparison. Given the minimal discrepancies between the sample metrics, we conclude that the model's reliability is acceptable.

TABLE 15.5
Reliability Evaluation of the Champion Model—Gradient Boosting 4

Metrics	Testing	Training	Validation
ACC	79.56%	80.42%	80.04%
MR	20.44%	19.58%	19.96%
Sensitivity or TPR	0.921	0.930	0.923
Specificity or TNR	0.655	0.666	0.665
F1	0.826	0.832	0.829
Gini	0.719	0.749	0.729
AUC (ROC Index)	0.859	0.874	0.864

VALIDITY OF THE CHAMPION MODEL

The next step is to evaluate the validity of the champion model by reviewing the complete model fit metrics for the validation sample, which is reserved for the final assessment. Table 15.6 shows the full metrics for this champion model and the evaluation of its performance.

- The prediction accuracy (ACC) is just above 80%, typically considered marginal.
- The sensitivity (true positive prediction) is quite good at 0.923.
- However, the lower ACC can be attributed to the poor specificity (true negative prediction) of 0.665, meaning we have a relatively high false positive rate.
- The F-1 score, which balances sensitivity and specificity, is considered acceptable.
- The Gini coefficient indicates a good fit, and the ROC index shows an acceptable fit.

Additionally, extracting the Lift value from the validation sample's results reveals a Lift score of 1.66 at a response rate of 20%. This implies that using 20% of the dataset, the model captures 33.2% of the positive cases, which is 1.66 times more than if no model were used.

Overall, while ACC and MR alone may suggest a marginal fit, looking deeper into the other metrics gives a more comprehensive view of the model's performance, indicating an acceptable model fit.

To further evaluate the performance of the gradient boosting model, I compared it with the best individual tree model, C&R Tree 2, identified in Chapter 8. With the exception of the sensitivity rate, Gradient Boosting 4 outperforms the individual tree model in most metrics.

TABLE 15.6
Validity Evaluation of the Champion Model—Gradient Boosting 4

Metrics	Validation	Evaluation	C&R Tree 2
ACC	80.04%	Marginal	79.5%
MR	19.96%	Marginal	20.5%
Sensitivity or TPR	0.923	Good	0.951
Specificity or TNR	0.665	Not good	0.623
F1	0.829	Acceptable	0.830
Gini	0.729	Good	0.673
AUC (ROC Index)	0.864	Acceptable	0.837

Additionally, Gradient Boosting 1 also surpasses C&R Tree 2 in most metrics. This comparison highlights that gradient boosting models perform better than individual tree models due to their ensemble approach.

Model Interpretation

Similar to Random Forest, Gradient Boosting is an ensemble of trees, meaning it consists of many decision trees combined in a boosting ensemble model. As a result, there is no single tree or specific hierarchy to explain, as the prediction process involves multiple trees working together. Instead, we will focus on the variable importance chart to identify the most significant predictors. This chart reveals which variables contribute the most to the model's predictions, helping us understand the factors driving the model's performance.

Variable Importance

Suppose we select Gradient Boosting 4 as the champion model in this example. We can retrieve the variable importance results from the model nugget. Figure 15.4 displays the variable importance. It is important to note that this variable importance plot is generated by the Python algorithm, and it uses the F score as the measure of importance rather than relative importance, which is often used with other nodes.

The chart reveals that the most important predictors are:

- Age (F16)
- Arrival delay (F6)
- Average number of flights per year (F5)

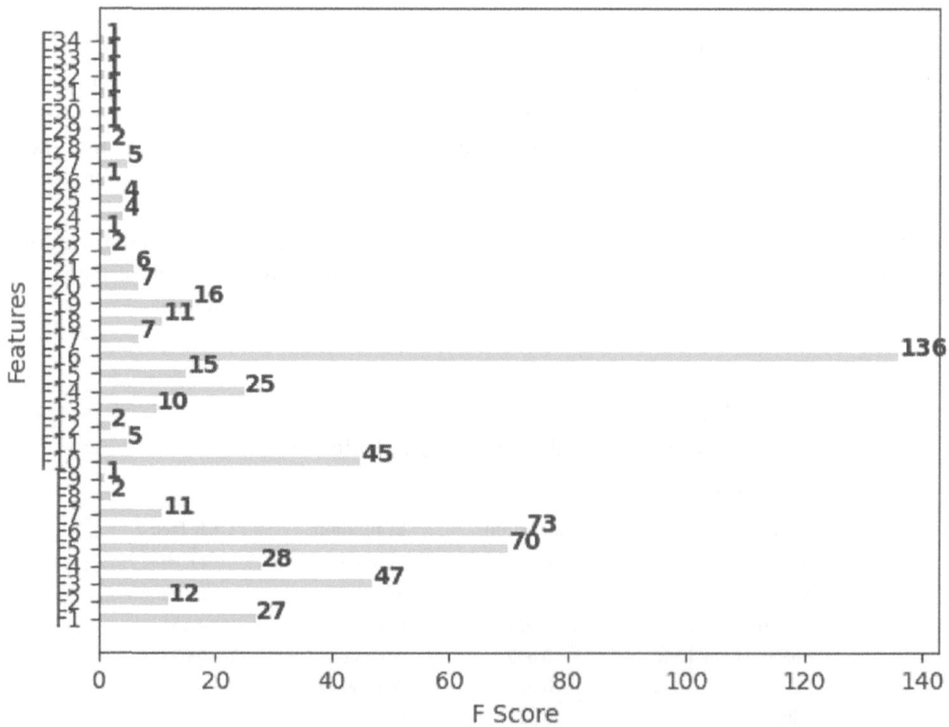

FIGURE 15.4 Variable importance plot.

These variables play a significant role in driving the predictions in the model.

> The variable importance chart for the XGBoost Tree node is generated by the Python algorithm, not by the SPSS Modeler. As a result, it uses the F-score to measure variable importance instead of the relative importance weights commonly seen in other SPSS nodes. Additionally, Python uses variable codes such as F1, F2, and so on instead of displaying field names or labels. Finally, due to these limitations, we are unable to download or reformat the data or the chart directly from this output.

CHAPTER SUMMARY

Gradient boosting is a powerful ensemble machine learning method that combines weak learners in an iterative process to build a strong predictive model. The method works by minimizing a specified loss function using gradient descent optimization. Each subsequent weak learner is trained to correct the errors made by the previous learners, with a focus on the cases that were misclassified in earlier iterations. This iterative process results in an ensemble model that can capture complex patterns and relationships within the data, leading to more accurate predictions. Gradient boosting is highly effective in both classification and regression tasks, making it a popular choice in machine learning.

One of the key advantages of gradient boosting is its ability to handle heterogeneous data, allowing it to incorporate various types of input, such as categorical and numerical variables, as well as missing values. Additionally, gradient boosting can employ different loss functions depending on the problem at hand, providing flexibility and adaptability across a wide range of scenarios. Another advantage is the interpretability of the model. By calculating variable importance, gradient boosting helps users understand which predictors contribute the most to the model's predictions, making it valuable for variable selection and gaining insights into data patterns. However, gradient boosting is also prone to overfitting if not properly configured. Therefore, tuning parameters and setting appropriate configurations are essential for achieving optimal performance and preventing overfitting in gradient boosting models.

CHAPTER DISCUSSION QUESTIONS

1. What is gradient boosting and how does it work?
2. How does gradient boosting differ from other ensemble methods?
3. What are the main advantages of using gradient boosting over other machine learning algorithms?
4. What are the key limitations of this method?
5. Define loss functions and explain their roles in gradient boosting? What are the common loss functions for gradient boosting and when to use them?
6. What is gradient descent? Why is it important in gradient boosting?
7. What are the key hyperparameters in gradient boosting, and how do they affect the model performance? How would you approach tuning these hyperparameters?
8. Gradient boosting is prone to overfitting, especially when the model becomes too complex. What are some common approaches that can be applied to mitigate overfitting in gradient boosting?
9. Can gradient boosting handle imbalanced datasets effectively? If yes, how does it address the issue of class imbalance?
10. What are the similarities and differences between traditional gradient boosting, Gradient Boosting Additive Models (GBAM), and Stochastic Gradient Boosting (SGB)?

11. Discuss the computational complexity of gradient boosting. Are there any techniques or optimizations that can be employed to improve its efficiency for large datasets?
12. What are some practical use cases or domains where gradient boosting has shown significant success? Provide examples and discuss the reasons behind its effectiveness in those scenarios.

CHAPTER EXERCISES

Exercise 1: Gradient Boosting Models for the Bank Dataset

1. Open the **bank** dataset in SPSS Modeler.
2. Sample the entire dataset. See the variable names and descriptions in Appendix 1.
3. Partition the data using a split ratio of 60–40 (train: 60%; test: 20%; validation: 20%)
4. Perform data modification as follows:
   - Impute missing values.
   - Normalize the data
5. Set **CRSCORE** (credit scores) as the target variable.
6. Configure five gradient boosting models for this dataset as follows.
   Gradient Boosting 1: Default
   - Basic options
     - Hyper-parameter optimization: **unchecked**
     - Tree method: **auto**
     - Number boost round: **10**
     - Max depth: **6**
     - Min child weight: **1**
   - Learning task
     - Objective: **reg:linear**
     - Early stopping: **unchecked**
   - Advanced options
     - Sub sample: **1.0**
     - Gamma: **0.0**

   Gradient Boosting 2
   - Basic options
     - Hyper-parameter optimization: **unchecked**
     - Tree method: **auto**
     - Number boost round: 100
     - Max depth: 3**0**
     - Min child weight: **1**
   - Learning task
     - Objective: **reg:linear**
     - Early stopping: **unchecked**
     - Advanced options
     - Sub sample: **1.0**
     - Gamma: **0.0**

   Gradient Boosting 3
   - Basic options
     - Hyper-parameter optimization: **unchecked**
     - Tree method: **auto**
     - Number boost round: 2**00**
     - Max depth: **50**
     - Min child weight: **1**

- Learning task
  - Objective: **reg:linear**
  - Early stopping: **unchecked**
- Advanced options
  - Sub sample: **1.0**
  - Gamma: **0.0**

Gradient Boosting 4 (hyper-parameter optimization)

- Basic options
  - Hyper-parameter optimization: **checked** (note: change the configurations below before checking this box).
  - Tree method: **auto**
  - Number boost round: **100**
  - Max depth: **50**
  - Min child weight: **1**
- Learning task
  - Objective: **reg:linear**
  - Early stopping: **unchecked**
- Advanced options
  - Sub sample: **1.0**
  - Gamma: **0.0**

Gradient Boosting 5 (hyper-parameter optimization)

- Basic options
  - Hyper-parameter optimization: **checked** (note: change the configurations below before checking this box).
  - Tree method: **auto**
  - Number boost round: 2**00**
  - Max depth: 1**0**
  - Min child weight: **1**
- Learning task
  - Objective: **reg:linear**
  - Early stopping: **unchecked**
- Advanced options
  - Sub sample: **1.0**
  - Gamma: **0.0**

7. Compare the model fit metrics between these four models using the testing results.
8. Select the champion model and justify this selection with the necessary evidence.
9. Present the results of the champion model.
10. Evaluate the reliability and validity of the champion model.
11. Present the variable importance chart and interpret the results.

Exercise 2: Gradient Boosting Models for the Passengers Dataset

1. Open the **passengers** dataset in SPSS Modeler.
2. Sample 20% of the full data. Set **return** (decision to return to the airline) as the target variable and change its role to binary. See the variable names and descriptions in Appendix 2.
3. Partition the data using a split ratio of 60-40 (train: 60%; test: 20%; validation: 20%)
4. Perform data modification as follows:
   a. Replace zero values for Likert-scale variables with missing values.
   b. Impute missing values.

5. Configure five gradient boosting models for this dataset as follows.

Gradient Boosting 1: Default

- Basic options
  - Hyper-parameter optimization: **unchecked**
  - Tree method: **auto**
  - Number boost round: **10**
  - Max depth: **6**
  - Min child weight: **1**
- Learning task
  - Objective: **binary:logistic**
  - Early stopping: **unchecked**
- Advanced options
  - Sub sample: **1.0**
  - Gamma: **0.0**

Gradient Boosting 2

- Basic options
  - Hyper-parameter optimization: **unchecked**
  - Tree method: **auto**
  - Number boost round: **100**
  - Max depth: **30**
  - Min child weight: **1**
- Learning task
  - Objective: **binary:logistic**
  - Early stopping: **unchecked**
- Advanced options
  - Sub sample: **1.0**
  - Gamma: **0.0**

Gradient Boosting 3

- Basic options
  - Hyper-parameter optimization: **unchecked**
  - Tree method: **auto**
  - Number boost round: **200**
  - Max depth: **50**
  - Min child weight: **1**
- Learning task
  - Objective: **binary:logistic**
  - Early stopping: **unchecked**
- Advanced options
  - Sub sample: **1.0**
  - Gamma: **0.0**

Gradient Boosting 4 (Hyper-parameter optimization)

- Basic options
  - Hyper-parameter optimization: **checked** (note: change the configurations below before checking this box).
  - Tree method: **auto**
  - Number boost round: **100**
  - Max depth: **50**
  - Min child weight: **1**
- Learning task
  - Objective: **binary:logistic**
  - Early stopping: **unchecked**

- Advanced options
  - Sub sample: **1.0**
  - Gamma: **0.0**

Gradient Boosting 5 (Hyper-parameter optimization)

- Basic options
  - Hyper-parameter optimization: **checked** (note: change the configurations below before checking this box).
  - Tree method: **auto**
  - Number boost round: **200**
  - Max depth: **10**
  - Min child weight: **1**
- Learning task
  - Objective: **binary:logistic**
  - Early stopping: **unchecked**
- Advanced options
  - Sub sample: **1.0**
  - Gamma: **0.0**

6. Compare the model fit metrics between these four models using the testing results.
7. Select the champion model and justify this selection with the necessary evidence.
8. Present the results of the champion model.
9. Evaluate the reliability and validity of the champion model.
10. Present the variable importance chart and interpret the results.

16 Bayesian Networks

LEARNING OUTCOMES

Upon the completion of this chapter, you will be able to

1. Describe the principles and concepts of Bayesian networks.
2. Explain how Bayesian networks capture probabilistic relationships among variables.
3. Discuss the advantages and disadvantages of Bayesian networks compared to other machine learning methods.
4. Describe the structure and components of Bayesian networks, including nodes, directed edges, and conditional probability tables.
5. Evaluate the Bayesian network modeling process.
6. Compare four types of Bayesian networks, Naive Bayes, Tree Augmented Naive Bayes (TAN), Bayesian Network Augmented Naive (BAN), and Markov Blanket.
7. Explore the algorithms for learning the structure of Bayesian networks from data, including constraint-based and score-based algorithms.
8. Interpret a Bayesian network structure, including the identification of causal relationships among variables.
9. Define causal data mining and explain the USELEI process.
10. Construct Bayesian network models using SPSS Modeler.

INTRODUCTION

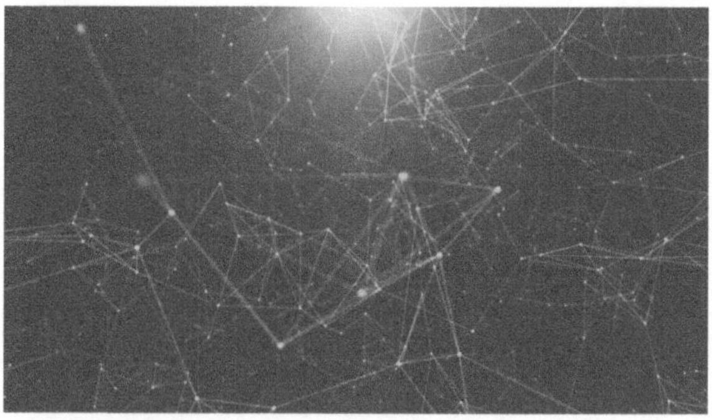

In this chapter, we explore one of the most robust and complex machine learning algorithms: the Bayesian network. Rooted in probabilistic reasoning and grounded in Bayes' theorem, this method provides a powerful approach to inferring future events based on existing evidence. Central to its functionality is the construction of a causal network—a directed acyclic graph (DAG) representing causal relationships between predictors and target variables, supported by a set of conditional probabilities. With its inherent capacity for causal inference, the Bayesian network is a formidable tool for uncovering cause-and-effect relationships within large datasets.

This chapter offers a comprehensive introduction to the application of Bayesian network methods in data mining. We cover various supervised learning algorithms, including Naive Bayes, Tree Augmented Naive Bayes (TAN), Bayesian Network Augmented Naive (BAN) Bayes Classifier, and Markov Blanket. This chapter also discusses the causal data mining process and provides real-world examples to illustrate the practical application of Bayesian network methods.

Given the complexity of the mathematical framework underlying Bayesian networks, our focus is on key concepts and the analytical process rather than detailed mathematical formulations. We deliberately omit in-depth mathematical details to prevent confusion. As Bayesian network modeling is often automated by software, the emphasis is placed on understanding the fundamental

DOI: 10.1201/9781003468196-19

principles, strengths, limitations, and interpretation of results. By presenting the probabilistic relationships between predictors and target variables in an accessible manner, we aim to improve comprehension of analysis outputs while maintaining the analytical rigor that is essential to the Bayesian approach.

WHAT IS BAYESIAN NETWORK?

CORRELATION VS. CAUSATION

Before defining a Bayesian network, it is important to understand the distinction between correlation and causation. Data mining has proven to be an effective method for detecting patterns in large, noisy datasets. In other words, it is an excellent tool for exploring vast amounts of data to uncover meaningful, previously unknown patterns that support decision-making processes in organizations. However, there are ongoing debates about the limitations of conventional data mining methods, such as regression. The most common criticism of these traditional methods is their inability to infer causation among variables (Cox, 2010).

While causation often implies correlation, correlation does not necessarily imply causation. To clarify this distinction, let's consider an example related to injury severity during a car accident (Figure 16.1). In this scenario, we examine the effects of two variables: age and failure to wear seatbelts. Suppose data analysis reveals a high correlation between age and injury severity. Although age may be highly correlated with injury severity, it is not the cause of the injury. The correlation simply suggests that people of different ages may experience varying levels of injury. For risk assessment purposes, age can serve as one of the predictors for injury severity, but it is not a factor that can be controlled to mitigate the risk of injury.

In contrast, failure to wear a seatbelt is a well-established cause of injury during car accidents. Without seatbelt protection, the likelihood of suffering severe injuries in an accident increases. Armed with this knowledge, authorities can reduce injury risk by enforcing seatbelt laws. This discussion highlights the importance of identifying causal relationships when using data mining for risk prediction.

HOW TO TEST CORRELATION AND CAUSATION?

Now that we understand the differences between correlation and causation, it's important to know how they are tested. There are distinct approaches to calculating correlation and causation.

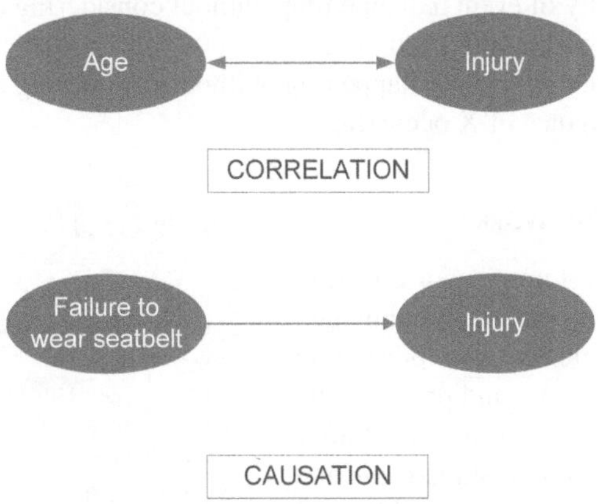

FIGURE 16.1 Correlation vs. causation.

Testing Correlation

The correlation between two variables, X and Y, can be determined by calculating the correlation coefficient. A positive correlation means that if X increases, Y also increases, while a negative correlation means that if X increases, Y decreases. The value of the correlation coefficient indicates the effect size or the strength of the correlation between the two variables. While correlation is commonly used in prediction, it is important to note that it cannot be used to infer causation. Additionally, data mining methods based on correlation are prone to overfitting, particularly with large sample sizes. The larger the sample, the more likely relationships between variables will appear statistically significant, even if they do not reflect causal connections.

> **DEFINITION**
>
> A probabilistic relationship refers to the existence of a connection or association between two or more variables, where the relationship is characterized by uncertainty and is expressed through probabilities.

Testing Causation

Causation is much harder to test because it requires isolating the variables of interest within a controlled environment. Causation occurs when variable X directly causes Y to occur. One common approach to testing causality is through probabilistic analysis, specifically by using conditional probabilities.

Unlike correlation, a probabilistic relationship describes the association between two or more variables in terms of uncertainty, and it is expressed using probabilities. A probabilistic relationship between variables X and Y can be determined using **Bayes' theorem**. This theorem describes the probability of event Y occurring given event X, based on the individual probabilities of X and Y, and the probability of observing X if Y is true. The formula below presents this probabilistic relationship. In this equation, the probabilistic relationship between X and Y is represented by the conditional probability P(Y|X), which takes into account the prior probabilities of Y and X, and the likelihood of X given Y.

$$P(Y|X) = \frac{P(X|Y).P(Y)}{P(X)}$$

where

- P(Y|X) represents the conditional probability of event Y happening, given that event X has already occurred. In other words, it tells us the chance of Y occurring when we know that X has happened.
- P(X|Y) is the conditional probability of event X happening, given that event Y has already occurred. It tells us the chance of X occurring when we know that Y has happened.
- P(Y) is the probability of event Y happening, without considering any other events. It represents the overall chance of Y occurring.
- P(X) is the probability of event X happening, without considering any other events. It represents the overall chance of X occurring.

Definition of Bayesian Network

A **Bayesian network**, also known as a Bayesian belief network or a probabilistic graphical model, is a powerful and popular method for representing and reasoning about uncertainty and probabilistic relationships between variables. It is named after Thomas Bayes, an 18th-century statistician and philosopher, and represents a set of variables and their conditional dependencies through a DAG.

> **DEFINITION**
>
> Bayesian network, also known as a Bayesian belief network or probabilistic graphical model, is a powerful and popular method for representing and reasoning about uncertainty and probabilistic relationships between variables. It is a graphical model that represents a set of variables and their conditional dependencies through a directed acyclic graph.

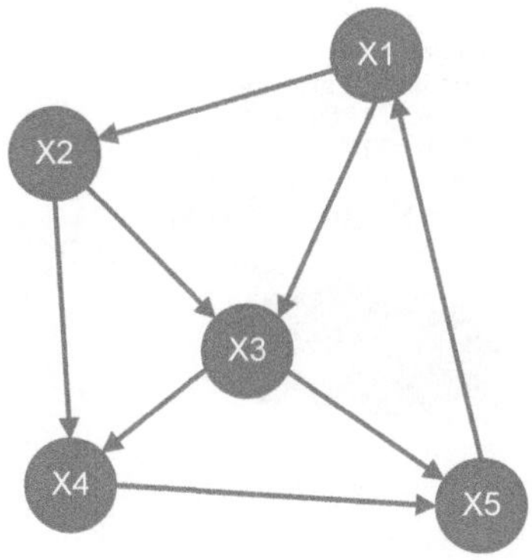

FIGURE 16.2 Annotated direct acyclic graph.

A Bayesian network consists of two main components: nodes and edges. Nodes represent variables of interest while edges represent probabilistic dependencies between those variables. The nodes can be either discrete or continuous, and they are connected by directed edges that indicate cause-and-effect relationships or conditional dependencies between the variables.

The structure of a Bayesian network is typically represented by a DAG, where each node corresponds to a variable, and the edges (or arcs) capture probabilistic dependencies. These edges represent conditional probabilities, showing how one variable influences another. Figure 16.2 illustrates a typical Bayesian network structure.

Conditional dependencies in a Bayesian network are defined using conditional probability tables (CPTs). Each node's CPT describes the probability distribution of that node, given its parent nodes in the graph. These CPTs allow for the quantification of uncertain relationships and enable probabilistic inference to be propagated throughout the network.

EXAMPLE OF BAYESIAN NETWORKS

A **Bayesian network** can be a complex method, so it's understandable if it feels confusing at first. Since its principles are rooted in the mathematical foundation of probability and Bayes' theorem, understanding how it works can be challenging.

To clarify the key concepts and components of a Bayesian network, let's walk through a simple example. In this case, we examine the factors that influence the injury level during accidents. We focus on four primary variables: **Traffic**, **Wearing a Seatbelt**, **Car Accident**, and **Injury**. The goal is to determine the probabilistic relationships between traffic conditions, wearing a seatbelt, the occurrence of a car accident, and the likelihood of sustaining an injury in an accident.

The Bayesian network for this example would consist of four nodes: Traffic, Wearing a Seatbelt, Car Accident, and Injury (as shown in Figure 16.3). The relationships between these variables are represented by directed edges: from Traffic to Car Accident, from Wearing a Seatbelt to Injury, and from Car Accident to Injury.

- **Traffic** has two possible states: **Light** or **Heavy**.
- **Car Accident** has two states: **True** or **False**.
- **Wearing a Seatbelt** has two states: **True** or **False**.
- **Injury** has two states: **True** or **False**.

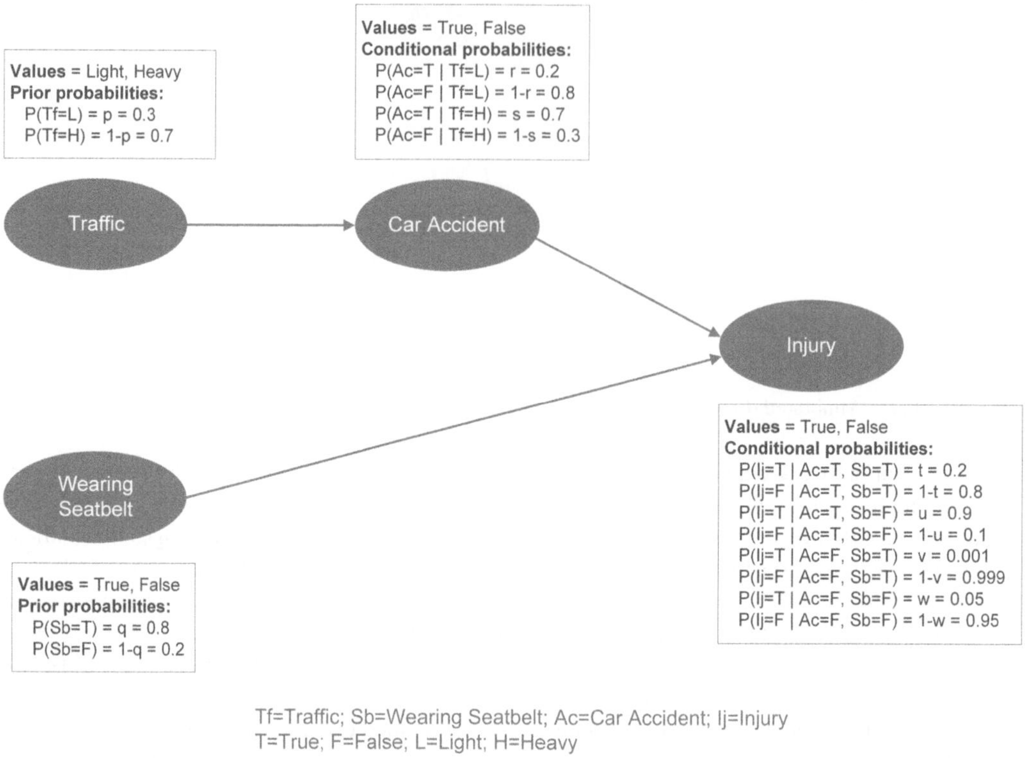

FIGURE 16.3 An example of Bayesian network structure for accident injuries.

We assume that heavy traffic may increase the likelihood of a car accident, and if an accident occurs, not wearing a seatbelt may increase the chances of sustaining an injury.

In this Bayesian network example, **Traffic** and **Wearing Seatbelt** are independent variables, meaning there is no direct causal relationship between them. The **Traffic** node serves as a parent to the **Car Accident** node because heavy traffic increases the likelihood of an accident. The **Car Accident** node and the **Wearing Seatbelt** node are both parents of the **Injury** node since the occurrence of an accident and failing to wear a seatbelt both contribute to the likelihood of injury.

The CPTs for each node define the probability distributions for the different states of these variables based on their relationships:

- Traffic and Wearing Seatbelt are precursors in this network, meaning we have prior probabilities for their events.
- The conditional probabilities for Car Accident and Injury depend on the values of Traffic and Wearing Seatbelt as parent variables.

Suppose, after running the analysis, we find the CPT for each node as follows.

Node 1—Traffic:

- **Variable:** Traffic
- **Values:** Light, Heavy
- **Prior probabilities:** P(Traffic=Light) = p; P(Traffic=Heavy) = 1 − p (where p is a known probability). In this case, we assume p = 0.3.

Node 2—Wearing Seatbelt:

- **Variable:** Wearing Seatbelt
- **Values:** True, False
- **Prior probabilities:** P(Wearing Seatbelt=True) = q, P(Wearing Seatbelt=False) = 1 − q (where q is a known probability). In this case, we assume p = 0.8.

Node 3—Car Accident: Influenced by Traffic

- **Variable:** Car Accident
- **Values:** True, False
- **Conditional probabilities:**
  - P(Accident=True|Traffic=Light) = r = 0.2 (interpretation: the probability of a car accident occurring when traffic is light is 0.2).
  - P(Accident=False|Traffic=Light) = 1 − r = 0.8 (interpretation: the probability of no car accident occurring when traffic is light is 0.8).
  - P(Accident=True|Traffic=Heavy) = s = 0.7 (interpretation: the probability of a car accident occurring when traffic is heavy is 0.7).
  - P(Accident=False|Traffic=Heavy) = 1 − s =0.3 (interpretation: the probability of no car accident occurring when traffic is heavy is 0.7).

Node 4—Injury: Influenced by Car Accident and Wearing Seatbelt

- **Variables:** Car Accident, Injury
- **Values:** True, False
- **Conditional probabilities:**
  - P(Injury=True|Accident=True, Wearing Seatbelt=True) = t = 0.2 (interpretation: the probability of injury when there is a car accident and the seatbelt is fastened is 0.2).
  - P(Injury=False|Accident=True, Wearing Seatbelt=True)= 1 − t =0.8 (interpretation: the probability of no injury when there is a car accident and the seatbelt is fastened is 0.8).
  - P(Injury=True|Accident=True, Wearing Seatbelt=False) = u = 0.9 (interpretation: the probability of injury when there is a car accident and the seatbelt is not fastened is 0.9).
  - P(Injury=False|Accident=True, Wearing Seatbelt=False) = 1 − u = 0.1 (interpretation: the probability of no injury when there is a car accident and the seatbelt is not fastened is 0.1).
  - P(Injury=True|Accident=False, Wearing Seatbelt=True) = v = 0.001 (interpretation: the probability of injury when there is no car accident and the seatbelt is fastened is 0.001).
  - P(Injury=False|Accident=False, Wearing Seatbelt=True) = 1 − v = 0.999 (interpretation: the probability of no injury when there is no car accident and the seatbelt is fastened is 0.2).
  - P(Injury=True|Accident=False, Wearing Seatbelt=False) = w = 0.05 (interpretation: the probability of injury when there is no car accident and the seatbelt is not fastened is 0.05).
  - P(Injury=False|Accident=False, Wearing Seatbelt=False) = 1 − w = 0.95 (interpretation: the probability of no injury when there is no car accident and the seatbelt is not fastened is 0.95).

The CPTs help quantify the likelihood of each event, given the conditions of their parent nodes. By using these probabilities, Bayesian networks enable reasoning about the likelihood of different outcomes based on observed evidence.

Using this Bayesian network, we can perform various types of inference, such as:

- Inferring the probability of a car accident, given heavy traffic:
  - If we know that the traffic is heavy, we can use the conditional probabilities in the CPT for the Car Accident node to compute the likelihood of a car accident. For example, based on the CPT:
    – P(Accident = True | Traffic = Heavy) = 0.7
- Inferring the probability of injury, given a car accident and whether the seatbelt was fastened:
  - If we have evidence that a car accident occurred and whether the driver wore a seatbelt, we can use the Injury node's CPT to compute the probability of injury. For example, if the driver was not wearing a seatbelt and a car accident occurred:
    – P(Injury = True | Accident = True, Wearing Seatbelt = False) = 0.9

This simple example illustrates how a Bayesian network captures the dependencies between variables and models probabilistic relationships. In practice, Bayesian networks can become significantly more complex by incorporating additional variables and more intricate causal structures. This complexity allows for sophisticated inference, prediction, and decision-making based on the network's probabilistic framework.

Refer to this example to understand how to read and interpret Conditional Probability Tables (CPTs) in a Bayesian network.

PRACTICAL APPLICATIONS OF BAYESIAN NETWORKS

Overall, the versatility of Bayesian networks lies in their ability to model complex dependencies, incorporate domain knowledge, handle uncertainty, and provide probabilistic reasoning and decision support.

They have practical applications in various domains. Some notable applications of Bayesian networks include the following:

1. **Medical diagnosis**: Bayesian networks can be used for medical diagnosis and decision support systems. They can model the relationships between symptoms, diseases, and test results to provide probabilistic assessments of different diagnoses, given observed symptoms and test outcomes.
2. **Risk assessment**: Bayesian networks can model the dependencies between various risk factors and provide probabilistic predictions of the likelihood of specific outcomes or events, such as financial risks, cybersecurity threats, or aviation incidents.
3. **Fraud detection**: This algorithm can be employed for fraud detection in various domains, such as credit card fraud, insurance fraud, or network intrusion detection. They capture the relationships between different variables and can identify patterns and anomalies indicative of fraudulent activities.
4. **Air traffic management**: Bayesian networks help optimize air traffic flow and manage congestion. They can take into account various factors, including weather conditions, aircraft performance, and air traffic control decisions, to make real-time predictions and decisions for efficient traffic management.
5. **Text classification**: This algorithm can model the relationships between words or features in text documents and classify new documents based on their probabilistic dependencies. Hence, it can be applied to text classification tasks, such as sentiment analysis or spam detection.
6. **Image and video processing**: Bayesian networks can capture the relationships between visual features, object appearances, and contextual information to enhance the accuracy

(ACC) of these tasks. Accordingly, they can be used for object recognition, scene understanding, and video analysis.

7. **Incident investigation**: When investigating aviation incidents or accidents, Bayesian networks can be used to analyze data and model the causal relationships between different factors. This can aid in identifying contributing factors and understanding the sequence of events leading to the incident.
8. **Predictive maintenance**: Bayesian networks can support predictive maintenance in industries such as air transportation, manufacturing, and energy. They model the dependencies between sensor data, equipment conditions, and failure events to predict and prevent equipment failures.

BASIC CONCEPTS OF BAYESIAN NETWORKS

Bayesian networks are an advanced and complex machine learning algorithm. They are graphical models that represent probabilistic relationships between variables. Hence, it is challenging to use and interpret these models effectively. Table 16.1 presents key concepts in Bayesian networks and their descriptions. By understanding these basic concepts, you can effectively construct, analyze, and perform probabilistic inference in Bayesian networks, making them valuable tools for handling uncertainty and making informed decisions in various domains.

TABLE 16.1
Basic Concepts of Bayesian Networks

Terms	Descriptions
Bayesian network	Bayesian network, also known as a Bayesian belief network or probabilistic graphical model, is a powerful and popular method for representing and reasoning about uncertainty and probabilistic relationships between variables. It is a graphical model that represents a set of variables and their conditional dependencies through a directed acyclic graph.
Bayes' Theorem	Bayes' theorem is a fundamental equation used in Bayesian networks to update probabilities based on new evidence. It calculates the posterior probability of an event, given prior knowledge and observed evidence.
Nodes	Nodes represent random variables or events in the domain being modeled. Each node corresponds to a specific quantity or attribute of interest.
Edges	Edges represent probabilistic dependencies between nodes. Directed edges indicate causal or direct influences between variables.
Directed Acyclic Graph (DAG)	The graphical structure of a Bayesian network is represented by a DAG, which is a graph without cycles (no path that starts and ends at the same node). The absence of cycles ensures that the network does not contain feedback loops and allows for efficient probabilistic inference.
Conditional Probability Tables (CPTs)	Each node in a Bayesian network has an associated CPT, which defines the conditional probability distribution of the node, given its parents' states. The CPT quantifies the probabilistic relationship between variables.
Parents and children	In a Bayesian network, the parents of a node are the nodes that directly influence it. The children of a node are the nodes that are directly influenced by it.
Probabilistic relationship	A probabilistic relationship refers to the existence of a connection or association between two or more variables, where the relationship is characterized by uncertainty and is described using probabilities.

(Continued)

TABLE 16.1 *(Continued)*
Basic Concepts of Bayesian Networks

Terms	Descriptions
Probabilistic inference	Bayesian networks allow for probabilistic inference, which means computing the probability of specific events or variables, given evidence (observed data or values) from other variables in the network. Inference is based on Bayes' theorem.
Prior probability	The prior probability represents the initial belief or knowledge about a variable before any evidence or data are observed. It is typically specified as part of the Bayesian network modeling process.
Posterior probability	The posterior probability represents the updated probability of a variable, given observed evidence or data. It is calculated by combining the prior probability with the likelihood of the observed data using Bayes' theorem.
Likelihood	The likelihood in Bayesian networks represents the probability of observing specific evidence, given the values of the variables in the network. It is derived from the conditional probability tables.
Evidence	Evidence refers to the observed values or states of certain variables in the Bayesian network. It is used to update the probabilities of other variables in the network.
Learning	Learning in Bayesian networks involves estimating the network structure and parameters from data. This can be done using approaches such as constraint-based methods or score-based methods.
Frequentist statistics	Frequentist statistics, also known as frequentist inference, is a statistical framework and approach to data analysis that focuses on the concept of repeated sampling and long-run frequencies of events.
Causal data mining	Causal data mining is a method of using causal machine learning algorithms to detect unknown patterns involving variables and their interrelationships in a causal network from data to predict outcomes
Score-based algorithm	A score-based algorithm is a Bayesian network structure learning approach that maximizes a scoring metric to identify the most probable network structure, given the data. The algorithm aims to find the network structure that best fits the data according to a specific scoring criterion.
Constraint-based algorithm	A constraint-based algorithm is an approach used to learn the network structure from data by discovering the probabilistic dependencies between variables without directly optimizing a scoring metric. It iteratively tests and evaluates conditional independence relationships among variables to determine the network's structure.

ADVANTAGES AND DISADVANTAGES

Like other methods, Bayesian networks have some important advantages and also disadvantages. It is important to consider these strengths and weaknesses when deciding whether to use Bayesian networks for a particular problem. The suitability of Bayesian networks depends on the specific application, available data, domain expertise, and the trade-off between model complexity and interpretability.

Advantages of Bayesian Networks

- **Probabilistic reasoning**: Bayesian networks provide a principled framework for probabilistic reasoning and uncertainty modeling. They can capture complex dependencies between variables and calculate probabilistic inferences, which allow us to gain unique insights from the data.

- **Causal modeling**: Unlike other machine learning methods, Bayesian networks represent causal relationships between variables. By explicitly modeling the causal structure, they enable causal reasoning and help uncover the underlying mechanisms driving the observed data.
- **Efficient inference**: This method provides us with efficient probabilistic inference. Once the network structure and parameters are defined, various inference algorithms can be employed to calculate posterior probabilities, make predictions, or answer queries about the variables in the network.
- **Handling missing data**: Bayesian networks handle missing data more effectively than some other methods. They can perform inference even when some values are missing, making them robust in dealing with incomplete datasets.
- **Modeling complex systems**: This method allows researchers to model complex systems with multiple variables and interactions. We can use Bayesian networks for problems where understanding the relationships between different components is essential.
- **Model transparency**: Bayesian networks offer interpretability and transparency. The graphical representation of dependencies between variables makes it easier to understand and communicate the relationships within the model, aiding in model interpretation and validation.

Disadvantages of Bayesian Networks

- **Model complexity**: Constructing Bayesian networks can be challenging, especially when dealing with a large number of variables or complex dependencies. Defining the network structure and estimating the parameters require domain expertise, data availability, and computational resources.
- **Dependency assumptions**: The ACC of Bayesian networks relies on correctly identifying the dependencies between variables. Incorrect assumptions or limitations in the available data may lead to biased or inaccurate models.
- **Data requirements**: We often need a substantial amount of data to accurately estimate the network structure and parameters. When data are scarce or limited, model performance and generalization may be compromised.
- **Computational complexity**: Inference in Bayesian networks can become computationally intensive, especially when dealing with large or complex networks. Exact inference algorithms may be infeasible, and we need to use approximate methods or sampling techniques.
- **Model uncertainty**: Although Bayesian networks handle uncertainty well, they can also introduce model uncertainty due to the many possible network structures. Determining the best structure and hyperparameters may require expert knowledge or additional model selection techniques.

BAYESIAN NETWORK PROCESS

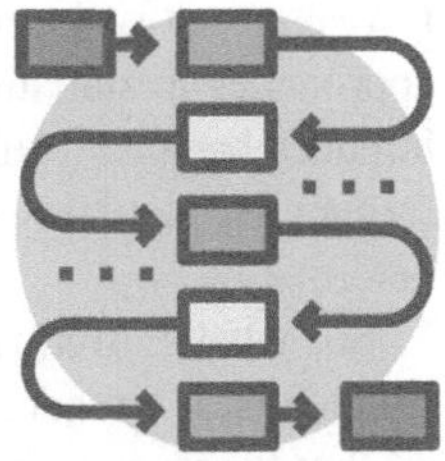

Bayesian network modeling requires a complex process. I provide below the important steps in constructing and validating Bayesian networks in data mining projects.

1. **Data Collection and Preprocessing**
   - Gather relevant data from various sources, ensuring that it represents the problem domain accurately.
   - Preprocess the data by handling missing values, dealing with outliers, and transforming variables if necessary.

2. **Variable Selection and Domain Knowledge**
   - Identify the variables of interest for building the Bayesian network. Domain knowledge and expertise play a crucial role in selecting relevant variables.
   - Understand the relationships and dependencies between variables, which can guide the network's structure.
3. **Structure Learning**
   - Determine the network's structure by learning the probabilistic dependencies between variables from the data.
   - Methods such as constraint-based algorithms or score-based algorithms can be used for structure learning.
4. **Parameter Estimation**
   - Estimate the CPTs for each node in the network. This involves calculating the probabilities of each variable, given its parent variables.
   - Maximum likelihood estimation (MLE) or Bayesian parameter estimation can be used for parameter estimation.
5. **Model Evaluation and Validation**
   - Split the data into training and testing sets for model evaluation.
   - Use appropriate model fit metrics to evaluate the model performance on the test set.
6. **Model Interpretation and Visualization**
   - Interpret the Bayesian network's structure to understand the relationships and dependencies between variables.
   - Visualize the network graphically to aid in model interpretation and communication.
7. **Inference and Predictions**
   - Perform probabilistic inference using Bayes' theorem to compute the probabilities of specific events or variables, given evidence (observed data).
   - Use the Bayesian network to make predictions or decisions based on the inferred probabilities.
8. **Model Refinement and Iteration**
   - Refine the Bayesian network by making adjustments to the structure or parameters based on feedback, domain knowledge, or additional data.
   - Iterate through the process to improve the model performance and ACC.

It's important to note that constructing a Bayesian network requires a combination of domain knowledge, data analysis, and expert judgment. The iterative nature of the process allows for continuous refinement and improvement of the model ACC and performance. I developed a more general process for the so-called causal mining. It is the USELEI process. I will present this process later in this chapter.

DIFFERENCES BETWEEN BAYESIAN NETWORK AND FREQUENTIST STATISTICS

Bayesian network methods and frequentist methods are two different approaches to statistical inference. It is important to understand these differences because you are likely much more familiar and experienced with frequentist statistics (even though you may not have heard of this term before).

So, what is frequentist statistics? Frequentist statistics, also known as frequentist inference, is a statistical framework and approach to data analysis that focuses on the concept of repeated sampling and long-run frequencies of events. The name comes from the method's emphasis on analyzing the frequency or proportion of events occurring in repeated experiments or observations. In frequentist statistics, the key idea is that probabilities are interpreted as limiting frequencies. It treats the parameters of a statistical model as fixed and unknown quantities that do not have probability distributions. Instead, the focus is on estimating these fixed parameters using observed data and making inferences based on the observed frequencies. Essentially, the statistics-based methods we covered earlier, including linear regression, logistic regression, cluster analysis, principal components, and other statistical methods you may be familiar with (such as t-tests, analysis of variance [ANOVA], multivariate analysis of variance [MANOVA], and discriminant analysis), are all frequentist statistics.

Since frequentist statistics focuses on frequencies rather than probability distributions, it differs from Bayesian networks. Table 16.2 highlights the key differences between these two methods.

TABLE 16.2
Differences between Bayesian Networks and Frequentist Statistics

Items	Bayesian Networks	Frequentist Statistics
Philosophy	Bayesian networks view probability as a measure of subjective belief or degree of uncertainty, incorporating prior knowledge and updating it based on observed data using Bayes' theorem.	Frequentist methods consider probability as the long-run frequency of an event occurring over repeated trials, focusing solely on the properties of the observed data.
Handling uncertainty	Bayesian networks explicitly model and reason uncertainty by assigning prior probabilities to uncertain quantities and updating them based on observed data.	Frequentist methods do not directly incorporate prior beliefs or quantify uncertainty but rely on the concept of sampling variability and estimate parameters based solely on observed data.
Parameter estimation	Bayesian networks estimate parameters by specifying prior probabilities and updating them with observed data to obtain posterior probabilities. The posterior distribution reflects the combined information from the prior knowledge and data.	Frequentist methods estimate parameters using techniques such as maximum likelihood estimation, which aims to find the parameter values that maximize the likelihood of the observed data.
Inference	Bayesian networks provide a framework for probabilistic inference, which enables the computation of various probabilities of interest, such as posterior probabilities, conditional probabilities, and predictions.	Frequentist methods focus on point estimation and hypothesis testing. They provide estimates and tests based on the observed data without quantifying probabilities directly.
Sample size	Bayesian networks can handle small sample sizes effectively by incorporating prior knowledge into the analysis. Prior information can act as a regularizer, helping to stabilize estimates and avoid overfitting.	Frequentist methods typically require larger sample sizes for reliable estimation and hypothesis testing due to their reliance on observed data.
Interpretability	Bayesian networks provide a natural framework for interpreting results in terms of probabilities, making it easier to communicate uncertainty and make informed decisions based on the posterior distribution.	Frequentist methods tend to focus more on correlation estimates and p-values, which can sometimes be misinterpreted or lead to binary decisions without explicit consideration of uncertainty.

BAYESIAN NETWORK AND CAUSAL INFERENCE

One common question when implementing the Bayesian network method is whether it allows for making causal inferences based on the results. As previously discussed, Bayesian networks provide valuable insights into the dependencies and conditional relationships between variables. They represent causal relationships through probabilistic relationships. However, it's important to understand that Bayesian networks alone cannot establish strict causality without prior domain knowledge or a theoretical foundation. While they do not explicitly prove causation, they can infer causal relationships based on the probabilistic dependencies modeled in the network. Let's explore this further.

Bayesian networks primarily capture statistical dependencies between variables based on observed data and assumptions. They model the conditional probabilities and directionality of the relationships between variables, which can be useful for predictions, probabilistic reasoning, and estimating the likelihood of events, given observed evidence. In a Bayesian network, the directed edges between nodes represent conditional dependencies or potential causal relationships. For example, if variable A is a parent of variable B in the network, it suggests that A has a direct causal influence on B. The structure of the network, along with the CPTs for each node, encodes these probabilistic relationships.

As mentioned earlier, it's important to note that Bayesian networks alone do not prove causation. To establish a true causal relationship, we must isolate input variable A and test its effect on output variable B while controlling for all other variables. This is typically done in a controlled environment, where it is possible to determine if A indeed causes B. However, such an approach is not practical in data mining projects, as it requires setting up experiments and usually only examines the causal effect of one or two variables at a time. In data mining, Bayesian network results can suggest causal relationships based on the observed data and network structure, but additional empirical evidence or domain knowledge is usually needed to definitively establish causality. For example, the probabilistic relationship derived from a Bayesian network may indicate that A has a direct causal influence on B, but further theoretical foundations, domain knowledge, and empirical evidence are required to confirm that A truly causes B.

Nonetheless, Bayesian networks serve as a useful tool for exploring and representing causal hypotheses. By integrating domain knowledge and causal assumptions into the network's structure and conditional probabilities, Bayesian networks provide a framework for reasoning about potential causal relationships and their implications. More specifically, Bayesian networks allow for probabilistic inference based on observed evidence, which can be used to analyze the potential causal relationships within the network.

Therefore, in this book, when we use terms like "causal inference" or "infer causality," we are referring to the causal relationships suggested by the Bayesian network structure through probabilistic inference, with the understanding that prior domain knowledge and causal assumptions are in place.

> Bayesian networks alone cannot establish causal relationships or infer causation in the strict sense without prior domain knowledge or a theoretical foundation. However, they can infer causal relationships based on the probabilistic dependencies modeled within the network.

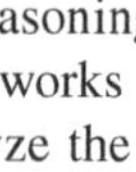

DIFFERENT TYPES OF BAYESIAN NETWORKS

Bayesian network is an advanced and powerful method that can be effectively applied in supervised learning when both the target variable and predictors are identified. The model captures dependencies between variables and uses learned probabilities to make predictions or classifications. This makes it highly effective for tasks such as classification, association, and probabilistic inference, especially in domains where labeled training data are available.

There are four primary types of Bayesian network structures: Naive Bayes, TAN, BAN, and Markov Blanket. Figure 16.4 presents these network structures in diagram form.

NAIVE BAYES

The Naive Bayes method is frequently used for constructing Bayesian networks, primarily due to its simplicity. In a Naive Bayesian network, the target variable serves as the parent node for all other nodes (predictors), resulting in a straightforward structure. There are no other connections

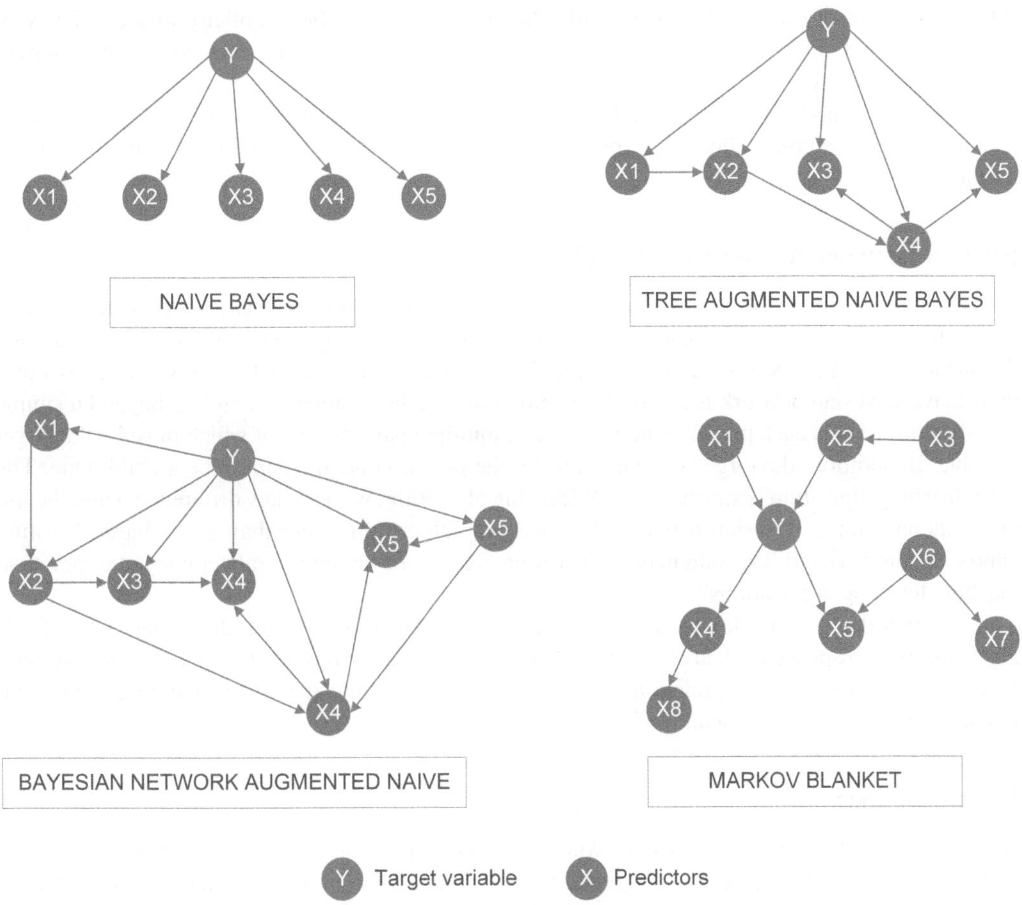

FIGURE 16.4 Four types of Bayesian Network learning algorithms.

between the predictors because the input variables are assumed to be conditionally independent of each other. This assumption allows for fast and efficient training and classification. However, Naive Bayes can be limited when the predictors are dependent on each other, as it ignores any dependencies among the variables.

This method is known for its simplicity and ease of construction, making it an efficient choice that demands a small training sample size. It is a commonly used supervised learning algorithm in data mining projects, allowing direct comparisons with other machine learning techniques. However, its applicability is constrained by the assumption of conditional independence, restricting its use to a simple network with direct relationships between predictors and the target variable. In more complex networks, where predictors are anticipated to exhibit causal dependencies, this assumption is typically violated.

Tree-Augmented Naive (TAN)

TAN extends Naive Bayes by adding a tree structure to capture dependencies between variables. The TAN Bayesian network connects the target variable to each predictor and connects the predictors in a tree structure. This algorithm constructs the network based on the maximum spanning tree algorithm. In a TAN, each predictor node has no more than two parent nodes, the target node, and at most one another predictor node. Thus, each predictor node can have only one augmenting edge pointing to it.

By relaxing the assumption of conditional independence, this method can outperform Naive Bayes while preserving its simplicity and efficiency. It retains the simplicity and efficiency of Naive Bayes but allows for more expressive models by incorporating limited variable dependencies. However, due to the tree structure's constraint, where each predictor node can have at most two parent nodes (the target node and another predictor node as an optional parent node), it doesn't permit a comprehensive exploration of the complete causal network structure among predictors.

Bayesian Network Augmented Naive (BAN)

BAN is an extension of TAN that incorporates a more general Bayesian network structure. It allows for arbitrary inter-variable dependencies by learning a complete Bayesian network structure among the variables. In a BAN, predictor nodes are direct children of the target node, yet a nearly comprehensive Bayesian network is established among the predictor nodes, extending beyond a simple tree structure. Thus, each predictor node can have multiple parents, one of which must be the target variable. In addition, the target variable must be the parent node and cannot be a child node. The BAN learning algorithm is similar to TAN learning algorithm, with the key difference being the use of an unconstrained Bayesian network (BN) learning algorithm rather than a tree-based learning approach. Unrestricted Bayesian network learning algorithms are commonly employed in practical machine learning applications.

BAN provides a more flexible modeling approach compared to Naive Bayes and TAN. BAN algorithm is appropriate for learning a causal network structure from big data in a complex system. It can capture more complex relationships among variables but at the cost of increased computational complexity for training and inference.

Markov Blanket

In the context of Bayesian networks, the Markov blanket of a node is a concept that defines a minimal set of nodes that renders the node conditionally independent of all other nodes in the network. In other words, given the Markov blanket nodes, the node is probabilistically independent of all other nodes in the network. A Markov blanket network establishes connections between the target variable and its predictors while also allowing connections between specific input variables. Only the children of the target variable are permitted to have an additional parent node, while all other variables remain conditionally independent of the target variable.

The Markov blanket of a target variable includes three types of nodes:

1. **Parents:** The parents of the target variable in the Bayesian network. These are the nodes directly connected to the target variable by incoming edges.
2. **Children:** The children of the target variable. These are the nodes directly connected to the target variable by outgoing edges.
3. **Spouses:** The parents of the children of the target variable. These are the nodes that are directly connected to the children of the target variable, excluding the target variable itself.

The Markov blanket is important because it represents a minimal set of variables that need to be observed or considered when predicting the target variable or performing probabilistic inference. Variables outside the Markov blanket are irrelevant to the target variable once the Markov blanket variables are known. The Markov blanket provides a way to identify and exploit the conditional independence relationships between variables in a Bayesian network. It helps reduce the computational complexity of probabilistic inference by focusing on the minimal set of relevant variables.

NETWORK STRUCTURE LEARNING ALGORITHMS

One important step in the Bayesian network process is learning the network structure from data. In data mining, learning algorithms are employed to accomplish this task. In the context of Bayesian networks, two primary types of learning algorithms are commonly used: score-based algorithms and constraint-based algorithms. In this section, we will gain a fundamental understanding of these algorithms and their differences.

Score-Based Algorithm

A score-based algorithm is a Bayesian network structure learning approach that aims to maximize a scoring metric to identify the most probable network structure, given the data. The algorithm searches through possible network structures and selects the one that best fits the data according to a specific scoring criterion. Commonly used scoring metrics include log-likelihood, Bayesian information criterion (BIC), and Akaike information criterion (AIC). The algorithm assigns a score to each structure, with higher scores indicating a better fit of the model to the data. The goal is to find the network structure that maximizes this score.

Constraint-Based Algorithm

A constraint-based algorithm learns the network structure by identifying the probabilistic dependencies between variables without directly optimizing a scoring metric. Instead of exhaustively searching all possible network structures, this algorithm tests and evaluates conditional independence relationships among the variables. By iteratively applying these tests, the algorithm constructs a network that captures the dependencies and independencies in the data, thus defining the structure.

TABLE 16.3
Differences between Score-based Algorithm and Constraint-based Algorithm

Item	Score-based Algorithm	Constraint-based Algorithm
Approach	Optimizes a scoring metric (e.g., log-likelihood, BIC, AIC) to find the network structure that best fits the data.	Identifies conditional independence relationships between variables to infer the network structure without directly optimizing a scoring metric.
Objective	Maximizes a scoring metric to balance model complexity and data fit.	Focuses on capturing direct conditional dependencies between variables to construct a more interpretable network.
Search space	Explores the search space by making local modifications to the network structure, such as adding, deleting, or reversing edges.	Utilizes statistical tests to identify conditional independence relationships and construct a skeleton graph, which is later refined into a partially directed acyclic graph.
Scoring metric	Directly uses a scoring metric to evaluate and compare different network structures.	Does not directly use a scoring metric, but the resulting network structure can be scored using a separate scoring method.
Strengths	• Efficient and scalable for large datasets. • Provides a direct measure of model fit. • Can handle both categorical and continuous variables.	• Robust to overfitting and requires fewer data samples. • Captures direct conditional dependencies, leading to a more interpretable network. • Handles mixed data types.
Limitations	• Sensitive to local optima and may not explore the entire search space thoroughly. • Requires a larger sample size for reliable parameter estimation.	• May not capture some complex relationships due to the focus on direct dependencies. • Can be computationally demanding when testing for conditional independence with many variables.
Interpretability	The resulting network may be complex and harder to interpret.	The network structure often reflects direct causal relationships and is more interpretable.
Common algorithms	Hill Climbing, Tabu Search, Greedy Equivalent Search (GES)	PC Algorithm (Peter-Clark), Fast Causal Inference (FCI)

Table 16.3 compares these two algorithms. Each of these approaches has its strengths and weaknesses, and the choice of the method depends on factors such as the size of the dataset, the complexity of the network, the availability of domain knowledge, and the desired interpretability of the resulting model.

CAUSAL DATA MINING PROCESS

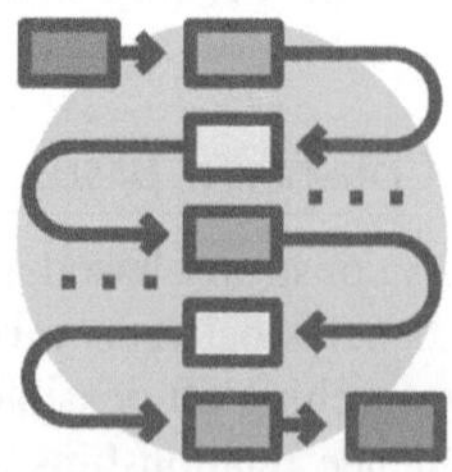

Given Bayesian networks' capability to capture probabilistic relationships among variables in a complex network, I want to introduce the concept of causal data mining and how Bayesian networks can be applied in causal data mining projects.

In traditional data mining, as discussed in previous chapters, the goal is often to use machine learning algorithms to detect unknown relationships within data that help predict a target variable. These projects generally result in a predictive model that accurately forecasts the target variable based on key predictors. However, the focus in such projects is not on understanding how the variables interact with and influence one another. The outcome users seek is the "champion model" that can make accurate predictions about the risk or event of interest.

But what if, in some cases, users want to delve deeper into the data, exploring how variables interact and influence one another? Or what if they want to know which variable has a direct causal effect on another, setting off a chain reaction that impacts further variables? Gaining insight into these causal relationships helps organizations make sense of their predictions and craft long-term strategies that are grounded in a deeper understanding of the data. In this scenario, traditional data mining techniques may not be sufficient. What we need here is a new type of data mining, which I call *causal data mining*.

DEFINITION

Causal data mining can be defined as a method of using causal machine learning algorithms to detect unknown patterns involving variables and their interrelationships in a causal network from data to predict outcomes (Truong, 2021b).

In my article published in the *Journal of Air Transport Management* in 2021, I define causal data mining as "a method of using causal machine learning algorithms to detect unknown patterns involving variables and their interrelationships in a causal network from data to predict outcomes." In projects of this nature, common processes such as SEMMA or CRISP-DM may not be the best fit, as they do not reflect the causal nature of the problem. To address this, I developed a six-step process tailored specifically for causal data mining: Understanding, Sampling, Exploring, Learning, Evaluating, and Inferring, abbreviated as **USELEI**.

In this section, I describe the USELEI process and explain how it can be applied to causal data mining projects to uncover deeper insights about how variables relate and influence each other.

Figure 16.5 presents the six steps of the USELEI process, and Table 16.4 provides the purpose and details of each step (Truong, 2021b).

Truong, D. (2021). Using Causal Machine Learning for Predicting the Risk of Flight Delays in Air Transportation. *Journal of Air Transport Management*, 91, 101993. https://doi.org/10.1016/j.jairtraman.2020.101993

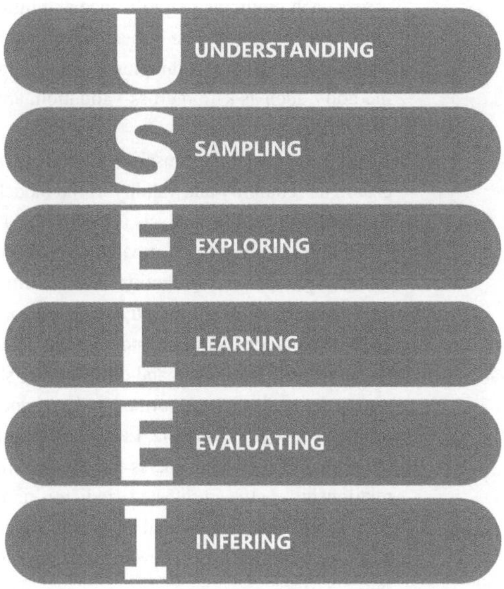

FIGURE 16.5 USELEI process.

TABLE 16.4
USELEI Causal Data Mining Process

Steps	Descriptions
STEP 1: UNDERSTANDING Purpose: Understanding the research problem and objectives	This step is dedicated to thoroughly understanding the research problem and objectives from a business perspective. Relevant information about the industry, market, stakeholders, systems, or organizations must be gathered, then translated into a clear problem statement and research questions that necessitate mining large datasets to predict an outcome. Researchers need to identify what they aim to achieve with causal data mining. The system's complexity and dynamics should be outlined, along with the potential variables to be included in the model.
STEP 2: SAMPLING Purpose: Selecting dataset, sampling, and partitioning	Data are collected and sampled from reliable sources. The sampling method should be determined based on the target population and research scope. The dataset should be large enough to capture relevant information but not so large as to hinder efficient modeling. Afterward, the data are divided into training and validation samples. The training sample is used to build the model, and the validation sample is used to test the model's performance. If cross-validation is applied, the dataset should be split into multiple subsamples.
STEP 3: EXPLORING Purpose: Exploring and preparing the data	The dataset is explored in detail to understand the variables, examine descriptive statistics, identify missing data, and detect outliers. Missing values and outliers should be handled appropriately to prevent information loss and reduce bias. A key step in data preparation for causal data mining is discretization, using algorithms such as equal distance, normalized equal distance, equal frequency, K-means, or tree-based methods.
STEP 4: LEARNING Purpose: Learning unsupervised and supervised causal networks and performing cross-validation	Initially, an unsupervised network structure is developed to explore preliminary causal relationships among variables without specifying a target variable. This network facilitates causal inferences between individual variables and identifies the most important variables in the entire system, but it does not indicate each variable's impact on the target variable. It serves as the foundation for constructing the full network. Next, a supervised network is built using causal learning algorithms with an identified target variable. This network helps predict the target variable and assess each predictor's significance and influence. The network structure is learned from the data, so it must be validated by comparing results from training and validation samples. For large datasets, more robust validation methods, such as k-fold cross-validation, are recommended to minimize overfitting.
STEP 5: EVALUATING Purpose: Evaluating the network performance and predictive precision	The performance of the causal network is assessed using model fit metrics to ensure its reliability and validity. If the model does not meet evaluation standards, a post-hoc analysis is performed. During this step, the model can be refined to improve performance by systematically removing variables with minimal predictive contribution, one at a time. This iterative process continues until the network achieves satisfactory performance.
STEP 6: INFERRING Purpose: Performing target analysis, sensitivity analysis, and causal inference	Key predictors are identified and reported, along with their effect sizes on the target variable. Both total and direct effects should be analyzed. Additionally, sensitivity analysis is performed to determine how changes in one predictor influence the target variable. Lastly, causal inference is conducted, based on the probabilistic dependencies modeled within the network, to explore causal relationships between selected predictors and the target variable under certain conditions. This analysis can be used to create "what-if" scenarios, which allow decision-makers to develop strategies to prepare for incidents or mitigate risks.

DIFFERENCES BETWEEN USELEI, SEMMA, AND CRISP-DM

You may ask: Why do we need another data mining process? And how does USELEI differ from the other two processes (SEMMA and CRISP-DM)?

Table 16.5 provides a summary and comparison of three data mining processes—USELEI, CRISP-DM, and SEMMA. These processes share similarities in various stages, such as sampling, data exploration, preparation, modeling, and model evaluation. However, when it comes to causal data mining, the USELEI process offers distinct advantages over the other two. USELEI places a stronger emphasis on constructing a causal network structure, performing causal inference, and conducting sensitivity analysis—key features that are particularly valuable for organizations in making informed business decisions.

Additionally, USELEI highlights the importance of understanding the research problem and identifying research objectives from a business perspective, which is not as prominently featured in SEMMA. These aspects allow for a more holistic approach to both predictive and causal modeling.

Overall, SEMMA and CRISP-DM are excellent processes for traditional data mining projects, where identifying causal relationships among variables is not a priority. In contrast, USELEI is specifically designed for causal data mining. Although it caters to a narrower audience, the USELEI process offers rigorous and comprehensive steps for conducting a causal data mining project.

It's important to note that Bayesian networks play a critical role in causal data mining, as this method enables the construction of a causal network from data, revealing probabilistic relationships among variables. The Bayesian network structure is used not only to predict the target variable but also to examine the causal relationships among variables within the network through probabilistic inference. While we won't be implementing the USELEI process in this book, I will demonstrate how to perform probabilistic inference using a Bayesian network structure in the example.

TABLE 16.5
Comparison between USELEI, CRISP-DM, and SEMMA

Items	USELEI	CRISP-DM	SEMMA
Steps	Understanding→Sampling → Exploring and preparing→Learning and validation→ Evaluating→Inferring	Business understanding →Data understanding →Data preparation →Modeling→Evaluation →Deployment	Sample→Explore →Modify→Model → Assess
Emphasis on understanding of business problems	Yes	Yes	No
Sampling process	Yes	Marginal	Yes
Data exploration and preparation	Yes	Yes	Yes
Modeling and Validation	Yes	Yes	Yes
Full causal network structure	Yes	No	No
Model evaluation	Yes	Yes	Yes
Decision support (causal inference)	Yes	No	No
Decision support (sensitivity analysis)	Yes	No	No

HYPERPARAMETER TUNING FOR BAYESIAN NETWORKS

Hyperparameter tuning for Bayesian networks typically involves finding the best structure and parameter settings that optimize the model performance. Unlike other machine learning algorithms that often have hyperparameters related to the optimization process, Bayesian networks have hyperparameters associated with the network structure and the parameter estimation process. Below are some key hyperparameters to consider when tuning Bayesian networks.

- **Network structure configuration**: Arrangement of nodes (variables) and edges (directed connections) in the Bayesian network. Different configurations of the network structure can result in different causal relationships and dependencies between variables.
- **Nodes and edges**: Configurations involving adding or deleting nodes and edges can alter the network structure. Adding new nodes can incorporate additional variables, while deleting nodes can simplify the model.
- **Maximum number of parents per node**: Maximum number of parent nodes each node can have in the network. Setting this parameter can control the complexity of the network and prevent overfitting.
- **Prior probabilities for CPTs**: Bayesian networks require estimating CPTs for each node. Configuring prior probabilities can influence the learning process and incorporate prior beliefs or knowledge about the variable relationships.
- **Scoring metric**: The choice of the scoring metric, such as log-likelihood, BIC, or AIC, influences the evaluation of different network structures during learning.
- **Search algorithm**: The search algorithm determines how the network structure is explored during learning. Score-based and constraint-based algorithms are common search algorithms to learn the network structure from data.
- **Prior knowledge**: Incorporating prior knowledge or constraints, such as known causal relationships, can be specified as part of the configuration to guide the learning process.

Remember that Bayesian network tuning is a complex task, as the number of possible network structures can grow exponentially with the number of nodes. It is essential to strike a balance between model complexity and performance. Additionally, domain knowledge and understanding of the data are critical for making informed decisions during the hyperparameter tuning process.

Fortunately, SPSS Modeler has a built-in automatic search algorithm that makes configuring Bayesian networks easier, especially for non-programmers. In addition, it still provides users with options for changing certain configurations. We will practice with those configurations in the example.

EXAMPLE: BAYESIAN NETWORKS

In this example, we will build predictive models for the **Airline Recommendation** dataset using the Bayesian network algorithm with various configurations.

- **Dataset:** AirlineRecommendation.sav
- **Target variable:** Recommendation (Flag: Yes/No)

SAMPLE-EXPLORE-MODIFY

Refer to the previous chapters for more details on these steps.

BAYESIAN NETWORK MODELING

In this example, we train five different Bayesian network models. Since Bayesian networks work better with complete datasets, we use the data with imputation.

Follow the steps below to set up the Bayesian Network models.

1. **Create a new stream**:
   - Open SPSS Modeler and create a new stream named "Airline_BayesianNetworks".
2. **Copy the existing stream**:
   - Open the stream "Airline_Tree".
   - Copy the section of the stream from the **data source**, **Filler node**, to the **Partition node**.
   - Paste these nodes into your new "Airline_BayesianNetworks" stream.
3. **Add five Bayes net nodes**:
   - From the **Modeling** group, add **four Bayes Net Nodes** to the stream.
   - Connect them to the **Partition node**.
   - Label the five XGBoost Tree nodes as:
     1. TAN 1
     2. TAN 2
     3. Markov Blanket 1
     4. Markov Blanket 2
     5. Markov Blanket FS

At this point, your stream diagram should look like the following:

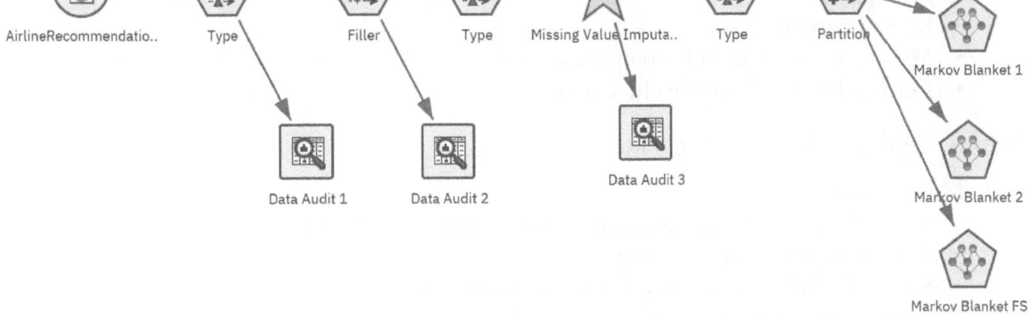

Then, use the details below to set the configurations for each Bayesian Network model. We focus on a few key configurations that are available under the Model options and Expert options in the Bayes Net node settings.

TAN 1 Model Configurations

- Model options
  - Select "Use partitioned data" and "Build model for each split."
  - Structure type: TAN
  - Parameter learning method: Maximum likelihood
- Expert options
  - Select "Expert"
  - Missing values: Use only complete records
  - Independent test: Likelihood ratio

TAN 2 Model Configurations

- Model options
  - Select "Use partitioned data" and "Build model for each split."
  - Structure type: TAN
  - Parameter learning method: Bayes adjustment for small cell counts
- Expert options
  - Select "Expert"
  - Missing values: Use only complete records
  - Independent test: Pearson chi-square

Markov Blanket 1 Model Configurations

- Model options
  - Select "Use partitioned data" and "Build model for each split."
  - Structure type: Markov Blanket
  - Parameter learning method: Maximum likelihood
- Expert options
  - Select "Expert"
  - Missing values: Use only complete records
  - Independent test: Likelihood ratio

Markov Blanket 2 Model Configurations

- Model options
  - Select "Use partitioned data" and "Build model for each split."
  - Structure type: Markov Blanket
  - Parameter learning method: Bayes adjustment for small cell counts
- Expert options
  - Select "Expert"
  - Missing values: Use only complete records
  - Independent test: Pearson chi-square

Markov Blanket FS Model Configurations

- Model options
  - Select "Use partitioned data" and "Build model for each split"
  - Structure type: Markov Blanket
  - Select "Include feature selection preprocessing step"
  - Parameter learning method: Maximum likelihood
- Expert options
  - Select "Expert"
  - Missing values: Use only complete records
  - Independent test: Likelihood ratio

To run each of the Bayes Net nodes separately and produce model nuggets, follow the same steps described in Chapter 7:

1. **Run each node separately**: After configuring the five Bayesian Network models (TAN 1, TAN 2, Markov Blanket 1, Markov Blanket 2, and Markov Blanket FS) with the specified settings, execute each of the nodes separately. This will generate the corresponding model nuggets for each.
2. **Connect model nuggets**: Since all models use the same input data with identical field names, you can use the approach described in Chapter 7 to connect all the model nuggets for comparison. Follow the same steps outlined in Chapter 7 under the section "How to run the analysis node with multiple models?"
3. **Add filter node**: Add a Filter node to the stream and connect it to the model nugget for Markov Blanket FS. Rename the model nuggets to the selected names (TAN 1, TAN 2, MB 1, etc.) (see the detailed instructions in Step 2 of the next process).
4. **Add analysis node**: After renaming the model nuggets, use an Analysis node to compare the performance of the models. The Analysis node will compare the models based on key performance metrics.
5. **Add evaluation graph node**: Use an Evaluation Graph node to produce ROC charts and Lift charts for all the models. This will provide a graphical representation of model performance.
6. **Run the diagram**: Run the entire diagram to generate the comparison results for each Bayesian Network model.

By following these steps, you'll have a diagram as below, including model comparison through the Analysis node and graphical evaluation with the ROC and Lift charts.

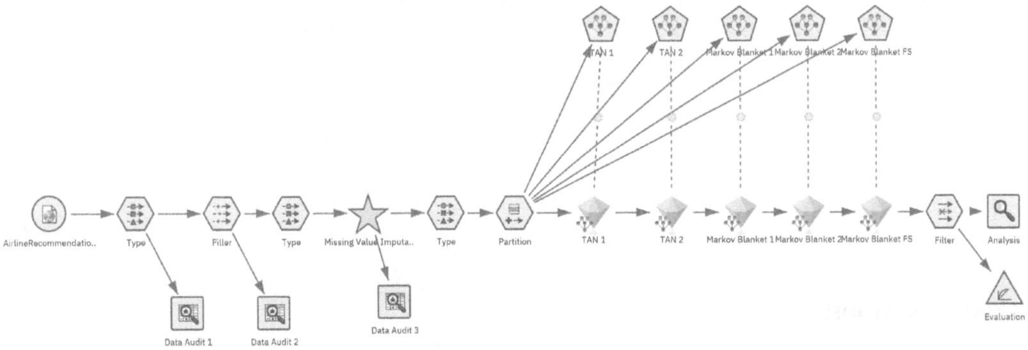

To configure the Filter node settings to change the model names for each of the Bayesian Network models, follow these steps:

1. **Open the Filter node**: After connecting the model nuggets (TAN 1, TAN 2, Markov Blanket 1, Markov Blanket 2, and Markov Blanket FS) to the Filter node, open the settings of the Filter node.
2. **Rename models**: Select the field to rename each model as follows:
   - Rename the field **$B-Recommend the airline** to **TAN 1**.
   - Rename the field **$B1-Recommend the airline** to **TAN 2**.
   - Rename the field **$B2-Recommend the airline** to **MB 1**.
   - Rename the field **$B3-Recommend the airline** to **MB 2**.
   - Rename the field **$B4-Recommend the airline** to **MB FS**.
3. **Apply the changes**: After setting the model names, click Apply to save the changes. The models should now appear with their new names in the stream.

This process ensures that each Bayesian Network model is clearly labeled in the final analysis, making it easier to differentiate between the results when comparing performance.

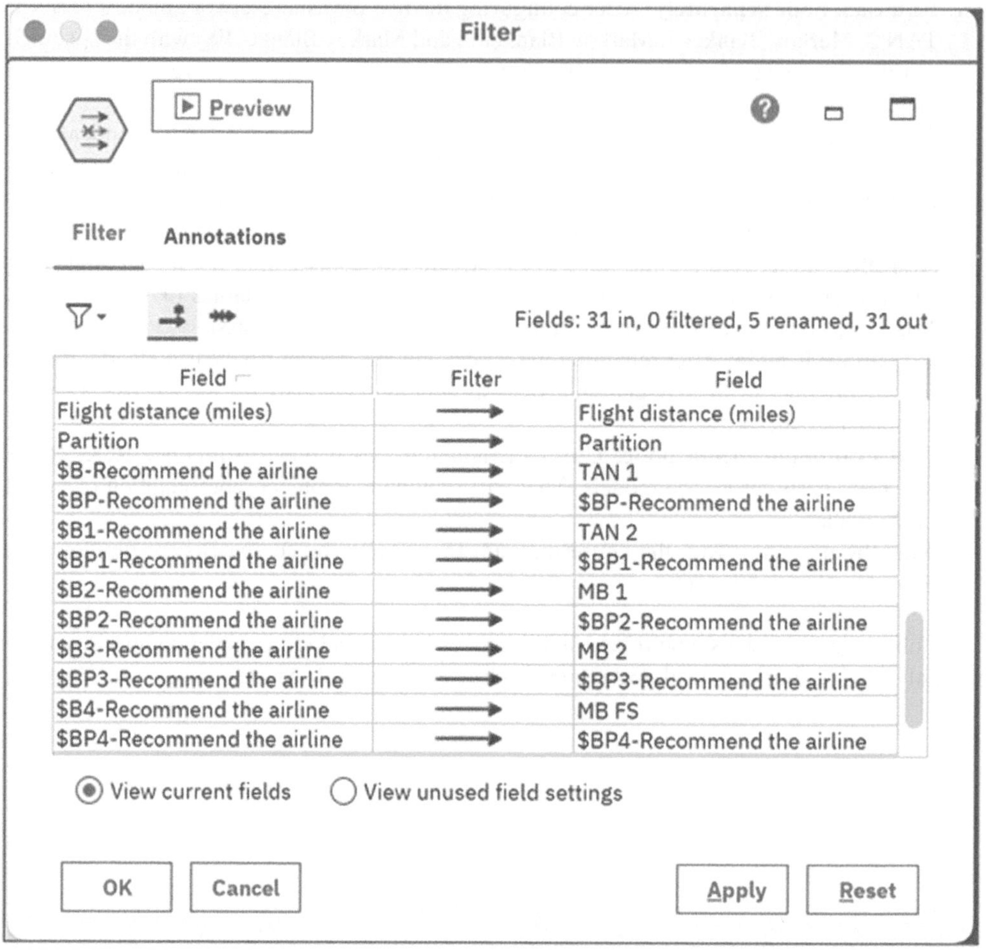

Champion Model

The model comparison results are shown in Table 16.6, using model fit metrics for the validation samples. Additionally, Figures 16.6 and 16.7 present the ROC comparison chart and Lift comparison chart, respectively.

- TAN 1 and TAN 2 models exhibit identical fit metrics.
- Overall, all models perform similarly, except Markov Blanket 2, which has the weakest performance.
- Markov Blanket FS outperforms the other models:
  - It has the highest ACC and F1 score.
  - It also achieves the highest Gini coefficient and ROC index.

The ROC comparison chart confirms this, showing Markov Blanket FS with the ROC curve closest to the top-left corner, although the differences between the curves are minimal. The Lift comparison chart further highlights Markov Blanket FS as having the best Lift curve.

Based on these results, we can conclude that Markov Blanket FS is the champion model.

TABLE 16.6
Model Comparison Results

Metrics	TAN 1	TAN 2	Markov Blanket 1	Markov Blanket 2	Markov Blanket FS
ACC	76.36%	76.36%	76.29%	76.06%	76.53%
MR	23.64%	23.64%	23.71%	23.94%	23.47%
Sensitivity or TPR	0.873	0.873	0.878	0.878	0.895
Specificity or TNR	0.641	0.641	0.634	0.630	0.620
F1	0.796	0.796	0.796	0.795	0.801
Gini	0.659	0.660	0.637	0.648	0.667
AUC (ROC Index)	0.829	0.830	0.818	0.82%	0.834

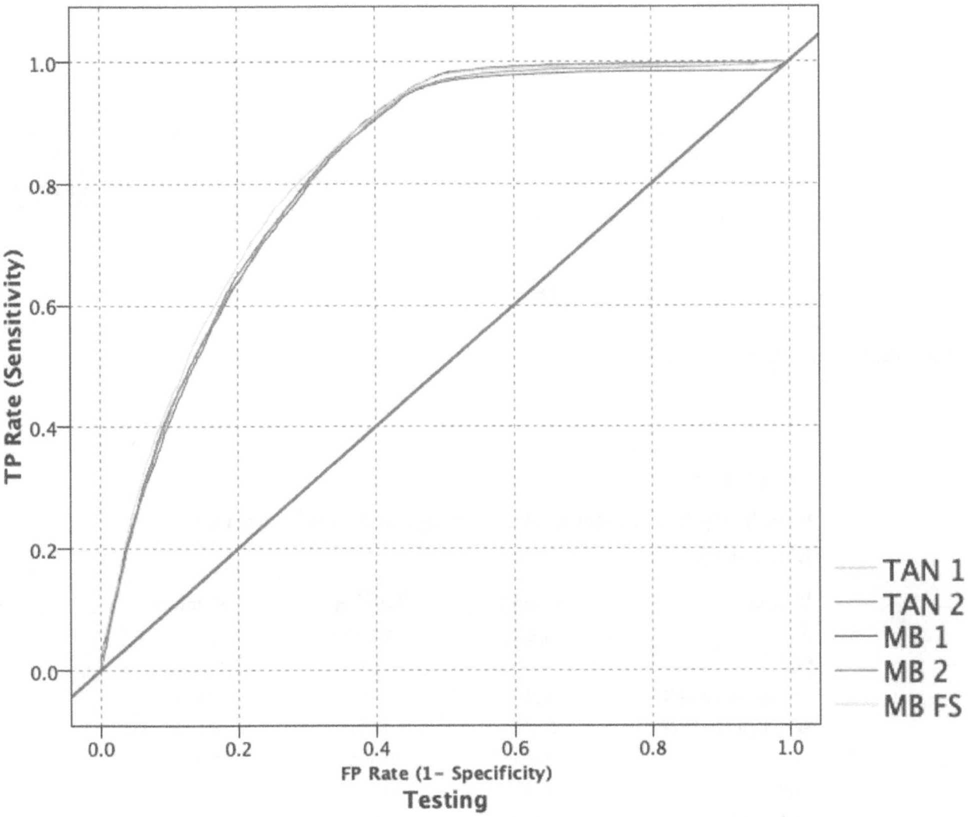

FIGURE 16.6 ROC comparison chart.

Reliability of the Champion Model

Table 16.7 compares the fit metrics across the three samples: testing, training, and validation. The results demonstrate consistent values, with differences lower than 4% between samples. This consistency suggests that the models are reliable. Given the minimal variation across the samples, we can conclude that the model reliability is satisfactory, and there is no evidence of overfitting.

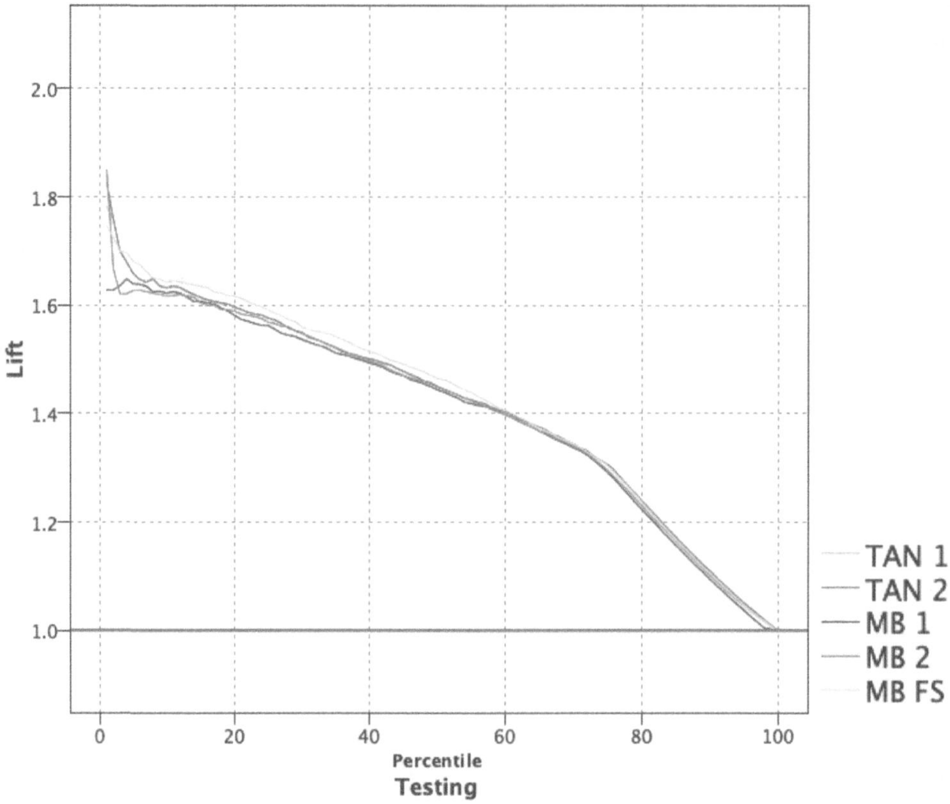

FIGURE 16.7 Lift comparison chart.

TABLE 16.7
Reliability Evaluation of the Champion Model—Markov Blanket FS

Metrics	Testing	Training	Validation
ACC	76.53%	77.22%	77.24%
MR	23.47%	22.78%	22.76%
Sensitivity or TPR	0.895	0.901	0.899
Specificity or TNR	0.620	0.631	0.632
F1	0.801	0.805	0.806
Gini	0.667	0.696	0.675
AUC (ROC Index)	0.834	0.848	0.837

Validity of the Champion Model

To evaluate the validity of the champion model, we present the model fit metrics for the validation sample, as shown in Table 16.8. The observed prediction ACC is just below 80%, which is generally considered marginal. However, the sensitivity (true positive prediction) performs well at 0.899, meaning the model identifies the majority of positive cases correctly.

TABLE 16.8
Validity Evaluation of the Champion Model—Markov Blanket FS

Metrics	Validation	Evaluation
ACC	77.24%	Not good
MR	22.76%	Not good
Sensitivity or TPR	0.899	Good
Specificity or TNR	0.632	Not good
F1	0.806	Good
Gini	0.675	Good
AUC (ROC Index)	0.837	Acceptable

The lower ACC can be attributed to the poor specificity (true negative prediction) of 0.632, indicating a higher false positive rate. The F1 score, which balances sensitivity and specificity, remains acceptable at 0.806. The Gini coefficient shows a good fit at 0.675, and the ROC index of 0.837 also indicates an acceptable fit.

Examining the Lift chart, we find that with 20% of the dataset, the Lift value is 1.63, meaning the model can capture 1.63 times more positive cases (32.6%) compared to random guessing.

Overall, based on these metrics, we conclude that the model demonstrates marginally acceptable predictive power.

Model Interpretation

The Results window for this node provides additional insights into the model, as illustrated by the Markov Blanket network structure in Figure 16.8. The figure reveals that we have a **Markov Blanket** network.

The Markov Blanket network in Figure 16.8 indicates ten predictors used in the model. A summary of the relationships and directions among these variables is as follows:

1. **Recommendation** is a child node of **airline status, price sensitivity, travel purpose, average number of flights per year, passenger class,** and **flight cancellation.** It is also a parent node to **age** and **gender.**
2. **Spending for eating and drinking at the airport** is the parent node of **gender.**
3. **Number of loyalty cards** is the parent node of **travel purpose** and **age.**
4. **Airline status** is the parent node of **recommendation, age, travel purpose, price sensitivity, average number of flights per year,** and **gender.**
5. **Passenger class** is the parent node of **recommendation, price sensitivity, average number of flights per year,** and **gender.**
6. **Flight cancellation** is the parent node of **recommendation** and **travel purpose.**
7. **Average number of flights per year** is the parent node of **recommendation** and **gender,** but it is the child node of **airline status, price sensitivity, travel purpose,** and **passenger class.**
8. **Travel purpose** is the parent node of **recommendation, age, gender,** and **average number of flights per year,** but it is the child node of **airline status, price sensitivity, number of loyalty cards,** and **flight cancellation.**
9. **Age** is the child node of **recommendation, airline status, price sensitivity, travel purpose,** and **number of loyalty cards.**

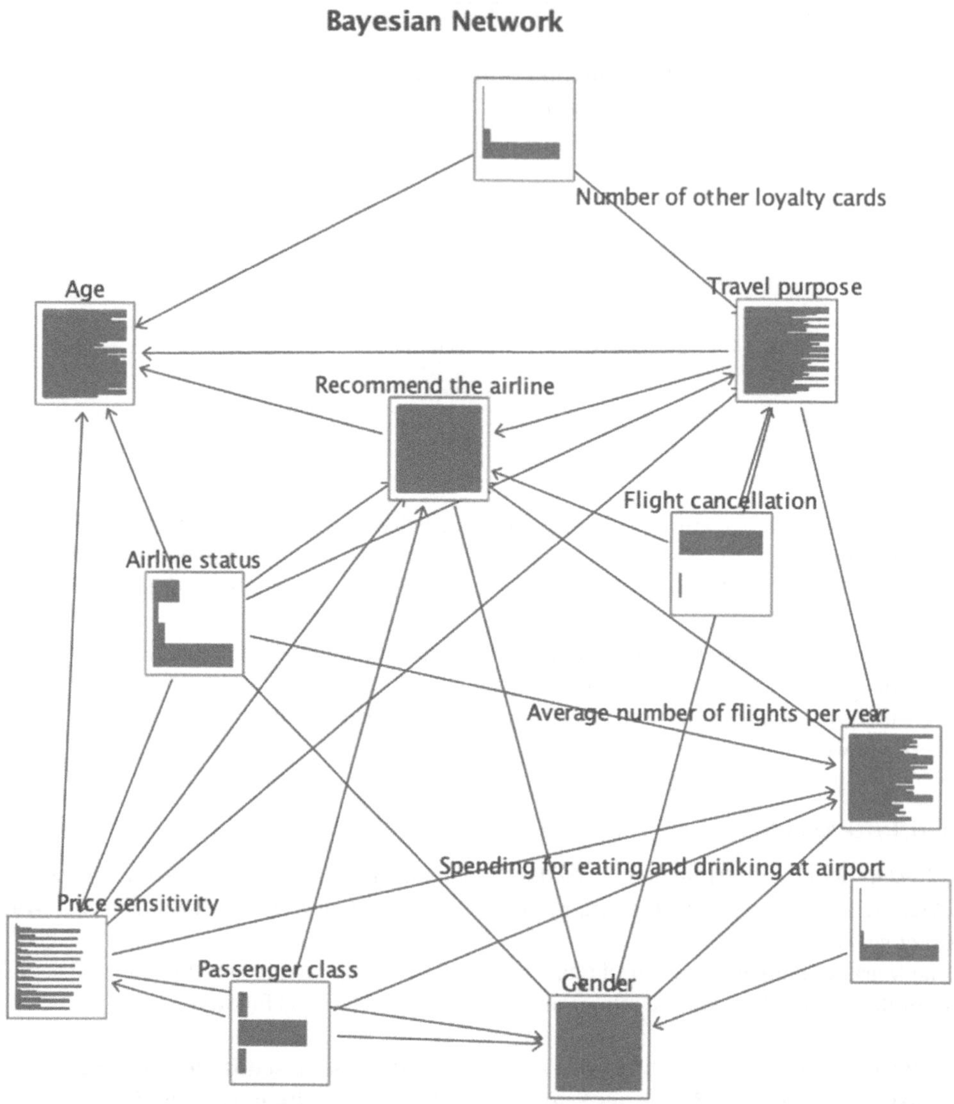

FIGURE 16.8 Markov Blanket network structure.

10. **Gender** is the child node of **recommendation**, **airline status**, **price sensitivity**, **average number of flights per year**, **spending for eating and drinking at the airport**, and **travel purpose**.
11. **Price sensitivity** is the parent node of **recommendation**, **average number of flights per year**, **travel purpose**, **gender**, and **age**, but it is the child node of **airline status** and **passenger class**.

As mentioned earlier, the effects of predictors on the target variable in Bayesian networks are not interpreted in the same way as in other machine learning methods. The key output is the CPT for each node. To access the CPTs in SPSS Modeler, navigate to the nugget window, select a node from the left panel, and choose Conditional Probabilities on the right panel. Prior and posterior probabilities for specific nodes are summarized below. Due to the length of the tables for nodes with parent nodes (such as recommendation, price sensitivity, travel purpose, average number of flights per year, age, and gender), only some conditions are demonstrated for clarity. The complete CPTs for these variables are displayed in Tables 16.9–16.19.

The CPT serves to encode conditional dependencies and probabilities between variables in the Bayesian network, enabling probabilistic reasoning, inference, and predictions. Each table provides the probabilities of a node assuming certain values, given the states of its parent nodes.

TABLE 16.9
Conditional Probabilities of Recommend the Airline

Parents						Probability	
Airline Status	Price Sensitivity	Average Number of Flights per Year	Travel Purpose	Passenger Class	Flight Cancellation	Yes	No
Blue	≤ 1	≤ 20	Business travel	Business	Yes	0.50	0.50
Blue	≤ 1	≤ 20	Business travel	Business	No	0.73	0.27
Blue	≤ 1	≤ 20	Business travel	Eco	Yes	0.41	0.59
Blue	≤ 1	≤ 20	Business travel	Eco	No	0.68	0.32
Blue	≤ 1	≤ 20	Business travel	Eco Plus	Yes	0.52	0.48
Blue	≤ 1	≤ 20	Business travel	Eco Plus	No	0.67	0.33
Blue	≤ 1	≤ 20	Mileage tickets	Business	Yes	1.00	0.00
Blue	≤ 1	≤ 20	Mileage tickets	Business	No	0.66	0.34
Blue	≤ 1	≤ 20	Mileage tickets	Eco	Yes	0.80	0.20
Blue	≤ 1	≤ 20	Mileage tickets	Eco	No	0.57	0.43
...							

TABLE 16.10
Conditional Probabilities of Passenger Class

Probability		
Business	Eco	Eco Plus
0.08	0.82	0.10

TABLE 16.11
Conditional Probabilities of Number of Other Loyalty Cards

Probability				
≤ 2.4	2.4 ~ 4.8	4.8 ~ 7.2	7.2 ~ 9.6	> 9.6
0.91	0.08	0.01	0.00	0.00

TABLE 16.12
Conditional Probabilities of Airline Status

Probability			
Blue	Gold	Platinum	Silver
0.67	0.09	0.03	0.21

TABLE 16.13
Conditional Probabilities of Spending for Eating and Drinking at Airport

Probability				
≤179	179 ~ 358	358 ~ 537	537 ~ 716	> 716
0.96	0.04	0.00	0.00	0.00

TABLE 16.14
Conditional Probabilities of Flight Cancellation

Probability	
Yes	No
0.02	0.98

TABLE 16.15
Conditional Probabilities of Price Sensitivity

Parents		Probability				
Airline Status	Passenger Class	≤ 1	1 ~ 2	2 ~ 3	3 ~ 4	> 4
Blue	Business	0.66	0.29	0.05	0.00	0.00
Blue	Eco	0.65	0.30	0.05	0.00	0.00
Blue	Eco Plus	0.70	0.24	0.06	0.00	0.00
Gold	Business	0.79	0.19	0.02	0.00	0.00
Gold	Eco	0.77	0.21	0.02	0.00	0.00
Gold	Eco Plus	0.80	0.17	0.03	0.00	0.00
Platinum	Business	0.72	0.23	0.06	0.00	0.00
Platinum	Eco	0.78	0.19	0.03	0.00	0.00
Platinum	Eco Plus	0.81	0.13	0.06	0.00	0.00
Silver	Business	0.74	0.23	0.03	0.00	0.00
Silver	Eco	0.74	0.22	0.03	0.00	0.00
Silver	Eco Plus	0.78	0.18	0.04	0.00	0.00

For example, to interpret the target variable recommendation, we focus on its parent nodes: airline status and travel purpose. From the CPT, we can interpret various scenarios:

- If airline status = BLUE, price sensitivity ≤ 1, average number of flights per year ≤ 20, travel purpose = Business travel, passenger class = Business, and flight cancellation = Yes, then the probability of recommendation = YES is 0.5.
- If airline status = BLUE, price sensitivity ≤ 1, average number of flights per year ≤ 20, travel purpose = Business travel, passenger class = Business, and flight cancellation = No, then the probability of recommendation = YES is 0.73.

TABLE 16.16
Conditional Probabilities of Travel Purpose

Parents				Probability		
Airline Status	**Price Sensitivity**	**Number of Other Loyalty Cards**	**Flight Cancellation**	**Business Travel**	**Mileage Tickets**	**Personal Travel**
Blue	≤ 1	≤ 2.4	Yes	0.43	0.09	0.47
Blue	≤ 1	≤ 2.4	No	0.60	0.07	0.32
Blue	≤ 1	2.4 ~ 4.8	Yes	0.50	0.13	0.37
Blue	≤ 1	2.4 ~ 4.8	No	0.67	0.09	0.24
Blue	≤ 1	4.8 ~ 7.2	Yes	0.75	0.00	0.25
Blue	≤ 1	4.8 ~ 7.2	No	0.60	0.11	0.30
Blue	≤ 1	7.2 ~ 9.6	No	0.33	0.11	0.56
Blue	≤ 1	> 9.6	No	1.00	0.00	0.00
Blue	1 ~ 2	≤ 2.4	Yes	0.34	0.12	0.54
Blue	1 ~ 2	≤ 2.4	No	0.51	0.10	0.39
Blue	1 ~ 2	2.4 ~ 4.8	Yes	0.38	0.19	0.42
Blue	1 ~ 2	2.4 ~ 4.8	No	0.58	0.11	0.32
Blue	1 ~ 2	4.8 ~ 7.2	Yes	0.00	0.00	1.00
...						

TABLE 16.17
Conditional Probabilities of Average Number of Flights per Year

Parents				Probability				
Airline Status	**Price Sensitivity**	**Travel Purpose**	**Passenger Class**	**≤ 20**	**20 ~ 40**	**40 ~ 60**	**60 ~ 80**	**> 80**
Blue	≤ 1	Business travel	Business	0.66	0.26	0.08	0.01	0.00
Blue	≤ 1	Business travel	Eco	0.67	0.26	0.06	0.01	0.00
Blue	≤ 1	Business travel	Eco Plus	0.75	0.23	0.03	0.00	0.00
Blue	≤ 1	Mileage tickets	Business	0.52	0.33	0.15	0.00	0.00
Blue	≤ 1	Mileage tickets	Eco	0.46	0.36	0.17	0.02	0.00
Blue	≤ 1	Mileage tickets	Eco Plus	0.57	0.36	0.07	0.00	0.00
Blue	≤ 1	Personal Travel	Business	0.34	0.40	0.22	0.04	0.00
Blue	≤ 1	Personal Travel	Eco	0.32	0.41	0.23	0.04	0.00
Blue	≤ 1	Personal Travel	Eco Plus	0.41	0.48	0.10	0.00	0.00
Blue	1 ~ 2	Business travel	Business	0.54	0.38	0.08	0.00	0.00
Blue	1 ~ 2	Business travel	Eco	0.54	0.38	0.08	0.00	0.00
...								

- If airline status = BLUE, price sensitivity = 1–2, average number of flights per year ≤ 20, travel purpose = Business travel, passenger class = Eco Plus, and flight cancellation = Yes, then the probability of recommendation = YES is 0.57.
- If airline status = BLUE, price sensitivity = 1–2, average number of flights per year = 40–60, travel purpose = Business travel, passenger class = Eco, and flight cancellation = No, then the probability of recommendation = YES is 0.61.
- And so on.

TABLE 16.18
Conditional Probabilities of Age

Parents					Probability				
Recommend the Airline	Airline Status	Price Sensitivity	Travel Purpose	Number of Other Loyalty Cards	≤ 31.4	31.4 ~ 44.8	44.8 ~ 58.2	58.2 ~ 71.6	> 71.6
Yes	Blue	≤ 1	Business travel	≤ 2.4	0.18	0.32	0.37	0.11	0.01
Yes	Blue	≤ 1	Business travel	2.4 ~ 4.8	0.22	0.64	0.13	0.01	0.00
Yes	Blue	≤ 1	Business travel	4.8 ~ 7.2	0.24	0.61	0.14	0.01	0.00
Yes	Blue	≤ 1	Mileage tickets	≤ 2.4	0.23	0.23	0.31	0.16	0.06
Yes	Blue	≤ 1	Mileage tickets	2.4 ~ 4.8	0.40	0.47	0.13	0.01	0.00
Yes	Blue	≤ 1	Mileage tickets	4.8 ~ 7.2	0.45	0.45	0.09	0.00	0.00
Yes	Blue	≤ 1	Mileage tickets	7.2 ~ 9.6	0.00	1.00	0.00	0.00	0.00
Yes	Blue	≤ 1	Personal Travel	≤ 2.4	0.20	0.34	0.32	0.12	0.01
Yes	Blue	≤ 1	Personal Travel	2.4 ~ 4.8	0.30	0.61	0.09	0.00	0.00
Yes	Blue	≤ 1	Personal Travel	4.8 ~ 7.2	0.00	0.67	0.33	0.00	0.00
Yes	Blue	1 ~ 2	Business travel	≤ 2.4	0.28	0.32	0.31	0.08	0.01
…									

TABLE 16.19
Conditional Probabilities of Gender

Parents							Probability	
Recommend the Airline	Airline Status	Price Sensitivity	Average Number of Flights per Year	Travel Purpose	Spending for Eating and Drinking at Airport	Passenger Class	Male	Female
Yes	Blue	≤ 1	≤ 20	Business travel	≤ 179	Business	0.53	0.47
Yes	Blue	≤ 1	≤ 20	Business travel	≤ 179	Eco	0.53	0.47
Yes	Blue	≤ 1	≤ 20	Business travel	≤ 179	Eco Plus	0.25	0.75
Yes	Blue	≤ 1	≤ 20	Business travel	179 ~ 358	Business	0.46	0.54
Yes	Blue	≤ 1	≤ 20	Business travel	179 ~ 358	Eco	0.61	0.39
Yes	Blue	≤ 1	≤ 20	Business travel	179 ~ 358	Eco Plus	0.30	0.70
Yes	Blue	≤ 1	≤ 20	Business travel	358 ~ 537	Business	1.00	0.00
Yes	Blue	≤ 1	≤ 20	Business travel	358 ~ 537	Eco	0.46	0.54
Yes	Blue	≤ 1	≤ 20	Business travel	358 ~ 537	Eco Plus	0.00	1.00
Yes	Blue	≤ 1	≤ 20	Business travel	537 ~ 716	Eco	0.67	0.33
…								

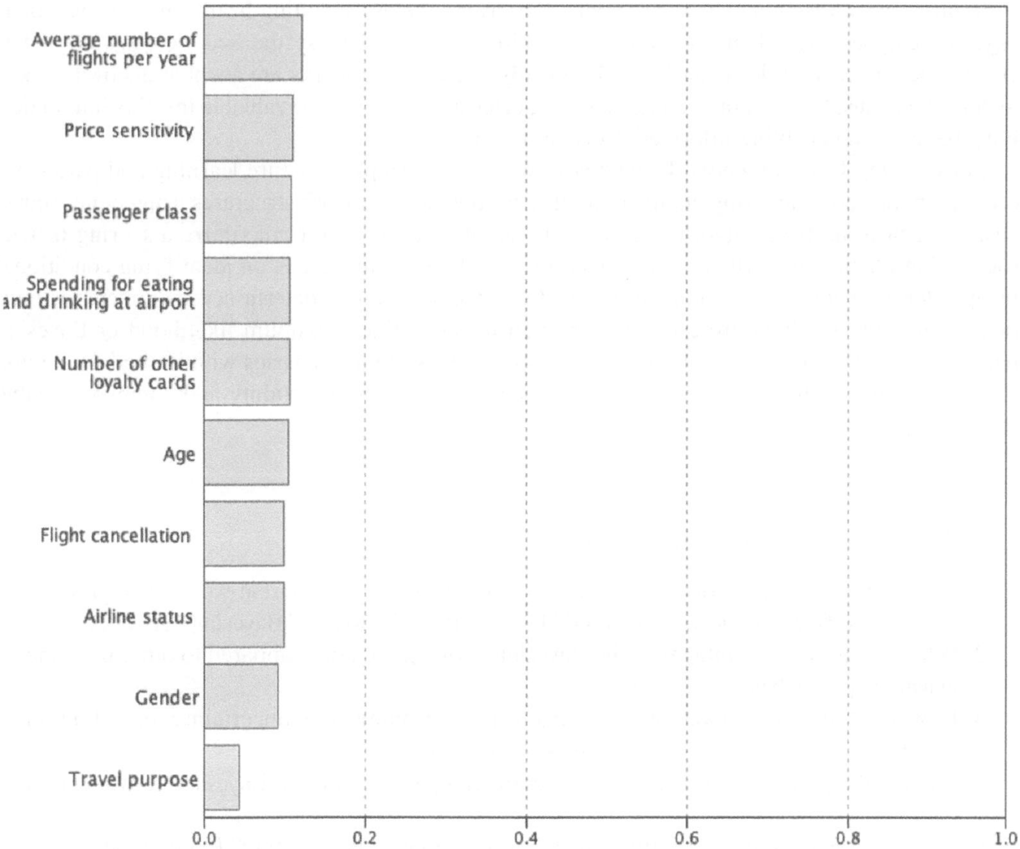

FIGURE 16.9 Variable importance plot.

We follow this same method to interpret the probabilistic relationships for other variables in the Bayesian network. Each CPT provides the necessary information to understand how the various predictors influence the target variable and how different factors interact within the model.

Variable Importance

Figure 16.9 presents the variable importance plot, highlighting the relative significance of each predictor. The results reveal that the average number of flights per year is the most important predictor, followed by price sensitivity, passenger class, spending for eating and drinking at the airport, and number of other loyalty cards. In contrast, travel purpose is shown to be the least important predictor.

CHAPTER SUMMARY

The Bayesian network method is a powerful algorithm within data mining, offering a rigorous approach for modeling and analyzing complex systems characterized by uncertain relationships between variables. This method uses probabilistic graphical models, represented by DAGs, to depict the dependencies among variables. Each node in a Bayesian network corresponds to a random variable, while the directed edges between nodes represent the causal relationships or dependencies between these variables. Leveraging Bayes' theorem, Bayesian networks enable probabilistic inference, allowing for the computation of posterior probabilities of variables, given observed data. This ability to handle uncertainty and provide interpretable causal relationships makes Bayesian networks valuable in various data mining applications.

In practice, Bayesian networks are applied in diverse fields, including healthcare, finance, biology, and engineering. Their robustness in handling large and noisy datasets demonstrates their effectiveness in real-world scenarios. Additionally, Bayesian networks are adept at uncovering key features and causal factors in complex systems, offering researchers invaluable insights into underlying patterns and enabling informed decision-making.

Constructing Bayesian networks involves two essential steps: structure learning and parameter estimation. Structure learning entails searching for potential network structures using either score-based or constraint-based algorithms. Score-based algorithms aim to maximize a scoring metric, such as log-likelihood or BIC, while constraint-based algorithms focus on identifying conditional independence relationships among variables. Once the structure is determined, parameter estimation follows, where CPTs are estimated through methods like maximum likelihood or Bayesian parameter estimation. Bayesian networks are especially useful in scenarios with limited data, making them an invaluable tool in data mining when dealing with uncertainty and complex variable relationships.

CHAPTER DISCUSSION QUESTIONS

1. Differentiate between correlation and causation. Provide specific real-world examples.
2. What is the Bayesian network method? Explain the principles of Bayesian networks.
3. What are the key advantages of the Bayesian network method compared to other machine learning approaches?
4. How does the incorporation of probabilistic reasoning and uncertainty modeling in Bayesian networks benefit decision-making and prediction tasks?
5. Explain the process of learning the structure and parameters of a Bayesian network from data.
6. What are the different algorithms for learning the network structure? Explain their similarities and differences.
7. Discuss the challenges or limitations of the Bayesian network method.
8. What are the four primary types of Bayesian network structures? Explain their similarities and differences.
9. What is frequentist statistics? Compare the Bayesian network method with the frequentist method.
10. What is causal inference? Can we infer causality using Bayesian networks? Why or why not?
11. Define causal data mining. Provide practical examples of causal data mining projects.
12. Describe the USELEI process. Compare it with SEMMA and CRISP-DM. When should we use the USELEI?

CHAPTER EXERCISES

Exercise 1: Bayesian Network Models for the Bank dataset

1. Open the **bank** dataset in SPSS Modeler.
2. Sample the entire dataset. See the variable names and descriptions in Appendix A.
3. Partition the data using a split ratio of 60–20–20 (train: 60%, test: 20%, and validation: 20%).
4. Perform data modification as follows:
   - Impute missing values.
   - Normalize the data.
5. Set **CRSCORE** (credit scores) as the target variable.

6. Configure five Bayesian Network models for this dataset as follows.
   TAN 1 Model Configurations
   - Model options
     - Select "Use partitioned data" and "Build model for each split".
     - Structure type: TAN.
     - Parameter learning method: Maximum likelihood.
   - Expert options
     - Select "Expert".
     - Missing values: Use only complete records.
     - Independent test: Likelihood ratio.

   TAN 2 Model Configurations
   - Model options
     - Select "Use partitioned data" and "Build model for each split".
     - Structure type: TAN.
     - Parameter learning method: Bayes adjustment for small cell counts.
   - Expert options
     - Select "Expert".
     - Missing values: Use only complete records.
     - Independent test: Pearson chi-square.

   Markov Blanket 1 Model Configurations
   - Model options
     - Select "Use partitioned data" and "Build model for each split".
     - Structure type: Markov Blanket.
     - Parameter learning method: Maximum likelihood.
   - Expert options
     - Select "Expert".
     - Missing values: Use only complete records.
     - Independent test: Likelihood ratio.

   Markov Blanket 2 Model Configurations
   - Model options
     - Select "Use partitioned data" and "Build model for each split".
     - Structure type: Markov Blanket.
     - Parameter learning method: Bayes adjustment for small cell counts.
   - Expert options
     - Select "Expert".
     - Missing values: Use only complete records.
     - Independent test: Pearson chi-square.

   Markov Blanket FS Model Configurations
   - Model options
     - Select "Use partitioned data" and "Build model for each split".
     - Structure type: Markov Blanket.
     - Select "Include feature selection preprocessing step".
     - Parameter learning method: Maximum likelihood.
   - Expert options
     - Select "Expert".
     - Missing values: Use only complete records.
     - Independent test: Likelihood ratio.
7. Compare the model fit metrics between these four models using the testing results.
8. Select the champion model and justify this selection with the necessary evidence.
9. Present the results of the champion model.
10. Evaluate the reliability and validity of the champion model.
11. Present the variable importance chart and interpret the results.

Exercise 2: Bayesian Network Models for the Passengers dataset

1. Open the **passengers** dataset in SPSS Modeler.
2. Sample 20% of the full data. Set **return** (decision to return to the airline) as the target variable and change its role to binary. See the variable names and descriptions in Appendix B.
3. Partition the data using a split ratio of 60–20–20 (train: 60%, test: 20%, and validation: 20%).
4. Perform data modification as follows:
   a. Replace zero values for Likert-scale variables with missing values.
   b. Impute missing values.
5. Configure five Bayesian Network models for this dataset as follows.

   TAN 1 Model Configurations
   - Model options
     - Select "Use partitioned data" and "Build model for each split".
     - Structure type: TAN.
     - Parameter learning method: Maximum likelihood.
   - Expert options
     - Select "Expert".
     - Missing values: Use only complete records.
     - Independent test: Likelihood ratio.

   TAN 2 Model Configurations
   - Model options
     - Select "Use partitioned data" and "Build model for each split".
     - Structure type: TAN.
     - Parameter learning method: Bayes adjustment for small cell counts.
   - Expert options
     - Select "Expert".
     - Missing values: Use only complete records.
     - Independent test: Pearson chi-square.

   Markov Blanket 1 Model Configurations
   - Model options
     - Select "Use partitioned data" and "Build model for each split".
     - Structure type: Markov Blanket.
     - Parameter learning method: Maximum likelihood.
   - Expert options
     - Select "Expert".
     - Missing values: Use only complete records.
     - Independent test: Likelihood ratio.

   Markov Blanket 2 Model Configurations
   - Model options
     - Select "Use partitioned data" and "Build model for each split".
     - Structure type: Markov Blanket.
     - Parameter learning method: Bayes adjustment for small cell counts.
   - Expert options
     - Select "Expert".
     - Missing values: Use only complete records.
     - Independent test: Pearson chi-square.

   Markov Blanket FS Model Configurations
   - Model options
     - Select "Use partitioned data" and "Build model for each split".
     - Structure type: Markov Blanket.
     - Select "Include feature selection preprocessing step".
     - Parameter learning method: Maximum likelihood.

- Expert options
  - Select "Expert".
  - Missing values: Use only complete records.
  - Independent test: Likelihood ratio.

6. Compare the model fit metrics between these four models using the testing results.
7. Select the champion model and justify this selection with the necessary evidence.
8. Present the results of the champion model.
9. Evaluate the reliability and validity of the champion model.
10. Present the variable importance chart and interpret the results.

Appendix A

Bank Customers Data—Variable Descriptions

Variable Name	Label/Description	Scale	Role
ACCTAGE	Age of Oldest Account	Continuous	Input
AGE	Age	Continuous	Input
ATM	ATM	Binary (Flag)	Input
ATMAMT	ATM Withdrawal Amount	Continuous	Input
BRANCH	Branch of Bank	Nominal	Input
CASHBK	Number Cash Back	Continuous	Input
CC	Credit Card	Binary (Flag)	Input
CCBAL	Credit Card Balance	Continuous	Input
CCPURC	Credit Card Purchases	Continuous	Input
CD	Certificate of Deposit	Binary (Flag)	Input
CDBAL	CD Balance	Continuous	Input
CHECKS	Number of Checks	Continuous	Input
DDA	Checking Account	Binary (Flag)	Input
DDABAL	Checking Balance	Continuous	Input
DEP	Checking Deposits	Continuous	Input
DEPAMT	Amount Deposited	Continuous	Input
DIRDEP	Direct Deposit	Binary (Flag)	Input
HMOWN	Owns Home	Binary (Flag)	Input
HMVAL	Home Value	Continuous	Input
ILS	Installment Loan	Binary (Flag)	Input
ILSBAL	Loan Balance	Continuous	Input
INAREA	Local Address	Binary (Flag)	Input
INCOME	Income	Continuous	Input
INV	Investment	Binary (Flag)	Input
INVBAL	Investment Balance	Continuous	Input
IRA	Retirement Account	Binary (Flag)	Input
IRABAL	IRA Balance	Continuous	Input
LOC	Line of Credit	Binary (Flag)	Input
LOCBAL	Line of Credit Balance	Continuous	Input
LORES	Length of Residence	Continuous	Input
MM	Money Market	Binary (Flag)	Input
MMBAL	Money Market Balance	Continuous	Input
MMCRED	Money Market Credits	Continuous	Input
MOVED	Recent Address Change	Binary (Flag)	Input
MTG	Mortgage	Binary (Flag)	Input
MTGBAL	Mortgage Balance	Continuous	Input
NSF	Number Insufficient Fund	Binary (Flag)	Input
NSFAMT	Amount NSF	Continuous	Input
PHONE	Number Telephone Banking	Continuous	Input
POS	Number Point of Sale	Continuous	Input
POSAMT	Amount Point of Sale	Continuous	Input
RES	Area Classification	Nominal	Input
SAV	Saving Account	Binary (Flag)	Input
SAVBAL	Saving Balance	Continuous	Input
SDB	Safety Deposit Box	Binary (Flag)	Input
TELLER	Teller Visits	Continuous	Input
CRSCORE	Credit Score	Continuous	Target

Appendix B

Airline Passengers Data—Variable Description

Variable Name	Description	Scale	Role
Gender	Gender	Binary (Flag)	Input
Customer_Type	Customer type	Nominal	Input
Age	Age	Continuous	Input
Type_of_Travel	Type of travel	Nominal	Input
Class	Ticket class	Nominal	Input
Flight_Distance	Flight distance	Continuous	Input
Inflight_wifi_service[a]	Passenger evaluation of inflight Wi-Fi service	Continuous	Input
Departure_Arrival_time_convenience[a]	Passenger evaluation of departure/arrival time convenience	Continuous	Input
Ease_of_Online_booking[a]	Passenger evaluation of ease of online booking	Continuous	Input
Gate_location[a]	Passenger evaluation of gate location	Continuous	Input
Food_and_drink[a]	Passenger evaluation of food and drink	Continuous	Input
Online_boarding[a]	Passenger evaluation of online boarding	Continuous	Input
Seat_comfort[a]	Passenger evaluation of seat comfort	Continuous	Input
Inflight_entertainment[a]	Passenger evaluation of inflight entertainment	Continuous	Input
On_board_service[a]	Passenger evaluation of on-board service	Continuous	Input
Leg_room_service[a]	Passenger evaluation of legroom service	Continuous	Input
Baggage_handling[a]	Passenger evaluation of baggage handling	Continuous	Input
Checkin_service[a]	Passenger evaluation of check-in service	Continuous	Input
Inflight_service[a]	Passenger evaluation of inflight service	Continuous	Input
Cleanliness[a]	Passenger evaluation of cleanliness	Continuous	Input
Departure_Delay_in_Minutes	Departure delay in minutes	Continuous	Input
Arrival_Delay_in_Minutes	Arrival delay in minutes	Continuous	Input
Return	Passenger decision in returning to the same airline in the next flight (0—No; 1—Yes)	Binary (Flag)	Target

[a] 5-point Likert scale, treated as continuous

Appendix C
Boston Housing Data—Variable Descriptions

Variable Name	Description	Scale	Role
CrimePC	Crime rate per capita	Continuous	Input
ZonePr	Proportion of residential land zoned for lots over 25,000 sq.ft.	Continuous	Input
IndustPr	Proportion of non-retail business acres.	Continuous	Input
Charles_River	Charles River dummy variable (1 if tract bounds river; 0 otherwise)	Continuous	Input
NiOxSQ	Nitric oxides concentration (parts per 10 million)	Continuous	Input
RmSQ	Average number of rooms per dwelling	Continuous	Input
AgePre40	Proportion of owner occupied units built prior to 1940	Continuous	Input
LogDisstance	Weighted distances to five Boston employment centers	Continuous	Input
LogAccess	Index of accessibility to radial highways	Continuous	Input
Tax	Full value property tax rate per $10,000	Continuous	Input
PtRatio	Pupil-teacher ratio	Continuous	Input
B	1,000(Bk – 0.63)^2 where Bk is the proportion of blacks by town	Continuous	Input
Lstat	Percentage of lower status of the population	Continuous	Input
Imv	Median value of owner-occupied homes in $1,000s	Continuous	Target

Term Definitions

Terms	Definitions
Activation function	An activation function is a function that determines the output of a neuron based on the weighted sum of its inputs. Common activation functions include sigmoid, ReLU (Rectified Linear Unit), and tanh (hyperbolic tangent).
Actual by predicted chart	The actual by predicted chart compares the actual values with the predicted values, showing us how well the trained model fits the data.
Adaptive stepwise	Adaptive stepwise is a forward stepwise method that allows removing a variable currently in the model but no longer significant. It is adaptive.
Aggregation methods	Ensemble models combine the predictions of base learners using aggregation methods. Common aggregation methods include averaging, voting, and weighted voting. The aggregated result is often more accurate and reliable than the predictions of individual models.
Artificial Intelligence (AI)	AI is the simulation of human intelligence in machines, enabling them to perform tasks that typically require human cognition, such as learning, reasoning, problem-solving, perception, and language understanding.
Association model	A predictive model with an interval target variable.
Backward propagation (or backpropagation)	Backward propagation is the process of adjusting the weights and biases in the network based on the prediction error. It involves calculating the gradients of the loss function and adjusting the parameters to minimize the error.
Backward stepwise	Backward stepwise is the method in which the model starts with all predictor variables, and at each step, the variable that contributes the least to the model's performance is removed until no further improvement is achieved or a predefined stopping criterion is met.
Bagging	Bagging is an ensemble method that involves training multiple base learners independently on different random subsets of the training data (sampling with replacement). The final prediction is typically an average or voting over the predictions of all the base learners.
Bar chart	A bar chart is a graphical representation of categorical data that use rectangular bars of varying lengths to visualize the frequency or count of each category.
Base learners (individual models)	Base learners are the individual models that constitute the ensemble. Base learners can be any machine learning algorithm, such as decision trees, support vector machines, neural networks, etc.
Bayes' theorem	Bayes' theorem is a fundamental equation used in Bayesian networks to update probabilities based on new evidence. It calculates the posterior probability of an event given prior knowledge and observed evidence.
Bayesian network	Bayesian network, also known as a Bayesian belief network or probabilistic graphical model, is a powerful and popular method for representing and reasoning about uncertainty and probabilistic relationships between variables. It is a graphical model that represents a set of variables and their conditional dependencies through a directed acyclic graph.
Bias	A bias term is added to the weighted sum of inputs in each neuron. It allows the network to learn a bias toward certain values and improves the accuracy of the model.
Big data	Big data is a term that describes the large volume of data—both structured and unstructured—that inundates a business on a day-to-day basis.
Boosting	Boosting is an iterative ensemble method where base learners are trained sequentially, and each learner focuses on correcting the mistakes made by its predecessors. Boosting assigns higher weights to misclassified instances, effectively giving more attention to challenging samples.
Box plot	A box plot (box-and-whisker plot) is a graphical summary of the distribution of numerical data that display the median, quartiles, and potential outliers, providing a concise view of the data's central tendency and spread.
Branch or sub-tree	A branch or a sub-tree is a sub-section of a decision tree.
Category	See "Class."

(Continued)

Terms	Definitions
Causal data mining	Causal data mining is a method of using causal machine learning algorithms to detect unknown patterns involving variables and their interrelationships in a causal network from data to predict outcomes
Centroids	Centroids are the mean values of the attributes for the objects in a cluster. They are used to represent the cluster and are often used as the basis for assigning new objects to a cluster.
Champion model	Champion model is the predictive model with the best model performance among the compared models in a data mining project. It is a relative best.
Chi-squared Automatic Interaction Detection (CHAID)	CHAID is a non-parametric method used for creating decision trees by recursively splitting the data into homogeneous subgroups based on the categorical predictor variables, using the chi-squared test to assess the independence between variables and find significant splits.
Class	A class refers to a distinct category or label that represents a specific group or type of data instances in a dataset, which is often the target.
Classification and Regression Tree (CART)	CART is a binary tree, which uses an algorithm to split a parent node into TWO child nodes repeatedly. CART can be used for both classification and association models.
Classification model	A predictive model with a binary or nominal target variable.
Cluster assignment	Cluster assignment is a process of assigning each object to a specific cluster based on its similarity to the cluster centroids. The assignment is usually done by comparing the distance or similarity of the object to the centroids.
Cluster or segment	A cluster, also known as a segment, refers to a group of data points or objects that are similar to each other based on certain characteristics or attributes.
Cluster profiling	Cluster profiling involves analyzing the characteristics of each cluster, such as the mean value of the attributes or the distribution of the objects within the cluster.
Cluster sampling	Cluster sampling is a sampling method where the population is divided into clusters or groups, and a random selection of entire clusters is chosen, followed by data collection from all the members within the selected clusters.
Cluster validation	Cluster validation is a process of assessing the quality and reliability of the clusters obtained. It involves evaluating the internal cohesion and separation of the clusters, as well as their stability and meaningfulness.
Clustering	A process of grouping objects or observations into clusters (segments) based on their similarity or proximity. The goal is to create clusters that are internally homogeneous (objects within the same cluster are similar) and externally heterogeneous (objects from different clusters are dissimilar).
Clustering algorithms	A clustering algorithm determines the number of clusters and assigns observations to clusters. Cluster algorithms can be categorized into hierarchical clustering and partitioning clustering.
Component rotation	Component rotation is used to transform the principal components to align those components with the original variables, which achieves a more interpretable and meaningful representation of the data.
Conditional Probability Tables (CPTs)	Each node in a Bayesian network has an associated CPT, which defines the conditional probability distribution of the node given its parents' states. The CPT quantifies the probabilistic relationship between variables.
Confidence level	The confidence level represents the degree of certainty or confidence we have in the estimates obtained from the sample.
Confusion matrix	Confusion matrix, also called the classification matrix, is the table that summarizes the performance of a classification algorithm.
Connection (edge) (in neural networks)	A connection connects neurons. Each connection has an associated weight that determines the level of the influence between connected neurons.
Constraint-based algorithm	A constraint-based algorithm is an approach used to learn the network structure from data by discovering the probabilistic dependencies between variables without directly optimizing a scoring metric. It iteratively tests and evaluates conditional independence relationships among variables to determine the network's structure.
Convenience sampling	Convenience sampling is a non-probabilistic sampling method where researchers select the most readily available individuals to be part of the sample, leading to a lack of randomization and potential bias.

(Continued)

Terms	Definitions
Cost-sensitive learning	Cost-sensitive learning is a method to evaluate and select a model based on the costs of misclassification.
Covariance	Covariance between two variables measures how much they vary together. If two variables have a high covariance, it means they are positively correlated, while a low covariance indicates no correlation. In PCA, we choose the principal components with the expectation that they are uncorrelated.
CRISP-DM process	Cross-Industry Standard Process for Data Mining
Data de-identification	Data de-identification is the process of removing or altering identifying information from a dataset to protect the privacy and anonymity of individuals.
Data mining	Data mining is a method of using machine learning algorithms to detect unknown patents involving relationships among variables within large datasets to predict outcomes of interest, which can lead to informed business decisions.
Data mining model	See "Predictive model."
Data mining process	Data mining process refers to steps involves discovering meaningful patterns, relationships, and insights from large datasets using various techniques and algorithms.
Data modification	Data modification involves preparing and transforming the raw data to make it suitable for training our predictive models.
Data science	Data science is a larger and multidisciplinary field, focusing on capturing and extracting knowledge from data and communicating the outcomes. Data science consists of data mining as an essential part of analyzing data and other components such as data collection and management, data treatment, data visualization, computer programming, and artificial intelligence applications.
Data scientist	Data scientist is a professional who does a task or a combination of tasks involving analytics, data collection and treatment, data mining, machine learning, and programming
Data visualization	Data visualization is the graphical representation of data and information through charts, graphs, and other visual elements to help understand patterns, trends, and insights in a more intuitive and accessible way.
Decision matrix	Decision matrix is the table that presents the costs of misclassifications, including costs of false positives and costs of false negatives.
Decision node	When a sub-node is divided into additional sub-nodes, it is referred to as a decision node.
Decision tree	Decision tree is a logical rule-based method that presents a hierarchical structure of variables, including the root node, parent nodes, and child nodes.
Deep learning	Deep learning is a complex neural network with many hidden layers. Deep learning breakthroughs lead to AI boom.
Dendrograms	A dendrogram is a tree-like diagram that shows the hierarchical relationship between clusters in hierarchical clustering. The height of each branch represents the distance between clusters at that level.
Density plot	A density plot is a graphical representation of the distribution of continuous data, providing an estimate of the underlying probability density function, often using smoothed curves.
Dependent variable	See "Target variable."
Descriptive statistics	Descriptive statistics is a branch of statistics that involves the summarization and presentation of data to provide a clear and concise understanding of its main characteristics, such as measures of central tendency, dispersion, and distributions.
Dimensionality reduction	Once the principal components have been identified, PCA can be used to transfer the high-dimensional data into a lower-dimensional data, while still retaining as much of the original variation as possible. The number of principal components retained determines the dimensionality of the new data.
Directed Acyclic Graph (DAG)	The graphical structure of a Bayesian network is represented by a DAG, which is a graph without cycles (no path that starts and ends at the same node). The absence of cycles ensures that the network does not contain feedback loops and allows for efficient probabilistic inference.
Distance or similarity measures	A distance, also known as a similarity measure, quantifies the similarity or dissimilarity between pairs of objects. These measures are used to determine the proximity between objects and form the basis for clustering algorithms. Common distance measures include Euclidean distance, Manhattan distance, and cosine similarity.

(Continued)

Terms	Definitions
Diversity	The primary strength of ensemble models lies in the diversity of their individual models. By using different learning algorithms or training on different subsets of the data, the ensemble captures various patterns and reduces the risk of overfitting.
Edges (in Bayesian networks)	Edges represent probabilistic dependencies between nodes. Directed edges indicate causal or direct influences between variables.
Eigenvectors and eigenvalues	PCA seeks to identify the directions of maximum variability in the data. Eigenvectors are the directions in the data that are not affected by a linear transformation, while eigenvalues indicate how much variance is captured by each eigenvector. In PCA, the principal components are the eigenvectors of the covariance matrix, sorted by their corresponding eigenvalues.
Ensemble	Ensemble is a machine learning method that involves combining the predictions of multiple models to produce a more accurate and robust prediction.
Ensemble size	The number of base learners in an ensemble is an important consideration. Increasing the ensemble size can lead to better performance up to a point, after which the returns diminish, and the model may become computationally expensive.
Error term	Error term (also known as the residual term) represents the discrepancy between the observed values of the target variable and the predicted values obtained from the linear regression model. It captures the part of the target variable that cannot be explained by the linear relationship with the predictors.
Ethical consideration	Ethical consideration refers to the principles, guidelines, and moral values that govern the responsible and respectful use of data throughout the entire data mining process. It is to ensure that we comply with any requirements and restrictions regarding data access to protect human subjects and avoid violating privacy.
Evaluation criteria	See "Model fit metrics."
Evaluation metrics	See "Model fit metrics."
Evidence (in Bayesian network)	Evidence refers to the observed values or states of certain variables in the Bayesian network. It is used to update the probabilities of other variables in the network.
Explained variance	The proportion of variance explained by each principal component is crucial for understanding the importance of each component in the data. It helps us determine how many principal components to retain for dimensionality reduction while preserving significant information.
Explanatory power (EP)	EP refers to the ability of a predictive model, especially association model, to explain or predict the variability observed in the data. It measures how well the model captures the underlying relationships between the input variables (predictors) and the target variable.
F-test	F-test is a test of the significance of the overall model. A significant F-test indicates that the regression model fits the data better than a model with no predictors. A significant F-test is expected for a well-fitting model.
F1 score	F1 score is the harmonic mean of the sensitivity (recall) and precision (positive predicted value); a high F1 score means low false positives (FP) and low false negatives (FN).
Factor analysis	See "Principal component analysis."
False negative (FN)	FN is the prediction that wrongly indicates the absence of the event.
False positive (FP)	FP is the prediction that wrongly indicates the presence of the event.
Feature importance	See "Variable importance."
Features	See "Predictors."
Forward propagation	Forward propagation is the process of transmitting input signals through the network, layer by layer, to produce an output. Each neuron in a layer receives input from the previous layer, performs computations using its activation function, and passes the output to the next layer.
Forward stepwise	Forward stepwise is the method in which the model starts with no predictor variables and iteratively adds one variable at a time, and at each step, the variable that improves the model's performance the most is added to the model until no further improvement is achieved or a predefined stopping criterion is met.
Frequentist statistics	Frequentist statistics, also known as frequentist inference, is a statistical framework and approach to data analysis that focuses on the concept of repeated sampling and long-run frequencies of events.
Gini coefficient	Gini coefficient is a measure of the model's discriminatory power, which is the effectiveness of the model in differentiating between positives and negatives. The Gini coefficient measures how equal the positive rates are across the attributes of a characteristic.

(Continued)

Terms	Definitions
Goodness of fit	See Model fit
Goodness-of-fit indices	See "Model fit metrics."
Gradient boosting	Gradient boosting is a machine learning ensemble method that sequentially combines weak learners to create a powerful predictive model by minimizing errors using gradient descent.
Gradient Boosting Additive Model	Gradient Boosting Additive Models (GBAM) is a variant of gradient boosting that extends the traditional algorithm to handle additive models. An additive model is a model that consists of a sum of functions of individual input variables, where each function depends only on a single input variable.
Gradient descent	Gradient boosting minimizes a specified loss function by using gradient descent optimization. It computes the gradients of the loss function and adjusts the subsequent weak learner to correct the errors made by previous learners. The learning process aims to iteratively reduce the loss and improve the overall model performance.
Hierarchical clustering	Hierarchical clustering builds a hierarchy of clusters, forming a tree-like structure called a dendrogram.
Histogram	A histogram is a graphical representation of continuous data that uses adjacent rectangles to display the distribution and frequency of data within specified intervals (bins).
Imbalanced dataset	An imbalanced dataset refers to a dataset where the number of samples in different classes significantly varies, resulting in unequal representation of the classes. In an imbalanced dataset, one class (the minority class) typically has significantly fewer instances than the other classes (majority classes).
Imputation	Imputation is a process of estimating or filling in missing values in a dataset with estimated values.
Independent variables	See "Predictors."
Input variables	See "Predictors."
Institutional Research Board (IRB)	An Institutional Review Board (IRB) is a committee responsible for upholding research ethics by reviewing proposed research methods to ensure they meet ethical standards.
Intercept	Intercept is the constant term in the regression equation, representing the predicted value of Y (target variable) when all predictors are zero.
K-means clustering	K-means is a partitioning clustering algorithm that partitions the data into k distinct clusters by iteratively updating the cluster centroids and assigning data points to the nearest centroid. The objective is to minimize the sum of squared distances between data points and their respective cluster centroids.
Label	See "Class."
Layers	Neurons are organized into layers in a neural network. The input layer receives the input data, the output layer produces the final output, and there can be one or more hidden layers in between.
Leaf node	Nodes that do not split are called leaf nodes. They represent the output or the predicted value/class for the specific subgroup of data that reaches that node.
Learning (in Bayesian networks)	Learning in Bayesian networks involves estimating the network structure and parameters from data. This can be done using approaches such as constraint-based methods or score-based methods.
Learning rate (in Gradient Boosting)	The learning rate is a hyperparameter that controls the step size in gradient descent. It determines how much each weak learner's contribution affects the final ensemble. A lower learning rate makes the algorithm converge slowly but can improve generalizability, while a higher learning rate leads to faster convergence but may increase the risk of overfitting.
Level	See "Class."
Level of measurement	See "Variable scale."
Lift chart	Lift is a measure of the effectiveness of a predictive model calculated as the ratio between the results obtained with and without the predictive model. The Lift chart, or Lift curve, presents the graphic representation between the lift values and response rates.
Likelihood (in Bayesian networks)	The likelihood in Bayesian networks represents the probability of observing specific evidence given the values of the variables in the network. It is derived from the conditional probability tables.
Line chart	A line chart is a graph that displays data points connected by straight lines, commonly used to visualize trends or changes over time or ordered categories.

(Continued)

Terms	Definitions
Linear regression	Linear regression is a regression method model in which the target variable is interval. Linear regression is an association model.
Logistic regression	Logistic regression is a regression method in which the target variable is categorical (binary or nominal). It is a classification model.
Loss function	A loss function measures the discrepancy between the predicted output of the neural network and the actual target values. It quantifies the error and guides the adjustment of the parameters during training.
Machine learning	Machine learning is data analytic methods that detect unknown patterns from large data to predict the outcomes of interest.
Machine learning model	See "Predictive model."
Machine learning process	See "Predictive modeling."
Margin of error	The margin of error is a measure of the uncertainty or variability in the estimates obtained from the sample.
Mean Absolute Percentage Error (MAPE)	MAPE is a common metric for association models. It measures the percentage difference between the predicted values and the actual values in a dataset.
Measurement of variable	See "Variable scale."
Measurement scale	See "Variable scale."
Misclassification Rate (MR)	MR is a measure of how often a model misclassifies (makes incorrect predictions) the target variable (including actual positives and actual negatives).
Model fit	Model fit measures how well the model fits the data.
Model fit metrics	Criteria used to evaluate model fit
Model performance	See "Model fit."
Model reliability	Model reliability refers to the internal consistency of a model. It means the model produces consistent results with different runs using different data sets.
Model selection and tuning	Ensemble models require careful selection and tuning of base learners and configurations. Different combinations of models and configurations can significantly impact the model performance.
Model validity	Model validity refers to the accuracy of the model. It means the model consistently produces an accurate prediction of the target variable.
Neuron (node) (in neural networks)	A neuron, also called a node or unit, is a basic building block of a neural network. It receives inputs, processes them using an activation function, and produces outputs.
Nodes (in Bayesian networks)	Nodes represent random variables or events in the domain being modeled. Each node corresponds to a specific quantity or attribute of interest.
Non-probability sampling	Non-probability sampling is the method of selecting individuals based on non-random criteria, and not every individual in the sampling frame has a chance of being included.
Normalization	Data normalization, also known as variable scaling, is a preprocessing step in machine learning that converts the variables of a dataset to a similar range.
Oblimin	Oblimin is an oblique rotation method that allows for correlation or obliqueness between the rotated components. The main objective of Oblimin rotation is to achieve a simpler structure by minimizing the number of variables with high loadings on a component.
Oblique rotation	Oblique rotation is the rotation method, in which the rotated components are allowed to be correlated with each other.
Odds ratio	Odds ratio represents the ratio of the odds of the event occurring for one group compared to another group. In the context of logistic regression, it measures how the odds of the binary outcome change with respect to a one-unit change in the predictor, while holding all other predictors constant.
Optimization algorithm (in neural networks)	The optimization algorithm is used during training to adjust the network's parameters based on the computed gradients. Gradient descent is a common optimization algorithm used for this purpose.
Orthogonal rotation	Orthogonal rotation is the rotation method, in which the rotated components are constrained to be orthogonal to each other, meaning they are independent and uncorrelated.

(Continued)

Terms	Definitions
Out-of-bag evaluation	Random forest utilizes an out-of-bag (OOB) evaluation technique. Since each decision tree is trained on a different subset of the training data, the samples not included in a tree's training subset can be used for evaluation. This provides an unbiased estimate of the model's performance without the need for a separate validation set.
Output	See "Target variable."
Overfitting	Overfitting is the situation in which the model is overtrained to the training sample and not generalized to other datasets. Hence, the model is invalid and unusable.
Oversampling	Oversampling is a technique that balances the data by incrementing the size of the minority (rare events) to the same size as the majority.
Parent node and child node (in decision trees)	These terms are relative in nature. Any node situated below another node is typically referred to as a child node or sub-node, while the node preceding these child nodes is commonly known as the parent node.
Parents and children (in Bayesian networks)	In a Bayesian network, the parents of a node are the nodes that directly influence it. The children of a node are the nodes that are directly influenced by it.
Partitioning clustering	Partitioning clustering divides the data into non-overlapping partitions or clusters. It directly assigns each data point to a specific cluster.
Pie chart	A pie chart is a circular chart that represents the proportions of different categories in a whole, with each category represented by a slice of the pie.
Population	Population is the primary target audience of which we want to examine the patterns of the data and make conclusions.
Posterior probability (in Bayesian networks)	The posterior probability represents the updated probability of a variable given observed evidence or data. It is calculated by combining the prior probability with the likelihood of the observed data using Bayes' theorem.
Prediction	A general term for a predicting act.
Prediction accuracy	Prediction accuracy is a measure of how accurate a model predicts the target variables (including actual positives and actual negatives).
Prediction errors	Errors measure the extent to which the trained model misfits the data. The lower the error, the more accurate the model.
Prediction model	See "Predictive model."
Predictive model	A model predicting either a continuous or categorical target variable.
Predictive modeling	A process of using machine learning algorithms to develop a model that predicts future or unseen events or outcomes based on data.
Predictive power	Predictive power is the ability of a model to accurately capture and represent the underlying patterns, relationships, or trends present in the data and generalize the results to other datasets.
Predictive research	Predictive strategy looks into the future. In other words, organizations analyze a large amount of data to predict new potential safety problems in the future. By mining big data, they can develop a predictive model that predicts an incident or accident before it happens.
Predictors	Inputs being used in predicting the output.
Principal components	Principal components are the eigenvectors that correspond to the highest eigenvalues of the covariance matrix. They are ordered in terms of their importance, where the first principal component explains the maximum variance in the dataset, followed by the second principal component, and so on.
Prior probability (in Bayesian networks)	The prior probability represents the initial belief or knowledge about a variable before any evidence or data is observed. It is typically specified as part of the Bayesian network modeling process.
Proactive research	Proactive research studies the present. In other words, organizations examine contributing factors to an incident/accident from various aspects of hazardous conditions and organizational processes and see how they are related to the incident or accident.
Probabilistic inference (in Bayesian networks)	Bayesian networks allow for probabilistic inference, which means computing the probability of specific events or variables given evidence (observed data or values) from other variables in the network. Inference is based on Bayes' theorem.
Probabilistic relationship (in Bayesian networks)	A probabilistic relationship refers to the existence of a connection or association between two or more variables, where the relationship is characterized by uncertainty and is described using probabilities.

(Continued)

Terms	Definitions
Probability sampling	Probability sampling is the method of selecting individuals randomly in such as way each individual in the sampling frame has an equal probability of being chosen.
Probability value (p-value)	The probability value (p-value) is a statistical measure for assessing the significance of the estimated coefficients (also known as the regression weights or beta values) for the predictors in the regression model. It determines whether the relationship between each predictor and the target variable is statistically significant or if it could have occurred by chance.
Promax	Promax is an oblique rotation method that extends the advantages of Varimax rotation while also accounting for possible correlations between the rotated components. It is considered a compromise between the simplicity of orthogonal rotation and the flexibility of oblique rotation.
Pruning	Removing the sub-nodes of a parent node is called pruning. It is a technique used to simplify the decision tree by removing nodes that do not significantly improve its performance on the validation set. A tree is grown through splitting and shrunk through pruning.
Pseudo R-squared	Pseudo R-squared is a metric used in logistic regression to assess the model fit. Unlike the traditional R-squared used in linear regression, which measures the proportion of variance explained by the model, pseudo R-squared measures the proportion of the deviance explained by the model.
Purposive sampling	Purposive sampling is a non-probabilistic sampling method where researchers deliberately select specific individuals or units based on pre-defined criteria to best represent certain characteristics or traits of interest.
R-squared or coefficient of determination	R-squared is a measure of the percentage of the variance in the target variable that is explained by variance in predictors, collectively.
Random forest	Random forest is a machine learning algorithm that combines the predictions of multiple decision trees to make more accurate and robust predictions. It falls under the category of ensemble learning, where multiple models are combined to form a more powerful predictive model.
Random subset selection	Each decision tree in the random forest is built using a random subset of the training data. This process, known as bootstrap aggregating or "bagging," involves randomly sampling data points with replacement to create diverse subsets for training each tree.
Random variable selection	The random forest algorithm selects a random subset of input variables for each decision tree. This technique, known as variable bagging or variable subsampling, introduces additional randomness and helps to reduce the dominance of any single variable in the ensemble.
Reactive research	Reactive research focuses on examining events that have already happened and identifying the root causes of a specific incident or accident. Based on the identified root causes, the organizations can form appropriate strategies to mitigate the risk.
Receiver Operating Characteristic (ROC) chart	The ROC chart displays the true positive rate (sensitivity) on the vertical axis and the false positive rate (sensitivity) on the horizontal axis, illustrating the trade-off between sensitivity and specificity.
Reconstruction	PCA allows the inverse transformation of the lower-dimensional data back into the original high-dimensional space. Although some information is lost during dimensionality reduction, the reconstruction can still be useful for interpretation or other downstream tasks.
Recursive partitioning	Recursive partitioning denotes an iterative procedure involving the division of data into partitions, followed by further splitting along each branch.
Regression coefficients	Regression coefficients are the coefficients assigned to each predictor (Xi), indicating the magnitude and direction of their influence on the target variable. A positive coefficient means an increase in the predictor leads to an increase in the target variable, and vice versa for a negative coefficient.
Regression method	Regression is a popular method that estimates the relationship between one dependent variable (target variable) and multiple independent variables (predictors).
Regression model	See "Association model."
Relative Squared Error (RSE)	RSE is a common metric for association models. It measures the relative improvement of the model's predictions compared to a simple baseline model that always predicts the mean of the actual values.

(Continued)

Terms	Definitions
Replacement	Replacement is a data modification technique to correct errors, reassign values, or remove incorrect information in the data.
Representative sample	Representative sample is a sample that accurately reflects the characteristics and diversity of the population.
Residual by predicted chart	The residual by predicted chart presents not only how close the actual values are to the predicted values but also indicates any patterns in the residuals.
Residuals	Residuals measure the difference between actual values and predicted values in a dataset.
Robustness	The algorithm is known for its robustness against overfitting, noisy data, and outliers. By aggregating multiple decision trees, it can capture complex relationships and handle noisy datasets more effectively.
Root mean squared error (RMSE)	RMSE is a common metric for association models. It measures the average difference between the predicted values and the actual values in a dataset, expressed in the same units as the predicted variable.
Root node	A root node is at the beginning of a tree. It represents the initial decision based on a specific feature that has the most significant impact on the target variable. Starting from the root node, the dataset is partitioned based on different features, and subsequently, these subgroups are further divided at each decision node located beneath the root node.
Sample	Sample is a subset of the population. We use the sample for data analysis and make inferences about the population.
Sample size	Sample size refers to the number of individuals or subjects included in the sample. It should be determined based on statistical considerations.
Sampling bias	Sampling bias occurs when the selected sample does not accurately represent the population, leading to skewed or inaccurate results.
Sampling frame	Sampling frame is the actual list of units or sources from which a sample is drawn.
Sampling method	Sampling method is the procedure that we use to select participants and collect the data.
Sampling with replacement	Sampling with replacement occurs when a randomly chosen unit from the population is put back into the population, and subsequently, another element is chosen at random. In this process, the population consistently retains all its original units, allowing for the possibility of selecting the same unit multiple times.
Scalability (in Random forest)	Random forest can be parallelized, as each decision tree in the ensemble can be built independently. This allows for efficient computation and scalability to large datasets.
Scale of measurement	See "Variable scale."
Scatter plot	A scatter plot is a graphical representation of paired data points on a two-dimensional plane, used to visualize the relationship and correlation between two variables.
Score-based algorithm	A score-based algorithm is a Bayesian network structure learning approach that maximizes a scoring metric to identify the most probable network structure given the data. The algorithm aims to find the network structure that best fits the data according to a specific scoring criterion.
Scree plot	A scree plot is a graphical tool used in PCA to visualize the eigenvalues associated with each principal component. It is a simple line segment plot that shows the eigenvalues for each individual principal component. The scree plot helps determine the number of principal components based on the amount of variance they explain.
SEMMA process	Sample-Explore-Modify-Model-Assess
Sensitivity, recall, or true positive rate (TPR)	Sensitivity is the probability of a positive result (event) or how well a model can predict true positives (events).
Significance level (alpha)	Significance level (alpha) refers to the significance level or the threshold used for hypothesis testing when evaluating the statistical significance of the estimated coefficients.
Simple random sampling	Simple random sampling is a sampling method that allows every member of the population to have an equal chance of being selected.
Snowball sampling	Snowball sampling is a non-probabilistic sampling method where initial participants are chosen based on specific criteria and then asked to refer other participants, creating a chain of referrals, commonly used for hard-to-reach or hidden populations.
Specificity, selectivity, or true negative rate (TNR)	Specificity is the probability of a negative result (non-event) or how well a model can predict true negatives (non-events).

(Continued)

Terms	Definitions
Splitting	Splitting is a process of dividing a node into two or more sub-nodes.
Splitting criterion	Splitting criterion is a measure used to evaluate and select the best predictor to split the data at each node of the tree. It is a measure of impurity or variance reduction.
Stacking	Stacking (stacked generalization) is a more complex ensemble technique that combines the predictions of multiple base learners by training a meta-model on their outputs. The meta-model learns to weigh the predictions of individual models effectively.
Standardized regression coefficient (beta)	Standardized regression coefficient is a measure used in linear regression to quantify the relationship between a predictor and the target variable after standardizing both variables to a common scale. Standardized coefficients range between 0 and 1 and are measured using the same scale in all cases via standard deviation.
Stepwise method	Stepwise method is an iterative variable selection method used in data mining to identify the most relevant subset of variables for a predictive model.
Stochastic Gradient Boosting	Stochastic Gradient Boosting (SGB) is a variant of the traditional gradient boosting algorithm that introduces randomness during the training process. It combines the concepts of gradient boosting and stochastic learning to enhance the model's generalizability and reduce the risk of overfitting.
Stratified sampling	Stratified sampling is a sampling method where the population is divided into distinct subgroups or strata, and a random sample is drawn from each stratum to ensure representation of different characteristics within the overall population.
Sum of squared error (SSE)	SSE measures unexplained variance, which is how far the data are from the model's predicted values.
Sum of squared regression (SSR)	SSE measures explained variance, which is how far the mean is from the model's predicted values.
Sum of squared total (SST)	SST measures total variance, which is how far the data are from the mean.
Supervised learning	A machine learning approach for structured data, in which a model is trained from labeled data to make predictions of a pre-defined target variable.
Systematic random sampling	Systematic sampling is a sampling method where every nth member is selected from the population after randomly choosing a starting point, ensuring equal and consistent spacing between selected samples.
Target	See "Target variable."
Target variable	The output being predicted.
Test sample	A sample of data used to evaluate and validate the model.
Training (in neural networks)	Training a neural network involves iteratively adjusting the weights and biases using a training dataset. The goal is to minimize the error or loss function and improve the network's prediction accuracy.
Training sample	A sample of data used to train the model.
Tree depth and number of trees	These are hyperparameters that control the complexity and size of the individual decision trees. The tree depth affects the model's capacity to capture complex relationships in the data, while the number of trees determines the boosting iterations and overall model complexity.
True negative (TN)	TN is the prediction that correctly indicates the absence of the event.
True positive (TP)	TP is the prediction that correctly indicates the presence of the event.
Undersampling	Undersampling is a technique that balances the data by reducing the size of the majority class to the same size as the minority class.
Unstandardized regression coefficient (B)	Unstandardized regression coefficient is a measure used in linear regression to quantify the relationship between a predictor and the target variable in its original units of measurement.
Unsupervised learning	A machine learning approach for unstructured data, in which a model is trained from unlabeled data without a specific target variable.
USELEI process	Understanding-Sampling-Exploring-Learning-Evaluating-Inferring
Validation sample	A sample of data used to perform the final assessment and validation of the model.
Variable discretization	Variable discretization is a process of converting a variable with continuous scale (interval variable) into discrete categories (a categorical variable).

(Continued)

Terms	Definitions
Variable importance	A measure of variable importance, indicating the relative contribution of each input variable in making predictions. It allows users to identify the most influential predictors and gain insights into the underlying data patterns.
Variable scale	A characteristic that describes the nature and properties of a variable; it indicates the type of values that a variable can take and the operations that can be performed on it.
Variance	Variance of a variable measures how much the values of that variable vary from their mean. In PCA, the goal is to find the directions in the data with the highest variance, as these directions contain the most information about the data.
Varimax	Varimax is an orthogonal rotation method that. aims to maximize the variance of the squared loadings on each component, leading to a simple and sparse structure where each variable has a high loading on one component and close to zero loadings on other components.
Voluntary response sampling	Voluntary response sampling is a non-probabilistic sampling method where participants self-select to be part of the sample, often in response to a survey or study invitation, which can lead to biased results due to the voluntary nature of participation.
Voting/averaging	During prediction, each decision tree independently generates a prediction, and the final prediction of the random forest is determined through voting (for classification models) or averaging (for association models) the individual tree predictions. This ensemble approach helps reduce bias and variance in the modeling and improves the overall accuracy and robustness of the model.
Weak learners	In the context of gradient boosting, weak learners are individual models that perform slightly better than random guessing but are not highly accurate on their own. Decision trees are commonly used as weak learners in gradient boosting, but other models can also be used.
Weight (in neural networks)	Each connection between neurons is assigned a weight, which represents the strength or importance of that connection. These weights determine how much influence the input of one neuron has on the output of another.

Bibliographies

Abramowitz, M., & Stegun, I. A. (1965). Handbook of Mathematical Functions: With Formulas, Graphs, and Mathematical Tables. Dover Publications.

Afendras, G., & Markatou, M. (2019). Optimality of training/test size and resampling effectiveness in cross-validation. Journal of Statistical Planning and Inference, 199, 286–301.

Aggarwal, C. C. (2015). Data Mining: The Textbook. Springer International Publishing.

Akaike, H. (1973). Maximum likelihood identification of Gaussian autoregressive moving average models. Biometrika, 60(2), 255–265.

Akaike, H. (1977). On the Entropy Maximization Principle, Applications of Statistics.

Alpaydin, E. (2020). Introduction to Machine Learning, 4th edition. MIT Press.

Barry de Ville. (2006). Decision Trees for Business Intelligence and Data Mining: Using SAS Enterprise Miner. SAS Institute.

Bengio, Y., Courville, A., & Vincent, P. (2013). Representation learning: A review and new perspectives. IEEE Transactions on Pattern Analysis and Machine Intelligence, 35(8), 1798–1828.

Bishop, C. M. (1996). Neural Networks and Pattern Recognition. Oxford University Press.

Bishop, C. M. (2006). Pattern Recognition and Machine Learning. Springer.

Boser, B. E., Guyon, I. M., & Vapnik, V. N. (1992). A training algorithm for optimal margin classifiers. In Proceedings of the Fifth Annual Workshop on Computational Learning Theory, July (144–152).

Box, G. E. P., & Cox, D. R. (1964). An Analysis of Transformation. Journal of the Royal Statistical Society Series B, 26, 211–252.

Breiman, L. (1984). Classification and Regression Trees. Routledge.

Breiman, L. (1996). Bagging predictors. Machine Learning, 24(2), 123–140.

Breiman, L. (1998). Arcing Classifiers. Annals of Statistics, 26, 801–824.

Breiman, L. (2001). Random forests. Machine Learning, 45(1), 5–32.

Breiman, L. (2003). Manual on Setting Up, Using, and Understanding Random Forests v3. 1. University of California.

Cai, T., & Dodd, L. (2006). Regression Analysis for the Partial Area Under the ROC Curve. Working Paper 36, Harvard University Biostatistics Working Paper Series.

Camastra, F., & Vinciarelli, A. (2015). Machine Learning for Audio, Image, and Video Analysis, 2nd edition. Springer.

Cao, L. (2017). Data Science: A Comprehensive Overview. ACM Computing Surveys (CSUR), 50(3), 1–42.

Caruana, R., Niculescu-Mizil, A., Crew, G., & Ksikes, A. (2004). Ensemble selection from libraries of models. In Proceedings of the Twenty-first International Conference on Machine Learning (p. 18).

Cattell, R. B. (1966). The scree test for the number of factors. Multivariate Behavioral Research, 1, 245–276.

Chawla, N., Bowyer, K., Hall, L. O., & Kegelmeyer, W. P. (2002). SMOTE: Synthetic minority over-sampling technique. Journal of Artificial Intelligence Research, 16, 321–357.

Chen, T., & Guestrin, C. (2016). Xgboost: A scalable tree boosting system. In Proceedings of the 22nd ACM SIGKDD International Conference on Knowledge Discovery and Data Mining (pp. 785–794).

Cheng, J., & Greiner, R. (1999). Comparing Bayesian network classifiers. Proceedings of the Fifteenth Conference on Uncertainty in Artificial Intelligence, July 30–August 01, 101–108. Stockholm, Sweden.

Cohen, J. (1987). Statistical power analysis for the behavioral sciences, 2nd edition. Laurence Erlbaum Associates.

Cohen, J., Cohen, P., West, S. G., & Aiken, L. S. (2003). Applied multiple regression/correlation analysis for the behavioral sciences, 3rd edition. Lawrence Earlbaum Associates.

Cover, T. M., & Hart, P. E. (1967). Nearest neighbor pattern classification. IEEE Transactions on Information Theory, 13(1), 21–27.

Cox, L. A., Jr. (2010). Regression versus causation, revisited. Risk Analysis, 30(4), 535–540.

Davenport, T. H., & Patil, D. J. (2012). Data scientist. Harvard Business Review, 90(5), 70–76.

DeVille, B., & Neville, P. (2013). Decision Trees for Analytics Using SAS Enterprise Miner. SAS Publishing.

Dhar, V. (2013). Data science and prediction. Communications of the ACM, 56(12), 64–73.

Dobbin, K. K., & Simon, R. M. (2011). Optimally splitting cases for training and testing high dimensional classifiers. BMC Medical Genomics, 4(31), 1–8.

Domingos, P., & Pazzani, M. (1997). On the optimality of the simple Bayesian classifier under zero-one loss. Machine Learning, 29(2–3), 103–130.

Dubbs, A. (2021). Test set sizing via random matrix theory. arXiv preprint arXiv:2112.05977. https://doi.org/10.48550/arXiv.2112.05977

Emerson, J. D., & Stoto, M. A. (1982). Exploratory Methods for Choosing Power Transformations. Journal of the American Statistical Association, 77, 103–108.

Federal Aviation Administration. (n.d.). Safety management system. Retrieved from https://www.faa.gov/about/initiatives/sms/explained/basis/

Freund, Y. (1995). Boosting a weak learning algorithm by majority. Information and Computation, 121, 256–285.

Freund, Y., & Schapire, R. E. (1996). Experiments with a new boosting algorithm. In Machine learning: Proceedings of the Thirteenth International Conference (Vol. 96, pp. 148–156).

Freund, Y., & Schapire, R. E. (1997). A decision-theoretic generalization of on-line learning and an application to boosting. Journal of Computer and System Sciences, 55(1), 119–139.

Freund, Y., Schapire, R. E., Abe, N., & Bello, P. (1999). A short introduction to boosting. Journal of Japanese Society for Artificial Intelligence, 14(5), 771–780.

Friedman, H. J., Hastie, T., & Tibshirani, R. (1998). Additive Logistic regression: A statistical view of boosting. Technical Report. Retrieved from ftp://stat.stanford.edu/pub/friedman/boost.ps.Z.

Friedman, J. H. (2001). Greedy function approximation: A gradient boosting machine. The Annals of Statistics, 29, 1189–1232.

Friedman, J. H. (2002). Stochastic gradient boosting. Computational Statistics & Data Analysis, 38, 367–378.

Friedman, N., Geiger, D., & Goldszmidt, M. (1997). Bayesian network classifiers. Machine Learning, 29, 131–161.

Gelman, A., Carlin, J. B., Stern, H. S., Dunson, D. B., Vehtari, A., & Rubin, D. B. (2013). Bayesian data analysis. CRC Press.

Goodall, C. (1983). M-Estimators of Location: An outline of Theory. Understanding Robust and Exploratory Data Analysis. Wiley.

Goodfellow, I., Bengio, Y., & Courville, A. (2016). Deep Learning. MIT Press.

Hair, J. F., Black, W. C., Babin, B. J., & Anderson, R. E. (2018). Multivariate Data Analysis, 8th edition. Pearson Education.

Han, J., Kamber, M. and Pei, J. (2011) Data Mining: Concepts and Techniques, 3rd Edition. Morgan Kaufmann Publishers.

Hand, D., & Till, R. (2001). A Simple Generalization of the Area Under the Curve for Multiple Class Classification Problems. Machine Learning, 45, 171–186.

Hastie, T., Tibshirani, R., & Friedman, J. H. (2009). The Elements of Statistical Learning: Data Mining, Inference, and Prediction, 2nd edition. Springer.

Hayashi, C. (1998). What Is Data Science? Fundamental concepts and a heuristic example. In Hayashi, C., Yajima, K., Bock, HH., Ohsumi, N., Tanaka, Y., Baba, Y. (Eds.), Data Science, Classification, and Related Methods. Studies in Classification, Data Analysis, and Knowledge Organization. Springer.

Heinrich Heine University Düsseldorf. (2022). G*Power. https://www.psychologie.hhu.de/arbeitsgruppen/allgemeine-psychologie-und-arbeitspsychologie/gpower

Hoaglin, D. C., Mosteller, F., & Tukey, J. W. (1983). Understanding Robust and Exploratory Data Analysis. John Wiley and Sons, Inc.

Jordan, M. I., & Mitchell, T. M. (2015). Machine learning: Trends, perspectives, and prospects. Science, 349(6245), 255–260.

Joseph, V. R. (2022). Optimal ratio for data splitting. Statistical Analysis and Data Mining: The ASA Data Science Journal, 15(4), 531–538.

Kass, G. V. (1980). An exploratory technique for investigating large quantities of categorical data. Journal of the Royal Statistical Society: Series C (Applied Statistics), 29(2), 119–127.

Kohavi, R. and Provost, F. (1998) Glossary of terms. Machine learning—Special issue on applications of machine learning and the knowledge discovery process. Machine Learning, 30, 271–274.

Kullback, S., & Leibler, R. (1951). On Information and sufficiency. Annals of Mathematical Statistics, 22, 79–86.

Langley, P., Simon, H. A., Bradshaw, G. L., & Zytkow, J. M. (1987). Scientific Discovery: Computational Explorations of the Creative Processes. MIT Press.

Larsen, J., & Goutte, C. (1999). On optimal data split for generalization estimation and model selection. In Neural Networks for Signal Processing IX: Proceedings of the 1999 IEEE Signal Processing Society Workshop (Cat. No. 98TH8468), 225–234.

Loh, W.-Y., & Shih, Y.-S. (1997). Split selection methods for classification trees. Statistica Sinica, 7(4), 815–840.

Mallows, C. (1973). Some Comments on Cp. Technometrics, 42, 87–94.
Masters, T. (1993). Practical Neural Network Recipes in C++. Academic Press, Inc.
Michalski, R. S., Carbonell, J. G., & Mitchell, T. M. (1983). Machine Learning: An Artificial Intelligence Approach. Springer.
Mitchell, T. M. (1980). The Need for Biases in Learning Generalizations. Department of Computer Science, Carnegie-Mellon University.
Mitchell, T. M. (1997). Machine Learning. McGraw Hill.
Mitchell, T. M., Keller, R. M., & Kedar-Cabelli, S. T. (1986). Explanation-based learning: A unified view. Machine Learning, 1(1), 47–80.
Mitchell, T. M., Shinkareva, S. V., Carlson, A., Chang, K. M., Malave, V. L., Mason, R. A., & Just, M. A. (2008). Predicting human brain activity associated with the meanings of nouns. Science, 320(5880), 1191–1195.
Mohamadi, A., Asghari, F., & Rashidian, A. (2014). Continuing review of ethics in clinical trials: a surveillance study in Iran. Journal of Medical Ethics and History of Medicine, 7, 22.
Murphy, K. P. (2012). Machine Learning: A Probabilistic Perspective. MIT Press.
National Institute of Health. (2019, June 25). Research Involving Private Information or Biospecimens. Retrieved from https://grants.nih.gov/grants/policy/hs/private-information-biospecimens-flowchart.pdf
Nguyen, Q. H., Ly, H. B., Ho, L. S., Al-Ansari, N., Le, H. V., Tran, V. Q., … Pham, B. T. (2021). Influence of data splitting on performance of machine learning models in prediction of shear strength of soil. Mathematical Problems in Engineering, 2021, 1–15.
Nisbet, R., Elder IV, J., and Miner, G. 2009. Handbook of Statistical Analysis and Data Mining Applications, 1st edition. Academic Press/Elsevier.
Pearl, J., & Russell, S. (2003). Bayesian analysis. In M. A. Arbib (Ed.), Handbook of Brain Theory and Neural Networks (pp. 157–160). MIT Press.
Picard, R. R., & Berk, K. N. (1990). Data splitting. The American Statistician, 44(2), 140–147. https://doi.org/10.2307/2684155
Provost, F., & Fawcett, T. (2013). Data Science for Business. O'Reilly Media.
Provost, F., Fawcett, T., & Kohavi, R. (1998). The Case against Accuracy Estimation for Comparing Induction Algorithms. In Proceedings of the 15th International Conference on Machine Learning (pp. 445–453), Madison, WI.
The Office of Human Research Protection. (n.d.). Institutional Review Board Guidebook: Chapter 3, Section A, Risk/Benefit Analysis. https://biotech.law.lsu.edu/research/fed/ohrp/gb/irb_chapter3.htm
Qualtrics. (2020, May 21). Sample size calculator & complete guide. https://www.qualtrics.com/blog/calculating-sample-size/
Quinlan, J. R. (1986). Induction of decision trees. Machine Learning, 1(1), 81–106.
Quinlan, J. R. (1993). C4.5: Programs for Machine Learning. Morgan Kaufmann Publishers, Inc.
Rawlings, J. O. (1988). Applied Regression Analysis: A Research Tool. Wadsworth and Brooks/Cole.
Rokach, L., & Maimon, O. (2014). Data Mining with Decision Trees: Theory and Applications, 2nd edition. World Scientific.
Sarma, K. S. (2013). Predictive Modeling with SAS Enterprise Miner—Practical Solutions for Business Applications, 2nd edition. SAS Press.
SAS. (n.d.). Big Data. Retrieved from https://www.sas.com/en_us/insights/big-data/what-is-big-data.html
SAS. (n.d.). Data Mining. Retrieved from https://www.sas.com/en_us/insights/analytics/data-mining.html
Schapire, R. E. (1990). The strength of weak learnability. Machine Learning, 5(2), 197–227.
Schechtman, E., & Schechtman, G. (2016). The relationship between Gini methodology and the ROC curve (SSRN Scholarly Paper No. ID 2739245). https://dx.doi.org/10.2139/ssrn.2739245
Scott, A. J., & Wild, C. J. (1989). Selection based on the response variable in logistic regression. In C. J. Skinner, D. Holt, & T. M. F. Smith (Eds.), Analysis of Complex Surveys (pp. 191–205). John Wiley & Sons.
Scutari, M., Vitolo, C. & Tucker, A. (2019). Learning Bayesian networks from big data with greedy search: computational complexity and efficient implementation. Statistics Computing, 29, 1095–1108.
Shearer, C. (2000). The CRISP-DM Model: The new blueprint for data mining. Journal of Data Warehousing, 5(4). 13–22.
Shmueli, G., Bruce, P. C., & Patel, N. R. (2016). Data Mining for Business Analytics: Concepts, Techniques, and Applications with XLMiner, 3rd edition. Wiley Publishing.
Somers, R. (1962). A new asymmetric measure of association for ordinal variables. American Sociological Review, 27(6), 799–811.
Soper, D. (n.d.). Sample size calculators. https://www.danielsoper.com/statcalc/

Soper, D. S. (2023). A-Priori Sample Size Calculator for Multiple Regression [Software]. https://www.danielsoper.com/statcalc

Tan, P. N., Steinbach, M., & Kumar, V. (2021). Introduction to Data Mining, 2nd edition. Pearson.

Tao, K. M. (1993). A closer look at the radial basis function (RBF) networks. In Conference Record of the Twenty-Seventh Asilomar Conference on Signals, Systems, and Computers (Vol. 1, pp. 401–405), Los Alamitos, CA.

Truong, D. (2021a). Estimating the impact of COVID-19 on air travel in the medium and long term using neural network and Monte Carlo simulation. Journal of Air Transport Management, 96, 102126.

Truong, D. (2021b). Using causal machine learning for predicting the risk of flight delays in air transportation. Journal of Air Transport Management, 91, 101993.

Truong, D. (2024). Data Science and Machine Learning for Non-Programmers: Using SAS Enterprise Miner. Chapman and Hall/CRC Press. ISBN: 9780367755386.

Truong, D., & Choi, W. (2020). Using machine learning algorithms to predict the risk of small unmanned aircraft system violations in the national airspace system. Journal of Air Transport Management, 86, 101822.

Truong, D., Friend, M, & Chen, H. (2018). Applications of business analytics in predicting flight on-time performance. Transportation Journal, 57(1), 24–52.

Truong, D., & Truong, M. D. (2022). How do customers change their purchasing behaviors during the COVID-19 pandemic? Journal of Retailing and Consumer Services, 67, 102963.

Truong, D., & Truong, M. D. (2023). Impacts of daily travel by distances on the spread of COVID-19: an artificial neural network model. Transportation Research Record, 2677(4), 934–945.

Tuffery, S., & Ebrary, I. (2011). Data Mining and Statistics for Decision Making. Wiley.

U.S. Department of Health & Human Service. (1979, April 18). The Belmont report—Ethical principles and guidelines for the protection of human subjects of research. U.S. Department of Health & Human Service. https://www.hhs.gov/ohrp/regulations-and-policy/belmont-report/read-the-belmont-report/index.html#xbasic

Vapnik, V. N. (1998). Statistical Learning Theory. Wiley.

Werntges, H. W. (1993). Partitions of Unity Improve Neural Function Approximation. In Proceedings of the IEEE International Conference on Neural Networks (Vol. 2, pp. 914–918), San Francisco, CA.

Witten, I. H., Frank, E., Hall, M. A., Pal, C. J. (2016). Data Mining: Practical Machine Learning Tools and Techniques, 4th edition. Morgan Kaufmann.

Zhang, Y., & Zhou, Z. H. (2014). A review on multi-label learning algorithms. IEEE Transactions on Knowledge and Data Engineering, 26(8), 1819–1837.

Index

Note: *Italic* and **Bold** pages refer to figures and tables, **respectively**.

E

N

O

P